The Chora of Metaponto 6
A Greek Settlement at Sant'Angelo Vecchio

Francesca Silvestrelli and Ingrid E. M. Edlund-Berry
Joseph Coleman Carter, Series Editor

The Chora of Metaponto 6
A Greek Settlement at Sant'Angelo Vecchio

Francesca Silvestrelli and Ingrid E. M. Edlund-Berry
Joseph Coleman Carter, Series Editor

Institute of Classical Archaeology
Packard Humanities Institute
University of Texas Press, Austin

Printed in the United States of America

First edition, 2016

♾ The paper used in this book meets the minimum requirements of ANSI/NISO Z39.48-1992 (R1997) (Permanence of Paper).

Library of Congress Cataloging-in-Publication Data

Names: Silvestrelli, Francesca, author. Edlund-Berry, Ingrid E. M., author. Carter, Joseph Coleman, editor. University of Texas at Austin. Institute of Classical Archaeology. Packard Humanities Institute.
Title: The chora of Metaponto 6, A Greek settlement at Sant'Angelo Vecchio / Francesca Silvestrelli and Ingrid E. M. Edlund-Berry ; Joseph Coleman Carter, series editor.
Description: First edition. Austin : University of Texas Press, 2016. "Institute of Classical Archaeology, Packard Humanities Institute." Includes bibliographical references and index.
Identifiers: LCCN 2016010364
ISBN 9781477309476 (cloth : alk. paper)
ISBN 9781477309681 (library e-book)
ISBN 9781477309698 (nonlibrary e-book)
Subjects: LCSH: Metapontum (Extinct city). Excavations (Archaeology)—Italy—Metapontum (Extinct city). Greeks—Italy—Metapontum (Extinct city). Agricultural colonies—Italy—Metapontum (Extinct city). Farmhouses—Italy—Metapontum (Extinct city). Kilns—Italy—Metapontum (Extinct city). Metaponto Region (Italy)—Antiquities.
Classification: LCC DG70.M52 S55 2016
DDC 937/.773—dc23
LC record available at http://lccn.loc.gov/2016010364

doi:10.7560/309476

For reasons of economy and speed, this volume has been printed from camera-ready copy furnished by ICA, which assumes full responsibility for its contents.

Title page illustration: Mold for relief-decorated loom weight with Eros. 4th–3rd century BC. (Ch. 31, "Loom Weights," Cat. No. LW 25) Photo: C. Williams/ICA.

A Rebecca.

Francesca Silvestrelli

To John, Dana, and Robert.

Ingrid E. M. Edlund-Berry

Special Acknowledgment

This volume and others in this series could not have been conceived,
much less carried through to publication, without the generous financial support of the
Packard Humanities Institute.
The Director and staff of the Institute of Classical Archaeology and its collaborators
express their profound gratitude.

Contents

Part III

Appendices

Reference Material

Acknowledgments

The excavations at Sant'Angelo Vecchio form part of the University of Texas Excavations at Metaponto, begun in 1974 under the direction of Professor Joseph C. Carter. With gratitude we acknowledge the permit granted by the Ministero per i Beni Culturali e Ambientali and the close collaboration with the Soprintendente Dott.ssa Elena Lattanzi, her predecessor, Prof. Dinu Adamesteanu, and the staff at Metaponto, Dr. Antonio De Siena, Sig. Antonio Indice, and Sig. Giuseppe Di Taranto. I am grateful to my colleague, Professor Joseph C. Carter for allowing me to direct the excavations at Sant'Angelo Vecchio in 1979 and for continuing the work himself there in 1980. The area of the kilns was studied separately in 1981. The publication of the finds is the result of the effort of all the staff members during the excavation season proper and during the period of study in preparation of the manuscript. I wish to acknowledge the splendid work of our photographer, Chris Williams, who took most of the field shots and documentary views of the objects for study and publication.

The project was supported by the University of Texas at Austin, the National Endowment for the Humanities, The James R. Dougherty Jr. Foundation, and many private individuals. Administrative support was provided by the Department of Classics at the University of Texas at Austin through its Chairman, Professor G. Karl Galinsky, and Administrative Assistant, Ms. Leoda Anderson. At Metaponto, the task of administration was handled by the staff of the Soprintendenza, Sig. Bruno Baccaro, and Sig. Bruno Chiartano. Facilities for workspace were granted by the Ente Irrigazioni Bari through Dott. Salvatore Dedono and Sig. Leonardo Torraco. Special thanks are due to the former landowner of Sant'Angelo Vecchio, Sig. Vincenzo Santoro and family, and to the present landowner, Alfredo Sarubbo.

The architects responsible for the site plan and drawings of sections were Michael Guarino (1979–1980), Ian Sutherland (1979), Gloria Howell (1981), Deena Berg (1981), and Professor Thomas Boyd (consultant 1981). Profile drawings of the pottery and drawings of the skeletons and votive plaques were executed by Ellen Simmons (1979–1981). Trenchmasters in the field were Mary Emerson (1979), Brian Aitken (1979), and Claire Lyons (1980).

Preliminary cataloguing and restoration of finds were carried out by Pamille Berg (1979), Jennifer Brehob (1981), Sondra Brady (1979), Anna Keys (1980), Anne Meyer (1980), John O'Connor (1979), Carole Pierce (1979), Francesca Quarato (1979–1980), Kerry Rodgers (1979), Gianni Scrinieri (1979), Johanna Smith (1979), Ginny Volterra (1979–1980), and Franco Zaccaro (1979).

Ingrid Edlund-Berry

The publication of Sant'Angelo Vecchio is, as that of any other excavation, the result of long, complex, and sometimes exhausting work, to which a large number of people gave their contribution. Ingrid Edlund-Berry has already acknowledged all those who were involved in the excavation and in the processing of the materials and of the results undertaken between 1979 and 1981. More names need to be added for the period 2009–2015, when the work was resumed.

The Soprintendenza per i Beni Archeologici della Basilicata has proved to be of the utmost importance for the completion of this project, as well as others with which I have been involved, providing constant support, and I wish to thank the administrative and technical staff of the National Museum of Metaponto (Enzo Arcuti, Nunzia Armento, Saverio Careccia, Giulia Gioia, Vita Quattromini, and Grazia Pinto). Special thanks are due to Antonio De Siena, former Superintendent of Basilicata and, for many years, Director of the National Museum at Metaponto, for the time and the knowledge he agreed to share with me.

The staff at the ICA both at Austin and at Pantanello was of invaluable help in every aspect of the job. Alberto Prieto took care of the translations of most of the contributions presented here, often offering important suggestions. Juan Carlo Fernández and Karl Strasen improved the graphic and photographic documentation needed for the publication. Jessica Trelogan helped me with her accurate maps in clarifying the site. Lauren Jackson provided her skill and her organizational capacity in editing the text, and much more. The insights both of them shared with me proved many times to be very helpful. To Keith Swift, always there to discuss, and to Esmeralda Moscatelli, who patiently and effectively fixed most of the problems we encountered in the realization of this volume, go my warmest thanks, and if I may, my friendship. On the Italian side, Tiziana Ria has guaranteed easy access to the library at Pantanello, and comfortable accommodations at the center. I am grateful to the ICA, which provided the best conditions possible to enhance the work, and to the Packard Humanities Institute for the role they played in making it possible.

Many colleagues provided insights and expertise that greatly assisted the research. I am glad to thank Annalisa Concilio, Liliana Giardino, Massimo Guastella, Elisa Lanza Catti, Erminia Lapadula, Mariastella Montemurro, Vincenzo Pugliese, Alessandro Quercia, Sveva Savelli, Thierry Van Compernolle, and Carlo Rescigno. For his constant support, I also wish to thank Mario Lombardo, former Director of the Department of Cultural Heritage at the University of Salento. I am also pleased to mention Giuseppe Di Taranto, foreman during the 1979 excavation, who in 2010 also completed the salvage excavation of Tombs 9 and 10, resuming his job at the site after 31 years.

Finally, my gratitude goes to Prof. Joseph Coleman Carter, Director of the ICA, who brought me onto the Metaponto Project in 2007. He offered me, with a freedom and generosity that I have learnt throughout the years to appreciate more than I can express, the possibility to be part of the study of the Metapontine chora in its variety and complexity, and to contribute with Sant'Angelo Vecchio to the wider picture his life-long research has created.

Francesca Silvestrelli

Donors

Dr. and Mrs. Leo Bini
Stefano Bini
The Brown Foundation
Mr. and Mrs. Joseph Coleman Carter III
Brandon Chenault
Judy Coker
Marian Davis
Ronald DeFord
Gilbert Denman, Jr.
Mary Patricia Dougherty
Sandra Heinrichs
Highland Resources, Inc.
The James R. Dougherty Jr. Foundation
Mr. and Mrs. John Allen King
Lois Ashton Larson
Vera Massaro
Barbara McCluer
Etelka McCluer
Benjamin D. Meritt
Lucy Shoe Meritt
Anne Meyer
Bert Henry Michelsen
Kathleen Monahan
The National Endowment for the Humanities
Mr. and Mrs. Ralph S. O'Connor
Packard Humanities Institute
Sheila Plotsky
Mr. and Mrs. Philip Schaefer
Johanna L. Smith
Texas Eastern
The University of Texas at Austin
Ben F. Vaughan, Jr.
Ben F. Vaughan, III
Mr. and Mrs. William Willis

Contributing Authors

Marshall Joseph Becker
Professor of Anthropology Emeritus
West Chester University

Philip Bes
Independent Roman pottery specialist

Anna Zsófia Biller
Archaeozoologist
Aquincum Museum of the Budapest History Museum, Hungary

Teresa Oda Calvaruso
Archaeologist

Joseph Coleman Carter
Director of the Institute of Classical Archaeology
University of Texas at Austin

Anna Cavallo
Archaeologist

Francisco da Conceição de Jesus Neto
Environmental Scientist

Annalisa Concilio
Archaeologist

Emanuela Conoci
Archaeologist

Lorenzo Costantini
Archaeobotanist
ISMEO—International Association of Mediterranean and Oriental Studies

Cesare D'Annibale
Archaeologist

Carlo De Mitri
Archaeologist

Antonietta Di Tursi
Archaeologist

Ingrid E. M. Edlund-Berry
Professor Emerita
University of Texas at Austin

Assunta Florenzano
Post-doctoral collaborator
Laboratorio di Palinologia e Paleobotanica
Università di Modena e Reggio Emilia

Lin Foxhall
Professor of Greek Archaeology and History
University of Leicester

Mauro Frattegiani
Forest Engineer
Dipartimento di chimica, biologia e biotecnologie
Università degli Studi di Perugia

Francesco Guizzi
Professor of Greek History and Epigraphy
La Sapienza–Università di Roma

Rachel Karlov
Geologist

Erminia Lapadula
Archaeologist

Massimo Limoncelli
Archaeologist

Marta Mazzoli
Archaeologist

Alessandro Montanari
Geologist
Osservatorio Geologico di Coldigioco

Francesco Perugino
Archaeologist

Fabrizio Pica
Archaeobotanist
Bioarchaeological Research Centre
National Museum of Oriental Art "Giuseppe Tucci"

Alessandro Quercia
Archaeological Officer
Soprintendenza Archeologia del Piemonte

Carlo Rescigno
Professore di seconda fascia in Archeologia Classica
Seconda Università di Napoli

Edward G. D. Robinson
Senior Lecturer in Archaeology
University of Sydney

Francesca Silvestrelli
Ricercatore Universitario
Università del Salento

Keith Swift
Institute of Classical Archaeology

Anna Lucia Tempesta
Conservatore Archeologo
Museo Archeologico Provinciale Sigismondo Castromediano

Mark Van Der Enden
Archaeologist
University of Leicester

Eloisa Vittoria
Archaeologist

Serena Viva
PhD student, Dipartimento di Civiltà e forme del sapere
Università degli Studi di Pisa

Eliana Vollaro
Archaeologist

Contributors

Site and Object Documentation
Nanita Ellen Barchi
Massimo Barretta
Pamille Berg
Jennifer Brehob
Saverio Careccia
Paolino Forino
Anna Keys
Kerry Rodgers
Ellen Simmons
Virginia Volterra

Site and Object Photography
Nunzia Armento
Maria Luisa Giuliani
Aldo La Capra
Cesare Raho
Chris Williams

Consultants
Marshall Joseph Becker
Lorenzo Costantini
Robert Folk
Carolyn Koehler
Sam Wolff
H. G. Wooldridge

Survey and Architecture
Brian Aitken
Enzo Arcuti
Deena Berg
Thomas Boyd
Michael Guarino
Gloria Howell
Carole Pierce
Alessandro Rizzo
Ian Sutherland

Conservation
Francesca Quarato
Gianni Scrinieri
Franco Zaccaro

Excavators

Supervisors
Brian Aitken
Joseph Coleman Carter
Ingrid E. M. Edlund-Berry
Mary Emerson
Claire L. Lyons
Francesca Silvestrelli

Participants
Donato Attanasio
Sondra Brady
Giuseppe Braico
Emanuela Conoci
Domenico D'Anzi
Giuseppe Di Taranto
Assunta Florenzano
Alfredo Gallitelli
Cosimo Goia
Giuseppe Marino
Vincenzo Matera
Hope McBride
Anne Meyer
Mimmo di Monte
Antonio Morelli
John O'Connor
Leonardo Pacciani
Geraldine Pease
Vittorio Plati
Alfredo Portarulo
Domenico Savoia
Roberto Sgraffeti
Johanna Smith

Production

Cartography
Jessica Trelogan

Translation
Alberto Prieto

Editing
Emily N. Harris
Lauren M. Jackson

Illustrations and Layout
Juan Carlo Fernández
Lauren M. Jackson
Karl Strasen

Collaborators

Soprintendenza per i Beni Archeologici della Basilicata
Dinu Adamesteanu
Bruno Baccaro
Bruno Chiartano
Antonio De Siena
Salvatore Dedono
Liliana Giardino
Antonio Indice
Elena Lattanzi

Museo Archeologico Nazionale di Metaponto
Enzo Arcuti
Nunzia Armento
Saverio Careccia
Giulia Gioia
Grazia Pinto
Vita Quattromini

Individuals
Emilio Caporale and family
Vincenzo Santoro and family
Leonardo Torraco and family

Institute of Classical Archaeology and Department of Classics
Leoda Anderson
Sondra Brady
Karl Galinsky
Elizabeth Graves
Robert King
Kathleen Monahan
Esmeralda Moscatelli
Tiziana Ria
Keith Swift

Illustration Credits

AF	Assunta Florenzano
ALT	Anna Lucia Tempesta
AM	Alessandro Montanari
CR	Carlo Rescigno
CW	Chris Williams
FS	Francesca Silvestrelli
JCC	Joseph Coleman Carter
JCF	Juan Carlo Fernández
JT	Jessica Trelogan
KJS	Karl Strasen
LC	Lorenzo Costantini
LMJ	Lauren M. Jackson
MF	Mauro Frattegiani
PF	Paolino Forino

Preface

The chora of Metaponto was filled with ancient sites from all periods in the history of this land, from the Neolithic to the Medieval, represented by a variety of structures. The period of the Greek settlement, of the chora proper, is by far the most present. Approximately 95% of the sites recorded in the Institute of Classical Archaeology's (ICA's) field survey of the chora (1981–2001) were Greek, though often with a prehistoric (usually Neolithic) component and, in a limited number of cases, an afterlife in the Roman period. We know all of this thanks to the foresight and energy of Dinu Adamesteanu, who organized the first systematic surveys and excavations in the 1960s. Adamesteanu's deep concern to preserve all of the history was manifest in his active stewardship (*tutela*) of the whole of the very large area—all of Basilicata—for which he, as its first Soprintendente, was responsible. In the 1970s, an ill-considered government-sponsored program of remodeling the landscape to create flatter fields was destroying ancient sites by the hundreds. The inspectors of the Soprintendenza were very busy.

I remember clearly that day in late May of 1979 when the assistant to the Soprintendente, Antonio Indice, took me to Sant'Angelo, where the remains of a large deposit of terracotta figurines and plaques was still visible along the road. At that time, Ingrid Edlund, with a group of professionals, students, volunteers, and highly skilled workmen, were preparing to continue the excavation of the Pantanello sanctuary. She was apprised of the urgency of the situation, and she and the group rose to the occasion. The site of Sant'Angelo Vecchio was excavated, essentially, that summer, with some follow-up in subsequent years.

The site proved unexpectedly rich. There was epigraphic evidence of a sanctuary here in the 6th century BC—one of a number of extra-urban sanctuaries that included Pantanello and San Biagio on the Basento River and the Tavole Palatine on the Bradano. There was a small farmhouse of the second half of the 6th century BC on the slope above the spring where the sanctuary would have been located; there was a small necropolis with burials of some importance. Among the grave goods were alabastra and red-figure vases, including one of the very first by the pioneer Italiote Greek vase painter, the Pisticci Painter. (Before the excavations in the chora revealed many burials with his work, it was mostly known from museum collections.) A ceramic production center flourished here from the 4th century BC to the early 1st century AD, thanks to the abundant good clay of the Basento valley and the presence of the spring, an essential element of all settlements in the countryside.

The site was safe because the landowners, the Santoro family, were not only friendly and helpful during the excavation, but took seriously their informal responsibility for the protection of the site. The current owners, the Sarubbo family, have continued to respect it.

Objects from the site were restored, cataloged, and documented photographically, and site plans and sections were drawn. Ingrid Edlund made a two-volume report in 1982 for the ICA archive, and there it may have continued to sit, had it not been for the generous support of the Packard Humanities Institute (PHI).

Sant'Angelo Vecchio and a number of other sites were saved for the historical record of Greek settlements in Italy by a program as foresighted and original as that of Adamesteanu. PHI made it possible to employ young, very well-prepared Italian scholars and other experts from around the globe in the study of unpublished excavations in the chora. This began in 2000 and received a big boost in 2007. The result is an extensive, ongoing publication series. The current volume is one of a half-dozen titles (10 separate volumes) completed so far,[1] with several more in the works, including one devoted exclusively to the abundant terracotta votive plaques found at Sant'Angelo Vecchio by Rebecca Ammerman. The scholars are based in the study center at Pantanello with its library and living quarters, while the editorial production staff works at ICA's offices in Austin.

[1] *Necropoleis*; Bökönyi and Gál 2010; *Survey*; *San Biagio*; *Fattoria Fabrizio*.

This is where Francesca Silvestrelli joined forces with ICA. An imaginative and meticulous scholar, she organized the study team and undertook a complete re-evaluation of the evidence from the original excavations at Sant'Angelo Vecchio, as well as that of some new excavation in 2010 and 2011. The publication follows the general outline of previous volumes in the series. There is a strong emphasis on the study of ceramics by top specialists in the field, including Francesca herself. This volume provides an invaluable tool for others working in Greek and Roman sites in southern Italy. There are chapters on little-known subjects like red ware. An unexpected Roman and Late Roman phase at the site came to light, based on a few unprepossessing fragments that could easily have been overlooked by a less attentive and prepared team. Sant'Angelo itself, in a quite limited area, illustrates with a wide range of materials the variety of activities that characterized the chora in the heyday of the Greek settlement and its successors.

The whole, however, is considerably greater than the sum of the parts. That is because Silvestrelli has seen connections with other sites and studies, and with the overarching picture that the survey results provide.[2] The archaeobotanical analysis, for example, re-creates the landscape surrounding the site, but it is part of a larger picture, one tessera in the mosaic of the ancient countryside that is slowly, piece by piece, coming into view at Metaponto—a historic reality that for now is unique in the Greek West. The authors are to be commended for the skill and persistence with which they have guided this impressive collaborative effort to completion.

Joseph Coleman Carter
Austin, Texas
September 2015

[2] See *Survey*.

Part I

1

The Many Lives of a Rural Site: Sant'Angelo Vecchio

Francesca Silvestrelli

Figure 1.1 View of Sant'Angelo Vecchio from the NE in 1979. The masseria is built at the edge of the terrace overlooking the Basento River valley. Further N, on the top of the hill, is a small abandoned rural building marking the position of the Late Archaic farmhouse. At the bottom is the gravel road, the construction of which led to the discovery of the archaeological site. (CW/ICA)

Introduction

The name Sant'Angelo Vecchio refers to an area dominated by a masseria (large farmstead) that consists of a majestic white building with arched openings in the front facade of the ground floor (*Fig. 1.1*). The masseria sits on a rise above the north bank of the Basento River, 8.2 km west of the ancient Greek city of Metaponto.[1] The area and the masseria owe their name to a chapel, now incorporated into the farmstead, dedicated to the archangel Michael, cited in documents as early as the beginning of the 12th century.[2] Now, as in the past, the masseria is clearly visible from the Basentana, the modern road that follows the Basento River from the coast inland toward Potenza in the region of Basilicata (*Fig. 1.2*) and traces the ancient route linking Metaponto to the inland region. The house and fields are set off from the main road, up the slope of the gently undulating hill (42 masl), bounded to the north and east by a tributary of the Basento, the Fosso Sant'Angelo, which extends to the north, beyond Masseria Sant'Angelo Nuovo toward Masseria Giamperduto.[3] The area between Sant'Angelo Vecchio and Sant'Angelo Grieco (Nuovo) forms a fairly level plateau, whereas further west the ground rises sharply toward the Cozzo della Guardia.

The land around the Masseria Sant'Angelo Vecchio was, at the time of the excavation (1979–1981), farmed quite extensively with alternating crops of grain, mel-

[1] As measured along a straight line from the western gate of Metaponto to the Kiln Complex of Sant'Angelo Vecchio.

[2] On the masseria, see Ch. 6, "Sant'Angelo Vecchio in the Medieval and Post-Medieval Periods."

[3] See *Carta d'Italia*, F201 Masseria S. Angelo II SO Sez. D, 1:10000.

Figure 1.2 Satellite image of Sant'Angelo Vecchio and its surroundings. (JT/ICA)

on, and tomato, watered by all-weather springs located on the lower eastern side of the hill, north of the masseria. In the last 20 years, the area has undergone dramatic transformations, and it is today cultivated only in part in olive groves and fruit orchards. Animal husbandry, which constituted one of the principal activities in past and present times and left deep marks on the countryside, is no longer practiced.[4]

The masseria shows evident signs of its abandonment, a tangible expression of the radical transformation in land use that this part of Italy has experienced

[4] On the changes in the landscape in recent times, see Ch. 10, "Observations on the Vegetation and Landscape Dynamics." On the importance of animal husbandry, see Ch. 8, "Archaeobotanical Analysis," and Ch. 10, "Observations on the Vegetation and Landscape Dynamics." The archaeological site was, and is still today, private property. Both the archaeological site and the masseria have recently been placed under a protective preservation order by the Soprintendenza Archeologica and the Soprintendenza ai Beni Artistici of Basilicata.

Figure 1.3 Plan of the site with indication of areas excavated year by year. NB: all maps of the excavation area are rotated (by 3.28° W of true N) to reflect the original site grid, oriented to magnetic N in 1979. (JT/ICA)

Figure 1.4 View of Sant'Angelo Vecchio from the NE in 2012. (FS/ICA)

in the last 40 years. No longer the thriving center of the economic management of the broad surrounding territory, the building today appears drab, while the coherence of the architecture and its interactive capacity with the surrounding countryside are threatened by a modern construction that has replaced it (*Fig. 1.1*). Extreme though these changes are, they represent only the most recent episode of the many lives that this site has experienced. Although Sant'Angelo Vecchio is not comparable in importance to the Pantanello complex or to the sanctuary at San Biagio,[5] its significance lies in the articulation of the evidence obtainable from it. These *disiecta membra*, often difficult to read and interpret, nonetheless bear witness to the various lives and transformations of one part of the Metapontino from prehistory through the current day, recounting and illustrating the extraordinary variety of settlements and uses of the territory throughout its history.

Discovery and Excavation

Although Sant'Angelo Vecchio appears in maps of the Metapontine chora made in the 1960s,[6] no secure information on archaeological evidence noted at the site was ever reported in contemporary publications. The history of research in this area seems to begin in spring 1979, when the leveling of a small gravel road (the so-called "modern road," *Fig. 1.1*) leading from the masseria toward the Fosso Sant'Angelo to the north led to the discovery, 140 m from the building, of a large quantity of fragments of votive terracotta plaques, pottery, and tiles. The density and quality of the evidence suggested that the site might have been part of a sanctuary complex, similar to the one found at San Biagio, discovered in 1964,[7] the topographical organization of which bears resemblances to Sant'Angelo Vecchio.

The preliminary collection of the materials revealed by the bulldozer (called the "Bulldozed Stipe"; see p. 51 in Ch. 2, "Assemblage 9") was undertaken by the Soprintendenza in 1979, and was followed, in the summer of the same year, by a four-week salvage excavation by the Institute of Classical Archaeology

[5] The Crossroads necropolis is published in *Necropoleis*. The sanctuary and farmhouse at Pantanello will be the subject of a forthcoming volume in this series, as will be the Late Republican ceramic workshop. See also Carter 2006, 7–8, 17–22, 150, 163–73, 246–47; *Survey*, 613–14, 666–69, 760–61, 842–43, 895–97, 910–12; and Carter 2012, 1108–14, with extensive bibliography. On the sanctuary and the general area of San Biagio, see Lapadula in *San Biagio* a, 19–20, and n. 2, with previous bibliography. See also Osanna/Pilo/Trombetti 2009 and De Stefano 2014.

[6] Adamesteanu 1967a, fig. 32.

[7] Adamesteanu 1965, 127–32.

Figure 1.5 First days of excavation at Sant'Angelo Vecchio in 1979: the crew starts cleaning the slopes. (CW/ICA)

(ICA) of the University of Texas at Austin.[8] Under the direction of Ingrid Edlund, the excavation aimed to define the stratigraphy and the extent of the architectural remains. Due to the heavy erosion and subsequent landfill that had taken place downslope and to the east of the modern road, efforts were at first concentrated on the area upslope and to the west, where the Bulldozed Stipe had been identified. Here, the recovery of lumps of fired clay, fragments of votive terracotta plaques, and roof tiles mixed with the plowed and eroded topsoil provided evidence not of a votive deposit, but rather of production of terracotta material. Excavations were subsequently extended west of the slope, bringing to light a small rural necropolis and the remains of a two-phase workshop with a set of kilns, three of them enclosed by a walled structure (Kiln Complex; *Fig. 1.3*).

Surveys were carried out in the surrounding area to detect other evidence of occupation. The recovery of Archaic pottery fragments on the top of the hill was suggestive, and subsequent investigation discovered the foundation walls of a rectangular building interpreted as an Archaic farmhouse (House on the Hill; *Fig. 1.3*).[9] A second trench, about 200 m west of the gridded area, yielded evidence of Hellenistic occupation (Westward Hummock; *Fig. 1.9*).[10]

Fieldwork resumed in 1980 to complete the analysis of the structures found in the Workshop Area and to establish the limits of the rural necropolis (squares H7, F7, and E6; *Fig. 1.3*). Sounding D, an east-west test trench 1.5 m wide and 8.60 m long, was excavated to natural soil east of the modern road, near the mouth of the all-weather spring (*Fig. 1.3*). It aimed to identify structures of the cult place connected to the spring, the existence of which was evidenced by the recovery of an inscribed stone interpreted as a *horos* (boundary stone) of a sanctuary in 1979,[11] but no such structure was identified. A further brief test was conducted in 1981 in the Kiln Complex, where a remaining baulk

[8] The permission was granted by the then Soprintendente alle Antichità of Basilicata, Dott.ssa Elena Lattanzi.

[9] See Ch. 2, "Site Phasing, Stratigraphy, and Site Assemblage."

[10] See Ch. 2, "Site Phasing, Stratigraphy, and Site Assemblage."

[11] Carter 1980a, 15. For the inscription, see p. 68 in Ch. 3, "The Inscribed Stone."

Figure 1.6 View of the Sant'Angelo Vecchio archaeological site from the E in 1979, showing the Workshop Area and the Slope Area with soundings. (CW/ICA)

just west of Sounding A was excavated.[12] Preliminary reports on the results were presented while the excavation was still ongoing and soon afterward.[13]

Methodology

Sant'Angelo Vecchio was excavated following a variation of the Wheelerian method, also adopted in other excavation projects carried out by ICA in the same period.[14] Fieldwork began with three trial soundings,[15] carried out to clarify the archaeological stratigraphy. The soundings were made inside the Kiln Complex, the eastern wall of which (Wall 9) was visible on the surface (Sounding A), and in the central and northern part of the Slope Area (Soundings B and C; see *Fig. 2.2*). The site was subsequently mapped on a 5-m Cartesian grid.[16] On the northern side (squares E–H 6–8), where the archaeological deposit was deep, the excavation preserved a 1-m-wide baulk to maintain vertical cross sections. In the southern area (squares E–H 1–5), the extensive soil erosion and agricultural activity had left only a shallow deposit covering the structures, and no baulks were kept. Section drawings of the baulks and cumulative sections of the entire excavation area record

[12] 1980, J. C. Carter; 1981, I. Edlund. They were assisted as site supervisors by Mary M. Emerson (1979), Brian Aitken (1979), and Claire L. Lyons (1980).

[13] Carter 1980a, 4; 11–12; 14–15. Carter 1980b, 31; Edlund 1979; Edlund 1982a; Edlund 1984, 287–90; Edlund 1986, 119–22; Edlund 1987, 99–100. The tombs, discussed in an unpublished report (Edlund 1982b), were first presented in *Necropoleis*, 443–47. On the sanctuary at Sant'Angelo: Carter 1994, 178. On the site more in general, see also Osanna 1992, 80–81 and Carter 2006, 19, 20 fig. 1.7A–B, 21, 41, 136 table 4.A, 137, 150, 174, 246.

[14] Lapadula in *San Biagio*, 21–23. See also *Fattoria Fabrizio*, ch. 1. Nonetheless the archaeological site, characterized by a complex stratigraphy, the presence of the slopes, and a large variety of contexts, required the development of strategies adapted to the evidence encountered.

[15] Trial trenches, although criticized even by Wheeler (1956, 80), were largely in use in contemporary excavations in southern Italy.

[16] This happened only two weeks after the beginning of the dig, once the harvest had been completed, and explains why material found in the earliest period of the excavation is labeled without a reference to the Cartesian grid (e.g., Slope 2, Level 1, battuta 1). The location of the slopes is indicated in *Figure 1.3*.

Strata	Description	Chronology
Stratum 1	Topsoil, in most cases disturbed, varied from a chalky whitish soil in the area of the kilns and the westernmost squares (especially in squares H6 and H7) to a yellowish brown soil ranging from pale to dark. The slope of the hill suggests that the darker soil represents fertile humus, whereas the chalky soil is equivalent to natural soil that has been exposed through erosion and, in modern times, leveling and earth-moving.	Modern
Stratum 2	Architectural features of the Kiln Complex, the foundation walls in the W part or the excavated area, and the kilns and deposits of votive plaques within the Slope Area to the E, toward the modern road. As demonstrated by the location of these features and the finds connected with them, this stratum includes material from different periods in terms of relative and absolute chronology.	4th c. BC–1st c. AD
Stratum 3	Features found at a level below those of Stratum 2 (Pottery Deposit in Sounding C, tombs in squares F6 and G6). For the most part these features and objects were placed within or rested immediately on natural soil. The natural soil is flaky and clayey and resembles the plowed topsoil in the kiln area, with the obvious difference that as topsoil it contained archaeological remains.	5th c. BC
Stratum 4	Identified in the interior of Tomb 3 (square F6), which was cut into the natural soil at a level lower than that of Tomb 2 (Ch. 4).	5th c. BC

Table 1.1 General stratigraphy and soil conditions for the site. (From Edlund 1982a)

the stratigraphic development of the site (see Ch. 2, "Site Phasing, Stratigraphy, and Site Assemblage").[17]

Each square was excavated by strata (originally called levels) divided into *battute* of 0.10–0.15 m. The strata were defined by the color and consistency of the soil or, alternatively, by archaeological features (*Table 1.1*). Although individual squares had different rates of progress, excavation removed one level at a time, so that it remained in phase. Plans of the architectural finds were made at the end of each season of fieldwork.

Work was recorded in a site notebook illustrated with sketches (usually drawn to scale) representing the most important features encountered. The sketches, along with many black-and-white and color photographs, offer a very detailed description of the excavation. The notebooks also include lists of the finds, each connected to the level and battuta where they were found and labeled with a progressive number.[18] Pottery fragments found in close association were grouped into so-called pottery lots (PL). All finds were collected, with the exception of roof tiles (for which examples of the main types encountered on the site, rather than all fragments, were kept). This allowed for a sound quantitative approach in the study of the material when the work was resumed in 2007.

The interaction between archaeology and other research fields pioneered in the Metapontino by ICA was also adopted at Sant'Angelo Vecchio (*Fig. 1.7*): the excavation and the study of the site involved many scholars of various disciplines.[19] The preliminary results of their work, although not always published,[20] are acknowledged and described within the text of this publication. This multidisciplinary approach continued also after study resumed in 2007. Geological, archaeobotanical, faunal, and landscape studies have been integrated into past research.

On the whole, the coherence expressed in the choice of the excavation methods, the variety of approaches used in the analysis of the site, and the overall quality of the recording system are noteworthy. An issue often undervalued in Metapontine excavations is the role played by the crew of hired workmen: trained in a series of digs in one of the most exciting seasons of Metapontine archaeology, they cooperated in a significant way by providing their expertise.[21]

Results of the 1979–1981 Excavations and Reevaluation of the Evidence

The fieldwork, encompassing a total area of 711.6 m^2, uncovered a complex stratification, the phases of which

[17] All elevations from the excavation were recorded relative to a local datum, with an arbitrary zero value located near the House on the Hill (per a local survey conducted in 1979). Our choice to leave these relative elevations was deliberate, because local sea level is variable and likewise arbitrary.

[18] E.g., SA 79-888P, where SA = Sant'Angelo Vecchio, 79 = year of excavation, and P = pottery. Other abbreviations used are T = Tiles and terracottas, M = Metals, and B = Bones.

[19] Specialized studies of aspects of the site were undertaken by Professor Robert L. Folk (University of Texas at Austin, 1979–1980), Dr. Lorenzo Costantini (Museo Nazionale d'Arte Orientale, Roma, 1979–1980), Dr. H. G. Wooldridge (University of Texas at Austin, 1979), Mr. Samuel Wolff (University of Chicago, 1979), Professor Carolyn Koehler (1979), and Professor Marshall Becker (West Chester State College, 1979–1980).

[20] With the exception of Becker (see p. 90 in Ch. 4, "Analysis of the Human Skeletal Material from the Classical Tombs at Sant'Angelo Vecchio") and Costantini (see Ch. 9, "Charred Plant Remains and Plant Impressions in Fired Clay Fragments from Sant'Angelo Vecchio").

[21] The crew of workmen in 1979 consisted of Alfredo Gallitelli, Vincenzo Matera, Leonardo Pacciani, Alfredo Portarulo, and Giuseppe Di Taranto; in 1980, of Alfredo Gallitelli, Cosimo Goia, Giuseppe Marino, Rocco Paradiso, Leonardo Pacciani, Roberto Sgraffeti, and Giuseppe Di Taranto. Student workers from a local cooperative were Donato Attanasio (1979), Giuseppe Braico (1979), Antonio Morelli (1979), Vittorio Plati (1979, 1981), and Domenico Savoia (1981).

Figure 1.7 Flotation at the spring in 1979. From left to right, Michael Guarino (front), Brian Aitken (back) and Ian Sutherland. (CW/ICA)

were summarized by Edlund as follows.[22] Occupation of the site began with the construction, on the highest point of the hill north of the Masseria Sant'Angelo Vecchio, of a farmhouse (*Fig. 1.2*) dated to the middle of the 6th century BC on the basis of the then-current chronology for Ionic cups. The construction of this "House on the Hill" inaugurated a period in which the area was inhabited by groups involved in animal husbandry and agricultural activities, lasting until the end of the 5th century BC. Although the occupation of the farmhouse was short-lived (possibly one generation), continued habitation in the Late Archaic period is documented by the remains of a Retaining (?) Wall in squares C4, C5, and D5, and by the Pottery Deposit identified in Sounding C, possibly connected to the nearby spring-sanctuary. West of the Retaining Wall (square E6 to H6), a cluster of burials suggests the presence of a moderately wealthy family of farmers living in the vicinity in the second half of the 5th century BC (*Fig. 1.3*).

Sometime in the 4th century BC, the function of the site changed quite abruptly. A workshop (square G7) and a set of kilns used for the production of votive terracotta plaques were built west of the Retaining Wall (squares D1–D5 and E1–E4). A very poorly preserved structure in squares G6 and F7 was interpreted as the residence of the artisan(s) in charge of the kilns. In the 3rd century BC, the set of kilns on the slope was decommissioned and replaced with a new complex constructed on the rise of the slope, comprising three kilns enclosed within a walled structure. This Kiln Complex was renovated in the 2nd century BC, when a workshop shed and a storage area for the clay were also built west of it (*Fig. 1.3*).

The study of the excavation data and material was resumed in 2007 as part of ICA's Metaponto project. Two cleanings of the site were undertaken to update the plan, in November 2010 (when, as a consequence of heavy rains, two new tombs of the Early Hellenistic period came to light) and July 2011.[23] The compre-

[22] Edlund 1979; 1982a; 1987, 99–100.

[23] The salvage excavation of the two tombs was carried out by ICA in

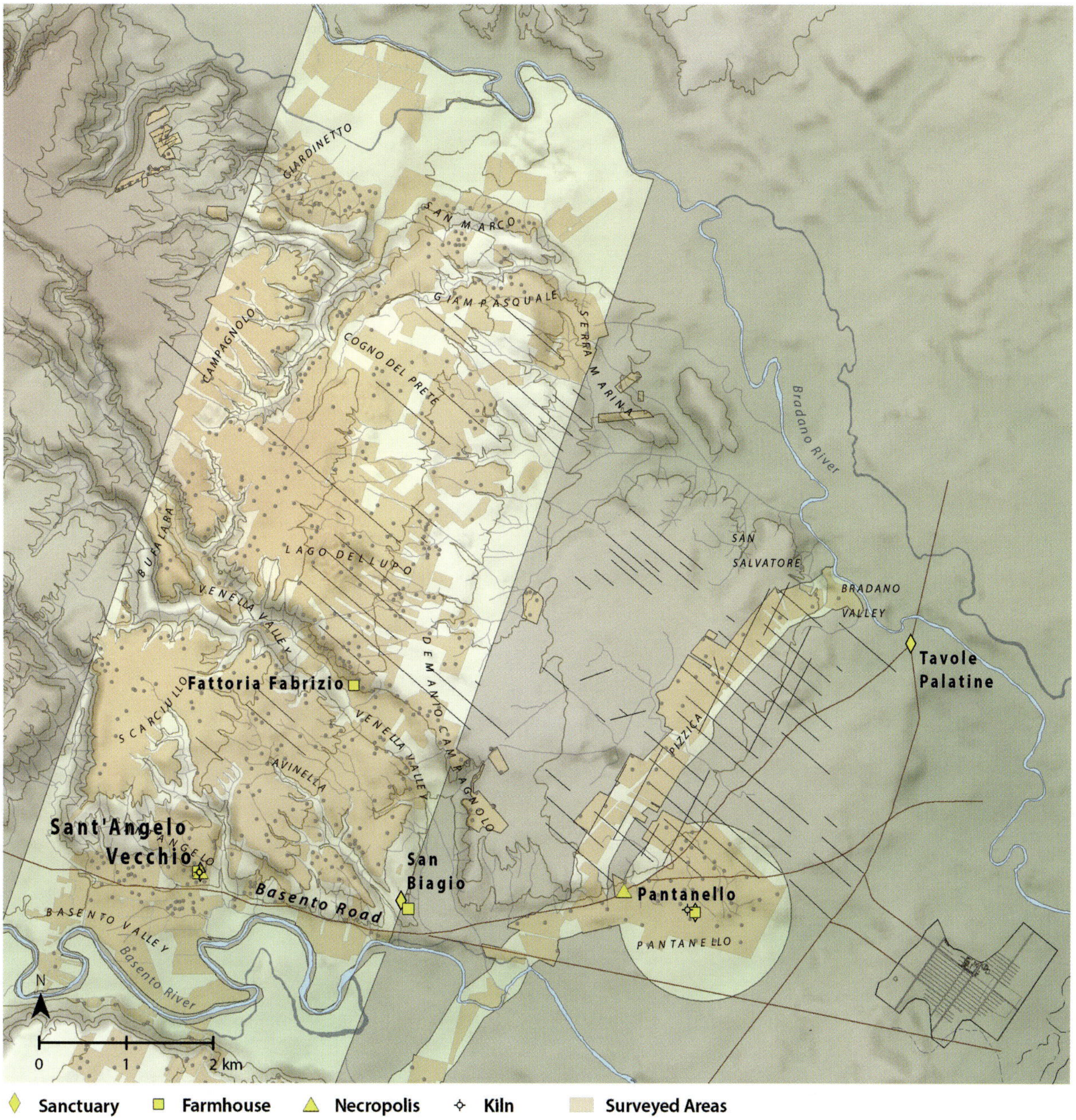

Figure 1.8 Map of the Metapontine chora between the Basento and the Bradano Rivers. (JT/ICA)

hensive results of this endeavor, which helped refine the general framework of the site phasing proposed by Edlund (*Table 1.1* and *Fig. 2.1*), are presented in this volume, though the essential source nonetheless remains the two unpublished reports written by Edlund in 1982.[24]

collaboration with the Soprintendenza della Basilicata. The new site plan was created from topographic data obtained with a total station operated by Emanuela Conoci, Antonio De Blasi, and Alessandro Rizzo.

[24] Edlund 1982a and 1982b.

Setting

Sant'Angelo Vecchio lies within the nearer chora of the Achaian colony of Metaponto.[25] It occupies a dominant position, which allows a view along both banks of the Basento River as far as the sea to the east and inland toward Pisticci and the mountains (*Fig. 1.8*). An array of advantageous features—including easy access to the all-weather spring, the relative fertility of the land, which was exploited for agriculture and animal

[25] *Survey*, 621–22, fig. 19.1.

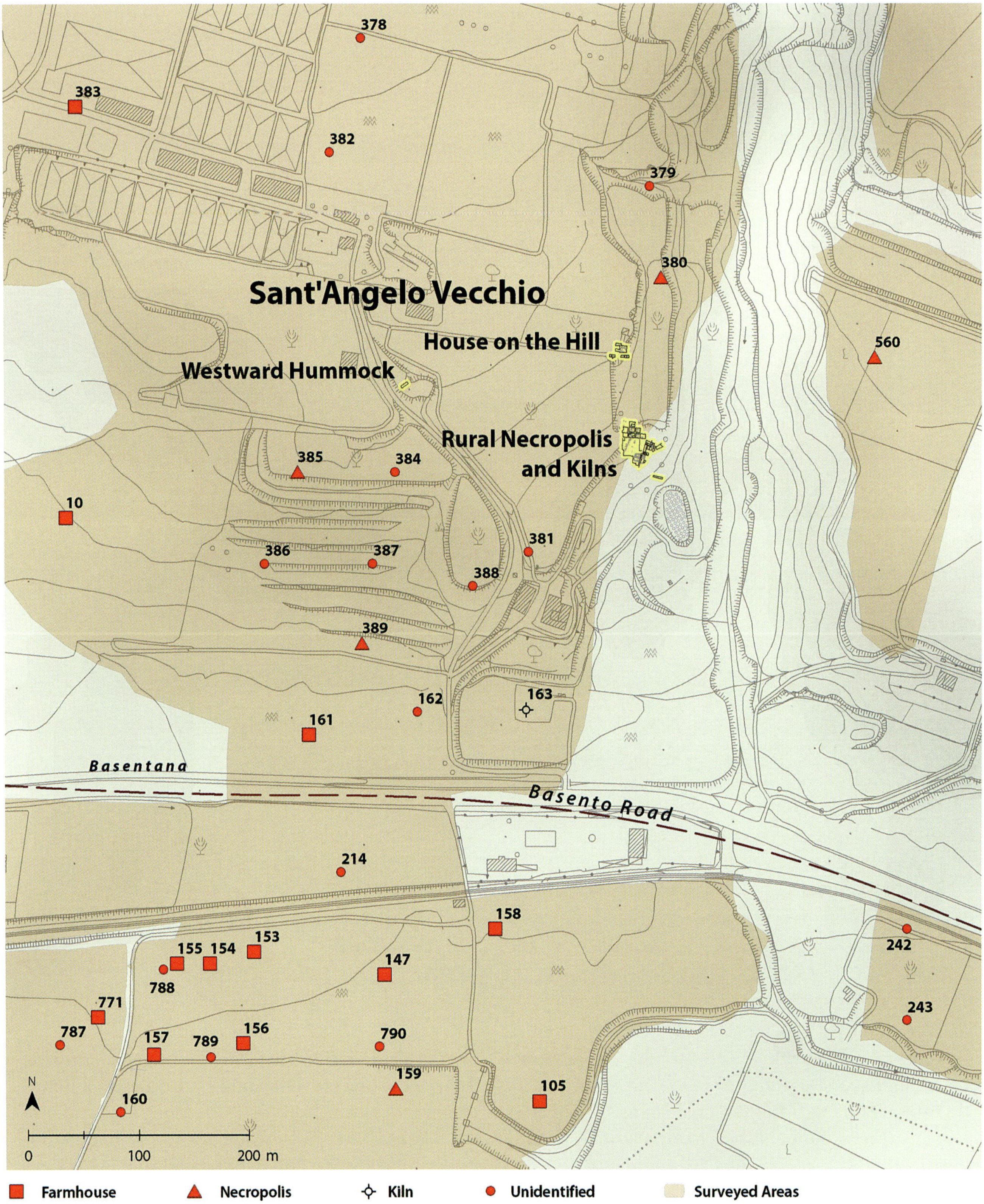

Figure 1.9 Map of the area of Sant'Angelo Vecchio comprising excavated areas (in yellow) and sites identified during the Bradano-Basento survey (in red). (JT/ICA)

husbandry both in antiquity and in modern times,[26] and the availability of Pliocene clays,[27] which facilitated the productive vocation of this area—provides an explanation for human presence, attested since the late Neolithic period.[28]

Road access was supplied by the Basento Road, the major north-south artery leading up the valley and connecting the city to the interior. The potential route of the road in this part of the chora might be suggested by the alignment of such sites as 154, 155, and 285 (*Fig. 1.9* and *Fig. 1.10*), identified by the Bradano-Basento field survey.[29] The ancient road, possibly also retraced by the medieval and modern *tratturo* (*Fig. 6.5*), was positioned approximately 500 m north of the north bank of the Basento River. The river itself was probably navigable in antiquity, and it has been suggested that the inlet of an old branch of the river near site 105, opposite of Sant'Angelo Vecchio, could have been used as a small port or landing serving the nearby sites.[30] Evidence for an east-west axis of communication leading from the Basento to the Metapontine plateau in the Classical and Roman periods is lacking, by contrast, and the scarcity of sites identified in the area does not permit any sound hypothesis.

Environs

The Late Archaic farmhouse and the workshops and kilns are part of a wider context partially disclosed by the field survey undertaken by ICA and now published.[31] One of the Early Village Clusters identified in the chora of Metaponto[32] is located in the area delimited by Cozzo della Guardia to the west, Fosso Sant'Angelo to the east, and the Basento River to the south. Here, a coherent system of individual dwellings and necropoleis was placed at regular intervals of 600 m and concentrated on the terraces facing the river valley. Noteworthy is the smaller number of survey sites detected on the hills overlooking the lower Fosso Sant'Angelo before the Late Republican period.[33] Although it is possible that the steep sides of the drainage proved unappealing for settlers, erosion and intensive agricultural activity performed especially after the bonifica could have seriously compromised the preservation of archaeological sites of the Classical and Roman periods. As suggested by the features discovered at Sant'Angelo Vecchio, these normally lay under a thin layer of topsoil.[34]

Prehistory

Evidence for the occupation of the area of Sant'Angelo dates back to the prehistoric period: a major Late Neolithic settlement (site 7/8) was identified 400 m to the west of Sant'Angelo Vecchio, around which gravitated sites of minor magnitude, such as 105 and 560 (*Fig. 13.1*). Human settlement in the area is almost certainly related to the Basento valley's role as an axis of communication connecting the interior to the coast.

To this evidence we can now also add residual impasto and figulina fragments and chipped stones found in close proximity of the all-weather spring of Sant'Angelo Vecchio. The evidence, broadly dated to the latter stages of the Neolithic and possibly the Eneolithic period (mid-5th to 3rd millennium BC),[35] suggests the frequentation of the source at Sant'Angelo Vecchio, placed at the entrance of a possible access linking the river valley to the plateau in the interior (*Fig. 13.1*).[36]

The Sant'Angelo Area in the 7th and 6th Centuries BC

The Basento valley floor maintained its importance also in the historical period, and thanks to its role as the route linking the coast to the interior along the north side of the river, the Greek colonists who founded Metaponto found their way to this part of the chora (*Fig. 1.10*). The sanctuary of San Biagio alla Venella, located 5.9 km from the city and 2 km east of Sant'Angelo Vecchio, was active at the beginning of the colonial phase, indicating that the territory already knew some form of occupation when the Achaian colony was founded. The area was probably also occupied by farmers, even if to a lesser extent than previously thought: the settlement pattern proposed in the publi-

[26] See Ch. 8, "Archaeobotanical Analysis," and Ch. 10, "Observations on the Vegetation and Landscape Dynamics."

[27] See Ch. 7, "Geologic Setting."

[28] See Ch. 13, "Prehistoric Artifacts."

[29] *Survey*, 649. On the Basento Road, partially excavated in the portion crossing the Pantanello necropolis, see *Necropoleis*, 33–39.

[30] *Survey*, 625; 788.

[31] The date bins utilized for the field survey (*Survey*, 104–6) have not been used for this study, in favor of more conventional dating by century and period.

[32] *Survey*, 562–63.

[33] For the scarcity of settlements not only in the lower Fosso Sant'Angelo but also in the upper course of the drainage, where only site 103 (active from ca. 725–225 BC) was found, see *Survey*, 824.

[34] See Ch. 2, "Site Phasing, Stratigraphy, and Site Assemblage."

[35] See Ch. 13, "Prehistoric Artifacts."

[36] *Survey*, 573 and Ch. 13, "Prehistoric Artifacts."

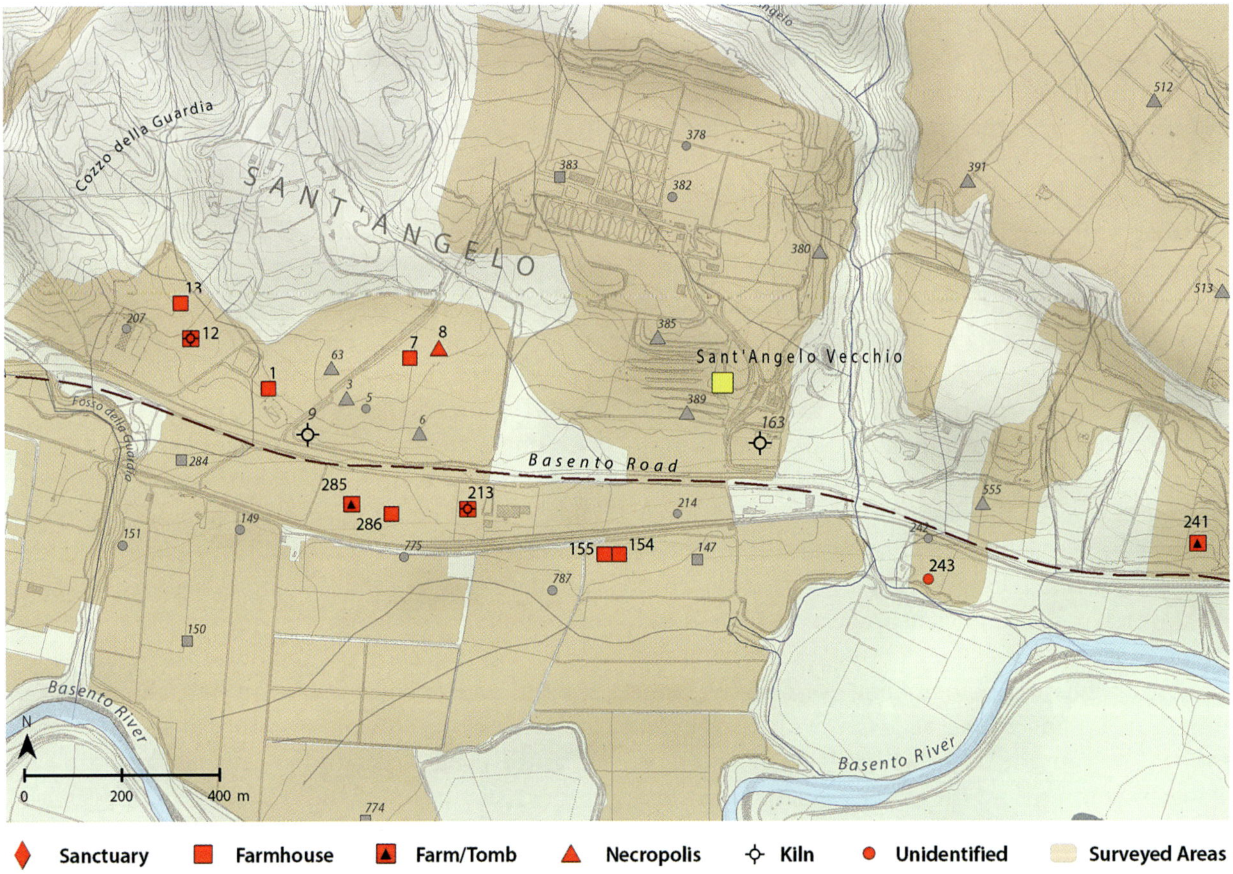

Figure 1.10 The Sant'Angelo area in the late 7th and 6th c. BC. (JT/ICA)

cation of the Bradano-Basento survey[37] for the period between the last quarter of the 7th and the first quarter of the 6th century BC has been recently revised. Only two (285, 800 m west of Sant'Angelo Vecchio, and 530,[38] at the confluence between the Venella drainage and the Basento River) of the nine farmhouses identified during the survey in the Basento valley attributed to this period yielded material datable to this phase.[39]

As a consequence, the existence of the Early Cluster Village of Sant'Angelo becomes evident only in the period between 575 and 525 BC, when a consistent population growth took place in the entire chora.[40] The village consisted of clustered farmhouses, placed at an average distance of ca. 600 m partly lined along the Basento Road (285, 286, 1, 154, and 155) or occupying the sloping valley walls, such as the isolated farmhouse 13, built on the scarp at the foot of the Cozzo della Guardia. Tombs were discovered at site 285,[41] while farmhouse 7 was associated with necropolis 8, located at a distance of 50 m.[42] Site 12, placed near a spring, yielded evidence of ceramic production.

The occupation of the terrace of Sant'Angelo Vecchio overlooking the lower Fosso Sant'Angelo probably took place within the second half of the 6th century BC, when the farmhouse was built on top of the hill.[43] Residual pottery of the first half of the 6th century BC found on the slope between the House on the Hill and the masseria, such as the rim of a Laconian amphora (SAV Amp 01), suggests that the spring in this part of the hill also may have been frequented at this time.

The crisis that affected both the asty (city) and the chora in the last quarter of the 6th and the first quarter of the 5th century BC and its consequent decrease in

[37] *Survey*, 641–76.

[38] *Survey Vol. IV*, map 3, F10.

[39] For the redefinition of the settlement pattern in the Basento River valley between the end of the 7th and the beginning of the 6th c. BC, see Carter 2012. Materials are in Tubelli in *Survey*, 158, no. 1; 159, nos. 3–6.

[40] *Survey*, 677–726.

[41] *Survey*, 682.

[42] Sites 13, 154, 155, 285, and 286: *Survey*, 648–50.

[43] Based on the preliminary survey of the material, Carter dated the farmhouse to the period 575–525 BC (*Survey*, 714). A thorough study of the material demonstrated that the principal moment of occupation is instead the Late Archaic period. See Ch. 2, "Site Phasing, Stratigraphy, and Site Assemblage."

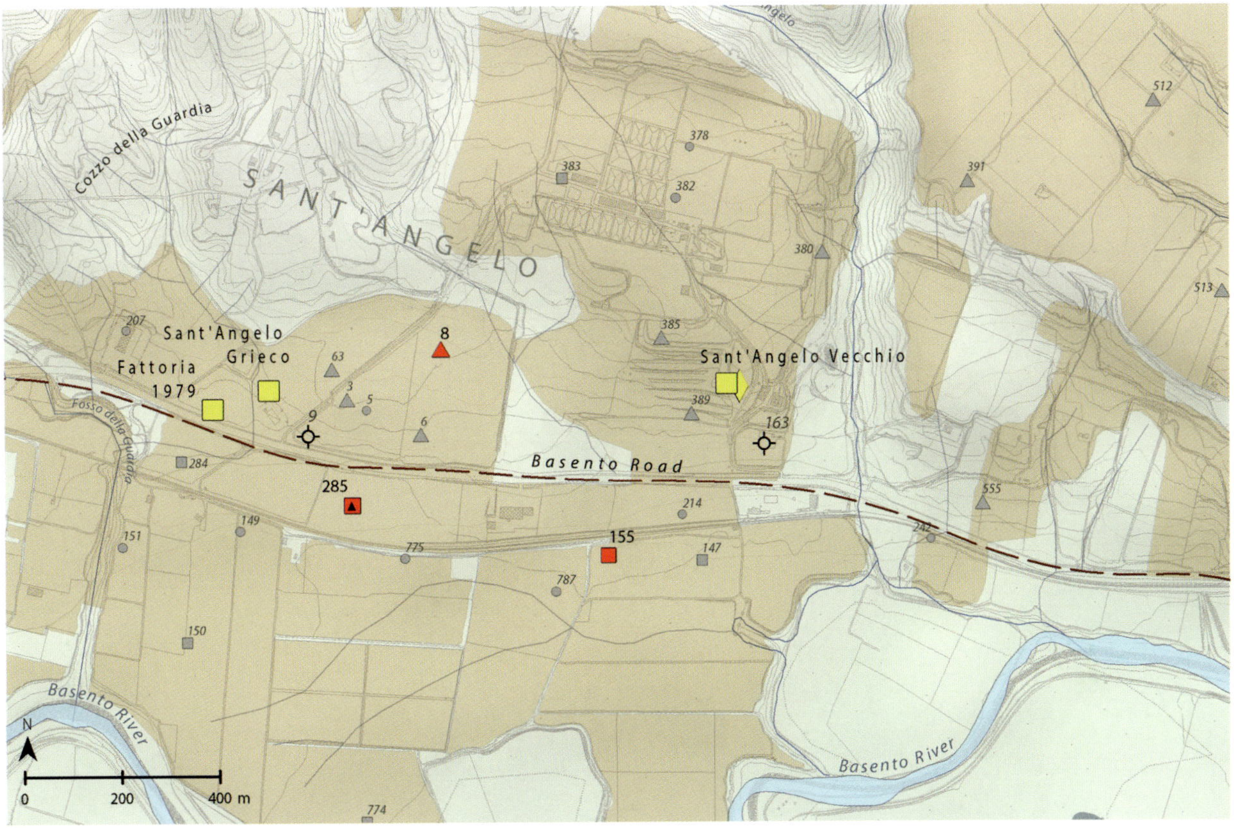

Figure 1.11 The Sant'Angelo area in the late 6th and early 5th c. BC. (JT/ICA)

population also influenced the cluster of Sant'Angelo, which experienced a drop in the number of sites (*Fig. 1.11*).[44] Nonetheless the House on the Hill was still active, and in this period, the existence of a sanctuary near the spring can be established with certainty.[45]

The Sant'Angelo Area in the 5th Century BC

The middle decades of the 5th century BC are characterized by an impressive development both in the asty and in the chora.[46] The number of sites increased in localities where the Division Lines are concentrated:[47] farmhouses were evidently built on the new lands made available by the creation of the canals. In the Sant'Angelo area, a net increase of sites is also evidenced (*Fig. 1.12*). Sant'Angelo Vecchio continued to be occupied by farmers, as indicated by the rural necropolis and by the so-called Pottery Deposit.[48]

Most of the sites identified in the northwest corner of the Sant'Angelo cluster (8, 11, 14, and 285) do not show signs of activity after the last quarter of the 5th century BC. There is, instead, evidence for a shift to the southeast, where site 105, placed near an inlet potentially used as a small port on the old stream of the Basento, became significant.[49] The necropolis 159, active only during the late 5th–early 4th century BC, was probably connected with this site. The following period (425–375 BC) was one of stability in the settlement pattern of the Sant'Angelo cluster,[50] though the Sant'Angelo Vecchio area in particular seems to have been rarely frequented if not abandoned (*Fig. 1.13*). The latest tombs of the rural necropolis date to 420 BC, and the pottery that can be soundly dated within the first half of the 4th century BC is scarce indeed.

The Sant'Angelo Area in the 4th and 3rd Centuries BC

The period between the last quarter of the 4th and the first quarter of the 3rd century BC is the moment of the greatest expansion in the chora.[51] After a long pe-

[44] *Survey*, 730.

[45] See p. 68 in Ch. 3, "The Inscribed Stone."

[46] *Survey*, 745–84.

[47] *Survey*, 769.

[48] See Ch. 2, "Site Phasing, Stratigraphy, and Site Assemblage."

[49] *Survey*, 786–88, fig. 24.4.

[50] *Survey*, 796.

[51] *Survey*, 809–68.

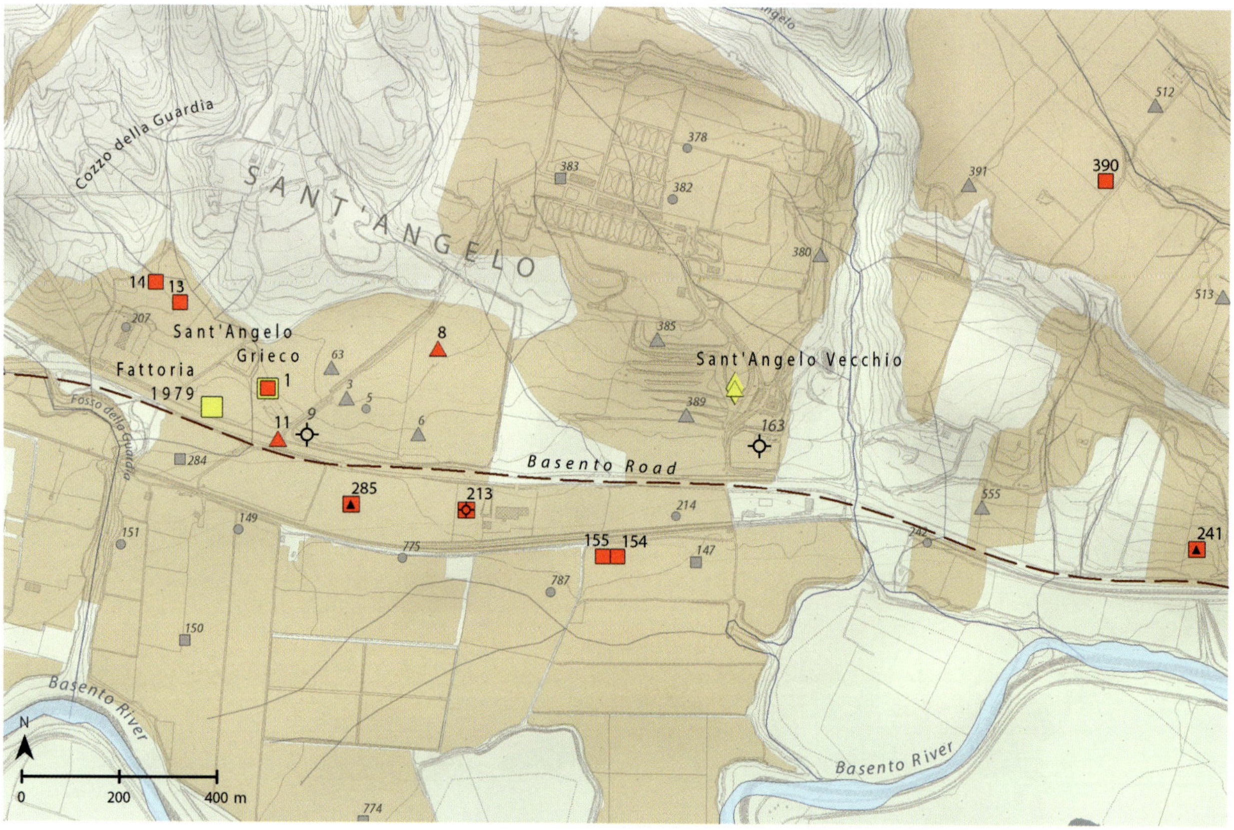

Figure 1.12 The Sant'Angelo area in the 5th c. BC. (JT/ICA)

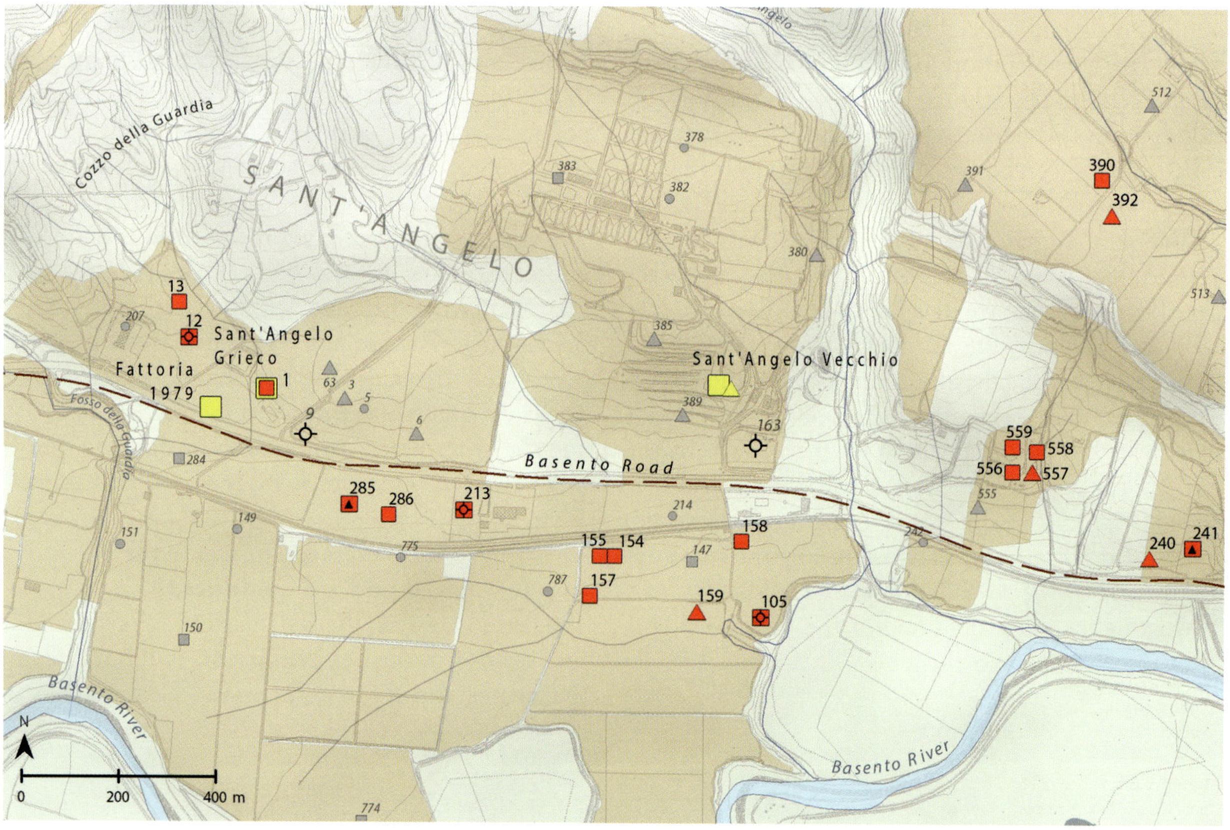

Figure 1.13 The Sant'Angelo area in the late 5th–early 4th c. BC. (JT/ICA)

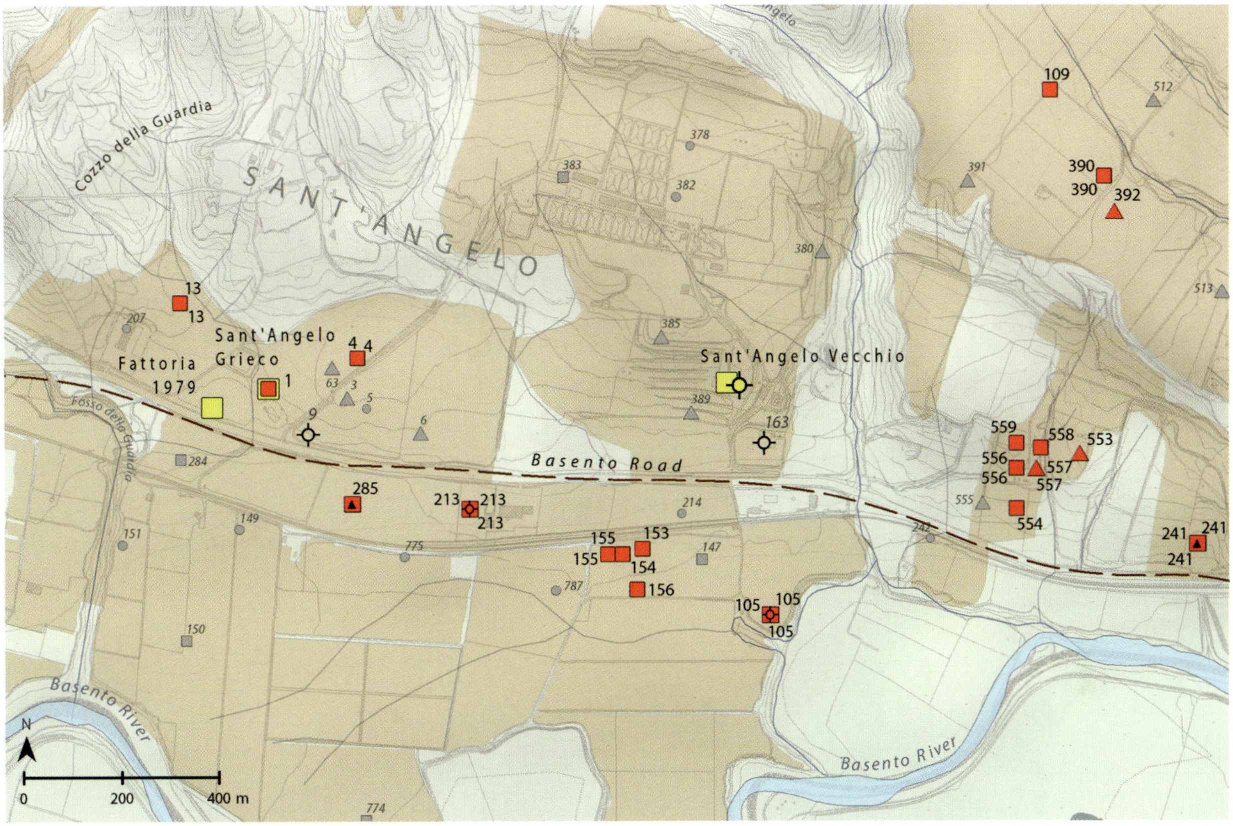

Figure 1.14 The Sant'Angelo area from 375 to 225 BC. (JT/ICA)

riod of stability marked by a tendency to decline, the Sant'Angelo Early Village Cluster gained several sites, but its population rate nevertheless remained lower than that in the period 575–525 BC (*Fig. 1.14*).[52] Sites 105 and 213 are among the most significant sites in this period.[53] Their relative prosperity is likely linked to ceramic production, evidence of which was found at both sites, and to the possible presence of the small port on the Basento, which would have allowed for the distribution of products. In particular, site 105 provided evidence for the production of black-gloss pottery in the late 4th–early 3rd century BC.[54] These two sites, along with Sant'Angelo Vecchio, which in this period revealed the activity of a small terracotta workshop possibly intended for the nearby sanctuary,[55] offer significant evidence for the prevalence of production in the local economy of the area. As in the previous periods, the scarcity of necropoleis associated with sites of this cluster, for which we lack evidence after the first quarter of the 4th century BC, is noteworthy.[56] The discovery of two tombs dated between the late 4th and first quarter of the 3rd century BC at Sant'Angelo Vecchio supports the idea that the people of Sant'Angelo disposed of their dead in as-yet undiscovered places.

Other Excavated Areas

The archaeological research conducted by the Soprintendenza has enhanced our knowledge of the area. Another salvage excavation, also conducted in 1979, brought to light a farmhouse, now destroyed (see Fattoria 1979 in *Fig. 1.14*). It was built on a scarp at the margin of the Basento valley, 30 m from the modern Basentana, and in antiquity was probably situated along the Basento Road.[57] The southern portion of

52 *Survey*, 813.

53 *Survey*, 813.

54 Lanza Catti et al. in *Survey*, 143–44; 213, no. 7.

55 See Ch. 2, "Site Phasing, Stratigraphy, and Site Assemblage."

56 *Survey*, 813.

57 "Nell'immediato entroterra di Metaponto occorre ricordare un intervento di urgenza che ha condotto il gruppo di lavoro di giovani assunti con la legge 285 in loc. Sant'Angelo Vecchio sulla Basentana, a circa 2 km prima del bivio per Bernalda. Sono stati individuati i resti di una fattoria antica, purtroppo manomessa da lavori agricoli, isolando un ambiente con tracce di bruciato, delimitato da strutture a secco, databile per la mancanza di materiali ceramici tra fine del IV e inizi del III a.C." (Lattanzi 1980, 406–7). The excavation was directed by Rosanna Ciriello (Soprintendenza alle Antichità della Basilicata) with the collaboration of Giulia Gioia, Saverio Careccia, Nunzia Armento, Enzo Arcuti, and Vittorio Plati from the cooperative of the Soprintendenza della Basilicata. Restoration of the material was carried out by Alessandro Pesare and Grazia Pinto. On the farmhouse, see also Amore 1992–1993, 86–88.

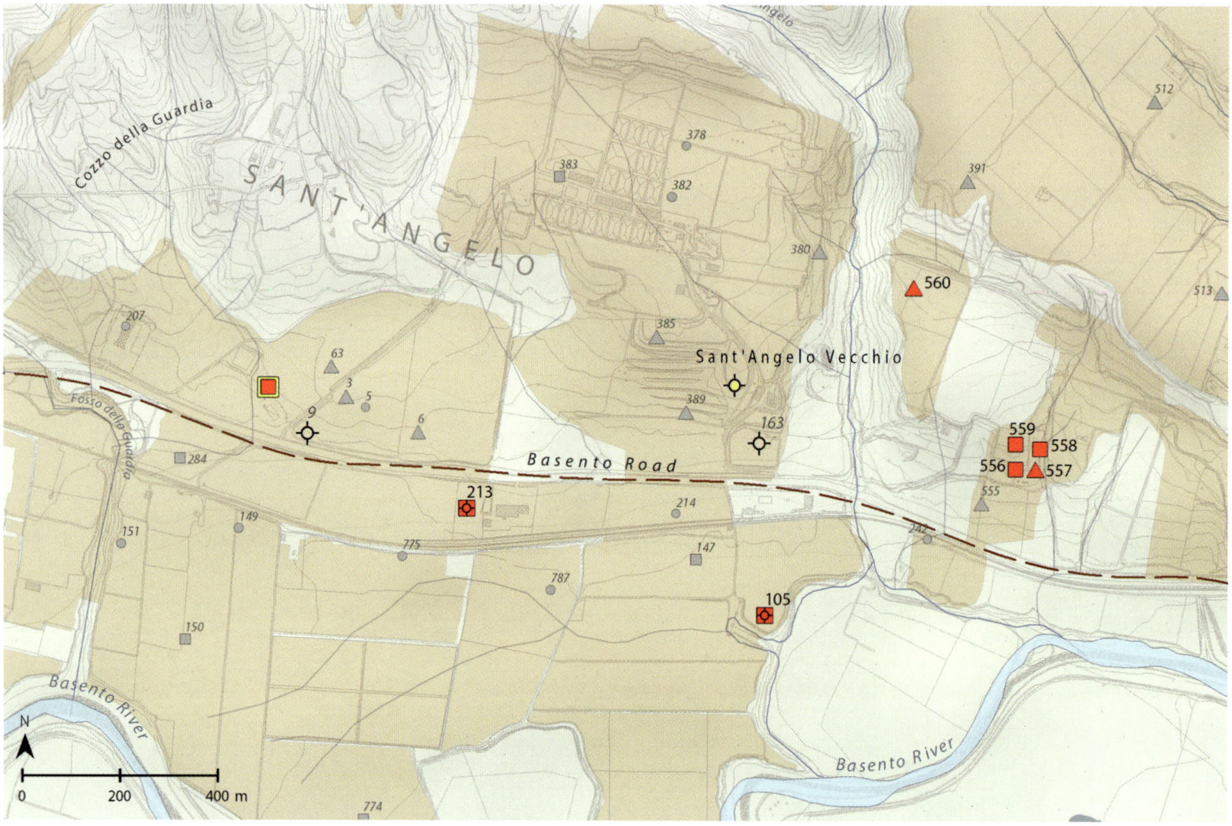

Figure 1.15 The Sant'Angelo area from 175 to 25 BC. (JT/ICA)

the farmhouse had collapsed at the time of the excavation due to landslide, and only an east-west wall delimiting three rooms, one with evidence of a hearth, remained. Not far from this farmhouse, the tomb of a child was discovered. Preliminary analysis of the pottery[58] shows that the farmhouse was occupied mainly in the second half of the 4th to first quarter of the 3rd century BC, and was abandoned within the first half of the 3rd century BC. Ionic-type cups, residual in the contexts, suggest the possibility of a Late Archaic occupation, also implied by the fortuitous nearby recovery of an Attic black-figure amphora of the 6th century BC.[59]

Further west, near the modern Stabilimento Troiano on the opposite side of the Basentana, an additional salvage project conducted in 1980 by the Soprintendenza led to the recovery of scattered material mainly dating to the second half of the 4th century BC. The function of this site remains uncertain.

The Sant'Angelo Area in the Late Republican and Roman Imperial Periods

The 3rd century BC saw a noticeable decrease in the number of settlements in the Sant'Angelo area (*Fig. 1.15*). Only sites 105 and 213, both significantly involved in ceramic production, show signs of activity,[60] and by the end of the 3rd century BC, the area seems to be virtually uninhabited. Occupation was resumed in the late 2nd and 1st century BC, a period in which the territory of Metaponto underwent a substantial change in its organization. The small farmhouses that were the most outstanding feature of the previous

[58] *Survey*, 852–53, fig. 25.30.

[59] The results of the excavation were presented in the exhibition "Archeologia e occupazione giovanile a Metaponto. Un anno di esperienza. Mostra documentaria a cura della cooperative 'Libertà nel lavoro e nella cultura'," held at the Archaeological Museum of Metaponto in 1980. According to a brochure written for the exhibition "lo scavo ha permesso l'identificazione dei resti di una fattoria in gran parte distrutta da lavori di scasso molto profondi. Della struttura originaria erano ancora riconoscibili tre ambienti, il focolare, un canale per il deflusso ed una tomba di bambino. Il materiale recuperato si riferisce essenzialmente a frammenti di ceramica e ha permesso di datare la costruzione della fattoria verso la metà del IV secolo a.C. e il suo abbandono agli inizi del III a.C. Tra la ceramica, le forme più rappresentate sono: ceramica da cucina, bacili, contenitori da dispensa, recipienti da tavola e ceramica fine da tavola, decorata a vernice nera." The author is grateful to Antonio De Siena and Nunzia Armento, who provided a copy of the brochure.

[60] *Survey*, 869–70, fig. 26.1.

periods, increasingly isolated in the 3rd century BC, disappeared by the end of the century. Settlements, now more spread out, changed in nature, reflecting the transformations in economy and land use that accompanied the development of animal husbandry and pastoralism.[61] Sizeable settlements or villas controlling a larger area than previously attested (nonetheless not real latifundia) populated the territory.[62]

In comparison with the previous periods, activity in the Sant'Angelo area shifted toward the southeastern portion in this period. In the west, the Late Republican complex at Sant'Angelo Grieco (1) was built,[63] but no other sites are attested nearby in the same period. The terraces and the river valley around the lower Fosso Sant'Angelo seem to have been more attractive for settlers, who were possibly searching for easy access to clay for pottery and tile production, good pastures, and water. Sites 556, east of Fosso Sant'Angelo and 105, placed near the old course of the Basento were active until the Early Imperial period.[64] These two sites and the workshop at Sant'Angelo Vecchio form a triangle with sides ca. 600 m long.[65] Between 125 and 75 BC, two small, short-lived sites (558 and 559) were added. They may form a small village with 556, which would have been the most significant site in the group. It is possible that this small community ought to be connected with site 105, or alternatively, with the workshop of Sant'Angelo Vecchio, where no evidence for a residence dating to this period was found.[66] A necropolis (560) was also active during this period on the plateau above Sant'Angelo Vecchio.

Pottery found at Sant'Angelo Vecchio suggests that the site may have been occupied as late as the beginning of the 1st century AD,[67] similar to nearby site 105. After this occupation, the area appears to have been abandoned, although evidence for rural dwellings dated to the Imperial period was found at sites 386–388.[68] The very small number of fragments from the sites prevents any sound interpretation.

A few fragments of Late Roman ware (see *Fig. 26.1*) dating to the 6th–7th centuries AD—among which are a possibly imported transport amphora (SAV LR 03)—indicate an occasional frequentation of Sant'Angelo Vecchio during this period. The activity may have been connected with the spring and the nearby farmhouse, site 388, occupied until the first half of the 6th century AD, and thus adds new evidence to the occupation of the territory from AD 300–600.[69]

[61] *Survey*, 914.
[62] Lapadula in *Survey*.
[63] *Survey*, 905–7. Lapadula in *Survey*, 1138. The site was previously occupied by a Greek farmhouse that was abandoned in the 3rd c. BC.
[64] Lapadula in *Survey*, 1137.
[65] *Survey*, 895.
[66] *Survey*, 896; 897–98.
[67] See Ch. 17, "Red Ware," and Ch. 26, "Roman and Late Roman Wares." See Schmiedt and Chevallier 1959, 349–70, and Adamesteanu 1974, 79–81.
[68] Lapadula in *Survey*, 1140, table 32.2.
[69] Lapadula in *Survey*, 1143–45, fig. 32.4.

2
Site Phasing, Stratigraphy, and Site Assemblage

Francesca Silvestrelli

The long occupation of Sant'Angelo Vecchio and its function determined a particular composition of the assemblages, often characterized by a small amount of pottery. Fine wares, the most useful tool in dating archaeological contexts and in establishing a chronological sequence of a site, are quite rare, and with uneven spatial and chronological distribution. With the exception of the vessels found in the rural necropolis,[1] figured pottery is virtually absent at the site. Ionic-type B2 cups and skyphoi with offset rim provide information for the Late Archaic period, and black-gloss ware occurs in Phase 3, mainly from the Pottery Deposit and from the tombs, but is rare in later contexts. For Phases 4–6, we rely mainly on plain and coarse wares and on cooking ware, characterized by a less precise chronology. Valuable information has been conveyed by tiles, usually not considered diagnostic. Nevertheless, the identification of two distinct layers of tile falls, each characterized by a different typology and by the presence of stamps has proved significant. Noteworthy also is the complete lack of coins. Occupation/frequentation phases of the site are summarized below.

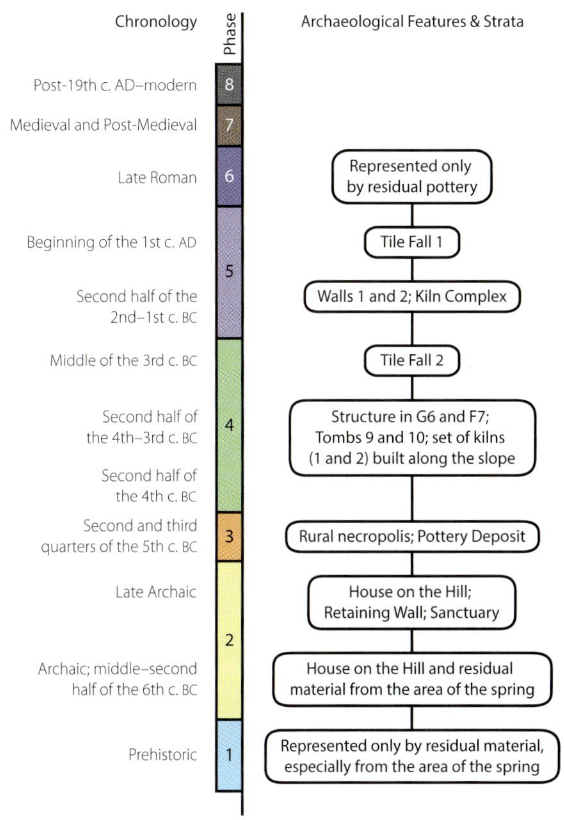

Figure 2.1 Sant'Angelo Vecchio site phases and chronology. (KJS/FS/ICA)

Phase 1: Prehistoric Period. The earliest traces of frequentation at the site go back to the prehistoric period, and consist of residual material,[2] with pottery from the Slope Areas broadly dated to the Neolithic (SAV Impasto 01, SAV Impasto 03, SAV Impasto 04, and SAV Impasto 06–SAV Impasto 08) and Eneolithic periods (SAV Impasto 02 and SAV Impasto 05). Since subsequent occupation probably removed prehistoric layers, the type and length of occupation known at the site in this phase cannot be hypothesized.

Phase 2: Archaic and Late Archaic Period, House on the Hill, Sanctuary, and Retaining Wall. Evidence for Phase 2 is comprised of the farmhouse identified on the hilltop, and by the sanctuary and the Retaining Wall in the Slope Area. The earliest datable fragment from the area of the House on the Hill is a Corinthian convex pyxis of the middle or second half of the 6th century BC (SAV Cor 01). Archaic fine wares of Metapontine production are the most useful markers for the occupation of the House on the Hill in the second half of the 6th and in the first quarter of the 5th century BC (see Assemblage 1, *Table 15.2*, and *Appendix A, Table A1*). Residual transport amphorae found in Sounding D (SAV Amp 01, dated to the first half of the 6th century BC) and from the Late Republican Kiln Complex (SAV Amp 10, second half of the 6th century BC) suggest the possible frequentation of

1 See Ch. 4, "The Tombs at Sant'Angelo Vecchio."

2 See Ch. 13, "Prehistoric Artifacts."

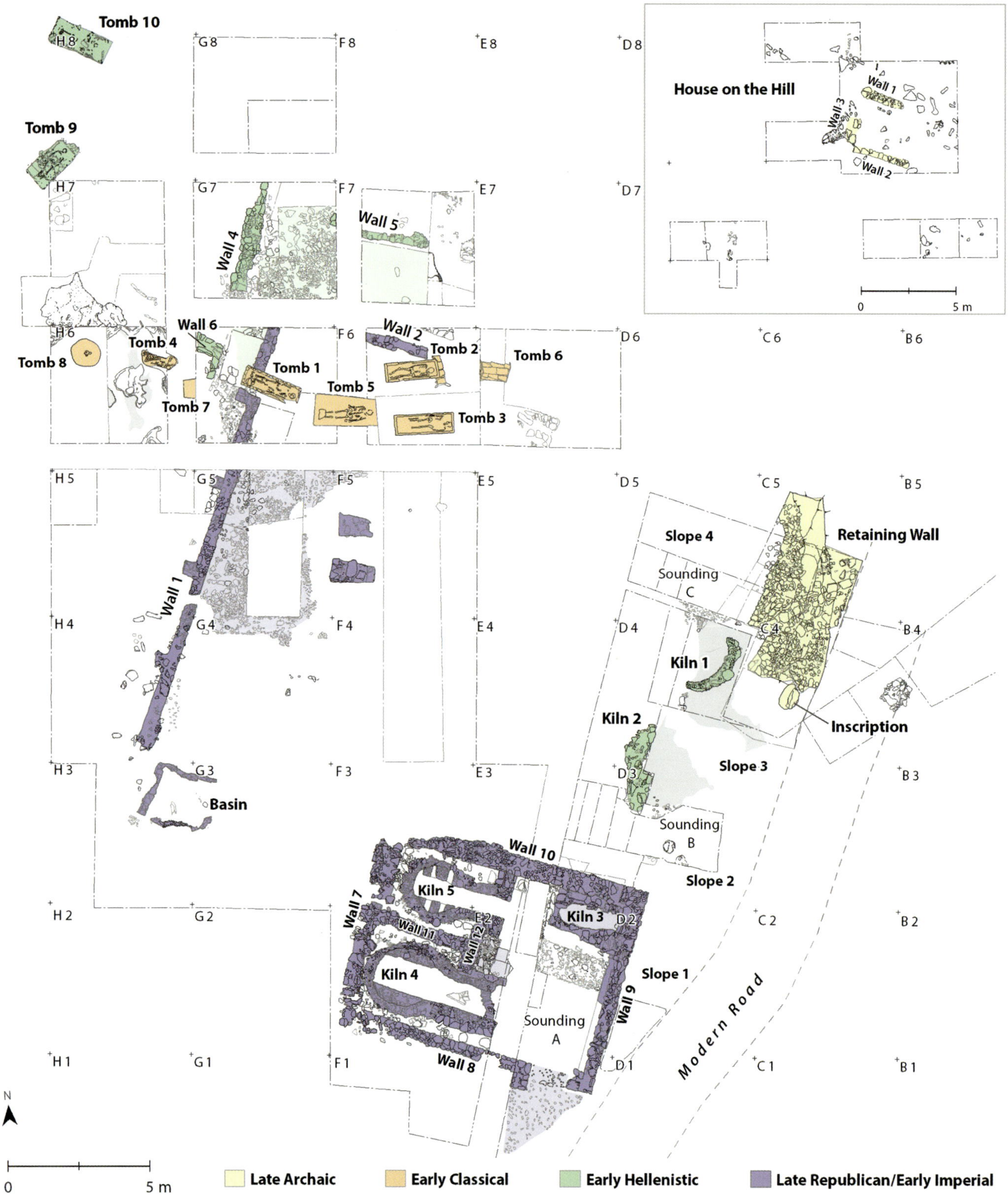

Figure 2.2 Plan of Sant'Angelo Vecchio by phase. (JT/ICA)

this part of the site in the mid- and second half of the 6th century BC, a period for which no archaeological contexts are known. Documentation becomes more frequent in the Late Archaic period, when the existence of the sanctuary can be established on the basis of epigraphic evidence.[3] In the Workshop Area, residual Archaic and Late Archaic material, attested by two antefix molds (SAV AT 05 and SAV AT 07), skyphoi with offset rim, and Ionic-type B2 cups (see *Fig. 2.20*, *Fig. 2.21*, and *Table 15.6*)—to which the hydria SAV BW 41 can also be added—were found, usually in levels overlaying the natural soil, in areas untouched by the Hellenistic workshop. Due to the lack of contexts datable to this phase, the nature of the frequentation cannot be established. The forms occurring could be consistent with funerary assemblages, as also confirmed by the presence of the amphora SAV BG 32, which could suggest the presence of Archaic tombs that have disappeared or are still undiscovered.

Phase 3: Rural Necropolis and Pottery Deposit. Although a few black-gloss fragments from the House on the Hill (SAV BG 34 and SAV BG 35) allow that the occupation of the farmhouse may have lasted into the second or beginning of the third quarter of the 5th century BC, evidence for the occupation of the site in this period is concentrated in the Workshop and Slope Areas, with the tombs ranging in date from 460 BC (Tomb 1) to 430–420 BC (Tomb 4 and Tomb Group 8),[4] and the Pottery Deposit, which yielded Late Archaic and especially Classical material (Assemblage 15; see *Appendix A, Table A2*). Residual pottery from the Workshop Area includes transport amphorae (SAV Amp 06 and SAV Amp 07) and black-gloss shouldered lekythoi and skyphoi with offset and concave rim of the first half of the 5th century BC (Assemblage 3; see *Appendix A, Table A9*), and the rim of a red-figure hydria (not included in the catalog because of its small size, Assemblage 16; see *Appendix A, Table A4*). All these belong to forms that usually occurred in funerary assemblages or in layers outside the tombs, and their presence can be understood in connection with the necropolis. The A-type skyphos SAV BG 21 and the bolsal SAV BG 28, dated to the last quarter of the 5th century BC, constitute the most recent evidence for the frequentation of the site in Phase 3.

Phase 4: Hellenistic Workshop, Tombs 9 and 10, and Kilns 1 and 2. Although some black-gloss vessels from the site belong to types whose chronology extends into the 4th century BC (see, for example, SAV BG 26 and the lamp SAV LA 44, both from the Pottery Deposit, Assemblage 15; see *Appendix A, Table A2*), they occur in Metapontine contexts already in the 5th century BC, when they would probably be more safely dated.[5] After a possible period of abandonment, the area seems to be reoccupied at the end of the 4th century BC, as suggested by pottery found in the Workshop and Slope Areas (especially SAV BG 40, and SAV Amp 21–SAV Amp 23). The lower chronological limit of the occupation can be fixed, based on the diagnostic pottery from the terracotta and tiles concentration in Slope 3 and Sounding C (Assemblage 13; see *Fig. 2.35* and *Appendix A, Table A8*), to the central decades of the 3rd century BC (SAV BG 36 and SAV BG 39; see also SAV BG 41). It should nonetheless be noted that the chronological seriation of pottery for the 3rd century BC still bears some uncertainties, and some of the cooking ware types encountered in the same assemblages and traditionally dated within the first half of the 3rd century BC also occur in contexts in which the chronology could extend into the second half of the 3rd century BC.[6]

Phase 5: Late Republican Workshop and Kiln Complex. After the abandonment of the Hellenistic workshop, the area apparently remained unused until occupation was resumed with the construction of a new workshop. Evidence for this phase consists of the shed in the Workshop Area and the Kiln Complex. Although no contexts related to the construction of these features was revealed, the presence of grey ware[7] and other pottery of the late 2nd and 1st centuries BC (see also the lagynoi SAV Lag 01–SAV Lag 04 and the Metapontine amphora SAV Amp 34, the latter dated to the 2nd century BC) suggests the moment in which the site was reoccupied, a chronology that finds indirect confirmation in the tiles related to the roofing of the structures of this phase. The stamp AVF found on the tiles connects them to the Aufidii, whose presence in the Metapontine territory could be referred

[3] See p. 68 in Ch. 3, "The Inscribed Stone."

[4] See Ch. 4, "The Tombs at Sant'Angelo Vecchio."

[5] See, for instance, the C-type skyphos from tomb 597, Proprietà Mutinati, found in association with a lebes gamikos by the Pisticci Painter and comparable to SAV BG 26: Osanna 2008, 928, pl. XX, 2.

[6] See, for example, SAV CkW 02 and SAV CkW 35 comparable to vessels from Bosco Andriace: Giardino 2015, 605, fig. 31.

[7] See Ch. 18, "Grey Ware."

to the same period.[8] Occupation of the site probably lasted until the beginning of the 1st century AD, as indicated by Early Imperial fragments found in the area (see *Fig. 26.1*).

Phase 6. The Late Roman period is represented by pottery fragments (*Table 26.2*) found in a secondary deposition in Sounding D and dated to the 4th–5th century AD (SAV LR 01), possibly extending into the Early Middle Ages (SAV LR 03, 5th–6th century AD). In the absence of structures and layers belonging to this phase, their presence suggests a frequentation of the area possibly in connection with the all-weather spring and of the occupation of the nearby sites 384, 387, and 388.[9]

Phases 7 and 8. Medieval and Post-Medieval frequentation of the site is connected to the presence of the church, the masseria, and on the top of the hill, of the 19th-century farmhouse.[10] The area occupied by the Greek and Roman evidence was probably, in these phases, exploited for cultivation or as pasture, and with the exception of the Post-Medieval green and yellow glazed basin, SAV PMP 01, no other datable finds were detected.

Stratigraphy and Phasing by Area: The Hillside

Most of the archaeological evidence from Sant'Angelo Vecchio is concentrated in the central part of the hillside, midway between the House on the Hill and Masseria Sant'Angelo. Here, the north-to-south slope drops to the east, gently at first, and then sharply, ending in the spring area. Excavations undertaken in 1979–1981 (see *Fig. 1.3*) exposed a rural necropolis of the Classical period,[11] overlaid by two ceramic workshops (the Workshop Area) with their sets of kilns at the edge of the slope (the Kiln Complex) and on its side (Slope Areas 1–4). Evidence for the presence of a sanctuary certainly functioning in the Late Archaic period was also found at the eastern edge of the Slope Area.[12]

The archaeological evidence is unevenly preserved due to erosion and the intensive agricultural activity that affected this part of the site before the beginning of the excavation. Further damage took place in 2001, when the slope was lowered for the construction of a path leading to the top of the hill, which resulted in the destruction of the excavated features in squares H3–H7 and in the southern half of G5 and F5 (*Fig. 2.4*).

The Workshop Area

Location and excavation. The Workshop Area lies west of the Kiln Complex and the Slope Area, on the highest elevation of this part of the hillside (*Fig. 2.3*, *Fig. 2.4*, and *Fig. 2.5*). The area, which extended for a total of 20 x 30 m, was excavated by grid squares; 1-m baulks were left on the east sides of H6 and H7, the north sides of G6–G8, and the north and south sides of F6 and F7, where a thicker deposit was preserved (see *Fig. 1.6*). Erosion accounts for the lack of archaeological deposits in the area south of square H3, where the sandy natural soil appears on the surface, and where, therefore, no excavation was undertaken. At the extreme north (G8, excavated to a depth of ca. 1.5 m), a sterile uniform dark yellowish brown layer (Munsell 10YR 3/4) of colluvial formation covered the natural soil, identified at -6.01 m, and sloped south; *Fig. 2.5*), suggesting that the evidence extended south of it.

Excavations uncovered features belonging to a Late Republican ceramic workshop comprised of a clay-settling basin (square H3) and a structure bordered to the east by the tile fall of the roof (squares H4–5, G5–6, and F5–6; Phase 5),[13] remains of a Hellenistic building with its collapsed roof and two tombs (G6, west of Wall 1, H7–F7, H8; Phase 4),[14] and, finally, a necropolis of seven tombs and a tomb group (H6–E6; Phase 3).[15]

Stratigraphy. The removal and disturbance of the stratigraphy has been observed in the western and in the southern parts of the area. In squares H6 and H7 (*Fig. 2.4* and *Fig. 2.21*), both excavated to a depth of ca. 1.5 m, the original topsoil had probably been leveled off in modern times and the deposit of pottery and bones from Tomb 4 and Tomb Group 8[16] (second half of the 5th century BC; *Fig. 2.4*) appeared on the surface, constituted by a sandy soil 0.1 to 0.9 m high overlying the irregular sandstone bedrock.

[8] See Ch. 32, "Roof Tiles and Bricks," 478ff.

[9] See Ch. 1, "The Many Lives of a Rural Site: Sant'Angelo Vecchio," and Ch. 26, "Roman and Late Roman Wares."

[10] See Ch. 6, "Sant'Angelo Vecchio in the Medieval and Post-Medieval Periods."

[11] See Ch. 4, "The Tombs at Sant'Angelo Vecchio."

[12] See Ch. 3, "The Structures at Sant'Angelo Vecchio."

[13] See p. 77 in Ch. 3, "The Workshop Shed."

[14] See p. 71 in Ch. 3, "The Early Hellenistic Workshop."

[15] See p. 85 in Ch. 4, "The Rural Necropolis of the Classical Period."

[16] See p. 85 in Ch. 4, "The Rural Necropolis of the Classical Period."

Figure 2.3 Part of the Workshop Area at Sant'Angelo Vecchio during excavation in 1979, looking W. In foreground, square F6; in middle ground, squares G6 (*center*) and G7 (*right*); and in background, square H6. (CW/ICA)

The natural layer was explored in a small quadrangular trench excavated at the center of the same square (H6, Level 1, batt. 6). At a depth of -6.99 m, the sandstone bedrock formed a wide basin, into which a darker sandy soil mixed with pebbles continued to filter (*Fig. 2.10*), allowing for natural drainage of the hill. With the exception of a few fragments found on the surface and immediately below dating to the Late Archaic, Hellenistic, and Late Republican periods (*Fig. 2.21*), the sandy soil proved to be sterile.

The same sandstone formation covered by sandy soil, which constitutes the natural soil of the area,[17] also stretched into square H7, as well as up- and downhill in the area west of the excavation. Here, it has been recently exposed to surface by the leveling of the topsoil that resulted from the planting of an olive grove west of the excavation.[18] Rills of flowing water from rainfalls in 2010 also uncovered Tombs 9 and 10, which date to the early 3rd century BC (squares H8–9 and I8; *Fig. 2.4* and *Fig. 2.21*).[19] The same layer appeared on the surface west of Wall 1, in squares H4–5 (where plow marks giving evidence of the disturbance of the area were observed in Level 1, batt. 2 (*Fig. 2.4* and *Fig. 2.21*), G5, and further south, to the east of Wall 1 in squares G3–4 (*Fig. 2.22*).

East of Wall 1 and in the northern area (squares G6–7, F6–7 and E6), in contrast, a deposit ranging in height from 0.9 to 1.2 m high separated the surface from the natural soil. This deposit preserved a complete stratigraphic sequence of the tile fall of the structure defined by Walls 1 and 2 (G5–6, F5–6), the destruction layer and tile fall of the feature defined by Walls 4 and 5 (squares G6–7, F7), and the natural soil in which the tombs of the rural necropolis were set (*Fig. 2.4*). Due to the complex stratigraphic sequence, it will be described by phase.

Phase 3. Evidence for the earliest occupation of this part of the hillside is concentrated in squares H6–E6. Here, seven tombs (1–7) and Tomb Group 8, aligned on a west-to-east slope, were exposed (*Fig. 2.12* and *Fig. 2.13*).[20] The tombs were all set in the natural soil, or just above it. With the exception of Tomb 1, which was covered by Tile Falls 1 and 2 and Wall 1 (*Fig.*

[17] For a detailed discussion of the geologic setting of Sant'Angelo, see Ch. 7, "Geologic Setting."

[18] See p. 154 in Ch. 7, "Geology of the Sant'Angelo Vecchio Archaeological Site," and *Fig. 10.3*, zone 3.

[19] See p. 91 in Ch. 4, "The Hellenistic Tombs."

[20] See p. 85 in Ch. 4, "The Rural Necropolis of the Classical Period."

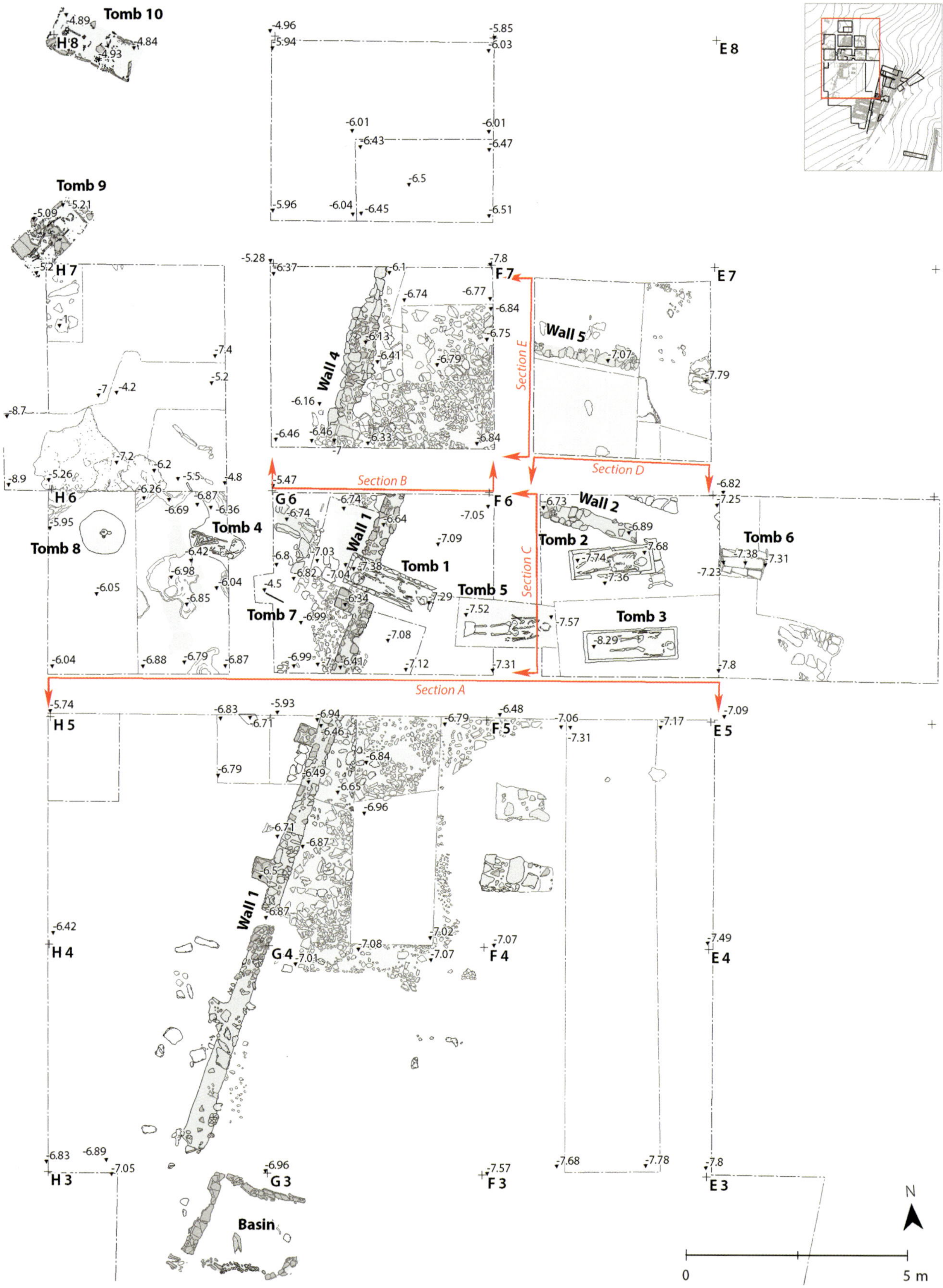

Figure 2.4 Detailed plan of the Workshop Area. (JT/ICA)

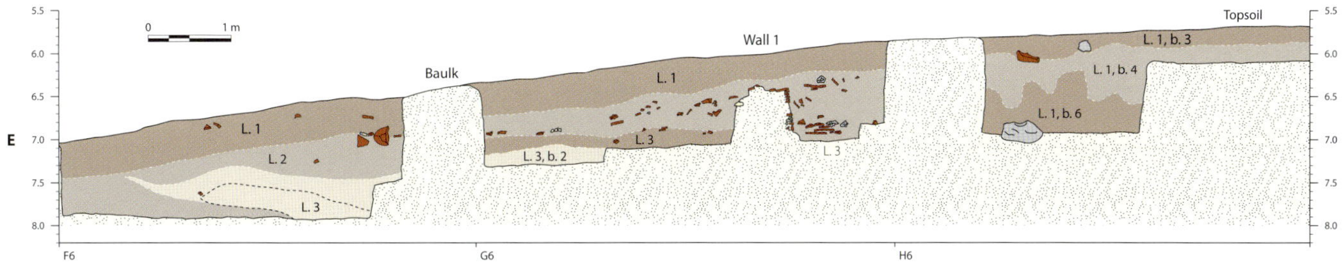

Figure 2.5 Section A, E-W section through S edge of squares H6, G6, and F6. (KJS/ICA)

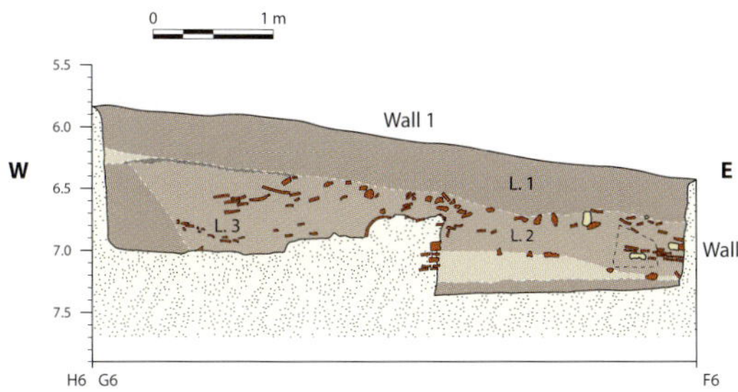

Figure 2.6 Section B, E-W section through N part of G6. (KJS/ICA)

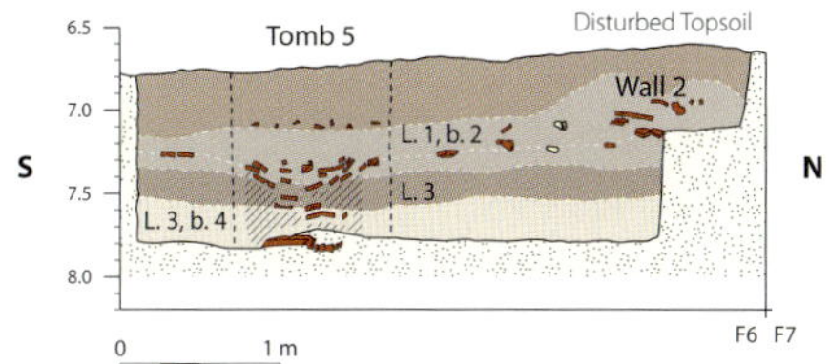

Figure 2.7 Section C, N-S section through W part of F6. (KJS/ICA)

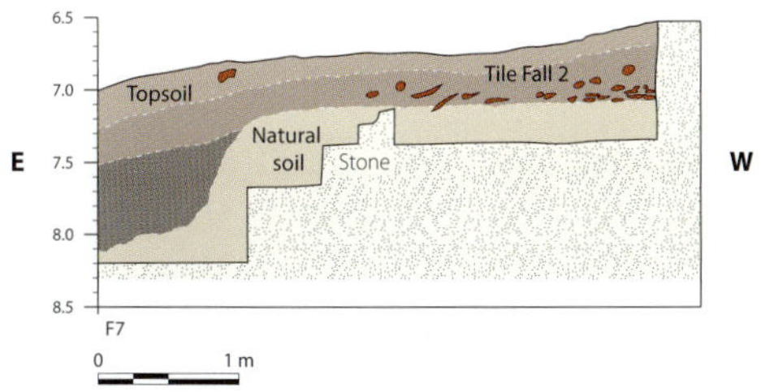

Figure 2.8 Section D, E-W section through S part of F7. (KJS/ICA)

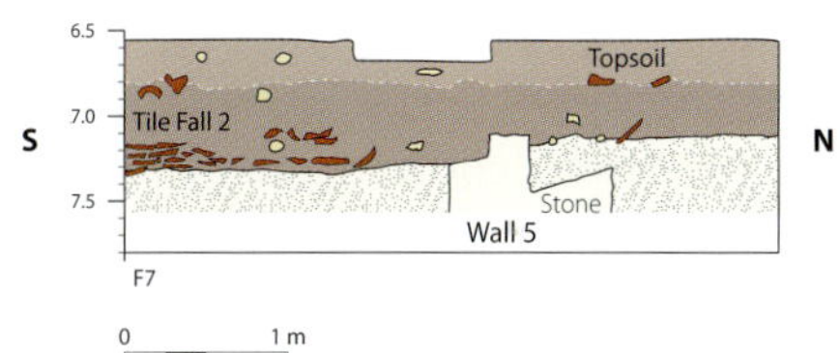

Figure 2.9 Section E, N-S section through W part of F7. (KJS/ICA)

2.12),[21] they appeared on the surface (Tomb 4 and Tomb Group 8) or beneath the topsoil and colluvial layers (Tombs 2, 3, and 6; F6, Level 2, batt. 3–4 and E6, lower half of Level 1, batt. 1; *Fig. 2.21* and *Fig. 2.23*), ranging in elevation from ca. -6.05/7.00 (Tomb Group 8 and Tomb 4), to ca. -7.20 m (Tomb 5), to -7.31 (Tomb 6).

Due to subsequent transformations, the original ground level of Phase 3 cannot be safely established. Nonetheless, if the stone slabs found in Tombs 5 (*Fig. 2.13*) and 7 are correctly interpreted as grave markers, their position can be used to estimate the original elevation of the soil covering the burials. In Tomb 7, the grave marker was at -6.45 m, ca. 0.4 m above the tomb floor. Further east, the uppermost part of the grave marker for Tomb 5 (which does not seem to be preserved to its original height) lay at ca. -7.20 m, ca. 0.5 m above the floor level of the tomb and 0.2 m above Tomb 3 (*Fig. 2.13*). Tombs 2, 3, 5, and 6 were evidently not visible at the time the Late Republican shed was constructed, since Wall 2 was built 0.4 m above the tombs (*Fig. 2.14*). The layers covering the burials in F6 yielded no material. Instead, 5th-century BC pottery was found on the surface of the natural soil, into which Tomb 6 was dug (E6, Level 2, batt. 3–4; *Fig. 2.23*). The tombs and their assemblages are described in detail in Chapter 4, "The Tombs at Sant'Angelo Vecchio."

Phase 4. The ubiquitous presence of Hellenistic material across the Workshop Area notwithstanding,[22] evidence for its occupation in Phase 4 is found only

[21] See below, p. 27, "Phase 4."

[22] See p. 71 in Ch. 3, "The Early Hellenistic Workshop," and p. 91 in Ch. 4, "The Hellenistic Tombs."

Figure 2.10 Natural drainage basin in square H6. (CW/ICA)

in a limited part: to the west of Wall 1 in G6, and outside the northwest corner of the Phase 5 shed in G7 and F7, where Late Republican frequentation was less intense (*Fig. 2.4*). Evidence of the Hellenistic Tile Fall 2 is lacking within the Late Republican workshop, likely due to the removal of the tiles for use in the construction of Walls 1 and 2.[23] The main feature for this phase is a poorly preserved structure defined to the west by Wall 4, revealed beneath topsoil and possibly continuing south as Wall 6, and to the north by Wall 5 (squares G7–F7; *Fig. 2.4* and *Fig. 3.11*). The junction between the walls was not preserved, and their connection, although likely, remains hypothetical. Evidence of erosion was observed south of G6 and prevents us from knowing whether and to what extent the Hellenistic workshop continued.

The western limit, however, was reached. It is likely that when the Hellenistic shed was constructed, the original ground of the hillside was leveled to create an even surface for the new building. The cut in the natural soil extended into the westernmost part of squares G6 (*Fig. 2.6*) and G7. This area to the west of Wall 4 consisted of a heavy dark yellowish soil (Munsell

Figure 2.11 Plow marks in squares G4–G5.

[23] See p. 77 in Ch. 3, "The Workshop Shed."

Figure 2.12 The Workshop Area, square G6 in 1979, looking E. The border of Tomb 1 appears on the surface below L. 1, b. 4, covered by Wall 1 and Tile Fall 2. In the background at the E end of the square, Tomb 5 prior to its excavation. (CW/ICA)

Figure 2.13 The Workshop Area, square F6, Tomb 3 in 1979, looking W. (CW/ICA)

10YR 5/6) with virtually no finds (Level 2, batt. 1–2 W of Wall 4; *Fig. 2.20*) covering natural soil, which appeared at a depth of -6.46 m, ca. 0.3 m beneath the preserved upper surface of the wall. The slope recovered its original profile in square G8. This earthwork explains both the location and the disturbance of Tomb 7 (Phase 3), found at -6.85 m (*Fig. 2.4*),[24] and the stratigraphic position of Tomb 1, found directly beneath the Hellenistic tile fall and the Late Republican Wall 1 (*Fig. 2.12* and *Fig. 2.22*).

A tile fall (Tile Fall 2, Assemblage 4) covered an area, defined by Walls 4 and 5, of approximately 8.5 x 6.0 m, mainly in square G7 but also extending for a maximum width of 1 m along and below the western edge of Wall 1 and directly above the western end of Tomb 1 (G6, Level 3, batt. 1; *Fig. 2.20*). It was deeper on its western side, as verified in G6, where it was first identified in the section of the north baulk at -6.87 m. It continued down to a level of -7.20 m, for a total depth of ca. 0.3 m. In F7, the tile fall was identified at -7.09 m and thinned to disappear at a lowest elevation of -7.25 m, about 1.5 m from the east edge of the square. No floor level was identified, and Tile Fall 2 was determined to be resting on the natural soil.

[24] See p. 106 in Ch. 4, "Tomb 7."

Figure 2.14 Square F6, Wall 2, and Tomb 2 in 1979, looking N. The wall, preserved only at the lowest foundation level, was set on the colluvial layer that covered Tomb 2, 0.4 m above it. (CW/ICA)

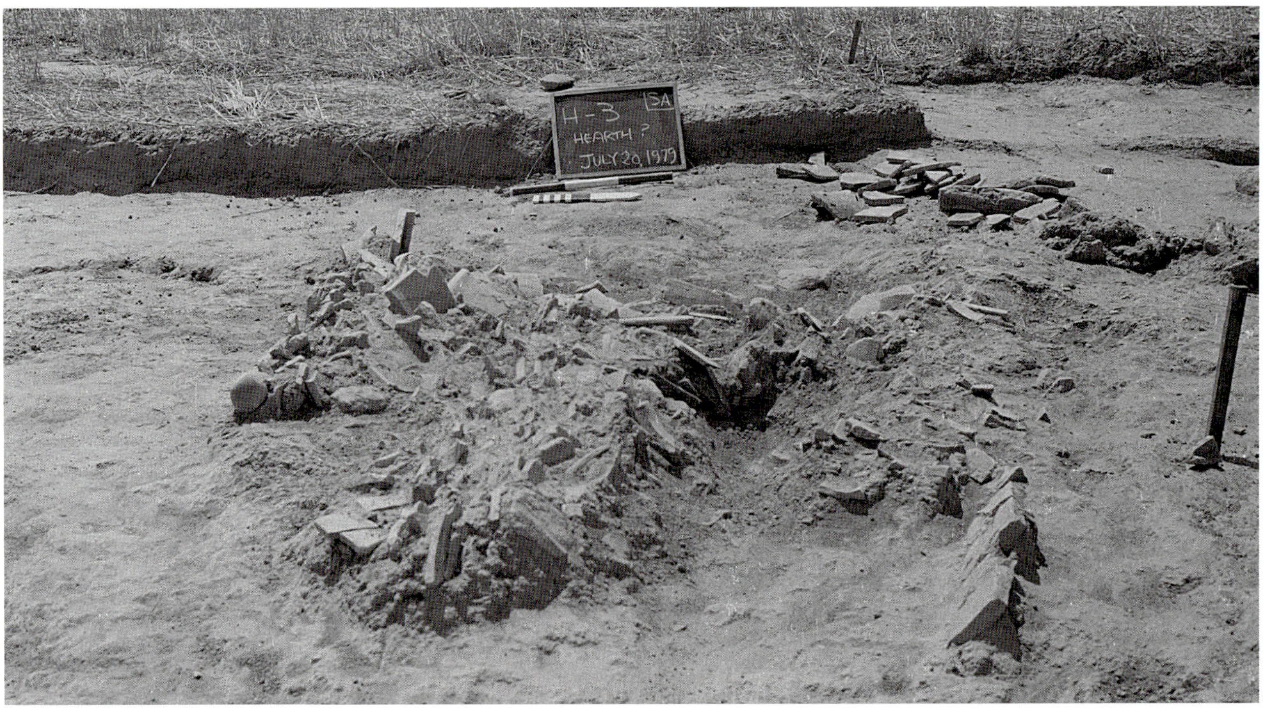

Figure 2.15 Square H3, the tile concentration in the clay-settling basin in 1979, looking W. (CW/ICA)

Figure 2.16 Square G5, Tile Fall 1 in 1979, looking N. (CW/ICA)

Figure 2.17 Wall 1 in square H4 in 1979, looking N. (CW/ICA)

To the north of Wall 5, in the area outside the building, only a few scattered stones and tiles appeared. The eastern side of the structure was likely built at the edge of the slope, as indicated by the natural sandy soil that dropped sharply to the east of Wall 5 to a depth of ca. -7.90 m (*Fig. 2.8*). The difference in elevation was filled by a layer of dark yellowish brown (Munsell 10YR 4/6) sandy soil mixed with small river pebbles and a cluster of rocks (Level 3, batt. 2–9). This layer might indicate a manmade terrace to level the ground surrounding the structure. It yielded pottery of the late 4th century BC together with residual fragments of the Late Archaic and Early Classical periods, connected to the active period of the Classical necropolis (*Fig. 2.20*).

To the west of Wall 4, remains of the destruction layer of this structure may have been identified. In squares H7, G6, and G7, beneath the topsoil—which here ranges in depth from -5.83 m to -6.09/6.17 m—a layer of reddish, burnt soil mixed with fired mudbricks was identified (Assemblage 16, *Fig. 2.6*, and *Fig. 2.20*). In G6 (where this layer was excavated as Level 1, batt. 4), it covered a tile concentration that abutted the sloping natural soil (Level 1, batt. 5, W of Wall 1), yielded only Hellenistic material, and lay atop Tile Fall 2 and Wall 6.

Phase 5. The key features of the area in Phase 5 are the structure defined to the north by Wall 2 (squares F6–G6), to the west by Wall 1 (squares H4–G6), and to the south by the clay-settling basin in square H3 (see *Fig. 3.19*), which marks the southwestern limit of the excavated area.[25] The northern end of the building lay over the Hellenistic shed, taking advantage of the leveled ground on which it had been constructed.

Wall 1 appeared beneath a layer of topsoil progressively decreasing from north to south to a depth ranging from ca. 0.4 m (G6, Level 1, batt. 1–3, W of Wall 1, and batt. 1–5, E of Wall 1; see *Fig. 2.6*) to 0.1 m (H4, Level 1, batt. 1; *Fig. 2.17*). It extended for a length of ca. 15.0 m, was preserved to a maximum depth of 0.5 m, and was mostly set in the natural soil. In square H5, the wall rested on the sandstone bedrock (here appearing at a higher elevation), while in square G6 it was built over and partly into the west

[25] See p. 77 in Ch. 3, "The Workshop Shed."

Figure 2.18 Square G7, Tile Fall 1, and Late Republican terracotta gutter pipes in 1979, looking N. (CW/ICA)

side of Tomb 1 (Phase 3),[26] and on top of the tile fall of the Early Hellenistic structure (Tile Fall 2; see *Fig. 2.12*). No foundation trenches were observed.

Perpendicular to Wall 1, Wall 2 was exposed in square F6 (*Fig. 2.14*), where it appeared beneath the topsoil at ca. -6.73 m (see *Fig. 2.7*). It continued to the west into square G6, where its southern face is visible in the north section of the baulk (see *Fig. 2.6*).

A heavy tile fall (Tile Fall 1, Assemblage 3) lay within the structure over a maximum area of 10.3 x 4.5 m. The tile fall was more dense in the northwest corner of G5 (*Fig. 2.16*) and in G6 (ca. 0.5 m, to a depth of ca -7.00 m), where it had been protected by a deeper layer of topsoil. The tile fall diminished in depth at its southern and eastern margins (G4 and F5, 0.1 to 0.3 m, -7.01), where it may have been removed by plowing or erosion.

The presence in square H4 of a pilaster meant to support the covering of the building suggests that Tile Fall 1 may have extended south of the preserved limit in G4. Square G4 was excavated to a depth of ca. 0.30 m, and plow marks were found on the surface of the remaining strip of Tile Fall 1 (see *Fig. 2.11*). This agricultural activity likely disturbed and removed the occupation layers of Phases 4 and 5. The theory is supported by the chronologically mixed material dating to the Early Imperial and Hellenistic periods (*Fig. 2.21*) that was yielded by the topsoil overlying and abutting the western edge of Wall 1 in square H4.

No floor levels were definitively recognized below Tile Fall 1. In G6 to the east of Wall 1, however, a thin layer of soil (Level 2, batt. 4; *Fig. 2.22*) that separated the tile fall from the covering of Tomb 1 and yielded the Late Republican pot SAV CkW 40, could potentially have been the remains of a floor.

In F5, the removal of the topsoil and of the strip of Tile Fall 1 exposed sandy natural soil in which a stone and tile cluster (*Fig. 2.4* and *Fig. 2.19*) was noted on the eastern margin. This cluster may have served as a foundation for a wooden post that supported the roof of Wall 1.[27] Evidence of plowing or earth removal was also observed.

A tile concentration directly over Tile Fall 2 was exposed to the north of Wall 2 in square G7 (Assem-

[26] See p. 71 in Ch. 3, "The Early Hellenistic Workshop," and p. 108 in Ch. 4, "Tomb 1."

[27] See Ch. 32, "Roof Tiles and Bricks."

Figure 2.19 Squares F5 and G5 with cluster of stones and tiles in 1979, looking W. (CW/ICA)

blage 17; *Fig. 2.18* and *Fig. 2.20*). At the southeastern corner, at ca. -7.00 m, the tile concentration formed a semicircle, and the soil immediately over these tiles appeared burnt, as if from a hearth. Fragments of gutter pipes appeared between this circle of stones and Wall 4.

In the southern limit of the excavated area (H3), which was dug to a depth of around 0.30 m, a thin layer of plowed topsoil (Level 1, batt. 1) covered a concentration of tiles (Level 1, batt. 3, Assemblage 2; *Fig. 2.15* and *Fig. 2.22*) that filled the interior of a 4.5 m^2 rectangular clay-settling basin made of upright tiles. The basin (only the north, south, and west sides of which were preserved) was set within the natural soil, which also appeared on the surface of square G3, where erosion and plowing activity compromised the preservation of archaeological deposits. A single battuta was undertaken in square G3 (Level 1, batt. 1), and neither features nor structures were detected.

Excavated Units and Assemblages from the Workshop Area

Assemblage 4
Tile Fall 2 (*Appendix A, Table A3; Fig. 2.20*)
Phase 4

Location and Lots: F7, Level 2, batt. 1, 4–6 (SA 80-6PL; SA 80-8PL; SA 80-21PL). G6, Level 3, batt. 1, W of Wall 1 (SA 79-1002PL, SA 79-1017PL). G7, Level 2, batt. 5–6, E of Wall 4 (SA 79-997PL).

Excavation and Stratigraphy: The tile fall was identified in G6, where it extended west of Wall 1, and in G7 and F7. In G7, Tile Fall 2 lay beneath the Late Republican tile concentration (Assemblage 17). Neither the existence of a layer dividing the two features nor its relation to Walls 4 (where part of the overlying tile concentration was left in situ) and 5 were mentioned. The tile fall extended west of Wall 4 for a length of ca. 3 m and to a depth of -7.05 m, with an estimated thickness of ca. 30 cm. In F7, Tile Fall 2 was exposed south of Wall 5 and extended to an area of ca. 2.0 x 2.30 m, with a maximum depth of ca. 0.15 m (ca. -7.07 m to -7.22 m). It lay beneath a

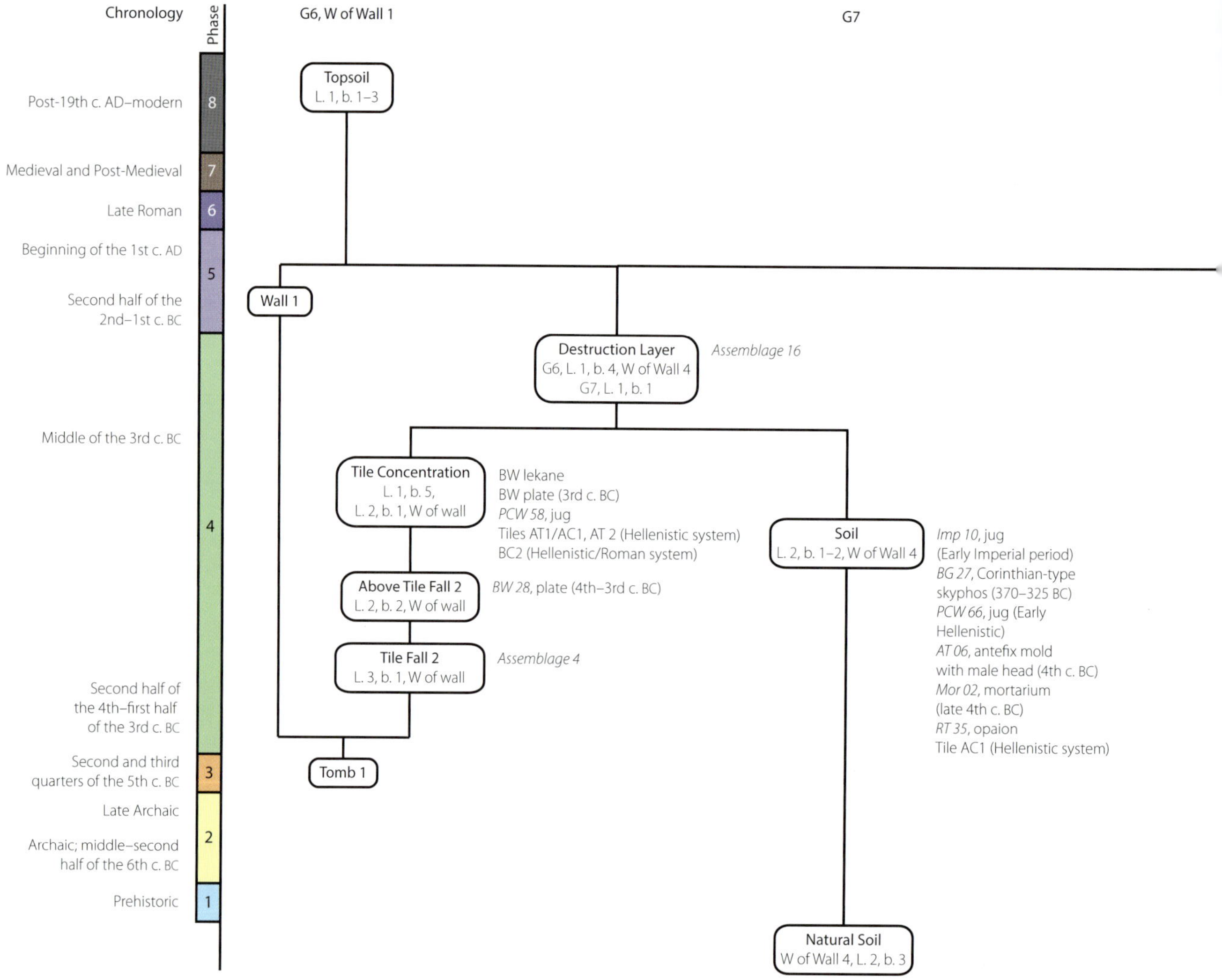

Figure 2.20 The Workshop Area to the W of Wall 1 (square G6) and E and W of Wall 4 (squares G7 and F7), phased matrix. NB: The phased matrices in this chapter (see also *Fig. 2.21*, *Fig. 2.22*, *Fig. 2.23*, *Fig. 2.30*, *Fig. 2.35*, *Fig. 2.39*, and *Fig. 2.42*) are the result of post-excavation reconstruction intended to give a visual representation of the stratigraphic sequence observed in the area. (KJS/FS/ICA)

colluvial layer (Level 1, batt. 1–3 and Level 2, batt. 2, Munsell 10YR 3/4, dark yellowish brown) ca 0.20 m deep, which yielded material of the 4th–3rd c. BC, as well as a Late Republican grey-ware fragment testifying to the frequentation of the area in Phase 5. No floor was recognized (although it may have been removed in Level 2, batt. 6, where the few pottery fragments from this level were concentrated) and, below the tile fall, natural soil appeared.

Assemblage Summary: Circa 779 tile fragments, all of the "curved type" (thus probably belonging to the Hellenistic system), were removed from squares G6 and G7 and discarded after the excavation. Only a few samples of roof tiles, all belonging to the AT1/AC1 type (Hellenistic system) were collected, among which is the AT1 tile SAV RT 02, bearing the *Herakas* stamp. Only a small amount of pottery was found in association. Diagnostic fragments include a ribbed kantharos of the second half of the 4th c. BC (SAV BG 40), an amphora (SAV Amp 24) dating to the first half of the 3rd c. BC, and an antefix (SAV AT 02) of the last quarter of the 4th c. BC. In square F7, terracotta relief plaques were found just above the tile fall (e.g., a standing Zeus, SA 80-9T) and mixed in with the tiles (e.g., an almost complete maenad-silenos plaque, SA 80-10T); a kiln firing support can also be dated to Phase 4.

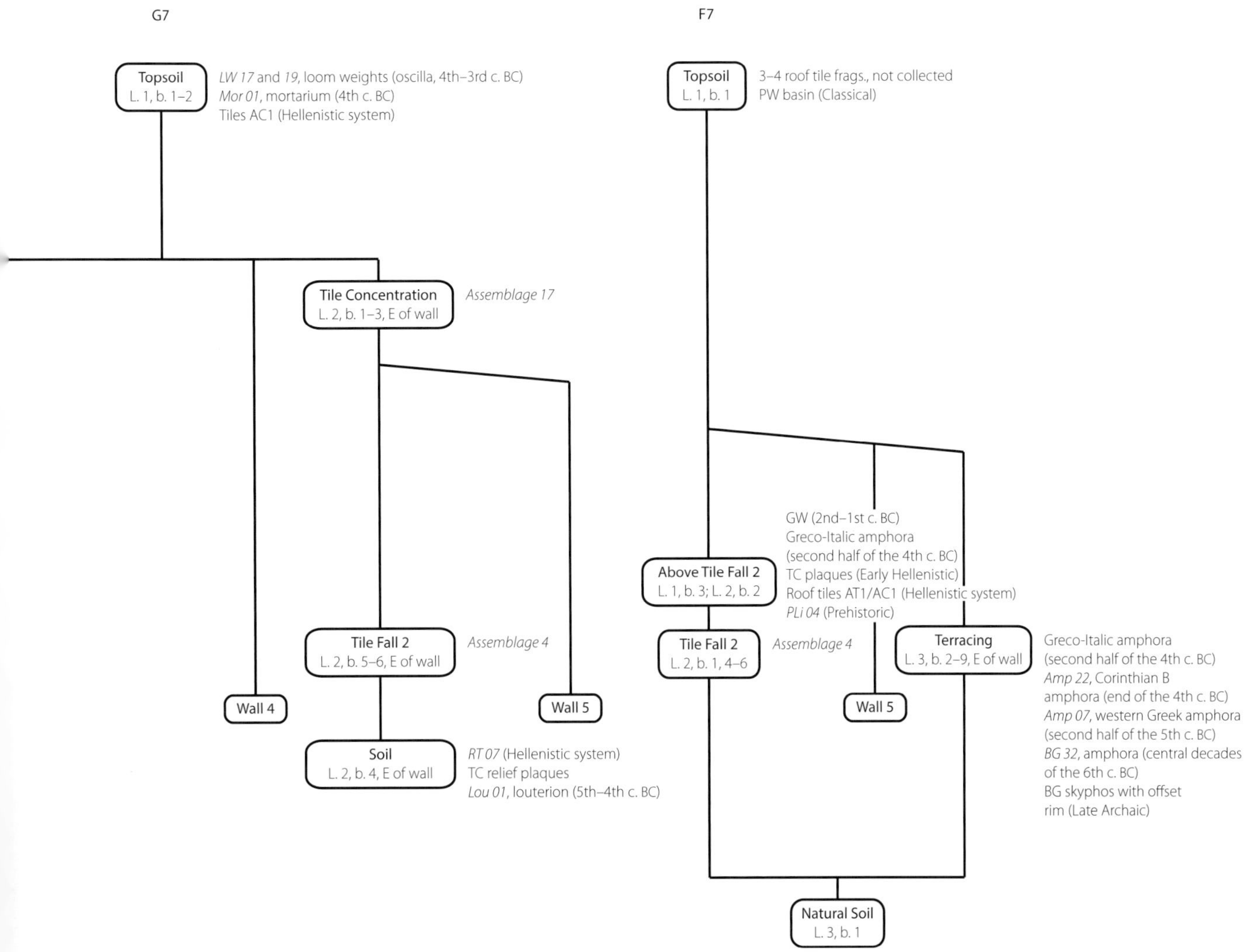

Assemblage 16

Destruction Layer (*Appendix A, Table A4; Fig. 2.20* and *Fig. 2.21*)

Phase 4

Location and Lots: G6, Level 1, batt. 4, (SA 79-678PL). G7, SW corner, Level 1, batt. 2 (SA 79-796PL). H7, Level 1, batt. 5.

Excavation and Stratigraphy: This layer of reddish soil (Munsell 10YR 5/4), which yielded a high concentration of carbon specks, extended for 4.0 x 1.50 m west of Wall 1 in square G6 (Level 1, batt. 4; *Fig. 2.6*) and continued in the southwestern corner of square G7 (Level 1, batt. 2), also protruding for ca. 0.5 m into the southeastern side of square H7 (Level 1, batt. 5). Its position in the stratigraphy can be determined in square G6, where it was visible in section at -6.17 m (at the same elevation as Wall 4), covered by the topsoil (Level 1, batt. 1) and by a colluvial layer composed of dark yellowish brown (Munsell 10YR 4/4) sandy soil (Level 1, batt. 2–3). It is probably the destruction layer of the Hellenistic structure and fired mudbricks, possibly from the elevation of the wall, observed in square H7. Although recognized during excavation, it was not excavated separately.

Assemblage Summary: Highly mixed assemblage, including also material from the topsoil. With the exception of a red-figure hydria rim of the second

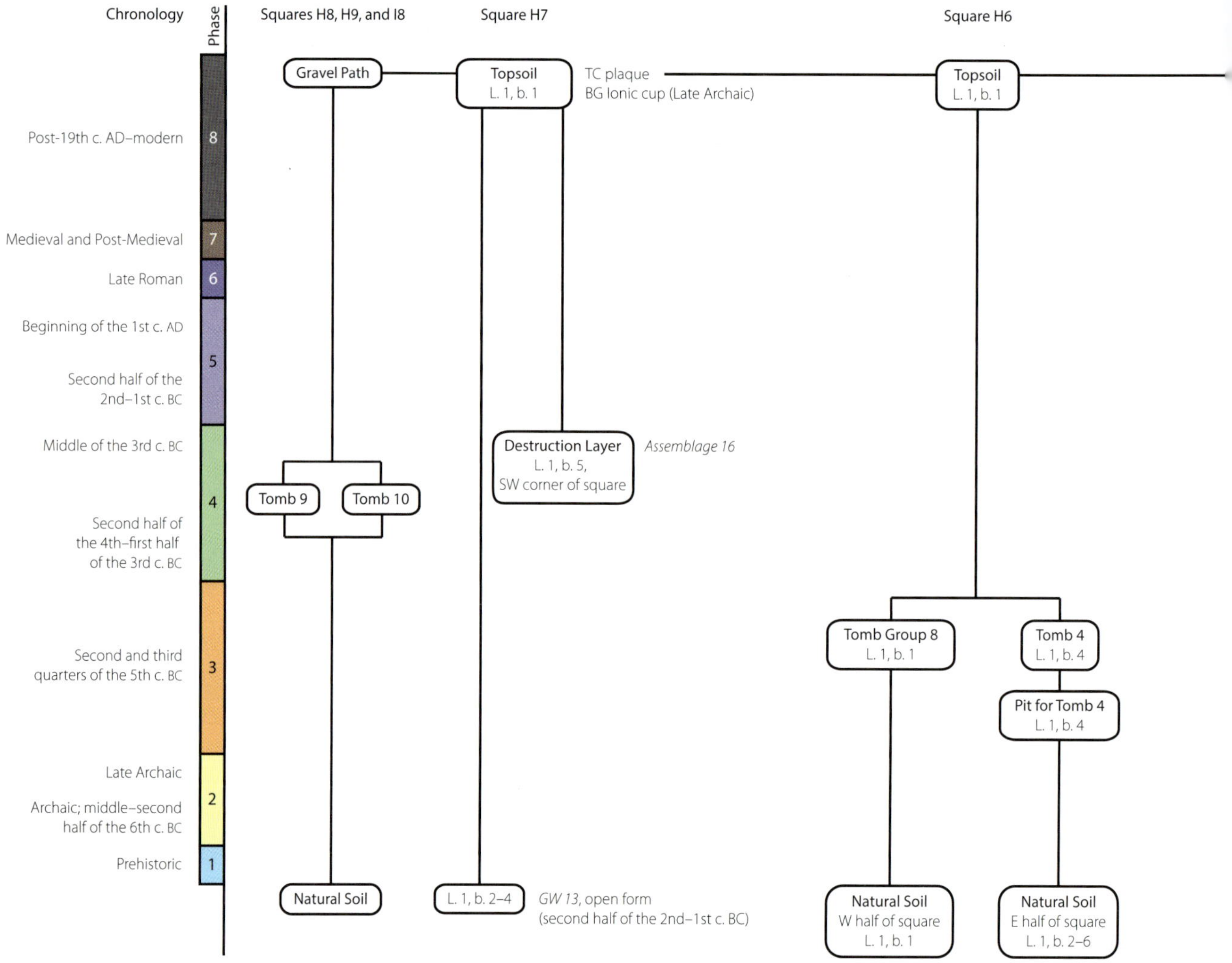

Figure 2.21 The Workshop Area to the W of Wall 1 (squares H4–9 and I8), phased matrix. (KJS/FS/ICA)

half of the 5th c. BC, all material dates to Phase 4. Diagnostic are the blackened rim of a type J4 jug (SA 79-678PL), which can be compared to SAV PCW 59 (Assemblage 13), and dated to the end of the 4th–3rd c. BC, and SAV CkW 24.

Assemblage 2
Tile Concentration inside the Clay-settling Basin (*Appendix A, Table A11; Fig. 2.22*)
Phase 5

Location and Lots: H3, Level 1, batt. 3 (SA 79-1024PL).

Excavation and Stratigraphy: The feature lay beneath a thin layer of plowed topsoil and filled a basin made of upright pan tiles, now destroyed.

Assemblage Summary: No samples of the tiles filling the basin were collected, except a fragment of an opaion 2 (SAV RT 104), attributed to the Roman system. A gutter pipe left in situ and now lost could be dated to the same period. The presence of chronologically mixed material dating to the Hellenistic and possibly Archaic period (SAV AT 07) confirms that the entire area had been disturbed. It is probable that the material yielded by the topsoil and in the area surrounding the basin were originally part of it (Level 1, batt. 2, Tiles BT3, SAV RT 55, 60 and 74), and that the basin was part of the Late Republican workshop.

Assemblage 3
Tile Fall 1 (*Appendix A, Table A9; Fig. 2.22*)
Phase 5

Location and Lots: G4, Level 2, batt. 1 (SA 79-544PL, SA 79-460PL). G5, Level 1, batt. 2 (SA 79-600PL), Level 2, batt. 1–2 (SA 79-636PL, SA 79-777PL).

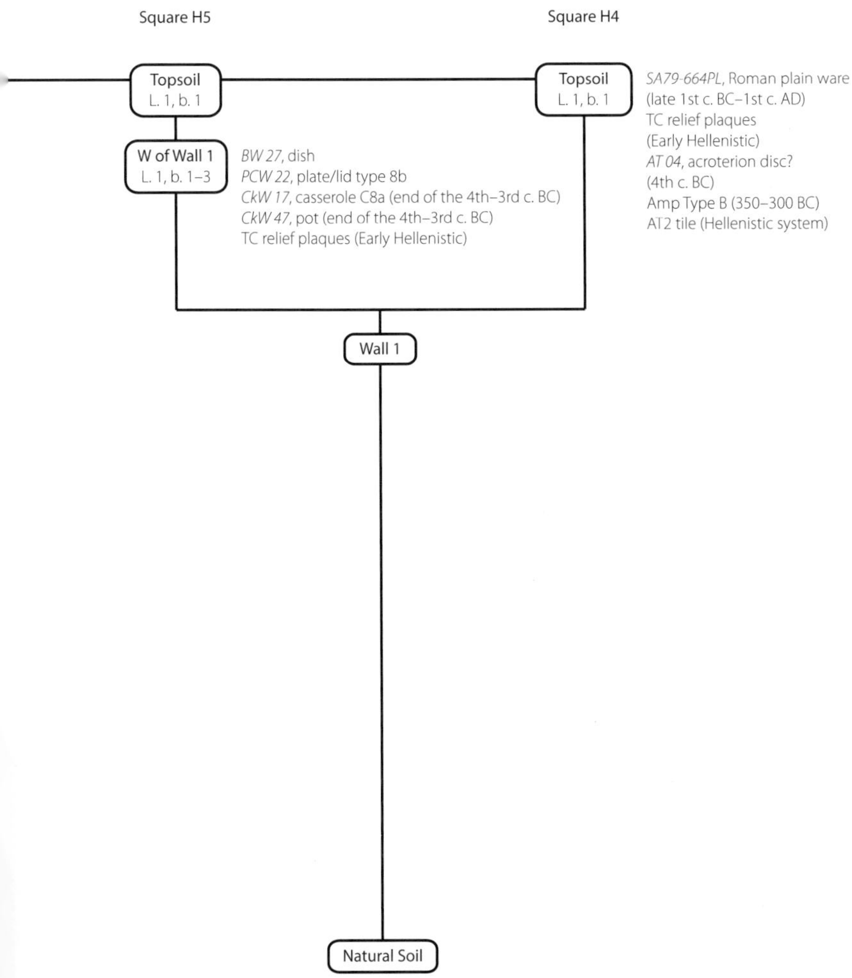

F5, Level 2, batt. 1. G6, Level 2, batt. 1. Cleaning. Level 2, batt. 2–3, E of Wall 1 (SA 79-739PL, SA 79-746PL, SA 79-986PL).

Excavation and Stratigraphy: Tile Fall 1 was identified in squares G6, G5, at the northern end of G4, and for 1.0 m in the northwest corner of F5. That a few tiles were observed in the west wall section of F6 just above Wall 2 (*Fig. 2.7*) could indicate that the tile fall extended this far. In square G4, where the tile fall was limited to the north end of the square, the integrity of the context was compromised by plowing activity. In square G5, the tile fall, abutting the east side of Wall 1, was excavated to its entire extent as Level 2, batt. 1. Excavation was then restricted to a 2.0 x 5.0 m trench, where the concentration continued to a depth of ca. -7.00 m (Level 2, batt. 2), at which point natural soil appeared. No floors or floor levels were identified beneath the feature, which lay on natural soil in squares G4, G5, and F5. In G6, in contrast, a thin layer of soil covering Tombs 1 and 5 (G6, Level 2, batt. 4), possibly part of a floor, was exposed beneath the Tile Fall.

Assemblage Summary: On the whole, ca. 1670 tiles were removed. Only a few samples were collected at the time of the excavation. In G4, G5, and F5, the discarded fragments were described as belonging to the "curved variety," and the samples collected all belong to AT1/AC1 type (Hellenistic system, SAV RT 05, SAV RT 13, and SAV RT 29), AT2 (SAV RT 45), and BT2 (SAV RT 50). Since in G5 the tiles were found mixed with stones, it is possible that the concentration here found included the collapse of Wall 1. This could be confirmed by a tile of type AT1 (SA 79-545T) bearing a mark similar to that encountered in SAV RT 05 and SAV RT 17. Encrustations on its surface could suggest that it was used as construction material for Wall 1. The presence of pan tiles, possibly part of the roof covering, was also noted in G6. Pottery found in association is limited and very fragmented, and diagnostic fragments include mainly vessels dated to Phase 4, such as a dish (SAV BW 27), a Corinthian A amphora (SAV Amp 18), and a type 8a casserole (SAV CkW 17). A skyphoid cup (SAV BW 15) belongs to a type popular especially in the 3rd c. BC. A few terracotta relief plaque fragments of the Early Hellenistic period (SA 79-549, 776, 783, and 857T) were also found, as well as loom weights (SAV LW 13–SAV LW 16 and SAV LW 18) and coarse ware pottery of similar chronology. Since in square G6 pottery is especially concentrated in the lower part of the tile fall, it is likely that floor level was partly excavated with the fall, explaining the presence of earlier material. The presence of grey ware (SAV GW 11), of a Late Republican pot (SAV CkW 41), and of BT3-type tiles confirms a chronology within Phase 5.

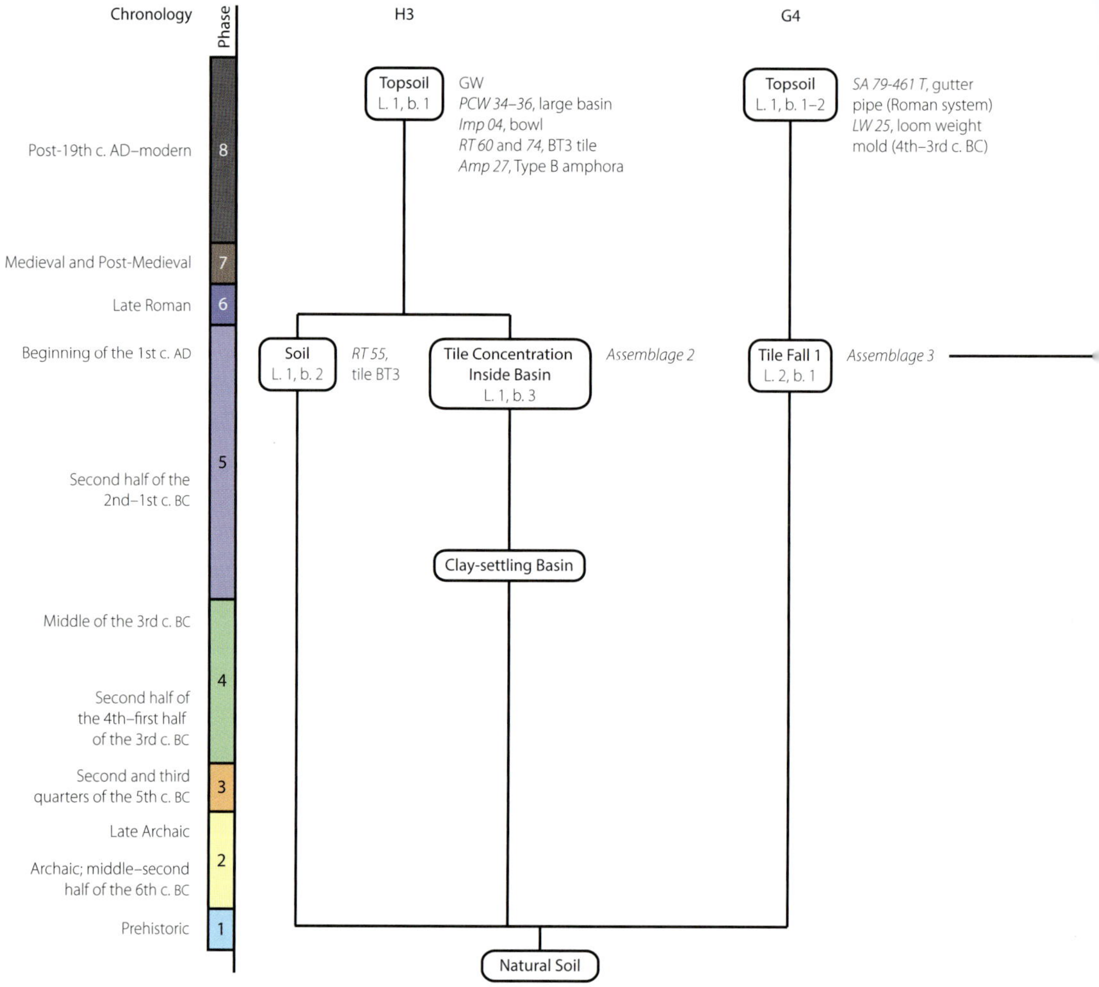

Figure 2.22 The Workshop Area to the E of Wall 1 (squares H3, F5, and G4–G6), phased matrix. (KJS/FS/ICA)

Assemblage 17

Tile Concentration in G7 (*Appendix A, Table A10; Fig. 2.20*)

Phase 5

Location and Lots: G7, E of Wall 4. Level 2, batt. 1–3. SA 79-802PL, SA 79-797PL.

Excavation and Stratigraphy: A tile concentration was also identified in square G7 (Level 2, batt. 1–3) beneath the topsoil (Level 1, batt. 1–2). Although the relationship was not specified, it probably abutted Wall 4, while Wall 5 lay beneath it. It extended for 3.20 m north-south with an irregular eastern limit forming a semicircle (*Fig. 2.18*) to a maximum east-west length of ca. 2.30 m. (Level 2, batt. 1–3, E of Wall 4; SA 79-802PL, SA 79-797PL).

Assemblage Summary: The concentration yielded 220 tiles, comprising "thick pan tiles" (possibly of the Roman system) and curved tiles. Few samples (all AC1 cover tiles), were collected. Mixed in with the tiles were five Roman gutter pipes (SAV RT 106 and SAV RT 107), possibly used in the drainage system of the area, for which no other evidence is preserved. The small amount of pottery found in association mostly dates to Phase 4, with a residual black-gloss skyphos with concave rim (SA 79-802PL) dated to 470–440 BC. Nevertheless, gutter pipes and the presence of "thick pan tiles," both related to the Roman system, suggest a chronology within Phase 5.

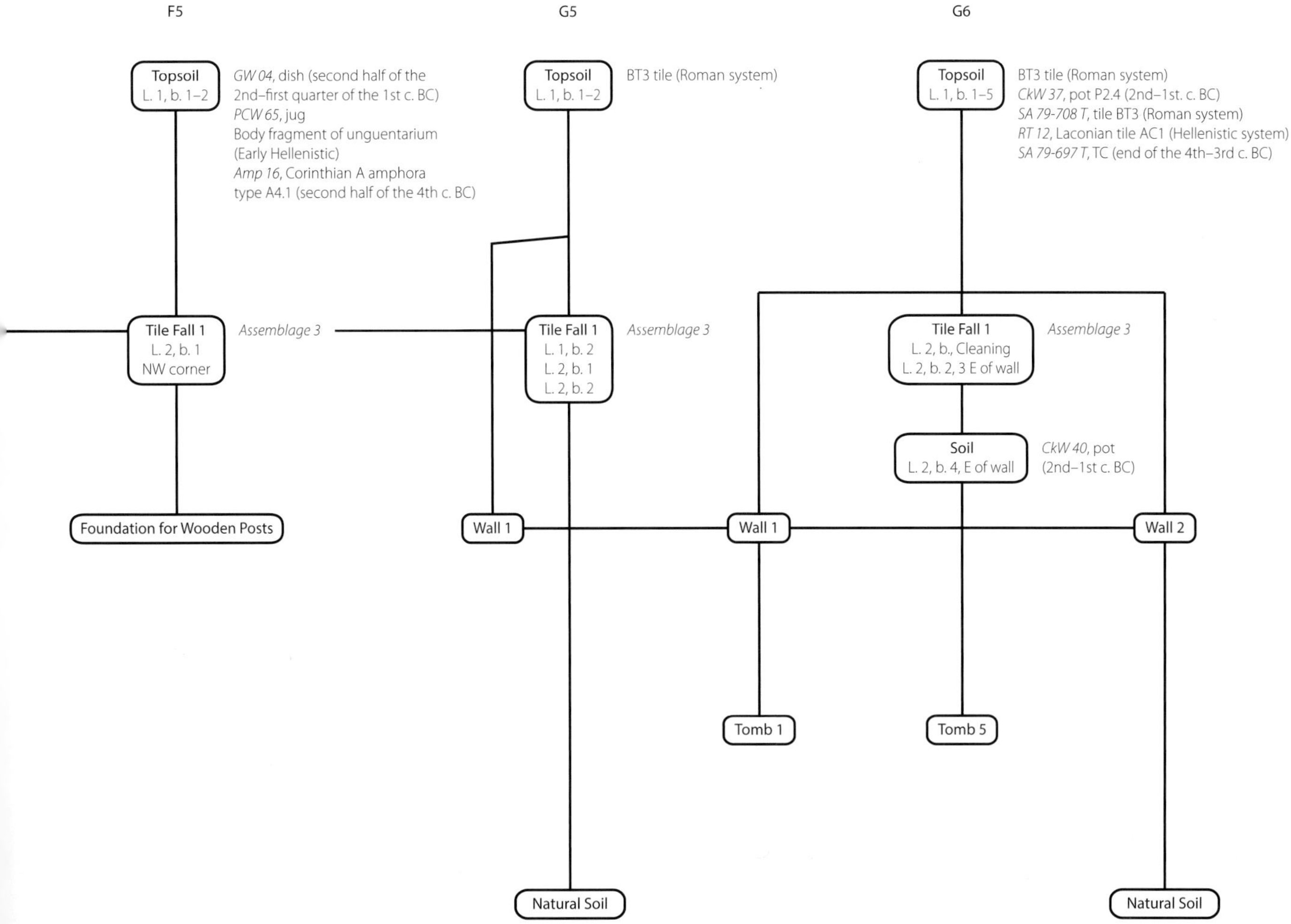

The Kiln Complex

Location and excavation. A rectangular Kiln Complex that encloses the remains of three kilns (Kilns 3–5) and a courtyard (*Fig. 2.25*) was exposed on the edge of the slope,[28] 5 m east of the clay-settling basin in H3. The structure was excavated in three separate instances. Sounding A (squares E1–E3) was undertaken in the courtyard on the southeastern side of the Kiln Complex (*Fig. 2.24* and *Fig. 2.27*) to determine the stratigraphic relationship between the kilns and Slope Areas. Here the ground sloped sharply between the west end (-8.93 m) and the level of Wall 9 (-9.61 m). Sounding A consisted of a square (2.5 m east-west x 7.5 m north-south) bounded to the south and east by Walls 8 and 9. Its northern extension (square E3, Sounding A-1, 1.0 x 2.4 m) uncovered Wall 10. The excavation was later extended west to squares F2 and F3, leaving a 1-m baulk between the two areas. The southern half of this baulk was removed in the third instance of excavation in 1981 (see *Fig. 1.3*).

Erosion and agricultural activity are likely responsible for the displacement and partial removal of the occupation and post-abandonment layers of the western half of the structure and outside it. Plow marks were detected near the northwestern corner of the complex while just south of the Kiln Complex, in square F1, natural soil appeared 0.10 to 0.15 m beneath topsoil. The only diagnostic fragments found in this area belong to a Post-Medieval basin (SAV PMP

[28] See Ch. 3, "The Structures at Sant'Angelo Vecchio."

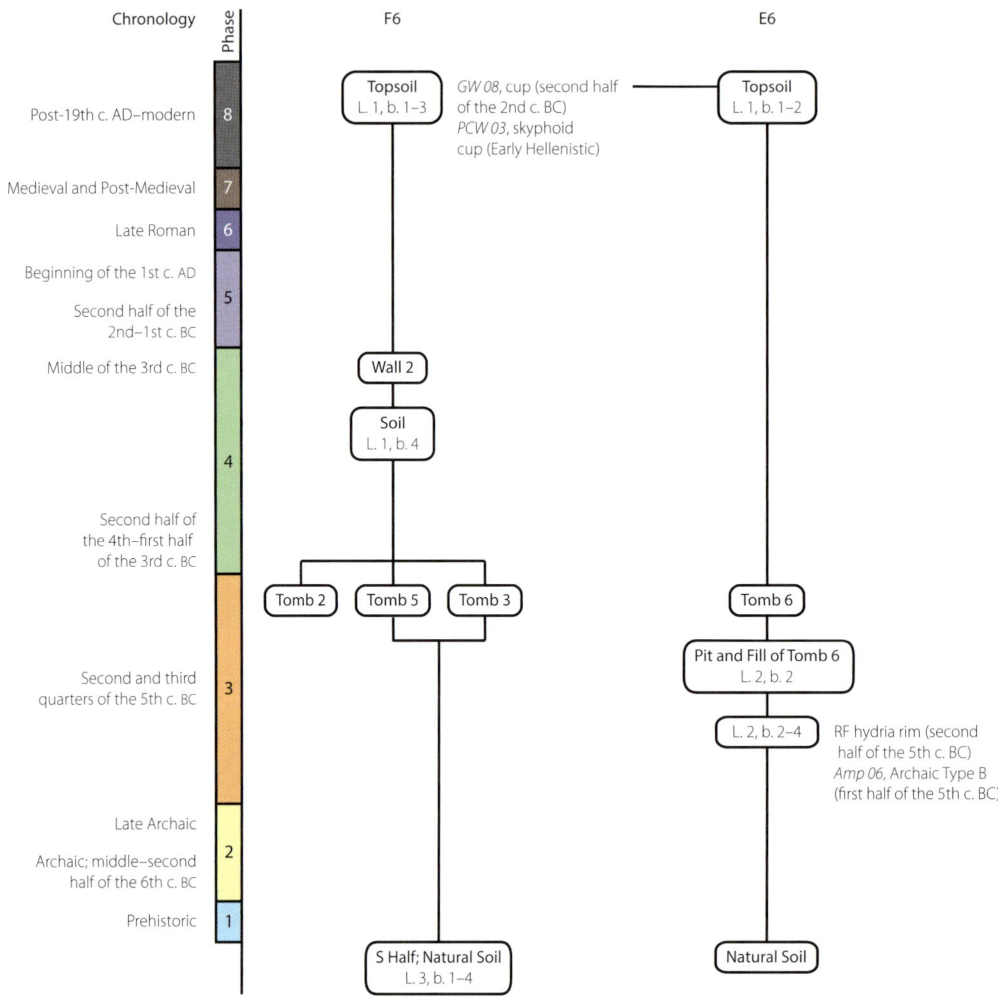

Figure 2.23 The Workshop Area (squares F6 and E6), phased matrix. (KJS/FS/ICA)

01), found together with an iron deadlock and a ring of the type used to secure animals (SA 79-827M), all probably dating to the 19th century AD. Stratigraphy was better preserved in the eastern half of the complex, which provided a complete sequence from tile fall to natural soil (*Fig. 2.30*).

Stratigraphy. Walls 7, 8, and 10 (preserved at the level of the foundations), the combustion chambers and the stocking channels of Kilns 4 and 5, and the tile fall found in the courtyard all lay beneath a thin layer of plowed topsoil (squares F2 and F3, Level 1, batt. 1; Sounding A, Level 1, batt. 1–4), 0.10 to 0.15 m deep (*Fig. 2.26* and *Fig. 3.21*). Wall 9 was already visible on the surface. Walls 7 and 8 were set in natural soil, which also constituted the surface outside the northern part of the complex (squares F2 and F3, Level 2, batt. 1, exterior

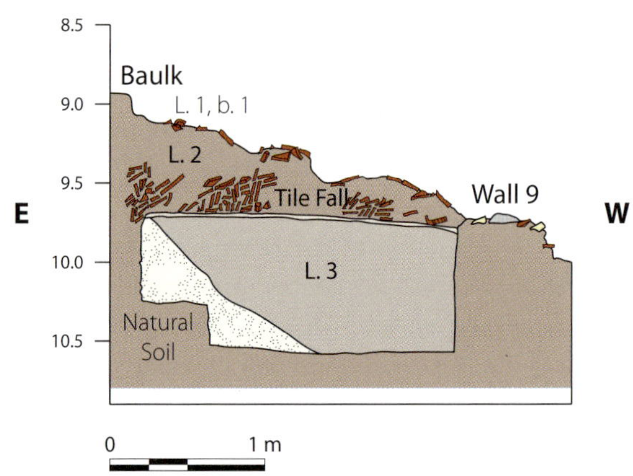

Figure 2.24 Section F, E-W section of the N wall of Sounding A in the courtyard of the Kiln Complex. (KJS/ICA)

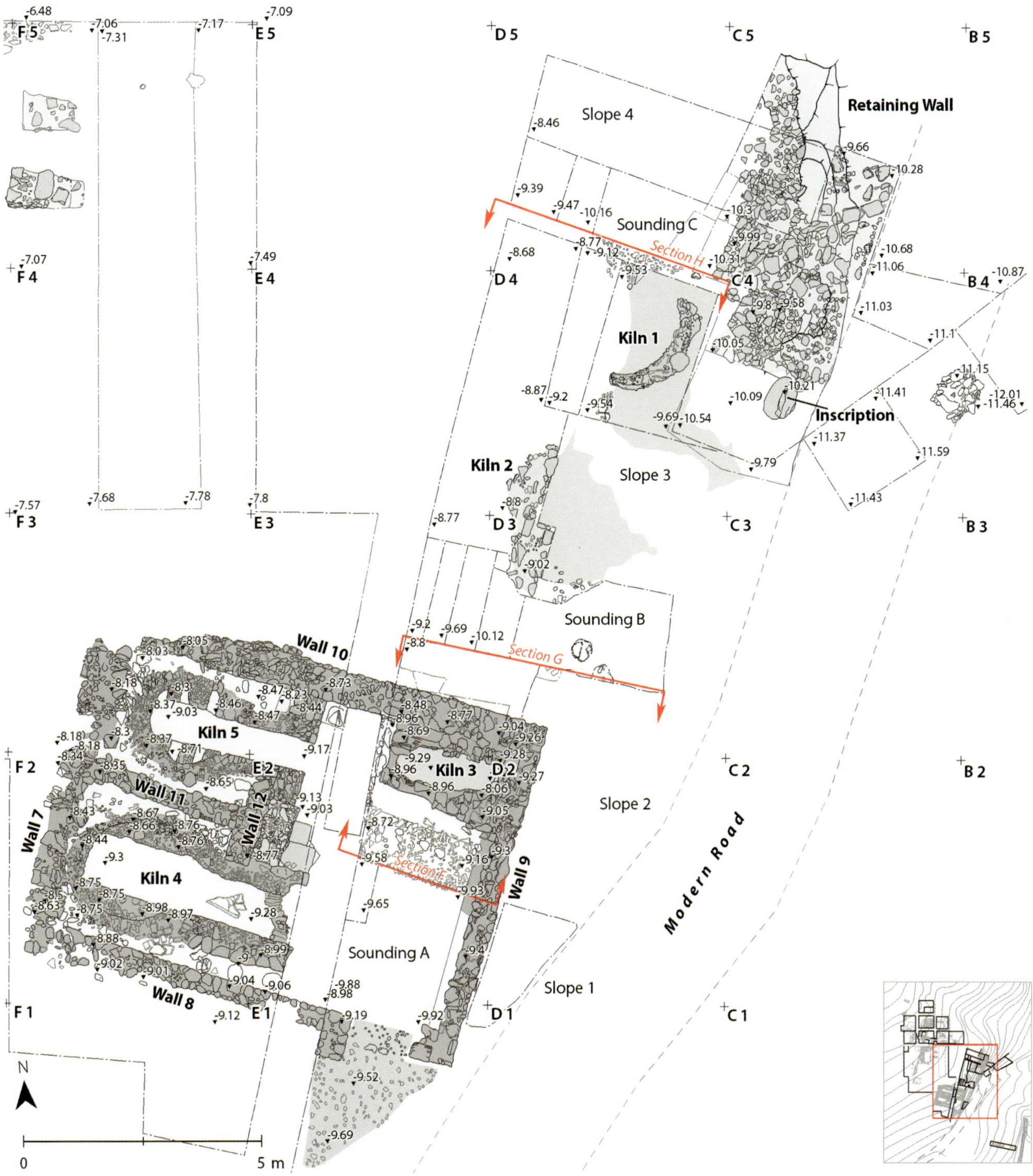

Figure 2.25 Detailed plan of the Kiln Complex and Slope Area. (JT/ICA)

of kilns). Wall 9 and the eastern half of Wall 10, on the other hand, rested on the slope covered with debris from the clay kilns and fragmentary votive plaques (Sounding A, Level 3, batt. 1–4). Both kilns were cut into the ground as far down as bedrock, with Kiln 4 (South Kiln) at a lower elevation than Kiln 5 (North Kiln).

The deposit found inside the combustion chamber of Kiln 4 (South Kiln, Level 2, batt. 1; Assemblage 7) yielded mainly tiles and can be interpreted as the collapse of the covering of the kiln (Phase 5), which had been left empty at the moment of its abandonment. Kiln 5 (North Kiln interior, Level 2, batt. 1–4,

Figure 2.26 Kiln Complex in 1979, looking N. Kiln 4 (South Kiln), Sounding A, and Wall 8 in foreground. (CW/ICA)

Assemblage 8), by contrast, was filled with structural debris from the collapse of the kiln covering (*Fig. 2.28*) as well as pottery that had been left in the kiln when it was abandoned.

The two kilns opened into a rectangular courtyard, partially occupied by a solid tile fall found beneath the plowed topsoil (Assemblage 5; *Fig. 2.27*). The tile fall, belonging to the courtyard's roof, extended across the entire northern half of Sounding A into the area of the entrance to the Kiln Complex (square E1) and between the openings of Kilns 4 and 5, but there is a distinct lack of tiles at the north end of the room. Its southern portion made a curve, leaving the southeastern corner of the courtyard empty. It reached its maximum depth of 0.6 m on the western side of the central part of the sounding and in front of the entrances to Kilns 4 and 5. The depth progressively reduced to 0.25 m at the eastern limit, where the bottom layer of the tile fall corresponded with the preserved upper surface of Wall 9, as it was in the northern extension of the sounding (A-1; *Fig. 2.27*). Apart from a 1-m baulk left in the central part of Sounding A, the tile fall was completely removed.

In the northeastern corner of the courtyard, the tile fall covered a small kiln (Kiln 3; see *Fig. 3.25*)[29] filled with a deposit (Sounding A-1, Level 2, batt. 3, Assemblage 6) that yielded a limited quantity of very fragmentary material, indicating that the feature was abandoned when it was almost empty. Its stratigraphical relation with the courtyard and Kiln 5 is not clear due to the presence of the unexcavated baulk.

In the remaining part of the courtyard, the tile fall rested on a floor constituted at the north and south ends by reddish soil laying directly over the bedrock and by a thin layer of loose ashy soil (Sounding A, Level 2, batt. 3). The bench between Kilns 4 and 5 sat on a pavement of heavily fired clay slabs, which was in turn placed on a thin layer of yellowish-orange soil immediately over the uneven surface of bedrock (*Fig. 2.29*). In front of Kiln 5, a few tiles were found in the soil immediately above bedrock. Two of these tile fragments seemed to be pressed into and underneath the pavement layer between the two kilns. Between the end of the south wall of Kiln 5 and the remaining part

[29] See Ch. 3, "The Structures at Sant'Angelo Vecchio."

Figure 2.27 Sounding A, looking E. To the N, a portion of the Courtyard Tile Fall is visible. (CW/ICA)

of the floor, a fragment of a grey-ware dish (SAV GW 06) appeared. At the northern end of the courtyard, a large pan tile formed a square or step (*Fig. 3.23*). A Late Republican cooking ware lid (SAV CkW 30) was found resting beneath the tile, on a surface of hard semi-fired soil that abutted Wall 10 and was covered by a layer of tiles, visible in the baulk east of Kiln 5.

In the southern half of Sounding A, the courtyard floor rested on a layer of dense, heavy clay (Level 3, batt. 1–4; *Fig. 2.30*), different in composition from the soil encountered on the slope. It is possible that the soil had been brought in to level the ground for the construction of the Kiln Complex. The soil was excavated down to a depth of -10.55 m, and natural soil appeared in a sloping line at the western edge of the section (*Fig. 2.24*). The finds include a type 1 kiln firing support (late 4th–3rd century BC) and a small amount of very fragmentary pottery (21 fragments), including some residual sherds of the Late Archaic period.

Excavated Units and Assemblages from the Kiln Complex and Soundings A and A-1

Assemblage 5

Courtyard Tile Fall (*Appendix A, Table A12; Fig. 2.30*)

Phase 5

Location and Lots: Baulk at entrance to N and S Kilns, Level 2, batt. 3. Cut in baulk between N and S Kilns, Level 2, batt. 1–2 (SA 81-07PL, SA 81-172PL, SA 81-12PL). E1, cleaning, Level 2 (SA 79-735PL). E2–E3, Sounding A-1, Level 2, batt. 1–2 (SA 79-798PL). E2-E3, N extension of the cut of the baulk, Level 2. E2, baulk at S Kiln, Level 2, batt. 1. E2, baulk at S Kiln, round vase and soil, Level 2, batt. 1. Sounding A, Level 2, batt. 1–2 (SA 79-16PL, SA 79-26PL). Sounding A, subdivision of Room 1 to S, Level 2, batt. 1. Sounding A, S sub-section of Room 1 down to wall, Level 2, batt. 1 (SA 79-18PL).

Excavation and Stratigraphy: Tile fall above the E part of the Kiln Complex courtyard floor.

Figure 2.28 Stones, mudbricks and tiles from the interior of Kiln 5 (North Kiln). (CW/ICA)

Figure 2.29 Clay slab pavement between Kilns 4 and 5 (South and North Kilns), looking W. (CW/ICA)

Assemblage Summary: In this large assemblage, most of the finds were roof tiles. Their exact number is unknown, but at least 3059 fragments were counted and discarded. A large number of the samples collected belong to BT3 pan tiles of the Late Republican period and some of the tiles bear the same stamps also found in Tile Fall 1 of the Workshop Area. They were probably part of the roofing of the courtyard, to which the carpentry nail SAV M 02 could also be related. Laconian tile SAV RT 120, on the other hand, has baked clay on the surface and was possibly reused in the construction of Kiln 4. This was also the case for the Archaic (antefix mold SAV AT 08 and the sima SAV RT 129), Late Archaic (Archaic Type B amphorae) and Classical material (louterion SAV Lou 01) from the tile fall. The pottery found in association (102 fragments, encompassing 27 individual vessels) is small in number and very fragmentary. The only exception is the pan SAV CkW 06, which was found at the lower end of the tile fall. It is likely that the pan was originally part of the courtyard floor assemblage and that there was some degree of mixture of material from the two contexts. The remaining diagnostic pottery ranges in date between the Late Republican and the Early Imperial periods (SAV GW 09, SAV Lag 03, SAV Lou 06, SAV RW 01, SAV RW 04, SAV Imp 06, SAV Imp 12, SAV Imp 13, SAV Imp 16, and SAV Imp 17). The Italian terra sigillata plate SAV Imp 01 and the bowl SAV Imp 02, dated to the 1st c. AD, provide the terminus post quem for the abandonment of the kiln.

Assemblage 6
Deposit within Kiln 3 (*Appendix A, Table A13; Fig. 2.30*)
Phase 5

Location and Lots: Level 2, batt. 2 (SA 79-768PL).

Excavation and Stratigraphy: Kiln 3 and its fill lay immediately beneath the northern strip of the Courtyard Tile Fall. It was excavated in a single battuta.

Assemblage Summary: The fill of Kiln 3 yielded a small assemblage. The diagnostic pieces are an Early Imperial closed form (SAV Imp 11) and a Late Republican BT3 pan tile (SA 79-770T) bearing a stamp representing a wreath (for the type, see *Fig. 32.10*), suggesting that Kiln 3 was abandoned at the same time as the other kilns. The remaining pottery includes tiny non-diagnostic fragments of black-gloss, banded, and cooking wares. Several chunks of fired clay, originally part of the coating of the kiln's inner walls, were also collected (see *Fig. 9.1*).

Assemblage 7
Kiln 4 Collapse (*Appendix A, Table A14; Fig. 2.30*)
Phase 5

Location and Lots: F2, S Kiln interior, Level 2, batt. 1–2. F2, within S Kiln, Level 2, batt. 1. F2, inside S Kiln, Level 2, batt. 1. F2, S Kiln interior, Level 2, batt. 1–2. F2, S Kiln interior, Level 2, batt. 4. E2, cleaning of S Kiln praefurnium, Level 2, batt. 2. SA 79-801PL, SA 79-860PL, SA 80-63PL, SA 80-65PL, SA 80-66PL.

Excavation and Stratigraphy: Exposed in the most battered area of the excavation, it is likely that only a portion of the collapse, ca. 40 cm deep and covering the clay floor of the kiln, was preserved.

Assemblage Summary: The assemblage is composed of the construction material of Kiln 4 that had collapsed within the combustion chamber, mainly comprising AT/AC1 and, to a lesser extent, BT3 type tiles, some with clay incrustation on the surface. Possibly used as construction material for the kiln was also the antefix mold SAV AT 04, found in the layer above the collapse. A 2nd-c. BC Greco-Italic amphora (SAV Amp 29) and one of the fragments of a red-ware jug (SAV RW 04, which has a joining fragment in Assemblage 5) of the late 2nd c. BC–early 1st c. AD are the only diagnostic pottery fragments found in the assemblage.

Assemblage 8
Kiln 5 Collapse (*Appendix A, Table A15; Fig. 2.30)*
Phase 5

Location and Lots: N Kiln interior: F3, N Kiln interior, Level 2, batt. 1. F3, N Kiln interior, Level 2, batt. 2. F3, N Kiln interior, Level 2, batt. 4. F3, SE corner of N Kiln near baulk, Level 2, batt. 1. F3, SE end of N Kiln near baulk, Level 2, batt. 4. F3, end of kiln, Level 2, batt. 1. F3, end of kiln, Level 2, batt. 4. F3, N Kiln, Level 2, batt. 1, 4. SA 79-880PL, SA 79.833PL, SA 79-873PL, SA 79-875PL, SA 79-883PL.

Excavation and Stratigraphy: The deposit, ca. 0.40 m deep and excavated in four battute, filled the interior

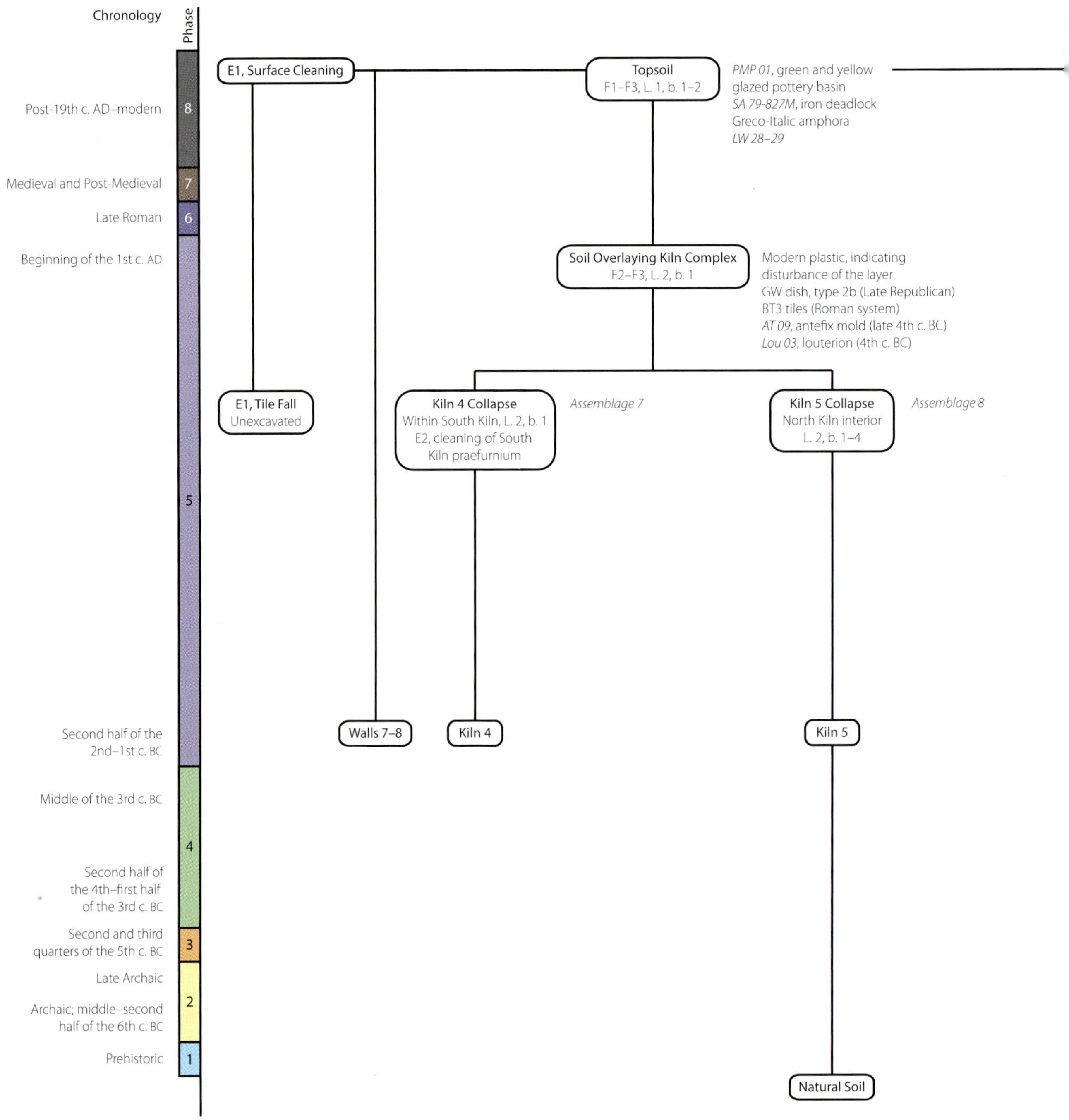

Figure 2.30 Kiln Complex, Sounding A, and Sounding A-1, phased matrix. (KJS/ICA)

of the combustion chamber of Kiln 5. No differences between the battute were noted, and both SAV PCW 32 and SAV CkW 43 are recomposed from fragments found in batt. 1 and 4, proving that it is a unitary fill consisting of the debris from the collapse of the kiln and pottery.

Assemblage Summary: This large assemblage consisted of structural debris from the collapse of the kiln, including AT1/AC1 (Hellenistic system) and BT3 (Roman system) tiles (often with clay mortar encrustations on their surface), bricks, stones, and mudbricks. Most of the pottery is very fragmentary. Since some vessels, such as SAV Amp 33, the large basin SAV PCW 32, and SAV CkW 43, have hard fired clay adhering to the surface, it is likely that they were probably recycled as fill in the struc-

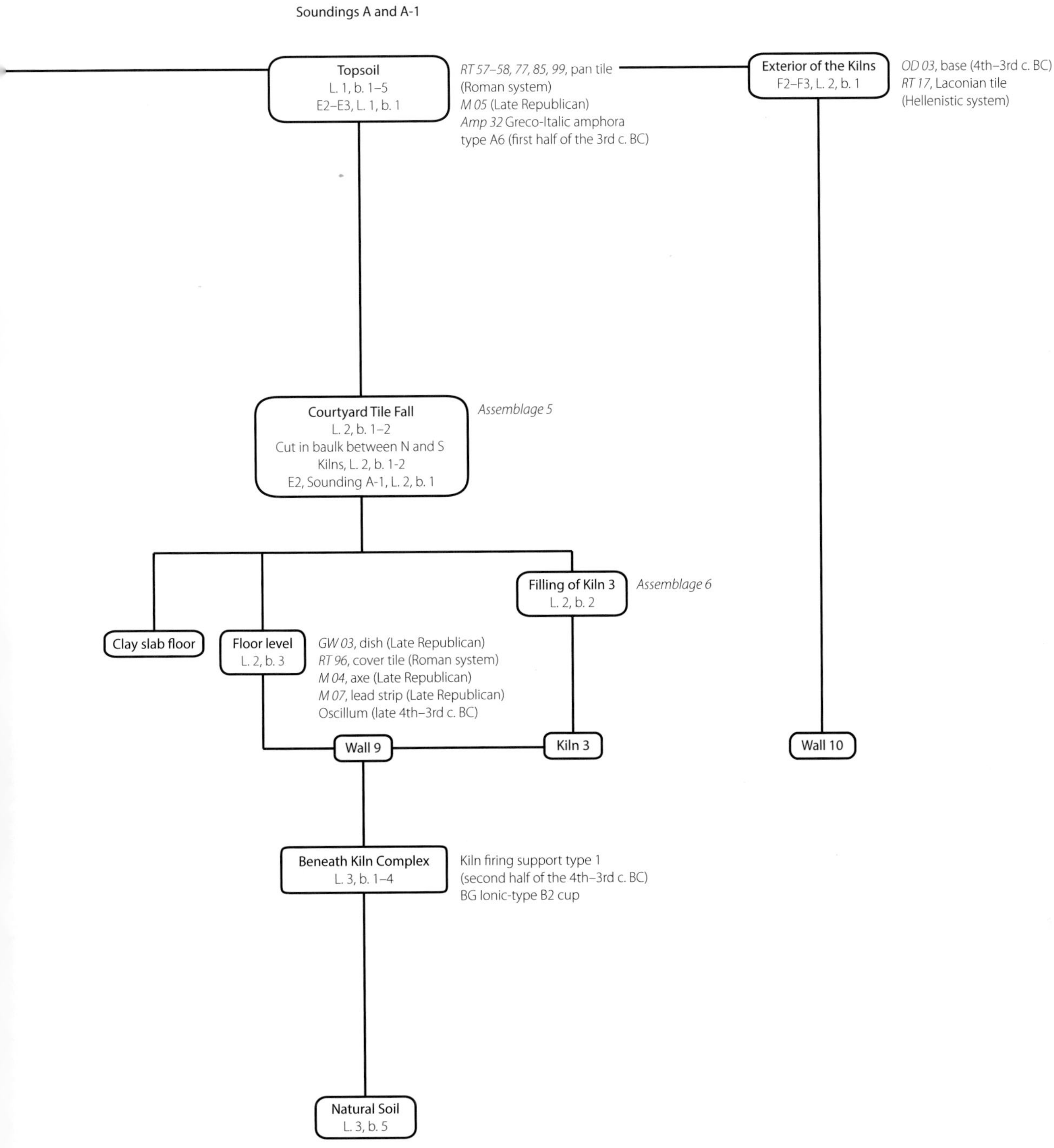

ture of the kiln. The finds from Level 2, batt. 2 differ from the remaining assemblage. Here, five almost complete Late Republican and Early Imperial vessels were found (SAV GW 01, SAV GW 02, SAV RW 02, SAV PCW 53, SAV Imp 09, SAV CkW 07, SAV CkW 25, SAV CkW 49, and a Metapontine amphora, not in catalog). None shows evidence of over-firing, and they were probably deposited or dumped in the kiln just before or sometime after its abandonment, and mixed with the debris from the collapse of the kiln. Key finds for the dating of this context are the two grey-ware mugs, SAV GW 01 and SAV GW 02 of the second half of the 1st c. BC, and the jug SAV Imp 09, dated to the 1st c. AD.

Figure 2.31 The Slope Area in 1979, looking W. Sounding B (left), Slope 3 (center-left), Kiln 1 and inscribed stone (center), Retaining Wall (center-right), and Slope 4 (right); modern road in foreground. (CW/ICA)

The Slope Area

Location and excavation. East of the Workshop Area and the Kiln Complex, the gently sloping hillside abruptly descends toward the area of the springs with an observable difference in height between the northern and southern parts of the slope (*Fig. 2.25*). The incline is cut in its central part by a modern gravel road heading north, now abandoned and almost completely hidden by the vegetation. Excavations concentrated on the area west of the path, where the so-called "Bulldozed Stipe" (Assemblage 9) was uncovered during the construction of the road and where material already appeared scattered on the entire surface of the section of slope located west of the road.

Since the grid was laid some time after the beginning of the excavations, the area was divided into four sections designated Slope Areas 1–4 (*Fig. 2.25*), all of them extending east of the modern road. Slope Area 1, at the lowest elevation, was a narrow strip ca. 6.0 m in length and ca. 1.5 m in width to the east of Wall 9 of the Kiln Complex, within square D2.

Slope Area 2, which extended over an area measuring ca. 5.0 m north-south and 5.5 m east-west, was defined to the south by the northeastern corner of the Kiln Complex. It ran parallel to the northern wall of the Kiln Complex (Wall 8) and perpendicular to the modern road. The northern half of Slope 2 was excavated separately as Sounding B, laid out within squares D3 and E3 and measuring 5.5 m east-west and 2.0 m north-south. The total depth of the sounding from the plowed topsoil to the horizontal base of natural soil was ca. 2.72 m (*Fig. 2.31* and *Fig. 2.32*). The stratification in the sounding consists of 3rd-century BC layers lying on the natural soil.

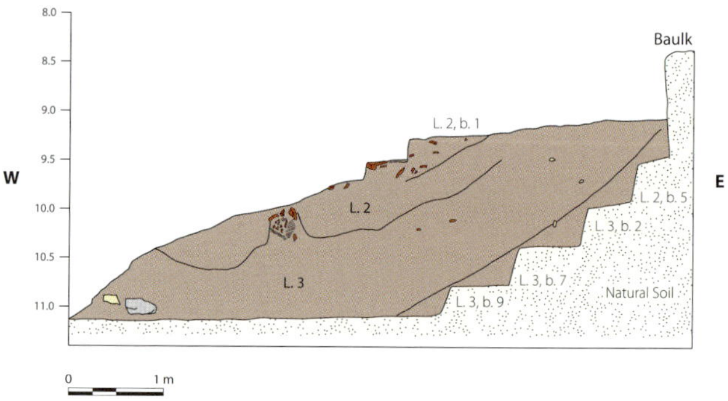

Figure 2.32 Section H, E-W section of Sounding B, S part. (KJS/ICA)

Slope Area 3, located in squares D3 and D4, measured ca. 6.6 m north-south and 6.5 m east-west. As in Slopes 1 and 2, Phase 4 levels and features found in Slope 3 lay directly on the natural soil.

Finally, Slope Area 4, measuring 3.5 m north-south and 7.8 m east-west, comprises the northernmost sector of the slope. The identification beneath the topsoil of the same concentration of rocks and a

Figure 2.33 The Slope Area in 1979, looking S. Sounding C and Kiln 1 in foreground. (CW/ICA)

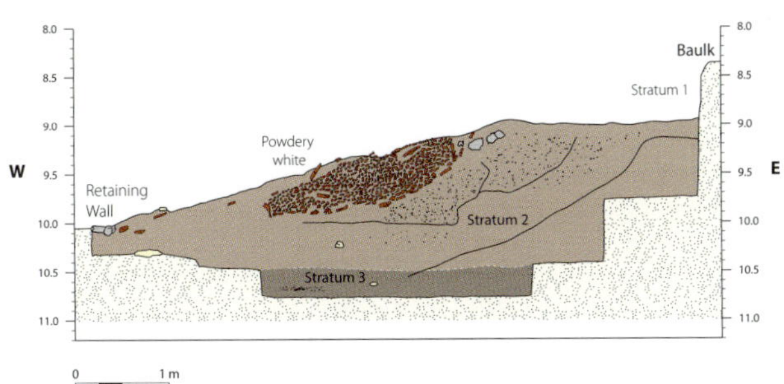

Figure 2.34 Section G, E-W section of Sounding C, S edge. (KJS/ICA)

cluster of greenish lumps of over-fired clay also observed in Slopes 2 and 3 led to the decision to excavate its southern part separately as Sounding C (*Fig. 2.33* and *Fig. 2.34*). This sounding, within square D5, measured 1.60 m north-south and 7.60 m east-west. The total depth of the sounding from the plowed topsoil to the natural soil was ca. 2.36 m. Sounding C provided a stratigraphic sequence ranging in date from Phase 3 to Phase 4 (*Fig. 2.35*). Further excavation was undertaken in the eastern half of Sounding C, where a north-south Retaining Wall was exposed,[30] and in squares D4 and C4, where a sounding 2.5 x 2.0 m was excavated to a depth of 0.85 m (-10.54; see *Fig. 3.9*).

The excavation of the archaeological deposits of the slope area was complicated by their position downslope. The technique adopted varied in the different areas. Both Soundings B and C were excavated mainly in horizontal cuts or passes (*battute*), which resulted in some (limited) mixing of material belonging to different contexts. In Slope 3, however, the natural inclination of the layers was respected in their removal, and the assemblages here are thus more reliably related. The entire eastern end of Slopes 1–4 had been subjected to modern earth-moving equipment, particularly invasive in Slope 4, as also indicated by the bulldozer tracks visible in the blocks of the Retaining Wall, resulting in partially compromised stratigraphy. Although excavated at different moments, the homogeneity of the stratigraphic sequence observed on the slopes allows us to describe it in a single narrative.

[30] See p. 71 in Ch. 3, "The Retaining Wall."

Stratigraphy. The difference in elevation between Slopes 1 and 4 accounts for the different conditions of topsoil preservation. In Slope 1 and the southern section of Slope 2, erosion explains the lack of topsoil. The finds (Slope 1, surface finds; Slope 2, Level 1, batt. 1; *Fig. 2.35*), found scattered on the surface, consisted of Hellenistic terracotta relief plaques, kiln firing supports, construction material of kilns (heavily fired lumps of clay with air bubbles and a crisp porcelain-like texture), and pottery that can be dated to Phase 4. Since natural soil appeared immediately below these fragments, only the surface area was excavated.

In the northern section of Slope 2 and Slopes 3 and 4, the surface consisted of plowed topsoil to a maximum depth of ca. 0.6–0.7 m along the western edge (Level 1, batt. 1–3; *Fig. 2.35*). With the exception of a few fragments dated to the Roman period (mainly roof tiles of type BT3, see *Fig. 32.7*, and the lid SAV Imp 08, *Fig. 26.1*), it yielded material of Phase 4 consistent with the finds of the archaeological deposits lying just beneath the topsoil. The Late Archaic and the Neolithic impasto (SAV Impasto 03, SAV Impasto 04, and SAV Impasto 06) fragments were probably collected at the eastern end of the slope, where stratigraphy had been highly compromised by mechanical earth-moving resulting from the construction of the road, with the partial removal of layers of previous phases. The removal of the topsoil revealed a concentration of material that extended irregularly north to south, from the northern half of Slope 2 up to square D5 (*Fig. 2.31*) and downslope toward the modern road, where its eastern end had already been partially cleared out as part of the Bulldozed Stipe.

Three clusters—two of them resting on the natural soil, the third covering the worn remains of Kiln 1—were identified and excavated separately. A circular concentration of terracotta relief plaques was detected at the southern end of Sounding B (Sounding B, S extension, Level 2, batt. 1–2; Assemblage 10). A second terracotta and tile concentration was found some 0.5 m north, extending across the northern end of Sounding B and the southern side of Slope 3 (Sounding B, Level 2, batt. 1–2; Slope 3, Level 2, batt. 1, Level 3, batt. 1–2; Assemblage 11, *Fig. 2.25* and *Fig. 2.35*). The presence of tile fragments with fired clay on the surface suggests that this concentration could correspond to the debris of a destroyed kiln (tentatively called Kiln 2).

Separated by an area of gray powdery soil ("Grey Stipe," *Assemblage 14*), a third terracotta and tile concentration was located in the northern half of Slope 3 and Sounding C (*Assemblage 13*). This cluster, a heavy concentration of tile fragments cut up into small squares and mixed in with lumps of burnt clay, measured as much as 0.5 m in depth (*Fig. 2.34* and *Fig. 2.35*) and can be interpreted as the debris from the collapse of Kiln 1. It covered the preserved portion of the combustion chamber of the kiln in D4. This concentration extended north into Sounding C (*Fig. 2.25* and *Fig. 2.34*), where the deposit was 0.87 m thick. Here, the same cubed tile fragments were concentrated in the central part of the sounding, just north of the partially preserved combustion chamber. To the west (where natural soil was identified at a depth of -9.78 m) and especially east of it, lumps of overfired clay and larger tile fragments were mixed in with pottery and terracotta relief plaques. The layer below this concentration, at a depth of -9.87 m to -10.44 m (Level 2, batt. 4–7), continued to produce tiles and relief plaque fragments, as did the soil surrounding and beneath these clusters in Slope 3 (Level 2, batt. 1, W extension, Level 2, batt. 1–5; *Fig. 2.35*).

Immediately to the west of the Retaining Wall, a concentration of ceramics, seemingly deposited in a pit ca. 0.3 m into the natural soil (*Fig. 2.34*), was found at a level of -10.44 m (Pottery Deposit, Assemblage 15). The ceramics were separated from the debris of the Early Hellenistic Kiln 1 in square D4 by a sterile layer as much as 0.5 m thick, which also covered the Retaining Wall and the inscription (*Fig. 2.35*). The total length of the cut in the natural soil was ca. 2.8 m, and the concentration of pottery was heavy throughout. It consisted mainly of black-gloss and banded wares, which can be mostly dated within the first half of the 5th century BC (*Appendix A, Table A2*).

The absence of cooking ware and the presence of some miniatures led the excavators to interpret the context as a votive deposit associated with the sanctuary. Several elements, however, invite caution. The deposit would have been located immediately outside of the sacred area, the limit of which was marked by the inscription. Furthermore, very similar material was identified along the whole length of the slope (as in Slope 4, Level 2, batt. 8–9, and C4, Level 1, batt. 4; *Fig. 2.35*), and in particular on the layer covered by the first course of the Retaining Wall, where the material was not concentrated as in a pit, but rather scattered along the slope (Stratigraphic Unit 40; see below). The characteristics of this material beg the question of whether it may actually have been detritus from the cleaning of the areas above the Retaining Wall, at the level of the upper section of the hill.

Cleaning of the Retaining Wall, 2011. In 2011, the area was cleaned to update and verify the plan of the Retaining Wall. This campaign removed two baulks that had been left during the original excavation, located immediately north and south of Sounding C. The removal of the baulks confirmed the stratigraphic sequence uncovered in 1979. Beneath a layer of debris and soil that had washed down after the end of the 1979 excavation (Stratigraphic Unit 34), the northern end of the terracotta and tile concentration found in Slope 3 and Sounding C (Assemblage 13, Stratigraphic Unit 31=36) was revealed. The concentration, here 0.2 m deep, lay atop a layer (US 35, approximately equivalent to Sounding C, Level 2, batt. 5, 7; *Fig. 2.35*) that yielded some construction material from Kiln 1 and a few diagnostic pottery fragments dated to Phase 4 (sherds of Greco-Italic amphorae, a cooking ware casserole C6.2, and AT1/AC1 tiles). The layer was separated from the upper surface of the Retaining Wall by a colluvial layer ca. 0.15–0.20 m deep with no material (Stratigraphic Unit 49).

Further cleaning was undertaken on the surface of the Retaining Wall, and the layer that had accumulated on top of the feature after the original excavation (Stratigraphic Unit 18) as well as the collapse of the feature at the southern margin (Stratigraphic Unit 9) were both removed. The cleaning revealed that the Retaining Wall was not connected to the *horos* stone, but ended 0.2 m north of it. In order to verify the stratigraphic relation of the Retaining Wall with the slope and to establish its chronology, a small portion of the feature was lifted (approximately 0.4 x 1.5 m; Stratigraphic Unit 39). From this sample, we learned that two rows of the Retaining Wall were preserved,[31] and it was built with unworked sandstone blocks that varied in dimension from 0.2 to 0.5 m and joined together with soil. Pottery dated to the end of the 6th and first quarter of the 5th c. BC (such as the krater SAV BG 16), Archaic type B amphorae, and skyphoi with offset and concave rims dated to the first half of the 5th century BC were mixed in with the sandstone blocks. The casserole SAV CkW 11, which probably came from the layer on which the Retaining Wall was set, can be dated to the same period. Beneath the Retaining Wall, a sloping layer (Stratigraphic Unit 40) was uncovered, which yielded a concentration of pottery with Late Archaic Ionic-type cups and a krater (SAV BG 02) dated to the second half of the 6th century BC. The other pottery (skyphoi with offset and concave rims, plain-ware basins similar to SAV PCW 07, SAV PCW 10, and SAV PCW 19 from the Pottery Deposit, and chytrai of types P1.4–1.5) can be dated to the first half of the 5th c. BC, when the construction of the Retaining Wall took place. As verified for the Pottery Deposit, this concentration of material rested on natural soil.

Excavated Units and Assemblages from Slope Areas 1–4

Assemblage 9
Bulldozed Stipe (*Appendix A, Table A5*)
Phase 4

Excavation and Stratigraphy: In spring 1979, during the completion of a gravel road leading from the masseria to the fields located far north of the building (*Fig. 1.2* and *Fig. 1.4*), fragments of terracotta relief plaques were brought to light along the western border of the road. The material found was gathered as part of a salvage effort, and there was no indication of stratigraphy, context, or precise location—it was retrieved possibly at the east end of Slopes 1–4.

Assemblage Summary: It is probable that just a selection of the material found was collected. Of the 1374 fragments recovered, most belong to terracotta relief plaques, with only 55 fragments of pottery, corresponding to a minimum of 14 individual vessels. The few diagnostic fragments, such as the dish SAV BG 37 and a cooking ware pan of type P3.1, can be dated to the end of the 4th–first half of the 3rd c. BC. The road also cut layers formed in the Late Archaic and Classical periods, thus explaining the presence of Ionic-type B2 cups and of the mug SAV BG 29, dated to the 5th c. BC. The terracotta plaque types found in this deposit (comprising the standing Zeus with thunderbolt in the left hand, SA 79-406T, and some well-preserved fragments of the maenad-satyr group such as SA 79-406T and SA 79-409T; *Fig. 3.14*) correspond to those found elsewhere on the slopes.

31 See p. 71 in Ch. 3, "The Retaining Wall."

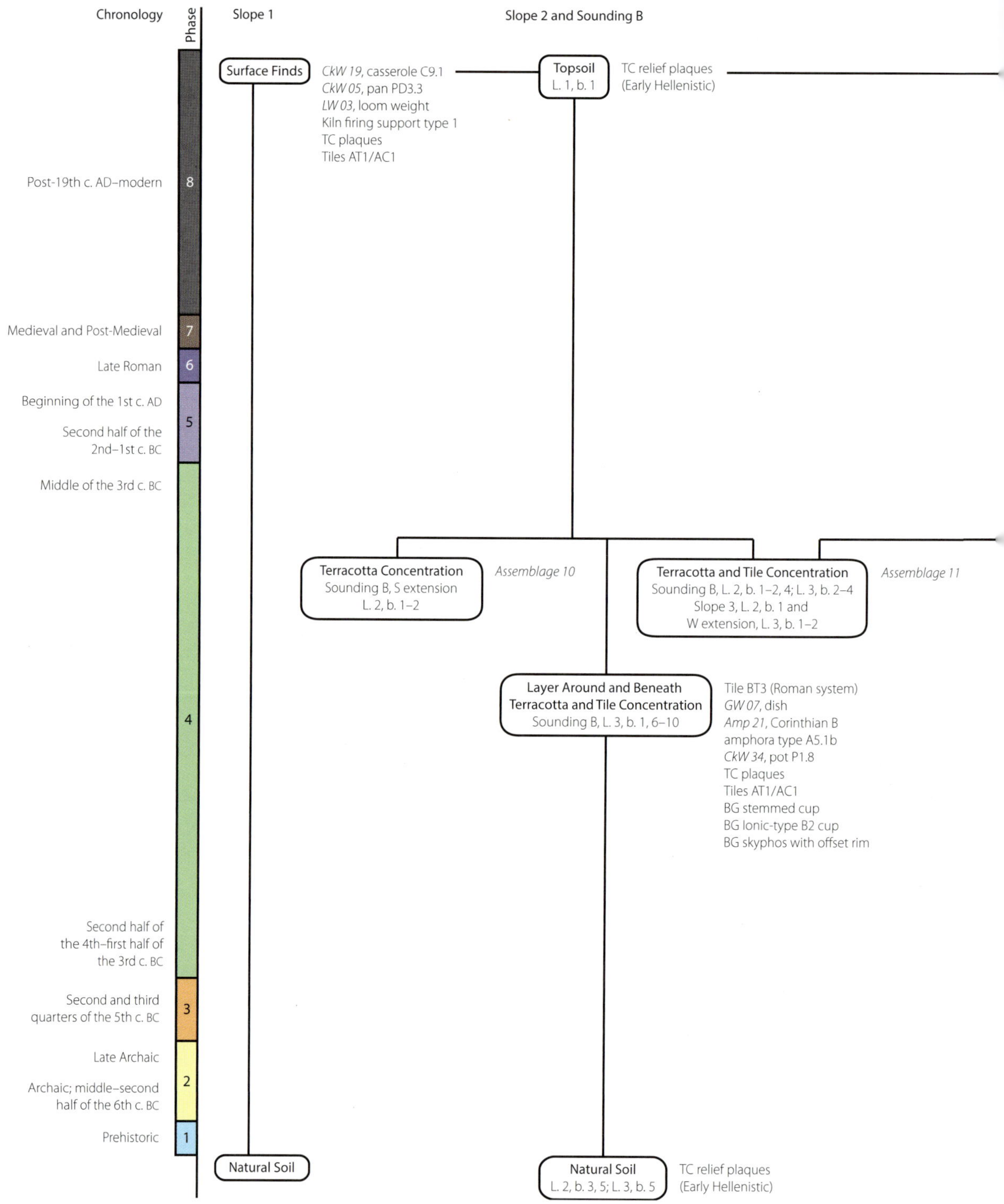

Figure 2.35 Slope Areas 1–4 and Soundings B and C, phased matrix. (KJS/ICA)

Assemblage 10

Terracotta Concentration in Sounding B, S Extension (*Fig. 2.35*)

Phase 4

Location and Lots: Sounding B, S extension, Level 2, batt. 1–2.

Excavation and Stratigraphy: Excavated in two battute, the deposit ca. 0.8 m thick was identified ca. 0.5 m south of the terracotta and tile concentration in Sounding B and Slope 3 (Assemblage 11).

Assemblage Summary: The deposit yielded mostly terracotta relief plaques (524 fragments) and a kiln firing

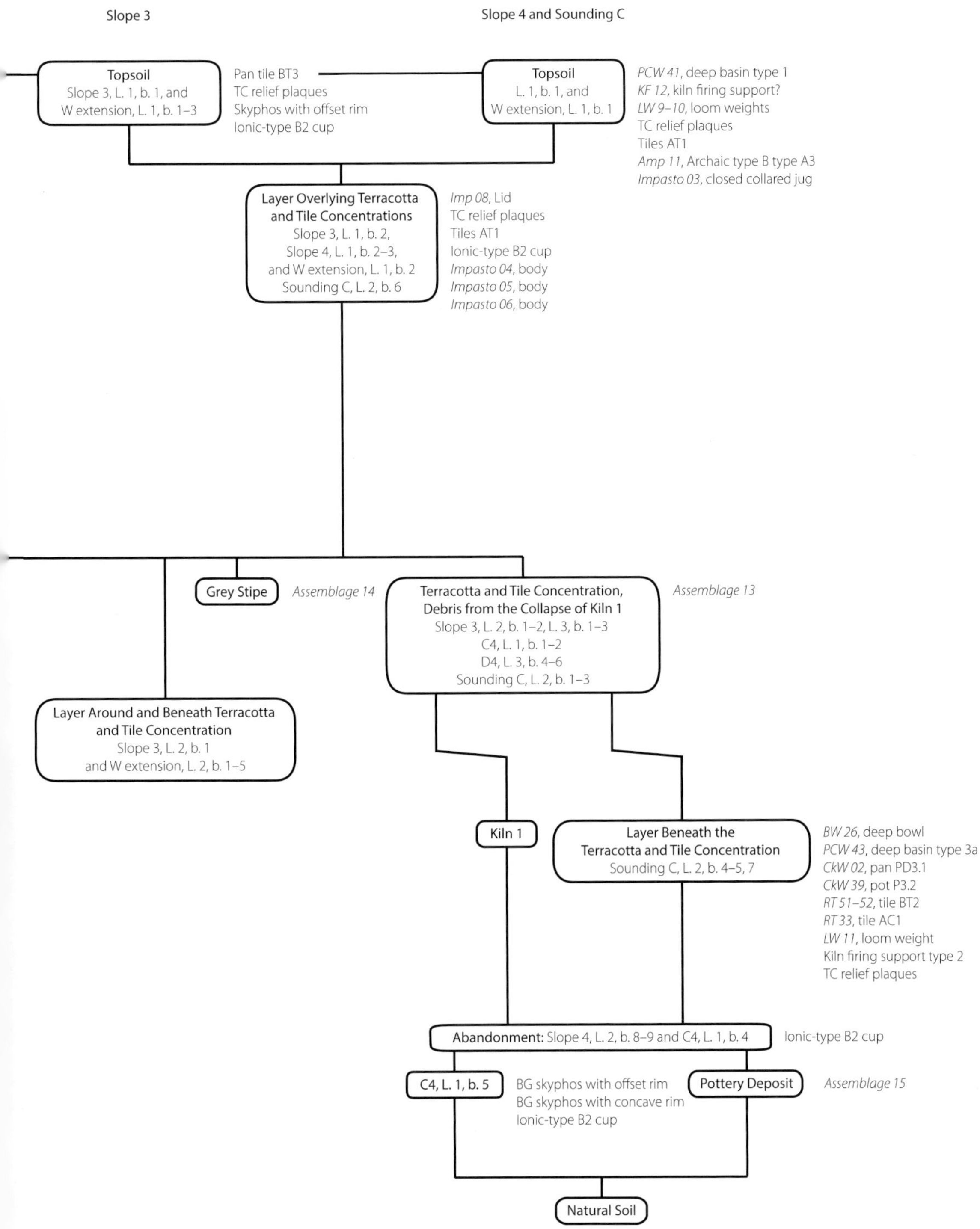

support of type 5. A single tile and a non-diagnostic cooking-ware body fragment were the only other finds associated with the terracotta relief plaques.

Assemblage 11

Terracotta and Tile Concentration, Sounding B and Slope 3 (*Appendix A, Table A7; Fig. 2.35*)

Phase 4

Location and Lots: Sounding B, Level 2, batt. 1–2, 4 (SA 79-309PL, SA 79-190PL, SA 79-295PL). Sounding B, Level 3, batt. 2–4 (SA 79-300PL, SA 79-305PL, SA 79-368PL). Slope 3, W extension, Level 2, batt. 1 (SA 79-132PL), Level 3, batt. 1–2 ("Stipe South," SA 79-376PL, SA 79-382PL, Stratigraphic Unit 31=36)

Excavation and Stratigraphy: This cluster was placed between Sounding B and the southern half of Slope 3 (*Fig. 2.35*). The soil here was clayey and dark yellowish brown (Munsell 10YR 4/6), and the concentration rested on the slope of the hill formed by natural soil. It is possible that the large tile fragments found in this cluster were the remains of a kiln (Kiln 2). The absolute level of this kiln material (ca. -9.94 m) corresponds with the level of Kiln 1, found at -9.87 m.

Assemblage Summary: The cluster yielded a conspicuous number of terracotta relief plaques mostly concentrated in the first battuta (W extension, Level 1, batt. 2). More than 350 roof tiles were found. The fragments collected can be attributed to AT1/AC1 type (Hellenistic system, SAV RT 20–SAV RT 28). Some examples (e.g., SA 79-306TA) had incrustations of heavily fired clay on the surface, confirming that they were used in the construction of a kiln together with the mudbricks (see *Fig. 9.1A*, Sample 1). Associated with the terracottas and tiles are also kiln firing supports and oscilla (SAV LW 06). The amount of pottery yielded by the assemblage was limited (102 fragments, for a minimum of 34 individual vessels) and mainly comprised of cooking ware dated to the end of the 4th–first half of the 3rd c. BC (SAV CkW 09, SAV CkW 20, and SAV CkW 23). The unguentarium SAV U 01, found just above the concentration, can be dated to the first half of the 3rd c. BC. Some Late Archaic Ionic-type B2 cup fragments and the rim of a banded lebes gamikos of the Classical period are residual. As in other areas of the slopes, a few roof tile fragments that probably washed down from the Kiln Complex can be attributed to the Roman system (SAV RT 102).

Assemblage 13

Terracotta and Tile Concentration, Debris from the Collapse of Kiln 1, Slope 3 and Sounding C (*Appendix A, Table A8; Fig. 2.35*)

Phase 4

Location and Lots: Slope 3, Level 2, batt. 1–2 (SA 79-131PL), Level 3, batt. 1–3 ("Stipe North," SA 79-385PL, SA 79.-23PL). C4, Level 1 batt. 1–2 (SA 79-1063PL). Slope 4, Sounding C, Level 2, batt. 1–3 (SA 79-325PL, SA 79-320PL, SA 79-322PL, SA 79-323PL). D4, Level 3, batt. 4–6 (SA 79-825PL, SA 79-896PL, SA 79-904PL, SA 79-907PL, SA 79-913PL)

Excavation and Stratigraphy: This concentration was exposed in the northern half of Slope 3, where the deposit was excavated as Level 2, batt. 1–2, Level 3, batt. 1–3 ("Stipe North"), and D4, Level 3, batt. 4–6 (the latter including the material found within and just immediately outside the preserved segment of Kiln 1's combustion chamber). It extended downslope to the modern road (C4, Level 1, batt. 1–2), and into Slope 4, covering an area of ca. 5.0 m in length. It was deeper in its central part, where it measured ca. 0.87 m and decreased downslope (*Fig. 2.34*). Some differences in the composition were observed during the excavation. The upper levels yielded mostly large tile fragments and terracotta relief plaques. A concentration of small tiles, mostly cut into cube-like shapes and measuring 0.03 to 0.05 m, was found just north of the preserved segment of Kiln 1 and excavated to a depth of 0.5 m (D4, Level 2, batt. 5). This extended into the central part of Sounding C (Level 2, batts. 1 and 3), while the terracotta relief plaques and large tiles were mainly found at the east end of the trench (Level 2, batt. 2). Nevertheless, the composition seems consistent all around, and that the concentration was a single feature composed of the construction material of Kiln 1 can be also confirmed by the presence of joins between different battute, as exemplified by SAV CkW 15. The presence of almost complete terracotta relief plaques (SA 79-419T, SA 79-841T, SA 79-935T) and the noteworthy concentration of kiln firing supports (27 individuals, mostly found within or just outside Kiln 1) suggests that the collapse of Kiln 1 merged with a discard deposit, possibly originally lying within the kiln itself or in its vicinity.

Assemblage Summary: This context yielded a large assemblage (1583 fragments), encompassing terracotta relief plaques (158), and especially tiles (most of them counted and discarded upon excavation) and other material (fired amorphous clay fragments, stones, and mudbricks) used in the construction of Kiln 1. Only some of the diagnostic tiles were collected; they belong mainly to the Hellenistic system.

Both the larger fragments and the small cube-like fragments were part of the construction of the kiln, and SAV RT 16, SAV RT 19, and SA 79-824T have heavily fired clay on the surface. A certain number of vessels (186 with 86 individuals identified) were recovered, mostly very fragmentary and possibly reused in the construction of the kiln (the surfaces of both SAV CkW 15 and SAV PCW 05 have incrustations). Joining fragments between the upper and lower levels of the concentration (as for the casserole, SAV CkW 15) were found. Diagnostic fragments are a type B amphora type A5.2c, SAV Amp 23 (end of the 4th c. BC), and especially the dish SAV BG 36 and the stemmed cup SAV BG 39, dated to the last quarter of the 4th–first half of the 3rd c. BC. Some residual material of the Late Archaic (such as the Ionic-type B2 cup SAV BG 05) and Classical periods (Archaic type B amphora type A3.1c, SAV Amp 05, first half of the 5th c. BC) was found. Noteworthy is the bolsal SAV BG 28 of the last quarter of the 5th c. BC, which provides the most recent evidence for the lower limit of the frequentation of the site in Phase 3.

Assemblage 14
Grey Stipe (*Appendix A, Table A6; Fig. 2.35*)
Phase 4

Location and Lots: Slope 3, W extension, Level 2, batt. 1 (SA 79-424PL).

Excavation and Stratigraphy: Between the two terracotta and tile concentrations lay a patch of powdery grayish soil with cinder (Munsell 10YR 5/3), possibly ash from the kiln, excavated separately in a single battuta.

Assemblage Summary: The context yielded a single fragment from a terracotta relief plaque and especially fragmentary tiles. The only noteworthy pottery fragment is the plate/lid SAV PCW 20, recomposed from fragments also found in the layers surrounding and beneath the terracotta and tile concentrations and in Assemblage 13.

Assemblage 15
Pottery Concentration in Sounding C, Pottery Deposit (*Appendix A, Table A2; Fig. 2.35*)
Phase 3

Location and Lots: Sounding C, Pottery Deposit, Level 3, batt. 1–4 (SA 79-357PL, SA 79-988PL).

Excavation and Stratigraphy: The so-called Pottery Deposit (Sounding C, Level 3, batt. 1–4) lay beneath a layer of soil that may have accumulated when the area was abandoned before its reoccupation in Phase 4 (Sounding C, Level 2, batt. 8–9). According to the excavators, the concentration of pottery, found at a level of -10.44 m, was placed in a cut in the natural soil 0.3 m deep and 2.8 m long.

Assemblage Summary: With 496 pottery fragments corresponding to a minimum of 107 vessels, this assemblage is set apart from any other context found at the site. The deposit yielded especially black-gloss (28.7%), banded (26.9%) and plain wares (33.3%). Transport amphorae and cooking ware are present, but with a small number of individual vessels (3.7%). The black-gloss, attested with a high number of reconstructable vessels, ranges in date from the Late Archaic period (Ionic-type cup SAV BG 04 and a skyphos with offset rim, SAV BG 08) to the end of the 5th c. BC (SAV BG 26, SAV BG 30, and SAV BG 33), with most of the vessels dated to the second and third quarters of the 5th c. BC. Few fragments from C-type skyphoi of the 4th c. BC are probably intrusive, such as the plain/coarse-ware dinos (SAV PCW 45). The Attic Castulo cup (SAV BG 01) is one of the few imports recognized at the site. The range of shapes includes mainly drinking vessels (see p. 221 in Ch. 15, "Shapes and Functions"). Banded, plain, and coarse wares include a large variety of jugs (e.g., SAV BW 38 and SAV PCW 47), lekanai (SAV BW 21, SAV BW 22, and SAV BW 23), basins (SAV PCW 07, SAV PCW 10, and SAV PCW 19), and one-handled bowls (SAV BW 01, SAV BW 04, SAV BW 06, SAV BW 10, SAV BW 11, SAV BW 13, and SAV BW 14).

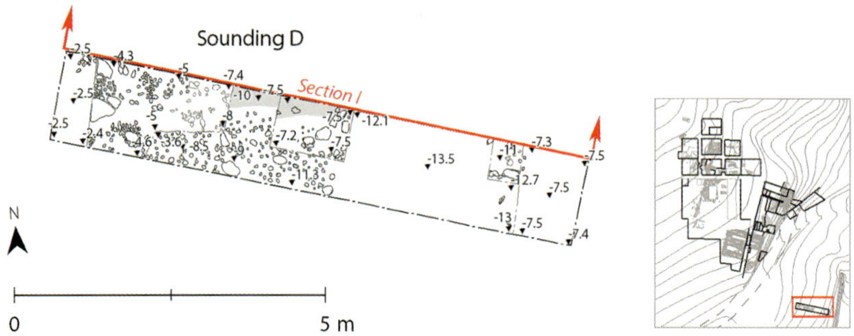

Figure 2.36 Detailed plan of Sounding D. Elevations for this sounding are relative to a different local datum than the rest of the excavation. (JT/ICA)

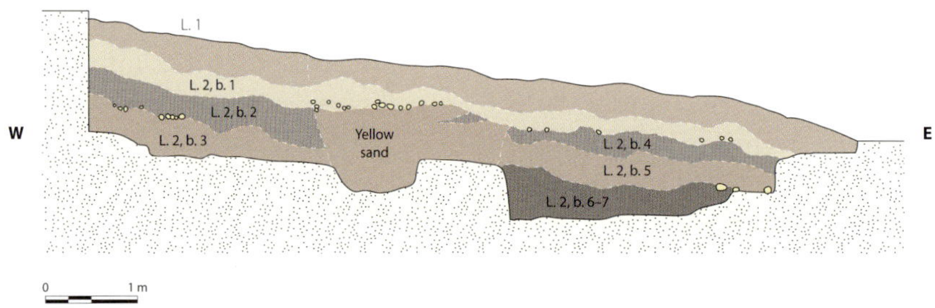

Figure 2.37 Section I, E-W section through the N edge of Sounding D. (KJS/ICA)

Sounding D

The surface of the slope southeast of the Kiln Complex (now hidden by vegetation) was, at the time of excavation, characterized by the presence of tiles and pottery that had washed down from the upper part of the slope. In 1980, excavation was extended into this area with Sounding D (8.40 x 1.5 m, *Fig. 2.36*), which cut the slope from east of the modern road down to the lower limit of the slope. Its aim was to explore the stratigraphy, organization, and function of this part of the site. The sounding helped define the extent of the production area, confirming that it was limited to the northern part of the slope.

The investigation showed that this part of the slope was an area where material washed down from above, and no architectural remains were found. The four sloping layers identified (*Fig. 2.37*) were interrupted by an area of yellow sand with no material, interpreted as a modern fill for a drainage system. Most of the pottery was found in the first two layers and ranged in date from the Late Roman to the Early Hellenistic periods, with an interesting concentration of prehistoric material in the western part of the sounding.

Stratigraphy and Phasing by Area: The House on the Hill

A survey undertaken on the top of the hill of Sant'Angelo Vecchio in June 1979 revealed a noticeable concentration of uneven stones, fragments of skyphoi with offset rims, cups of Ionic type, and banded lekanai, which suggested the presence of a Late Archaic structure lying below. Three test trenches, oriented east to west, were therefore opened in order to determine the nature of the context (*Fig. 2.38*), which was subsequently called the "House on the Hill."[32]

Excavation and stratigraphy. Test Trench 1 (2.0 x 7.0 m) and Test Trench 2 (2.0 x 5.0 m, with a rectangular extension to the south of 0.8 x 1.4 m) were both laid north of the gravel road (*Fig. 2.38* and *Fig. 3.1*). In

[32] Preliminary reports on the House on the Hill are in Edlund 1979, 1986 and Carter 2006, 137. See also Amore 1992, 85 and *Survey*, 714, table 21.9.

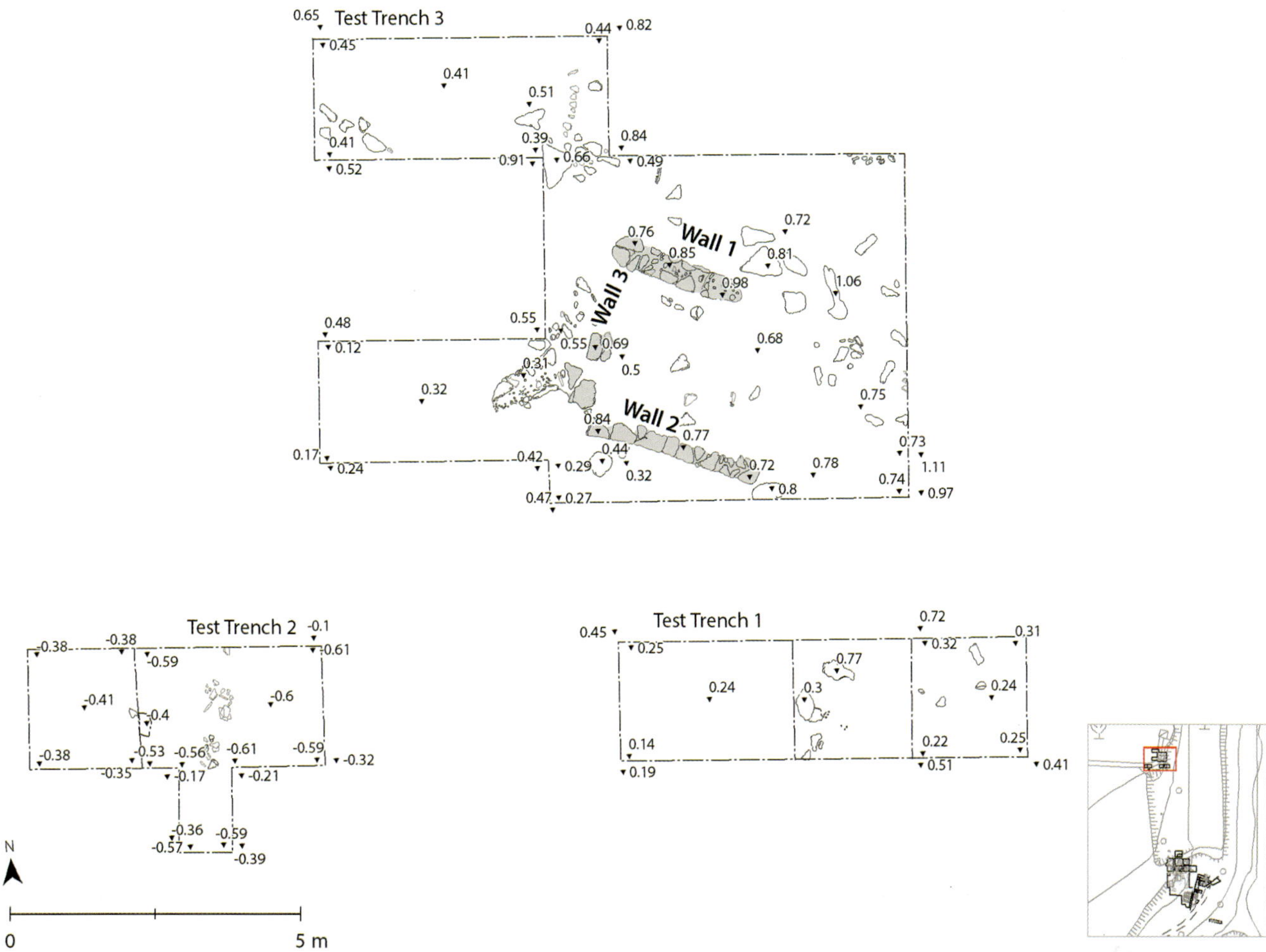

Figure 2.38 Detailed plan of the House on the Hill. (JT/ICA)

Test Trench 1, the only identifiable feature was a cluster of irregular chunks of local sandstone mixed with the topsoil and resting directly on natural soil. Similar irregular scattered rocks resting on natural soil were also present in Test Trench 2 (laid out 5 m to the west of the first), where some 120 roof tile fragments were found mixed with the soil. Test Trench 3 (5.60 x 6.15 m, with two extensions to the west, for a total length of 10.10 m east to west) was laid out on the highest point of the hill between the modern farmhouse and Test Trenches 1 and 2. It eventually led to the discovery of the foundation walls of a structure.[33] The building and the surrounding area lay beneath a shallow layer of topsoil (0.2–0.3 m thick; *Fig. 2.39*), resting directly on natural soil and the sandstone formation. Although the telltale traces of plowing were nowhere to be seen, the slipping of the stones that originally formed the walls and the presence of joins between fragments found in significantly different contexts (e.g., between fragments found in Test Trenches 1 and 2, and some found in Test Trench 3, Level 1, batt. 2 and the surface) provide indirect evidence for disturbance, possibly due to agricultural activity and/or erosion. The area defined by the walls, equally disturbed, was nevertheless excavated separately (Level 2, batt. 1; *Fig. 2.39*).

33 See p. 61 in Ch. 3, "The House on the Hill."

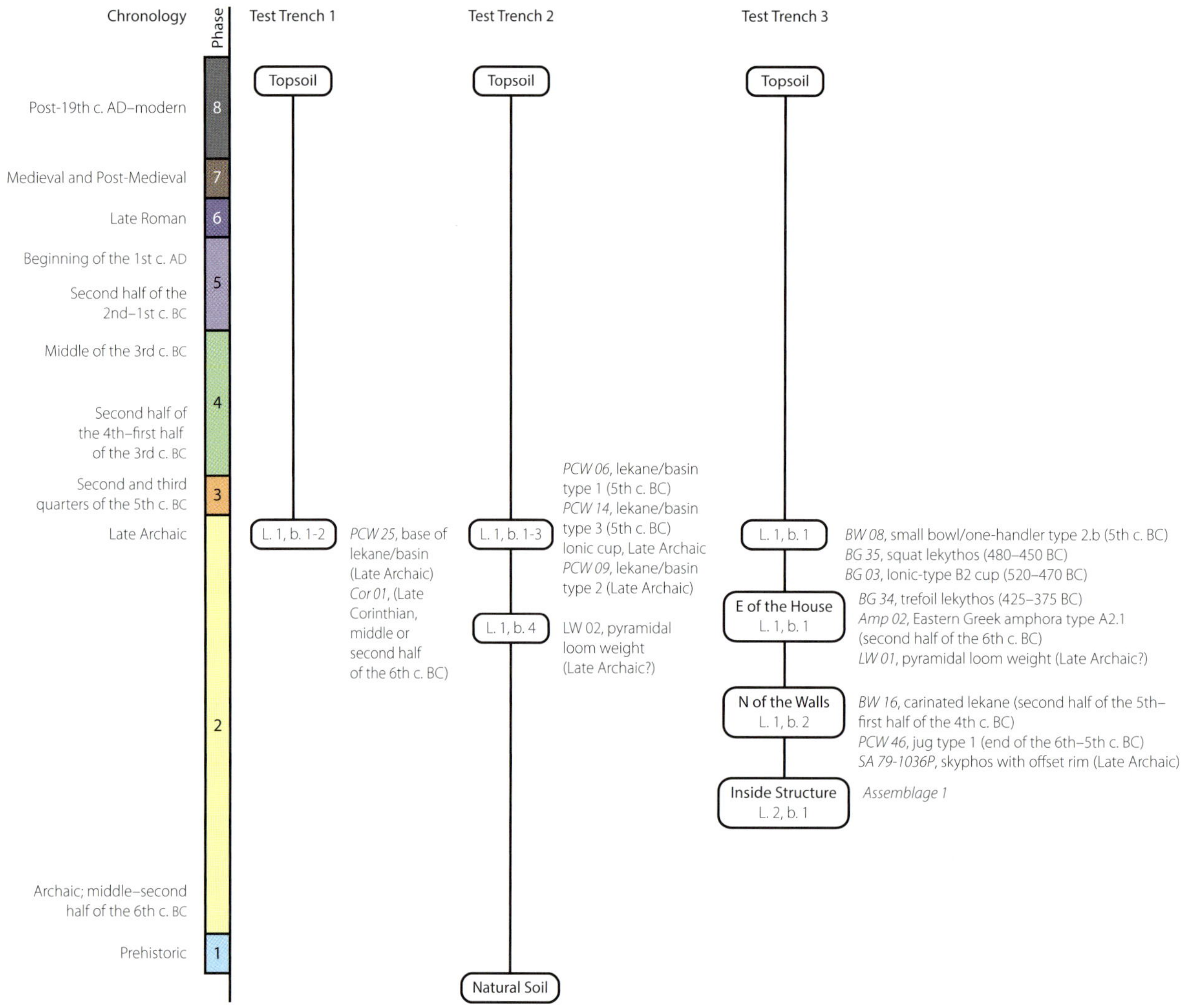

Figure 2.39 House on the Hill, phased matrix. (KJS/FS/ICA)

Excavated Units and Assemblages from the House on the Hill

Assemblage 1

Test Trench 3, within the structure (*Appendix A, Table A1*)

Phase 2

Location and Lots: Level 2, batt. 1. Lot: SA 79-1062PL.

Excavation and Stratigraphy: Excavated in a single battuta, the soil within the structure defined by Walls 1–3 consisted of a uniform layer formed by agricultural processes and therefore disturbed. It lay on the natural soil.

Assemblage Summary: The assemblage comprises mostly pottery (561 fragments, for 100 individual vessels) with a range of forms encompassing table and storage wares, with some cooking ware bearing traces of use as well as transport amphorae. SAV Li 07 and SAV Li 08 could be interpreted as tools used in the preparation of food. The very fragmentary material ranges in date from the second half of the 6th (SAV BG 07) to the first half of the 5th century BC (SAV BG 18), especially concentrated in the Late Archaic period with a high number of Ionic-type B2 cups and skyphoi with offset rim. See also p. 61 in Ch. 3, "The House on the Hill."

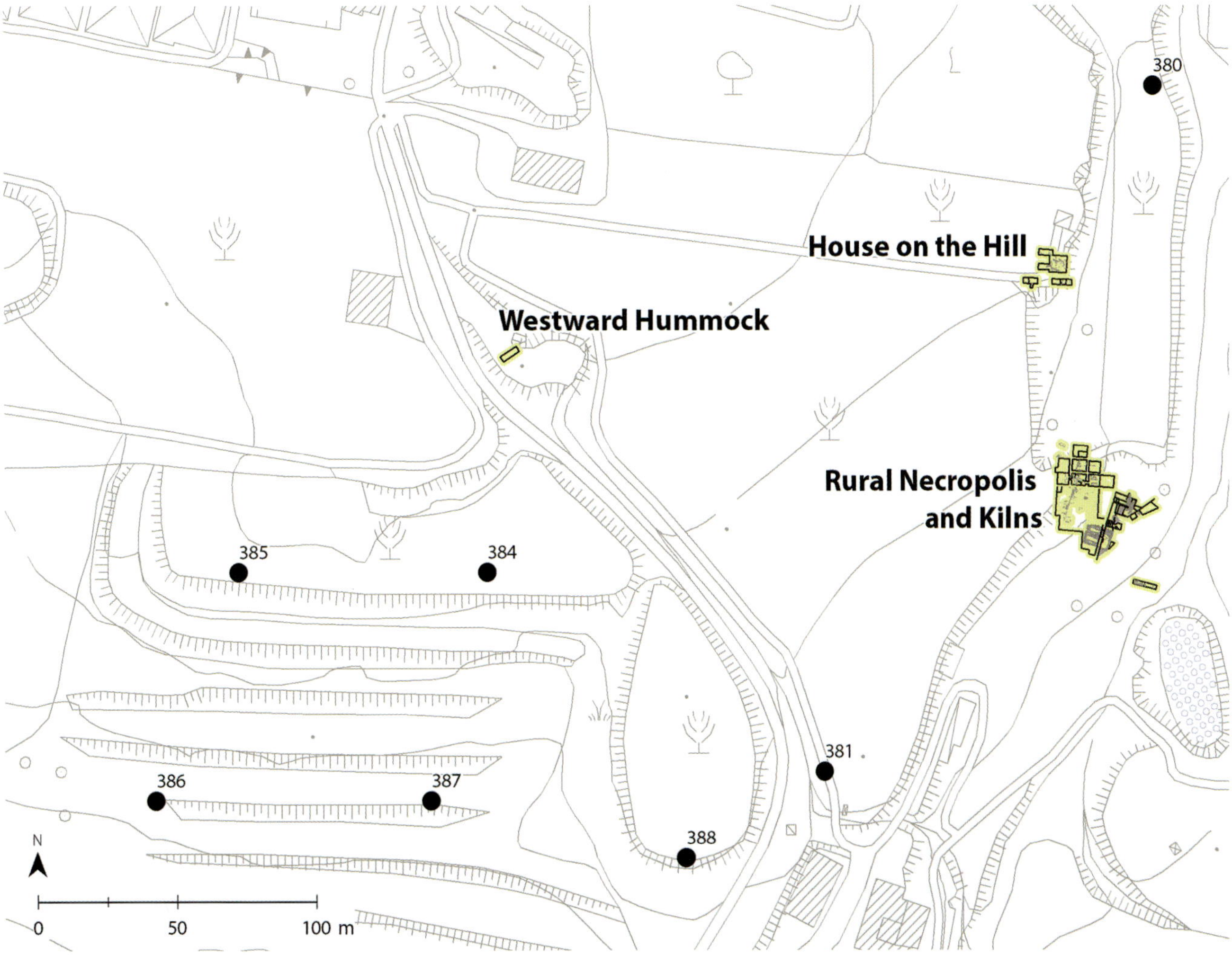

Figure 2.40 Location of the House on the Hill and the Westward Hummock in relation to the hillside area at Sant'Angelo Vecchio. (JT/ICA)

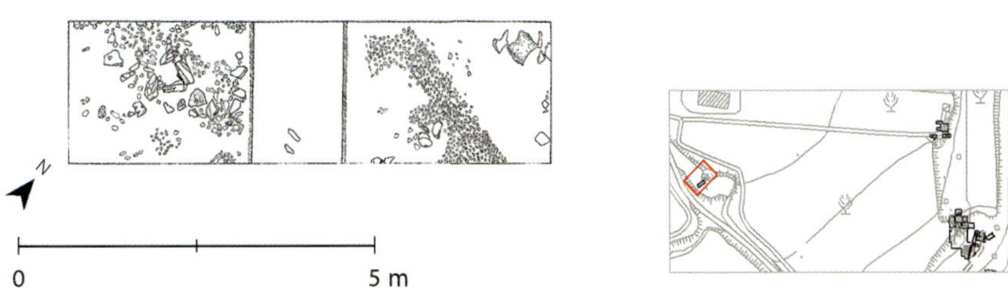

Figure 2.41 Detailed plan of the Westward Hummock. (JT/ICA)

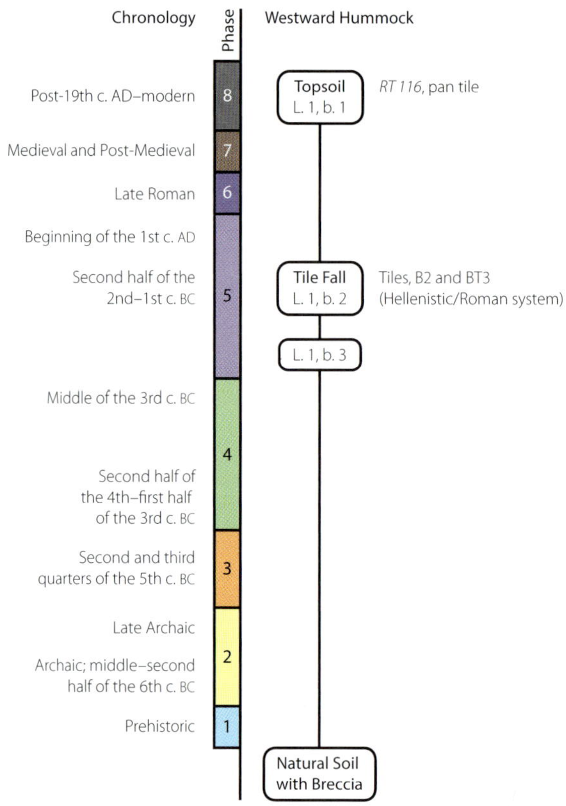

Figure 2.42 Westward Hummock, phased matrix. (KJS/FS/ICA)

Stratigraphy and Phasing by Area: The Westward Hummock

Evidence of occupation was also found 200 m west of the gridded area (*Fig. 2.40*), where tiles were noted during the construction of a tarred road leading to the top of the hill. The trench (2.0 x 7.0 m; *Fig. 2.41*), oriented northeast to southwest, lay just east of the road on the hummock, and brought to light a tile fall concentrated in the northwestern part of the excavated area. No walls were identified, although scattered limestone blocks were found. The few plain ware fragments found in association (SA 79-1004PL) are not diagnostic. The tile fall, resting on the natural soil, which also appeared on the surface in the eastern part of the trench, yielded cover tiles of type BC2 (Hellenistic/Roman system) and a flat pan tile of type BT3 (Roman system),[34] suggesting a Hellenistic or Late Republican chronology for the site, otherwise destroyed.

[34] See Ch. 32, "Roof Tiles and Bricks."

3

The Structures at Sant'Angelo Vecchio

Francesca Silvestrelli and Ingrid E. M. Edlund-Berry
With a contribution by Francesco Guizzi

Figure 3.1 The House on the Hill from the W in 1979. (CW/ICA)

The House on the Hill

Overview

The occupation of the area of Sant'Angelo Vecchio appears to have begun in the second half of the 6th century BC on the highest point of the hill, 214 m north of the masseria. The place is now marked by a small, deserted, and nearly collapsed farmhouse dating to the early part of the 20th century (*Fig. 3.1*). At the time of excavation, the surface immediately to the south of the southern wall of the modern farmhouse sloped gently downward, and sandstone formations, covered by a thin layer of topsoil, were visible in places. The profile of the hill changed approximately 20 years ago, when the entire area was bulldozed for the planting of a fruit orchard (since removed), causing the complete destruction of the archaeological evidence (*Fig. 3.1*).

Structure and Plan

The area was identified as potentially interesting after the recovery of Archaic pottery,[1] and subsequent excavation in 1979 uncovered the remains of a structure defined by two parallel east-west walls (Walls 1 and 2) and traces of a connecting north-south cross wall (Wall 3; *Fig. 3.2*), though no evidence of a fourth wall delimiting the southern side was found. Constructed from unconsolidated cobbles and sandstone with occasional conglomerate chunks, the walls were preserved to just one course at the lower level of the foundations and rested directly on natural soil and bedrock (*Table 3.1*). No foundation trenches were observed. The wall construction technique differs from that normally used in Archaic structures in the extensive use of sandstone chunks. The foundations of Archaic structures were

[1] See Ch. 2, "Site Phasing, Stratigraphy, and Site Assemblage."

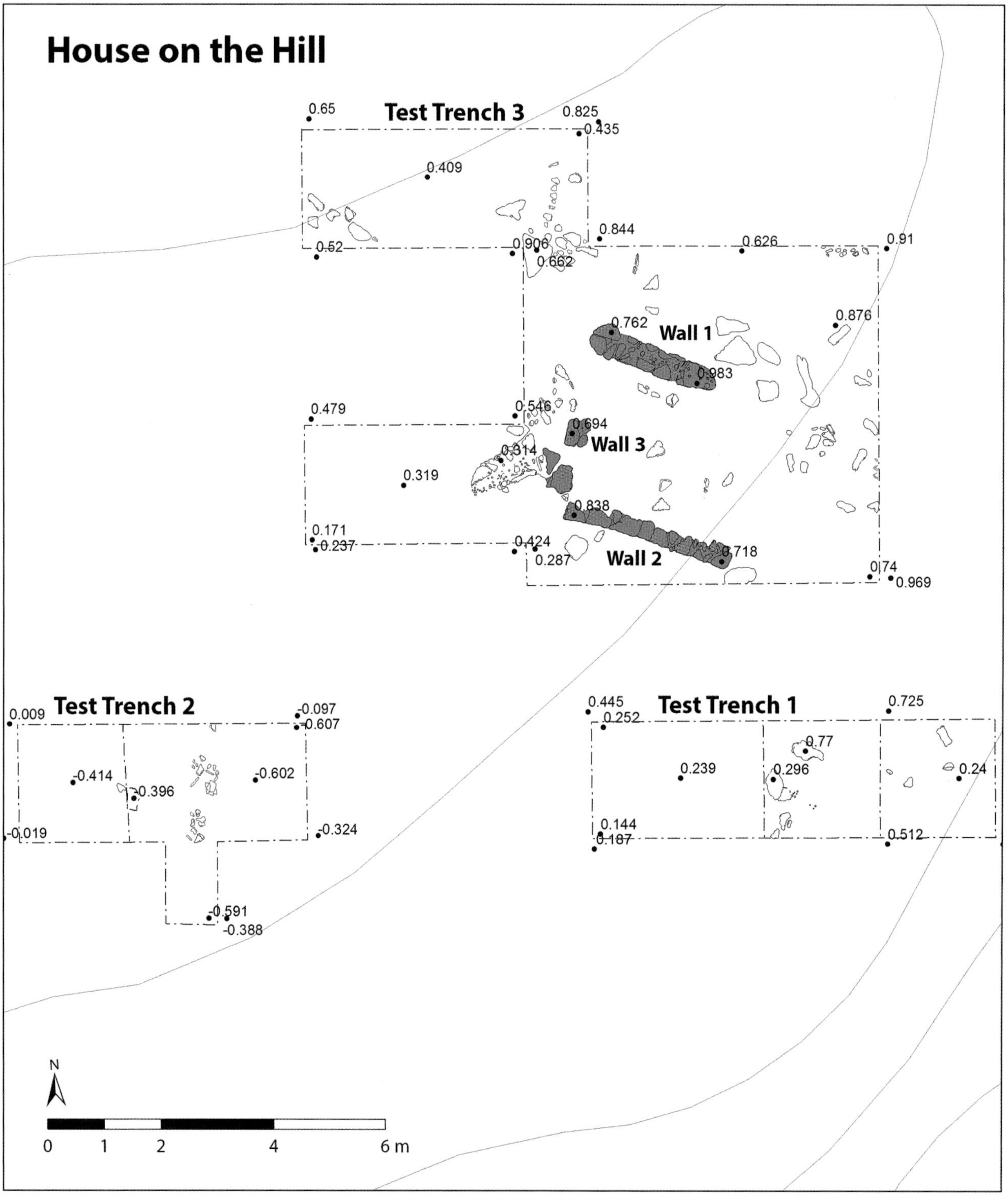

Figure 3.2 The House on the Hill: plan of the excavations. (JT/ICA)

usually built with unconsolidated river cobbles and occasionally small stones.[2] The peculiarity of the technique is likely due to the local availability of sandstone,[3] also employed in the construction of the Retaining Wall (see below).

Wall 1, the southern wall, preserved to a length of ca. 3.0 m, was placed at a distance of ca. 2.4 m from Wall 2, the northern wall, which was 2.4 m in length. The two walls vary in width from 0.35 m (Wall 1) to 0.5 m (Wall 2). A few scattered blocks (Wall 3) probably marked the western limit of the structure, which had a projected length of at least 3.5 m and a preserved outer width of slightly more than 0.3 m.

No indication of the superstructure was identified during the excavation. The socle may have been surmounted by mudbrick, commonly used in walls of the Archaic and Classical periods at Metaponto and elsewhere.[4] The presence of wooden posts propping up the tile roof cannot be ruled out, but no trace was found among the archaeological evidence.

Although no proper tile fall was discovered, several curved roof tile fragments (SAV RT 14 and SAV RT 15) and an opaion (smoke hole tile, SAV RT 36) of the Laconian A system were found scattered in Test Trenches 1 and 3 and, to a larger extent, in Test Trench 2 (see *Fig. 32.1*). These fragments may be the remains of an original collapse that was disturbed and perhaps partially dragged by the plow to the lower part of the slope, along with chunks of sandstone that may have belonged to the upper courses of the foundation

Wall	Location	Construction Technique	Photo
1	House on the Hill, S limit of the room	One course of the unconsolidated foundation wall built with worked blocks of local sandstone, one block of conglomerate.	 
2	House on the Hill, N limit of the room	Single course of the unconsolidated foundation wall; W portion at the probable conjunction with Wall 3 built with 2 sandstone blocks (ca. 30 cm). Faces built with sandstone blocks and inner space filled with river cobbles. Conglomerate chunk at the E end, possibly to reinforce the structure.	
3	House on the Hill, W limit of the room	Two isolated large sandstone blocks probably pertaining to the east-west cross wall.	

Table 3.1 The House on the Hill: construction techniques of the walls.

[2] Uggeri 1969, 61 (Fattoria Musillo in contrada Saldone); Adamesteanu 1974, 82–85 (Cugno del Pero); *Fattoria Fabrizio* (Wall 2).

[3] See Ch. 7, "Geologic Setting."

[4] Fragments of mudbrick and denuded soil from them were identified in room 2 at Fattoria Fabrizio (see *Fattoria Fabrizio*). For the technique, see A. Liseno 2007, 78–79, with extensive bibliography.

Figure 3.3 The House on the Hill at the end of excavation in 1979. (CW/ICA)

walls. Nevertheless, the highly fragmentary preservation of the tiles prevents either a reconstruction of the roofing system or a sound chronology for the recovered examples, which may belong to the Archaic or the Hellenistic system.[5] A pan tile of type BT1, SAV RT 47, can be more specifically attributed to the Archaic/Classical system.

No trace of the original floor, usually laid at the same level of the foundations or just above it, was identified. The three preserved walls delimited a single room of at least 10.5 m^2, but unfortunately, their state of preservation at the time of the excavation impedes a reliable reconstruction of the plan. Whether the structure consisted of a single room, or rather (more probably, judging from the few available comparanda) formed part of a larger building is impossible to determine based upon the available evidence. If part of a larger building, it may have originally extended into the area occupied by the modern farmhouse, and since disappeared due to erosion and the later frequentation and utilization of the site.[6]

Most of the pottery was concentrated in the layer excavated within the walls (Assemblage 1). Although from a uniform layer formed by agricultural processes and therefore disturbed, the material recovered inside the room and in its immediate surroundings contributes some useful information about the chronology of the building's occupation and its interpretation. The

[5] See Ch. 32, "Roof Tiles and Bricks."

[6] Although many of the excavated farmhouses yielded evidence of Archaic occupation (Fattoria Stefan: Adamesteanu 1973, 58; Adamesteanu 1974, 85–87; Fattoria Fabrizio: *Fattoria Fabrizio*; Fattorie Pacciano, Musillo, and Pascucci-Torraca in contrada Saldone: Uggeri 1969; Adamesteanu 1974; Amore 1992–1993, 64; Fattoria La Cappella all'Incoronata: Amore 1992–1993, 82–84; Pantanello, Proprietà Morlino: Nava 2002, 733–35; see also Ch. 1, "The Many Lives of a Rural Site: Sant'Angelo Vecchio"), the only rural building known to a certain extent is the farmhouse at Cugno del Pero, in Contrada Bufalara (Adamesteanu 1974, 82–85; Amore 1992–1993, 79–81; Carter 2006, 135–36). This farmhouse, dated to the second half of the 6th c. BC and occupying an area of ca. 225 m^2, comprised four rooms with foundation walls in river cobbles and a Laconian roofing system. A comprehensive census of rural buildings in the Metapontine chora can be found in *Fattoria Fabrizio*, appendix B.

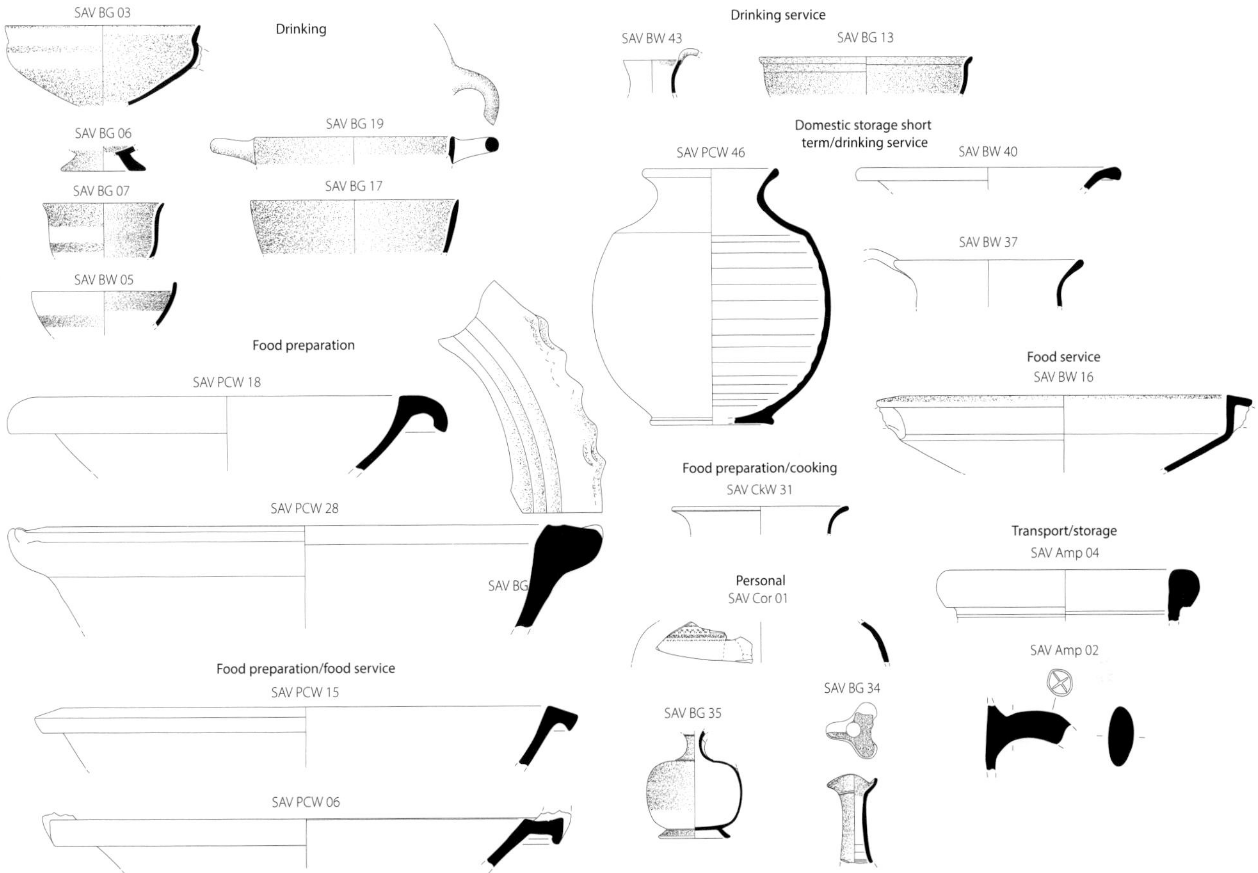

Figure 3.4 Pottery from the House on the Hill. (KJS/ICA)

upper chronological limit is provided by a Corinthian fragment of the middle of the 6th century BC (SAV Cor 01), while the black-gloss and transport amphorae indicate that occupation continued into the middle decades of the 5th century BC (SAV Amp 04, SAV BG 34, and SAV BG 35; *Fig. 3.4*).

The range of forms consisted primarily of table and storage wares with some cooking ware bearing traces of use and transport amphorae types normally utilized for wine and oil (*Fig. 3.4*). The scope is consistent with ceramic assemblages from other domestic contexts, and suggests that the room probably belonged to a farmhouse. Evidence for activities that took place at a farmhouse is, however, extremely meager, leaving room for few tentative hypotheses. The assemblage can be entirely ascribed to the daily life of the occupant family. Further evidence comes from palaeobotanical analysis, which has demonstrated that the farmhouse was surrounded by an open, predominantly grassy environment with occasional areas of Mediterranean scrub, pastures, and limited cultivation areas.[7] The high concentration and diversity of fungi that live on excrement, as well as the presence of parasite eggs, indicate the presence of herbivores, suggesting that animal husbandry constituted an important part of the local economy.

A small number of wasters found in secondary depositions, including a one-handled bowl of the late 6th–5th century BC, may indicate limited pottery production. If that is indeed the case, production activities at the site would have begun even before the kilns on the slopes were built in the late 4th century BC, as also suggested by the antefix molds SAV AT 05, SAV AT 07, and SAV AT 08.[8]

[7] See Ch. 8, "Archaeobotanical Analysis."

[8] See Ch. 5, "Ceramic Production at Sant'Angelo Vecchio and in the Metapontine Chora," and Ch. 29, "The Architectural Terracottas: Some Considerations on Production."

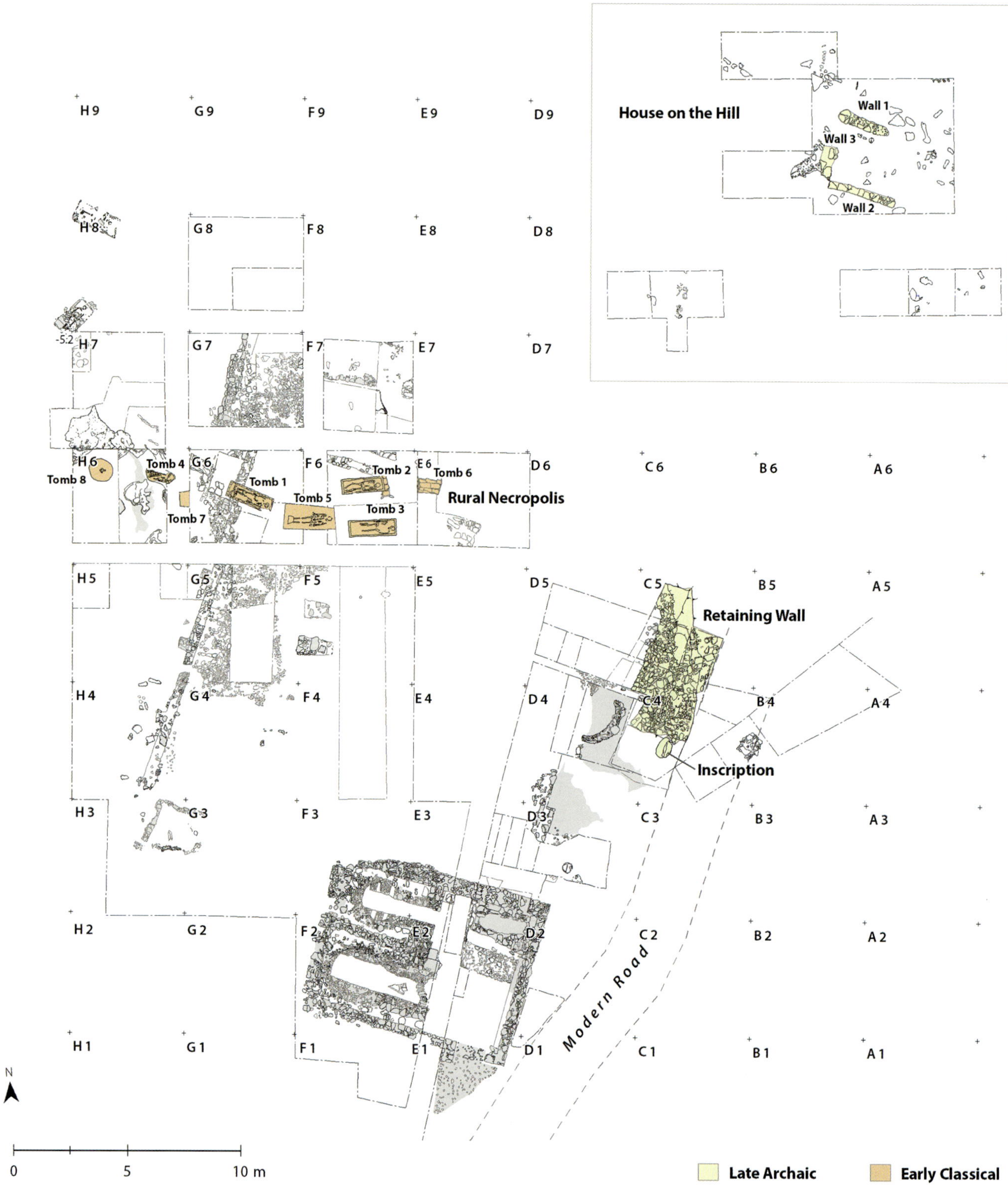

Figure 3.5 Sant'Angelo Vecchio in the Late Archaic and Early Classical periods. (JT/ICA)

The Late Archaic and Early Classical Periods

During the Late Archaic and Early Classical periods, occupation at Sant'Angelo Vecchio expanded to include a sacred area by the spring, a terrace Retaining Wall, and a rural necropolis (*Fig. 3.5*).[9]

The Sanctuary

A sacred area was probably located southeast of the Late Archaic House on the Hill, at the foot of the slope just above the springs.[10] Here, the excavations brought to light a boundary marker (*horos*) in primary deposition bearing the inscription τō hιϙρο ("of the sanctuary") in the Late Archaic Achaian alphabet carved into the western side of an apparently unworked block of sandstone (*Fig. 3.7* and *Fig. 3.8*).[11] It was securely situated in the natural soil (*Fig. 3.6*). The *horos* marks the limit of a sanctuary that probably extended immediately east of it, incorporating the area that still today includes the perennial springs supplying water to the site.

Due to the occupational history of Sant'Angelo Vecchio, a reconstruction of the structure and shape of this sacred area is not feasible. The presence of the masseria and the intense agricultural and pastoral exploitation of the surrounding territory led to consistent transformation of the landscape, compromising its archaeological legibility. The slope immediately to the east of the modern road, which was probably part of the sacred area, has been affected over the last 20 years by accelerated erosion processes, while the absence of cultivation has led to the growth of invasive tree species that completely obscure the surface.[12] The springs, used for irrigation and probably also for watering animals, were channeled by a conduit that fed the water into a series of basins placed at regular distances along the slope of the hill (see *Fig. 1.7*). These basins and the water that fed them may have destroyed evidence of other structures connected to the sanctuary. In recent times, the water from the springs was channeled into a sort of pond more than 3 m deep (see *Fig. 1.2*). Whatever evidence there is, as a consequence, has been hidden or has disappeared, and nothing certain can be said about the appearance of this sanctuary. Still, the recovery of an Archaic channeled sima that was reused inside the Late Republican kilns (SAV RT 129) could be interpreted as part of the roofing system of a building of some importance, which may have been part of the sacred area.

Figure 3.6 The *horos* boundary marker from the E in 2010; in the foreground, the cut of the modern road. (FS/ICA)

According to ancient sources,[13] every place equipped with a spring was sacred, and this was all the more true in Basilicata, where the model of the sanctuary connected to the spring is well known in both the Greek and the indigenous worlds.[14] With its proximity to the Basento Road and the connection with an all-weather spring, the placement of this sanctuary is consistent with what we know from other sacred areas.[15] The general organization of the area in fact clearly evokes two of the most important sanctuaries discovered in the territory of Metaponto,

[9] For a detailed discussion of the necropolis, see p. 85 in Ch. 4, "The Rural Necropolis of the Classical Period."

[10] On the sanctuary: Edlund 1984, 287–90; 1986, 119–22; 1987, 99–100; Osanna 1992; Leone 1998, 122–23.

[11] See below, p. 68, "The Inscribed Stone."

[12] See Ch. 10, "Observations on the Vegetation and Landscape Dynamics."

[13] Servius ad Aen. 7.84.

[14] A. Russo 1999.

[15] *Survey*, 706–7.

at Pantanello[16] and San Biagio,[17] both organized around a spring that is monumentalized to varying degrees and serves as the center of cult life.

There is no indication for either the date of the sanctuary's creation or for the identity of the cult practiced there. Residual pottery shows that the site was frequented at least from the central decades of the 6th century BC onward, as suggested by the Laconian amphora (mid-6th century BC; SAV Amp 01) and the fragments of black-gloss column kraters (second half of the 6th century BC; SAV BG 02 and SAV BG 16), possibly the earliest fragments found in the Slope Area.

There is also insufficient information for determining the date of the sanctuary's abandonment. The inscription appears to have been obscured by a layer of sterile soil,[18] in turn covered by the deposits created by the terracotta production facility that was active at the site by the late 4th century BC. This fact could suggest that the sacred area was abandoned sometime after the second half of the 5th century BC, when the House on the Hill was abandoned (as has been observed in other minor sanctuaries in the chora),[19] or that it underwent a radical transformation.

Figure 3.7 Drawing of the inscription; scale 1:10. (KJS/ICA)

The Inscribed Stone

Francesco Guizzi

A massive living rock that is almost certainly in its original position (*Fig. 3.7* and *Fig. 3.8*) was used to demarcate the boundary of a sacred area at Sant'Angelo Vecchio.[20] The sacred area very likely included a spring, the presence of which is suggested by a small pool located downslope from the structures. Both the near total absence of signs of carving on the epigraphic medium and the stone's unsuitability for carving suggest that this was a natural element used as a boundary marker as seen in Homeric tradition. In *Iliad* XX.403–405, Athena assails Ares with a rock that she has picked up from the ground, using it as an offensive weapon and depriving it of its primary function as a *horos*.[21]

Inscribed *cippi* are attested at Metaponto, but they were found around the urban temples and were therefore urban dedications. The *cippus* bearing the name

[16] Carter 2006, 163–73, with previous bibliography. The sanctuary at Pantanello will be the subject of a future title in this series.

[17] De Stefano 2014, with previous bibliography.

[18] See Ch. 2, "Site Phasing, Stratigraphy, and Site Assemblage."

[19] *Survey*, 663.

[20] The inscription was first interpreted by L. Moretti (Edlund 1982a, 8).

[21] On the ambiguity (or semantic complexity) inherent in the absence of inscriptions on *horoi*, see Ober 1995, which takes into account speech-act theory and analyzes the Homeric passage in detail (96–100).

Figure 3.8 Detail of the inscription on the boundary stone in 1979. (CW/ICA)

of Zeus Aglaos from San Biagio alla Venella likewise almost certainly comes from the urban sanctuary.[22] Although the function of these artifacts appears to be the same as that of the stone from Sant'Angelo Vecchio, there are differences in the way the medium was worked and in the type of inscription. The numerous aniconic votives in the urban center[23] are an altogether different matter, and cannot be compared to the object and text found at Sant'Angelo Vecchio, which at present has no parallel anywhere in the Metapontino.

The sandstone *cippus* (*Fig. 3.7*) is in the shape of a wedge on its axis, with a stone divider (max. 89 x 69 x 33 cm). On the upper surface, there is a hole of unclear origin (whether natural or artificial). The inscription is on the west side, which also preserves a small step-like projection that was left blank. A lighter band located lower down (h ca. 15 cm) probably marks the point above which the block was visible in antiquity.

Size (cm)	Letter
3.8	Initial tau and squared omicron
4.5	Terminal rhomboid omicron (type 2 in *LSAG*², 248), or squared omicron, as above, but rotated 45°
6.0	H (open aspiration; type 2 in *LSAG*², 248)
8.5 (pres.)	Three-stroke sinistrorse iota (type 1 in *LSAG*², 248), rho (type 2 in *LSAG*², 248)

Table 3.2 Comparison of letter sizes on the inscription.

The size of the letters varies widely (*Table 3.2*). The text (*Fig. 3.8*) runs from left to right on a single line. The lettering is marked and irregular, adapting to the variable roughness of the surface. The first three letters are aligned, but those inscribed after the projection, apparently left untouched, start from a higher point and tend to deviate toward the bottom, following the

22 For the two *cippi* of Zeus Aglaos, see Giacometti 2005, 118–19, nos. 2–3 (one of which she considers a votive offering), with bibliography. Guarducci (1978, 49) considered the San Biagio inscription—the only one known at the time—a boundary marker.

23 On Metapontine votive offerings and their various forms in the broader context of southern Italy and Sicily, see Doepner 2002, 19–81, with catalog (195–223).

Figure 3.9 The inscription and the Retaining Wall in 1979 (from the E). (CW/ICA)

orientation of the stone. Except for the last one, they are also larger. Between the iota and the rho, traces of two oblique strokes (the first one starts beneath the two upper strokes of the iota) are visible, while the smooth surface of the stone suggests the presence of a horizontal stroke joining the two oblique ones.

TO HIẠPO
τō hιạρō
Of the sanctuary

The alphabet is Archaic Achaian. The lone diagnostic letter is the three-stroke iota oriented opposite the rest of the text. There are examples of sinistrorse iotas in otherwise normal (left-to-right) Achaian inscriptions on blocks reused in the walls of Poseidonia-Paestum.[24] The rho appears to lack the stroke added beneath the eye, a characteristic of the western alphabets (Achaian and Chalcidean). The upper right section of the eye, however, seems to have been lost through wear.

The text may represent a simplified format that implies the name of both the object and the divinity; this example preserves the article, which is usually absent elsewhere.[25] Just as the presence of the inscription unambiguously reveals the function of the block, so the orientation of the text marks with equal certainty the boundary of the sacred area, indicating that it extended to the east from the site of the stone. In fact, the texts of boundary markers[26] always faced away from the areas they served to demarcate, even when they were made on natural features like walls or stumps of living rock.[27] This particular example does not appear to

[24] Dubois 2002, 69–72, no. 27. In only one case is the name clearly Greek: *Kikon* or *Kigon*. For the etymology and possible parallels of this example, see the bibliography in Dubois 2002, 71.

[25] Guarducci (1978, 46–73), in a broad-ranging discussion of sacred boundary markers, cites a mid-5th c. BC Athenian *cippus* with the same format of inscription as the current example (Guarducci 1978, 64), first published by Lalonde (1968, 122–33, esp. 127–28; 1991, 23, H8).

[26] For inscriptions on *horoi* and the functions and formats of their supports, as well as their historic significance, see Lalonde 1991, 5–6. The use of living rock as a support for *horoi* with the addition of appropriate inscriptions is widely attested in Attica, especially at the deme boundaries; see Langdon 1999, 481–508. For the Archaic *horoi* of the Sanctuary of Zeus near the Athenian Agora, see Lalonde 2006b.

[27] See, e.g., Lalonde 2006a, 83–119, according to which "horos texts usually face the reader outward from the domain to which they refer" (91).

Figure 3.10 The Retaining Wall from the N in 2011. (FS/ICA)

contradict that rule. The inscription also served to advertise the sacred function of the bounded space. The lettering and context suggest a date between the end of the 6th and the beginning of the 5th century BC.

The Retaining Wall

Next to the inscription, at a level of ca. -9.95 to 10.25 m, is a north-south wall following the lower edge of the slope (squares C4–5 and D4–5, *Fig. 3.10*). It was built with medium-sized irregular sandstone blocks placed side by side, tightly packed, and filled with earth. The wall, which ends near the inscription,[28] is only preserved for about 6.4 m, with a preserved width at its larger point of 2.4 m. Because it respects the position of the inscription, it was probably built after the *horos* had been set in place, as also confirmed by pottery fragments found within and below the structure, dating to the first half of the 5th century BC. The stratigraphic relationship with the surrounding area was completely compromised by the construction of the modern road that started the investigation, given that it used to pass right next to this structure. Further damage was inflicted after excavation, when a large portion of the northern section was removed by a bulldozer. Because of the massive construction of the wall and its location, it can best be interpreted as a retaining wall, built for the purpose of securing the topsoil of the fields further up the slope.

The Early Hellenistic Workshop

Occupation of the site resumed in the late 4th century BC, concurrent with a period of great expansion in the chora. At the end of the century, a workshop with a set of kilns was built on the terrace and the slopes above the spring. Two tombs discovered to the northwest of the workshop and earlier, Classical tombs, were also dated to the Hellenistic period.[29] With its proximity to clay sources, availability of water, and favorable location on transportation and trade routes, Sant'Angelo Vecchio is ideally situated for production, and it was clearly utilized for that purpose (see Ch. 5).

The Workshop Structure

Evidence for this phase is very poorly preserved and highly compromised by later activity, but it is nonetheless possible to recognize the limited remains of foundation walls. Built with blocks of sandstone, the foundations outline a roofed structure of uncertain plan, probably used by the artisans for their everyday activities. The proximity of the structure to the 5th-century BC tombs (1–8)[30] confirms that there must have been a change in the identity of the people living on the site. North-south Wall 4 in square G7 (*Fig. 3.11*) was 3.5 m long and 0.5–0.6 m thick at the widest. It was crudely built of irregular stones that caused its rough edges. Tiles appeared around the wall and on top of it, but did not form part of the wall construction. At ca. 3.5 m from the southern edge of the square, the wall seems to end quite abruptly, except for a line of scattered stones that continues toward the north, possibly part of its collapse. A few stones remaining to the west of the tomb in square G6, called Wall 6, may belong to the original line of this earlier structure.

East-west Wall 5, at a lower elevation, was identified in square G7 and continued in square F7 (*Fig. 3.11* and *Fig. 3.12*). The wall was built rather flimsily of sandstone blocks with a preserved total length of

[28] See above, p. 68, "The Inscribed Stone."

[29] In 2010, two Hellenistic tombs came to light a short distance from the workshop after heavy rain; see p. 91 in Ch. 4, "The Hellenistic Tombs."

[30] See p. 85 in Ch. 4, "The Rural Necropolis of the Classical Period."

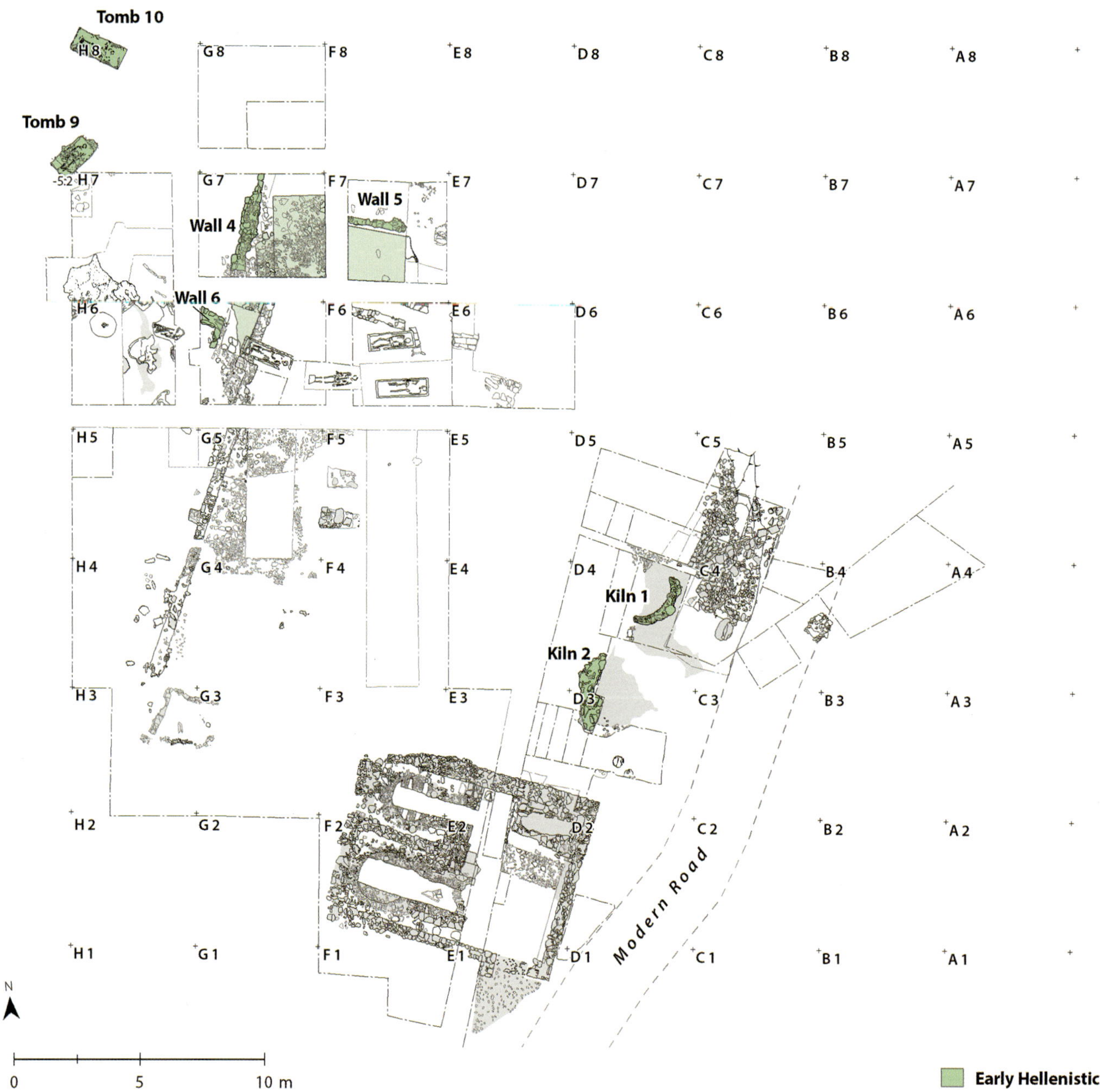

Figure 3.11 The Early Hellenistic Workshop and tombs. (JT/ICA)

almost 6.0 m, measured from the point where it intersects with Wall 4 in square G7. The preserved width may be estimated to a maximum of 0.5 m. Both outer walls were constructed poorly, using a technique consisting of stones of different sizes placed at random and without the use of tiles as reinforcement.

The two walls formed part of a shed that extended as far south as the southern baulk of square G6, as confirmed by the distribution of the roof collapse of the structure, made up of Laconian tiles that bear the *Herakas* stamp.[31] That Wall 4 is preserved at a level considerably higher than Wall 5 presents a note of caution in considering Walls 4 and 5 as part of a single structure. It is possible that the workshop shed was built on the slope, and—although some earthwork appears to have been undertaken to create a flat surface[32]—the foundation of the western wall remained elevated in relation to the floor level of the structure.

[31] See Ch. 32, "Roof Tiles and Bricks."

[32] See Ch. 2, "Site Phasing, Stratigraphy, and Site Assemblage."

Figure 3.12 Square F7, Wall 5 in 1980, looking S. (CLL/CW/ICA)

The workshop may have been destroyed by fire, as evidenced by a burnt layer and reddish soil preserved to the west of Wall 1 in square G6 (see p. 35 in Ch. 2, "Assemblage 16"). For all the evidence described above, a late-4th-century BC date can be suggested.

The Firing Area on the Slopes

A firing area occupied the Slope Area just above the spring, 15 m east of the workshop, extending for ca. 20.0 x 5.5–7.5 m.[33] The surface level throughout this part of the excavation had been disturbed, but gave general indications of the presence of two kilns built against the slope (Kilns 1 and 2).

The Kilns in the Slope Area

Kiln 1 is located at the northern end of the excavated area (*Fig. 3.13*). Only a clay segment ca. 2.10 m long and 0.35 m wide of the apparently pear-shaped combustion chamber remains in situ. Part of the collapsed dome, constructed with curved tiles of the Hellenistic system bound together with clay, was found in the surrounding area (Assemblage 13). Another part of its collapse was also identified north of the combustion chamber, where the construction material of the kiln consisted of tile fragments cut up into small squares and mixed in with lumps of burnt clay. The kiln would have been entered from the south. Its total size can be estimated from the extent of its collapse, and would originally have measured at least 3.0 m in length with a width of probably 2.5 m. Its dimensions are quite consistent with kilns of the Hellenistic period, which ranged from 1.0 to 3.0 m in diameter.[34] Because of the fragmentary walls of the kiln, it is not possible to estimate the interior construction with any degree of certainty.

The large tile fragments found in sizable quantities in the terracotta and tile concentration of Sounding B and in Slope 3 (Assemblage 11) can be interpreted as the remains of a second kiln, Kiln 2 (*Fig. 3.11*), resting on the natural soil of the slope. The absolute level of this kiln material (ca. -9.94 m) corresponds with the

[33] See Ch. 2, "Site Phasing, Stratigraphy, and Site Assemblage."

[34] Hasaki 2002, 155.

Figure 3.13 Square D4, Kiln 1 at the time of excavation in 1979, looking E. (CW/ICA)

Figure 3.14 Terracotta plaques SA 79-406T (*left*) and SA 79-409T (*right*) from the Bulldozed Stipe; scale 1:3. (CW/ICA)

level of Kiln 1, found at -9.87 m. Judging from the material found in association and in the nearby area (the terracotta concentration in Sounding B, Assemblage 10, and the Bulldozed Stipe, Assemblage 9), it is likely that these kilns were used primarily for the production of terracotta relief plaques.[35]

Because of erosion and the later development of the Slope Area, it cannot be determined whether more than two kilns were built in the area. The time span within which these kilns functioned is determined by the stratigraphy of the area and the finds. If the Pottery Deposit and Retaining Wall in squares C1–C5 and Sounding C provide a clear terminus post quem for the kilns and plaques, the stylistic date of the plaques points to the end of the 4th–first half of the 3rd century BC. Of the pottery, the latest datable material belongs in the middle decades of the 3rd century BC.[36]

The exploitation of a slope for the construction of kilns is not unusual. This location allowed the artisans better control over the various activities involved in production. Building the kiln against a slope eliminated the need for an entire pit on flat ground, and part of the combustion chamber may have been dug directly into the hillside. Furthermore, the slope provided good heat insulation and greater stability against thermal shocks during firing.[37] The presence of water and the relative distance from the workshop alleviated safety concerns.

[35] See Ch. 5, "Ceramic Production at Sant'Angelo Vecchio and in the Metapontine Chora." The extensive assemblage of terracotta relief fragments is in course of study, and will be published separately by R. M. Ammerman.

[36] See Ch. 2, "Site Phasing, Stratigraphy, and Site Assemblage."

[37] Hasaki 2002, 73.

Figure 3.15 Sant'Angelo Vecchio in the Late Republican period. (JT/ICA)

The extensive use of tiles in the construction of these kilns is noteworthy. Although few contemporary kilns are published well enough to make a sound comparison, those known from the urban *kerameikos* were built using a completely different technique. The urban examples made use of mudbrick produced specifically for the different parts of the kiln.[38]

[38] D'Andria 1980a. Nevertheless, the use of tiles in the construction of the dome of the Early Classical rectangular kiln at Torretta di Pisticci has been reported (see *Appendix B, Table B2*, no. 46; Lecce 2010–2011, 17).

The Late Republican and Early Imperial Periods

No sound indicators of the second half of the 3rd or first half of the 2nd century BC have been identified, and Sant'Angelo Vecchio seems to have been abandoned in this period. Sometime in the late 2nd or first half of the 1st century BC, the site was reoccupied. Three new kilns were built as the Kiln Complex, and to the west of the kilns, a new shed was built.

Figure 3.16 Wall 1 looking S (*left*) and Wall 1 pilaster built over Tomb 1 looking N (*right*) in 1979. (CW/ICA)

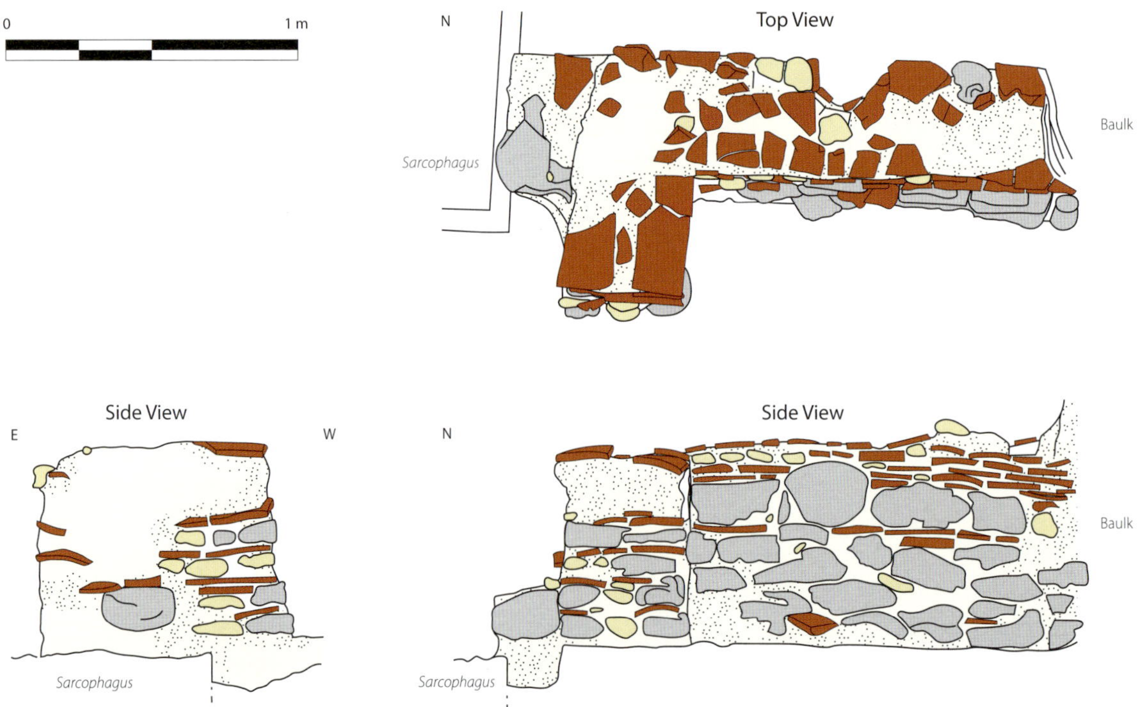

Figure 3.17 Section of Wall 1 in square G6. (KJS/ICA)

Figure 3.18 Wall 1 in square G6 in 1979, looking W. (CW/ICA)

Intensive agricultural activity and erosion processes compromised the layers of this period, which, in the southern portion of the area, had disappeared prior to excavation.[39] The structures of this phase, lying just a few centimeters below the surface, are only partially preserved due to post-depositional factors.

The Workshop Shed

The area within squares H3–5 and G4–6 was crossed by a long north-south wall (Wall 1) with a preserved length of ca. 15.0 m (*Fig. 3.15* and *Fig. 3.16*). The southern end was badly damaged, but the few remaining scattered stones indicate that it might have continued into square H3 to the south, beyond the basin (see below).

The best-preserved portion of the wall was found in squares G5, H5, and G6. It consists of tightly packed small sandstone blocks held in place by the tiles that form the outer edges (*Fig. 3.16*, *Fig. 3.17*, and *Fig. 3.18*). The wall's maximum preserved depth of 0.5 m occurs at the point where the wall crosses Tomb 1 in square G6 and at the pilaster in square G5. Its preserved width is about 0.4 m. At regular intervals of 3.1 m, pilasters were placed on the west side of the wall (*Fig. 3.16* and *Fig. 3.17*). Flat pan tiles packed in tight layers define the straight sides of the pilasters, and between the tiles are layers of small sandstone blocks. Over the upper tile layer of each pilaster lay a large conglomerate block to reinforce its construction. The pilasters measure ca. 0.4 m x 0.4 m. Two pilasters remain in square G5, at a depth of ca. -6.75 m, and one is preserved at a lower level in square H4 (-7.11 m). A fourth appeared at the southern edge of Tomb 1 in square G6 at a depth of -6.66 m.

An east-west cross wall, Wall 2, built in the same technique, was found in the northwest corner of square F6 at a depth of ca. -7.00 m. With its preserved length of ca. 2.0 m and width of ca. 0.45 m, the wall constitutes the northern end of the structure. The construction technique of Walls 1 and 2 is different from the one observed in the previous phase. Builders in this

[39] See Ch. 2, "Site Phasing, Stratigraphy, and Site Assemblage."

Figure 3.19 The basin in square H3 in 1979, looking W. (CW/ICA)

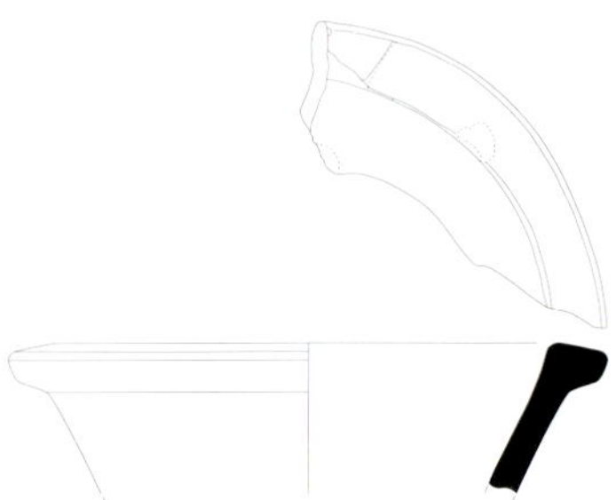

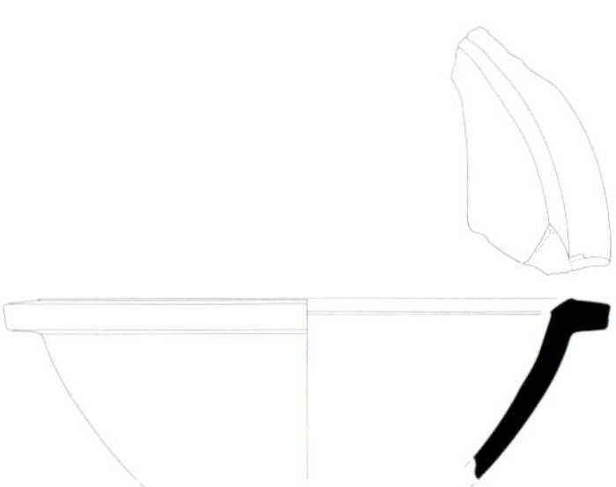

Figure 3.20 Large basins SAV PCW 35 and SAV PCW 36 from square H3 (*left*); a modern large basin used for the decoration of cooking ware in a pottery workshop at Lucugnano (Lecce, Italy) in 2014 (*right*). (FS/ICA)

Figure 3.21 The Kiln Complex at the beginning of the excavation in 1979, looking SE. (CW/ICA)

period made extensive use of tiles partly recycled from the tile fall of the Early Hellenistic building.

Although later activity on site partially compromised the abandonment layers at the southern end of the shed, the stratigraphy is well preserved in the central and northern areas, yielding precious information about the covering of this structure. The tiles found in the roof collapse belong to System Bc (type BT3, Tile Fall 1, Assemblage 3), which can be dated to the Late Republican period due to the presence of the stamp bearing the name of the Aufidii, the owners of the property, whose presence at Metaponto is recorded by textual sources.[40]

More or less aligned to the eastern limit of Wall 1 and to the northern pilaster in G5 is a cluster of stones noted in square F5. It may have served as a foundation for a wooden post that supported the roof. The southern end of the building is unfortunately not preserved, and therefore the precise dimensions of this structure cannot be determined with certainty. Together, the walls can be interpreted as the foundations of a shed open to the east (possibly a portico?) with a preserved area of 84 m^2.

Of some interest is a facility identified at the southern end of the workshop structure, and unfortunately now destroyed. At the southern end of the workshop shed in square H3, set within the clayey and sandy natural soil that here appeared on the surface, was a rectangular structure that consisted of flat tiles placed at an angle so as to form a shallow basin (*Fig. 3.19*). The preserved exterior dimensions are ca. 2.0 m north-south and 2.25 m east-west, and the interior space measured ca. 1.6 m north-south by 1.9 east-west. There was no visible difference between the soil inside the basin and the surrounding area due to the exposure of the whole square to the surface through plowing. It is possible that the tile-lined space served as a clay-settling basin or a storage area for the clay to be used in production. The discovery of fragments of large basins (SAV PCW 33–SAV PCW 36, *Fig. 3.20*) and gutter pipes in the surrounding area confirms its function.[41]

[40] See Ch. 32, "Roof Tiles and Bricks." For the reconstruction of the roof, see *Fig. 32.7* (flat tile system Bc).

[41] Parallels for the feature are also known from Late Republican and

Figure 3.22 Kiln Complex, interior of Kiln 4 (South Kiln) in 1979, looking E. (CW/ICA)

Figure 3.23 Kiln Complex, area in front of Kiln 5 (North Kiln) in 1981, looking N. (CW/ICA)

The Kiln Complex

A later firing area[42] was located at the edge of the sloping hill (squares E1–D3/E3, F2–F3; *Fig. 3.15* and *Fig. 3.21*), where three kilns (Kilns 3–5) opened onto a courtyard and defined a rectangular space (7.0 m north-south and slightly over 9.0 m east-west, occupying an area of 63 m^2).[43] The foundations of three of the enclosure walls (7, 8, and 9) measure between 0.4–0.5 m in width, while Wall 10 is wider, ca. 0.7 m. Only the first course of Walls 7 and 8 (the former damaged by a plow) were exposed. They were constructed from river cobbles and chunks of local sandstone, though conglomerate was also observed (see *Fig. 2.27*). Wall 9 preserves two courses, the lower of unconsolidated spherical river cobbles,[44] above which are blocks of local sandstone mixed up with cobbles, especially concentrated in the inner part of the wall. The construction technique can be better appreciated in Wall 10, the internal face of which was exposed to a height of ca. 0.58 m (*Fig. 3.23*). It was constructed from tiles and chunks of local sandstone of irregular shape and dimensions. The former, only occasionally present in the lower part of the wall, make up proper courses in the upper preserved portion, resembling the technique used in Wall 1.

An entrance to the complex measuring ca. 1.25 m is located at the southeastern corner of Wall 8. A short anta wall marks it on the west side, whereas the east side consists of a break in the line of the wall proper. The entrance opens into a small courtyard (excavated in its eastern portion as Sounding A) between Kiln 3 and the entrance to Kiln 4 that continues as a narrow corridor toward Kiln 5.

Imperial ceramic workshops; see, for example, the basins found at Pesaro (Cerri/Ciuccarelli/Lani 2012, 56–57, figs. 2 and 7.

42 During excavation and for much of the time afterward, the Kiln Complex was thought to have an earlier, Hellenistic phase, remodeled in the Late Republican period. Due, however, to the scarcity of available evidence to support this interpretation, the Kiln Complex is presented here as belonging to a single phase.

43 See Ch. 2, "Site Phasing, Stratigraphy, and Site Assemblage."

44 See *Fattoria Fabrizio*, wall 7.

Figure 3.24 Kiln Complex, interior of Kiln 5 (North Kiln) in 1979, looking E. (CW/ICA)

The combustion chamber of Kiln 4 (the South Kiln, *Fig. 3.22*), the larger of the two kilns, was built as a circle of packed tiles lined with clay with an estimated diameter of 1.9–2.0 m. Two shelves built of pan tiles held together with clay defined an ash pit at the center of the combustion chamber. They may have been surmounted by arches to support the perforated floor, as in Kiln 5 (the North Kiln). The corridor of the stocking channel, lined with carefully placed pan tiles covered with a thin layer of clay and fired to a hard greenish-gray fabric, had a maximum preserved length of ca. 2.7 m. At the eastern end of the kiln, large blocks of terracotta bricks had fallen on the floor of the kiln in such a way that they preserved the original arched entrance to the stocking channel (see *Fig. 2.29*).

Kiln 5, preserved to a length of ca. 3.35 m, had a narrow entry (0.6 m) from the east, formed by two side walls that created a façade (*Fig. 3.23*). The corridor of the stocking channel was lined with tightly packed thin pan tiles and clay in a line that continues to a round chamber with a diameter of at least 1.5 m. Two shelves similar to those encountered in Kiln 4 (*Fig. 3.24*) defined the ash pit, and four rectangular niches (two on either side) measuring ca. 0.4 m in length that projected from the wall were likely the springs of the arches that originally supported the perforated floor (not preserved).

The two kilns, which belong to Cuomo di Caprio's Type Id,[45] have another parallel: the Late Republican large kiln at Pantanello.[46] Both kilns were cut into the shallow layer of clayey natural soil, and the area around them was packed with stones and rubble. Aligned with Walls 8 and 10, a T-shaped wall (Wall 11) separates the two kilns and forms a rectangular space between the openings. This wall was built in the same technique as the outer walls, with rounded fieldstones and smaller rocks. The east-west line is well

[45] Cuomo di Caprio 1971–1972, 1979.

[46] Carter 1983a, 426–31.

Figure 3.25 Kiln 3 in 1979, looking E. (CW/ICA)

defined and measures ca. 0.35–0.45 m in width, with a preserved length of ca. 3.5 m. The cross wall (Wall 12) measures 0.5 m wide and ca. 0.6 m long on the south section, stretching toward Kiln 4. The north section, on the other hand, is less well defined, and is narrower, 0.4–0.5 m, extending as far as the south edge of Kiln 5, ca. 0.9 m. The area between the enclosure and partition walls is packed with rubble. The elevation of this wall and of the enclosure walls cannot be evaluated. Since they were probably built as an insulating device to limit the loss of heat—a practice also attested in other cases[47]—it is likely that they were the same height as the kilns. It should also be noted that the position of the Kiln Complex at the edge of a slope exposed the kilns to the winds, which may have affected the firing process.

[47] Naxos, Kiln 3 (Pelagatti 1972, 214, figs. 1, 6). Metaponto, *kerameikos*, Archaic kiln and Kiln C (D'Andria 1980a). Locri Epizefirii (*Locri III*, 27ff). Crete, Priniàs, 7th c. BC (Rizza/Palermo/Tomasello 1992, 43); Hasaki 2002, 340–41, nos. 31–36. Vamvouri, Ammoudia, Thasos (Garlan 1986). Phari, Thasos, 525–475 BC (Blondé/Perreault/Péristéri 1992, 12, fig. 1; Hasaki 2002, 336–37, nos. 25–26).

Both kilns open to the east; this orientation may have been determined by the direction of the slope.[48] Kiln 4, left empty when the site was abandoned, was found filled with the collapse of its combustion chamber dome.[49] More complex is the composition of the inner deposit found in Kiln 5,[50] which does not seem to show the same careful closure. A number of nearly complete vases was found mixed in with the debris from the collapse of the dome (SAV CkW 07, SAV CkW 25, SAV CkW 43, SAV CkW 48, SAV CkW 49, and SAV Imp 09).

The area between the entrances to Kilns 4 and 5 contains a rectangular bench or platform preserved to the same level as the foundation course of the kilns (see *Fig. 2.29*). It is made up of tightly packed tiles set in a chalky crumbly soil, and its border is formed by vertical fragments. The platform rests on a pavement of heavily fired terracotta slabs, cracked and broken at the time of excavation, which presumably extended all the way in front of the platform.

The northeastern corner of the Kiln Complex was occupied by a third kiln (Kiln 3) with a different orientation (*Fig. 3.25*). Dug into the soil, it preserved a narrow stocking channel (ca. 0.25 m wide), open to the west and marked by tiles placed upright with terracotta bricks on either side. The pear-shaped combustion chamber[51]—constructed from courses of curved tiles placed on an angle, bonded with clay, and lined with clay and tiles—is of relatively uniform width (0.7–0.8 m). The preserved total length of the kiln is ca. 2.0 m. No trace of either an internal column supporting a perforated floor or of the combustion chamber was found, and the original construction and height of the kiln cannot be estimated. It seems likely, however, that it was built up with an arched dome, similar to that of Kiln 1 in square D4. The small dimensions of the kiln (the average diameter of the type normally ranges from 1.2–1.5 m)[52] are not unusual, especially in workshops that also had larger kilns.

The three kilns opened into a courtyard excavated as Sounding A.[53] The courtyard yielded a tile fall layer that covered the whole eastern half of the Kiln Complex with a preserved depth of at least 0.5–0.6 m (see

[48] Hasaki 2002, 81–82.

[49] See Ch. 2, "Site Phasing, Stratigraphy, and Site Assemblage," *Assemblage 7*.

[50] See Ch. 2, "Site Phasing, Stratigraphy, and Site Assemblage," *Assemblage 8*.

[51] On the pear-shaped kiln type and its relation to circular kilns, see Hasaki 2002, 153.

[52] Hasaki 2002, 152.

[53] See p. 39 in Ch. 2, "The Kiln Complex."

Fig. 2.24, Assemblage 3). This tile fall was concentrated in the central part of the courtyard and in the area within and in front of the entrance to the Kiln Complex. There was, by contrast, a distinct lack of tile at the north end, where the paving consisted of uneven natural rock, partially covered by reddish hard semi-fired soil. The tiles consist of pan tiles of type BT3 (Roman system), also used in the covering of the Late Republican shed in the Workshop Area, some of which have the stamp AVF (see below). They originally belonged to the roofing of the courtyard.

Due to the post-depositional history of the site, the evidence to establish the type of production performed in this period is limited. The only definitive evidence is a grey-ware waster (SAV GW 07).

The roofing system in the Late Republican workshop shed and courtyard of the Kiln Complex provides the most reliable evidence to establish the moment of Sant'Angelo Vecchio's reoccupation. The stamp AVF, present on some of the tiles, can be connected to the Roman family of the Aufidii, whose presence in the territory of Metaponto is attested by the sources between the end of the 2nd and first decades of the 1st century BC.[54] Grey ware, found in limited quantities both in the area of the workshop shed and reused in Kilns 4 and 5, confirms this chronology. Occupation of the site ended with the beginning of the 1st century AD, as suggested by the pottery found in abandonment layers (i.e., the tile fall) of the Kiln Complex and the Workshop Area (see *Fig. 26.1*). Under this point of view, Sant'Angelo finds comparison with other sites of the Metapontine chora, such as Troyli and Sant'Angelo Grieco, also abandoned in the same period, when an economic downturn appears to have taken place.[55]

---

[54] For a detailed discussion of the stamp and the Aufidii, see Ch. 32, "Roof Tiles and Bricks."

[55] Giardino in *San Biagio*, 7.

4
The Tombs at Sant'Angelo Vecchio

Marshall Joseph Becker, Ingrid E. M. Edlund-Berry, Massimo Limoncelli,
Edward G. D. Robinson, Francesca Silvestrelli, and Serena Viva
With contributions by Cesare D'Annibale, Marta Mazzoli, and Francesco Perugino

Figure 4.1 Detailed plan of the burial ground. (JT/ICA)

THE RURAL NECROPOLIS OF THE CLASSICAL PERIOD
Topography and Excavation

Ingrid E. M. Edlund-Berry and Francesca Silvestrelli

A necropolis with the remains of seven tombs (*Fig. 4.1*) documents the 5th-century BC occupation at Sant'Angelo Vecchio. The preserved burials were found unexpectedly in 1979, and their exploration was completed the following year. They are all concentrated in an area measuring less than 20 m by 5 m (squares E6–H6). The tombs have an east-west orientation following the configuration of the slope, and were found at different elevations on or within natural soil. The two most recent contexts (Tomb Group 8 and Tomb 4, dated to the second half of the 5th century BC) were located on the highest point (square E6), while Tomb 6 (square H6) was laid at the lowest elevation, just above the gully ending in the spring. These differences also explain their dissimilar conditions of preservation. Tombs 1 and 2 were partly compromised by the later occupation of the area, while Tombs 3, 5, and 6 were undisturbed.

Figure 4.2 Three-dimensional virtual reconstruction of the burial ground at Sant'Angelo Vecchio: Tombs 1, 5, 2, and 3; for details on virtual reconstruction techniques, see below, p. 95, "3D Reconstruction of the Sant'Angelo Vecchio Necropolis." (ML/ICA)

Figure 4.3 Reconstruction of the burial ground at Sant'Angelo Vecchio: Tomb Group 8, Tombs 4, 7, and 1. (ML/ICA)

The limits of this cluster of burials, which may have been part of a relatively larger nucleus,[1] could not be defined with certainty, since the area surrounding the necropolis underwent many significant anthropic and natural transformations throughout the site's long period of occupation. As previously stated,[2] Tomb Group 8 and Tomb 4 appeared on the surface, here constituted by natural soil, in an area where archaeological deposits had been leveled by modern plowing (see *Fig. 2.21*). The precise limits of the cluster also cannot be established for the area south of it: no excavation was undertaken in squares E4 and E5, while in F4 and F5, natural soil appeared beneath 0.3 m of topsoil. The chronologically mixed material, mostly concentrated on the surface and ranging in date from the Archaic (type B amphora, SA 79-457PL) to the Hellenistic (SAV Amp 16), Late

[1] According to farmers living near the archaeological site, other tombs of unknown chronology were found some 50 m northeast of the excavated cluster in 2000, when the area was prepared for the planting of an olive grove. The information, however, could not be verified.

[2] See Ch. 2, "Site Phasing, Stratigraphy, and Site Assemblage."

Figure 4.4 Tomb 3 in 1979, looking E. (CW/ICA)

Republican (SAV GW 04, SAV GW 05, and SAV RT 56), and Late Roman periods (SAV LR 03), confirms that the archaeological deposit had been compromised.

The area to the east of Tomb 6 was eroded. A 10-cm-deep layer of topsoil, which lay atop the natural soil, yielded 30 pottery fragments, mostly small, undiagnostic sherds from plain-ware vessels (see *Fig. 2.23*). The rim of a Late Archaic type B amphora (SAV Amp 06) and the rim of a hydria of the second half of the 5th century BC (SA 80-61PL) may have washed down from the layers originally covering the tombs, which are lost across the entire area.

An analysis of the funerary assemblages indicates that the necropolis was in use between the end of the second quarter and the beginning of last quarter of the 5th century BC. Following common practice in the chora of Metaponto,[3] the farmhouse with which these burials were associated was likely located nearby. Tombs were a very common landscape feature in agrarian societies, and certainly in the Metapontine chora. Many burials were placed near the edges of

Figure 4.5 Tomb 6 in 1979, looking E. (CW/ICA)

properties, often—especially in the second half of the 4th century BC—with highly visible ceramic or stone markers to reinforce boundary lines, but a large number of the identified necropoleis in the chora of Metaponto were located within 25–30 m of a farmhouse. No evidence of such a dwelling was detected in the lower area of Sant'Angelo Vecchio, but it is nonetheless possible that the tombs might be connected to the family living in the House on the Hill, where the most recent material found dates to the middle decades of the 5th century BC.[4] The amphora SAV BG 32 of the second half of the 6th century BC, found in a secondary deposition in square F7 (see *Fig. 2.20*), belongs to a type that occurs, although rarely, in funerary assemblages and it could hint at the presence of tombs of the Archaic period, now lost.

Residual black-gloss pottery found scattered in the area of the tombs confirms that the occupation lasted until the end of the 5th or the very beginning of the 4th century BC.[5] After this moment, Sant'Angelo Vecchio seems to have been abandoned. Occupation was certainly resumed in the Early Hellenistic period, when we have evidence for the production of votive plaques.[6]

[3] *Survey.*

[4] See Ch. 2, "Site Phasing, Stratigraphy, and Site Assemblage," and Ch. 3, "The Structures at Sant'Angelo Vecchio."

[5] See Ch. 2, "Site Phasing, Stratigraphy, and Site Assemblage."

[6] See Ch. 2, "Site Phasing, Stratigraphy, and Site Assemblage."

Figure 4.6 Tomb Group 8 in 1979. (CW/ICA)

Nonetheless, the builders of the Hellenistic workshop apparently saw no need to respect the burial area, the existence of which had probably been forgotten, indicating a change in the people occupying the site.

A ninth burial, Tomb A, is similar in date, though separate from the tomb cluster at Sant'Angelo Vecchio.[7] The sarcophagus tomb was identified in 1981 during the construction of a road 500 m southeast of the site, on the slope facing the Basento valley (site 555).[8] This burial, which can be dated to the middle decades of the 5th century BC, is the only feature identified on the site, and no farmhouses active in this period have been recognized during the survey. The evidence nonetheless enhances the general picture of the population of this part of the chora, where necropoleis are, in this period, quite rare.

[7] The tomb was excavated by the Soprintendenza in collaboration with the Institute of Classical Archaeology. A preliminary note is in *Necropoleis*, 447.

[8] *Survey IV (Atlas)*, 82.

Burial Rites and Tomb Types

Edward G. D. Robinson

The small nucleus of 5th-century BC tombs at Sant'Angelo Vecchio contains all of the major tomb-types found at Metaponto and in its chora (Pantanello, Saldone, San Salvatore), with the exception of the cist burial type. The three sarcophagus tombs (Tombs 1, 2, and 3; *Fig. 2.12*, *Fig. 2.13*, *Fig. 2.14*, and *Fig. 4.1*) had tiled lids, and all contained the bodies of males. At the Pantanello necropolis (a distance of 5.5 km to the east), sarcophagi with tiled lids account for only 25% of the sarcophagus burials found; the type does not appear in that cemetery after 425 BC, a date which accords well with the burials of Sant'Angelo Vecchio.[9] These containers must have been expensive, since the *carparo* stone came from at least 20 km away, and possibly from as far as the Salentine peninsula.[10] Given the expense of the sarcophagi, it is perhaps surprising that their grave goods are so modest, but this charac-

[9] Carter in *Necropoleis* a, 61.

[10] Carter in *Necropoleis* a, 112–13, appendix 3A.1

Figure 4.7 Tomb 4 in 1979, looking E. (CW/ICA)

teristic is true of sarcophagus burials at Pantanello in general, and of tile-covered sarcophagi in particular.[11]

Tomb 5 (see *Fig. 2.12*) is the only example at Sant'Angelo Vecchio of the *a cappuccina* tomb type, constructed of Corinthian pan tiles. While curved tiles apparently did not cover the central ridge, pan tiles were used to make a floor for the tomb. This is a relatively rare feature at Pantanello, appearing for the first time around the middle of the 5th century BC (e.g., T 230, T 311).

Tomb 6 (*Fig. 4.5*) is another *a cappuccina* tomb, but it uses curved Laconian tiles and is unusual in several respects. It had a cover tile in addition to the two tiles that normally made up such a burial, and was closed at the ends by flat pan tiles rather than by sections of Laconian tiles. Both features are documented at Pantanello, but are rare at that site.[12] The use of stones to outline the burial is another feature seen rarely at Pantanello, and only in fossa tombs (e.g., T 56, a child's burial of the mid-5th century BC).[13] But the use of this tomb type for a child burial accords perfectly with its usage at Pantanello. Around one-third of Laconian tile tombs at Pantanello contained no grave goods. Tomb 7 is less well documented, but also appears to have been a Laconian tile tomb; it contained the remains of an adolescent. At Pantanello, this tomb type was reserved almost exclusively for very young children, but the burials of adolescents (T 295) and even adults (T 338) are also documented.

The two fossa tombs are both problematic. Tomb Group 8 (*Fig. 4.6*), apparently containing the bones of a child, was disturbed, and it is not clear whether it should be seen as a tomb or a ceramic deposit, a type of deposit found frequently at Pantanello and generally interpreted as evidence for the cleaning up of earlier burials. It may even be connected with Tomb 4, just 2 m distant, a fossa burial that contained two skeletons.

The depositional history of Tomb 4 (*Fig. 4.7*) is of interest, since at least one of the skeletons may represent a secondary burial. Double burials are extremely rare at Pantanello and, with the exception of some late 4th-century BC monumental cist tombs, are generally restricted to females and children.[14] Secondary deposition is much more common in the indigenous world, especially in Messapia,[15] but also at sites to the north of Metaponto. Since Tomb 4 contained a number of highly unusual grave offerings of types more often found in indigenous burials (shell, glass, amber, and bronze), it is possible that either of these mature women (one 40+, the other 60+ years old) had noteworthy connections with the Italic world. Though the material culture and habits on display in the Sant'Angelo Vecchio necropolis were overwhelmingly Greek in nature, its location was certainly prime for frequent interaction with Italic people: the major Italic center of Pisticci was clearly visible from the site.

Tombs 5 and 7 both had vertical slabs of stone placed at the feet of the deceased (*Fig. 2.13* and *Fig. 4.8*), presumably as grave markers. Such markers are unknown at Pantanello and very rare elsewhere in the Metapontino.[16] Ceramic vessels are much more commonly used at Greek sites in southern Italy, and are widely attested at Metaponto and Taranto.[17]

11 Carter in *Necropoleis* a, 76.
12 Carter in *Necropoleis* a, 97ff.

13 Carter and Hall in *Necropoleis*, 270.
14 Carter in *Necropoleis* a, 108–10.
15 E.g., Giannotta 1994, 90–91.
16 Carter in *Necropoleis* b, 124.
17 Carter in *Necropoleis* b, 125ff.; Hoffmann 2002, 15; Lippolis 1994b.

Figure 4.8 Tomb 5 in 1979, looking E. (CW/ICA)

Tomb no.	Estimates from photographs alone	Conclusions from direct observation
1 (G6)	Adult, possibly male (?)	Adult male, possibly 40–50 years
2 (F6-N)	Adult, female (?), age ca. 30–35	Young adult male, age ca. 25 years
3 (F6-S)	Adult, male (?), age ca. 40	Male (?), at least 30 years of age
4 (H6)	Adult	Two adult females, age 60+ and 40+
5 (F6/G6)	Adult, possibly female (?)	Young adult male, age 22–23 years
6	Not evaluated	Child, age 3–5 years
7	Not evaluated	Child, age 8–12 years

Table 4.1 Comparison of initial observations made from photographs and final conclusions following direct observation.

The Sant'Angelo Vecchio necropolis fits comfortably within the patterns seen elsewhere in the region (e.g., Pantanello, Saldone, and Metaponto itself) in terms of tomb types, burial rites, and the numbers, types, and associations of grave goods. Table amphorae are, as expected, common in 5th-century BC tombs. At Pantanello, the role of the amphorae was completely subsumed by pelikai from c. 425 BC onward.[18] In the 5th-century BC tombs, mugs are overrepresented, skyphoi are rare, and the preponderance of male burials stands in contrast to Pantanello, where female burials are much more common.[19] But all such observations must be tempered by the very small size of the sample from Sant'Angelo Vecchio. Relatively few types of object at Pantanello were found to be strongly correlated with male or female burials, but the beads, pin, and ring in Tomb 4 (a double female burial) conform to gender expectations. The hydria and alabastron in the same tomb might also be suspected of being typical of a female burial, but at Pantanello no such correlation was found.[20]

Analysis of the Human Skeletal Material from the Classical Tombs at Sant'Angelo Vecchio

Marshall Joseph Becker

During June of 1980, the human skeletal remains recovered earlier from seven 5th-century BC tombs from Sant'Angelo Vecchio near Metaponto were subjected to osteological analysis. The tombs were discovered during the 1979 excavations by the University of Texas at Austin, but only after the author had completed an earlier program of skeletal analysis at the site[21] and had left the area.[22] In five of these tombs, the human remains were cleared with great care and the exposed bones photographed *in situ*. Field measurements and these photographs provided the basis for a preliminary report[23] until the tombs could be directly examined. Two children's tombs were left until the following season.[24]

These inhumations probably do not reflect the total population of this small cemetery, which may have

[18] Hall in *Necropoleis*, 566.

[19] Henneberg and Henneberg in *Necropoleis*, 509–12.

[20] Hall in *Necropoleis*, 585.

[21] See Becker 1980a.

[22] Becker 1980b.

[23] Becker 1979.

[24] Becker 1981. Cf. also Becker 1996–1997.

served a rural farm. The tombs were found in an area less than 20 m long by 5 m broad. The long axis of the area runs east-west in the direction of the slope: Tomb 4 on the western side is at the highest point, and Tomb 6 is at the lower side. As would be expected in this situation, the damage to the skeletons prior to excavation, probably from plowing, was considerable.

The overall dental disease rate (decay plus loss) appears high for this population, but can be shown to be extremely variable. Tomb 2 (age 25) has a 6% rate and Tomb 5 (age 23) no disease, yet Tombs 1 (40–50 years) and 4B (age 40+) show 71% and 56% respectively. Thus dental disease and loss, as expected, increase with age. A low rate of third molar eruption is suggested.

Strong third trochanters are evident on the males of Tombs 1, 2, and 5, and traces are found on the male from Tomb 3 and at least one female from Tomb 4. The lineae asperae of all of these people seem large, with a strong protrusion in the central zone. Wide femoral necks also appear to characterize these people. Small femur head diameters are present on the males except the one in Tomb 5, who happens to be the largest person in this population.

Stature can be calculated for only five people. The three males are 180, 164, and 160 cm tall (an average of 168 cm), and the tallest male is exceptionally so. The two females, both much older than the males, are 150.6 and 146.7 cm tall (an average of 148.6 cm). A test of the Manouvrier formula for Italian populations would require a larger sample and one with more accurate *in situ* measurements.[25] The data on stature should be tested against the Musgrave and Harneja (1978) formulae.

Locations of perinatals and children below the age of 5.5 years cannot be sought due to the small size of this population.[26]

The Hellenistic Tombs

Topography and Excavation

Francesca Silvestrelli

The area west of the structure in square G7 and the Late Republican workshop shed in squares H4–G6 underwent consistent transformations after the completion of excavation in 1981. This part of the slope, previously planted with wheat, was in 2001 leveled and transformed into an olive grove. The creation of a new path leading to the hilltop resulted in a substantial lowering of the original level (possibly 1 m), which caused the removal of the topsoil in the non-excavated area and the destruction of the excavated archaeological evidence in squares H3–H5 and in the western half of squares H6 and H7. In November 2010, runoff rills from the rains revealed two tombs lying 0.10 m below the surface (*Fig. 4.1*), which were immediately excavated.[27]

The two burials, located ca. 7 m west of the structure in square G7, both date to the Early Hellenistic period and can probably be connected to the artisans living on the site. The position of the tombs near the workshop is less unusual than it might appear at first sight. Evidence from the survey shows that burials are usually within 25 m of farmhouse scatters. The proximity to the rural necropolis of the Classical period notwithstanding, the two nuclei are probably independent, as suggested by the differences in elevation and orientation. The tombs of the Classical period were likely no longer visible and their existence could have been ignored.

Burial Rites and Tomb Types

Edward G. D. Robinson

The two burials of the Hellenistic period are both of well-known types. Tomb 9 is an *a cappuccina* burial using two pan tiles on each long side with a further tile placed vertically at either end to close the tomb (one is lost). This type had been common in the Metapontino since the second quarter of the 5th century BC.[28] Tomb 10, by contrast, is of a type that seems to be restricted to the Hellenistic period in the region: the so-called vault tomb, in which the body was covered by curved terracotta slabs with a semicircular profile, with the vault closed by pan tiles at the ends. Such elements were used to line well shafts and examples are known from wells in sanctuaries of the Late Classical and Early Hellenistic periods at Pantanello and Herakleia.[29] In the Pantanello necropolis some of these well liners may have been reused functional elements, but others were clearly made for the grave.[30] At Pan-

[25] See Becker 1999.

[26] See Becker 2007, 2011, 2012.

[27] The salvage excavation was conducted by the Institute of Classical Archaeology in cooperation with the Soprintendenza Archeologica della Basilicata. Excavation participants included Saverio Careccia (Soprintendenza Archeologica della Basilicata), Emanuela Conoci, Assunta Florenzano (University of Modena), Francesca Silvestrelli (Salento University), and Giuseppe Di Taranto (working crew).

[28] Carter in *Necropoleis*, 90–97.

[29] Carter in *Necropoleis*, 100, fn 120; Osanna/Prandi/Siciliano 2008, 84–86.

[30] Carter in *Necropoleis*, 60, 100, fig. 3.59.

Figure 4.9 Right hip bone (T10A), scale 1:2. (CW/ICA)

	SAV		Henneberg in *Necropoleis*		Safont 2000	
	T9	T10	M	F	T9	T10
			m ± std	*m ± std*	*sex*	*sex*
(FMC)	85.0	80.0	89.6 ± 6.8	80.1 ± 5.8	M	F
(HMC)		55.0	62.8 ± 6.8	58.8 ± 4.3		F
(TMC)		68.0	84.4 ± 7.6	72.7 ± 4.4		F
(RMC)		37.0	44.5 ± 4.1	40.3 ± 3.2		F
(UMC)		32.0	39.3 ± 2.6	38.0 ± 3.1		F

Table 4.2 Circumference at the mid-point of the femur and minimum circumference of the humerus, tibia, radius, and ulna.

	SAV	Henneberg in *Necropoleis*		Trotter and Gleser 1957
	T10A	M	F	Stature
Ulna (max l)	235.0 mm	259.0 ± 16.1 mm	242.6 ± 15.4 mm	158.1 cm

Table 4.3 Maximum length of the left ulna of T10A compared to the averages for men and women at Metaponto, along with estimation of stature.

tanello, the vault tomb was almost entirely restricted to female burials (15 of 16) and strongly correlated with advanced age there,[31] so the female skeleton (40–60 years old) recovered from Tomb 10 corresponds well to the pattern. The tomb was later opened and the body of a child inserted; among the eight double burials at Pantanello, the female/child combination was the most common.[32]

The one Hellenistic tomb that contained grave goods (Tomb 10) was typical of its time in that it contained no large ceramic forms. One might be tempted to connect the miniature vase from the tomb, found separately from the cluster of three full-sized vessels, as belonging to the later interment of a child, but at Pantanello miniature vases were not correlated with child burials.[33]

Anthropological Analysis of Three Skeletons from the Hellenistic Tombs at Sant'Angelo Vecchio

Serena Viva

The subject of this study is the osteological material from Tombs 9 and 10 (T9 and T10) discovered in November 2010 at Sant'Angelo Vecchio. The bones' state of preservation is heavily compromised by several factors. In Tomb 9, which yielded the remains of an adult, the poor state of preservation can be attributed to the low pH of the soil. Increased soil acidity, often caused by certain types of cultivation (in particular the olive), is the primary cause of the dissolution of the calcium that constitutes the majority of bone tissue.[34] This poor preservation sharply limited the amount of information that could be retrieved about the individual buried in Tomb 9.

Tomb 10 contained two individuals, an adult (T10A) and a child (T10B). The adult's state of preservation is markedly better than that of Tomb 9, but the integrity of the osteological material is nevertheless compromised by disturbances that presumably followed a reopening of the tomb in antiquity. The child's state of preservation is very poor due to the fragility of the immature bones: even the teeth show notable demineralization.

Sex, age at death, and stature. Determination of the sex was successful for both of the adults. In the case of T10A, determination of the sex was based on morphological analysis of the hip bones and the cranial features.[35] The hip bones contain a particularly wide greater sciatic notch and a preauricular sulcus (*Fig. 4.9*). On the cranium, the glabella and the brow bone are very weak, the supraorbital margin has a sharp edge, and the mastoid process is very small. The mandible, although incomplete, appears slender. These features suggest identification of the individual as a female.

The availability of a large quantity of metrical data from Metapontine necropoleis[36] allowed comparison of the minimal circumference of the humerus (HMC),

[31] Carter in *Necropoleis*, 101.
[32] Carter in *Necropoleis*, 110.
[33] Hall in *Necropoleis*, 586.
[34] Carter in *Necropoleis*.

[35] Acsádi and Nemeskéri 1970.
[36] Henneberg and Henneberg in *Necropoleis*.

Figure 4.10 Detail of the right femur with enthesophytes (T9), scale 1:1. (CW/ICA)

	Martin No.	d
HUMERUS		
Max diam at mid-point	5	20.0
Min diam at mid-point	6	17.5
FEMUR		
Diam A/p at mid-point	6	26.0
Diam m/l at mid-point	7	28.0
Perimeter at mid-point	8	85.0
Transverse diam sub troch.	9	31.0
Sagittal diam sub troch.	10	26.0

Table 4.4 Measurements of T9.

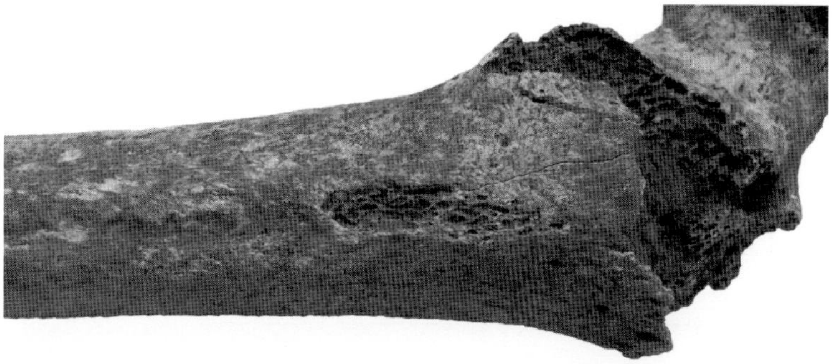

Figure 4.11 Detail of right femur with bone erosion (T10A), scale 1:1. (CW/ICA)

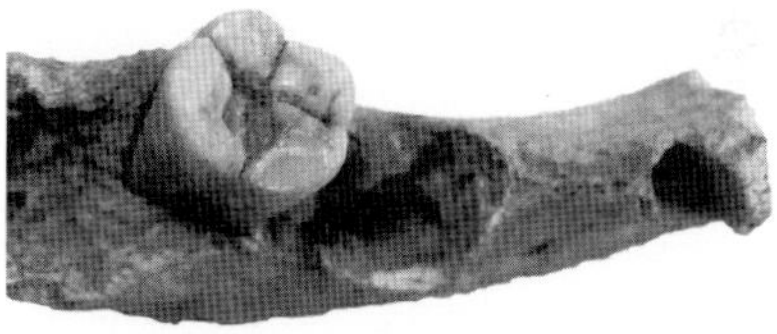

Figure 4.12 Part of the right hemi-mandible (T10A), scale 1:1. (CW/ICA)

radius (RMC), ulna (UMC), and tibia (TMC), and the circumference at the mid-point of the femur (FMC) between the individuals of Tombs 9 and 10 at Sant'Angelo Vecchio and individuals throughout the chora of Metaponto (*Table 4.2*). The measurements follow Martin and Saller's numeration and methodology.[37] This comparison confirms that the individual in T10A was a female.

In the absence of the cranium and pelvis, T9 was identified as a probable male by the circumference at the mid-point of the right femur (FMC). These measurements were calculated with Safont's equation for determining the sex from the circumference of the long bones,[38] providing additional confirmation of the results (*Table 4.2*).

[37] Martin and Saller 1957.

[38] Safont/Malgosa/Subirà 2000.

In terms of the age at death, an age range could not be determined for T9, who can be described only as an adult, while for T10A a very wide age range (40–60 years) can be proposed based on the observation of changes in the auricular surface of the ilium.[39] The degree of tooth wear is not particularly high,[40] but the loss of teeth during life suggests an advanced age. The child's age at death was determined to be approximately 3.5 years based on the development and eruption of the teeth.[41]

[39] Buckberry and Chamberlain 2002.

[40] Molnar 1971.

[41] AlQahtani 2009.

	Martin No.	d	s
MANDIBLE			
H mental foramen	69 (1)	23.0	
Th mental foramen	69 (3)	10.0	
HUMERUS			
Max l	1		
Total l	2		
W upper epiphysis	3	43.0	
Max diam at mid-point	5	19.0	19.0
Min diam at mid-point	6	17.0	16.0
Min perimeter	7		55.0
Transverse diam head	9	36.0	
Sagittal diam head	10	40.0	
RADIUS			
Min perimeter	3	37.0	
ULNA			
Max l	1		235.0
Physiological l	2		208.0
Min perimeter	3		32.0
W olecranon	6	17.0	17.5
Upper transverse diam	13	17.5	18.0
Diam upper A/p	14	19.0	20.0
FEMUR			
Diam A/p at mid-point	6	26.0	
Diam m/l at mid-point	7	25.0	
Perimeter at mid-point	8	80.0	
Transverse diam sub troch.	9	30.0	
Sagittal diam sub troch.	10	25.0	
Vertical diam head	18	41.5	
Horizontal diam head	19	41.0	

Table 4.5 Measurements of T10A.

Stature was estimated only for the female (T10A) using Trotter and Gleser's equations[42] and the only intact long bone, the left ulna, which yielded a maximum length of 235.0 mm. The estimated stature of T10A, 158.1 cm, is an average figure for the female population of Metaponto (*Table 4.3*).[43]

Pathologies. All of the osteological material was analyzed macroscopically for evidence of pathologies. T9

	Martin No.	d	s
TIBIA			
W upper epiphysis	3	62.0	
Diam a/p lower epiphysis	7		33.0
Max diam at mid-point	8	29.0	27.0
Transverse diam at mid-point	9	21.0	20.0
Sagittal diam foramen nutricium	8a	31.0	
Transverse diam foramen nutricium	9a	23.0	
Perimeter at mid-point	10	78.0	74.0
Min perimeter	10b		68.0
PELVIS			
Max diam acetabulum	22	47.0	
H greater sciatic notch (Sauter '55)		37.0	
W cotylo sciatic (Sauter '55)		32.5	

Table 4.5 cont. Measurements of T10A.

has an enthesopathy of the psoas major muscle insertion on the right femur (*Fig. 4.10*). During intense and repeated movements, the entheses (the insertion points of the tendons that attach the muscles to the bones) are subjected to biomechanical stress caused by microlacerations of the tendons and muscle tears. Enthesophytes are formed as a result of the repair process.[44] This pathology indicates repeated flexion, adduction, and extrarotation of the thigh.

T10A shows a bone erosion corresponding to the gluteal tuberosity, in the area of the gluteus maximus muscle insertion (*Fig. 4.11*). This loss of substance is evidence of intense and repeated external rotation and extension of the thigh, analogous to the enthesopathy of T9.

Of the teeth from T10A, only the first upper right incisor (I20r), the second lower right incisor (I2r), the first lower right premolar (P1r), and the second lower right molar (M2r) are preserved. A caries can be observed on the occlusal surface of the M2r (*Fig. 4.12*). Of the four preserved tooth sockets (mandibular alveoli), a strong alveolar bone retraction can be observed corresponding to P1r and M2r, as well as evidence for the loss of P2r during life with consequent alveolar bone substance loss, or resorption (*Fig. 4.12*).

[42] Trotter and Gleser 1957.

[43] Henneberg and Henneberg in *Necropoleis*.

[44] Ortner and Putschar 1981.

Taphonomy. The excavation photos demonstrate with certainty that T9 was a single, primary burial of the *a cappuccina* type. The body was laid with the head oriented northwest. The deceased was buried in a supine (dorsal decubitus) position with the upper and lower limbs extended and roughly parallel to the body (on the median sagittal plane). Despite the absence of anatomical connections caused by the destruction of almost all of the articulated surfaces and the epiphyses of the long bones, the anatomical arrangement is nevertheless consistent. Evidence for determining the exact mode of decomposition is lacking, apart from the evident opening of the pelvis. Nevertheless, given the *a cappuccina* container, decomposition probably occurred in an empty space.

The excavation photos of Tomb 10 show that the tomb contained two individuals, an adult (T10A) and a child (T10B). The possibility of a double burial, in which the two cadavers were deposited simultaneously,[45] can be excluded, first and foremost because the adult burial was subjected to a disturbance, which may have occurred at the time of the child's burial or independently of it. The tomb was presumably reopened before becoming completely filled with earth and after the decomposition of the adult T10A. Nevertheless in this case, too, decomposition could have occurred in an empty space. A large part of the skeleton was rearranged within the confines of the tomb, in the western half; this appears to have been an anthropic action, judging from the nearly parallel arrangement of the long bones. The only bones apparently left in their original position were the left fibula and several bones belonging to the left foot, elements that demonstrate the northwest-southeast orientation of the skeleton.

The remains of the child's skeleton, T10B, were located west of the rearranged remains of T10A and occupied one-third of the length of the tomb. The photos hint at the profile of the cranium and, immediately to the east, parts of ribs and long bones belonging to the upper limbs. This arrangement demonstrates that T10B shared the orientation of T10A.

The reopening of an *a cappuccina* tomb, the rearrangement of the skeleton, and the subsequent burial of another individual in the same tomb were apparently fairly rare events which currently lack precise parallels in this historical period.

[45] Duday 2006.

3D Reconstruction of the Sant'Angelo Vecchio Necropolis

Massimo Limoncelli

Virtual archaeology deals with the reconstruction of artifacts, buildings, and landscapes of the past. These may be extant or vanished, but they are known through archaeological, historical, or artistic sources. The use of three-dimensional models, hypertexts, and multimedia solutions—in addition to other digital technologies—makes a notable communicative and educational impact, presenting high quality science and dense cultural content.[46] Virtual archaeology therefore takes on a didactic and popularizing role for the broader field. It aims to communicate information about ancient areas, spaces, and objects (monuments, cities, and territories) via reconstructions in an effective, rapid, and repeatable way. The discipline places cultural patrimony in its proper context in order to allow an accurate interpretation of the past in ways that facilitate understanding in the wider public.

Virtual restoration shares many common traits with virtual archaeology,[47] but it is nevertheless distinct in its principles, methods, and techniques of reconstruction. Virtual restoration must be viewed in the context of conservation science—and the set of disciplines and professional competencies (e.g., in evaluation, project planning, verification, and hypothetical reconstructions) that cooperate therein—rather than in the context of restoration *per se*.[48]

This study integrates the two approaches with the aim of reconstructing the specific context of the necropolis at Sant'Angelo Vecchio (see *Fig. 4.2* and *Fig. 4.3*). The act of "reconstructing" entails a methodological and scientific path directed toward the most realistic possible re-creation of the context under study. The process of elaborating the information that leads to the final result requires what may perhaps be considered the most important aspect of the reconstruction project: a cognitive process. In fact, the virtual restoration of any item—whether a building, a fresco, a sculpture, or, as in this case, several tombs with grave goods—makes sense only if the reconstruction provides new or supplementary information that could not be obtained from the traditional methods of scientific research. Virtual reconstruction therefore should be understood as an act that starts with a "fragmentary (or partial) reality" and concludes by defin-

[46] Niccolucci 2002; Moscati 2009; Antinucci 2007.

[47] Limoncelli 2012.

[48] Giannini and Roani 2000, 186.

ing a "possible reconstructed (or complete) form." The most important phase of research during this act poses all of the questions useful in clarifying the function, relationships, and quality[49] that the archaeological reconstruction must address.

The creation of a virtual reconstruction follows a path opposite to that of the item itself during its use-life: conception, creation, duration, and disuse (whether abandonment, destruction, etc.). Starting with evidence from the material data (sources), an attempt is made to reconstruct the original appearance (mainly shape and colors, etc.), with the aim of re-inserting it into the context for which it was conceived and created, and in which it was placed. Different types of sources may be considered, including "direct" sources, which are retrieved from the documentation of the material evidence; "indirect" sources,[50] or information that can no longer be observed directly on the item but can be deduced from, for example, written or iconographic attestations;[51] the so-called "rule" (or laws)[52] valid for the historical-archaeological context of the evidence under consideration; and, finally, the "techniques of execution" used in the creation of the various categories of items. In the specific case of the reconstruction of the necropolis, the sources employed were mainly the data retrieved from the excavation documentation, including plans of the tombs and drawings and photographs of the finds.

The first step was the reconstruction of the individual tombs within a digital terrain model. The nearly rectangular pits into which the tombs were inserted were re-created using the "subdivision surface" organic modeling technique. The method allows for the creation of complex forms using the same methods as for the modeling of polygonal surfaces, but with the advantage of being able to operate on objects defined by a low number of faces, and with the possibility of acting on the curvature of the surfaces by assigning greater or lesser weight to the points of the 3D mesh structure.[53]

The skeletons of the individuals found inside the tombs were also created with the same modeling technique. The dimensions and the positions of the individual bones within each grave reflect their actual locations as documented by the detailed plan of the tomb with elevations. The models of the skeletons include only the bones actually found during the excavation, even though the individual bodies were obviously complete at the moment of burial. The structure of each tomb was re-created using a specific modeling technique (polygonal, NURBS, and subdivision surface), depending on the specific morphological features of the tomb type (a fossa, *a cappuccina*, and stone sarcophagus).

Finally, all of the grave goods found within each tomb were placed in their correct positions (see below). In this case, too, the reconstruction of each individual item was based on the drawings of the item in the excavation documentation, which were used as the reference source for the 3D model, and on the digital photographs needed for the creation of the textures.

The resulting reconstruction of the tombs is the most accurate possible visualization of the archaeological context. The expected result of virtual archaeology is the transformation of information into a perceptible form via various methods of scientific visualization: video, multimedia platforms, or, as in the present case, a printed image. Even the mere creation of a reconstructed image represents a description of an inductive and deductive reasoning process followed in the cognitive path associated with the act of reconstruction. The pairing of knowledge derived from the psychology of perception on the one hand, and from the techniques of graphic design on the other, compensates for the limit imposed by a simple written description by providing a hypothetical reconstruction.

The creation of reconstructions using virtual three-dimensional models represents knowledge, study, and verification of the material evidence, as well as an effective method of visualization and interpretive synthesis. The evidence can be interpreted at different levels, with the result that the reliability of the reconstruction is variable, but it increases proportionally in relation to both the quantity of information incorporated and the degree of analysis of the underlying archaeological evidence.

[49] Medri 2000, 134.
[50] Fancelli 2001, E13.
[51] Limoncelli 2012, 187.
[52] Medri 2000, 134.
[53] Della Calce 2004, 161–63; Catmull and Clark 1978; Kundert Gibbs and Lee 2004, 412–13; Caraccia 2005, 79.

Burial Descriptions: Classical Tombs

Tomb Group 8

For catalog conventions and abbreviations, *see p. 209, "Catalog Abbreviations."*

Location: Square H6, Level 1
Type: Fossa (?)
Orientation: NA
Terminus date range: 430–400 BC
Weighted date range: 432–416 BC
Basis for date: Red-figure and black-gloss pottery
Sex: Unknown
Age: Unknown
Dimensions: diam. 1.0 m
Ceramic objects: T08-01, red-figure hydria (SA 79-970P); T08-02, red-figure owl-skyphos (SA 79-971P); T08-03, black-gloss mug (SA 79-972P, inv. no. 319859); T08-04, black-gloss lebes gamikos (SA 79-973P, inv. no. 319860)
Comments: About 2 m from Tomb 4, four vases were carefully arranged and placed together. Skeletal remains of a child were found but were not available for final analysis. On a slab of bedrock near the vases were modern plow marks. It may be significant that a red-figure hydria and black-gloss lebes gamikos were found here, as with Tomb 4. Although it is possible that one of the skeletons of Tomb 4 belonged here, it is more likely that the skeleton of this burial was removed by modern plowing, although it would be remarkable that the vases remained unscathed as they did. It is difficult to know whether to regard this as a burial or a ceramic deposit.
Excavated: 12 July 1979.

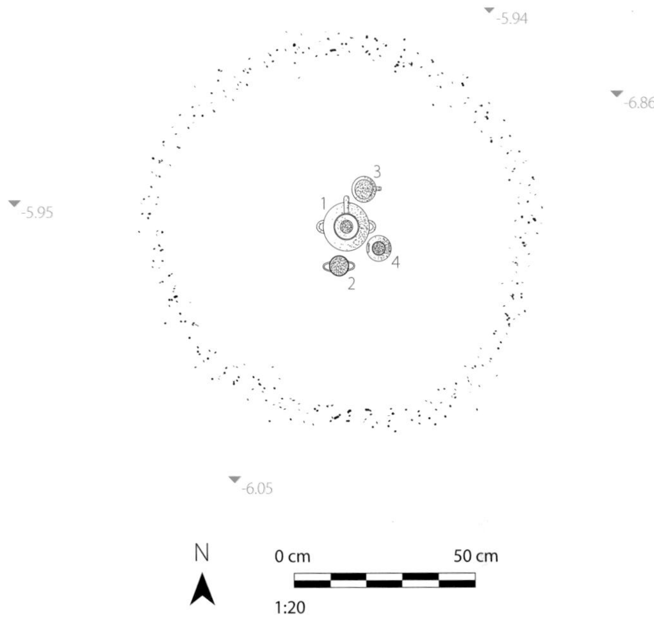

T08-01 Red-figure hydria
Provenance: SA 79-970P. Inv. no. 319857.
Dimensions: rim diam. 7.2, foot diam. 6.7, h 18.0, max body diam. 13.0 cm.
Condition: Complete. Recomposed from 12 fragments.
Description: Clay Munsell 7.5YR 7/4 (pink). Thick black gloss. Relief line used for inner details of the figures. Upper surface of the rim, outer lower margin, and underside of the foot reserved. Rim molded in two degrees, ogee foot. Short neck, slightly flat shoulder, and plump body. Horizontal lateral handles curved upward. On the rim, looped egg pattern. Eggs and dots on the neck. Below the scene, running key pattern. Draped woman with outstretched right arm rising from a *diphros* and addressing a second draped woman moving to left and looking back. The presence of the *klismos* suggests that the scene is set in an inner space.
Comments: For the Pisticci Painter's hydriai, see T04-03. Subsidiary decoration, with the double register of eggs on both rim and neck and the running key pattern, is unusual; the looped egg pattern on the rim also occurs on the hydria *LCS* Suppl. 3, 1/24b (Athens, Benaki Museum) and on an unpublished vase from Montescaglioso, Edificio Scolastico, tomb 7 (Matera, Museo Archeologico Nazionale D. Ridola). The other two patterns are not paralleled in vases by the Pisticci Painter. The presence of egg patterns both on the rim and on the neck is highly unusual in South Italian pottery but is frequently adopted in hydriai by the Washing Painter and his workshop.
Attribution and date: Pisticci Painter. The hydria is attributed by Trendall to the Early Phase of the painter (*LCS* Suppl. 3, 4–5). For the treatment of the drapery there is a parallel with the Louvre bell-krater (*LCS* 1/48 and *CVA* Louvre 25, pl. 1), which belongs to the Developed Style. As a consequence, the hydria could be dated to 430–420 BC.
Bibliography: Edlund 1979, p. 9 fig. 22. *LCS* Suppl. III, 1/7a, pl. I, 1. *Necropoleis*, 447. *CVA* Louvre 25, 16 (M. Denoyelle). Söldner 2007, 177, n. 1346.

(FS)

Tomb Group 8 cont.

T08-02 Red-figure owl-skyphos

Provenance: SA 79-971P. Inv. no. 319858.

Dimensions: h 4.8, rim diam. 6.0, foot diam. 3.2 cm.

Condition: Restored from four joining fragments. Rim slightly chipped.

Description: Clay Munsell 7.5YR 7/4 (pink). Black-to-brown gloss. No traces of *miltos* are preserved on the reserved areas. Lower side of foot glossed, with a central dot. On both sides, owl between olive branches with interior line.

Comments: South Italian owl-skyphoi, derived from similar vessels produced in Athens especially in the second and third quarter of the 5th c. BC, are another sign of the Atticizing character of the Metapontine red-figure production in the second half of the 5th c. BC. Owls decorating skyphoi of types A, B, and C and bolsals appeared within the workshop of the Pisticci Painter. The success encountered by this decoration is also attested by its utilization in other shapes (as in the hydria and the pelike found in association with a vase by the Pisticci Painter: Nava 2003a, 667, pl. XXXIII, Metaponto, site M, tomb 55). While Type B skyphoi were dismissed soon after their introduction, production of the other types of owl-cups and skyphoi continued within the Amykos Painter's workshop (to whom the skyphos from Pantanello T87 can possibly be attributed: Burn in *Necropoleis*, 616, T87-2) and lasted at least until the first quarter of the 4th c. BC, when owl-skyphoi are part of the production of the Dolon-Creusa workshop, as testified by the fragments yielded by deposit no. 1 (inv. no. 29346) and 4 (inv. nos. 29188+146484, D'Andria 1980a, 434, no. 346), and by the vessel found in association with a pelike by the Creusa Painter at Pantanello, T219 (*Necropoleis*, 341). As already stated by Trendall, the Sant'Angelo Vecchio skyphos can be attributed to the Pisticci Painter himself. The owls of the pelike in tomb 55, cited above, and the bird on the hydria *LCS* Suppl., 1/55a are also similar. For the shape of the skyphos, see Elliott in *Necropoleis*, 681, S34 (from Pantanello, T209, third quarter of the 5th c. BC, found in a context that also yielded a pelike by the Pisticci Painter: Burn in *Necropoleis*, 620, T209-7).

Date: 430–420 BC

Bibliography: *LCS* Suppl. 3, 9, OS3 (Pisticci Painter). Burn in *Necropoleis*, 640, no. SAV 8-2.

(FS)

T08-03 Black-gloss mug

Provenance: SA 79-972P. Inv. no. 319859.

Dimensions: h 5.3, base diam. 5.4, rim diam. 7.1 cm.

Condition: Intact.

Description: Strap-handle and ring base; entirely glossed except for the underside, which has a glossed circle toward its center and another on the inside surface of the ring base. Clay Munsell 5YR 7/4 (pink).

Comparisons and comment: See comments for T01-02. The ring base is a highly unusual feature for mugs in southern Italy, although it was common for Attic mugs after the middle of the 5th c. BC (*Agora XII*, 250–51). The small size and narrow base (in comparison to the rim) are noteworthy. Pantanello T206-14 is the closest comparison from that site, but it has a disk base (Elliott in *Necropoleis*, 668, M9).

Date: Second half of the 5th c. BC.

(EGDR)

T08-04 Black-gloss lebes gamikos with lid

Provenance: SA 79-973P. Inv. no. 319860.

Dimensions: rim h 6.5, handle h 8.3, h with lid 11.5, base diam. 4.3, rim diam. 4.5 cm.

Condition: Intact.

Description: Thick vertical handles, round in section, set on the shoulder. High ring base, almost vertical on its outside face, with a slight convex profile at the top of the foot; a thick rounded molding at the junction of base and body. The lid has a tall, complex knob that is best understood by referring to the drawing. Black gloss covers the exterior; underside reserved. Clay Munsell 5YR 7/6 (reddish yellow).

Comparisons and comment: This small vessel, part of Elliott's group of "plain" lebetes gamikoi of the second half of the 5th c. BC at Pantanello (Elliott in *Necropoleis*, 660), has thick walls and handles. There is no close parallel at Pantanello for the shape of the foot with the thick molding at its top, and neither for the tall, complex knob of the lid. The best parallel for the form of the lid is a lebes gamikos from Gravina, Botromagno tomb 2 (*Arte e artigianato*, 407, no. 20). The tomb contains almost a dozen proto-Italiote vases, including a number from the Pisticci-Amykos workshop, and is dated 430–400 BC. The black-gloss lebes gamikos, not far in shape and size from the Sant'Angelo Vecchio example, is probably Metapontine.

Date: 440–400 BC.

(EGDR)

1:2

T08-01

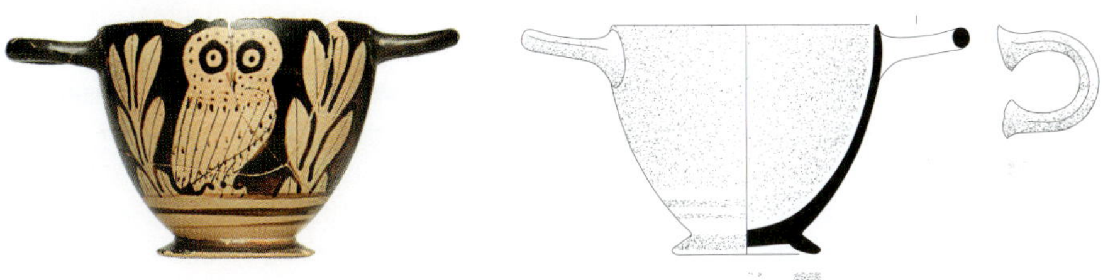

T08-02

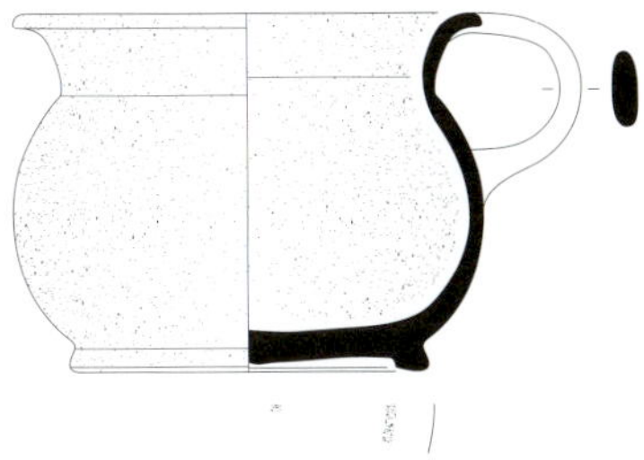

T08-03

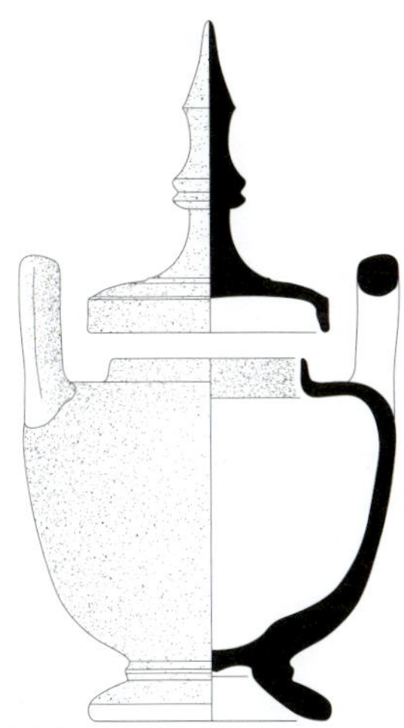

T08-04

Tomb 4

Location: Square H6, Level 1, batt. 4.
Type: Fossa; two skeletons, A and B.
Orientation: 68°
Skull (A): SE
Skull (B): Unknown
Terminus date range: 440–400 BC
Weighted date range: 440–420 BC
Basis for date: Red-figure and black-gloss pottery
Sex (A): Female
Age (A): 70–x
Sex (B): Male
Age (B): 50–60-x
Dimensions: extant l 1.2 m, w 0.6 m
Bibliography: Edlund 1979, figs. 2.a–b.
Ceramic objects: T04-01, black-gloss lebes gamikos (SA 79-959P, inv. no. 319854); T04-02, black-gloss lebes gamikos (SA 79-966P, inv. no. 319852), T04-03, red-figure hydria (SA 79-967P, inv. no. 319855)
Non-ceramic objects: T04-04, alabastron (SA 79-958P, inv. no. 319853); T04-05, iron pin (SA 79-960M); T04-06, bronze ring (SA 79-961M); T04-07, bronze bead (SA 79-962M); T04-08, glass bead (SA 70-963G); T04-09, amber (bead?) (SA 79-964SS); T04-10, seashell (SA 79-965SS); T04-11, shell (SA 79-968SS)
Comments: This tomb consisted of two burials. Individual A was placed in an extended, supine position; it was partially disturbed. Individual B was completely disarticulated: the cranium and jaw were missing, and the mandible was at the opposite (N) end of the burial from the skull of Individual A. It is possible that both individuals were originally buried elsewhere then brought to the present site fairly soon afterward (especially Individual A, which remained somewhat articulated). Other scenarios are possible, the most simple that skeleton A was a later burial that disturbed the earlier burial of Individual B. In this case, the grave goods should be associated solely with Individual A (see also notes on Tomb Group 8, the vase collection of which may provide a connection between the two burial sites). Note that there seem to be no examples of secondary burials at Pantanello. The two skeletons were found in a cluster within sandy natural soil. The grave goods consist of a hydria (T04-03) found apart from the other items. The two lebetes gamikoi T04-01 and T04-02 were found with the lid inverted. Six of the grave goods (T04-06–T04-11) were found near lebes T04-02, and were probably originally preserved inside the vase. The alabastron T04-04 and the iron pin T04-05 were placed at the edge of the cluster of bones.
Excavated: 17 July 1979.

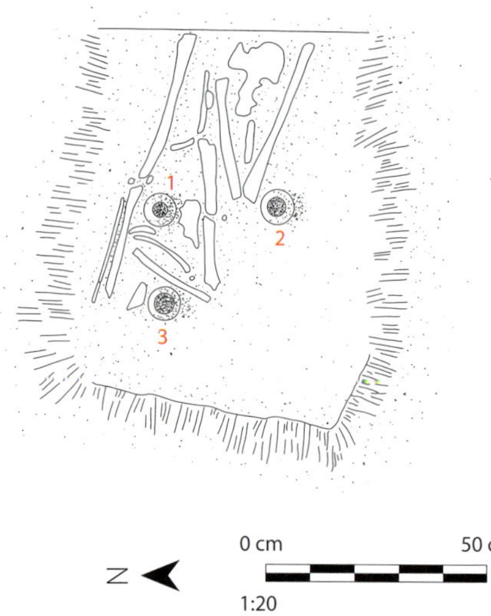

Skeletal Material

Location: SA 79-1040B.
Sex: (A) Female; (B) Female
Age: (A) 60+ (smaller); (B) Mature adult, at least 40 (larger)
Condition: The area of the third trochanters are not intact on the damaged femurs recovered with one exception. The relatively intact section evidences a third trochanter, suggesting probable presence on all. The lineae asperae are faint on A and strong on B, but both are raised in the central area like other individuals in this population. The age of the smaller woman has been evaluated from the right pubic symphysis (Krogman 1962, 94).
Stature: (A) 146.7; (B) 150.4. Long bone lengths: (A) r. femur, ca. 37.5, (B) l. femur, ca. 39.0 (Trotter and Gleser 1958, table 13).
Skull: The skull that appears to relate to this burial (Individual A) was recovered from the baulk where it was found. The crushed condition appears to be the result of normal post-interment processes. The maximum length (reconstructed) is 188 mm and the breadth is estimated at 140 mm. The small mastoids suggest that this is a female. The sutures are all fully closed.
Dentition: A mandible (of Individual B?) was located with nine of the 16 teeth (50%) lost antemortem. At least one had been lost within a year prior to death. This mandible was found at the opposite end of the grave from the skulls, but no mandible was recovered with the other skull.
Pathology: A vertebral body with some anterior lipping belongs to one of these females. A distorted (diseased?) metatarsal also was found.

Excavation: The materials recovered from Tomb 4 were recognized as problematic in location by the excavators. The photograph taken prior to the removal of the material suggested that some bones were in articulation and others were not, and that at least two people were represented by this collection of bone. Recognizing the difficulties involved, the excavators took several photographs in the field and then removed a large section of the earth enveloping the bones of the tomb for specialized study. This provided a series of bones that were, in effect, *in situ*, as well as a second collection of miscellaneous fragments that were recovered from the grave. The bones were all labeled and drawn in situ before being taken from the earth matrix. When studied they were found to represent two people. The bones then were separated by size to define the two individuals (not more) who are represented. Lists of the two bone sets (by individual) and of those bones that could not be specifically placed with either person are listed elsewhere (Becker 1980b). A description of the detailed excavation of this mass of earth also appears in these field notes.

Analysis: Two problems confront the analyst. One is to determine how many individuals are represented by the remains piled together in this locus. This was achieved with ease, and the two people identified were determined to be females. The second task is to determine how or why these bones came to be placed into this situation, which is quite different from what would be expected under normal circumstances. Since the archaeological data reveals that a large number of vessels were present in the grave, the problem is even more peculiar. After evaluating the remains as being those of two females, I was informed that the vessels were in sufficient number to provide the grave goods for two female burials, but they might represent a single wealthy woman's equipment. The problem of skeletal placement or arrangement therefore becomes an important means by which the problem of which artifacts belong with each person could be resolved.

Conclusions: This series of tombs, and presumably the two women of Tomb 4 all date from the 5th century BC. Individual A, the smaller and older woman, appears to be in an extended position, supine, with her head to the east in the pattern typical for the population in this cemetery. Individual A must have been disturbed at some point in time, as the full skeleton is not present. Individual B, on the other hand, has been completely disarticulated and must have either been disturbed at this locus, being the first person interred here, or brought from another grave. Three completely different possibilities exist regarding the presence of these remains and their states of articulation or disarticulation. Only the archaeological evidence may serve to support one of the following three possible series of events, A–C.

A. (1) Individual B may have been buried at this locus (no direct evidence found). (2) The interment of Individual A resulted in the removal of the bones of B, which were later replaced over the body of A. (3) The bones of both A and B are disturbed by subsequent activities. Note that Tomb 4 is at the highest elevation of this group and thereby was subject to the most disruption by erosion or plowing.

B. (1) Individual A was buried at this locus. (2) Individual B, who may have been buried somewhere else, either before or after A, was removed from another location after A was buried. (3) The disarticulated remains of B were intruded into the Tomb 4 locus, dislodging some of the bones of Individual A.

C. (1) Both A and B were buried elsewhere. (2) Both A and B were disinterred and brought to this locus. (3) Individual A, probably recently buried, was somewhat articulated and was placed in Tomb A in a "normal" position. (4) The remains of B were simply added to the grave.

Although the first of these scenarios had seemed most probable, the fact that the skull of A and mandible of B were found at opposite ends of the grave and that the other cranium and jaw did not appear at all causes us to shift interest to the third possibility.

(MJB)

Tomb 4 cont.

T04-01 Black-gloss lebes gamikos
Provenance: SA 79-959. Inv. no. 319851.
Dimensions: rim h 7.4, handle h 9.7, base diam. 4.6, rim diam. 4.6 cm.
Condition: Intact, one handle missing.
Description: Vertical handles, round in section, with small plastic knobs on the shoulder, one to either side of each handle. Ring base with two broad concave moldings on the exterior surface. At the lower part of the body, a groove marks a sharp carination making a short stem down to the foot, the junction of which is marked by a small circular molding. Ribbed decoration from the top of the shoulder, with the gap between each groove closed at the top by a semicircular groove; on one side of the vase, at the bottom of each vertical groove a stamped seven-leaved palmette, facing downward. The palmettes are absent on the other side of the vase. Underside reserved. Clay Munsell 5YR 7/6 (reddish yellow).
Comparisons and comment: Stamped patterns were first applied to black-gloss vases in Athens a little before 450 BC (*Agora XII*, 22ff.) and the practice was adopted in Metaponto, with impressed decoration being used on much the same shapes as in Athens: mainly drinking cups and squat lekythoi (see, e.g., D'Andria 1980a, 406ff.). In Metaponto, stamps were occasionally applied to other shapes as well, and at Pantanello a total of 14 small lebetes gamikoi with palmettes stamped on their shoulders have been found (Elliott in *Necropoleis*, 660–61). Other lebetes gamikoi from Pantanello have ribbed bodies, but ribbing is never combined with stamped decoration in the way we see it on the two lebetes gamikoi from Sant'Angelo Vecchio Tomb 4. The sharp carination toward the bottom of the body, creating a short stem for the vase, finds no parallel at Pantanello nor elsewhere in southern Italy, although this detail is not always easy to see in a photograph.
Date: 440–400 BC.

(EGDR)

T04-02 Black-gloss lebes gamikos with lid
Provenance: SA 79-966P A+B. Inv. no. 319854.
Dimensions: rim h 8.8, handle h 11.3, h with lid 14.7, base diam. 5.1, rim diam. 5.5 cm.
Description: Vertical handles, round in section, with small plastic knobs on the shoulder, one to either side of each handle. Ring base with three rounded moldings separated by two deep grooves; the top of the foot is convex, and a small rounded molding marks the junction with the body. The lid has two narrow concentric moldings toward its center with a tall conical knob, grooved at its base. Entirely glossed on the exterior; below a horizontal groove at the shoulder, incised ribbed decoration on the upper part of the body. The vertical grooves which make up the ribbing are joined at the top and bottom by semicircular grooves; at the bottom of each vertical groove, a downward-facing seven-leaved palmette. Underside reserved. Clay Munsell 2.5YR 5/8 (red).
Condition: Intact.
Comparisons and comment: For general comments, see above, T04-01. This vessel can be compared with Pantanello tomb 350-32, with a weighted date range of 458–438 BC (Carter and Hall in *Necropoleis*, 386–87). Ribs which have been "closed-off" at the bottom by semicircular grooves are extremely rare. In this case, a semicircular stamp has been used, as perhaps one would expect: the semicircles meet the vertical grooves at the top, but overlap at the bottom, where the diameter of the vessel has diminished. Hanging palmettes are unusual; the Metapontine potters probably got the idea from Attic sessile kantharoi (e.g., *Agora XII*, 281, no. 635, 450–425 BC), and seem to have produced sessile kantharoi that used the scheme, e.g., Pisticci, Matina Soprano tomb 1 (the tomb, which should date to the third quarter of the 5th c. BC, is described in *Popoli anellenici in Basilicata*, 21–23, but the sessile kantharos, inv. no. 20124, surely of Metapontine manufacture, was not placed in the exhibition and hence is not listed).
Date: 440–400 BC.

(EGDR)

1:3

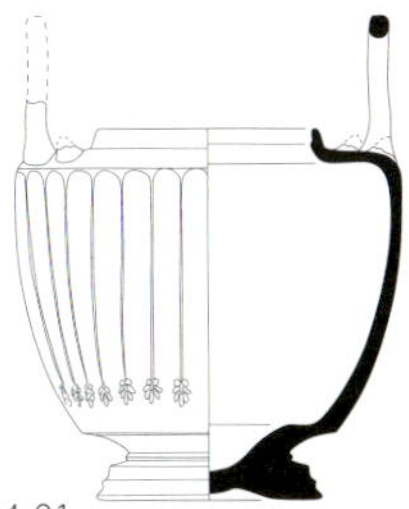

T04-01

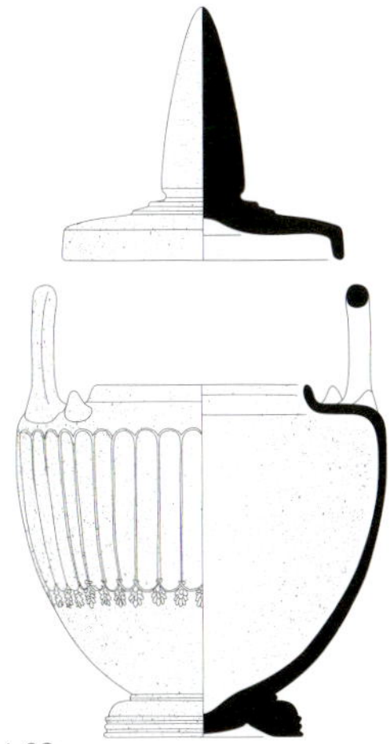

T04-02

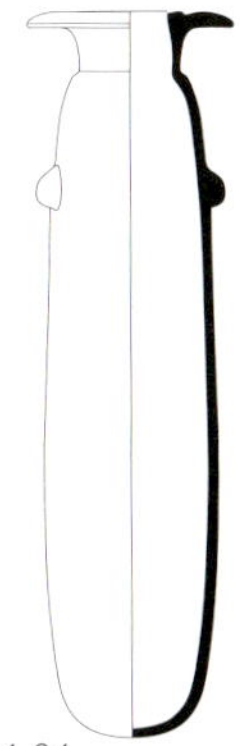

T04-04

T04-03

1:2

T04-10

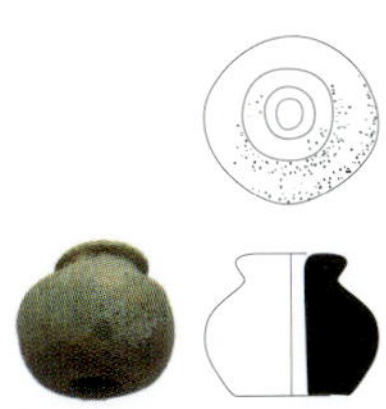

T04-07

T04-09

T04-08

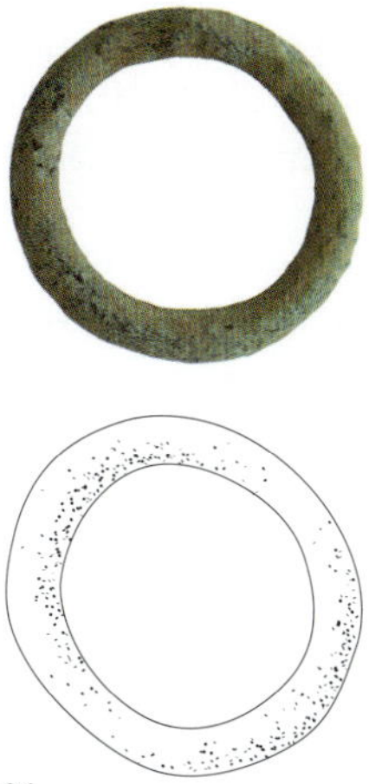

T04-06

T04-11

Tomb 4 cont.

T04-03 Red-figure hydria
Provenance: SA 79-967P. Inv. no. 319855.
Dimensions: rim diam. 9.2, foot diam. 7.9, h 22.8.
Condition: Complete. Recomposed from 13 fragments. Small portion of the rim missing.
Description: Clay Munsell 7.5YR 7/4 (pink), thick black gloss. Miltos on reserved area.
Shape: Rim molded in two degrees, ogee foot. Slightly rounded shoulder and plump body. Horizontal lateral handles curved upward.
Decoration: On the rim, vertical lines. On the neck, laurel wreath facing right. At the bottom of the scene, a band with esses. Upper surface of the rim and lower part of the foot reserved. Youth with himation pursuing a woman with wide-sleeved chiton and himation wearing diadem and moving to right. Between them, tendril.
Comment: The Pisticci Painter decorated 27 known hydriai, ranging in height from almost miniature examples of around 10 cm (Nava 2003a, 667, pl. XXXIII, 1 from Metaponto, site M, tomb 55) to a single vessel of 40 cm (*LCS* 22, no. 55), with the majority in the 20s and 30s. The picture is on the body and comprises two or (in vases of large dimensions) three figures; pursuits and gynaeceum scenes are the subjects represented. The variety of heights is matched by a certain homogeneity in the shape, which indicates a common workshop tradition. The hydria from Tomb 4 can be paralleled for its shape to the vase at the Benaki Museum in Athens (*LCS* Suppl. 3, 6, no. 24b, Tzannes 2012) and to the hydria from Pantanello, tomb 88 (Burn in *Necropoleis*, 616). Subsidiary decoration is normally characterized by an egg band on the rim, a laurel wreath on the neck, and a meander frieze with three to six meanders broken in the middle and squares with uprights or saltire crosses. Reserved areas are present on the upper surface of the rim, on the lower part of the foot, and below the handles. The decorative pattern of the hydria from tomb 4, on the contrary, is unparalleled; the rare band with esses also occurs on a pelike at Athens (Andreiomenou 1971, 561–62, no. 8, pl. 568γ, attributed to the Early Phase of the Pisticci Painter by Trendall in his unpublished fourth supplement on Lucanian pottery), on a shouldered lekythos from Rutigliano, tomb 1978/11 (*Arte e artigianato,* 408, no. 348.27), and on some black-gloss squat lekythoi possibly produced within the same workshop. The same decoration is also used on the neck of the volute-krater by the Sisyphus Painter (B.M.F 178, *RVAp I*, 1/90). Vertical lines on the rim are also quite unusual in the Pisticci Painter's hydriai (see the vase from the market *Art in the Ancient World XXI*, 64, no. 163), while they occur more frequently in the Amykos Painter's vases (see *CVA* Louvre 38, pl. 8, attributed by M. Denoyelle to the Early phase; hydria from Pantanello, tomb 87: Burn in *Necropoleis*, 615; Taranto, Collezione Rotondo: *LCS* Suppl. 1, 9, no. 232a). The youth is a variant of the Eros frequently used by the painter (*LCS* 10, i); for the woman, see hydriai from Saldone (*LCS* Suppl. 3, 6, 25a, pl. I,3) and at St. Petersburg (*LCS*, 18, no. 26).
Attribution and date: Pisticci Painter, developed style, early phase (Trendall), 440–430 BC.
Bibliography: Edlund 1979, 1, figs. 2a–b; 9, fig. 21; *LCS* Suppl. 3, 6, no. 24a; Denoyelle 2002, 106–7; Colucci 2002, 162–63, no. 147; Carter 2006, 19–20, fig. 1.7B.

(FS)

T04-04 Alabastron
Provenance: SA 79-958P.
Dimensions: h 14.5, rim diam. 4.3.
Condition: Recomposed from 15 fragments with some parts of the body missing.
Description: Horizontal lip rim, short neck, long narrow body with two small lug handles.
Comparisons and date: Alabastra are quite common in burials in and around Metaponto; this is a Type A alabastron in the typology formulated for Pantanello, where 29 were recovered (Carter and Parmly Toxey in *Necropoleis* a). They are not easy to date on the basis of their form. While at Pantanello the width-to-height ratio seems to be indicative of chronology, the figure for this alabastron (0.29) fits into the small area of overlap between the alabastra dated 460–435 BC (which range from 0.27–0.34) and those dated 445–370 (which range from 0.23–0.30). One clear correlation for alabastra at Pantanello was their appearance in the more expensive burial forms for the period: sarcophagi and stone cists (Carter and Parmly Toxey in *Necropoleis*, 760). The alabastron at Sant'Angelo Vecchio comes from a fossa tomb, which may support the proposition that one or both of the skeletons in the grave were secondary burials.
Date: 460–370 BC.

(EGDR)

T04-05 Iron pin
Provenance: SA 79-960M
Dimensions: l 11.5, w 1.7.
Description: Part of the shaft.
Condition: Fragmentary.
Date: Based on tomb context.

T04-06 Bronze ring
Provenance: SA 79-961M
Dimensions: diam. 2.6/2.8, w 0.4.
Condition: Intact.
Description: Bronze ring with oval section.
Date: Based on tomb context.

T04-07 Bronze bead
Provenance: SA 79-962M
Condition: Intact.
Dimensions: h 1.8, max diam. 2.3, w of central hole 0.35.
Comparisons and date: No similar object has been encountered. In form, the bead has something in common with beads and pendants that were widespread in the indigenous world, mainly in amber, representing vases such as aryballoi, alabastra, and particularly amphorae (e.g., Braida di Vaglio tomb 102, Bottini and Setari 2003, 40, pl. 44–45; for Rutigliano, *Museo Taranto II.2*, 601–4). Those beads and pendants, though, were generally pierced through near the "neck," not laterally as here.

(EGDR)

T04-08 Glass bead
Provenance: SA 79-963G
Dimensions: diam. 1.1.
Condition: Intact.
Description: Thick disc-shaped green bead with a large central hole.
Comparisons and date: Only two tombs at Pantanello contained glass beads, and both had the same spherical type with ribbed sides. Glass beads similar to that from Sant'Angelo Vecchio appear in the indigenous world, often alongside amber beads of the same shape (e.g., Rutigliano tomb 83, *Museo Taranto II.2*, 83–84).

(EGDR)

T04-09 Amber (bead?)
Provenance: SA 79-964S
Dimensions: h 1.4, max diam. 1.6, w of perforation 0.25.
Condition: Intact.
Description: Pierced spherical bead with flattened ends.
Comparisons and date: Not a single object made out of amber was discovered in the Pantanello necropolis, but the substance had been in wide use in the Italic communities of South Italy since the Iron Age. There are some very spectacular collections of amber in elite tombs at indigenous sites in Basilicata, most famously at the "princely" tombs of Serra di Vaglio of the late 6th and early 5th c. BC, where a number of amber beads of sub-spherical shape appear amongst the necklaces from tombs 102 and 106 (Bottini and Setari 2003, pl. 44–47). Taken together with the bronze and glass beads listed above, and the bronze ring and shell from the same tomb, there is a distinctly Italic flavor to the ornaments from this burial.

(EGDR)

T04-10 *Monetaria annulus* (Ring cowrie or gold ring cowrie)
Provenance: SA 79-965SS; Tomb 4
Condition: Complete, very worn.
Dimensions: l 2.3, w 1.7.
Comment: The domed back of the shell has been cut away to reveal the central whorl. The golden yellow band worn and intermittently visible. Found together with two *Chlamys glabra* valves (SAV Sh 04–05) and one *Tapes decussates* valve (SAV Sh 19).
Date: 441–429 BC (weighted date range based on ceramics).
See also: SAV Sh 01.

(CDA)

T04-11 *Chlamys glabra* (Smooth scallop)
Provenance: SA 79-968SS; Tomb 4, Level 1, Batt. 4.
Condition: Complete.
Dimensions: l 4.1, w 4.3.
Comment: Found together with another *Chlamys glabra* valve (SAV Sh 05) and a *Tapes decussates* valve (SAV Sh 19).
Date: 441–429 BC (weighted date range).
See also: SAV Sh 04.

(CDA)

Tomb 7

Location: Square G6, Level 2, batt. 1 (Stratum 2A).
Type: A cappuccina with Laconian tile
Orientation: 95°
Date range: Mid-5th c. BC
Basis for date: Banded ware
Sex: Unknown
Age: Child (8–12 years)
Dimensions: l 0.4 m, w 0.6 m
Ceramic summary: T07-01, banded squat lekythos (SA 79-952P, inv. no. 319856)
Comments: This burial was severely damaged by the overlaying tile fall, and only one bone of the skeleton remained. The location of the skull is unknown. The short ends of the burial were defined by the stone slabs, of which only the eastern one is preserved. Curved tiles (exact number not noted) were placed over the body. A large stone slab, probably a tomb marker, stood vertically at the west end (cf., Tomb 5). The squat lekythos T07-01 was found in the area of the tomb and is possibly originally part of the burial, disturbed during the construction of the workshop. Because of its extremely poor condition, no illustration of this tomb exists.
Excavated: 2 July 1979.

Skeletal Material
Location: Baulk in square H6
Sex: Unknown
Age: Child, 8–12 years
Condition and Excavation: This burial had been cut through in 1979, leaving only part of the western (foot) end in the baulk. The author decided that careful recovery of these remains despite absence of the dentition might provide information by which an estimate of age could be made. Fragments of the left (?) tibia and fibula were recovered, indicating that everything above the knees had been lost. As in the case of Tomb 6, we can assume that the Tomb 7 child was lying supine in its grave. Comparative evaluation of bone size leads me to suggest an age of about 8 to 12 years for this child.

(MJB)

T07-01 Banded squat lekythos
Provenance: SA 79-952. Inv. no. 319856.
Dimensions: h 10.0, base diam. 7.5, rim diam. 3.3.
Condition: Broken into several pieces and repaired; weathered surfaces; the upper and lower bands of gloss have largely disappeared.
Description: Thickened rim with slightly concave inner side; narrow neck; sloping shoulder; saggy body with greatest diameter just above the base, which is flat and slightly concave underneath. Handles, roughly oval in section, attach to the shoulder and to a broad section encompassing the bottom of the rim and the entirety of the shoulder. A broad band of gloss covers the upper part of the interior and exterior of the rim and its top surface; two further broad bands are below the level of the handle, and another just above the base; exterior surface of handle painted; reserved underneath. Clay Munsell 5YR 7/4 (pink).
Comparisons and comment: Banded squat lekythoi are rare at Pantanello: there are only five in total and two of them, with ring bases, are not properly comparable (Elliott in *Necropoleis*, 704–5). The other three have the same flat, slightly convex bases as the Sant'Angelo Vecchio lekythos, and to them others can now be added from excavations around Metaponto (e.g., propr. Bonaventura tomb 295: De Siena 2007b). Such vessels are scattered thinly but broadly in the indigenous world, and occur in various 5th-c. BC tombs, often together with other material that can plausibly be associated with Metaponto, at sites such as Matera (from tombs in loc. Rifeccia: Lo Porto 1973, 80–81, pl. 70:1), Ceglie del Campo (tomb FXX, Marin et al. 1982, 150ff., pl. 22), Torre di Satriano (Colangelo 2009, 16, fig. 9, tomb 33) and Ordona (tomb LX/92, *Ordona VII/2*, pl. 36, no. 40, apparently undecorated but of identical shape). Similar squat lekythoi appear also to have been made at Poseidonia, again in both ring-base and flat-base types (e.g., Cipriani 1989, fig. 11, where both types appear in tomb 341). Poseidonia is the probable source of examples found in indigenous sites in Campania such as Fratte (e.g. *Fratte*, fig. 427, tomb 30/1972) and San Brancato (Donnarumma and Tomay 1999, pl. 23, tomb 21), and may even have been the origin for lekythoi found further afield, in Sicily (e.g. Selinus, Manicalunga cemetery, tomb 47: Kustermann Graf 2002, no. 935). Versions from Metaponto and Poseidonia are distinguishable, it seems, on the basis of their shapes (Campanian versions have a baggier body, with a shoulder that slopes more steeply) and their decoration (Metapontine examples almost invariably have a thin band of gloss above a thick band on the shoulder, while the Campanian versions tend to have bands of equal thickness). On both of these grounds the lekythos from Sant'Angelo Vecchio more closely resembles the Poseidonian type. A Poseidonian import so close to Metaponto may seem implausible, but the two Achaian colonies are thought to have been in close contact by the overland route that passed by Sant'Angelo Vecchio, and it is also worth recalling that this vase was probably sold for its contents rather than for itself.
Date: Mid-5th c. BC.

(EGDR)

1:2

T07-01

Tomb 1

Location: Square G6, Level 2/Level 3, batt. 1.
Type: Carparo sarcophagus with tile lid
Orientation: 68°
Skull: SE, facing W.
Date range: 460 BC
Basis for date: Black-gloss pottery and tomb type
Sex: Male
Age: 25–30
Dimensions: l 2.1 m; w 0.7 m; extant h 0.3 m; wall th 0.06 m.
Ceramic objects: T01-01, black-gloss amphora (SA 79-937P, inv. no. 319845); T01-02, black-gloss mug (SA 79-938P, inv. no. 319846)
Bibliography: Edlund 1979, fig. 17.
Comments: The NW end of the tomb was disturbed by the construction of the wall of the Late Republican workshop shed. Consequently, the upper edges of the sarcophagus were uneven, and the skull and the lower end of the body were crushed. The precise number of tiles used for the lid is unknown, but two of them were preserved at the middle of the tomb. The cover of the sarcophagus uses the same type of tile as Tombs 2, 3, and 5. On one of the tiles (SA 79-1074T, *Fig. 32.11*), only one of the two flanges is in the usual thick, low, and squat style, while the flange on the opposite side has the same quarter-circle section but is smaller and less squat. For tomb type, cf. Tombs 2 and 3. *(FP)*
Excavated: 11 July 1979.

Skeletal Material

Location: SA 79-1039B. Square G6.
Sex: Male
Age: Adult, possibly a mature adult of 40–50 years.
Condition: Tomb 1, originally identified as Burial G6, was in extremely poor condition. Not one long bone was intact, although the remains appear massive. The right humerus midshaft diameter (lateral) is 2.3 cm. The pelvic angle is acute and both mastoids large, but no supraorbital crests are seen and the frontal bone generally appears gracile, as is common in Italic populations (Becker and Salvadei 1992).
Stature: Estimated at 164 cm. Reconstruction of the right femur provides a basis upon which stature may be calculated. The maximum length of 42.8 cm can be used to compute stature, especially when linked with a right tibia length of 34.4 cm. Estimates of stature are 164.4 ±3.74 (Trotter and Gleser 1958, 120) or 163.0 cm according to the Manouvrier table (Krogman 1962, 155), an impressive congruence. Since we have an in situ measurement of 155 to 156 cm, these calculations suggest that the supposedly more accurate measurement in situ should be evaluated in light of the caution noted earlier (Becker 1979) regarding the measurement of skeletal material within open tombs where skulls tend to become dislocated. The left femur has a bicondylar length of 42.3 cm. and a head diameter of 4.6 cm. Other data are recorded in note form (Becker 1980b). Of note with this individual is the presence of a large third trochanter. The *linea aspera* (left) differs from all the others in this population in that it is quite large and bifurcate at the distal end. A slight bifurcation is evident on the proximal end. Although the right tibia can be reconstructed, the left is completely smashed. A left patella is intact.
Skull: The skull was crushed within the earth, but some restoration might be possible with considerable time commitment.

Dentition: Dentition is very poor. The maxillary area was destroyed, but two molars found in the fill may have been from the maxilla. Both are moderately worn, but reflect uneven wear such as would be expected if opposing teeth had been lost. Only four of the 14 (or 16) mandibular teeth were not lost, confirming the observations derived from the molars. Resorbtion of the gums is extreme and tooth loss must have occurred over several years and was in process at death. I would estimate, conservatively, a decay-loss rate of 20/28 or 71%.

Pathology: The discovery of a seriously deformed right mandibular condyle may reflect alteration due to tooth loss, but could be an entirely independent pathology. A slight gonial flare is evident, as are robust gonial tubercles. Measurements and photographs of all these features were taken.

(MJB)

T01-01 Black-gloss amphora

Provenance: SA 79-937P. Inv. no. 319845.

Dimensions: h 37.7, base diam. 12.3, rim diam. 17.2.

Condition: Broken in 22 fragments and repaired at the neck and shoulder with small fragments missing.

Description: Straight base, glossed on the upper surface; dipped decoration with a relatively small reserved band above the foot. Brown and opaque gloss with large oxidation stain. Traces of a double overpainted band. Ridge at the junction of body and neck, triple-reeded handles, rim slightly convex on its exterior surface. Internal surface of the neck is glossed. Clay Munsell 7.5YR 6/6 (reddish yellow).

Comparison and comment: While similar in dipped decoration and in size to many of the black-gloss amphorae of the 5th c. BC at Pantanello, this Nolan amphora differs from them in a number of details, most notably the profile of the rim and foot and the triple-reeded handles. Handles of this type are characteristic of the earliest phase of the production of Attic red-figure Nolan amphorae from 490 BC, but according to Euwe's study of the shape, only single-ridged or convex handles are used after 460–455 BC (Euwe 1985, 144). Attic red-figured Nolan amphorae certainly reached Basilicata in the early 5th c. BC (e.g., Guardia Perticara Tomb 192; Macrì 2009, fig. 13; Nava 1998, 133, 219) and were available for imitation by Metapontine potters. While never common in Metaponto or its hinterland, a version with ribbed decoration on the shoulder was apparently produced in Metaponto just after the middle of the 5th c. BC (third deposition of a tomb found in via Rinaldi at Pisticci in 1937, Lo Porto 1973, 173, no. 30, pl. 22: 3). An amphora with triple-reeded handles of almost precisely similar dimensions was found in the chora of Metaponto in 1938 (Sestieri 1940, 115, no. 2, fig. 54, with a reserved zone at the top of the shoulder, painted with rays). The closest parallel from Pantanello is a considerably smaller amphora from Tomb 295 (no. 2, h 20 cm, with flat handles; the tomb is given a weighted date range of 472–445 BC: Carter and Hall in *Necropoleis*, 316).

Date: 460 BC.

(EGDR)

Tomb 1 cont.

T01-02 Black-gloss mug

Provenance: Square G6, Level 3, batt. 1. SA 79-938P + SA 79-1009PL. Inv. no. 319846.

Dimensions: h 8.1, base diam. 8.0, rim diam. 9.25.

Condition: Broken into many pieces and repaired, with fragments missing from the shoulder, neck, rim, and handle.

Description: Flaring rim, concave neck, strap handles, flat disk base. Entirely glossed on the exterior and interior, with the black gloss misfired to red over most of the surface. Underside reserved, with one glossed circle near the center of the base and another at the edge of the base. Clay Munsell 7.5YR 6/6 (reddish yellow).

Comparisons and comment: The form is the so-called Pheidias mug, apparently invented in Athens in the second quarter of the 5th c. BC (*Agora XII*, 72–73) and immediately adopted in Metaponto (Elliott in *Necropoleis*, 667). This example has the flat disk base decorated with circles and strap-handle of the earliest Attic mugs; double handles soon took over in Athens, but the fashion found little favor in Metaponto. Mugs of this shape are rare in the Pantanello necropolis, where they occur in only five tombs of the 5th c. BC, but they are present in three graves in the small Sant'Angelo Vecchio necropolis. Black-gloss mugs may have been produced in Metaponto principally for an indigenous market, since vessels of this shape are extremely common from the second quarter of the 5th c. BC in Peucetia and eastern Basilicata, where small round-mouthed jugs/mugs of local manufacture were already popular: see, e.g., Rutigliano tomb 36 (1978) for a tomb of the last quarter of the 6th c. BC with a range of small one-handled, round-mouthed vessels in matt-painted, red-gloss, and plain fabrics (*Museo Taranto II.2*, 109ff.). Tomb 79 (1978) is more typical of the second quarter of the 5th c. BC, with a black-gloss mug (*Museo Taranto II.2*, 218–19). Some of the mugs in indigenous graves are undoubtedly Attic, but the majority are probably Metapontine; Taranto is an unlikely source since the form is extremely rare there. The relative popularity of mugs in the Sant'Angelo Vecchio necropolis may possibly be an indication of the influence of indigenous practices on the inhabitants of the chora of Metaponto; it is worth recalling that Pisticci, where mugs are frequent among the (unpublished) tombs of the middle decades of the 5th c. BC, is clearly visible from Sant'Angelo Vecchio.

Date: 460 BC.

(EGDR)

1:2

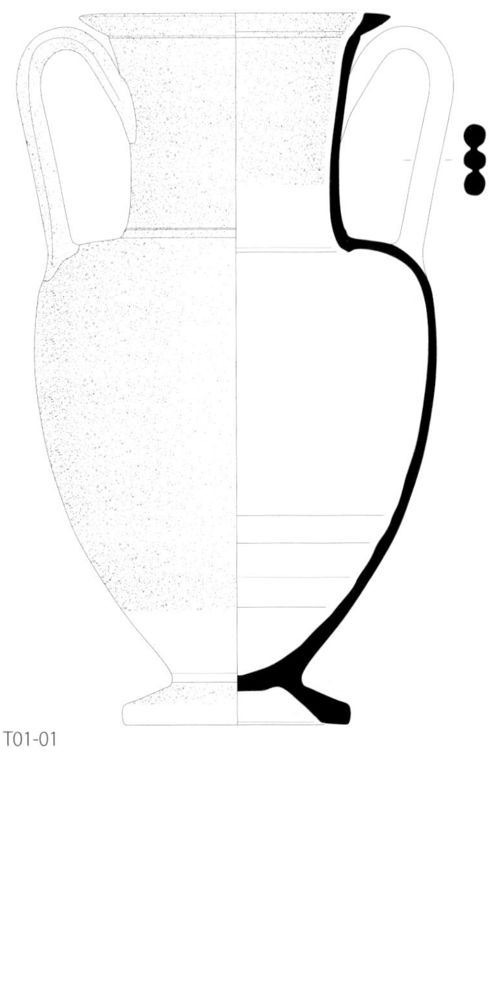

T01-01

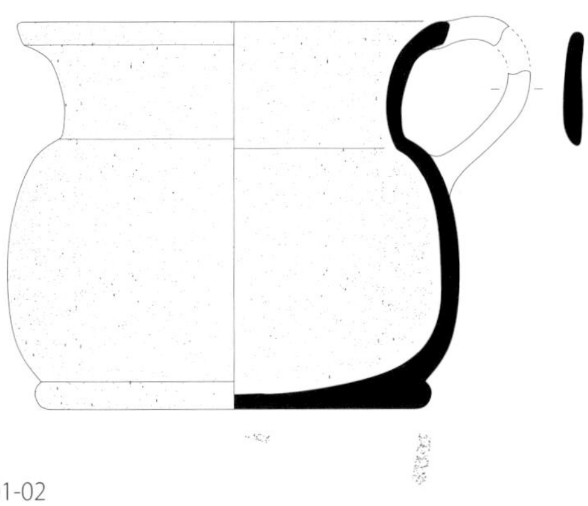

T01-02

Tomb 5

Location: Squares F6/G6
Type: A cappuccina
Orientation: 74°
Skull: E
Date range: 470–300 BC
Basis for date: Black-gloss pottery
Sex: Male
Age: 25–30
Dimensions: l 1.95 m, w 0.65 m
Ceramic objects: T05-01, black-gloss lekythos (SA 79-1054P)
Comments: A large stone slab, presumably a marker, stood vertically in the ground at the tomb's eastern end. Around the stone slab a large number of tiles were found. These tiles, which collapsed over the body, were originally arranged in the *a cappuccina* style. Their exact number is unknown. The burial uses the same type of tiles as Tombs 2–3. One of the tiles from this burial (SA 79-1028T), of the B1 type, has (beside the squat flange) a raised band just a few millimeters high running diagonally between the flanges along the rear edge or a short distance from it. SA 79-1028T was used to reconstruct the standard dimensions of Type B1 (*Fig. 32.12*). *(FP)*
Excavated: 19 July 1979.

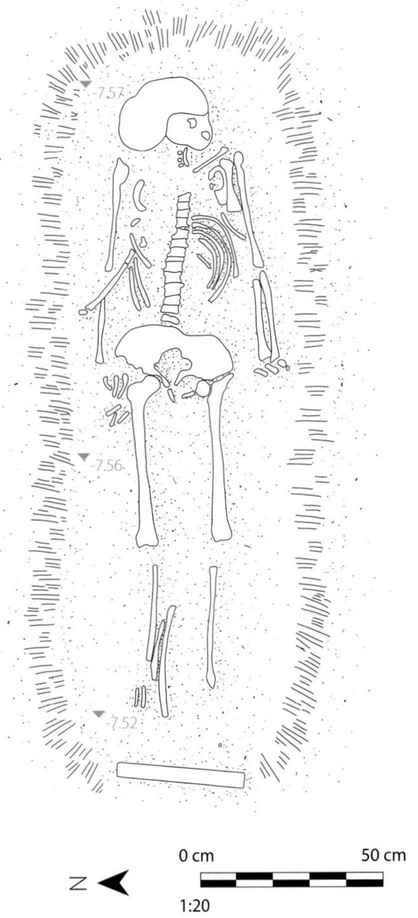

Skeletal Material

Location: SA 79-1042B. Baulk between squares F6 and C6
Sex: Male
Age: Young adult, 22–23 years
Condition: The age of this young man is clearly indicated by dentition, the lingering evidence of fusion lines of the iliac crests, and some long bones and incomplete sacral fusion (only S3 and S4 are fully fused). The massive and powerful bone structure and huge ribs and clavicles clearly reflect a powerfully built male. Although the bones were badly damaged before excavation, some pieces provide useful information. Third trochanters are strongly evident and the breadth of the femoral neck is so great as to appear deformed. The lineae asperae are quite massive in the center of the bone, as is the case in this population. Slight gonial eversion is noted.
Stature: ca. 180 cm. The maximum diameter of the left femoral head is 50 mm, reflecting the massive size of this person. The left radius is 26.9 cm long, which calculates to a stature of 181.37 cm (Trotter and Gleser 1958, 120).
Dentition: Twenty-seven teeth were found, most with slight wear. No caries are present on any of the 27 teeth found and the five not found appear to have been lost postmortem (0/32 disease rate). Individual tooth data appears in the field notes (Becker 1980b). An auxiliary buccal cusp appears to be present on the right maxillary second molar. Although all four maxillary incisors are present, only the left lateral (maxillary) incisor shows possible shoveling. A slight overbite (malocclusion) is evident in the buccal-lingual wear pattern that was developing when this young man died.

(MJB)

Tomb 5 cont.

T05-01 Black-gloss lekythos mouth
Provenance: SA 79-1054P. Inv. no. 319845.
Dimensions: Preserved h 5.9, rim diam. 5.5.
Condition: Broken into a number of fragments.
Description: Tall, narrow lekythos mouth, thickened on the exterior; streaky black gloss on the interior and exterior of the mouth; reserved in the lower part.
Comparison and comment: Mouth from a squat lekythos, or conceivably an epichysis. Mouths from such vessels at Pantanello are quite varied in their shape (Elliott in *Necropoleis*, 683–85, 688), and without the body of this vessel, no close dating is possible.
Date: 5th/4th c. BC.

(EGDR)

1:1

T05-01

Tomb 2

Location: Square F6, N end. Level 3, batt. 1. Long N side of the tomb partly covered by later foundation wall from the workshop.
Type: Carparo sarcophagus with tile lid
Orientation: 95°
Skull: E
Date range: 460–450 BC
Basis for date: Black-gloss pottery
Sex: Female
Age: 22–25
Dimensions: l 1.90 m, w 0.68 m, h 0.46 m, wall th 0.06 m
Bibliography: Edlund 1979, fig. 18.
Ceramic objects: T02-01, black-gloss amphora (SA 79-969P, inv. no. 319847)
Non-ceramic objects: T02-02, pebble (SA 79-1029S)
Tiles: SA 79-921T A (l 80, w 53, th 2.5); SA 79-921T B (l 79, w 53, th 2.5); SA 79-921T C (l 32, w 27, th 2.5)
Comments: The lid of the sarcophagus, badly preserved due to the later disturbance of the foundation wall of the workshop partly covering the N side of the sarcophagus, was made with three pan tiles (SA 79-921T A-C) of type BT1 (see Ch. 32, "Roof Tiles and Bricks"), with low and squat flanges, semicircular profile, and overlapping assembly. The type was long-lived, attested from the Archaic through the Roman periods. No evidence of curved cover tiles were found. *(FP)* The black-gloss amphora T02-01 was found at right foot of skeleton.
Excavated: 12 July 1979.

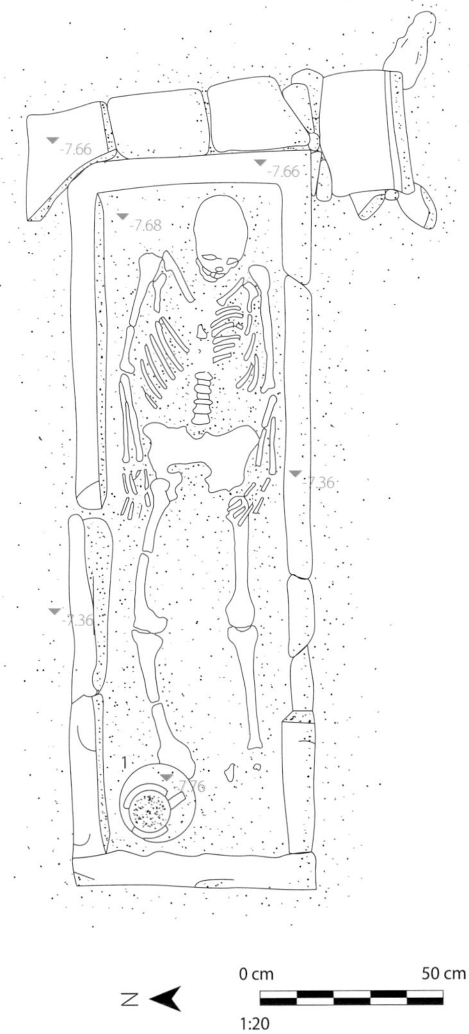

Skeletal Material

Location: SA 79-1029 B. Square F6, N tomb (Tomb 2).
Sex: Male. This individual is clearly a male based on pelvic configuration and large mastoid size.
Age: Young adult. The degree of tooth wear and lack of suture closure suggests an age of about 25 years.
Condition: The young man was buried with his left palm down, with the thumb below the acetabulum of the pelvis, and thus below the head of the femur. The fourth metacarpal appears deformed, probably through having been broken and imperfectly healed. Both patellae are unusually long (4.5 and 4.5 cm) and thin (breadth 4.3 and 4.2 cm). The clavicles are rugged and very stocky. The medial aspect of the left clavicle is thick and appears either deformed or altered due to articulation with the ribs, producing a large sinus. This could be a bone retreat, but it has the appearance of a tubercular lesion.
Stature: 160 cm. Both femora were reconstructed and full measurements taken. The head diameters of 4.64 and 4.5 cm also suggest a small male. Long bone maximum lengths in cm: l. femur, 39.2; r. ulna, 24.2; r. tibia, 32.0?; r. radius, 22.0?; r. humerus, 30.2. These can be calculated to provide a stature of 156.8 ±3.74 cm (fem. + tib.) or 162.97 ±4.3 cm. (hum. + rad.). I suggest 160 cm as an average (Trotter and Gleser 1958, 120).
Skull: Maximum length 180 mm; breadth 138 mm. Damage to the temporal and facial region limited further observations. Gonials slightly everted, gracile skull, no wormian bones of any type, and a lack of parietal foramina were noted. This skull merits further attention.

Tomb 2 cont.

Dentition: All of the teeth but the upper left third molar had erupted. The upper left second molar was decayed to the root stubs. One other tooth was found decayed, and some have been lost postmortem. Thus two of 24 (probably two of 31) teeth were decayed in this young man (6%). Large central incisors are present (genetic recessive trait), as is a considerable overbite. Other problems were also noted (Becker 1980b).

Pathology: In addition to possible broken metacarpal this man had a severe infection (mastoiditis?) of the right mastoid.

(MJB)

T02-01 Black-gloss amphora

Provenance: Square F6, N tomb (Tomb 2). SA 79-969P. Inv. no. 319847.

Dimensions: h 29.7, base diam. 10.5, rim diam. 13.7.

Condition: Intact, with chipped rim.

Description: The base has a convex profile; the rim a concave profile, and the handles are flat with elliptical sections. The exterior is entirely covered with brown-to-black gloss except for a reserved narrow band on top of the foot and a broader band at the bottom of the body. The interior surface of the neck is glossed. Clay Munsell 7.5YR 7/6 (reddish yellow).

Comparison and comment: This is an example of the most common type of glossed amphora in the Pantanello necropolis, where the type was assigned a date of 460–440 BC (Elliott in *Necropoleis*, 648–49, Group iii). The amphora from Tomb 2 is considerably larger than most of those from Pantanello (which range in height from 16–27 cm), its foot is articulated in a more complex fashion, and its form is slightly closer to a neck-amphora than the majority of amphorae from Pantanello, where the closest comparison is Pantanello T29-3 (Carter and Hall in *Necropoleis*, 280, dated 460–440 BC).

Date: The size and articulation of this amphora should date it around 460 BC, or a little before.

(EGDR)

T02-02 Pebble

Provenance: Square F6, N tomb (Tomb 2); SA 79-1029S.

Material: Sandstone.

Dimensions and technical features: l 2.8, w 2.2, th 2.2, weight 17 g.

Condition: Complete.

Description: Spherical pebble, smoothed all around.

Comparison and comment: Although the pebble was slightly larger, parallels for this occurrence are to be seen at field survey burial site 635 (*Survey, Vol. IV*, 83). Pebbles in burial contexts can be seen from a late-4th to mid-3rd-c. BC burial at Muro Tenente where five pebbles were found (Tetteroo and Waagen 2010, 130–40). See p. 552 in Ch. 36, "Spherical Pebbles," for further references.

See also: SAV Li 06

(CDA)

1:2

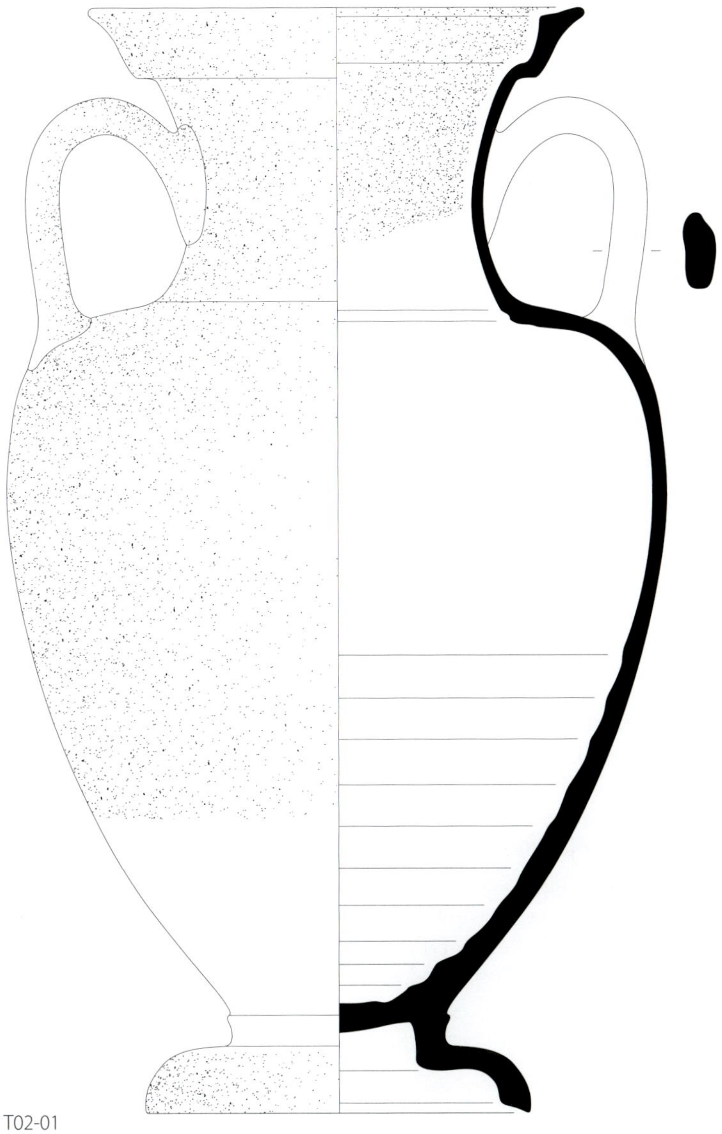

T02-01

T02-02

Tomb 3

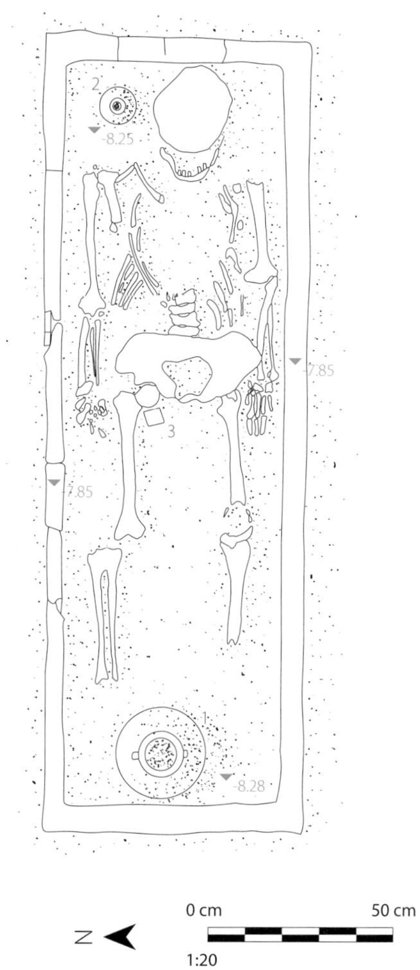

Location: Square F6, Level 3, batt. 4.
Type: Carparo sarcophagus with tile lid.
Orientation: 93°
Skull: E
Date range: 460–440 BC
Basis for date: Black-gloss pottery
Sex: Male
Age: 60–?
Dimensions: l 2.10 m, w 0.70 m, h 0.47 m, wall th 0.06 m
Bibliography: Edlund 1979, figs. 18–19.
Ceramic objects: T03-01, black-gloss lebes gamikos (SA 79-967 A-B, inv. no. 319849); T03-02, amphora (SA 79-975P, inv. no. 319848)
Non-ceramic objects: T03-03, strigil
Pan tiles (BT1): SA 79-981T A (l 79, w 52, th 2.5, flange h 4.1); SA 79-981T B (l 79, w 53.5, th 2.4, rim h 4.2); SA 79-981T C (l 80, w 53.5, th 2.5, rim h 4.3); SA 79-981T D (l 79, w 52, th 2.2, rim h 3.9)
Cover tiles (BC1): SA 79-1020T A (l 79, w 19, th 1–3); SA 79-1020T B (l 79, w 16, th 1–3); SA 79-1020T C (l 78, w 14, th 1–3)
Comments: The amphora T03-02 was found left of the feet of the skeleton. The lebes gamikos T03-01 was placed at the eastern end of the tomb. The strigil T03-03 was found in the vicinity of the pelvis. As in Tomb 2, the stone sarcophagus was covered with four tiles (SA 79-981T A–D) of Type BT1, with low and squat flanges, semicircular profile, and overlapping assembly. The flanges were covered with three cover tiles (SA 79-1020T A–C) of Type BC1, narrow and horseshoe-shaped. SA 79-981 A–D and SA 79-1020T were used to reconstruct the standard dimensions of Types BT1 and BC1 (see *Fig. 32.11* and *Fig. 32.12*). *(FP)*
Excavated: 13 July 1979.

Skeletal Material
Location: SA 79-1041 B. Square F6, S tomb (Tomb 3).
Sex: Male?
Age: At least 30 years, possibly older
Condition: Traces of third trochanters and wide femoral necks are evident, as are the unusual configurations of the lineae asperae. No left patella was recovered but the right is "notched" or manifests dual centers or ossification. This condition has not been noted previously in this area but exists in relatively high percentage in some parts of the world. The closest location where this author has seen this is at Ikiztepe on the Turkish Black Sea, in a proto-Hittite population, which is far removed in time and space (Becker 1988). Two metacarpals were located. Five teeth and numerous hand and foot bones were recovered by water washing (SA 79-864 B). A single middle phalanx was recovered from inside a black painted amphora from within the tomb. Slight gonial eversion is noted. The shallow acetabulum suggest the possibility that this person might be female. The margins of the acetabulum are extremely rounded, almost flat. Unfortunately, very little is left of the pelvis.
Stature: Unfortunately, no measurements were taken of the metacarpals, which might have permitted an evaluation of stature (Musgrave and Harneja 1978). This person appears smaller than the male of Tomb 5, perhaps sug-

gesting that Tomb 3 may have contained a female. Long bone maximum lengths in cm: r. femur 38.5, r. tibia 32.5?, l. ulna 24.3, r. ulna 24.4, l. radius, 22.5.

Skull: Although crushed by ground pressure, badly distorted estimates of length (184 mm) and breadth (148+) were made. Slight brow ridges suggest that this was a male.

Dentition: The mandibular third molars never erupted. Of the 10 known tooth spaces, only one had been lost antemortem and none were decayed (1/10=10%), although wear was extreme.

(MJB)

T03-01 Black-gloss lebes gamikos

Provenance: SA 79-976P. Inv. no. 319849.

Dimensions: h 7.1, base diam. 4.5, rim diam. 5.8.

Condition: Broken and restored from 22 fragments, with many large fragments missing; large irregular hole in the center of the base.

Description: Ring base, convergent with a straight profile and flat resting-surface; short offset rim; handles on the shoulder canted outward. Clay Munsell 7.5YR 7/4 (pink). Inside of rim and body glossed, except for a wide reserved band on the lower body; exterior surface of foot glossed, underside reserved. A thin band of red/purple paint is visible on the top of the shoulder and at the lower edge of the glossed zone.

Comparison and comment: This lebes gamikos is unusual in a number of respects. The shoulder is less flat than is normal, giving the body a piriform shape, and the canted handles are set lower on the shoulder than on other comparable lebetes gamikoi. The vessel appears to have been first dipped, and afterward brushed with black gloss. Whether accidentally or by design, the central section of the body was not brushed, and as a result the surface here is more matte than elsewhere, and lighter in color. Some splashes of gloss are visible in this dipped-only zone, and also on the reserved lower part of the body. The vessel was evidently not fired for long enough and/or at a high enough temperature, since the reserved zones on the exterior are corroded. The fabric body was protected by the gloss in the upper part, but the reserved lower part of the vessel has weakened and detached. It is unclear whether the large, irregular hole in the center of the base is the result of weathering or of intentional perforation. Thin bands of added red paint are quite common on South Italian black-gloss wares of the late 6th and first half of the 5th c. BC, as indeed they are in Athens (*Agora XII*, 18–19), but the practice seems to have been rare at Metaponto, and it is not attested on any lebes gamikos from Pantanello. The vase belongs amongst Elliott's "plain" lebetes gamikoi from Pantanello (Elliott in *Necropoleis*, 660). The closest parallel from Pantanello is T350-4 (Carter and Hall in *Necropoleis*, 387; Elliot in *Necropoleis*, 661–63). The tomb has a weighted date-range of 458–438 BC and contains six lebetes gamikoi, one of which has a perforated base.

Date: 440 BC.

(EGDR)

T03-02 Undecorated amphora

Provenance: SA 79-975P. Inv. no. 319848.

Dimensions: h 33.5, base diam. 9.0, rim diam. 14.5.

Condition: Intact; rim and foot chipped.

Description: Thickened rim with a convex profile on the exterior, cylindrical neck, globular body, ring foot, handles with oval profile. Clay Munsell 5YR 7/4 (pink).

Comparison and comment: This vessel is without any close comparison at Metaponto, or indeed elsewhere in southern Italy. The rim profile, cylindrical neck, globular body, and handles are remarkably reminiscent of Corinthian B transport amphorae, which, while rare, certainly existed in the Pantanello necropolis around the middle of the 5th c. BC (e.g., Morter and Leonard in *Necropoleis*, 735, fig. 17.4, T 256-1). Metapontine potters were not averse to experimenting with the Corinthian B amphora shape in this period: a black-gloss version of the form appears in T207 at Pantanello (Carter and Hall in *Necropoleis*, 406, where the tomb is dated 460–440 BC; Morter and Leonard in *Necropoleis*, 736, fig. 17.6, with discussion). Further afield, a type of undecorated amphora of similar size and shape was produced in Corinth in the same period (Boulter 1953, pl. 34, no. 106, h 36 cm, mid-5th c. BC) and examples have been found in the Athenian Agora (*Agora XII*, pl. 62, no. 1480, h 36 cm, 460–440 BC). The example from Sant'Angelo Vecchio is different enough in the details (taller rim and more flattened, less spherical body) that a Corinthian import seems unlikely.

Date: 460–440 BC.

(EGDR)

Tomb 3 cont.

T03-03 Strigil
Provenance: Tomb 3. Square F6. Level 4, batt. 1. Lot: SA 79-977M. Inv. no. 319850.
Dimensions: l 4.3, w 2.8, th 0.6, weight 16.0 g.
Condition: Incomplete; oxidized.
Description: Fragment of slightly curved strip with rectangular section. Identifiable as a part of a strigil handle.
Comparison and comment: Carter in *Necropoleis*, 444–45; Prohàszka in *Necropoleis*, 799–800.
Date: 460–440 BC (from context).

(MM)

1:2

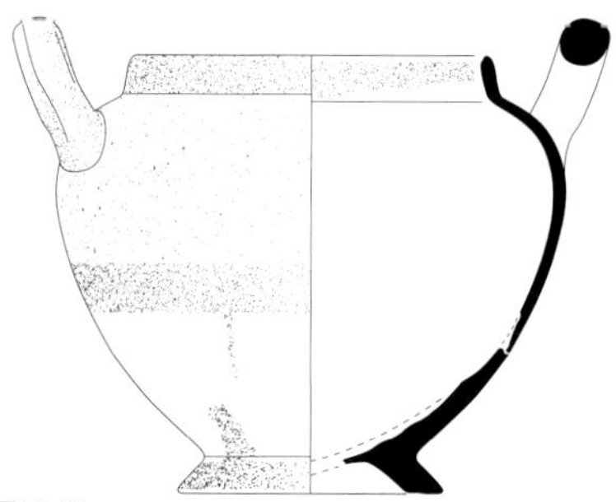

T03-01

T03-03

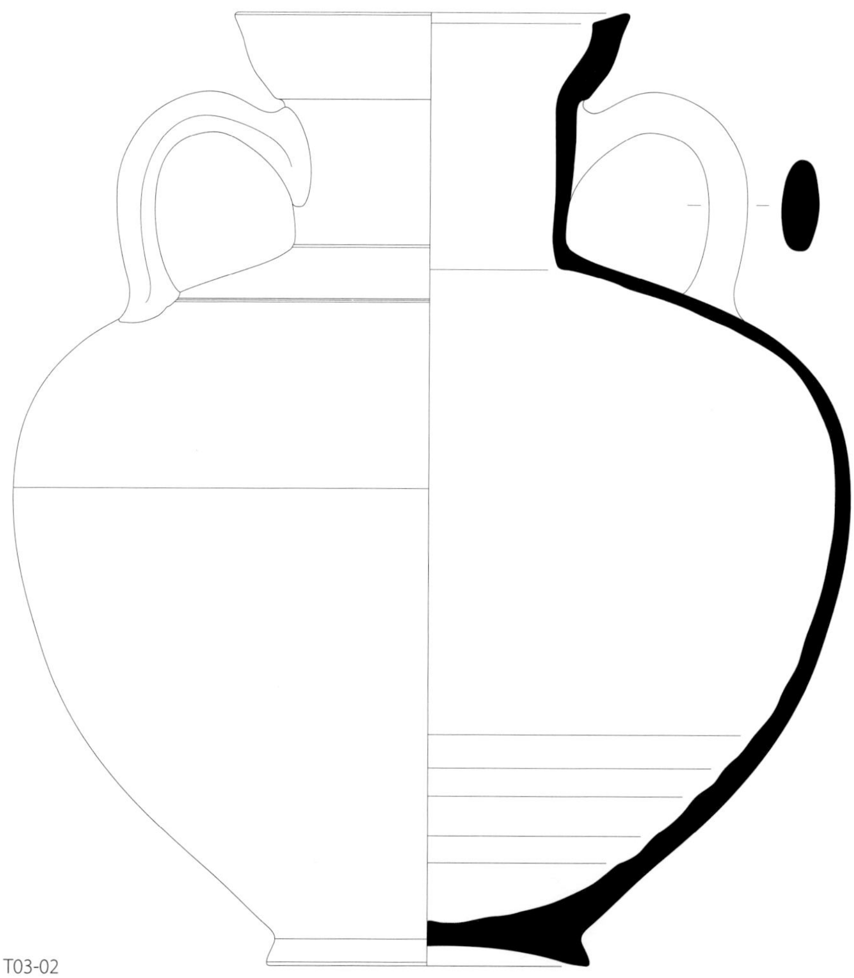

T03-02

Tomb 6

Location: Square F6/E6
Type: Laconian tile
Orientation: 74°
Date range: 475–275 BC
Basis for date: Tomb type
Sex: Unknown
Age: Child (3–5 years old)
Dimensions: l 0.94 m, w 0.57 m
Comments: The tomb was outlined by a rectangle of stones and bordered at each end by an upright pan tile. A single Laconian tile broken into three pieces was placed over the body. A single bone of the skeleton remained, and no grave goods were found. The pit dug for the tomb was filled with olive-colored sand.
Excavated: 11 June 1980.

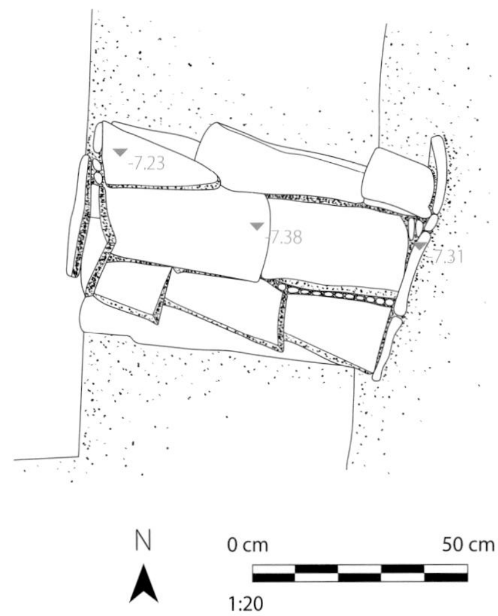

Skeletal Material
Location: SA 80. Square E6.
Sex: Unknown
Age: Child, 3–5 years?
Condition and Excavation: This small tomb, covered with large roof tiles set *a cappuccina* or like a tent over the bones, was excavated by a workman who lacked the ability to identify the small bones and tiny teeth of the child. That problem plus the high rate of bone deterioration from this situation of alternating wet and dry environment resulted in the recovery of only one small piece of bone from the west of the center of the chamber area. This appears to be a femur, and if so, would represent a child of age 3–5 years. The fragment, however, was identified more by location than by morphology, and by inference that only the largest bones would leave any trace in this situation. If it is a tibia fragment, a lesser possibility, then the child may be as old as 8 or 9 years. Since we have no data on the position of the body I suppose that it had been supine. Because no teeth were seen *in situ*, I requested that a sample of earth be collected. When sieved this might produce additional evidence by which the age of the child could be evaluated. The absence of perinatals and children in this small collection is unfortunate.

(MJB)

Burial Descriptions: Hellenistic Tombs

Edward G. D. Robinson

With contributions by Francesco Perugino

Tomb 9

Location: Squares I7–H8
Type: A cappuccina with tile lid
Skull: NW
Basis for date: Tomb type
Sex: M
Age: Adult
Dimensions: l 1.95 m, w 0.87, h 0.17 m.
Comments: The cover, composed of pan tiles, is only partially preserved. There are fresh breaks on the tiles, and it is therefore probable that the cover had been damaged recently by a tractor. The vertical tile at the head is preserved, while the tile at the opposite (SW) end is lost. All of the tiles of both the cover and the ends belong to a single type, BT2 (see Ch. 32, "Roof Tiles and Bricks"), which has flanges with square section and overlapping assembly (for this type of tomb, see *Necropoleis*, 90–96). The type was used from the Archaic through the Roman period; at Sant'Angelo Vecchio similar tiles have been referred to the Hellenistic and Roman periods (see *Table 32.1*). Two of the examples belonging to the cover were partially reconstructed. The first tile, reassembled from six fragments, preserves the two side edges with the flanges (flange h 4.5) and the complete rear edge. The standard width was reconstructed as 56 cm, while the length was only partially reconstructed as 50 cm. The second tile was partially reconstructed from four fragments (max. pres. l 37, max. pres. w 36.5) and preserves part of the flange. *(FP)* No distinction could be made between the dirt fill (Munsell 10YR 5/8, yellowish brown, sandy) and the dirt covering the tomb.
Excavated: 25 November 2010.

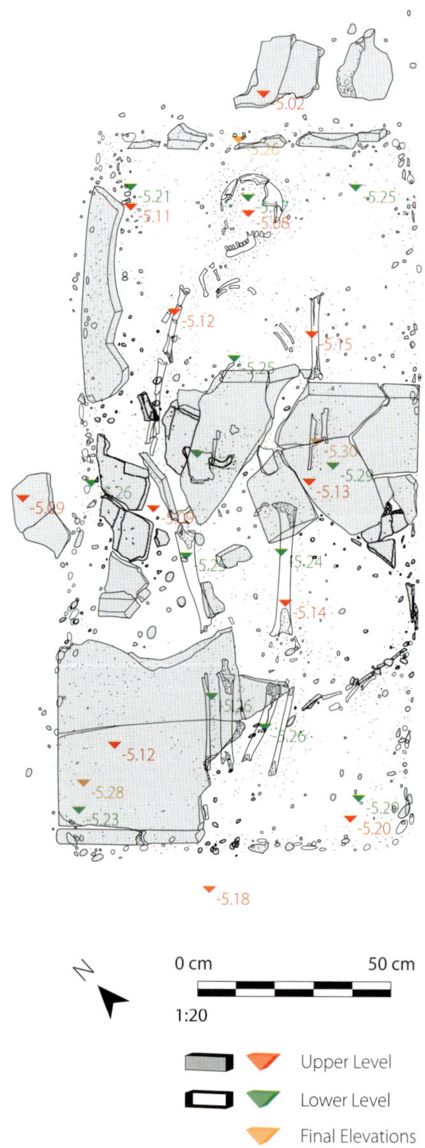

Tomb 10

Location: Squares H8–H9
Type: Vault tomb
Skull: T10A, W
Date range: End of the 4th–early 3rd c. BC.
Sex: T10A, Female; T10B, Child
Age: T10A, 40–60; T10B, 3.5
Dimensions: l 2 m, w 0.90, h. 0.19.
Excavated: 26 November 2010
Ceramic Objects: T10-01, black-gloss ring-handled cup; T10-02, black-gloss small bowl (found inside the ring-handled cup T10-01); T10-03, black-gloss miniature; T10-04, banded-ware olpe
Non-ceramic Objects: T10-05, shell SAV Sh 02; T10-06, shell SAV Sh 03
Comments: The cover of the tomb (partially damaged immediately before the excavation) consisted of semicylindrical terracotta well-liners, while the head end was made with type BT2 pan tiles, of which only part of a flat surface with flange was preserved. Four well-liners were partially reconstructed, yielding the full standard length of 59–60 cm, while the width could be only partially re-created as 17–25 cm. The front and rear edges are flat and well-modeled with oblique profiles at the extremities, while the flat area tends to become thinner toward the center, reaching a minimum thickness of 1.5 cm. Terracotta well-liners placed side-by-side and overlapping to form a complete circumference were used primarily to line the walls of wells. The front and rear edges increase in thickness (3.5–4.0 cm) to facilitate the assembly by overlapping. These examples could belong to the type of wells lacking a *peristomion* (Sconfienza 1996, 29), with the first cylinder projecting above the surface. They began to circulate in the Archaic period and are known at Metaponto and other sites in Magna Grecia (see the Pantanello necropolis, where these items were always reused as parts of tomb covers, *Necropoleis*, 60, 100. All are dated by grave goods to 320–280 BC. For the possible use of wells in the chora, see *Survey*, 839). Terracotta cylinders similar to the examples from Tomb 10 were used to line a Hellenistic well in the Sanctuary of Demeter at Herakleia (Sconfienza 1996, 29 n. 23, fig. 11; De Siena 1980, 88–90, pl. IV.3; Nava 2001, 953, pl. XXXIII). *(FP)* Grave goods T10-01, 02, and 04 were placed at the feet of the adult female (T10A); the miniature T10-03 was found just above. The dirt fill (Munsell 10YR 4/4, dark yellowish brown, clayey) was identical to the dirt covering the tomb. The two shells were found at the level of the skull and the thoracic cavity.

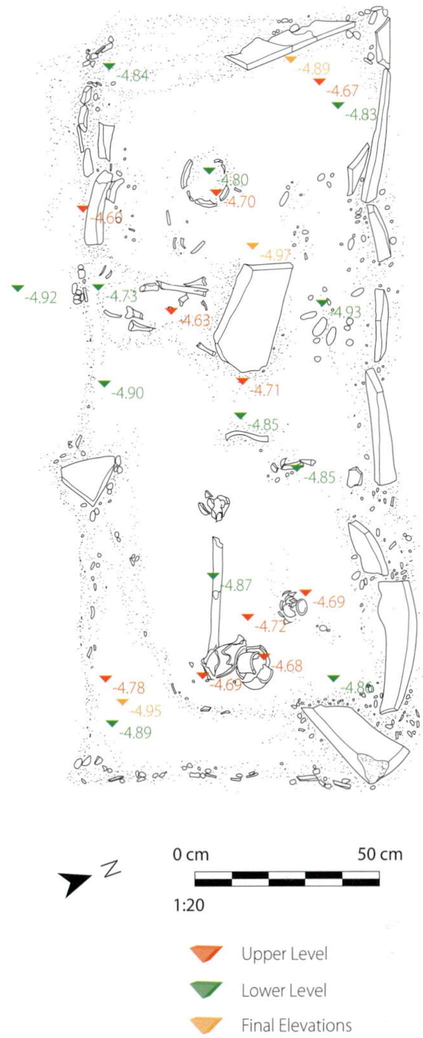

Tomb 10 cont.

T10-01 Black-gloss ring-handled cup
Dimensions: h 7.6, base diam. 3.3, rim diam. 9.2.
Condition: Broken and repaired; one handle missing.
Description: Convex body profile with slightly out-turned lip. Dilute matte black slip, worn in places, with unslipped lower exterior and underside.
Comparison and comment: Ring-handled cups are common in Hellenistic Apulia and Basilicata, particularly in and around Metaponto. Nineteen such vessels were found in the Pantanello necropolis, some of which are virtually identical to the example from Sant'Angelo Vecchio (e.g., tomb 80-6, Elliott in *Necropoleis*, 651). No clear shape development could be detected at Pantanello, and the cups were given a general date of the early 3rd c. BC. The history and distribution of the shape has recently been studied in exhaustive fashion (Lanza Catti et al. in *Survey*, 189–91; its first appearance at Metaponto has been pushed back into the late 4th c. BC, and the form is made into the late 3rd c. BC and beyond).
Date: Late 4th/early 3rd c. BC.

(EGDR)

T10-02 Black-gloss small bowl
Dimensions: h 3.3, base diam. 4.0, rim diam. 7.2.
Condition: Intact.
Description: Thick convex body with interned rim and base ring. Consistent matte black slip, with unslipped lower exterior and underside. Found within the ring-handled cup T10-01.
Comparison and comment: Small black-gloss bowls with convex profiles, ultimately of Attic origin, occur sporadically in the Pantanello necropolis but are abundant amongst the surface finds from sites in the chora; their small bases are likely to survive and remain identifiable on the surface (Lanza Catti et al. in *Survey*, 55–56). At Pantanello, small bowls are concentrated in the final period of the necropolis (late 4th/early 3rd c. BC), and are closely comparable to the example from Sant'Angelo Vecchio (e.g. T189-06 and T193-12: Elliott in *Necropoleis*, 682). Finds from the *kerameikos* show that the form was certainly made in Metaponto (D'Andria 1980a, fig. 67, no. 323), but no doubt it was also produced in many other parts of southern Italy, and the form continues to be found in contexts of the later 3rd and 2nd c. BC. A full discussion of the date and distribution of the shape appears in Lanza Catti et al. in *Survey*, 55–56.
Date: Late 4th/early 3rd c. BC (although a date later in the 3rd c. BC cannot be excluded).

(EGDR)

T10-03 Black-gloss miniature
Dimensions: h 3.3, base diam. 1.8, rim diam. 2.0.
Condition: Slightly worn on the exterior of the rim; intact.
Description: Small, thick-walled vessel with baggy body, slightly constricted neck and out-turned rim; solid disk base. Carelessly manufactured with uneven surface in places. Dilute matte black slip over the body and inside of the neck, with an unslipped underside and lower exterior (in places the slip has run down over the foot).
Comparison and comment: Miniature vases are an infrequent but consistent feature of the Pantanello necropolis from the 5th to the early 3rd c. BC, made in black-gloss, banded ware, and unglazed ceramic (Elliott in *Necropoleis*, 691; Carter and Parmly Toxey in *Necropoleis* b, 715–17; Parmly Toxey and Carter in *Necropoleis*, 729–30). These vessels, and a number of others from urban and rural necropoleis and votive deposits around Metaponto, tend to be recognizable versions of full-sized shapes, but this is not the case for the miniature vessel from Tomb 10. The best parallel is a black-gloss vase from a female tomb at nearby Pizzica, dated by its Gnathia pottery to the early 3rd c. BC (Nava 2002, pl. 71: 1). A miniature form, possibly related to the example from Sant'Angelo Vecchio, was in widespread use at Taranto from the early 3rd c. BC (e.g., Viale Virgilio, chamber tomb beta of 1981–1982: *Museo Taranto III.1*, 48, pl. 4 and 101ff.; Graepler 1997, 102, no. 626 "Becher"). A range of similar miniature "cups" from the south necropolis at Herakleia may also be connected (e.g., Pianu 1990, pl. 36: tomb 72.2 and pl. 69: tomb 202.3).
Date: Uncertain, but a date in the early 3rd c. BC is plausible.

(EGDR)

1:2

T10-01

T10-04

T10-02

T1

T10-03

T10-04 Banded-ware olpe
Dimensions: h 11.8, base diam. 3.8, rim diam. 3.3.
Condition: Intact.
Description: Piriform vessel with a flaring rim, flat, slightly concave underside and one strap handle from shoulder to neck; a thin band of consistent, matte black slip below the handle and another on top of the rim with the upper part of the exterior of the handle also glazed.
Comparison and comment: This form is descended from the tall, narrow Attic banded-ware olpe of the 5th c. BC (e.g., *Agora XII*, pl. 12, no. 259). Metaponto is presumably the source of the colonial imitations that appeared in tombs at indigenous sites in the second half of the 5th c. BC (e.g., Rutigliano tomb 24: Riccardi 2007, 361, with further bibliography). But the form is absent from the Pantanello necropolis, and rare elsewhere in the Metapontino in the 5th c. BC (one example comes from tomb 19 in loc. Pizzica, site 736: Nava 2000, pl. 53:2). The piriform version of the shape appears in a few Metapontine contexts of the late 4th and early 3rd c. BC (e.g., Sestieri 1940, 66, fig. 15) and a closely-related vessel was found in Pantanello tomb 193, which is dated 310–280 BC (Carter and Parmly Toxey in *Necropoleis*, 701).
Date: Early 3rd c. BC.

(EGDR)

T10-05 *Tonna galea*
Provenance: Tomb 10, SAV 11.
Condition: Incomplete.
Dimensions: l 5.7, w 4.5.
Comment: Found together with a *Chlamys glabra* valve (SAV Sh 03).
Date: End of the 4th–first quarter of the 3rd c. BC.
See also: SAV Sh 02

T10-06 *Chlamys glabra*
Provenance: Tomb 10; SAV 11.
Condition: Complete, very worn.
Dimensions: l 4.5, w 4.8.
Comment: Found together with a *Tonna galea* shell (SAV Sh 02).
Date: End of the 4th–first quarter of the 3rd c. BC.
See also: SAV Sh 03.

Burial Descriptions: Other
Tomb A

Type: Sarcophagus
Orientation: N-S (approximate)
Skull: S
Terminus date range: 460–420 BC
Weighted date range: 460–440 BC
Basis for date: Black-gloss pottery
Sex: Unknown
Age: Unknown
Dimensions: l 1.8 m, w N/A
Ceramic objects: TA-01, black-gloss lekythos; TA-02, black-gloss lekythos; TA-03, black-gloss mug; TA-04, black-gloss mug; TA-05, black-gloss amphora
Non-ceramic objects: TA-06, bronze strigil
Comments: The tomb, located 500 m SE of the Sant'Angelo Vecchio site along the SS Basentana highway, was recovered during a road-building operation in 1981. No drawings or photos of the tomb were made and even its precise location is unknown.
Excavated: 1981.

TA-01 Black-gloss lekythos
Provenance: SAV 81. Inv. no. 319861.
Dimensions: h 12.8, base diam. 3.2, rim diam. 2.9.
Condition: Smashed and restored from 16 pieces, with a large section from the top of the body, the shoulder, and the neck replaced with plaster. The gloss, particularly in the central area of the body, is misfired to brown/red and flaky.
Description: The underside, exterior of the foot, shoulder, neck, and interior of the handle are reserved; the top of the foot, body, mouth, and exterior of the handle are glossed. On the shoulder two rows of tongues, the lower one larger and better-formed. Clay Munsell 7.5YR 6/6 (reddish yellow).
Comparisons and comment: The lekythoi of Pantanello are extremely varied in terms of size, body shape, foot profile and decoration, but one is very close in all respects to this small lekythos from Sant'Angelo Vecchio: T230-8 (Carter and Hall in *Necropoleis*, 379; weighted date range 463–444 BC). It forms part of Elliott's Group ii (Elliott in *Necropoleis*, 664–65, L25).
Date: 460–440 BC.

(EGDR)

TA-02 Black-gloss lekythos
Provenance: SAV 81. Inv. no. 319865.
Dimensions: h 13.5, base diam. 4.3, rim diam. 3.1.
Condition: Recomposed from seven fragments. Neck broken and reattached; gloss has flaked off many parts of the body, particularly in the lower part.
Description: The exterior of the foot, a band at the top of the body, and the neck are reserved, as is the underside; the rest of the vessel is glossed, including the interior of the mouth, to a short distance down the neck. Two rows of painted tongues on the shoulder, and a simple meander pattern between two thin bands at the top of the body; the patterns do not continue under the handle-zone. Misfired with just a few patches of shiny black gloss. Clay Munsell 5YR 7/6 (reddish yellow).
Condition: Recomposed from seven fragments. Neck broken and reattached; gloss has flaked off many parts of the body, particularly in the lower part.
Comparisons and comment: This lekythos can be compared with Pantanello T256-2 and T350-22 (Carter and Hall in *Necropoleis*, 321, 386). The dates of these tombs, and of the Attic red-figure lekythoi from which the shape and subsidiary ornament were borrowed by Metapontine potters, are all similar: mid-5th c. BC.
Date: 460–440 BC.

(EGDR)

TA-03 Black-gloss mug
Provenance: SAV 81. Inv. no. 319864.
Dimensions: h 8.2, base diam. 8.4, rim diam. 9.4.
Condition: Intact, except for a large fragment missing from the rim.
Description: Strap handle and concave disk base; entirely glossed (misfired over much of the surface) except for the underside, which is reserved and has a thick glossed circle near its center and a thinner circle on the exterior edge (the resting surface).
Comparisons and comment: For general comments on the form, see T01-02 above; this mug is as tall as any from the Pantanello necropolis, where the best parallel is T206-14, which was found with Attic red-figure lekythoi by the Berlin Painter (490–480 BC) and close to the Painter of BM London E 342 (Burn in *Necropoleis*, 610–11).
Date: 460 BC.

(EGDR)

1:2

TA-01

TA-02

TA-03

TA-04

Tomb A cont.

TA-04 Black-gloss mug
Provenance: SAV 81. Inv. no. 318863.
Dimensions: h 6.0, base diam. 6.9, rim diam. 6.9.
Condition: Intact.
Description: Strap handle and concave disk base; entirely glossed except for the underside, which is reserved and has a thick glossed circle near its center and a thinner circle on the exterior edge (the resting surface). Clay Munsell 5YR 7/6 (reddish yellow).
Comparisons and comment: For general comments on the form, see T01-02 above. This mug has somewhat eccentric proportions, with a low body and relatively high, straight neck. Pantanello T295-4 is very similar (Carter and Hall in *Necropoleis*, 316). Despite the dramatic difference in size and proportions to the other mug in this tomb, the painted circles on the underside are identical.
Date: 460–420 BC (the generic date of squat mugs at Pantanello).

(EGDR)

TA-05 Black-gloss amphora
Provenance: SAV 81. Inv. no. 319866.
Dimensions: h 21.4, base diam. 8.3, rim diam. 10.4.
Condition: Intact, recomposed from five fragments.
Description: Ring base, slightly convex lip rim, flat handles. Body entirely coated with a thin gloss (apparently dipped) ranging in color from brown to gray, save for the lowest 2.5 cm of the body; top of foot glossed and exterior surface reserved. Traces of an overpainted line below the handles.
Comparison and comment: This is an example of the most common type of glossed amphora in the Pantanello necropolis, where the type was assigned a date of 460–440 BC (Elliott in *Necropoleis*, 648–49, Group iii). The best parallel from Pantanello is T256-2 (Carter and Hall in *Necropoleis*, 321); that tomb is dated 460–440 BC and contains a patterned black-gloss lekythos (remarkably similar to TA-02 above) and a bronze strigil. There are also similarities between this amphora and the slightly earlier amphora T339-2 (Carter and Hall in *Necropoleis*, 321, profile drawing Elliott in *Necropoleis*, 649, tomb dated 480–460 BC).
Date: 460 BC.

(EGDR)

TA-06 Bronze/copper alloy strigil
Provenance: Tomb A. Rep. 2. Lot: SA 81-T1. Inv. no. 319862.
Dimensions: l 24.8, w 12.2, th 0.14, weight 59.0 g.
Material: Bronze/copper alloy.
Condition: Complete.
Description: Blade with curved section forming an approximate quarter-circle; flat, rectangular handle folded over the back and terminating in a lanceolate leaf. Curved between blade and handle, with notable angle where the blade attaches to the handle.
Comparisons and comment: See Carter and Hall in *Necropoleis*, 447; Tarditi 2007, 569. It seems to fall within the third group of examples found at Corinth (*Corinth XIII*, 92) dated to the mid-5th c. BC and continuing through the 4th c. BC.

(MM)

1:2

TA-05

TA-06

5

Ceramic Production at Sant'Angelo Vecchio and in the Metapontine Chora

Francesca Silvestrelli

Although the number of studies conducted on the territories of Greek cities is steadily increasing, the manufacturing resources and methods of Greek chorai are still unclear. The subject is of a certain importance. The presence of farmers permanently established in the territory and the activities they were performing made necessary the use of tools and other objects needed in their everyday work in the fields or for the animals, which could have been purchased in the city, but also may have been locally produced.[1] Despite a clear demand, there is a dearth of evidence of the activity of artisans who made these tools, and the complexity and variety of the types and modes of productions that rural life implies and needs still escapes us.

The crucial evidence yielded by the chora of Metaponto can enrich our understanding of the rural perspective of some aspects of production. Both excavations and field survey have revealed considerable evidence for ceramic manufacturing, identifying 37 production sites that range in date from the Archaic to the Roman period. The large number of sites identified in the chora, the necessity of amphorae and pithoi to transport and store agricultural products, the early adoption of the tile roof in most buildings from the earliest occupation phase, and the everyday needs of the people living (and being buried) in the chora may explain why kilns were a very common feature of the territory of Metaponto from the beginning of the history of its occupation. Although much of the evidence from the chora is still largely unpublished, even precluding detailed studies and interpretations, there are enough data to attempt a preliminary outline of the general character of ceramic production in the chora, and to compare it to what is known within the city of Metaponto.

Production at Sant'Angelo Vecchio

Sant'Angelo Vecchio is well equipped for the production of pottery and other clay objects. Clay beds are easily accessible near the spring,[2] which would also have provided water necessary for production. No direct information about the type of fuel used in the kilns has been recovered, but evidence of plant and pollen remains may provide some clue. Archaeobotanical analysis suggests that tree cover of Sant'Angelo Vecchio was low, but the Mediterranean shrubland (*macchia*, or maquis) that characterized the area could have provided at least part of the fuel needed. Some of the plants whose presence has been recognized on the site (*Olea*, *Quercus ilex* type, *Pistacia* cf. *lentiscus*, *Phyllirea*, and *Rhamnus*)[3] were also used as a source of fuel in the kilns of the so-called urban *kerameikos*.[4] The proximity of the site to the Basento Road also deems the area suitable for this type of activity, since it guarantees easy communications.[5]

The Archaic Period (Phase 2)

Owing to the scanty remains preserved at Sant'Angelo Vecchio, the reconstruction of its history as a production site and the identification of the types of production are not easy tasks. Its identity as a manufacturing site could potentially date as far back as the 6th century BC. Although no kilns of this period were identified, three Archaic antefix molds, all found in secondary depositions, (SAV AT 05, SAV AT 07, and SAV AT 08; see *Fig. 30.1*) open up the possibility of a local production of architectural terracottas in the earliest years of the site's occupation. Two of them are preserved enough to recognize the type, the first representing a palmette (SAV AT 05) and the second a silenos head (SAV AT 08). While no cast of the silenos head has ever been identified at Metaponto or in its territory, the palmette antefix possibly belongs to a type also used in the nearby sanctuary of San Biagio.[6]

---

[1] See Amouretti 2000 for a detailed discussion of the evidence provided by literary sources. The meager evidence for Greece is discussed in Sanidas 2013, 237–38.

[2] See Ch. 7, "Geologic Setting."

[3] See Ch. 8, "Archaeobotanical Analysis."

[4] Nuzzo 2003. The same species occur also at Giancola (Brindisi): Manacorda 2012, 99–101.

[5] See Ch. 1, "The Many Lives of a Rural Site: Sant'Angelo Vecchio."

[6] See Ch. 29, "The Architectural Terracottas: Some Considerations on Production," and Ch. 30, "The Molds from Sant'Angelo Vecchio."

The lack of evidence makes a clear understanding of the organization of this production difficult. It may have been a permanent production site that manufactured items to be used in other sites,[7] or alternately it could have been a short-term production completed for the decoration of a specific building, possibly located within the sanctuary that was certainly active in this period or in the nearby area. Sound evidence for the presence of such a building is lacking, but some pieces permit grounded speculation. A sima with drip channel (SAV RT 129) found in Sounding A may have belonged to the roof of a structure, to which the positive of the silenos-type antefix mold SAV AT 08 might also be attributed (see *Fig. 32.2*).

Sant'Angelo Vecchio is one of the few South Italian rural settlements where some evidence for Archaic production of architectural terracottas has been recognized. The asty (city) of Metaponto, however, yielded evidence of a second Archaic workshop: two still unpublished quadrangular kilns were found at the western end of the Sanctuary of Apollo Lykaios. The two kilns have been interpreted as temporary installations used to produce the architectural terracottas for the temples; wasters of the terracottas were found inside and nearby the two kilns (*Fig. 5.4*, no. 7).[8] They could constitute an example of temporary facilities built in a sanctuary for the on-site production of building material for the construction of the temples, of which other examples in southern Italy and in Greece are also known.[9] A landscape of mobile artisans who took up residence for limited periods in response to specific commissions has been hypothesized for the Archaic period. Permanent workshops of architectural terracottas appear elsewhere only in the 5th century BC, when many western cities, such as Croton,[10] were involved in intense building activity.

A small number of wasters and overfired fragments found in the House on the Hill, including a one-handled bowl (SA 79-1037P) and a lekane (SA 79-1062PL) of the late 6th–5th century BC,[11] may also indicate the existence of limited pottery production, possibly supplying the everyday needs of farmers living on the site.

[7] See Ch. 30, "The Molds from Sant'Angelo Vecchio."

[8] *Appendix B, Table B1*, no. 7, with bibliography. Rectangular kilns of similar types seem to be attested in Greece especially from the Classical period onward, with rare and uncertain evidence for the previous period: Hasaki 2002, 165–66, 170.

[9] Billot 2000, 211–12. Sanidas 2013, 226–27. Pisani 2012 (Leontinoi).

[10] Aversa 2012, 33–35.

[11] See Ch. 34, "Evidence for Pottery Production: Fabric Analysis."

The Early Hellenistic Period (Phase 4)

In the Early Hellenistic period, a workshop with multiple kilns covering a minimum area of 200 m^2 was active at the site. Although evidence for this phase is poorly preserved and highly compromised by the later activity that took place at the site, it is possible to recognize the meager remains of the foundation walls of a roofed structure, probably used by the artisans for their everyday activities,[12] and of a firing area, located to the east, where a set of kilns was built just above the spring.[13] The presence of the two tombs, discovered in 2010, suggests that the artisans responsible for the production were living on the site.[14]

The main evidence for this workshop is a large collection of terracotta relief plaques. More than 2700 fragments attributed to ca. 118 individual plaques were found during the excavations.[15] Most of the fragments were discovered in the Slope Area, and half of the total number from the so-called Bulldozed Stipe (Assemblage 9), a deposit destroyed during the construction of the modern road that led to the identification of the site.[16] That these were produced locally at Sant'Angelo Vecchio is confirmed by the presence of two molds (see *Fig. 30.1*),[17] by the restricted number of mold series represented, and by the lack of variants and different generations in the casts. The types occurring at the site (standing female with cross-torch, basket of fruit, and oinochoe; reclining male banqueter; standing Zeus with lightning bolt and scepter; Nymph and Pan; and Artemis Bendis, *Fig. 5.1*) were also present in the asty of Metaponto and in its chora[18] in contexts of the last quarter of the 4th to the first quarter of the 3rd century BC. The relief plaques from Sant'Angelo Vecchio were found associated with pottery (a few sherds of black-gloss, and especially cooking and plain wares) as well as loom weights and transport amphorae, which can for the most part be dated to the late 4th–first half of the 3rd century BC.[19]

In the Greek world, and most likely at Metaponto as well,[20] the production of terracotta figurines and

[12] See Ch. 3, "The Structures at Sant'Angelo Vecchio."

[13] See Ch. 3, "The Structures at Sant'Angelo Vecchio."

[14] See p. 91 in Ch. 4, "The Hellenistic Tombs."

[15] Ammerman 2011a. The terracotta plaques from Sant'Angelo Vecchio will be published by R. Ammerman in a future volume in this series.

[16] See Ch. 1, "The Many Lives of a Rural Site: Sant'Angelo Vecchio."

[17] See Ch. 30, "The Molds from Sant'Angelo Vecchio."

[18] Ammerman 2011a; in *Survey*, 494–97, 502 (reclining banqueter), 498–500, 503 (Nymph and Pan).

[19] *Appendix A, Table A5–Table A8.*

[20] As in the *kerameikos*, deposit no. 2. See below, p. 136, "Production in the City."

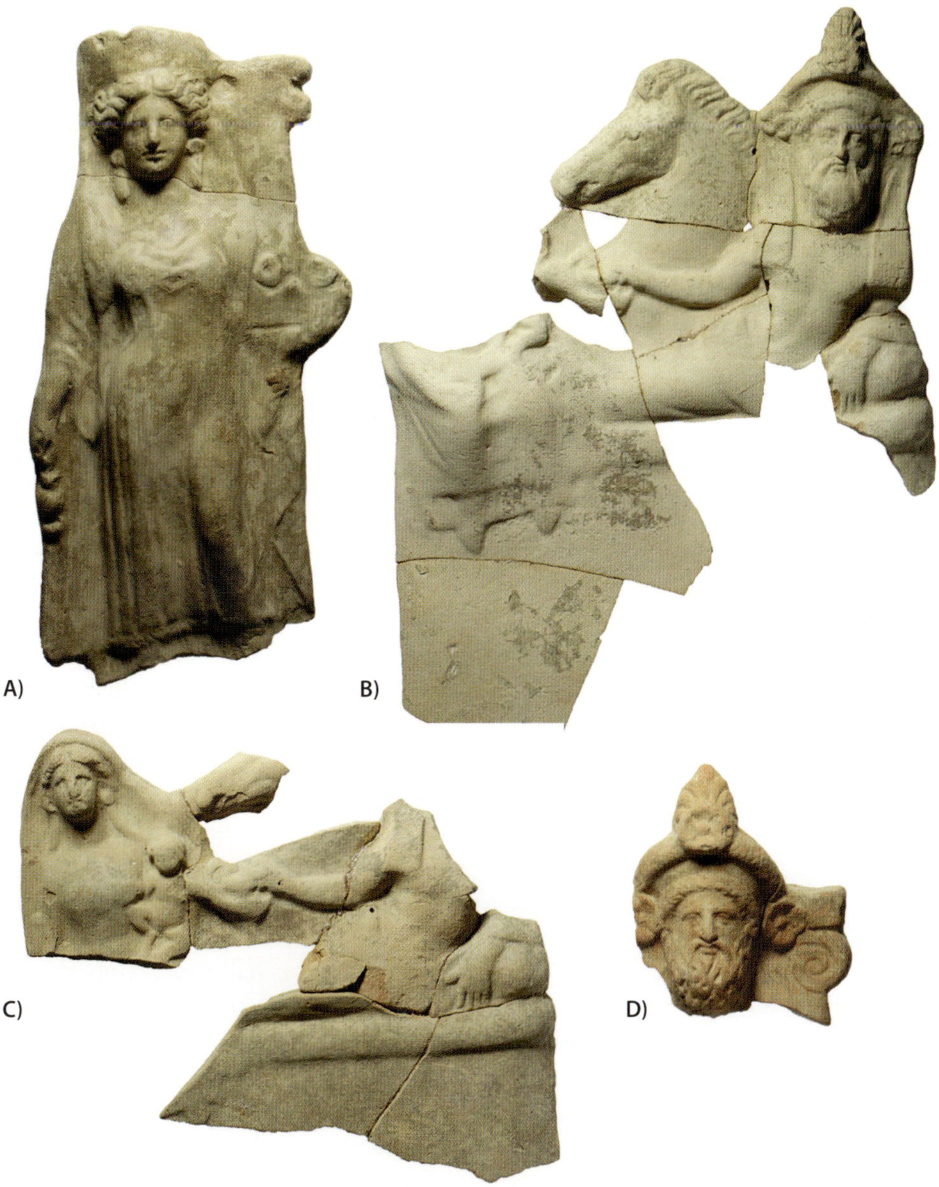

Figure 5.1 Terracotta plaques: (A) SA 79-411; (B) SA 79-219; (C) SA 79-250; and (D) SA 79-510. (CW/ICA)

molded relief plaques often takes place in the same workshops where vases were also manufactured.[21] At Sant'Angelo Vecchio, evidence for pottery production in Phase 4 (the Early Hellenistic period) is ambiguous. The only waster identified (no. 2 in Ch. 34) is a body sherd from an unidentified closed form found in the topsoil covering Tile Fall 1 in square G5. Most of the 21 overfired fragments of plain and coarse wares recognized are not diagnostic, although they were found in contexts of Phase 4. The plain and coarse wares from the discard deposits of the slopes occur mainly as small fragments and, as a consequence, the evidence at our disposal is not conclusive. No wasters or misfired fragments of cooking ware were identified. Nonetheless, in light of the large variety of forms and types attested and the high degree of preservation of many of the vessels from the deposits of Phase 4, which normally do not present traces of fire-exposed blackening, the possibility of production at the site cannot be ruled out.

Adding to the evidence in favor of ceramic production at the site during this period, a mold for disc-shaped loom weights (*oscilla*) with the figure of Eros (SAV LW 25, see *Fig. 30.1*) was found in square G4. Owing to the small number of loom weights of this type found in the Workshop Area and in the discard deposits of the slopes (25 examples), and the absence of wasters, they should be regarded as part of a subsidiary production. Many of the examples, however, are very worn, and Foxhall and Quercia suggest that they could have been recycled as kiln spacers.[22] Evidence for the production of loom weights is rare. Wasters and molds can occur in domestic contexts,[23] but at production sites they seem to be a subsidiary production associated with tiles,[24]

[21] Muller 2000, 98–99.

[22] See Ch. 31, "Loom Weights."

[23] Monte San Fratello, Messina: Bonanno 2008b, 57–58, 61, nos. 20–21, pl. 34.5. Megara Hyblaea: Gras/Tréziny/Broise 2004, 83, fig. 89. Taranto: Ferrandini Troisi/Buccoliero/Ventrelli 2012, nos. 11 and 88. Trypitos (Crete): Sophianou 2011, 427–28, fig. 36.7.

[24] Cetamura del Chianti, 300–100 BC: Meyers 2013, 266. Coulàdere (Haute-Garonne), 1st c. AD: Maniére 1971, 191–93.

transport amphorae,[25] coroplastic works,[26] or pottery.[27] This is also the case at Metaponto, where 18 examples of the pinched and discoid type, most of them over-fired, were found in deposit no. 7 of the *kerameikos* (first half of the 4th century BC).[28]

Noteworthy also is a mold decorated with a Medusa head (SA 79-1019T, see *Fig. 30.1*).[29] The type and dimensions are consistent with molds known from the urban *kerameikos*, where they were used for medallions of black-gloss gutti.[30] Sant'Angelo Vecchio's contexts of Phase 4 did not produce evidence for black-gloss pottery production and, although no example with a similar decoration has been identified, its use for loom weights is plausible.

Kiln spacers from contexts of Phase 4 provide indirect confirmation of the absence of fine ware production at Sant'Angelo Vecchio, which instead occurs at the nearby site 105 (*Fig. 5.2*, no. 46).[31] Thirty-six of the 68 examples belong to the wedge-shaped support, a type quite widespread both in Greece (Athens, Corinth, Thasos, Elis, and Locris) and southern Italy (Croton,[32] Taranto, Locri, Entella, and some southern Lucanian sites such as Montescaglioso, Pisticci, and Gravina di Puglia[33]) from the Archaic through the Hellenistic, Roman, and Byzantine periods. At Metaponto, spacers of similar types are attested at the production site of Torretta di Pisticci in the chora (*Fig. 5.2*),[34] (first half of the 5th century BC) where pottery (especially black-gloss and banded wares) was also produced. Evidence from the *kerameikos*, however, demonstrates that cylinder- and ring-shaped kiln spacers were used in fine ware production beginning from the second half of the 5th century BC onward.[35] These types, of Attic origin and probably introduced at Metaponto along with the red-figure technique, were adopted by Metapontine artisans not only in the city but also in the chora, as confirmed by the finds from Fattoria Casamassima, Fattoria Venezia, and Pantanello (*Fig. 5.2*, nos. 20, 21, and 29).[36]

The wedge-shaped spacers from Sant'Angelo Vecchio could have been used for the stacking of tiles, the production of which could be suggested by the presence of a small number of deformed tile fragments, possibly to be interpreted as wasters attributed to the roofing system bearing the stamp *Herakas* (SAV RT 126 and SAV RT 127). This stamp, a nominal mark in genitive that likely indicates the owner of the workshop,[37] is also known from other sites within the Metapontino (see *Fig. 32.9*). The variation in the stamps attested in several contexts opens up two possible explanations for the workshop at Sant'Angelo Vecchio. In one scenario, the site could have been used by itinerant artisans specializing in roof tile production, who moved around from a single workshop for specific commissions—in this case the construction of the workshop. Alternatively, Sant'Angelo Vecchio may have been a branch of this *ergasterion* established in the countryside and serving the surrounding area.

It is also possible that architectural terracottas were produced at the site. An antefix mold representing Artemis Bendis (SAV AT 09, see *Fig. 30.1*) and dating between the end of the 4th and the first half of the 3rd century BC was found in the collapse of Kiln 5 (North Kiln), where it had probably been reused. An overfired antefix with Dionysos (SAV AT 01) was found in the roof collapse of the late-4th- to 3rd-century BC workshop.

The Early Hellenistic workshop of Sant'Angelo Vecchio seems to be a polyvalent atelier supplying ceramic materials and especially votive terracotta relief plaques. For whom, though, was this production intended? Due to the highly disturbed condition of the spring area, no sound evidence for the continuity of use of the Archaic sanctuary into the Early Hellenistic period has been found,[38] but the hypothesis that the sanctuary could have been still active and that they could eventually have been used there cannot be ruled out. Examining nearby sanctuaries, however, indicates a pattern. The sanctuaries at both San Biagio

[25] Giancola (Brindisi), 2nd–1st c. BC: Firmati 2012, 206, pl. XXVI.

[26] Phere, Thessaly, end of the 3rd–beginning of the 1st c. BC: Doulgeri-Intzesiloglou 1997, 304. This workshop also produced molded pottery.

[27] Sallèles d'Aude: Laubenheimer/Widemann/Attas 1978–1979, 115–16. For a detailed analysis on the evidence of production, see Quercia and Foxhall forthcoming b.

[28] D'Andria 1980a, 441–43, nos. 400–1. Silvestrelli 2000, 253.

[29] See Ch. 30, "The Molds from Sant'Angelo Vecchio."

[30] Silvestrelli 2000.

[31] *Appendix B, Table B2*, no. 35.

[32] Verbicaro 2010, 232, fig. 16.22 (end of the 4th–beginning of the 3rd c. BC, associated with pottery production).

[33] See Ch. 33, "Kiln and Workshop Furniture."

[34] Cracolici 2003, 49, type VI, nos. 337–338, fig. 12. Lecce 2010–2011. *Appendix B, Table B2*, no. 46.

[35] Cracolici 2003. Only a ring-shaped support was identified at Sant'Angelo: SAV KF 08.

[36] *Appendix B, Table B2*, nos. 20, 21, and 29.

[37] See Ch. 32, "Roof Tiles and Bricks."

[38] See Ch. 2, "Site Phasing, Stratigraphy, and Site Assemblage," and Ch. 3, "The Structures at Sant'Angelo Vecchio." If found, such evidence could help in understanding the location and products of this workshop, which may have been devoted to the worshippers attending the sanctuary.

and Pantanello were abandoned in the 3rd century BC giving way to farmhouses. This phenomenon has been interpreted as one of the consequences of the crisis Metaponto underwent in this period, with, among other effects, sacred places in the chora losing their importance, and an inability by the city to prevent the private takeover of formerly public spaces,[39] as it also happens at Masseria Durante, where a farmhouse of the end of the 3rd century BC was built on a section of the urban necropolis.[40] It is possible that the workshop at Sant'Angelo Vecchio could instead be a sign of a new occupation and an existence separate from the cult place of previous periods. The problem must, as a consequence, remain open.

The Late Republican and Early Roman Periods (Phase 5)

The occupation of the site resumed in the late 2nd century BC, when the workshop shed and the Kiln Complex were built on the ruins of the late-4th-century BC structure. Both structures had coverings made with tiles that bear the Latin stamp AVF, possibly indicating that the area of Sant'Angelo Vecchio was property of the Aufidii family.[41]

Erosion on the one hand and agricultural activity on the other destroyed the layers related to this period, and therefore the evidence for the Late Republican and Early Imperial occupation and abandonment of the site is scarce.

The workshop, extending for a minimum area of 350 m^2, comprises a working space (with a roofed workshop shed and clay-settling basin) and the firing area of the Kiln Complex. The discard areas for rejected pieces were not identified and, as a consequence, identification of the range of material produced at the site is complex and relies only on the small number of wasters found.

Although possible, no sound evidence for the manufacture of tiles was found. Production of pottery on the site can be inferred from the presence of misfired cooking ware (SAV CkW 07 and SAV CkW 25) and a grey-ware waster datable within the 1st century BC (SAV GW 07). The small dimensions of the kilns, measuring 2 m in diameter, testify to a limited output, possibly serving the needs of the people in the settlements of the river valley around Fosso Sant'Angelo (see *Fig. 1.2*).

Production in the Chora of Metaponto

Identifying ceramic workshops in the chora of Metaponto depends on three types of archaeological indicators characterized by a variable degree of reliability, and found in both excavation and survey. Permanent features such as kilns, the soundest indicators of production activity, were identified at 12 sites;[42] with the exceptions of Pantanello and Torretta di Pisticci,[43] they are unpublished and offer only limited information. Six other sites have been recognized as possible workshops due to a less conclusive but equally important type of evidence: the equipment used in ceramic production, such as molds or kiln spacers.[44] The refuse of the production process, ceramic wasters and slag, comprise the third and least decisive (unless properly studied)[45] body of evidence that has been interpreted to suggest the possible presence of workshops.

The 37 production sites identified by these indicators are widespread throughout the entire area of the chora (*Fig. 5.2*). The highest concentration can be observed along the Basento valley,[46] and, to a lesser extent, along the Bradano[47] and Cavone valleys.[48] These locations put the sites close to the raw material offered by fluvial clay-beds and to major arteries of distribution, such as the Basento Road. Most (23) of the production sites can be probably dated to the Greek period of the chora's occupation,[49] six are active in the Late Republican period,[50] and the chronology of the other sites is uncertain.[51]

Production in the chora was a reality already in the Archaic period. Indicators of ceramic production (slag) were found at Sant'Angelo, site 12 (middle of the 6th century BC?), and at site 530.[52] The presence of wasters of Ionic-type cups were reported from Musillo farmhouse at contrada Saldone, built in the 6th and

[39] De Siena 2005, 442–46; 458. Giardino 2015 (San Biagio). Carter in *Survey*, 880–82 (Pantanello and San Biagio).

[40] Giardino in *San Biagio*, 6.

[41] See Ch. 32, "Roof Tiles and Bricks."

[42] *Appendix B, Table B2*, nos. 16, 18, 24, 43, 46, and 47 (Greek period); nos. 30, 31, 34, and 42 (Roman period); nos. 32 and 39 (uncertain chronology).

[43] *Appendix B, Table B2*, nos. 30 and 46.

[44] *Appendix B, Table B2*, nos. 20, 21, 26, 28, 29, and 35 (Greek period).

[45] See Ch. 34, "Evidence for Pottery Production: Fabric Analysis," *Appendix B, Table B2*, nos. 12, 15, 23, and 38 (Greek period).

[46] *Appendix B, Table B2*, nos. 29–41.

[47] *Appendix B, Table B2*, nos. 13 and 14.

[48] *Appendix B, Table B2*, no. 46.

[49] *Appendix B, Table B2*, nos. 12, 13, 18, 20–29, 33, 35–38, 40, 41, 43, 46, and 47.

[50] *Appendix B, Table B2*, nos. 19, 30, 31, 34, 35 (uncertain), and 42.

[51] *Appendix B, Table B2*, nos. 14, 16, 17, 32, 39, 44, and 45.

[52] *Appendix B, Table B2*, nos. 33 and 41.

Figure 5.2 Production sites in the Metapontine chora, Late Archaic to Roman periods. (JT/ICA)

reoccupied in the 4th century BC.[53] An *ergasterion* dated to the Late Archaic and Early Classical period was found at Torretta di Pisticci on the Cavone River,[54] where a rectangular kiln produced Ionic-type cups, skyphoi with offset rim, and other forms in both black-gloss and banded ware during the first and sec-

[53] *Appendix B, Table B2*, no. 12. The farmhouse was excavated by G. Uggeri (Uggeri 1969, 60; Amore 1993, 60–62; Van Compernolle 1996, 299). Amore (1993, 107) reports the presence of wasters also from the Pacciano farmhouse, located nearby. Evidence of the production of Ionic-type cups at the farmhouse of Cugno del Pero (*Appendix B, Table B2*, no. 25) is only based on the type of clay, and it is highly uncertain.

[54] *Appendix B, Table B2*, no. 46.

ond quarters of the 5th century BC. The kiln is probably part of a significant rural settlement that was occupied from the second half of the 6th to the end of the 5th century BC. The range of shapes produced is consistent with material found in funerary assemblages and votive deposits of the surrounding area, and it is possible that the production was meant not only for local consumption but also for limited export.

The number of production sites identified as well as the type of wares manufactured increased in the 4th and 3rd centuries BC. A black-gloss production dating to the end of the 4th–early 3rd century BC was present at site 105, located a short distance from Sant'Angelo Vecchio.[55] Cylindrical kiln spacers were found at the Casamassima farmhouse in Demanio Campagnolo,[56] occupied in the 4th and 3rd centuries BC, and at the 4th-century BC Venezia farmhouse in contrada Lago del Lupo.[57] A kiln of uncertain date with a long stocking channel, a circular combustion chamber 3 m in diameter, and a free-standing wall at the center crossed by three perpendicular sections was found near the farmhouse at Cappa D'Amore.[58] Its large dimensions, more common from the Hellenistic period onward, might suggest production with a consistent output, potentially reaching beyond the everyday needs of the people living nearby. Due to the scarcity of information, it is not possible to evaluate the modes of production active in these sites.

Medium-scale production sites, in various ways comparable to Sant'Angelo Vecchio, are also known. Now unfortunately gone, one of those sites was excavated by M. Lacava at the end of the nineteenth century at Bufalara, near Cugno del Pero, an area characterized by a large number of rural settlements.[59] Here, three kilns "of shape and dimensions similar to the features used nowadays by our potters"[60] were excavated. The deposits found inside the kilns yielded a mold of a clay disc (diam. 17.1 cm) with a handle and decorated by an olive or laurel branch, a lyre, a patera, a bird, an altar with vases, and other symbolic objects that were not recognized.[61] A mold bearing a similar decoration dated to the 3rd century BC was found at Herakleia.[62] Architectural terracottas may have also been produced at the site, as suggested by the recovery of a second mold with a lion's head. Lacava referred to vase fragments (possibly transport amphorae?) with impressed stamps bearing the names of ΣΟΤΕΛΕΣ and ΑΓΑΘΙΝΟΣ. A fragment possibly from a cover tile bore the stamp ΔΑΜΑΤΡΙΟΣ, which also occurred in the body fragment of a vase (probably a transport amphora). The same stamp is attested in tiles from the urban Sanctuary of Apollo at Metaponto,[63] suggesting that he also sold his products in the city, as did Herakas at Sant'Angelo Vecchio in the same period.

In the 4th century BC, architectural terracottas may also have been produced at Bufalara, site 489,[64] and a mold for terracotta relief plaques of the reclining banqueter type was found at Avinella, site 404.[65]

The Bradano-Basento survey yielded evidence for the production of transport amphorae. As suggested by discards, Metapontine Type B and Metapontine Greco-Italic amphorae along with plain wares and possibly terracottas, were produced at Cogno del Prete, site 266.[66] Wasters of Metapontine Greco-Italic amphorae were also found at site 153, at the northern margin of the Basento valley,[67] and at Pantanello.[68] Evidence for the manufacture of transport amphorae in the city is debatable, and only known for the Late Republican period.[69] The recently published evidence from the chora fills a gap, confirming that the production of these vessels was probably connected mainly to the agricultural activity in the countryside.

Although the evidence at our disposal is scarce and unevenly published, making it difficult to establish types and modes of production of the identified workshops, it is evident that the production system active in the Greek-period chora was complex, multifaceted, and lively. The production of fine wares seems to be mainly concentrated in urban workshops, but wasters and kilns demonstrate that Ionic-type cups in the Archaic period and black-gloss in the Early Classical and Early Hellenistic periods were also produced in the rural chora. At least three of these *ergasteria* (Torretta di Pisticci in the Late Archaic period, and Sant'An-

55 *Appendix B, Table B2*, no. 35.

56 *Appendix B, Table B2*, no. 20.

57 *Appendix B, Table B2*, no. 21.

58 *Appendix B, Table B2*, no. 18. The sections corresponded to pilasters projecting from the circular wall, probably designed to support the perforated clay floor.

59 *Appendix B, Table B2*, no. 24.

60 Lacava 1891, 327.

61 Lacava 1891, 327–28.

62 Giardino 1996, 39. Giardino 2015, fig. 7.

63 *Metaponto I*, 59, fig. 46f.

64 *Appendix B, Table B2*, no. 26.

65 *Appendix B, Table B2*, no. 28.

66 *Appendix B, Table B2*, no. 23.

67 *Appendix B, Table B2*, no. 38.

68 Swift in *Survey*, 466. The waster of a transport amphora of unknown type was also found at Cozzo Presepe: *Appendix B, Table B2*, no. 15.

69 Giannotta 1980, 62.

gelo Vecchio and site 105 in the Early Hellenistic period) could be considered permanent production sites active for one or more generations and probably serving the surrounding territory.

Ceramic production is one of the functions performed in the Early Cluster Villages, as is the case of site 530 in the Basento valley.[70] Kilns have also been found in connection with farmhouses. The sites of Pantanello, where pottery production is attested already in the 4th century BC,[71] and San Biagio (cited with caution),[72] where a kiln of uncertain date was found some 50 m from the sanctuary, yield evidence of the presence of workshops possibly established to fulfill the needs of sanctuaries. The model was well known in central Italy,[73] but rarely attested in Greece[74] or in Greek colonies of southern Italy.[75]

The Late Republican period witnessed a great transformation in the organization of the territory and new forms of settlement and production are attested. Alongside the traditional cultivation of cereals, animal husbandry became increasingly important, leading to more areas allotted to grazing and the emergence of rural complexes (comprising small to medium-sized farmhouses and large villas) administrating extensive holdings. The complexes were usually built along or in a commanding position of major roads, and in proximity to water sources.[76] Ceramic kilns have been found in connection with some of those complexes. The products of these kilns were meant both to serve the daily needs of the inhabitants and to store agricultural products that were either consumed on-site or traded locally. A kiln was found near the Destra Basento villa,[77] tiles, dolia, and transport amphorae are said to have been fired in two kilns found near the residential structure at La Cappella,[78] and site 105 may have still been active as a production site in this period.[79]

Pantanello is the only production site of the period that has undergone intensive study for thorough publication. The workshop, which could be part of the pars rustica of a yet-undiscovered villa,[80] extends over an area of at least 1000 m^2, and comprises a three-wing covered structure defining a rectangular courtyard. The west wing of the building, probably used as a drying shed for the ceramics manufactured at the site, opens into a complex with two kilns, one with a diameter of 3.9 m, the second measuring 1.5 m. Near the kilns, a deposit yielding large quantities of grey ware, molded bowls, plain ware, Greco-Italic transport amphorae and tiles was found.[81] The wide range of products attested suggests that the output was probably meant for local consumption but also traded elsewhere.

Production in the City

Archaeological investigations undertaken in the asty of Metaponto in the last few decades have served to clarify the great wealth of local craftmanship, characterized by a variety of high quality types of production, and the importance of the city as a center for pottery manufacturing is now evident. As in the chora, the identification of workshops relies on the discovery of permanent features such as kilns, of the equipment used in the process, and of rubbish left over. Some production workshops have been identified and excavated, many of which were located in the area known as the *kerameikos*,[82] near the northern side of the fortification walls of the city.[83] The so-called *kerameikos* occupies a triangular area of around 2080 m^2, defined to the south by the stenopos that originates just east of the Sanctuary of Apollo Lykeios,[84] and to the north

70 Carter in *Survey*, 717.

71 *Appendix B, Table B2*, no. 29.

72 *Appendix B, Table B2*, no. 32. According to Adamesteanu (1965, 131), the kiln was used in the production of Archaic terracotta statuettes meant for the sanctuary. Nonetheless, the kiln was recently dated to the Hellenistic-Roman period: Leone 1998. The kiln was recently cleaned, but no useful information for its chronology was found (A. De Siena, pers. comm., 2012).

73 Di Giuseppe 2012.

74 Muller 1999, 281 n. 13. Hasaki 2002, 289–91. More frequent is the presence of evidence for metalworking: Risberg 1992 and 1998.

75 See, for instance, Locri, Piani Caruso: Barra Bagnasco 1997, 213. The two kilns found here produced miniatures and terracotta figurines that could have been connected to the sanctuary of Grotta Caruso.

76 Lapadula in *Survey*, 1137–40. Giardino in *San Biagio*, 4–9.

77 *Appendix B, Table B2*, no. 31.

78 *Appendix B, Table B2*, no. 34.

79 *Appendix B, Table B2*, no. 35.

80 Carter in *Survey*, 910–12.

81 *Appendix B, Table B2*, no. 30. The study of the Pantanello workshop is in progress and will be published in a future volume in this series.

82 The workshops were identified at the end of the 1960s by D. Adamesteanu and were excavated between 1973 and 1977 by F. D'Andria. Further campaigns were held in 2003 and 2005 (D'Andria, Silvestrelli). For the history of the archaeological research undertaken on the site, see Silvestrelli 2005, 113. On the materials, D'Andria 1980a and 1980b, Silvestrelli 2000, and Mannino 1993 and 1998 (terracotta coins), Signore 1996 (terracotta plaques from deposit no. 2), Silvestrelli 1996 and 2005 (red-figure pottery), Quercia 2003 (cooking ware from deposit no. 1), and Cracolici 2001 and 2003 (kiln spacers). See also Robinson 2012 (archaeometric analysis); and D'Andria 1997a and 1997b (fingerprints).

83 The walls can be dated to the second half of the 4th c. BC (Sconfienza 2005, 122, with bibliography). Evidence for the presence of earlier fortifications with a similar orientation were identified west of the 4th-c. BC walls (De Siena 1998, 145–51; D'Andria 1977a, tav. XL).

84 Mertens 1985, fig. 1a. The stenopos running near the *kerameikos* was excavated in 2005, when the road pavement used in the first decades of the 3rd c. BC was brought to light. It is not possible to establish when it was created, but it is nonetheless important to underline that the Archaic workshop, as all the other structures found in the area, seems to follow its orientation.

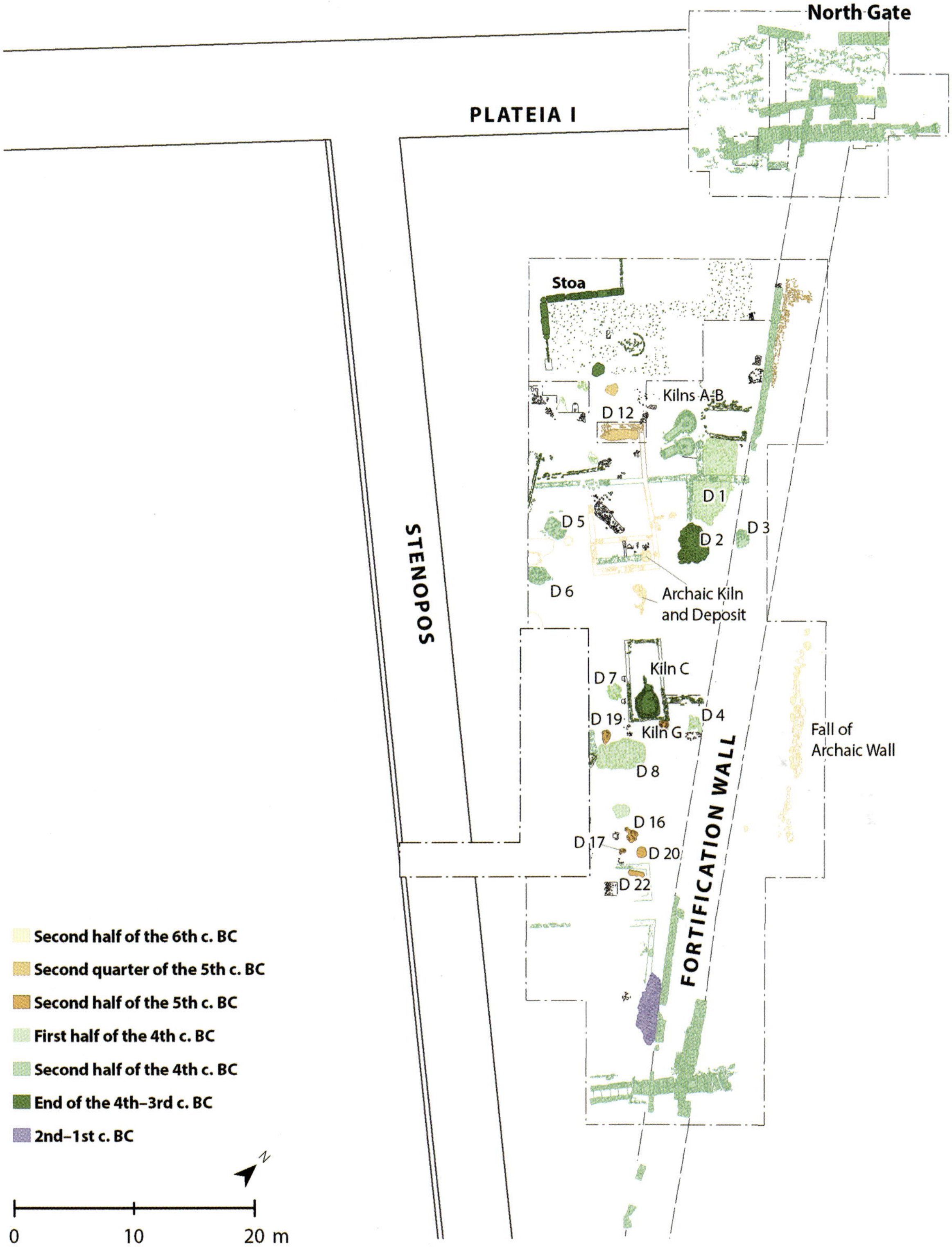

Figure 5.3 *Kerameikos* of Metaponto. (From Silvestrelli 2005, additions by JT/ICA)

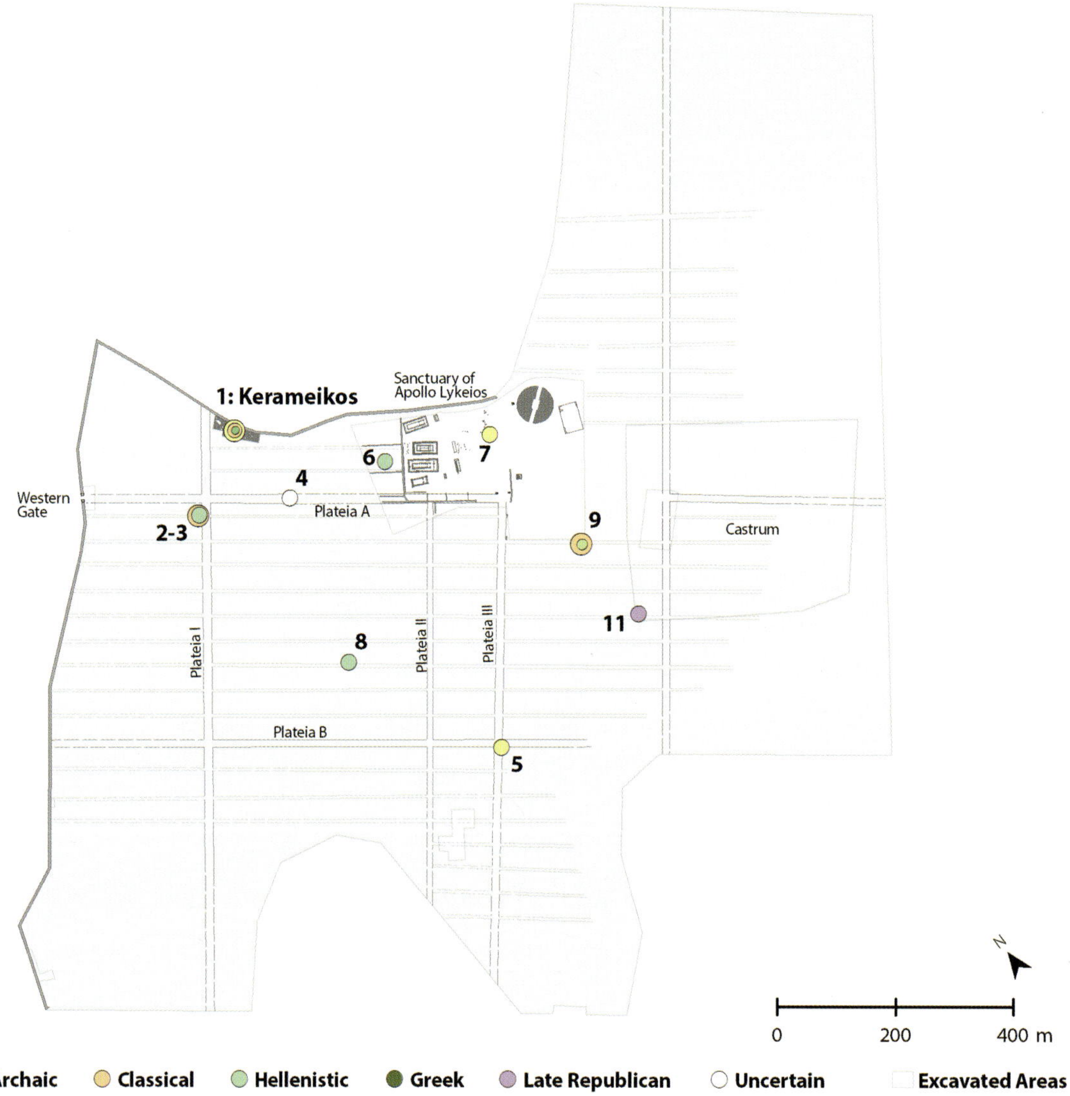

Figure 5.4 Workshops at Metaponto, Late Archaic to Late Republican periods (JT/ICA, street grid based on Mertens 2006, fig. 276, and Osanna 2008, tav. XXIV).

by Plateia I (*Fig. 5.3*).[85] The area is marginal to the city center, but its close proximity to two major roads ensures easy connections with the city center and the chora. The excavations, which covered an area of around 2032 m2, brought to light kilns, structures, and discard deposits ranging in date from the second half of the 6th to the 3rd century BC.[86] Evidence for the distinct phases of occupation identified is, however, uneven. The area seems to have been used for production for the entire period of its occupation. Some evidence, such as the secondary deposition of slag from metal production, allows us to hypothesize that pottery production was not, at least in some periods, the only activity performed at the site.

This area, included from the beginning in the city limits,[87] does not seem to have been occupied before the second half of the 6th century BC, the date of the earliest identified workshop. Archaic contexts, however, yielded residual fragments of imitation Proto-Corinthian cups and three fragments of Bradano Subgeometric vessels, all dated to the 7th century BC.[88] The Archaic workshop occupied an area of at least 150 m^2, divided between outdoor and cov-

[85] Although the northern section of the plateia seems to date to the second half of the 4th c. BC (D'Andria 1980a, 355–56), the path should have been in use at least from the end of the 6th c. BC (Mertens 1999, 249 and 259).
[86] D'Andria 1977a; 1978; 1980a; 1980b; 1980c.
[87] De Siena 1998, 145–51.
[88] D'Andria 1978, 409.

ered areas (*Fig. 5.3*). As attested by the pottery found in a discard deposit, the production was comprised of skyphoi with offset rims, Ionic-type B2 and B3b cups, and banded lekanai and one-handled cups.

Evidence for the activity in the second quarter of the 5th century BC is very limited, and was recovered only in an area of about 68 m^2 (4% of the total area excavated). The workshop active in this period was located in the complex formerly occupied by the Archaic *ergasterion*, which was enlarged. No kilns were detected, but a deposit (*Fig. 5.3*, no. 12) was excavated within the complex. The dump[89] yielded banded (41.4%), plain (13.2%), and cooking wares (16.7%). Fine wares are attested in lesser quantities (17.6% and 1.7%, respectively). Its composition, when compared to other deposits found in the same area, presents some peculiarities. Only a few wasters were identified, and a large number of transport amphorae (5.4% of the total) were found. The widened spectrum of classes (which now also included cooking ware) and forms manufactured suggests that a transformation of some significance occurred in production organization. Noteworthy are the palmette lekythoi and black-gloss vessels, which prove that Metapontine potters of this period had mastered the firing technique required for their manufacture—a technological background was necessary for the production of red-figure vessels. Wasters confirm that this workshop produced black-gloss pottery characterized by shapes partially derived from the local tradition (cups and skyphoi with offset rims) and partially borrowed from the Attic morphological repertoire (A- and C-type skyphoi and Type C cups).

The *kerameikos* area did not yield evidence of the middle decades of the 5th century BC, when the production of red-figure pottery was introduced at Metaponto by the Pisticci Painter and his workshop. Nevertheless, discards of vases by the Amykos Painter (last quarter of the 5th century BC), associated with a kiln that has disappeared, were found in a salvage excavation held in 1966 some 200 m south of the North Gate along Plateia I (*Fig. 5.4*, no. 2).

In the *kerameikos*, four deposits (nos. 19, 22, 16 and 17; *Fig. 5.3*)[90] of the end of the third quarter to the last quarter of the 5th century BC fill the gap between deposit no. 12 and the workshop in which the Dolon Painter worked. These dumps did not yield Metapontine red-figure pottery. Instead, similar to what was observed in later dumps, the best attested classes were black-gloss ware and kiln spacers, with percentages ranging between 30 and 40 percent of the total.

The contexts of the end of the 5th and the first half of the 4th century BC yielded the largest amount of pottery, even though layers dated to this phase were uncovered only in limited areas of the excavation and are still partly obliterated by later structures (Kiln G, probably in use in this period, is partially cut off by Room E, which dates to the end of the 4th–3rd century BC).[91] Dump areas were identified along the fortifications, which, though built in the second half of the 4th century BC, probably followed the alignment of the earlier defensive structures. The most important context is deposit no. 1,[92] the protected position (pottery was thrown inside a pit) of which ensured its partial isolation from the surrounding area, thus limiting post-depositional interventions. Two other dumps (deposit no. 8 and, perhaps, no. 4) are probably related to the same workshop. The assemblage yielded by deposit no. 1, which is comprised of kiln spacers,[93] test pieces,[94] vessels of the Xenon Group, and black-gloss, banded, plain, and cooking wares (all attested by wasters), confirms the variety of the atelier's types of production. But it is especially the presence of red-figure pottery, attributed to the Dolon Painter and, to a lesser extent, to the Creusa Painter, that determined the significance of this context,[95] which also differs from contemporary deposits in the amount and quality of the figured material yielded.[96]

The production of red-figure pottery continued into the second half of the 4th century BC, as confirmed by the material found in four deposits (nos. 3, 5, 6, and 18; *Fig. 5.3*) located in close proximity of the two kilns (A and B) active in this period.[97] Though not quantitatively comparable to those of the previous phase, the deposits also yielded Gnathia, black-gloss,

89 For the shapes found in the deposit, see D'Andria 1980a.

90 D'Andria 1977a, 857, pl. CXXXIX,1; Cracolici 2003, 81–84. Silvestrelli 2000.

91 Silvestrelli 2005, 114.

92 D'Andria 1980a, 375–418. For fortifications, see above, note 83.

93 Cracolici 2003, 64–74, pl. I–XII.

94 D'Andria 1980a, 376; 394 no. 111, fig. 43; 397, nos. 129–31, fig. 396 and no. 142, fig. 46. D'Andria 1980c, 44. Cracolici 2003, nos. 134–48 and pl. XII. Papadopoulos 2003, 262, nos. M7 and M9; 263, nos. M13–M15; 264, no. M20. Silvestrelli 2004, 114.

95 D'Andria 1980a, 379–406, nos. 52–193, figs. 33–53; Silvestrelli 1996 and 2005.

96 Deposit no. 1 yielded some 1310 red-figure fragments; deposit nos. 8, 72. With the exception of a bell-krater attributed to the Anabates Painter (D'Andria 1980a, 443, no. 402, figs. 74–75), the evidence is comprised of small fragments, each normally representing an individual vessel.

97 D'Andria 1980a, 362–64 (kilns); 422–34 (deposit no. 3); 435–40 (deposits 5 and 6). Silvestrelli 2000.

banded, plain, and cooking wares. Banded and plain wares, however, are poorly represented.

Between the end of the 4th and the first half of the 3rd century BC, the types of production were altered, as attested by deposit no. 2.[98] This deposit yielded terracotta plaques and molds,[99] a type of production not previously attested in this part of the city. The black-gloss ware found in association can be dated to the end of the 4th and first half of the 3rd century BC. The production of terracotta plaques is not the only change: the recovery of tile wasters in levels of this period is also a new development.

Later in the 3rd century BC, after more than two centuries of activity, the area near the North Gate was no longer as dominated by production workshops, and seemed in fact to be less frequented in general. Other areas in the northwestern part of the city saw a similar decline.[100] The abandonment of the *kerameikos* must nonetheless have been gradual: Kiln C, for example, was found empty (possibly indicating maintenance activity prior to its complete disposal) and filled with debris from the collapse of the dome.[101] An abandonment layer gradually accumulated atop production structures and dumps of the *kerameikos*; along with much residual material, it yielded grey ware and molded bowls.[102]

In terms of craft specialization, the *ergasteria* excavated in the *kerameikos* come closest to the category of "nucleated workshops."[103] Their position could be interpreted as a confirmation of the preferred location for workshops at the margin of a city, and the *kerameikos* has often been indicated as an example of a potters' quarter—a specialized area earmarked for pottery production. No evidence for the presence of houses was recovered, suggesting that the area served only for production. The existence of specialized quarters, though, has recently been questioned.[104] In this understanding, the *kerameikos* might be better interpreted as a number of *ergasteria* clustering in proximity of the road system, as the locations of other workshops seem to confirm.

According to the distribution map of production sites within the city of Metaponto (*Fig. 5.4*), the presence of workshops seems, in any period, far from being concentrated only in this sector of the city. Production discards of unspecified chronology have been observed in the area between the *kerameikos* and the sanctuary of Apollo.[105] Evidence for what seems to have been the earliest workshop found at Metaponto and dated to the first half of the 6th century BC was recovered in a sounding excavated at the intersection between Plateia B and III (*Fig. 5.4*, no. 5).[106] In the Late Archaic period, pottery workshops could have been active near the agora, as suggested by the recovery of vitrified clay and Ionic-type cup wasters from a road pavement fill of the stenopos that defined the southwestern side of the agora (*Fig. 5.4*, no. 9).[107] As discussed above, the Amykos Painter's workshop may have been located along Plateia I.[108]

The same dispersion can be noted for the Hellenistic period (*Fig. 5.4*, nos. 3, 6, and 8). A pottery workshop with two kilns was excavated in Insula I, just west of the sanctuary.[109] The insula is excavated only in part and, with the exception of a plan, remains unpublished.[110] Adamesteanu mentioned the discovery of a discard deposit with unguentaria,[111] and it is likely that the workshop can be dated to the 3rd century BC, when the insulae west of the sanctuary began to be occupied.[112] Kiln spacers, mud-bricks, and wasters dated to the second half of the 3rd century BC were found southwest of the *kerameikos*.[113] Several examples of the same type of kiln spacers were collected in the area where the red-figure wasters by the Amykos Painter were found.[114] This evidence confirms that production in the city was still quite active in the second half of the 3rd century BC, as has been confirmed by recent studies.[115]

In the Late Republican period, evidence for production activity is found in the area of the castrum

98 D'Andria 1980a, 418–22.

99 Signore 1996.

100 On the transformation occurring in the 3rd c. BC in the city, see Mertens 1999, 284–92 and De Siena 1999, 242–45, with previous bibliography. See also Giardino 2005, 406–10, and Carter in *Survey*, 884–91.

101 D'Andria 1980a, 367–70, figs. 14–20.

102 D'Andria 1980a, 361, no. 25, fig. 25.

103 Peacock 1982.

104 Esposito and Sanidas 2012.

105 A. De Siena, pers. comm., 2008.

106 *Appendix B, Table B1*, no. 5.

107 *Appendix B, Table B1*, no. 9.

108 *Appendix B, Table B1*, no. 2.

109 *Appendix B, Table B1*, no. 6.

110 Mertens 1985, fig. 1a.

111 *Metaponto I*, 59 note 168.

112 Mertens 1999, 286.

113 *Appendix B, Table B1*, no. 8. Cracolici 2003, 97, fig. 52.

114 *Appendix B, Table B1*, no. 3. D'Andria 1980a, 448–52. The kiln spacers belong to types II-C2 and II-D2: Cracolici 2003, 37.

115 Giardino 2015, figs. 22–37.

and along Plateia A.[116] West of the castrum, [117] an area yielding mudbricks, spacers, transport amphorae, and grey-ware wasters was found during agricultural activity. Molds for molded bowls from the same area confirms the Metapontine production of this type of ware.[118] Slags also indicate the likely presence of metal workshops. A Late Republican metal workshop (2nd–1st century BC) partially occupying Plateia A was excavated just west of the sanctuary.[119] The topographical distribution of workshops and production indicators is no coincidence. In this period, a large part of the city lay abandoned and the settlement clustered in the areas of the castrum, the fluvial harbor, and along Plateia A.[120]

With the exception of the *kerameikos*, the workshops excavated at Metaponto are unpublished, and the other evidence is of limited value. This precludes the possibility to understand in detail the choices made in the placement of the workshops and their modes of production. What can be confidently observed is that the locations of the ateliers were probably chosen following principles of ceramic ecology. On the other hand, though, it is equally true that workshops could clearly be present in multifunctional insulae, as in other cities (e.g., Taranto, Locri, and Herakleia), and their location choices may have been connected with the roads.

Regarding the type of wares manufactured in these workshops, the city yields evidence for fine ware, which, at least in the *kerameikos*, is the best represented. Plain and coarse wares as well as cooking ware seem to have been produced in the same workshops where fine wares were manufactured. This is noteworthy especially for the cooking ware. As far as is currently known, the two types of production were, for technical reasons, normally separated.[121] The shapes produced were meant for local consumption and are the same found in tombs and sanctuaries, though little can be said about domestic consumption. At least part of this production was certainly meant for export, evidenced by the B2 Ionic-type cups in the Archaic period, and red-figure and black-gloss wares in the second half of the 5th and the beginning of the 4th century BC. Evidence for terracotta relief plaques has also been found. As in other contexts, such as at Thasos, the production appeared to take place within pottery workshops. Unfortunately, the evidence at our disposal is limited to the Early Hellenistic period, but this may be due to the casualties of archaeological research. Tiles must have been produced in the city, although the only evidence comes from random Early Hellenistic fragments from the *kerameikos*.

The extensive presence of production sites in the chora is no surprise and fleshes out what is already known from literary sources referring to other areas.[122] The widespread presence of workshops allows for the chora's substantial self-sufficiency in terms of the vessels needed for everyday life. Comparing production in the chora and in the city, an overlap of shapes and classes, and possibly also techniques, is evident, suggesting the existence of a common background. The remarkable opportunity to compare two workshops active in the same period, Torretta di Pisticci[123] and deposit no. 12 from the *kerameikos*, demonstrates that the same tradition occurs in both the city and the chora, with workshops near the places where their products are consumed. This same model can be witnessed for the terracotta relief plaques, with similar types being produced both in the city and in the countryside.

At present, evidence for the production of transport amphorae is, at least in the Classical and Early Hellenistic periods, limited to the chora. The existence of urban production could nonetheless be suggested by misfired pieces of the 3rd century BC found in a deposit excavated near the Eastern Gate,[124] and by the waster found in the area nearby the castrum.[125]

As far as we know, red-figure ware is produced only in the city. In consideration of the great diffusion of this type of ware in the territory, this might be surprising. This may simply be a coincidence of the archaeological research, but it also might be true that the production of this class was confined to the city, due to the highly specialized skill set its production required.

Notwithstanding the difficulties in reconstructing the production pattern of Sant'Angelo Vecchio, it seems evident that the site is not an isolated phenomenon but part of a wider network of production.

[116] Giardino 2005, 422–23. In a sounding undertaken in Plateia A, the most recent road pavement (possibly dating to the 2nd c. BC) was built with construction elements from destroyed kilns (*Fig. 5.4*, no. 4): *Metaponto I*, 297, fig. 312.

[117] *Appendix B, Table B1*, no. 10.

[118] They have also been found in the abandonment layers of the *kerameikos*. See also D'Andria 1980a, 361.

[119] Mertens 1985, 648, fig. 2. De Siena 1993, 120; 1999, 237, fig. 2.

[120] Giardino 2005, 423–24. Carter in *Survey*, 919–22.

[121] This is the case at Thasos: Blondè and Picon 2000, 15.

[122] Amouretti 2000.

[123] *Appendix B, Table B2*, no. 46.

[124] Giardino 2015, fig. 28a–b.

[125] *Appendix B, Table B1*, no. 10.

6

Sant'Angelo Vecchio in the Medieval and Post-Medieval Periods

Erminia Lapadula

Figure 6.1 General view of the masseria at Sant'Angelo Vecchio, 2012. (EL/ICA)

The Medieval occupation phase of the site of Sant'Angelo Vecchio has always been a subject of debate and discussion. The Post-Medieval phase, on the other hand, is represented by the homonymous masseria that occupies the site today (*Fig. 6.1–Fig. 6.3*), built in several discrete sections that include a church. On a wall above the church's altar, a fresco, datable to the 17th century and attributable to a South Italian artist, depicts the Archangel Michael in the act of transfixing Satan (*Fig. 6.4*).[1]

Situated on a modest rise on the northeast bank of the Basento River between the Agnone S. Angelo and Avinella/Venella localities, the site formed part of a sparsely populated Medieval and Post-Medieval landscape.[2]

At present, the only known archaeological evidence comes from the intensive field survey between the Bradano and Basento Rivers conducted by the Institute of Classical Archaeology over the course of 20 years, the results of which have recently been published.[3] The survey documented several surface scatters of modest size containing Roman artifacts (sites 384, 386, 387, and 388) and Medieval artifacts (sites 381 and 388), which have been interpreted as agrarian structures (see *Fig. 1.9*).[4]

Medieval sources document the existence of a *casale* (village) at Sant'Angelo di Avena, property of the ab-

---

[1] The author wishes to thank the Superintendency for the Historical, Artistic, and Ethnoanthropological Heritage of Basilicata in Matera for the information about, and the photograph of, the fresco, in particular the art-historical and technical report written by Dott.ssa M. Francione. Additional thanks go to M. Falla Castelfranchi, M. Guastella, V. Pugliese, and G. Scarcia for the information pertaining to its date.

[2] On the Metapontine landscape in the Medieval and Post-Medieval periods, see Lapadula in *Survey*.

[3] See *Survey*.

[4] For analysis of the data, see Lapadula in *Survey*. See also maps 15–16 and the gazetteer of sites in Survey, Vol. IV.

Figure 6.2 General view of the masseria at Sant'Angelo Vecchio from the southeast, 1979. (CW/ICA)

bey of Santa Maria del Casale at Pisticci, at the beginning of the 12th century.[5] Despite many doubts and uncertainties surrounding the authenticity of several *chartae*, a document that places Rodolfo Maccabeo's donation of the church of San Michele Arcangelo to Giovanni, abbot of Santa Maria del Casale at Pisticci, in the year 1100 nevertheless appears to be credible.[6] The document describes the location of the church thus: "que sita est iuxa flumen Basenti cum casale et hominibus in eo habitantibus et cum omnibus pertinentiis, acquis, pascuis etc."[7]

An earlier document of dubious authenticity[8] records in 1083 the donation by Umfredo, Count of Montescaglioso, "del locum quae vocaturar Avena ubi olim erat casale ad habitationem hominum,"[9] and the church of San Giovanni on the road "que venit de Appio,"[10] to the abbot of San Michele Arcangelo at Montescaglioso.[11] Another controversial document is a record of King Ruggero from 1133 or 1135 confirming the donation of the church of San Michele Arcangelo di Avena with *casale* (village), the Trinità monastery at Monte Finesio, and the church of Sant'Angelo at Barco (Varco), with all of their appurtenances, to Gaudio the abbot of Santa Maria del Casale at Pisticci.[12]

The fief of San Michele Arcangelo di Avena remained in the possession of the abbey at Pisticci until 1451, when the Benedictine monastery was suppressed by papal bull and its property annexed to the Certosa di San Lorenzo at Padula.[13] Sound management by the Certosa to a large extent restored the property's economic prosperity, as is demonstrated by the *platea* (map) created in 1678 by Angelo Grammatico as a detailed catalog of the Certosa's possessions (*Fig. 6.5*).[14]

[5] Di Meo 1801, 81; Pedio 1990, 85; Lafratta 1996a.
[6] Sacco 1914, 220–21; Pedio 1998, 131.
[7] "which is located near the Basento River with the village and the people who live there and with every appurtenance, streams, pastures, etc."
[8] Pedio 1998, 116–17.
[9] "...of the place which was called Avena where once there was a village for human habitation."
[10] "which came from the [Via] Appia."
[11] Tansi 1746, 134; Pedio 1998, 116–17.
[12] Spinello 1968; Pedio 1998, 157–58.
[13] D'Angella 2006, 58.
[14] Sacco 1914.

Figure 6.3 General view of the masseria at Sant'Angelo Vecchio from the southwest, 1979. (CW/ICA)

The fief at San Michele Arcangelo di Avena is indicated on the map with the name Sant'Angelo di Bernalda or Santa Maria dell'Incoronata e dell'Avinella (*Fig. 6.5*).[15] The property S. Angelo in Bernalda is described as a very large estate with six boundary markers bearing the Certosa's emblem to advertise its ownership.[16] The detailed and precise description is not matched by a relevant position on the map, which indicates only the church and the fountain, while the *difesa* (a property enclosed by a wall or otherwise physically bounded) is crossed from west to east "… dal Tratturo publico, che va a Torre di mare, et in altri luoghi."[17] The map contains a short description of the church, locating it near the *tratturo* (route), flanked by a house used as lodging for the monks, lodgings for the keepers of the animals, and *fabbriche* (buildings) used to shelter the livestock.[18]

The documentary sources indicate that the settlement at Sant'Angelo formed part of the network of Benedictine properties spread around the Metapontino belonging to the monasteries of San Michele Arcangelo at Montescaglioso and Santa Maria del Casale at Pisticci. This network also included S. Maria del Corno (Incoronata), S. Biagio, S. Giovanni dell'Avinella, S. Salvatore, S. Maria del Vetrano, and S. Basilio.[19] Within these rural settlements, the site of Torre di Mare (*Civitas Sanctae Trinitatis*) is prominent as a fortified site with a long life, from the 11th to the 17th century.[20] The significance of this focal point in the territory is clear from the important road connecting the coast and the interior. The road was lined with *casali*, including the one at Sant'Angelo Vecchio, for which Torre di Mare served as the arrival and departure point for goods, people, and ideas.

The Medieval *casali* often are found at sites previously occupied by Greek and Roman structures. The

15 D'Angella 2006, 78.

16 Sacco 1914, 260.

17 "from the public route which goes to Torre di Mare, and in other places."

18 Sacco 1914, 261.

19 Caputo and Bubbico 1983, 81–83.

20 For the history and archaeology of Torre di Mare, see Bertelli and Roubis 2002.

Figure 6.4 Wall painting in the chapel of Masseria Sant'Angelo Vecchio. (Courtesy of the Soprintendenza ai Beni Artistici della Basilicata)

continuous occupation of these places over time is due to their particularly favorable positions along major roads and their geomorphology. They are located on terraces and ridges well-suited for agro-pastoral activities and, especially, free from the risks of malaria, which infested the low-lying areas of the Bradano and Basento valleys due to the presence of stagnant water. Many of these settlements were sparsely inhabited, abandoned, or destroyed between the end of the 14th and the 15th centuries.[21]

In the 16th century under Spanish domination, the territory was subjected to a new process of agrarian

[21] Pedio 1990. On the causes behind the abandoned villages phenomenon, see most recently Milanese 2006 with related bibliography.

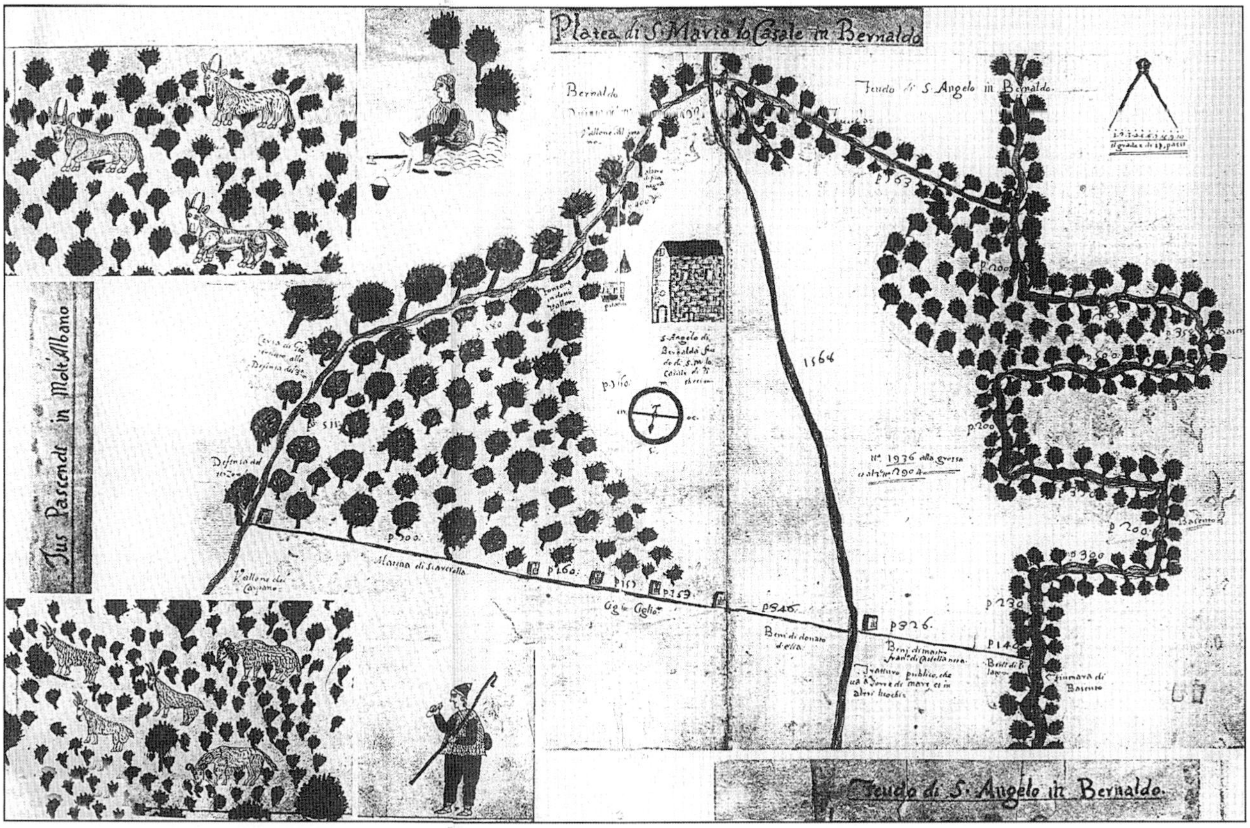

Figure 6.5 Excerpt of the platea by Angelo Grammatico (1678), showing the large estate at Sant'Angelo. (From Sacco 1914, Courtesy of Arte Grafica Boccia S.p.A.)

feudalization, the masseria. The large estates and their methods of production were reorganized by incorporating vast tracts of land partly left uncultivated and intended for pastoral activities. An autonomous farm covering a large area and engaged in agriculture or pastoralism, a masseria generally consisted of a complex of rural buildings with different uses, including lodging or production-related activities, forming a single operational unit. The origin of the masseria in its very basic form stretches back to the earliest examples of farmhouses scattered throughout the chora of the Greek colony. The farm's later evolution in the Imperial period approached the enormous areas of land that characterize the rural administrative and landed entities of the *massae* or the *villae rusticae*. Between Late Antiquity and the early Middle Ages, the *villae*, together with the rural agglomerations known as *pagi* and *vici*, formed the backbone of residence and production in the rural landscape. They were replaced during the Middle Ages by new inhabited nuclei, *castra* and *casali*, which were built for agricultural, defensive, and religious functions. Many *casali* served as the core around which masserias later developed. In the Post-Medieval period, masserias were amalgamated from existing structures and land tracts, creating an agricultural system that continued into the 19th century.

In recent years, a large body of archaeological data generated by research and studies conducted in Basilicata and Apulia has reinforced this understanding of the transformation of the ancient landscape, proposed by scholars several decades earlier.[22] Among the most important studies are those performed in the Salento and eastern Basilicata.[23] As far as the Metapontino (including Sant'Angelo Vecchio) is concerned, the archaeological and architectural evidence documenting the Medieval phase is visible at the continuously occupied masserias at San Salvatore, Pantanello, Torre di Mare, and San Basilio.[24]

22 Lepre 1981; Mongiello 1984.

23 For the Salentine peninsula, see Arthur 1996; Arthur and Bruno 2009; Tinelli 2006. For Basilicata, see most recently the example of Vagnari (Small 2011, 34–36).

24 San Salvatore: Lapadula 2009. Pantanello: Carter 1983a. Torre Mare: Bertelli and Roubis 2002. San Basilio: Tommaselli 1984, 366–72; 1986, 178–81; Lafratta 1996b, 180–82.

The decline of the two abbeys at Montescaglioso and Pisticci during the 18th century led to the abandonment of their rural possessions, which were rented or leased in perpetuity. The organization of the landscape began to change, and the first documented changes occurred in the early 19th century, following the suppression of the religious orders and the feudal system under Murat.[25] The large ecclesiastical and feudal estates, broken up and subdivided, were replaced by masseria-villages and their related holdings. This began a new agricultural regime based on an increase in the extent of fields for sowable and tree crops after large-scale deforestation and recuperation of large humid areas.[26] This organization remained stable until the bonifica and the agrarian reform of the 20th century transformed the socioeconomic and geographic conditions by breaking up both the large estates and the unitary production model of the masserias.[27] The new economic situation and heavily mechanized production methods introduced as part of the reform created the conditions for several types of settlement that replaced the masseria in the countryside.[28]

[25] Among the many military, political, and administrative reforms initiated by Joachim Murat, king of Naples, beginning in 1809, the law that provided for the abolition of the feudal system and the division of the land, which favored the rise of the landed bourgeoisie of southern Italy, is particularly important in this context. So, too, is the decree that suppressed the religious orders in the kingdom of Naples, and the confiscation of their property for other (often military) uses that followed. For detailed discussion, see Woolf 1973 and S. Russo 2007.

[26] On the masseria system, see Sereni 1961 and Licinio 1981, 1998.

[27] On the agrarian reform, see De Leo 2008.

[28] Tommaselli 1986, 23–25.

Part II

7

Geologic Setting

Alessandro Montanari, Francisco da Conceição de Jesus Neto, and Rachel Karlov

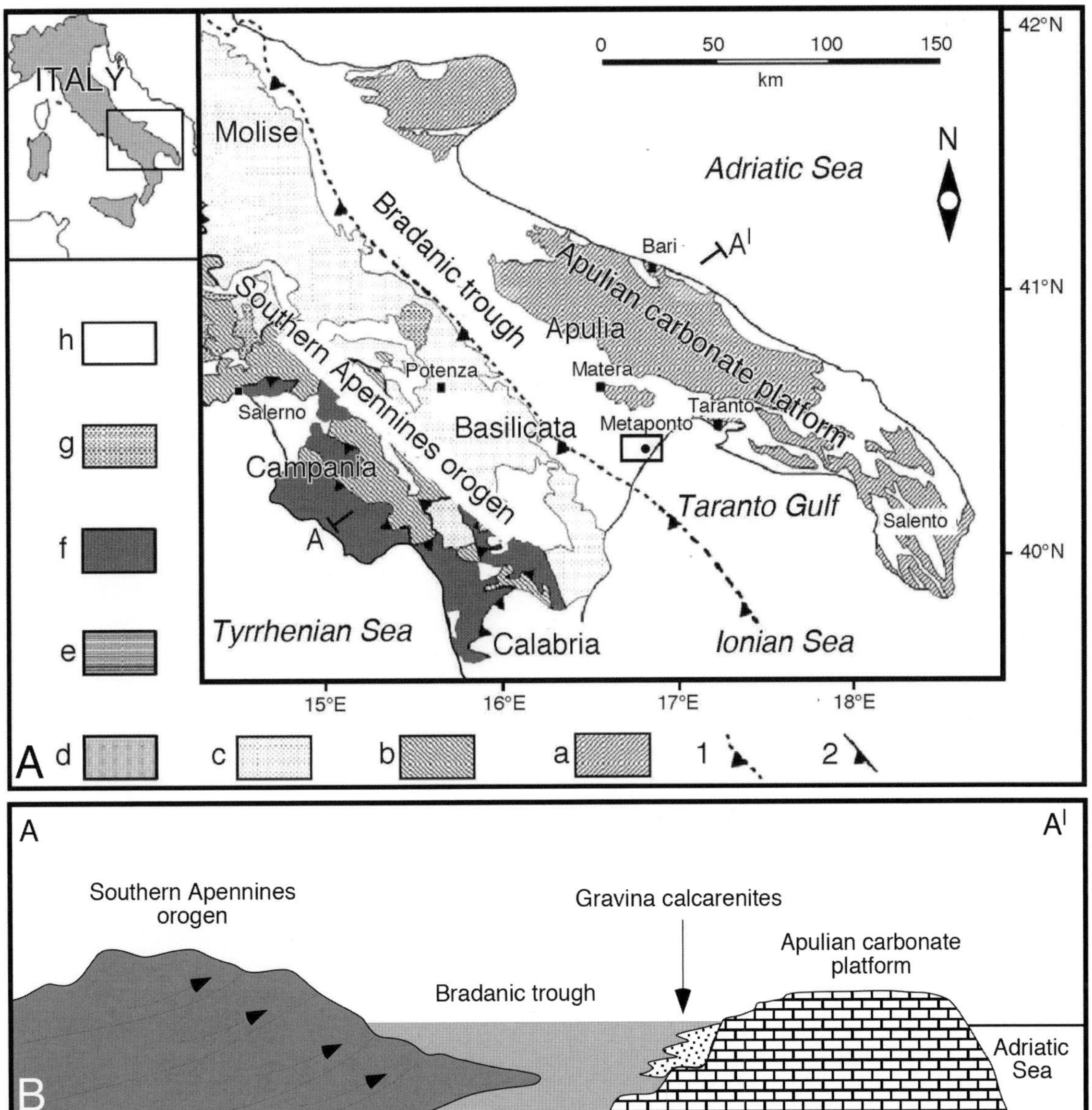

Figure 7.1 (A) Generalized and simplified geologic map of the southern Apennines: (a) Apulian foreland, (b) Apennine carbonate platform, (c) Molise-Sannio-Lagonegro pelagic basin, (d) Laga Fm, (e) Umbria Marche pelagic basin, (f) Liguride units, (g) Magmatic units, (h) Pliocene and Quaternary deposits, (1) front of the southern Apennine orogenic wedge, (2) main thrust front, A–A^{1} trace of cross-section (redrawn from Sauer et al. 2010, 50, fig. 1). (B) Simplified structural cross-section model across the southern Apennines (modified from Folk in *Survey*, 3, fig. 1.1).

Regional Geologic Background

Traversing the five southernmost regions on the Italian peninsula, the southern Apennine Mountains are a prime example of an orogenic fold-and-thrust belt (*Fig. 7.1A*). The range resulted from the closure, in Oligocene time (i.e., between 34 and 24 million years ago), of the so-called Ligurian Ocean (a western branch of the ancient Tethys Ocean), and consequent

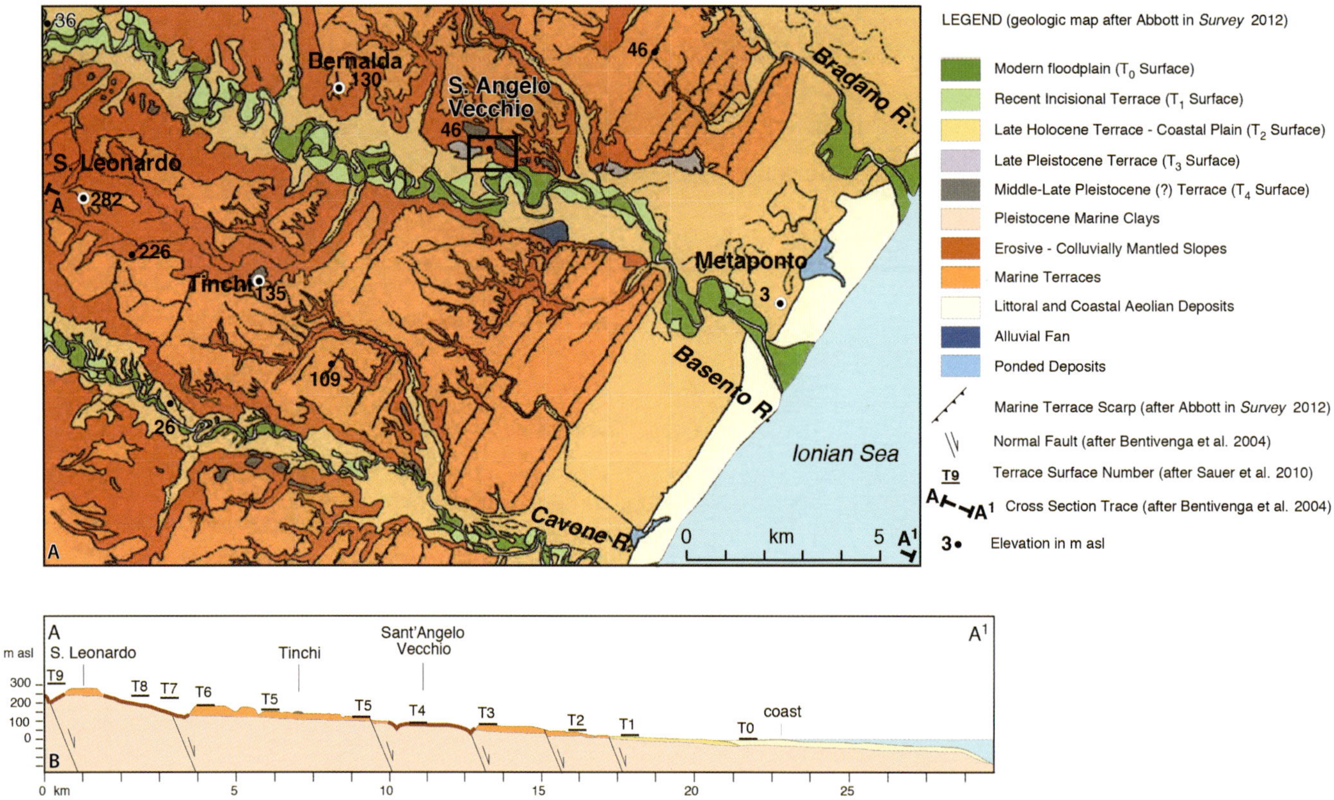

Figure 7.2 (A) Geomorphic map of the Metapontino with location of the Sant'Angelo Vecchio archaeological area (from Abbott in *Survey*, 33, fig. 2.2). (B) Cross-section of the Metapontino area from San Leonardo to the offshore of the Ionian Sea (redrawn from Bentivenga et al. 2004, 398, fig. 9b), with numbered marine terrace surfaces (according to Sauer et al. 2010, 54, tab. 2).

southwest-northeast contraction and deformation of an adjacent marine epeiric basin, which was previously formed on a stretching and subsiding continental crust.[1] Notwithstanding the geological, stratigraphic, sedimentological, structural, and tectonic complexities of this region, which have been the subjects of innumerable studies published in the past half-century or so,[2] the southern Apennines can be subdivided into three main tectono-stratigraphic domains: (1) a southwestern Apennines orogenic complex consisting of an accretionary wedge, (2) a central Bradanic foredeep trough, and (3) a northeastern, essentially undeformed and stable foreland Apulian carbonate platform (see schematic model in *Fig. 7.1B*).

The Bradanic trough hosts a 3.4-km-thick Plio-cene-Pleistocene sedimentary succession, which covers unconformably the Cretaceous limestones of the Apulian carbonate platform to the northeast, and is overthrusted by the imbricated units of the Apennines accretionary wedge to the southwest.[3] The oldest sediments of this foredeep fill are marine, hemipelagic marls and marly clays of Early Pliocene age, which pass upward to thick sequences of siliciclastic turbidites and clays of Late Pliocene to Middle Pleistocene age. Along the northeastern margin of the Bradanic trough, these sediments interfinger with calcarenitic bodies of the Gravina Formation,[4] which were derived from the dismantlement of the adjacent Apulian carbonate platform's limestones.

While the lower part of the Bradanic sedimentary succession is involved in the compressional tectonic deformation along the advancing front of the Apennines accretionary wedge, the Middle Pleistocene sediments, which constitute a shallowing-up succession of marine clay and silt, cover both the allochthonous units and the older foredeep deposits.[5] This indicates that the orogenic compressional phase and conse-

[1] D'Argenio and Alvarez 1980.
[2] E.g., Casero et al. 1988; Pescatore/Renda/Tramutoli 1988; Sella/Turci/Riva 1988; Bigi et al. 1996; Monaco/Tortorici/Paltrinieri 1998; Calamita et al. 1999; Patacca and Scandone 2001; and references therein.
[3] Tropeano and Sabato 2000.
[4] Pomar and Tropeano 2001.
[5] Ciaranfi/Pieri/Ricchetti 1988; Ciaranfi et al. 1996.

quent subsidence of the Bradanic foredeep ceased in the Middle Pleistocene, followed by a phase of passive filling and, ultimately, by uplifting and emersion. According to Pieri et al.,[6] the process of filling and emersion dates back to the Early Pleistocene (~1.8 Ma) in the northern sector of the Bradanic trough, whereas in the southern sector marine sediments making up regressive successions are radioisotopically dated to the Middle Pleistocene (~0.7 Ma).[7]

The Recent Geologic History of the Metapontino

The quaternary geological history of the Bradanic trough has gained, since the beginning of the last century, special attention from a number of earth scientists.[8] These scholars were particularly attracted by the arrangement of marine terraces making up the stepped topography that characterizes the area of the Metapontino, the hinterland of the western littoral of the Gulf of Taranto between the Cavone and the Bradano Rivers (*Fig. 7.2*). The stepped marine terraces are the result of a ca. 700,000-year history of constant tectonic uplifting punctuated by glacio-eustatic fluctuations of the sea level, which are controlled by cyclic glacial and interglacial episodes. During a glaciation, the sea level drops and, with respect to the coastline, the sea regresses. With the regression of the sea, offshore pelitic (clay) sediments deposited at depths of tens to a few hundred meters become covered by more proximal sublittoral sediments, such as sand and silt, and as the regression proceeds, by even coarser and shallower foreshore sediments, such as coarse sands and gravels. Locally, foreshore marine sediments are covered by fluvial and colluvial continental sediments. The ultimate sedimentary expression of such a regression is a coarsening-up succession of neritic marine sediments (i.e., from clay to sand to gravel; *Fig. 7.3*).

As the glaciation comes to an end, and a climatic inversion opens the way to an interglacial (warm and wet) period, the sea level rises again, transgressing over the coast. Thus, the sediments of the regressive succession get progressively submerged by the sea and covered by ever deeper and finer marine neritic sediments, making up a transgressive sedimentary succession. During this cyclic regression-transgression process,

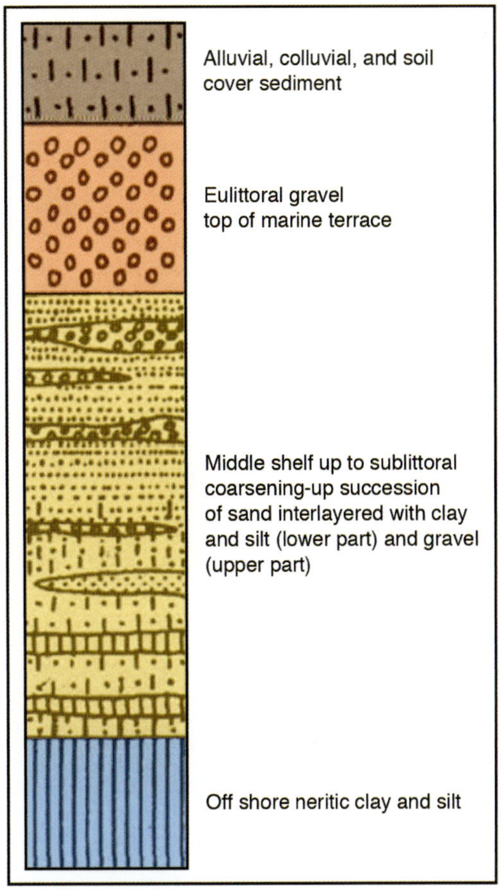

Figure 7.3 Generalized sedimentary stratigraphy of the marine terrace units in the Gulf of Taranto. (Adapted from Sauer et al. 2010, 51, fig. 3)

however, tectonic uplifting raises the marine sediments, which, at the end of a renewed sea transgression, will be spared from submersion. In essence, this is how marine terraces form as flat-topped, elongated sedimentary bodies running parallel to the coast. The higher the elevation is, the older the age of the terrace. After emersion, these terraces fall under the action of morphogenic processes such as erosion, denudation, soil formation, land sliding, accretion, and incision, which shape the Metapontino in a sort of stepped topography made up of the remains of these marine sedimentary deposits.

The number of terraces has been strongly debated in literature, ranging from a record maximum of 11 terraces to just one.[9] Over a general decrease of elevation toward the coast (and thus a general decrease of the terrace ages),[10] terraces are delimited by small escarpments but, in many cases, they seem to

[6] Pieri/Sabato/Tropeano 1996.

[7] De Marco 1990.

[8] E.g. Gignoux 1913; Gigout 1960; Selli 1962; Cotecchia and Magri 1967; Vezzani 1967; Boenzi et al. 1971; Neboit 1982; Dai Pra and Stearn 1977; Dai Pra and Hearty 1988; Hearty and Dai Pra 1992; Brückner 1980; Fuchs 1980; Amato 2000; Bentivenga et al. 2004; Sauer et al. 2010.

[9] Brückner (1980) claims 11 terraces, and Bentivenga et al. (2004) cite one.

[10] Sauer et al. 2010; Abbott in *Survey*.

Figure 7.4 Panoramic view of the Sant'Angelo Vecchio archaeological site. (AM/ICA)

split or join laterally into a single terrace, leading to this range of interpretations of their number. In their interpretation resulting from a detailed geomorphologic, sedimentologic, and structural study of the area, Bentivenga et al.[11] envision essentially one terrace made up of a single yet diachronous regressive sedimentary succession, which is cut by a series of small displacement normal faults trending south-southwest–north-northeast in correspondence to what was previously interpreted as marine terrace escarpments (see *Fig. 7.2B*).

Geology of the Sant'Angelo Vecchio Archaeological Site

Apart from this difference in the interpretation of the number of terraces, the geology of the Metapontino is by now extremely well resolved and mostly understood in all its geological, sedimentological, geochronological, geomorphological, and pedological aspects, as thoroughly documented in the recent works by Folk and Abbott.[12] Both of these authors place a strong emphasis on the geoarchaeology of this area. On the basis of a detailed multidisciplinary study of dozens of key sites, such as exemplary outcrops of the marine sedimentary succession, soil profiles, and natural vs. anthropogenic geomorphic features and processes (stream patterns, erosional and depositional processes, modification of the topography, etc.), they give a wide range of detailed information to the archaeologists who are interested in why and how a particular site has been originally occupied and developed by a human community. As for the case of the Sant'Angelo Vecchio archaeological site, the simple question is why Greek colonists decided to establish a small ceramic workshop at that particular location. A detailed geologic ground-truthing of the immediate surroundings of Sant'Angelo Vecchio is sufficient for answering these archaeological questions.

The archaeological site of Sant'Angelo Vecchio (see *Fig. 7.2* for location), comprising a farmhouse, a small necropolis, a ceramic workshop, and a retaining wall (*Fig. 7.4*), is located on the gentle southwestern slope of the Sant'Angelo Vecchio creek gully, which flows for a few kilometers in a north-to-south direction, deeply incising the north slope of the Basento River valley. The very few outcrops exposing bedrock

[11] Bentivenga et al. 2004.

[12] Folk in *Survey*; Abbott in *Survey*.

Figure 7.5 (A) The bluff exposing the sand unit of the marine terrace at Sample Site 2, about 300 m SE of Sant'Angelo Vecchio. Note cross-bedding and protruding hard layers of well-cemented sandstone. (B) Outcrop of the gravel unit of the marine terrace exposed at Sample Site 5, about 400 m NW of Sant'Angelo Vecchio. (AM/ICA)

in the area immediately surrounding the archaeological site reveal, nevertheless, the marine terrace's regressive succession (*Fig. 7.3*), from neritic clay to sand (*Fig. 7.5A*) to gravel (*Fig. 7.5B*).

From these few outcrops, it is possible to trace the contacts between these units over the area, following the contour lines on a 1:5000-scale topographic base map, and assuming essentially horizontal bedding (*Fig. 7.6*). The marine terrace, with a flat top at an elevation of about 60 m above sea level, would correspond to Abbott's surface T4,[13] and is covered by variously thick and compositionally different loamy soils (i.e., more or less pebbly, sandy, or clayey), depending on the parent rock type.

[13] Abbott in *Survey*.

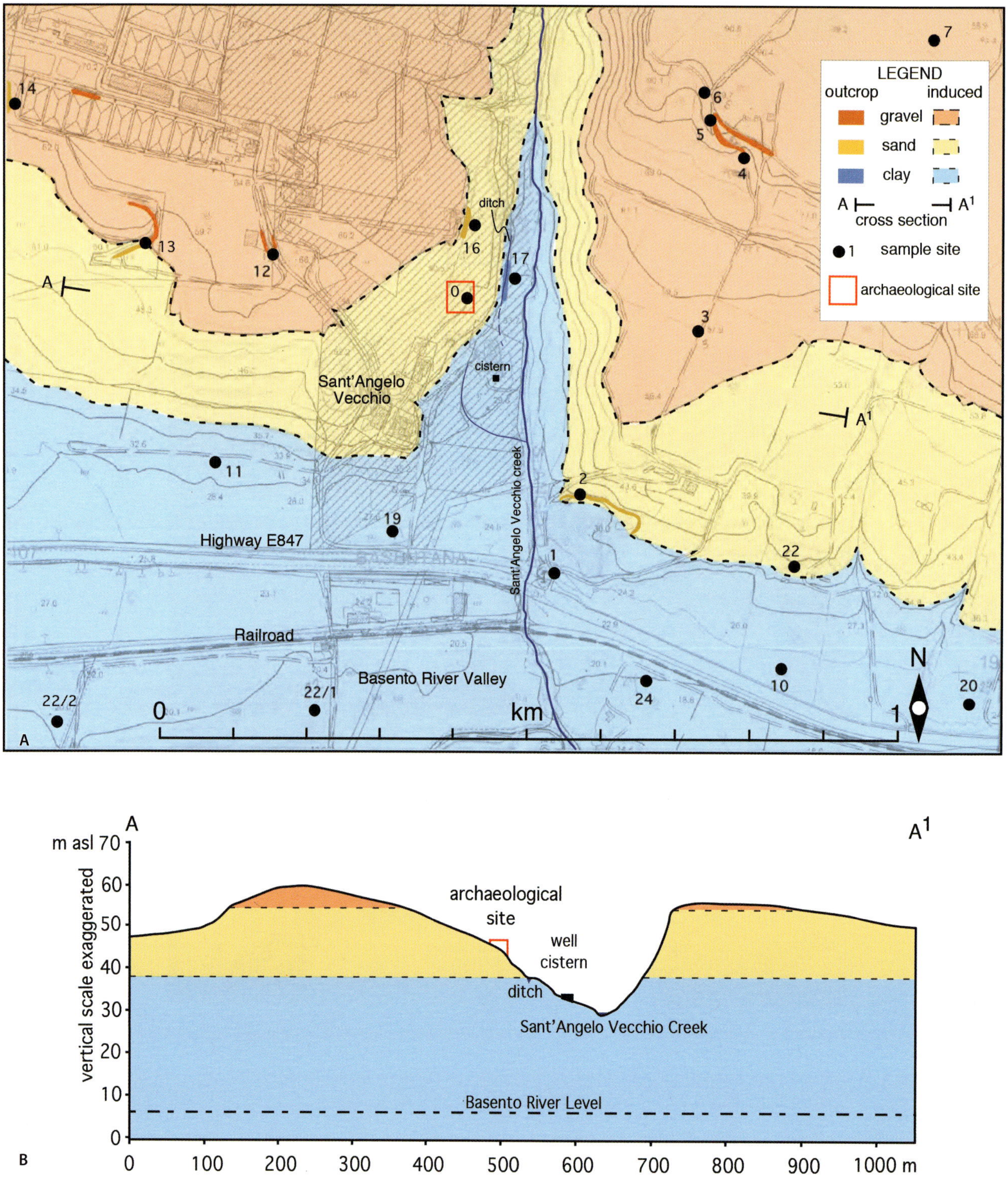

Figure 7.6 (A) Geologic map of the area surrounding Sant'Angelo Vecchio. (B) Cross-section A-A[1] across Sant'Angelo Vecchio. (AM/ICA)

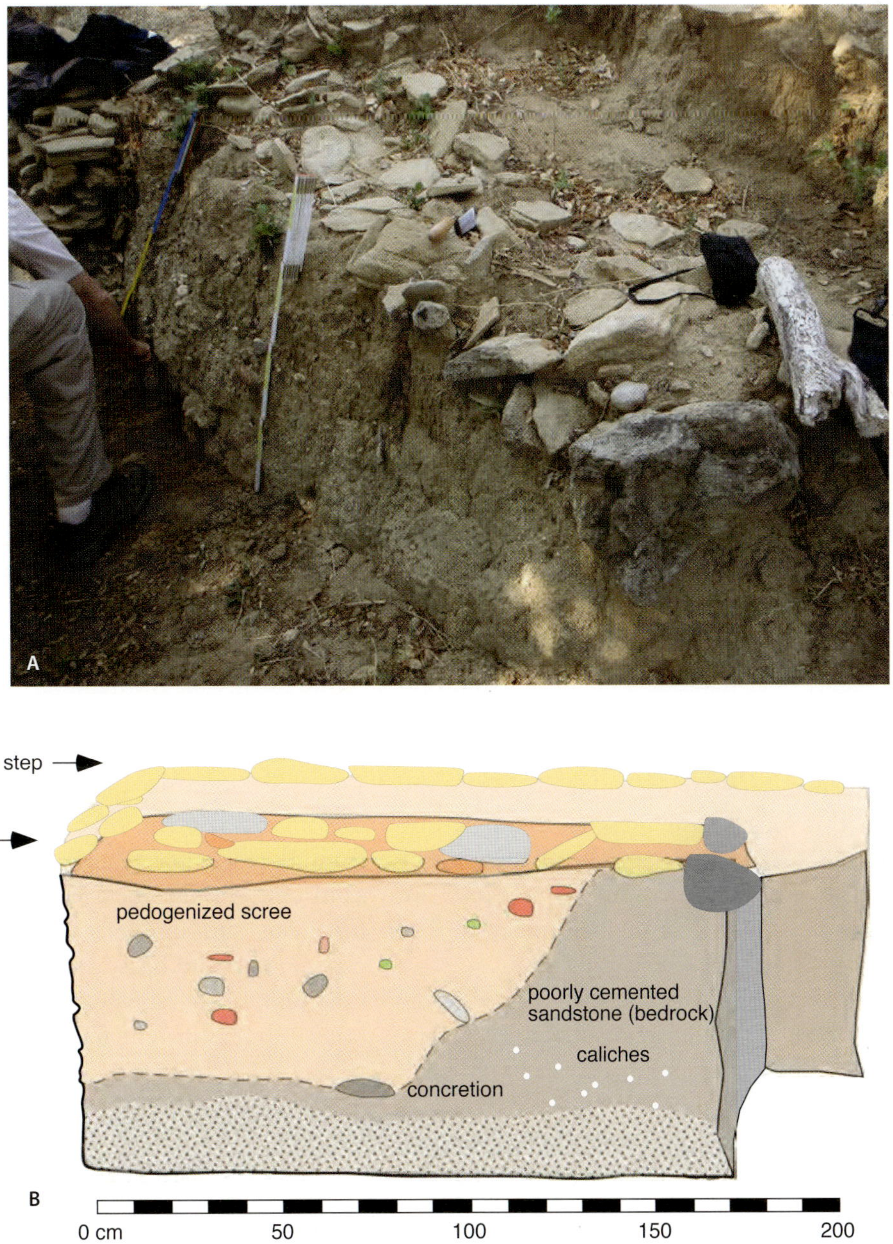

Figure 7.7 (A) View of the excavated lower paved step at the archaeological site of Sant'Angelo Vecchio. (B) Schematic drawing of the sediment units exposed in the front of the excavation. (AM/ICA)

A closer look at the excavated steps (Sample Site 0 in *Fig. 7.6*, and *Fig. 7.7A*) reveals the sediment stratigraphy underlying the steps. The sand unit making up the bedrock below the lower step is incised by a channel filled with pedogenized scree, a sandy loam containing small pebbles of various silicic rocks (mostly metaquartzite and chert; see *Fig. 7.7B*). The steps and the rest of the archaeological area are covered by a rather homogeneous, 40- to 60-cm-thick sandy loamy soil lacking a distinct stratigraphy (see *Fig. 7.4*). Sand grains in the soil residue are mostly subangular and silicic (i.e., quartz, some feldspar, and silicic lithoclasts), whereas pebbles and cobbles in the gravel fraction are mostly made of well-rounded metaquartzite and chert clasts. No carbonate clasts were found in the several test samples collected in the area, except for some fossil fragments of shallow water mollusks found in the few outcrops of the marine terrace's sand unit. Locally the sand unit exhibits well-cemented horizons, turning the otherwise loose or poorly cemented sand into hard sandstone. The same localized cementation occurs also in gravel deposits.

As for the clay unit at the base of the regressive marine terrace succession, it is never exposed in an outcrop section, such as a road cut, a bluff, or a river scarp. The closest to an outcrop found in the area around Sant'Angelo Vecchio is a narrow ditch that runs along the foot of a ca. 10-m-high forested scarp on the western slope of the Sant'Angelo Vecchio creek, about 5 m above the actual creek channel and about 60 m northwest of the Sant'Angelo Vecchio archaeological site (see Sample Site 17 in *Fig. 7.6*). The ditch delimits a ca. 200-m-long, ca. 20 m flat and horizontal field, with a recently planted olive grove. The soil is clayey, and samples collected with an auger in the ditch and in the olive grove next to it revealed pure gray-blue clay in the immediate subsurface. The half-meter-wide ditch, which is probably man-made as a device for draining the olive field, was filled with water at the end of July 2012, and also in mid-October of the same year when this site was surveyed after an exceptionally long summer drought. This suggests the presence of a perennial water seepage, which happens to be at the contact between the (impermeable) marine clay unit and the overlying (porous) sand unit of the marine terrace cut by the Sant'Angelo Vecchio creek gully.

The clay deposit and the associated freshwater spring represent, for the early Greek settlers, a precious resource, ideal for the manufacture of ceramics. The settlement and the kiln itself are found at a short distance from this clay deposit, on a gentle soil-mantled slope over the sand unit. This site is the closest to the clay and the spring; the settlers could have chosen to build edifices on stable and well-drained ground. Apparently, houses were built with unburnt clay bricks or adobe walls erected on foundations consisting of trenches filled with loose gravel, which would have ensured efficient drainage. These building materials, the clay from the spring area and the gravel from the top unit of the marine terrace, are abundantly available in the immediate reaches of the settlement. Much rarer are sizable stones that could be used as building material. In fact, the whole Metapontino lacks exploitable rock sources, such as exposed limestone or sandstone formations. The sarcophagi found in the small Sant'Angelo Vecchio necropolis are made of a calcarenite likely imported from some quarry locality 20–30 km up north in the Matera region—where, in fact, the Gravina calcarenite formation is extensively exposed. The dry stone wall and the paved steps unearthed in the lower part of the archaeological site, near the edge of the forested scarp leading to the spring and clay deposit, are made of slabs and boulders of sandstone, which were derived from the cemented horizons of the marine terrace sand unit. Similar rock pieces can be found sparse in the fields surrounding the Sant'Angelo Vecchio archaeological site on soil-mantled slopes covering the sand unit; in any case not enough material for building houses, but sufficient for erecting dry stone walls or paving small areas.

8

Archaeobotanical Analysis

Assunta Florenzano

In the framework of the multidisciplinary analysis carried out at Sant'Angelo Vecchio, the archaeobotanical investigation contributes to a better understanding of the economy of the site and helps to reconstruct the plant landscape of the area during the Hellenistic period.

The study of pollen, or palynology, on archaeological sites has been utilized for many decades,[1] and is an important part of archaeobotany. Pollen is constantly present in the air as part of the biological aerosol, and may fly and fall down to the soil surface as "pollen rain." In the past, as in the present, pollen produced by plants mirrors the flora and vegetation of a certain region. In archaeological sites, plants are frequently collected outside and transported onto the site by humans. In this latter case, pollen may be transported with plants, and therefore its recovery can be an indication of human behaviors and activities. The large number of studies on archaeopalynology continually adds details to archaeological reconstructions, and has increasing chance of success when research is carried out within a multidisciplinary approach. Pollen data from Sant'Angelo Vecchio, together with the information presented by the NPPs (Non-Pollen Palynomorphs), provide detailed descriptions of the plant landscape and the related cultural context of the site.

Material and Methods

Sampling

In November 2010, a total of 10 pollen sequences were taken from stratigraphic sequences of the site from trenches within the squares G6 (Sequences 1, 2, and 3), G7 (Sequences 4 and 5), and F7 (Sequence 6) in the area of the rural necropolis (*Fig. 8.1*). Samples were also taken from the squares F2 (Sequence 7; kilns area), D4 (Sequences 8 and 9; Slope 3), and C4 (Sequence 10; Retaining Wall area). The samples in all sequences were collected at 5–10 cm intervals from about 10 cm to 100–115 cm in depth, taking into account both stratigraphy and archaeological phases. One pollen sample was also collected from the surface soil, intended as a control to check the composition of modern pollen rain near the site. The samples for pollen analysis were stored in a refrigerator, protected from contamination, before being submitted to laboratory treatments.

Extraction and Counts

Twenty-eight samples from four pollen sequences (2, 8, 9, and 10) and one surface soil sample were selected for pollen analysis in accordance with the archaeological contexts (*Table 8.1*). The 29 pollen samples have been treated according to the routine method in use in the laboratory at the University of Modena.[2]

About 5–6 g of dry matter for each sample was treated in 10% tetra-sodium-pyrophosphate to deflocculate the sediment matrix, 10% hydrochloric acid to eliminate carbonates, acetolysis to eliminate the organic matter different than pollen walls (made by sporopollenins), separation with heavy liquid (sodium-metatungstate hydrate) to encourage pollen flotation, 40% hydrofluoric acid to eliminate silicates, and ethanol. The residues were dried on a stove. Permanent pollen slides were mounted in glycerol jelly. *Lycopodium* spores were added to calculate concentrations, which were expressed as pollen per gram (= p/g). The method has provided sufficient amounts of pollen even when pollen concentrations were very low, and therefore it proved to be very useful for pollen extraction from archaeological samples.

Pollen and NPPs were counted in the same samples. Pollen identification was made at 400x and at 1000x magnification with the help of keys, atlases,[3] and the reference pollen collection of Modena. Pollen percentages were calculated on a Pollen Sum, including all pollen grains except those forming a cluster constituted by a number of >10 pollen grains, which were considered as a unit.

[1] Dimbleby 1985; Hevly 1981; Faegri/Kaland/Krzywinski 1989.

[2] Florenzano et al. 2012. The method was imported and adapted from the Institute of Earth Sciences, Vrije Universiteit Amsterdam in the 1990s.

[3] Moore/Webb/Collins 1991; Reille 1999.

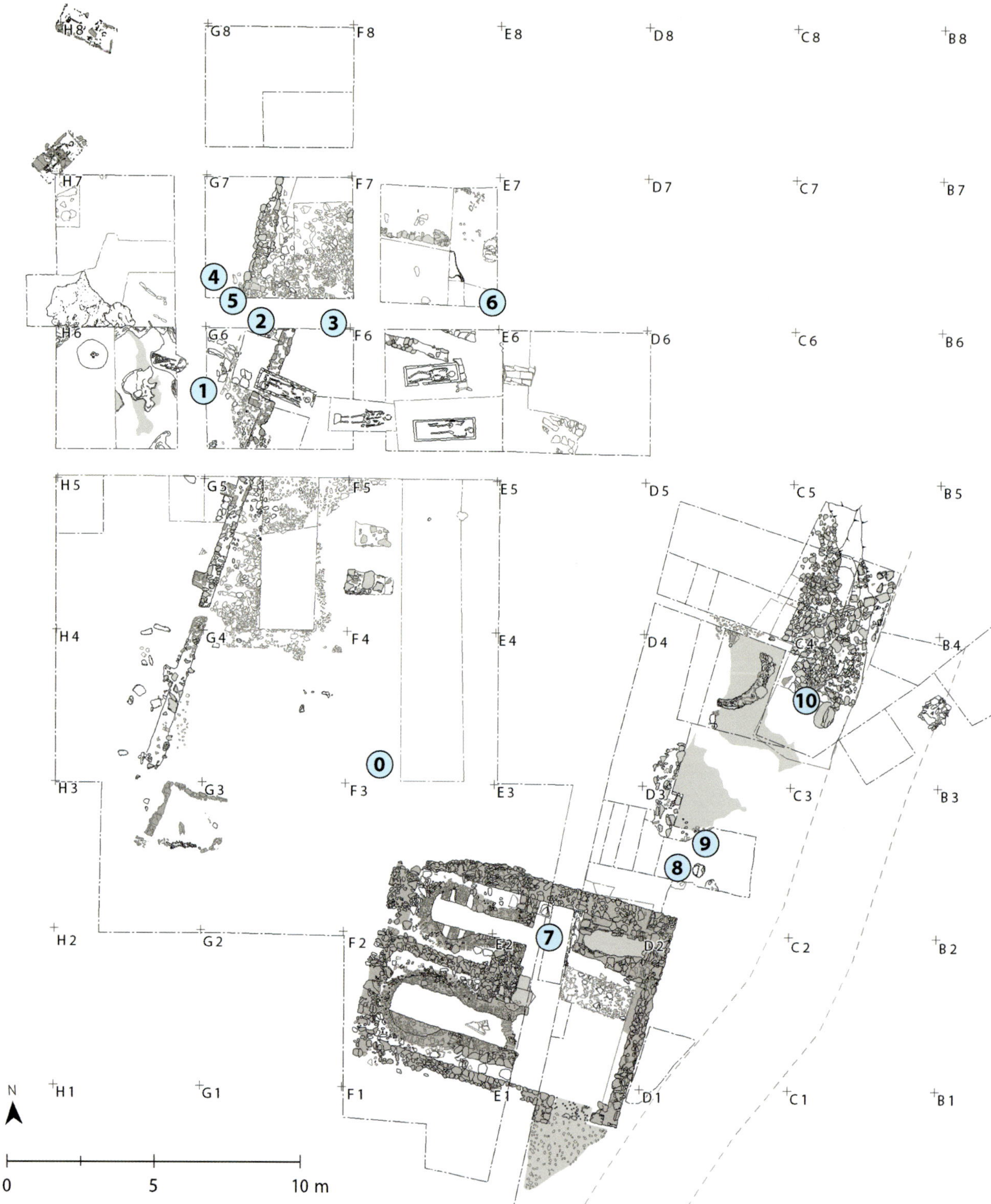

Figure 8.1 Plan of the site of Sant'Angelo Vecchio showing the location of the pollen sampling from the stratigraphic sequences of different squares of the excavation. (JT/ICA)

<table>
<tr><th>Area—Square</th><th>Pollen Sequence</th><th>Sample</th><th>Depth (cm)</th><th>Chronology</th><th>Description</th></tr>
<tr><td></td><td></td><td>P0</td><td>0</td><td>Recent</td><td>Surface soil</td></tr>
<tr><td rowspan="12">Burial ground—G6</td><td rowspan="12">2</td><td>P1</td><td>12</td><td>Post-1st c. BC</td><td>Abandonment level</td></tr>
<tr><td>P2</td><td>18</td><td>3rd c. BC</td><td>Concotto level</td></tr>
<tr><td>P3</td><td>27</td><td>3rd c. BC</td><td>Combust soil</td></tr>
<tr><td>P4</td><td>34</td><td>3rd c. BC</td><td rowspan="3">Tile concentration</td></tr>
<tr><td>P5</td><td>40</td><td>3rd c. BC</td></tr>
<tr><td>P6</td><td>50</td><td>3rd c. BC</td></tr>
<tr><td>P7</td><td>58</td><td rowspan="3">Second half of the 3rd c. BC</td><td rowspan="3">Soil covering the collapse layer 2</td></tr>
<tr><td>P8</td><td>68</td></tr>
<tr><td>P9</td><td>76</td></tr>
<tr><td>P10</td><td>83</td><td>4th–3rd c. BC</td><td>Collapse layer 2</td></tr>
<tr><td>P11</td><td>90</td><td rowspan="2">5th–4th c. BC</td><td rowspan="2">Stratum covering the graves of the 5th c. BC</td></tr>
<tr><td>P12</td><td>100</td></tr>
<tr><td rowspan="6">Retaining Wall—C4</td><td rowspan="6">10</td><td>P13</td><td>8</td><td rowspan="6">6th–5th c. BC</td><td rowspan="6">Phase of frequentation of the sacred area and the necropolis</td></tr>
<tr><td>P14</td><td>15</td></tr>
<tr><td>P15</td><td>20</td></tr>
<tr><td>P16</td><td>25</td></tr>
<tr><td>P17</td><td>35</td></tr>
<tr><td>P18</td><td>42</td></tr>
<tr><td rowspan="10">Slope 3—D4</td><td rowspan="5">8</td><td>P19</td><td>30</td><td rowspan="3">4th–3rd c. BC / 2nd–1st c. BC</td><td rowspan="3"></td></tr>
<tr><td>P20</td><td>50</td></tr>
<tr><td>P21</td><td>70</td></tr>
<tr><td>P22</td><td>90</td><td rowspan="2">End of the 4th–mid-3rd c. BC</td><td rowspan="2"></td></tr>
<tr><td>P23</td><td>115</td></tr>
<tr><td rowspan="5">9</td><td>P24</td><td>30</td><td rowspan="3">4th–3rd c. BC / 2nd–1st c. BC</td><td rowspan="3"></td></tr>
<tr><td>P25</td><td>50</td></tr>
<tr><td>P26</td><td>70</td></tr>
<tr><td>P27</td><td>90</td><td rowspan="2">End of the 4th–mid-3rd c. BC</td><td rowspan="2"></td></tr>
<tr><td>P28</td><td>115</td></tr>
</table>

Table 8.1 List of pollen samples collected from trenches and from recent surface sediments of the site of Sant'Angelo Vecchio.

The sums of Local Pastoral Pollen Indicators (LPPI), Regional Human Activities Pollen Indicators (RHAPI), and wild Anthropogenic Indicators (wAI) were calculated in all samples.[4] Non-Pollen Palynomorphs (NPPs) were identified according to van Geel.[5] NPP values are expressed in concentration per gram of dry sediment (npp/g).

Results

A total of 11,489 pollen grains (ca. 400 p/sample) were counted (*Appendix C, Table C1* and *Table C2*). Pollen was found in all archaeological samples in fairly variable concentrations, ranging from 1537 p/g (P22) to 18,914 p/g (P1). In general, the pollen sequences show quite a good presence of pollen (ca. 5500 p/g, on average), suggesting some organic matter in the layers along with good conditions of preservation. Interestingly, clumps

[4] Mazier 2007.

[5] van Geel 1978, 2001.

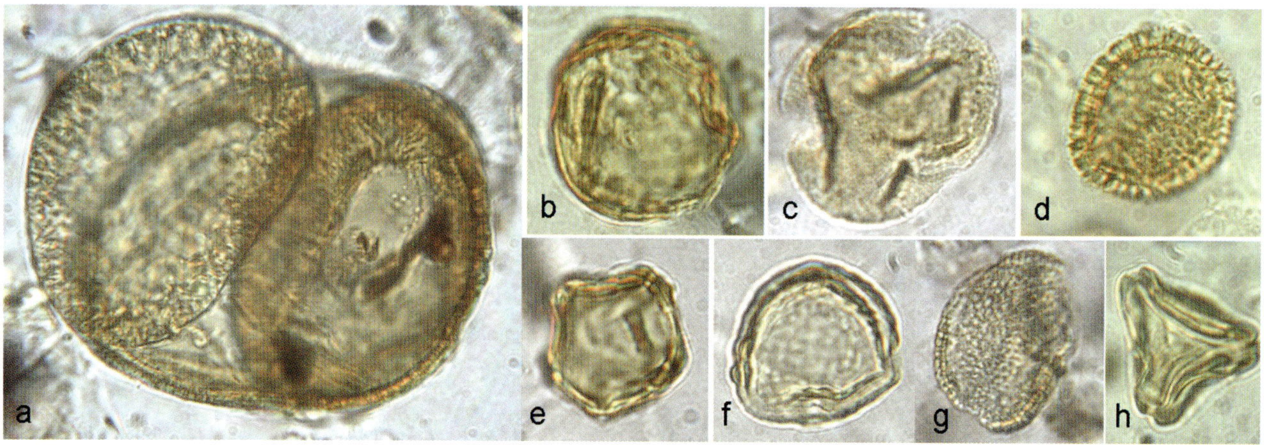

Figure 8.2 Pollen from Sant'Angelo Vecchio: main woody taxa. (a) *Pinus* (diam. max.: 110 µm); (b) deciduous *Quercus* (33 µm); (c) *Quercus cerris* (40 µm); (d) *Daphne* cf. *gnidium* (29 µm); (e) *Alnus* (21 µm); (f) *Ulmus* (28 µm); (g) *Fraxinus* (45 µm); (h) *Myrtus* (24 µm). (AF/ICA)

of pollen are observed in almost all samples; they are both single-type clumps (remains of anther fragments) and mixed-type clumps (including more than one pollen type, compatible with fecal pellets).

The number of taxa is high, consisting of 152 different types, and it varies from 22 (P21) to 69 (P12) taxa per sample. Trees, shrubs, and lianas are represented by 35 taxa, almost one third of those of herbaceous plants (117 taxa). The ratio of the percentages of the Arboreal, or woody pollen (AP) to the Non Arboreal, or prevalently herbaceous pollen (NAP) totaled 12/88 as an average of the spectra. The maximum value of the AP/NAP ratio is 19/81 (mean of the Sequence 10), while the lowest one is 9/91 (Sequences 8 and 9). In pollen spectra, the dominant woody pollen grains (*Fig. 8.2*) are the trees oak *Quercus* (mean percentage of all samples = 2.2%), olive *Olea* (2.1%), and pine *Pinus* (1.7%), as well as the shrub flax-leaved daphne *Daphne* cf. *gnidium* (Thymeleaceae; 2.4%). Cichorieae (21%) and Poaceae wild grass group (14%) prevail among the herbaceous pollen (*Fig. 8.3*), and Asteraceae (10.5%), Brassicaceae (5.4%), Fabaceae (4.8%), Urticaceae (4.7%), Cyperaceae (4.6%), and Scrophulariaceae (2.9%) are also well represented.

Among cultivated plants, cereals are present in all samples (*Avena/Triticum* group; 0.7%). Ubiquitous also is *Olea*, a fruit tree widespread at Sant'Angelo Vecchio in the past (2.1%), but also during the present day (5.1%). Other "economic plants" (walnut *Juglans*, grape *Vitis*, and hemp *Cannabis*) are only sporadic in the spectra (0.3%—P16, P17, 0.8%—P5, 0.2%—P8, P9, respectively; *Fig. 8.4* and *Appendix C, Table C3* and *Table C4*).

Non-Pollen Palynomorphs (NPPs) were observed in all samples: 34 morphotypes of spores of fungi, algae elements, and parasite eggs were identified (*Fig. 8.5*; *Appendix C, Table C5* and *Table C6*). Among the 24 observed fungal spores, the mycorrhizal fungus *Glomus*, an index of soil erosion, is the prevalent type along with coprophilous fungi *Sordaria* type and *Sporormiella* type, which are indices of the presence of the dung of grazing animals.[6] The most frequent microfossil algal is HdV-181,[7] a recurrent finding in stagnant shallow open water. An interesting record found in only one sample per sequence (P4, P18, P21) is the parasite egg of *Dicrocoelium*, a parasite of ruminants.

Detailed results for each pollen sequence are given below. In particular, analytical data referring to concentration, state of preservation, pollen richness, AP/NAP ratio, and flora composition with specific indication on cultivated and grazing areas are reported, along with the record of main identified NPPs. Pollen Sequences 8 and 9 are discussed together because they were collected from the same levels of two facing sections obtained after removing part of Slope 3. The same aspects are discussed for the surface soil sample.

Discussion

Features of the pollen samples. The pollen record obtained from the site of Sant'Angelo Vecchio has some peculiarities that have resulted in excellent find-

[6] O. Davis 1987; van Geel et al. 2003.

[7] NPPs are often identified by the "HdV type," that indicates the Hugo de Vries Laboratory (Department of Palynology and Palaeoecology, University of Amsterdam), in which the palynomorph was observed and defined for the first time. For the ecological meaning of HdV-181, see van Geel/Hallewas/Pals 1983.

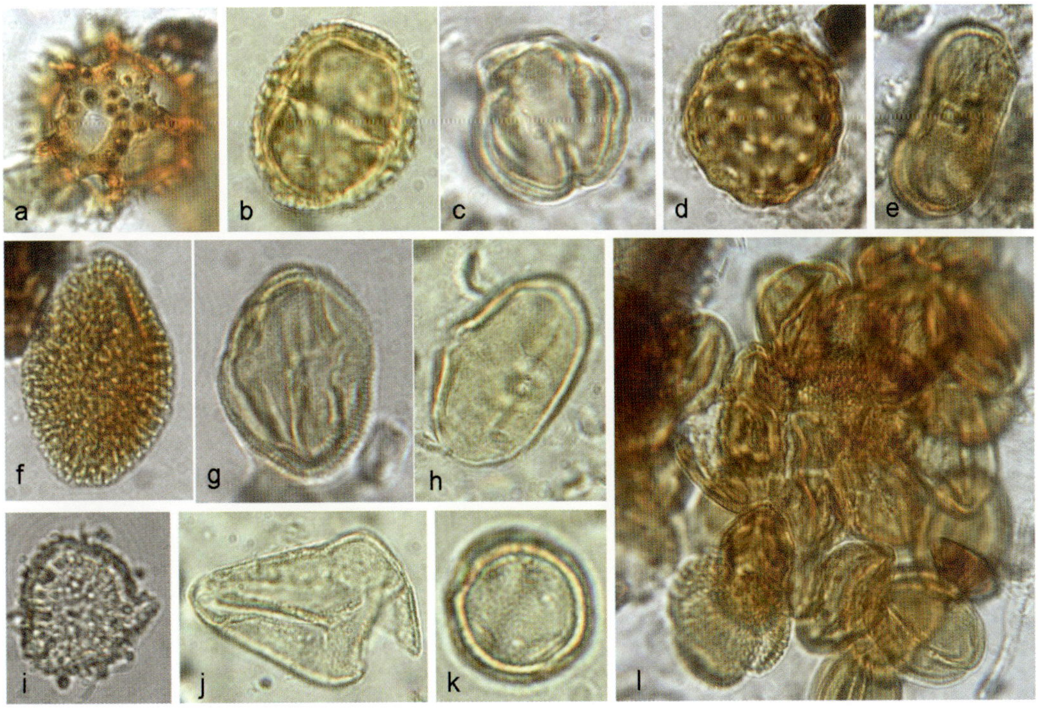

Figure 8.3 Pollen from Sant'Angelo Vecchio: main herbaceous taxa. (a) Cichorieae (diam. max.: 35 μm); (b) *Centaurea nigra* type (29 μm); (c) *Artemisia* (24 μm); (d) *Chenopodium* type (28 μm); (e) Apiaceae (38 μm); (f) *Brassica* type (35 μm); (g) Scrophulariaceae (28 μm); (h) Fabaceae (34 μm); (i) *Nymphaea* (26 μm); (j) Cyperaceae (47 μm); (k) *Urtica membranacea* (15 μm); (l) mixed-type pollen clump. (AF/ICA)

ings for palaeoenvironmental reconstructions. First, pollen showed a state of preservation that was better than is normally observed in archaeological layers. The chance of pollen recovery is, in fact, fairly difficult in sediments that were upset by trampling or similar human disturbance, and in calcareous sediments that are common in contexts from southern Italy.[8] At this site, however, a good quantity of pollen was found, and it was identifiable in all samples. Pollen clumps were also often found, and their typologies may easily be associated to the presence of fecal remains or coprolites in the layers. This is confirmed by coprophilous fungi, which were common and most abundant.

Second, the list of flora was quite long (152 taxa), the longest that was observed in sites from the Basilicata region.[9] A comparable pollen richness was only observed at Fattoria Fabrizio (117 taxa),[10] which is also located in the Metapontine area and fairly similar for chronology. A notable richness was also observed for NPPs (34 taxa). The list of fungal remains was particularly long (24 taxa), followed by that of algae.

[8] Dimbleby 1985; Accorsi/Bandini Mazzanti/Mercuri 1995.

[9] Florenzano et al. 2009; Mercuri et al. 2010.

[10] Florenzano in *Fattoria Fabrizio*.

Pollen assemblage and palaeoenvironmental reconstruction. The pollen spectra from the series show some signs of natural environments and considerable evidence of the presence of houses/walls, cultivation, and breeding/pastoral activities near the site.

Natural vegetation. Woody plants were only minimally represented in these spectra (AP/NAP=12/88), and the tree cover especially was very low. This suggests that in the site and its surroundings there was an open, herbaceous plant–dominated environment. This accords both with the assumption that forest cover must be lower in the settlement/site rather than in the surroundings, and also that the area was largely treeless in the past. The comparison with the modern surface samples shows that it was even more treeless in the 6th–1st centuries BC than in current times.

Mesophilous woods with oaks and isolated pines were distributed at a distance from Sant'Angelo Vecchio. The striking trait of the vegetation cover corresponded with a Mediterranean shrubland, with *Quercus ilex* type, *Juniperus* type, *Helianthemum*, *Myrtus*, *Pistacia cf. lentiscus*, *Cistus*, *Phillyrea*, *Erica*, *Pinus* cf. *halepensis*, and *Rhamnus*. This was, however, mostly a

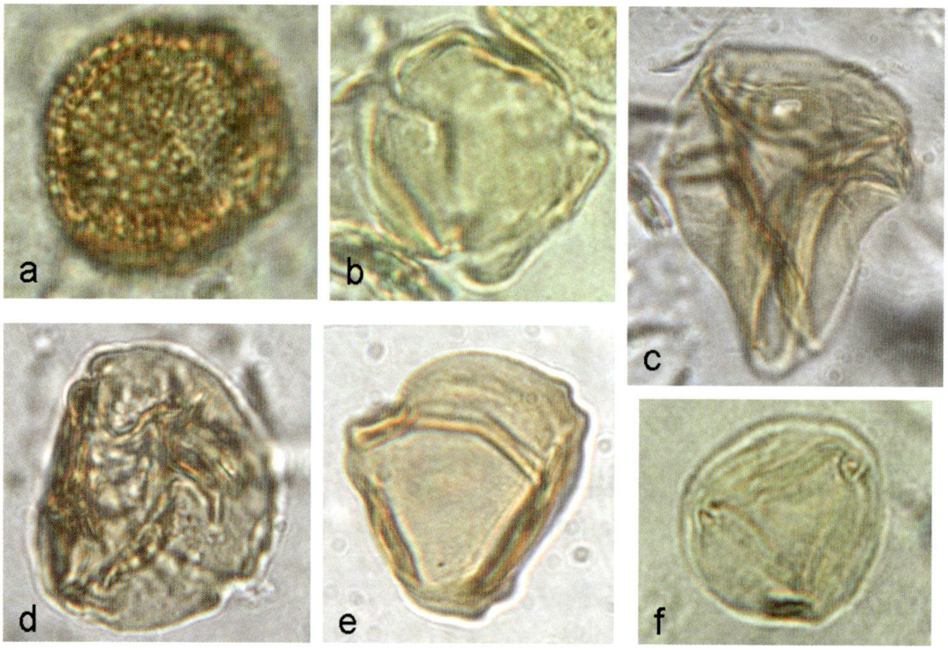

Figure 8.4 Pollen from Sant'Angelo Vecchio: cultivated/cultivable plants and weeds. (a) *Olea* (diam. max.: 20 µm); (b) *Vitis* (22 µm); (c) *Avena/Triticum* group (53 µm); (d) very crumpled *Juglans* (38 µm); (e) *Corylus* (30 µm); (f) *Cannabis* (20 µm). (AF/ICA)

degraded environment rather than natural vegetation. In fact, *Olea* (2.1%) may have been cultivated, and the shrub *Daphne* cf. *gnidium* (Thymeleaceae; 2.4%) may have developed under continued grazing pressure. The "wild" vegetation, therefore, may also have been the result of several human activities (see below).

Wet environments were present and fairly extended or diversified including different small habitats. In fact, there were low amounts of aquatic and hygrophilous plants that match the presence of the alga HdV-181. This evidence suggests that there were some circumscribed springs or stagnant shallow water not far from the site.

Crops and fields. Cultivated plants are represented by *Olea* and traces of *Juglans* and *Vitis* among woody plants. Their total amount is so low that we must hypothesize that they were grown fairly far from the site. Olive groves, for example, are found to produce pollen in fairly high quantities, up to about 60%, but the pollen rapidly falls down within a short range. *Olea* pollen decreases at approximately 87% in the first few hundred meters from the sources, passing from about 58% in the olive grove to 8% at 500 m.[11]

Cereals are mainly represented by the oat/wheat type, suggesting that wheats rather than other cereals were mainly cultivated in the fields. Weeds, such as scarlet pimpernel *Anagallis*, wormwood *Artemisia*, hemp *Cannabis, Convolvulaceae,* bedstraw *Galium*, Papaveraceae, knotgrass *Polygonum*, and sorrel *Rumex*, are also an indication of crop practices. Overall, the low amount of pollen from fields suggests that these cultivations were not carried out near the site, and that cereals were not transported into the site for plant processing.

Grazing and pasturelands. In all samples, plants and other evidence from grazing environments are particularly evident. The general composition of spectra is indicative of dry pastures: there is an abundance of Cichorieae, together with Poaceae wild grass group, Asteraceae, Chenopodiaceae, and Brassicaceae. The sum of LPPI has higher percentages in the archaeological samples (32%) than in the modern surface soil sample (20%), suggesting that there were more breeding activities in the past than today.

The coprophilous fungi *Sordaria* type and *Sporormiella* type are indices of the presence of dung, and therefore associated with pastoral/breeding activities. The parasite egg of *Dicrocoelium* is also indicative of the presence of herbivores at the site. Some soil ero-

[11] Florenzano/Benassi/Mercuri 2012.

Figure 8.5 Main NPPs found in the pollen samples from Sant'Angelo Vecchio. (a) *Glomus* (diam. max. chlamydospores 50 μm, and micelium); (b) *Sordaria* type (27 μm); (c) *Sporormiella* type (14 μm); (d) *Delitschia* type (32 μm); (e) *Podospora* type (45 μm); (f) HdV-20 (48 μm); (g) *Chaetomium* (14 μm); (h) *Coniochaeta* (16 μm); (i) *Valsaria variospora* type (65 μm); (j) HdV-181 (21 μm); (k) HdV-221 (37 μm); (l) insect head with mandible (58 μm); (m) *Dicrocoelium* (30 μm). HdV = Hugo de Vries Laboratory, University of Amsterdam, The Netherlands. (AF/ICA)

sion, possibly linked to human disturbance, is evident from the mycorrhizal fungus *Glomus*.

Pollen Sequence 2 (12 samples: 5th/4th–3rd century BC). The sequence shows good content (ca. 5,000 p/g) and richness (133 taxa) of pollen and NPPs (33 taxa). Grazing plants are the best represented in spectra, while olive trees are the only (possibly) cultivated trees found in a significant amount. Crop cultivation is represented by only a few records, mainly of wheat, testifying that they were present but not very near to the site. Interestingly, *Vitis* was found in traces (0.8% in P5) in a layer in square G6. Though this is thin evidence, the absence or low amount of *Vitis* pollen in archaeological sites is not surprising because the plant is a low pollen production species. For this reason, we interpret the recovery of this pollen as strong evidence of local presence of grapevine in this location within the site.

Regarding pollen spectra, plants from pastures include edible herbs (Fabaceae, Poaceae, Brassicaceae) and inedible/toxic herbs (Cichorieae and Chenopodiaceae). Coprophilous fungi (*Sordaria* type, *Podospora* type, *Sporormiella* type, *Cercophora* and *Delitschia* type) confirmed the presence of herbivores that dropped fecal remains. Moreover, the parasite *Dicrocoelium* is indicative of the presence of infected ruminants, such as cows and sheep.

This sequence suggests that the economy of the site was mainly based on pastoral/breeding activities, rather than on cultivation. In general, the site was already largely influenced by human presence as the pollen from nitrophilous (Urticaceae) or trampled environments (Plantaginaceae) made clear. These two are typical wild anthropogenic pollens that are known to spread according to the frequentation of sites by animals and humans.[12]

Oak woodlands and some isolated pine specimens are the main elements of the natural vegetation situated at a significant distance from the site. In the vicinity of the site, however, the Mediterranean evergreen woods and maquis are represented by *Quercus ilex* type, *Pistacia* cf. *lentiscus*, *Phillyrea*, and probably some wild olive. The shrubs would have been a prevalent element of the

[12] Behre 1981; Mercuri et al. 2012.

vegetation cover near the site, including *Helianthemum*, and especially *Daphne* cf. *gnidium* and *Juniperus* type that developed after continuous grazing pressure. The high frequency of *Glomus* is evidence of soil disturbance.

Rivers or permanent wet environments were not in the vicinity of this site. They likely would have been bordered by trees and shrubs growing at the margins, while aquatics lived on the water surface. There were, however, eutrophic stagnant ponds where the fungus *Valsaria variospora* type and the Zygnemataceae algae lived. Probably some springs, hosting the sedges and cattails, were present.

Pollen Sequence 10 (6 samples: end of the 6th–5th century BC). The sequence has a content of pollen that was higher, nearly double that of Sequence 2 (ca. 9,000 p/g), but the richness of pollen (105 taxa) and NPPs (25 taxa) was lower. This suggests that the organic matter would have been more present in this sequence than in the previous one. The earlier age of this series may have corresponded to a different depositional consistency that formed before those of the other series, and near a sacred area, possibly with less trampling *in loco*.

The main characteristics of the plant landscape are similar to those described above for Sequence 2, describing a prevalently human environment with pastures and restricted cultivation areas. The presence of herbivores is again confirmed by the high concentration and diversity of coprophilous fungi among which *Delitschia* type is prevalent, as well as the parasite *Dicrocoelium*. Pollen evidence of fields is less than in the previous sequence. Some features, however, may be evidence of very local differences in the distribution of vegetation in the vicinity of the different points of sampling.

In this sequence, the Mediterranean shrubland is better represented by *Daphne* cf. *gnidium* and *Juniperus* type (7% vs. 1% of Sequence 2). These two plants are evidence of growing shrubs as a vegetation response to grazing. However, the other shrubs are lower than in series 2 (0.4% vs. 0.9%), though they include the same diversified list as *Pistacia* cf. *lentiscus*, *Helianthemum*, *Erica*, *Myrtus*, *Phillyrea*, and *Rhamnus*. Thus, we can hypothesize that this vegetation was mainly locally developed near the point of sampling of Sequence 10.

Moreover, wet environments should have been more developed in this sector as suggested by the higher percentage of their pollen sum (8% vs. 6%) and the lesser diversity of plants of wet environments found in these samples. In fact, only *Potamogeton* is exclusive of this series with respect to series 2, while many other taxa disappeared in this series (*Nuphar*, *Ranunculus* cf. *fluitans*, *Sagittaria*, and *Ulmus*). Contrarily, the diversity of algae looks larger than that in Pollen Sequence 2.

Evidence for the walnut tree is of interest: the pollen found was so scanty that its cultivation cannot be hypothesized, but this important tree would have been among the plants providing edible fruits in the area.

The traces of cereals are the lowest on the site (0.4%), and though they are not so evident in all the site, it is possible that cereal fields were nearer to the other places, also belonging to more recent chronologies where their values are a bit higher (1.3% in Pollen Sequence 2; 1.0% in Sequences 8 and 9). Actually, the presence of fields would have been farther from the site in this earlier phase, when it is documented by archaeological data that this area was devoted to sacred practices and a necropolis.

Pollen Sequences 8 and 9 (10 samples: end of the 4th/3rd–2nd/1st century BC). These sequences had the lowest concentrations observed in the samples (ca. 4,000 p/g) that parallel the shortest list of flora at the site (92 taxa). The diversity of NPPs, however, is comparable to previous ones (29 taxa). Woodlands were even more thinned from these spectra (9%), while wet environments (7%) are comparable to the previous ones. The plants floating on the water surface are less represented than in Pollen Sequence 10. Mediterranean shrubland with *Daphne* cf. *gnidium* and *Juniperus* type were comparable to the mean values among those of the previous sequences (2.5%).

Traces of cereal fields are similar to those observed in Sequence 2 (0.7% vs. 0.9%). All the plants produced by human environments, including pastures, are well represented as previously described in the other sequences. Pastures are also documented by the *Dicrocoelium* parasite. Coprophilous fungi, however, among which the *Sordaria* type is prevalent, are less diversified and less numerous with respect to other sequences.

Surface Soil Sample P0. The pollen concentration of the modern sample is four times more (ca. 20,000 p/g) than that found in archaeological layers. This corresponds to the current depositional situation indicated by soils less (or not) disturbed by trampling and intense frequentation as in the past. The floral richness, however, is decidedly lower (67 vs. 110). This may be an effect of minor land use or a less diversified land use that occurs today in the area as compared to the past.

Interestingly, the forest cover is to some extent higher today than previously (21% vs. 12%), and pines have a major role in the forest composition. Mediterranean maquis is near the site, as confirmed by direct observation, and the representation of *Olea*, largely coming from olive groves, is 5.1%.

Wet environments (6.2%) show fairly similar percentages but the amount of telmatophytes (wet-meadow plants) is more significant that previously. The data agree with Frattegiani, who reports that common marsh reed (*Phragmites australis*) is the prevalent hygrophilous herb in damp environments, and that there are not habitats with stagnant water.[13] Today, a small river and some springs are present around 50–100 m from the site. This suggests that currently the wet areas are less extended and less permanent than in ancient times.

The cereal fields, which were grown at about 0.5 km from the site at the time of archaeological excavation, have values comparable to those obtained from the archaeological samples.

Conclusions

The current landscape, and especially the plant-agrarian landscape, is characterized by intense agricultural activity, including sowable (mostly cereals) and orchard crops (including olive groves and vineyards).[14] The archaeopalynological data suggest that these kinds of land use had begun as long ago as the prehistoric period at Sant'Angelo Vecchio.

The palynological research was carried out on a high number of samples taken from layers that were deposited between the 6th and the 1st centuries BC. Analyses have shown an intense use of the territory surrounding the site, especially for pastoral or breeding activities. Cultivations of cereal fields would have been made fairly far from the site, while woody plants were only sparingly cultivated. Among them, only some unextended olive groves were present at that time, while vineyards were possibly cultivated near some houses.

The most striking feature of the past economy of the site—pastoral practices—has already transformed the landscape into a shrubland where Mediterranean shrubs were well developed after continuous grazing action by herbivores.

The palaeoenvironmental reconstruction leads to three main conclusions. The present environment is not as wet as it has been in the past, probably as an effect of the increasing dryness of climate and of the reclamation practices that have occurred since the latter half of the 20th century in the Metapontine plain. Additionally, from its earliest history, the territory of the Metapontino has been a human environment, deeply transformed by human presence and activities, with open spaces and low forested areas distributed far from the site. Finally, the exploitation of the territory was more diversified in the past than at present, where the plant diversity seems to have been reduced by a less intense and less varied typology of exploitation. In the past, the more frequent human presence allowed the landscape to be a mosaic of habitats, as fragmentation is among the effects of human action.

To sum up, this territory was richer in water and species in the past. It has been more intensively exploited in the past than in the present, and it is possible that continuous human actions have locally diminished both soils and natural resources at the site. The archaeopalynological research, therefore, has permitted scholars to discover features of past exploitation, and has allowed us to conclude that the current landscape may be truly considered the result of the more intense exploitation that occurred in the past.

[13] See p. 181 in Ch. 10, "Description of the Current Environment."

[14] See Ch. 10, "Observations on the Vegetation and Landscape Dynamics."

Pollen Sequences

Pollen Sequence 2 (*Appendix C, Table C1*)

Concentration: 5,300 p/g on average, ranging from ca. 2 to 19 x 10^3 p/g (P9 and P1, respectively).

State of preservation: Pollen grains are generally found in a fairly good state of preservation. Some pollen grains were observed crumpled or with folded exines due to post-depositional disturbances. In each sample, some small pollen clumps (< 10 pollen grains) have been found, including one or more (3–4) pollen types. In some cases, the single-type clumps presented an incomplete sculpture of exine, and they were identified as anther fragments (included in counts as a unity). Sometimes, the mixed-type pollen of the bigger clumps (15–20 grains; not included in pollen sum) was observed broken, suggesting that it had been subjected to browsing.

Pollen richness: The number of taxa is high (133), ranging from 46 (P10) to 69 (P12) taxa per sample. The woody taxa number 29, almost one third that of herbaceous plants (104).

AP/NAP: The mean percentage ratio of woody/herbaceous plants is 11/89 as an average of the spectra, with great uniformity in the samples. This suggests a very low forest cover of the site and its surroundings.

Flora composition: The main elements of woodlands are deciduous *Quercus* (1.9%) and *Pinus* (1.3%). Mediterranean evergreen woods and maquis (mean: 5.5%; max.: 15.1%—P2; min.: 0.7%—P4) are represented by *Olea* (3.3%), *Daphne* cf. *gnidium* (0.8%), *Phillyrea*, and *Helianthemum* (0.3% each), together with holm oak *Q. ilex* type, juniper *Juniperus* type and mastic tree *Pistacia* cf. *lentiscus* (0.2% each). Wet environments are represented by the hygrophilous trees poplar *Populus* and willow *Salix* (0.2% each), alder *Alnus* (0.1%) and elm *Ulmus* (0.1%), and by hydrophytes Cyperaceae (4.1%), Nymphaeaceae (0.8%), and Typhaceae (0.3%). In the spectra, herbaceous taxa prevailed, with Asteraceae (34.4%, mainly Cichorieae at 22.3%), and Poaceae (16.2%, mainly wild grass group at 15%). In all samples, they were found with Brassicaceae (5.7%), Fabaceae (5.5%), Scrophulariaceae (3.5%), Ranunculaceae (2.7%), Urticaceae (2%), Plantaginaceae (1.9%), and Chenopodiaceae (1%).

Cultivated and grazing areas (Appendix C, Table C3): The cultivated woody plants are represented by *Olea* (3.3%; max.: 13.1%—P2; min.: 0.4%—P8) and *Vitis* (0.8%—P5). *Cannabis* is present in traces (0.2%—P8, P9). Cereals are generally few, and include oat/wheat (*Avena/Triticum*, 0.9%), found in all samples, and barley (*Hordeum* group, 0.4%—P4). Plants from grazing environments, in the LPPI sum, are well represented. Among them, the plants strictly correlated to local pastoral activities (Asteraceae including Cichorieae, *Galium* type, cinquefoil *Potentilla* type, Ranunculaceae) were found in all samples in good quantities (LPPI = 37.2% on average; max.: 51.5%—P8; min.: 15.3%—P12). Pollen from plants correlated to various human activities at the regional scale is added up to the RHAPI sum. It includes wormwood *Artemisia*, Chenopodiaceae, cereals, ribwort plantain *Plantago lanceolata* type, sorrel *Rumex*, and Urticaceae that were ubiquitous and fairly well represented (RHAPI = 6.4% on average; max.: 8.5%—P12, min.: 3.2%—P2). The category of wAI groups the wild plants correlated to human presence. The sum includes several taxa (and excludes Cichorieae) in different quantities for each sample (mean: 5.7%; max.: 7.6%—P6; min.: 2.4%—P2).

Non-pollen palynomorphs (Appendix C, Table C5): The NPPs were prevalently algae and spores of fungi, and an average of 17 types were identified. Fungal spores (24 morphotypes plus some undifferentiated types; 1656 npp/g on average) belong to several groups of fungi, among which *Glomus* is prevalent (580 npp/g; max.: 1293 npp/g—P1; min.: 154 npp/g—P7). Coprophilous fungi are also quite well represented (493 npp/g); in all samples *Sordaria* type (350 npp/g), *Podospora* type (81 npp/g), *Sporormiella* type (49 npp/g), and traces of *Cercophora* and *Delitschia* type (8 and 5 npp/g, respectively) were found. Other interesting fungal spores observed were ascospores of *Chaetomium* (cellulose-decomposing fungi occurring on, for example, plant remains, fiber, wallpaper, and dung; 110 npp/g) and *Valsaria variospora* type (occurring in deposits formed under eutrophic wet conditions; 45 npp/g). Spores and algal elements (8 morphotypes; 288 npp/g on average) are recurrent findings. The most frequent algal trace is HdV-181 (229 npp/g), which occurs in all samples in association with Zygnemataceae (40 npp/g), suggesting the presence of stagnant water. Small amounts of zoomicroremains were found (e.g., mandibles and hair of insects, and other invertebrate fragments). The most interesting record was the parasite egg belonging to *Dicrocoelium* (P4), which is produced by a tapeworm that usually infects ruminants, such as cows and sheep (Le Bailly and Bouchet 2010; Florenzano et al. 2012).

Pollen Sequence 10 (*Appendix C, Table C2*)

Concentration: 8,990 p/g on average, ranging from ca. 4.7 to 14 x10^3 p/g (P18 and P14, respectively).

State of preservation: Pollen grains were recorded with different degrees of preservation, from good to bad, in the same archaeological sample. Small amounts of little pollen clumps (< 10 pollen grains), both single- and mixed-type, were found.

Pollen richness: The number of taxa is high, consisting of 105 different types, ranging from 48 (P14) to 64 (P13) taxa per sample. The woody plants are represented by 28 taxa, less than one-third of those of herbaceous plants (77 taxa).

AP/NAP: The mean percentage ratio of woody/herbaceous plants is 19/81 as average of the spectra, indicating an open landscape around the site.

Flora composition: The best represented woody plants are the Mediterranean shrub daphne *Daphne* cf. *gnidium* (5.8%), and the trees pine *Pinus* (3.8%), ash *Fraxinus* (1.9%), and deciduous oak *Quercus* (1.7%). Mediterranean plants are recurring in all samples (mean: 9%; max.: 15.5%—P17; min.: 3.9%—P14). Besides the overrepresented *Daphne* (max.: 12.8%—P17; min.: 2.3%—P14), there are *Olea* (1.2%) and *Juniperus* type (1%), while *Cistus*, *Helianthemum*, *Myrtus*, *Phillyrea*, *Pistacia* cf. *lentiscus*, and *Q. ilex* type are present in traces (their sum totaled < 1%). All samples showed good evidence of wet environments (mean: 8.3%; max.: 10.2%—P18; min.: 6.5%—P13), with the hygrophilous trees *Populus* (0.4%), *Alnus* (0.3%), and *Salix* (0.3%), and hydrophytes Cyperaceae (5.4%), Typhaceae (0.7%), common marsh reed *Phragmites australis* cf. (0.5%), water lily *Nymphaea* (0.4%), and traces of grass rush *Butomus*, common duckweed *Lemna*, and pondweed *Potamogeton* (0.1%, for each of them). The spectra are dominated by herbs, with Poaceae (15.7%, mainly wild grass group at 14.8%) and Asteraceae (13.8%, mainly Cichorieae at 7.6%). Other recurring herbaceous plants are Urticaceae (10.5%; nettle *Urtica membranacea*: 9.2%, with the highest value 31.7% in P14), Brassicaceae (8.3%), Fabaceae (4.9%), Scrophulariaceae (3.3%), Ranunculaceae (2.8%), Chenopodiaceae (2.1%), and Plantaginaceae (1.3%).

Cultivated and grazing areas (Appendix C, Table C4): There are *Juglans* (0.1), found only in this sequence, and *Olea* (1.2%; max.: 3.1%—P13; min.: 0.3%—P18). In almost all samples cereals are recorded, but always in very low amounts (*Avena/Triticum*: 0.4%). In contrast with the other sequences, taxa from grazing environments are low, and the percentage of plants strictly correlated to local pastoral activities is almost half of those found in the other series (LPPI = 16.8% on average; max.: 20.1%—P16; min.: 13.3%—P14). Pollen from plants correlated to different human activities at the regional scale is fairly well represented (RHAPI = 16% on average; max.: 37.2%—P14; min.: 8.1%—P18), as well as pollen from wild plants correlated to human presence (wAI = 14.3%; max.: 36.6%—P14; min.: 7.1%—P18).

Non-pollen palynomorphs (Appendix C, Table C6): The number of identified NPPs types per sample is 15 on average. Among fungal elements (18 morphotypes plus some undifferentiated types; 3104 npp/g on average). The mycorrhizal fungus *Glomus* is the most frequent type (902 npp/g; max.: 1695 npp/g—P14; min.: 418 npp/g—P18). Coprophilous fungi (1145 npp/g) include spores of *Delitschia* type (734 npp/g), *Sordaria* type (318 npp/g), *Sporormiella* type (35 npp/g), *Podospora* type (33 npp/g), and *Cercophora* type (25 npp/g). Other recurrent fungal remains were spores from decomposing fungi, such as *Pleospora* (235 npp/g), *Urocystis* (169 npp/g), *Chaetomium* (141 npp/g), and *Coniochaeta* cf. *lignaria* (61 npp/g). Spores of algae (7 morphotypes; 320 npp/g on average) are quite spread in all samples. The prevalent type is HdV-181 (174 npp/g), along with Zygnemataceae (78 npp/g) and Desmidiales (37 npp/g). Small amounts of *Concentricystes* and *Spirogyra* (10 npp/g each) were also observed. In sample P18, one parasite egg of *Dicrocoelium* was found.

Pollen Sequences 8 and 9 (*Appendix C, Table C2*)

Concentration: 3,632 p/g on average, ranging from ca. 1.5 to 12 x 10^3 p/g (P22 and P19, respectively).

State of preservation: Most pollen types were recorded with mediocre/bad state of preservation, with folding of the grains and exine thinning. Very few clumps of 4–5 pollen grains, both single- and mixed-type, and one of >10 grains of *Daphne* (P28), were observed.

Pollen richness: The number of taxa is high (92), ranging from 22 (P21) to 46 (P19) taxa per sample. The woody taxa are represented by 21 taxa, about one-third of those of herb plants (71 taxa).

AP/NAP: The ratio of woody/herbaceous plants was 9/91 as average of the spectra.

Flora composition: Only four woody plants reach mean values ≥ 1%: *Daphne* cf. *gnidium* (2.2%, on average), deciduous *Quercus* (1.8%, on average), *Olea* (1.1%), and *Pinus* (1%). Mediterranean plants (mean: 4.8%; max.: 15%—P28; min.: 1%—P25; absent in P21) are represented by the above-mentioned *Daphne* and *Olea*, along with *Myrtus* (0.5%), *Q. ilex* type (0.4%), *Juniperus* type (0.3%), and traces of *Pistacia* cf. *lentiscus* (0.1%). Plants of wet environments (mean: 6.7%; max.: 8.8%—P26; min.: 4.7%—P21, P23) are the hygrophilous trees *Populus* (0.5%), *Alnus* (0.2%), and *Salix* (0.1%), and the hydrophytes Cyperaceae (4.6%), *Phragmites australis* cf. (0.6%), Typhaceae (0.4%), *Nymphaea* (0.2%), and *Lemna* (0.1%). Asteraceae (38.9%), mainly represented by Cichorieae (27.6%), and Poaceae (14.9%), including mainly wild grass group (13.7%), prevail in the spectra. Other herbaceous plants are Urticaceae (4.4%; *Urtica membranacea*: 3.7%, with the highest value 34.1% in P23), Fabaceae (3.8%), Brassicaceae (3.3%), Ranunculaceae (3.1%), Chenopodiaceae (2.1%), Scrophulariaceae (1.9%), and Plantaginaceae (1.6%).

Cultivated and grazing areas (Appendix C, Table C4): Olea (1.1%; max.: 3.5%—P28; min.: 0.3%—P23; absent in P19, P21) is the only cultivated woody plant in the spectra. Cereals are present only in traces (*Avena/Triticum*: 0.6%; *Hordeum* group: 0.4% in P27). Plants from grazing environments are abundant. Pollen from plants strictly correlated to local pastoral activities reaches the highest values (LPPI = 42.3% on average; max.: 58.3%—P24; min.: 21.5%—P23). The RHAPI percentages totaled 9.6% on average (max.: 37.3%—P23; min.: 2.8%—P24) while the wAI shows fairly good values (mean: 7.6%; max.: 37.3%—P23; min.: 1.6%—P21).

Non-pollen palynomorphs (Appendix C, Table C6): The number of identified NPPs types per sample is 14 on average. Fungi (22 morphotypes plus some undifferentiated types; 1599 npp/g on average) are mainly represented by spores of HdV-20 (383 npp/g) and *Glomus* (317 npp/g). Spores of coprophilous fungi were found in quite low quantities (215 npp/g), with a dominance of *Sordaria* type (123 npp/g) and *Podospora* type (44 npp/g). In sample P19, very good amounts of spores of *Chaetomium* (1038 npp/g) and *Coniochaeta* cf. *lignaria* (807 npp/g) were found, while they are quite absent in the other samples. *Urocystis* (267 npp/g) recurs in almost all samples, along with HdV-18 (72 npp/g) and *Valsaria viariospora* type (62 npp/g). Spores of algae (7 morphotypes; 138 npp/g on average) are few. The best-represented type was HdV-181 (91 npp/g), in association with Zygnemataceae (29 npp/g) and small amounts of *Spirogyra* (11 npp/g). One parasite egg of *Dicrocoelium* was observed in sample P21 (series 8).

Surface Soil Samples

Surface Soil Sample P0 (*Appendix C, Table C1*)

Concentration: 19,871 p/g, decidedly higher than those observed in the archaeological samples.

State of preservation: Pollen grains are found in a good state of preservation. Few single-type pollen clumps (of 4–5 grains of *Olea* and *Aster* type) were observed.

Pollen richness: The number of taxa is fairly high (67), with a floristic diversity less varied than in the archaeological samples (110 on average). The woody plants are represented by 15 taxa, about a quarter of those of herbaceous plants (67 taxa).

AP/NAP: The mean percentage ratio of woody/herbaceous plants was 21/79, suggesting a low forest cover at Sant'Angelo Vecchio. In fact, nowadays grasslands and herbaceous cultivations cover the surrounding area and an olive grove grows just near the site.

Flora composition: Woody plants are relatively low (mean: 21%; about 4,243 p/g), with *Pinus* (6.8%), *Olea* (5.1%), *Quercus ilex* type (2.7%), deciduous *Quercus* (2.6%), and *Daphne* cf. *gnidium* (1.3%). In pollen spectra, Mediterranean taxa prevailed (10.6% on average): along with the above-mentioned olive (the site lies in a olive grove), *Q. ilex* type, and *Daphne*, Mediterranean plants are represented by *Cistus* (0.5%), *Pinus halepensis* (0.4%), *Phillyrea*, *Pistacia* cf. *lentiscus*, and buckthorn *Rhamnus* (0.2% each). Plants of wet environments are poorly represented (6.2%) by the hygrophilous trees *Alnus* (0.4%), *Salix* (0.4%), and *Populus* (0.2%), and the hydrophytes Cyperaceae (4.6%), *Phragmites australis* cf. (0.5%), and *Nymphaea* (0.2%). Pollen spectra are dominated by herbs (mean: 79%; about 15,628 p/g), with Asteraceae (15.3%; Cichorieae: 2.9%), Poaceae (12.4%; Poaceae-wild grass group: 11.1%), Fabaceae (10.2%), Brassicaceae (8.4%), Scrophulariaceae (4.6%), Ranunculaceae (4.4%), Chenopodiaceae, Plantaginaceae (2.9% each), Rosaceae (2.2%), Caryophyllaceae (1.6%), and Urticaceae (0.9%).

Cultivated and grazing areas (Appendix C, Table C3): Not surprisingly, *Olea* (5.1%) is the main cultivated plant found in the spectra from the recent sample (it was collected close to an olive grove). Cereals (*Avena/Triticum* group 0.4%; *Hordeum* group 0.3%) and a low amount of Papaveraceae (weeds of wheat fields, 0.5%) are recorded. Elements attesting grazing environments are less represented in the modern sediments than in archaeological samples. In fact, the LPPI sum totaled 19.7%, almost a half of percentages calculated in the pollen sequences. The RHAPI sum shows lower percentages in the modern surface samples (8.8%) than in the archaeological samples (10.7% on average). Moreover, the wAI percentages (P0: 9.7%) are of similar value to those observed in the samples of the past (P1–P28: 9.2%).

Non-pollen palynomorphs (Appendix C, Table C5): A list of 11 morphotypes of fungal spores and elements of algae was identified. Among fungi (7 morphotypes plus some undifferentiated types; 2357 npp/g), the most frequent type was *Glomus* (1813 npp/g). Coprophilous fungi (218 npp/g) were represented by *Sordaria* type (145 npp/g), *Podospora* type, and *Sporormiella* type (36 npp/g each). Other fungal spores found in the surface soil sample were HdV-18 (218 npp/g), *Coniochaeta* cf. *lignaria* (73 npp/g), and *Pleospora* (36 npp/g). Spores of algae (4 morphotypes; 580 npp/g) are fairly well represented. The most recurring type was HdV-181 (326 npp/g), along with Zygnemataceae (145 npp/g), HdV-221 (73 npp/g), and Desmidiales (36 npp/g).

9

Charred Plant Remains and Plant Impressions in Fired Clay Fragments from Sant'Angelo Vecchio

Lorenzo Costantini and Fabrizio Pica

Summary

Charred seeds and plant impressions in fired clay fragments from the site of Sant'Angelo Vecchio (Matera), located about 8.2 km from the asty (city) of Metaponto along the course of the Basento River, document the cultivation of cereals and legumes in a period dated between the Late Archaic and Early Imperial periods. Flotation samples provided limited evidence of emmer (*Triticum dicoccum*), free-threshing wheat (*Triticum durum/aestivum*), and six-row barley (*Hordeum vulgare*) together with bitter vetch (*Vicia ervilia*), vetch/vetchling (*Vicia/Lathyrus* sp.), garden/field pea (*Pisum* sp.), and oat (*Avena* sp.).

An analysis of plant impressions (mainly including straw, chaff, seeds, etc.) revealed the presence of emmer (*Triticum dicoccum*), free-threshing wheat (*Triticum durum/aestivum*), two-row barley (*Hordeum distichum*), and six-row barley (*Hordeum vulgare*) among the crop byproducts used for tempering clay in mudbrick production. The occasional presence of seeds of chickpea (*Cicer arietinum*), darnel (*Lolium* cf. *temulentum*), and sun spurge (*Euphorbia elioscopia*) were also detected, providing important evidence of the plant economy at the site.

Introduction

In 1979, during the excavation performed by the University of Texas at Austin at Sant'Angelo Vecchio, flotation of soil samples from several contexts (tombs, buildings connected with the ceramic workshops, kilns, etc.) was carried out. Samples of fired clay were also collected from three different areas of the site for plant impression analysis, as a complement to the archaeobotanical investigation. The initial examinations of both soil samples and fired clay fragments showed that some of them contained archaeobotanical remains in the forms of charred seeds and plant impressions.[1] The material provided a good chance to improve the limited knowledge of farming practices and the exploitation of cereal-threshing byproducts in the manufacturing of bricks. Moreover, the archaeobotanical evidence offered an opportunity to discuss the economic features of the settlement. Therefore a more detailed analysis of the samples from Sant'Angelo Vecchio has been performed.

[1] Costantini 1980.

Materials and Methods

A total of 55 soil samples, for a total volume of about 180 liters, and five samples of fired clay fragments were collected for the archaeobotanical investigation, as summarized in *Table 9.1*. Soil samples from cultural layers were processed at the site using a flotation method and sieves with a mesh size of approximately 2 mm. Smaller seeds, thus, are presumably greatly underrepresented in the samples. The floated material was sorted, cleaned, and prepared for identification at the Bioarchaeological Research Unit of the National Museum of Oriental Art (Rome).[2] Plant material was found only in seven of the 55 samples processed, and seeds and grains were analyzed using a binocular stereomicroscope at a magnification ranging from x10 to x50.

Fired clay fragments were sampled from three contexts in the area of the kilns, where remains of structural material (bricks) were accumulated as a result of the abandonment of some of these areas. During assessment, 16 small fragments were selected because the clay was tempered with plant material (chaff and straw). These fragments were grouped into five samples (*Fig. 9.1*). The fragments were characterized using a binocular microscope at a magnification of x15 to x25 for identification during recording.

Identification of plant macroremains and plant impressions was carried out by comparison with material of a reference collection and various reference works.[3]

Results

A total of 30 macroremains were recovered in the seven samples collected from three different contexts (cf.,

[2] The authors wish to thank G. Morgione for assistance and collaboration in registering and documenting the plant material.

[3] E.g., Bertsch 1940; Berggren 1969; Cappers/Bekker/Jans 2006; Bojňanský and Fargašová 2007.

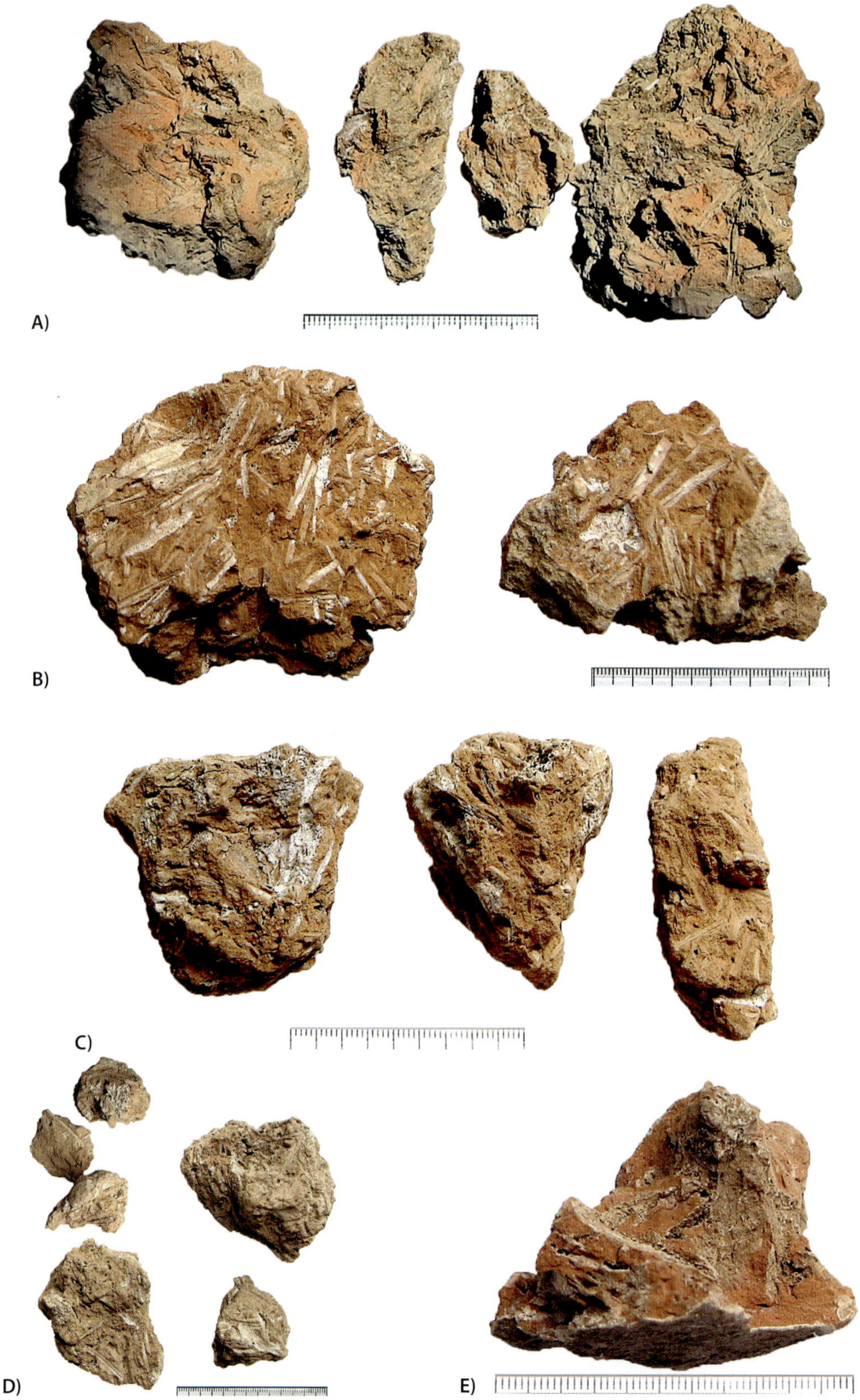

Figure 9.1 Fired clay fragments examined for plant impressions. (A) Sample no. 1; (B) Sample no. 3; (C) Sample no. 4; (D) Sample no. 2; (E) Sample no. 5 (see *Table 9.3*). (LC/ICA)

Chronology	Phase	Soil Samples	%	Vol. (l)	%	Fired Clay Samples
Second–third quarters of 5th c. BC	3	11	20.00	23	12.78	0
Second half of 4th–3rd c. BC	4	24	43.64	89	49.44	5
Second half of the 2nd c. BC–beginning of 1st c. AD	5	14	25.45	55	30.56	0
Unknown	Top-soil	6	10.91	13	7.22	0
Totals		**55**	**100.0**	**180**	**100.0**	**5**

Table 9.1 Counts of the examined soil samples and fired clay fragments from Sant'Angelo Vecchio.

Area	G6 Level 2	G7 Level 2	F3 Kiln
Period	**Phase 5**	**Phase 5**	**Phase 5**
CEREALS			
Triticum dicoccum	4		
Triticum durum/aestivum	3	2	
Triticum sp.			3
Hordeum vulgare		2	
Hordeum sp.	3	2	4
PULSES			
Pisum sp.	1		
Vicia ervilia		3	
WEEDS			
Avena sp.		1	
Vicia/Lathyrus sp.	1	1	
Total	**12**	**11**	**7**

Table 9.2 List of the species identified by carbonized grains and seeds from Sant'Angelo Vecchio.

Table 9.2). All the material was preserved by charring, which occurred because the plant material was exposed to fire. A number of plant remains, even if reduced to fragments, have retained enough characteristics for identification at genus level. The main categories of plant material were cereals (24 grains), pulses (4), and weed seeds (3), as listed in *Table 9.2*.

The analysis of the fired clay fragments allowed for the selection of plant impressions, which were produced in large part by cereal chaff and straw used as temper in the production of the clay bricks. Single impressions of barley grains, chickpea, ryegrass/darnel, and spurge were also found (*Table 9.3*). Due to the limited size of the fragments and the difficulty in identifying some impressions, the number of the impressions in each fragment was not uniform, varying from 2 to 10.

Phase	Phase 4	Phase 5			
Sample	**1**	**2**	**3**	**4**	**5**
Clay fragments	4	6	2	3	1
Area	Sounding B L. 2, b. 1	E2/E3, kiln area L. 2, b. 1–2	E2/E3, kiln area L. 2, b. 1–2	E2/E3, kiln area L. 2, b. 1–2	E2/E3, kiln area L. 2, b. 1–2
Triticum dicoccum	x	x	x	x	
Triticum durum/aestivum		x			
Triticum sp.	x	x	x		
Hordeum distichum	x	x			
Hordeum vulgare			x	x	x
Hordeum sp.		x		x	
Triticum/Hordeum chaff	x	x	x	x	x
Cereals straw	x	x	x	x	x
Cicer arietinum	x				
Euphorbia elioscopia		x			
Lolium cf. *temulentum*		x			

Table 9.3 List of the species identified in plant impressions from Sant'Angelo Vecchio.

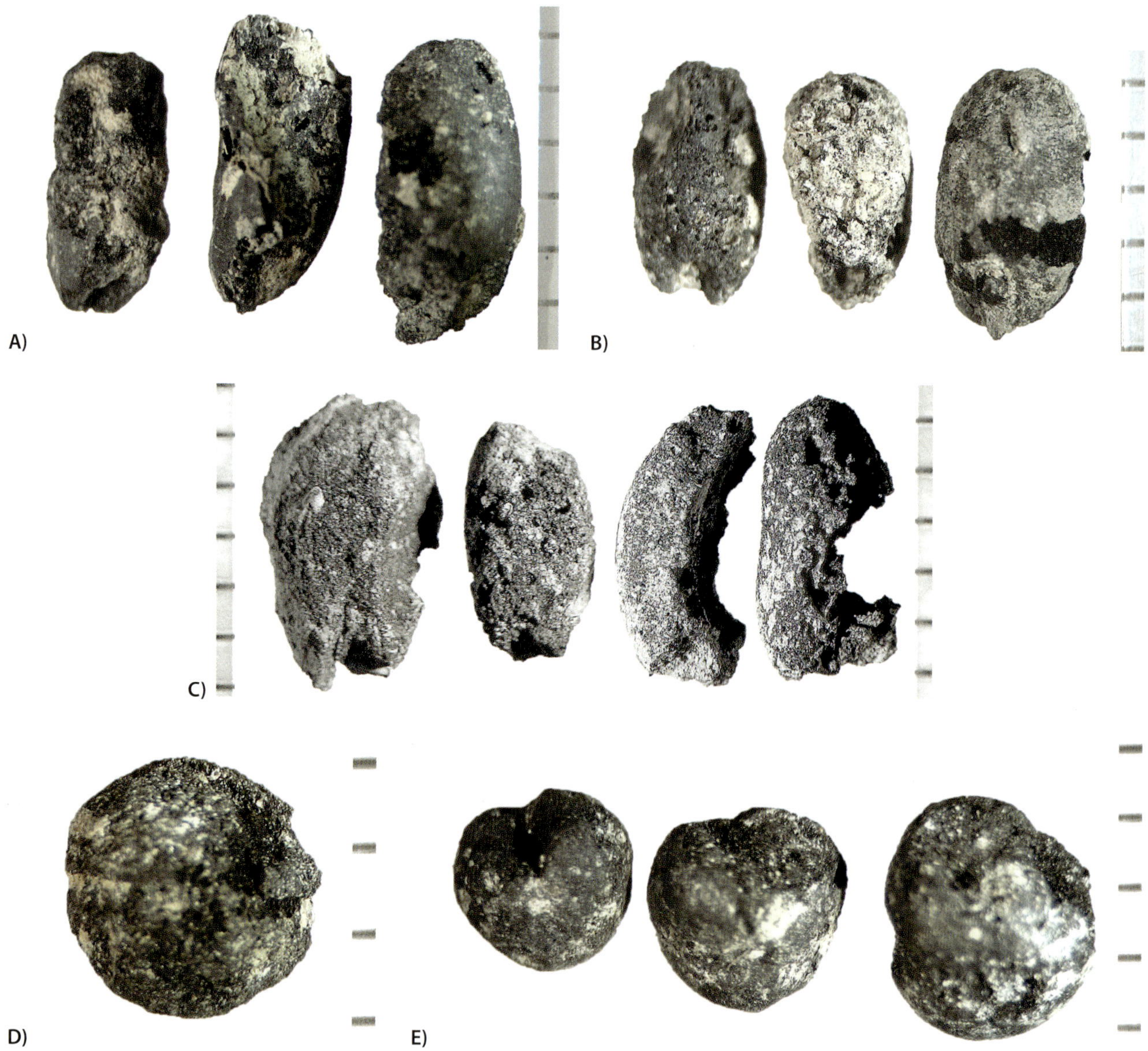

Figure 9.2 Charred plant material from Sant'Angelo Vecchio. (A) Grains of emmer wheat (*Triticum dicoccum*; G6, L. 2); (B) Grains of free-threshing wheat (*Triticum durum/aestivum*; G6, L. 2); (C) Grains of barley (*Hordeum vulgare* and *Hordeum* sp.; G7, L. 2); (D) Seed of pea (*Pisum sativum/arvense*; G6, L. 2); and (E) Seeds of bitter vetch (*Vicia ervilia*; G6, L. 2). (LC/ICA)

Cereals

Cereal macroremains were made up of 12 wheat grains—including both glume-based grains (*Triticum dicoccum*) and free-threshing wheats (*Triticum durum/aestivum*)—and 11 badly preserved barley grains. The crop processing (threshing and winnowing) residues were the main component of the plant inclusions in the examined fired clay fragments, including stems, leaf fragments, rachis fragments, spikelet bases, glumes, and a few grains.

Emmer Wheat

Four grains of *Triticum dicoccum* (*Fig. 9.2A*) recovered in a sample from the deposit of a building connected with the workshop shed and five impressions (spikelets, spikelet bases and a grain) (*Fig. 9.3*) found in the examined clay fragments documented the cultivation of this important crop in phase 5. Emmer, the *far* of the Latin agronomists, was the most widely cultivated glume wheat in ancient Italy and the oldest and most important grain crop grown by the Latins (*primus antiquis Latii cibus*).[4] Grains of emmer have been record-

[4] Pliny *NH*, XVIII.xix.83.

Figure 9.3 Imprints of emmer wheat (*Triticum dicoccum*): *left*, spikelet, sample no. 4, and *right*, grain, sample no. 2. (LC/ICA)

ed in the deposits of Pantanello (Metaponto),[5] Monte Irsi,[6] Torre di Satriano,[7] San Giovanni di Ruoti[8] and in many other archaeological sites of central and southern Italy.[9]

Durum/Bread Wheat

Grains of free-threshing wheat (*Triticum durum/aestivum*; *Fig. 9.2B*) were found in two samples collected in the Kiln Complex area (Phase 5). The grains may belong either to tetraploid hard wheat (*Triticum durum*) or to bread wheat (*Triticum aestivum* sl). Some impressions (*Fig. 9.4*), identified as rachis fragments of free-threshing wheat, were also found in the fired clay slabs, but their study could not clarify if they were produced by tetraploid or hexaploid wheat-threshing residues.

A number of wheat species (*Tritici genera*) were known by Pliny, who believed that no other kind of wheat could equal that of Italy either for whiteness or weight, qualities for which it was more particularly distinguished.[10] Pliny, to prove the quality of the Italian wheat, translated a line of Sophocles' *Triptolemos*: *et fortunatam Italiam frumento canere candido*. Whiteness was one of the peculiar merits of the Italian wheat and Pliny was very surprised to note that none of the Greek writers had made any reference to it.

Figure 9.4 Imprint of rachis internodes of free-threshing wheat (*Triticum durum/aestivum*), sample no. 2. (LC/ICA)

Evidence of free-threshing wheat was found at Pantanello (Metaponto),[11] Pomarico Vecchio,[12] and San Giovanni di Ruoti,[13] and the presence of free-threshing wheat remains has been recorded in other archaeological deposits from the Neolithic period onward.[14]

[5] Costantini and Costantini Biasini 2003.

[6] Hjelmqvist 1977.

[7] Osanna 2007; Novellis 2009.

[8] Costantini 1983.

[9] Costantini 2002.

[10] *NH* XVIII.xii.63.

[11] Costantini and Costantini Biasini 2003.

[12] Caramiello and Siniscalco 1997.

[13] Costantini 1983; Monckton 2002.

[14] Costantini 2002.

Figure 9.5 Imprint of a two-row barley spikelet (*Hordeum distichum*) with a fresh spikelet for comparison, sample no. 1. (LC/ICA)

Figure 9.6 Imprint of a six-row barley grain (*Hordeum vulgare*), sample no. 4. (LC/ICA)

Wheat

Three poorly preserved grains could not be assigned to glume or free-threshing wheat, and were thus classified at genus level as *Triticum* sp.

Barley

Barley grains were found in all five samples that yielded plant material for a total of 11 items, two of which were classified as hulled six-row barley (*Hordeum vulgare*). The other nine fragmentary grains could be identified only at genus level as *Hordeum* sp. (*Fig. 9.2C*). Imprints of spikelets recognized as two-row barley (*Hordeum distichum*; *Fig. 9.5*), six-row barley (*Hordeum vulgare*; *Fig. 9.6* and *Fig. 9.7*), and *Hordeum* sp. (*Fig. 9.8*) were also detected in the examined fired clay fragments.

Pliny, on the authority of Menander, said that barley was a most ancient aliment of mankind, as evidenced by a custom of the Athenians and also by the name of *hordearii*, given to the gladiators.[15] According to Pliny, barley bread was extensively used by the ancients, but during his time it had fallen into universal disrepute, and barley was mostly used as a food for cattle only.

Barley remains were the most common grain found in Italian archaeological deposits, being recorded in nearly all the investigated sites in central and southern Italy.[16] In Basilicata, barley grains were found at Pantanello (Metaponto),[17] Pomarico Vecchio,[18] and San Giovanni di Ruoti.[19]

[15] *NH* XVIII.xiv.72.

[16] Costantini 2002.

[17] Costantini and Costantini Biasini 2003.

[18] Caramiello and Siniscalco 1997.

[19] Costantini 1983; Monckton 2002.

Pulses

Four leguminous seeds and one impression of pulses were counted among the charred remains and the plant impressions, allowing the identification of three cultivated species of food legumes: chickpea (*Cicer arietinum*), garden/field pea (*Pisum sativum/arvense*), and bitter vetch (*Vicia ervilia*).

Chickpea

A single globose impression, found in a quite large clay fragment of sample no. 1, has been classified as a chickpea seed (*Cicer arietinum*; *Fig. 9.9*). According to Pliny,[20] several different kinds of chickpea were known, (*differentiae plures, magnitudine, colore, figura, sapore*), and its pods were described as round, fitting with the shape of the seed, whereas in other food legumes they are long and broad. One variety, with white and black seeds, resembles in shape a ram's head, and for this reason it has been called *arietinum* (*est enim arietino capiti simile, unde ita appellatur, album nigrumque*). Pliny suggested that it should be sown after it had been steeped in water for a day, because chickpea is naturally salty.

Seeds of chickpea are rarely attested in the Italian archaeobotanical record, though in Basilicata the presence of chickpea has been documented at Monte Irsi[21] and at Pantanello.[22]

[20] *NH* XVIII.xxxii.124–xxxiii.125.

[21] Hjelmqvist 1977.

[22] Costantini and Costantini Biasini 2003.

A)

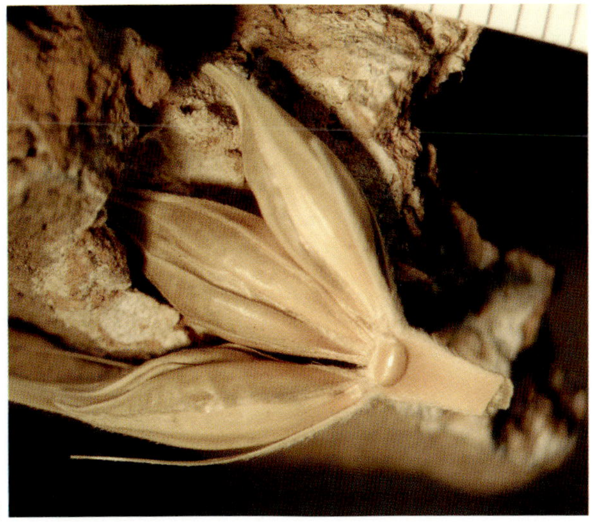

B)

Figure 9.7 (A) Imprint of a triplet of six-rowed barley (*Hordeum vulgare*), including a rachis segment, part of the ventral surface of the central grain and part of a lateral grain, sample no. 5; (B) the same with a triplet of six-row barley from the reference collection over the impression. (LC/ICA)

Figure 9.8 Imprint of a barley grain (*Hordeum vulgare*), sample no. 3. (LC/ICA)

Garden/Field Pea

A single rounded seed, classified as garden/field pea (*Pisum sativum/arvense*; *Fig. 9.2D*) by comparison to reference collection material, has been collected from a sample assigned to Phase 5. Seeds of this pulse have been previously found at Pantanello (Metaponto),[23] and San Giovanni di Ruoti.[24]

[23] Costantini and Costantini Biasini 2003.

[24] Costantini 1983; Monckton 2002.

Bitter Vetch

Three seeds, triangular in cross-section with the hilum at the margin of a triangular end, found in a sample of the Phase 5 deposit have been identified as bitter vetch (*Vicia ervilia*; *Fig. 9.2E*). *Ervum*, the classical name for bitter vetch or fitch, was regarded more as medicine than a food (*Cibis idem hominis alienum*) because it caused intestinal trouble. As with chickpea, limited evidence documented the utilization of this pulse in ancient southern Italy.[25]

Weeds

One grain of oat (*Avena* sp.), thought to represent a weed (*Avena fatua*, *Avena sterilis*) rather than a cultivated crop (*Avena sativa*), and two seeds of wild vetch (*Vicia/Lathyrus* sp.) have been grouped as wild plants. The poor preservation did not allow a more precise identification, limiting their ecological interpretation. Both the taxa are commonly considered potential arable weeds, and they are the most common recorded weeds in the charred plant assemblage. Imprints of a darnel grain (*Lolium temulentum* type; *Fig. 9.10*) and of a sun spurge seed (*Euphorbia elioscopia*; *Fig. 9.11*) were also identified in the fired clay fragments of sample no. 2. Darnel is regarded as a weed of cereal fields. Classical writers pointed out its poisonous properties, and it was called *temulentum* for its intoxicating

[25] Hjelmqvist 1977; Costantini 2002; Costantini and Costantini Biasini 2003.

Figure 9.9 Imprint of a chickpea seed (*Cicer arietinum*), sample no. 1. *Left*, impression recomposed from two fired clay fragments; *middle*, the two fired clay fragments; *right*, part of the impression with a chickpea seed from the reference collection. (LC/ICA)

Figure 9.10 *Left*, imprint of a darnel grain, sample no. 2; *right*, grain of *Lolium temulentum* from the reference collection. (LC/ICA)

Figure 9.11 Imprint of a sun spurge seed (*Euphorbia elioscopia*), sample no. 2. (LC/ICA)

or narcotic powers when taken alone or mixed with wheat. For this deleterious property it was termed by Virgil *infelix lolium*.[26] Columella reported *cribro viciario vel etiam loliario*,[27] vetch and darnel sieves used by the Romans to part weed seeds from cereal grains.

Remarks

The plant assemblages found at Sant'Angelo Vecchio consist principally of charred remains and plant impressions, the latter likely originating from threshing remains. In spite of the generally poor preservation of the plant material, a good range of species has been identified. Cereals, as emmer wheat (*Triticum dicoccum*), free-threshing wheat (*Triticum durum/aestivum*), two-row barley (*Hordeum distichum*), six-row barley (*Hordeum vulgare*), and food legumes, as chickpea (*Cicer arietinum*), garden/field pea (*Pisum sativum/arvense*), and bitter vetch (*Vicia ervilia*), were the main crops documented by the archaeobotanical investigation. The arable weeds, even if of insignificant amounts, seem to be linked with the cereal cleaning processes, as they were present among the charred material and the chaff and straw residues used as temper in the production of clay bricks.

The identified species are consistent with the plant economy understood for the period under study, as documented by the plant evidence collected in other contemporary sites of the region. The lack of any fruit remains, like those of grape and olive, seems to be related more to the limited archaeobotanical research than to a real absence of the fruit plant cultivation in the territory. In point of fact, published archaeobotanical data confirm the relevance of the two fruit plants for the agricultural economy of the coastal and inland regions of southeastern Italy.

[26] *Buc.* V.36–37; *Georg.* I.154.
[27] *RR* VIII.5.16.

10

Observations on the Vegetation and Landscape Dynamics

Mauro Frattegiani

According to the European Landscape Convention,[1] a landscape is the result of the combination of geographic location, environmental conditions, and the historical and current uses of a place. It is a zone or territory, perceived as such by local inhabitants or visitors, and its nature and character derive from the action of natural and/or cultural (i.e., anthropic) factors. This definition takes into account that landscapes evolve over time due to the effects of natural forces and of human beings. It likewise emphasizes the concept that a landscape consists of an inseparable set of factors, the natural and cultural elements of which are considered simultaneously.

A landscape is therefore anything but static, instead changing noticeably over time. Every modification of the landscape results in a change in its current state and consequently determines the direction of the future evolution of the landscape. Analyses of these evolutionary processes, performed on past events or future projections, are the aim of studies of landscape dynamics, the techniques of which were developed and refined beginning in the last decade of the 20th century. The spread of Geographic Information Systems (GIS) has brought about a notable increase in the capacity to analyze and manipulate data pertinent to these subjects, accompanied by a deeper understanding of the processes that drive the evolution of ecosystems.[2]

Aims of This Study

The goal of this chapter is to describe the current state of the landscape surrounding the archaeological site at Sant'Angelo Vecchio in the Comune di Bernalda (MT), as well as to analyze the principal modifications that have affected the landscape over the last 150 years, concluding with a hypothetical analysis of its probable future evolution.

Analysis of the landscape dynamics will enable a better interpretation of the processes that have disturbed the archaeological site in the last few decades, as well as a correlation of the changes in the modern landscape with the results of the archaeobotanical and archaeological studies. This chapter is an initial contribution to such a research project, which in the future can be furthered and directed toward more specific problems.

Materials and Methods

The analyses were conducted using 1:25,000-scale topographic maps created by the Istituto Geografico Militare (IGM) between 1873 and 1956, aerial photographs taken in 1947 and 1954, and digital orthophotos created in 1997, 2000, and 2006.

The aerial photographs were acquired from the archives of the IGM, while the orthophotos were analyzed in a GIS environment (Quantum GIS) utilizing the online geospatial data of the Ministero per l'Ambiente, la Tutela del Territorio e del Mare. The aerial photographs were orthorectified and georeferenced by identifying ground control points visible also in the orthophoto made in 2000, used as the reference image. Warping was performed with third degree polynomial algorithms. The scanned topographic maps were georeferenced using the same procedure.

The study of the current state was performed primarily via visits to the area and examination of photographs, utilizing portable GPS receivers lacking differential correction but equipped with SIRF III correction to locate the differences.

Description of the Current Environment

The area under study is located in the Comune di Bernalda in Basilicata, at an elevation of approximately 45 masl. The archaeological site lies approximately 10 km from the Ionian coast and less than 1 km from the Basento valley floor.

The climatic conditions can be evaluated using the temperature and precipitation data from the Metaponto meteorological station (25 masl): the mean annual precipitation is 550 mm, with typical distribution in the fall and winter. The maximum monthly value

[1] *Consiglio d'Europa, 2000.*

[2] Particularly important and emblematic of this trend is, for example, Shugart's work on the evolution of forest ecosystems (1984).

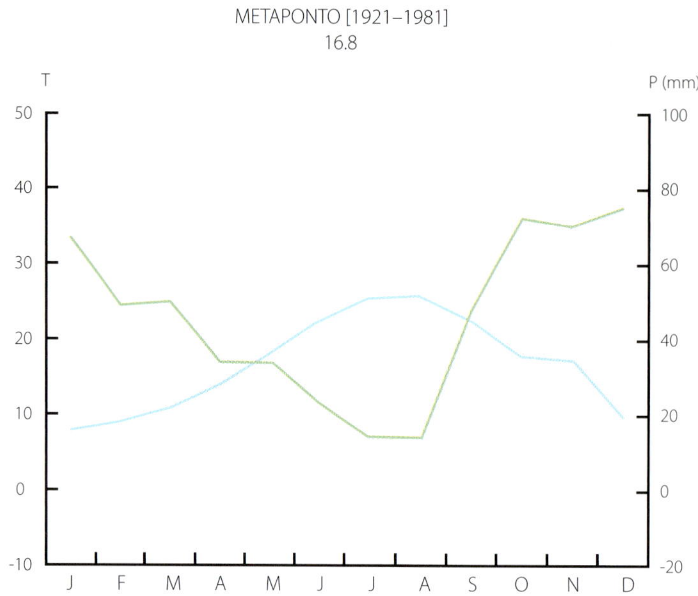

Figure 10.1 Annual temperature (blue) and precipitation (green) as measured at the Metaponto meteorological station. (After Walter and Lieth 1967)

is reached in December (75 mm), the minimum in July and August (14 mm in both months). The mean annual temperature is 16.8°C with higher monthly averages in July and August (25.6–25.7°C), while the minimum temperature occurs in January (8.1°C). When interpreted using Walter and Lieth's diagram (*Fig. 10.1*),[3] the meteorological data demonstrates a very broad water deficit period beginning in May and ending in September.

Thornthwaite's climatic formula places the zone in a semi-arid climate (D) with aridity index of 41.7. Pavari's phytoclimatic classification places it in the Lauretum area, hot sub-zone, type II with summer drought.[4]

The soils mostly have moderate proportions of clay (with more sand in the areas closer to the Basento valley floor) and clearly originate in sedimentary rocks. The soils map of Basilicata classifies the soils in the Basento valley floor as part of the pedologic region known as "soils of the alluvial plains."[5] The area under examination belongs to the "soils of the alluvial valley floors" pedologic unit lying between the oldest terraces (or the hillsides) and the areas most prone to flooding alongside the streams.

These types of soil are encountered in the incised parts of the valleys and in the valley floors of the main tributary rivers of the Ionian Sea (Sarmento, Sinni, Agri, Cavone, Basento, and Bradano), in level or nearly level areas characterized by alluvial deposits with variable soil texture encompassing recent alluvial surfaces that are often slightly terraced, alluvial cones, and colluvial fans at the bases of the slopes. The sediments from which they originate vary in nature and composition, coming from the inundations of the main river and more local contributions, such as torrents and gullies that converge in the valley from the surrounding hillsides and colluvial material eroded from the slopes.

Away from the river and above 40 masl, the soils belong to the "soils of the western interior highlands" pedologic region, and more specifically to the "marine terraces of the Ionian hinterland" pedologic unit. These soils consist primarily of sediments (loose sand and gravel) deposited in layers up to several dozen meters in thickness over the clay marine terraces formed by the oscillations in sea level.[6]

On the surfaces that are best preserved in relation to their original condition, the soils of the marine terraces are very evolved, given that the pedogenic processes were able to act for a long period of time due to the prolonged geomorphologic stability. The types of vegetation and land use available in these places are strongly associated with the characteristics of the soils,

[3] Walter and Lieth 1967.

[4] For Pavari's classification, see De Philippis 1937.

[5] Cassi and Viviano 2006.

[6] Folk in *Survey*.

Ecosystem Type	Status	Main Characteristics	Photograph
Grassland	Actively cultivated	Absence of trees and shrubs; cut or burned grass (less than 50 cm tall); compact turfgrass and polycultural vegetation. (Photo: MF/ICA)	
	Recently abandoned	Absence of trees and shrubs; high grass (more than 50 cm tall); compact turfgrass marked by differential development depending on the zone. (Photo: MF/ICA)	
	Abandoned at length	Presence of trees and/or shrubs, as well as polycultural herbaceous vegetation of different sizes; absence of grazing. (Photo: MF/ICA)	
Olive grove	Actively cultivated	Harrowed or mown surface; absence of other species of tree or shrub. (Photo: FS/ICA)	
	Abandoned	Naturalized olive plants within a shrubby matrix of typical species of Mediterranean *maquis*. (Photo: MF/ICA)	

Figure 10.2 Synthetic description of the principal ecosystem types in the territory around Sant'Angelo Vecchio. (MF/FS/ICA)

Ecosystem Type	Status	Main Characteristics	Photograph
Orange grove and vineyard	Actively cultivated	Harrowed or mown surface; absence of other species of tree or shrub. (Photo: MF/ICA)	
	Under conversion	Areas of activity, with movement of earth and holes of max. 1 m depth; litter of discarded materials (irrigation tubes, training struts, fence sections) over the surface, with large piles of deadwood at the edges. (Photo: MF/ICA)	
Mediterranean maquis	Growing	In abandoned fields and pastures or denuded areas subject to recolonization. Predominance of low and heliophilous shrubs such as heath (*Erica* spp.) and mitnan (*Thymelaea hirsuta*), which nevertheless do not cover the entire surface. (Photo: MF/ICA)	
	Consolidated	Predominance of low and heliophilous shrubs along with some larger shrubs and tree-like plants. The shrubby vegetation covers the soil in a practically continuous manner. (Photo: MF/ICA)	
	Stable	Complete, dense, and practically impenetrable coverage. The mean height of the vegetation crown exceeds 3 m and is homogeneous across the entire area. (Photo: MF/ICA)	

Figure 10.2 cont. Synthetic description of the principal ecosystem types in the territory around Sant'Angelo Vecchio. (MF/FS/ICA)

Ecosystem Type	Status	Main Characteristics	Photograph
Moist areas	Reed stands	Periodic submersion or elevated moisture, along the banks and on the floor of the stream, just as on the first alluvial terraces. (Photo: MF/ICA)	
	Poplar stands	Patches of white poplar (*Populus alba*), intentionally planted in valley floor areas, standing out clearly for the color of their leaves and bark and their tall stature. (Photo: MF/ICA)	
	Riparian thickets	Primarily broadleaf deciduous trees, both autochthonous (field elm, willow, field maple) and allochthonous (black locust, which is the predominant species in these areas). (Photo: FS/ICA)	
Anthropic areas	Greenhouses and other structures	Buildings and structures, in particular greenhouses and (most recently) solar panel farms, which impede the absorption of runoff water and often cause the destruction of the natural vegetation. (Photo: MF/ICA)	
	Sowable fields	Cultivation predominantly in wheat, barley, or corn. (Photo: FS/ICA)	

Figure 10.2 cont. Synthetic description of the principal ecosystem types in the territory around Sant'Angelo Vecchio. (MF/FS/ICA)

Figure 10.3 Positions of the areas surrounding the excavation zone described in the text.

which can be generally classified as soils of good fertility, at least in the flat areas. The territory is therefore characterized by intense agricultural activity including sowable (mostly cereals) and orchard crops, especially olive and citrus as well as vine and peach.

The areas with greater slope instead feature primarily shrubby vegetation, which results in the formation of Mediterranean *maquis* in different stages of development. The streams and the lowest areas are characterized by a range of vegetation types, especially hygrophilous species. The dampest areas have patches of vegetation dominated by white poplar (*Populus alba*) or common marsh reed (*Phragmites australis*), while the riparian zones are dominated by field elm (*Ulmus minor*) and/or black locust (*Robinia pseudoacacia*), species that thrive in cool and moist areas but do not tolerate stagnant water beyond very limited periods. *Figure 10.2* lists the principal ecosystem types in the territory with their essential features.

Notes on Visits to the Areas Surrounding the Excavation

Visits to the site in May 2011 and March 2012 revealed several signs of the modifications made to the areas surrounding the excavation zone in the past few decades, providing a level of detail impossible to obtain from examination of the photographs and topographic maps. This section describes several specific situations relevant to interpreting the changes in the area. *Figure 10.3* shows the positions of the areas described.

The zone that has undergone the most radical changes, even recently, is the newly planted olive grove located on the second terrace on the (hydrographic) right side of the stream (zone 1, *Fig. 10.3*). Local informants indicated that the grove was planted in 2001 after a deep-digging operation (depth greater than 70 cm with inversion of the soil horizons) and marked changes to the local topography, with excavation and filling operations exceeding 5 m in certain areas (*Fig. 10.4A*).

According to the analysis of the vegetation present on the residual section of the original scarp located uphill, the area occupied today by the olive grove was once covered with a shrubby maquis vegetation spread out along a natural slope. The scarps created by the digging operations using the backhoe and deep digger are still free of grass, leaving the area subject to erosion, which will likely become marked in the near future.

A similar situation is encountered also in the olive grove planted on the lower terrace (zone 2), but here the impact on the local topography is less pronounced

Figure 10.4 (A) Scarp above the olive grove planted in zone 1; (B) Olive grove on the plateau located above the excavation, in zone 3; (C) Willows near the river bed in zone 5; (D) Zone 7, south of the excavation, beneath the spring. (MF/FS/ICA)

due to more limited excavation and filling operations. In this zone the greatest impact is associated with the reduction of the reed stand, which consists particularly of swamp reed (*Phragmites australis*). This species, however, has a strongly invasive behavior, given its notable capacity for agamic reproduction via rhizomes (horizontal subterranean stems), and is already partly re-colonizing the olive grove.

Another recently planted olive grove is located on the upper terrace (zone 3), where in 1979 a field cultivated in grain had stood (*Fig. 10.4B*). The creation of this grove in 2001 nevertheless did not include remodeling of the terrain, although here, too, a deep digging operation was probably performed.

Among the areas covered by spontaneous vegetation, there is a greater dynamism immediately beneath the excavation zone toward the area of the springs (zone 4), previously used for horticulture and currently characterized by the pervasive presence of brambles (*Rubus* spp.), which notably exceed the brambles in the surrounding areas in height and vigor. The presence of brambles and their stature are, in fact, correlated with the greater presence of nitrogen in the soil, a consequence of the repeated application of manure in the garden. The anthropogenic vegetal formations will therefore be rapidly transformed into formations dominated by swamp reed and/or riparian thickets.

In the areas under study, the riparian thickets are present along the course of the Fosso Sant'Angelo. These formations are characterized by several different situations: in the areas where the stream becomes wider and forms small floodplains (zone 5, *Fig. 10.4C*) there are small nuclei dominated by poplar (*Populus alba*) or white willow (*Salix alba*), while in the areas where the stream becomes narrower the scarps are dominated by elm (*Ulmus minor*) and, especially, black locust (*Robinia pseudoacacia*).

The origin of the poplar and/or willow nuclei is unclear, inasmuch as they could have been planted during land-reclamation activities designed to make the areas closest to the stream cultivable. The stumps appear to be very vigorous, although the buds seem relatively young.

Figure 10.5 Sant'Angelo Vecchio in the 1873 IGM topographic map (original scale 1:50,000). (Istituto Geografico Militare)

The woody formations dominated by elm and black locust feature plants of varying age, predominantly young, and primarily agamic in origin (i.e., the plants do not grow from seed, but instead sprout directly from the pre-existing stumps). The structure of the thicket probably derives from the use of modest quantities, at short intervals, of a part of the plants present on each stump (*taglio della formica*). Black locust is an exotic species from North America, introduced into Europe at the beginning of the 17th century. It is invasive, propagating very rapidly by seed and root buds, and it probably reached the study area by natural diffusion, beginning with nuclei, individual trees, and rows of trees planted along the SS407 Basentana to stabilize the soil along the scarps. Black locust adapts very well to moist environments, although it is able to spread and grow even in drier conditions.

The area below the Fosso Sant'Angelo (zone 6) preserves more developed types of ecosystems, with vestiges of riparian-plain forest. The wood is nevertheless limited in extent and concentrated on the scarps along the streambed, where there are century-old elms and a reduced presence of black locust. This is a stable strip of wood, which has certainly been present in this area for several centuries, and there is also a remarkably developed shrubby stratum featuring typical prickly species including buckthorn (*Rhamnus catharticus*), a fairly rare autochthonous shrub.

The area beneath the spring (zone 7), continuously characterized by the presence of artificial terraces, is dominated by recently planted fruit trees. The northern border of the orchard still has a scarp featuring much less diffused exotic plants (*Fig. 10.4D*), such as false pepper (*Echinus mollis*) and cape gum (*Acacia horrida*). This latter species, originating in Senegal and east-central Africa, is not common in Italy generally or in Basilicata, but it is more common in Apulia. The plants observed were last cut approximately 8–10 years ago, while the stumps are probably even older.

The least transformed area is the eastern side of the Fosso Sant'Angelo (zone 8), characterized by areas of Mediterranean maquis in varying stages of development. The predominant species are olive (*Olea europea*), phyllirea (*Phyllirea latifolia*), and myrtle (*Myrtus communis*) in the upper shrubby stratum, while among the less-developed shrubs the most common species is mastic (*Pistacia lentiscus*), along with a diffuse presence of Mediterranean heather, besom heath, and mitnan (*Thymalaea hirsuta*) in the less densely covered areas.

The different development in the maquis (denser and taller in the central area, thinner and lower in the north and south areas, with a greater proportion of small shrubs) appears to be associated with the use of these areas for pasturing domestic animals, especially sheep. Nevertheless, the presence of a nearly continuous covering over the entire surface reveals that the

Figure 10.6 Air photo of Sant'Angelo taken in 1947. The red circle indicates the area containing the most significant differences with respect to the 1873 topographic map as described in the text. (Istituto Geografico Militare)

area had not been used for pasture for several decades, except in a very light form.

Analysis of the Changes to the Landscape, 1873–2012

Comparison of the 1873 topographic map (*Fig. 10.5*) and the 1947 air photos (*Fig. 10.6*) produced no significant results, mainly due to the paucity of useful information in the map about such a limited area. Nevertheless, some significant changes can be observed, such as the expansion of the olive grove below the masseria and the creation of a road for wheeled traffic in the eastern part of the new olive grove. The cultivated areas to the north (Giamperduto locality), furthermore, were probably expanded.

More information, on the other hand, can be obtained from analysis of the air photos and orthophotos. *Figure 10.6* and *Figure 10.7* illustrate the main changes detected using a sequence of imagery covering the period from 1947 to 2006.

The information derived from this analysis broadly confirms the general picture created by the visits to the site to document the changes in the landscape that have occurred in the last few decades. The analysis highlights the periods of greatest change, in particular the remarkable transformation of the area between 2000 and 2006, when a more intensive form of agriculture was introduced and a large quantity of agricultural land and woods was given over to the development of infrastructure. The images also confirm the use of large-scale digging and reshaping of the ground surface to create new terraces and expand the existing ones.

Reflections on the Changes in the Landscape and the Evolutionary Trends

The results of the study were compared with results extrapolated from the studies conducted by the Region of Basilicata for the creation of its Regional Ecological Network, a synthesis of which appears in the following table (*Table 10.1*).[7]

The studies performed for the creation of the Regional Ecological Network analyzed the land-cover dynamics in the period 1960–2000. The analysis of the

[7] *Regione Basilicata, 2009.*

1954
1—Maquis converted to sowable fields
2—Increased density of orchards

(This photograph, positive/negative: 6622/21314, from the mission flown by the Gruppo Aereo Italiano in 1954–55, is reproduced courtesy of L'Aerofototeca Nazionale, Rome.)

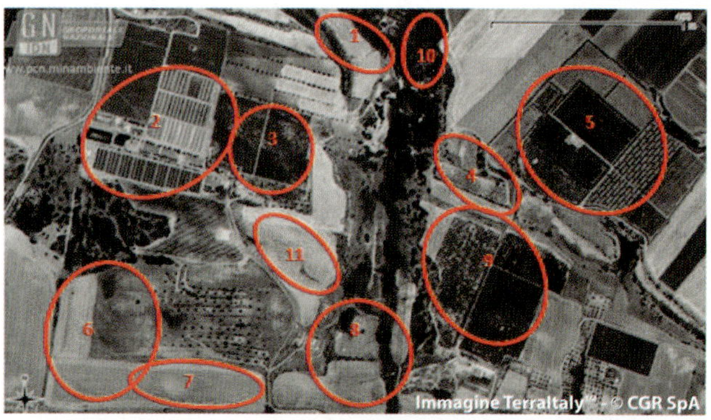

1997
1, 4—Maquis converted to sowable fields
2—Sowable fields transformed with construction of greenhouses
3, 5, 9—Sowable fields converted to orchards
6, 7, 8—Orchards converted to sowable fields
10—Pasture developed to maquis
11—Elimination of isolated rows and trees

(Courtesy Immagini Terraltaly™ - © CGR SpA - Parma www.terraitaly.it)

2000
1—Sowable fields transformed with construction of greenhouses

(Courtesy Immagini Terraltaly™ - © CGR SpA - Parma www.terraitaly.it)

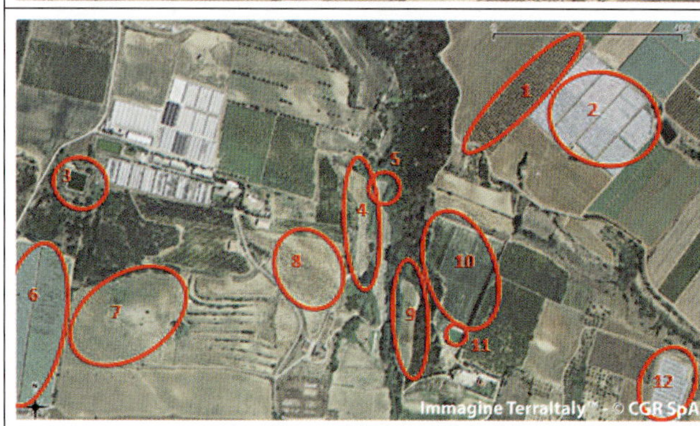

2006
1, 9—Sowable fields converted to orchards
2, 6, 12—Sowable fields and orchards transformed with construction of greenhouses
3, 11—Sowable fields transformed with construction of irrigation basins
4, 5—Maquis converted to orchards
7, 8—Orchards converted to sowable fields
10—Orchards converted to sowable fields and vice versa

(Courtesy Immagini Terraltaly™ - © CGR SpA - Parma www.terraitaly.it)

Figure 10.7 Principal changes in the landscape, detected using air photos and orthophotos.

		Area (ha)	Percentage	
Continuity	Agricultural continuity (agricultural to agricultural)	210.66	66.8%	78.0%
	Pasturable continuity (pasture to pasture)	16.25	5.2%	
	Water bodies (water body to water body)	18.96	6.0%	
Change	Agricultural tillage (pasture to agricultural)	44.94	14.3%	22.0%
	Extensive exploitation of pasturable land (agricultural to pasture)	0.04	0.0%	
	Forest management of pastures (pasture to woods)	4.56	1.4%	
	Forest management of agricultural areas (agricultural to woods)	19.89	6.3%	

Table 10.1 Comparison of land use in the study area between 1960 and 2000 (1-km radius from the excavation area). (Data source: Sistema Ecologico Funzionale Territoriale; Regione Basilicata, 2009)

changes in land cover was performed in a GIS environment via overlay mapping of the following cartographic datasets:

1. the 1:200,000-scale Map of Land Use in Italy created by the Center for Studies in Economic Geography of the National Research Council (CNR) in collaboration with the General Directorate of the Land Register and the Cartographic Office of the Touring Club Italiano, published between 1956 and 1960;[8] and
2. the CORINE Land Cover database published in 2000.[9]

The data demonstrate that the changes in the area under examination (within a radius of 1 km from the archaeological site) between 1960 and 2000 were limited: 78 percent of the surface area maintained the conditions documented in 1954. The most significant changes were a reduction in pasture, especially in favor of agriculture, and a much more limited recolonization of pasture and agricultural land by forest vegetation.

These changes are very different from those documented broadly across the territory of the region of Basilicata, where the surface area dedicated to woods registered a much greater increase, affecting more than 25 percent of the total territory. Such a result can be attributed to the greater importance of agriculture for this area, which was affected only minimally by the marginalization of agriculture.

It is, however, interesting that the changes in the landscape documented by the studies for the Ecological Network are rather different from those documented by this study. This is the result of three factors. First, the temporal span of the Ecological Network studies is much broader, and in several areas multiple changes were documented in this period which are not considered in a comparison of the conditions in only two years (1960 and 2000). Second, the CORINE land-cover classification scheme is less articulated than that utilized in this study, and therefore certain transformations are not documented (for example, from orchard crops to sowable crops and vice versa). Finally, the Ecological Network studies were conducted at a larger scale than that utilized for this study, and therefore changes across limited areas are not documented.

As for the evolutionary trends, in the absence of anthropic use the natural processes of colonization by shrubs and trees are relatively rapid. Colonization by invasive allochthonous species is even more rapid, in particular black locust, although the study also documented small areas recently colonized by tree of heaven (*Ailanthus altissima*), a much more aggressive Asian species which in the course of just a few years could spread broadly across the territory.

Conclusions

The analyses described in this contribution reveal the transformation of the landscape of Sant'Angelo Vecchio as the result of anthropic activities, which certainly took place many centuries previously. Comparison of the analyses described above with the results of the archaeobotanical investigations[10] reveals a substantial correspondence in the types of landscape, with the obvious exceptions of the spread of the allochthonous species from overseas and the increase in surface area occupied by infrastructure.

[8] *CNR, Direzione Generale del Catasto, 1956–1960.*

[9] *European Environment Agency, 2004.*

[10] See Ch. 8, "Archaeobotanical Analysis."

Figure 10.8 Areas subject to slumping after mechanical reshaping of the surface profile, in various stages of recolonization and stabilization. (MF/ICA)

The main result is the limited spread, both currently and in the past, of the typical species of forest and maquis environments, suggesting that the conversion of these areas to pasture, and subsequently to agriculture, happened a very long time ago, and therefore the agrarian character of the landscape at Sant'Angelo Vecchio has significantly modified the natural setting.

Nevertheless, this transformation is in no way consolidated, and even in recent times significant changes have been documented in the use and morphology of the land. Particularly important is the reshaping of the surface profile for the creation of new terraces to host new orchard crops (especially olives), which has caused large-scale erosion and slumping, as illustrated in the photographic documentation of *Figure 10.8*. The analyses identify the principal types of ecosystem present in the territory (*Fig. 10.9*), which can be classified by the degree of anthropic transformation and seasonal conditions.

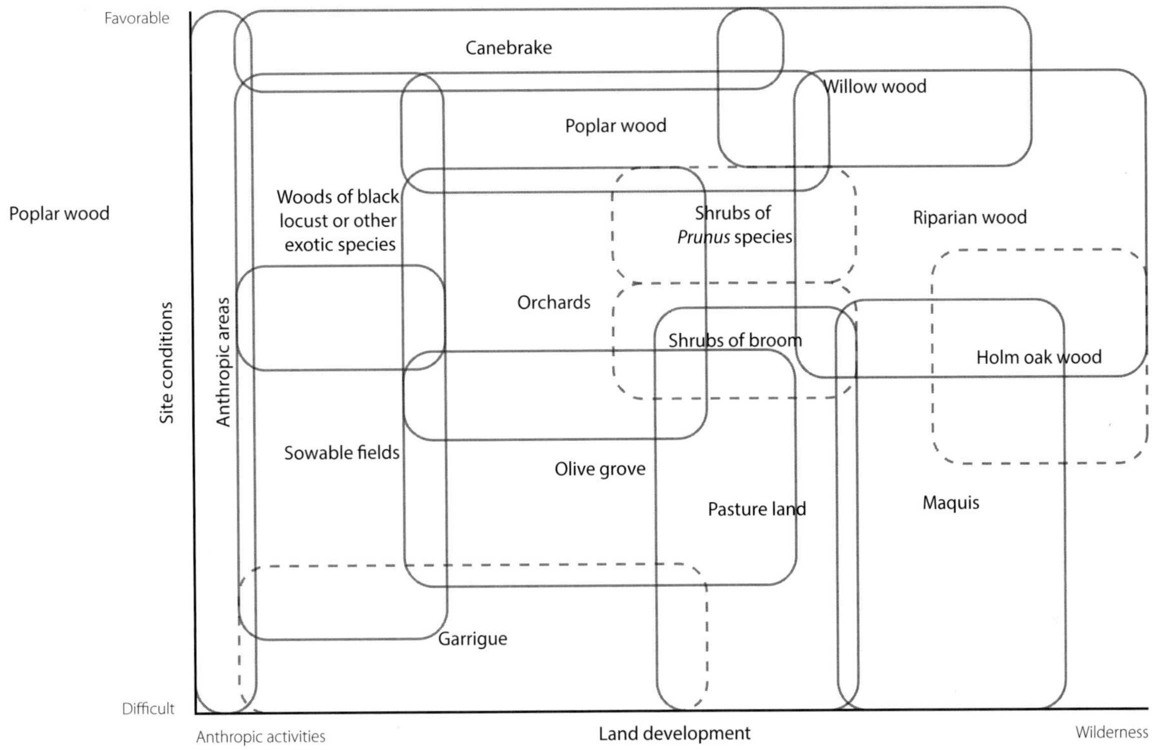

Figure 10.9 Graphic representation of the principal ecosystem types present (solid outline) or potentially present (dotted outline) in the territory in relation to local levels of moisture and development. (KJS/MF/ICA)

According to Liebig's Law,[11] the ecological potential of the territory under examination is limited primarily by the moisture and the water resources available to the vegetation. *Figure 10.9* is therefore a synoptic illustration of the principal ecosystem types present or potentially present in the territory in relation to the available water resources and the degree of anthropic transformation. *Figure 10.9* broadly identifies the features of the individual places and their potential evolution in the absence of anthropic disturbance (the transformation into the ecosystem types is located to the right of the departure condition).

This study should be considered a first step in analysis of the modifications to the landscape in recent times. Further studies could involve quantification of the changes with specific GIS techniques utilizing the main landscape metrics and multitemporal analysis.

[11] Liebig's Law or the Law of the Minimum, one of the fundamental laws of ecology as a discipline, states that the growth of an organism is controlled not by the sum total of the available natural resources, but by the availability of the scarcest resource. The law was subsequently refined by others and extended from single organisms to ecosystems in their entirety without, however, invalidating the basic principle established by Liebig (1848).

11
Faunal Analysis

Anna Zsófia Biller

Area	PN	PL	Species	Body Part	Length (mm)	Total
HoH	SA 79-714B			unknown		1
Kilns	SA 79-1033B		*Bos*	femur		1
	SA 79-30B	SA 79-29PL	Small ungulate?	unknown	35	1
	SA 79-442B	SA 79-438PL	*Sus*	scapula		2
	SA 79-804B		Large ungulate?	unknown	70	1
	SA 79-815B (A–B)			unknown		2
	SA 79-882B		Large ungulate?	unknown	80	1
Slope 3	SA 79-994B (A–C)		Small ungulate?	unknown	70	3
		SA 79-81PL	Large ungulate*	long bone	36	1
Slopes—Sounding D		SA 80-54PL	*Bos**	femur	70	1
Workshop Area	SA 79-1032B		Large ungulate?	unknown	160	3
	SA 79-535B		*Sus*	radius	55	1
	SA 79-537B (A–F)		*Canidae*	mandibula		6
	SA 79-543B		*Sus*	mandibula		1
	SA 79-548B (A–C)		*Equus*	upper tooth (P^2)	60	3
	SA 79-672B (A–C)		Small ungulate?	unknown	57	3
	SA 79-703B		*Sus*	lower tooth (I)		3
	SA 79-713B (A–C)		*Sus*	ulna	80	3
	SA 79-734B	SA 79-733PL	Large ungulate?	unknown	55	1
	SA 79-805B		Large ungulate?	flat bones	37	2
	SA 79-1007B		Large ungulate?	unknown	22	1
Total						**41**

Table 11.1 The list of animal remains. Specimens studied physically are marked by asterisks.

The animal bone material yielded by the Sant'Angelo Vecchio excavation is very limited, consisting of 41 bone fragments mainly from topsoil layers or found within contexts of phases 4 and 5.[1] A single specimen was found in the Late Archaic House on the Hill (*Table 11.1*). Dr. H. G. Wooldridge (Austin, Texas) completed the preliminary study of the material.[2] Subsequent research has had to rely on this earlier analysis and the excavation database because many of the examples are unavailable for firsthand observation.

Of the 41 specimens, the author was able to observe only three bone fragments. They include two smaller fragments of a cattle femur (SA 80-54PL) and a large ungulate long bone fragment (SA 79-81PL; marked by asterisks in *Table 11.1*). Where fragments themselves were not available, tentative species identification was attempted based upon accessible drawings and photographs. Bearing in mind the unreli-

[1] The author is grateful to László Bartosiewicz, whose translation assistance and ideas have been invaluable.

[2] Preliminary results of the faunal analysis based on Wooldridge 1980 were first reported in Carter 1980a, 12, fig. 8, and were also discussed in Edlund 1982a.

Figure 11.1 Large ungulate long bone fragment showing poor surface preservation; scale 1:1. (CW/ICA)

Statistics	mm
Mean	63.4
Standard deviation	33.0
Median	58.5
Minimum	22.0
Maximum	160.0

Table 11.2 Univariate statistics of the length of 14 bone fragments available for study.

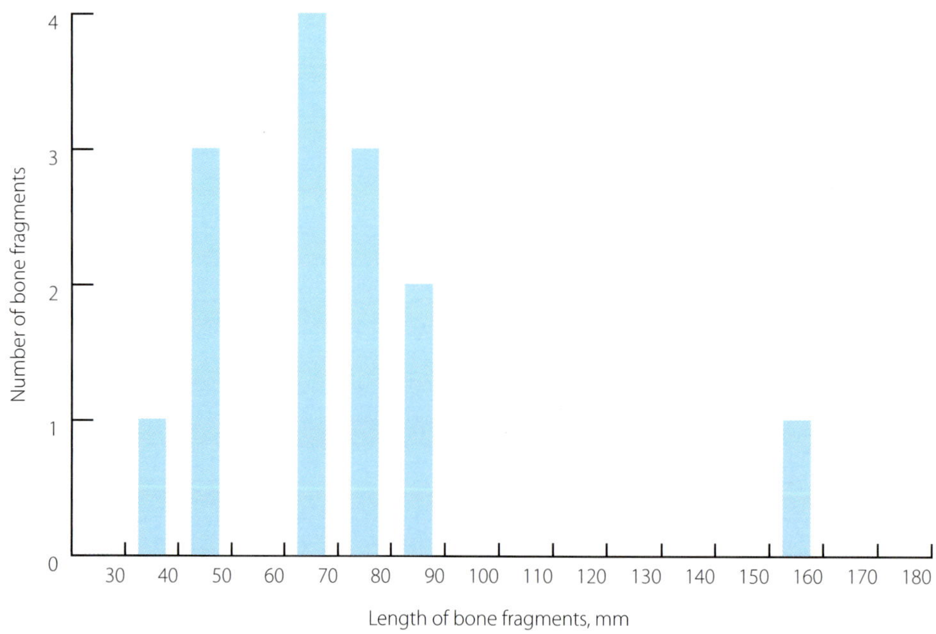

Figure 11.2 The size distribution of 14 bone fragments available for study.

ability and limitations of identifying species based on two-dimensional documentation, the following conclusions may be drawn.

There were two cattle bones (*Bos taurus*, Linnaeus 1758), 10 pig (*Sus domesticus*, Erxleben 1777), three horse (*Equus caballus*, Linnaeus 1758), six canine (dog: *Canis familiaris*, Linnaeus 1758; or wolf: *Canis lupus*, Linnaeus 1758), 10 large ungulate (cattle: Bovinae subfamily; or horse/donkey/mule: Equidae family), seven small ruminant (sheep/goat: Caprinae subfamily) or pig (Suidae family), and three totally unidentifiable bones. The presence of bones from atypical food animals in this sample (Equids and especially canine) may be indicative of a redeposited general disposal. If the canine bones were dog bones, they belonged to a bigger young adult dog, about the size of a German shepherd. The pig radius (SA 79-535B) and ulna (SA 79-713B [A–C]) probably belonged to one adult pig, and the horse tooth to an adult horse.

The bones were heavily eroded showing poor surface preservation attributable to soil conditions (*Fig. 11.1*). The material was also very fragmented: the length of 14 bone fragments of known sizes was on average 63 mm, but this value (as well as the standard deviation of the sample) is increased by the unusually large long bone fragment (SA 79-1032B) found in square F6 (*Table 11.2*). This tendency is clearly shown in the distribution presented in *Figure 11.2*. Most of the fragments fall between 20 and 80 mm.

The composite nature of the archaeological site suggests that many of the bones were probably found in a secondary deposition, possibly following trampling, exposure to weathering, and multiple redepositions. This would account for the bad taphonomic condition of the finds, which made species-level zoological identification very difficult in several cases.

12
Marine Shells

Cesare D'Annibale

Species	Tomb 4	Tomb 10	Work-shop Area	Kiln Com-plex	Slope Area	House on Hill	Count
Bivalves							
Pectinidae				3	9		12
Chlamys glabra	2	1					3
Aequipecten opercularis					3		3
Pseudamussium clavatum					2		2
Glycymeris sp.					1		1
Ostrea sp.			2		1	2	5
Ostrea edulis			1	1	2	2	6
Tapes decussates	1						1
Gastropods							
Monetaria annulus	1						1
Tonna galea		1					1
Total	4	2	3	4	18		**35**

Table 12.1 Quantification of marine shells from Sant'Angelo Vecchio.

The shell assemblage from Sant'Angelo Vecchio consists of 35 marine mollusks: 29 from the site, and six from the tombs. All of the mollusks from the site are bivalves, whereas the tombs produced two gastropods as well.[1] The shells, primarily the bivalves, could have been harvested on the nearby Ionian coast. The Metapontine littoral is characterized by long stretches of beaches with sandy substrates and muddy gravelly bottoms at the mouths of the large rivers such as the Bradano, Basento, and Cavone—the typical habitat for most of these mollusks. A geologic origin for some shells is also possible as the Metapontine basin consists of a series of marine terraces laid down during the Pleistocene, many of which contain strata with fossilized mollusks.

The low frequency of shells at Sant'Angelo Vecchio could be attributed to off-site disposal or perhaps poor preservation, as half of the bivalve remains are fragmented. Despite Metaponto's proximity to the sea, most farmhouse sites in its chora produce few marine shell remains,[2] a pattern that reflects a minimal dependency on mollusks as a staple food source, and instead indicates at best a supplementary role for mollusk consumption. The few examples point to a sporadic procurement, suggesting that fresh shellfish may have been only an occasional addition to a meal. If mollusks did form part of the regular diet, they may have been shelled at a coastal production center and transported in long-term storage containers either dried, salted, or in brine. The low frequency of shells may also suggest they had functions other than food, such as medicinal or curative elements, utensils, ornaments, or simply collected curiosities—all activities that preclude a large number of examples.

In contrast, the mollusks from the tombs were likely procured as standard merchandise for inclusion in funerary assemblages, evidence for the customary association of shell artifacts with gender, ethnicity, and as a reflection of belief systems. The shells may represent items of a more personal nature as individual belongings or keepsakes buried with the person as part of their ritual attire. The ritual aspect that shells take on in this context can also be proposed as domestic cult activity, in which shells are integral elements as either implements or votive offerings.

[1] I would like to express my gratitude to Lindsey Groves of the Natural History Museum of Los Angeles County for his verification of some shell species. For general identification of shell species, see Poppe and Goto 1993.

[2] The pattern is repeated at several farmhouse sites excavated by the University of Texas at Austin, such as Sant'Angelo Grieco, San Biagio, and Fattoria Fabrizio (D'Annibale in *Fattoria Fabrizio*).

Catalog of Shells

Cypraeidae

Cypraeidae species are gastropods that are generally not eaten but characteristically recovered as funerary goods in southern Italian archaeological contexts. Some recoveries of *Cypraea* sp. in tomb contexts include Lavello—Alleanza Contadina, tomb 196, an infant burial dated to the last quarter of the 6th c. BC (Osanna and Russo 1988, 100); Cavallino—Fondo Villa tomb 1 (Arias 1979, 225, fig. 92, nos. 10–11); Alianello—Cazzaiola tomb 315 and tomb 309, female burials, both dating to the end of 7th/early 6th c. BC (Tempesta 1996a, 152, fig. 2.12; 155–57, fig. p. 113); Minervino Murge tomb 6, a female burial dating to the end of the 5th/early 4th c. BC, which produced a *Cypraea* shell in association with a rich complement of bronze and amber adornments (Corrente 1997, 162); Taranto, tomb 9 dated to mid- to last quarter of the 2nd c. BC (Colivicchi 2001, 109, no. 9.28). The example from Tomb 4 at Sant'Angelo Vecchio, *Monetaria annulus*, or ring cowrie shell, is regarded today as a species foreign to the Mediterranean (SAV Sh 01). Its natural distribution is localized to the Indo-Pacific region, with the Red Sea its closest habitat to the Mediterranean. The presence of this species at Sant'Angelo Vecchio is perhaps one of the most enigmatic and exotic of recoveries from the site. Cowries are endowed with a long tradition of amuletic associations shared by many cultural groups spanning a wide area, extending their distribution well beyond their natural habitat.

The *Monetaria annulus* shell was recovered in a double adult female burial dating to the second half of the 5th century BC (Tomb 4), in association with glass and bronze beads. In the same burial, three other shells (two scallop, SAV Sh 04 and SAV Sh 05, and one carpet shell, SAV Sh 19) were also found. These three examples represent a distinct grouping of funerary goods belonging to the other individual, as these appear to be cosmetic in nature and may be from a reburial episode. One of the burials appears to be an episode of reburial, indicating that there is no direct association between the shells in question (for the description on the nature of the tomb, see p. 100 in Ch. 4, "Tomb 4"). The back of the Sant'Angelo Vecchio cowrie has been cut away just above the natural golden yellow band around the base of the dorsum. The aperture is likely for threading or suspension, making it an object of personal ornament to be attached or strung on a band. Many cowrie shells are found as decorative elements in composite belts, necklaces, and headdresses (at Alianello on a belt and headdress in combination with other decorative items including glass, amber, and bone beads, Tempesta 1996a; at Cavallino on a necklace with similar beads and also bronze ring, Arias 1979). This association would seem to be applicable to the assemblage from Tomb 4, where glass and bronze beads were also found. Although lacking the evidence for a specific jewelry item, the combination of these various elements could imply a similar type of composite decorative piece. Nevertheless, these shells do sometimes occur as single ornamental artifacts (at Lavello, Osanna and Russo 1988; at Taranto, Colivicchi 2001).

In ethnographic and anthropological contexts, cowrie shells have a long history of powerful associations that range from objects of commercial value to charms and amulets associated with religious and magical or superstitious belief systems. The amuletic qualities generally bestowed on *Cypraea* shells is further validated by their reproduction in prestigious material such as amber. The *Cypraea* form in amber is particularly prominent in indigenous burials from Guardia Perticara where it is found as a pendant in three female burials dating to the first half of the 6th c. BC (Bianco 2012, 220, 232, 252). The presence of an exotic cowrie in Tomb 4 at Sant'Angelo Vecchio invites allusions to the potent nature of this artifact that transcend its use as a simple decorative element. The gender and fertility connotations commonly attributed to cowries impart perhaps a more powerful symbolism to this example. Cowries were widely distributed in prehistoric societies as an element of decoration perhaps due its resemblance to female genitalia. The aperture on the back of the shell opens to reveal the central whorl, in itself offering other allusions to fertility—possibly to male genitalia, representing the dual nature of gender, or as an abstraction of the womb with fetus. This longstanding tradition was likely known during Greek times as most cowries are found in female burials—as was also the case here.

Although *Monetaria annulus* is a rarity in southern Italian archaeology, several examples sharing an identical cut-away dorsum are found in the ancient Greek world. A similar assemblage of funerary goods consisting of the combination of cowrie shell and glass beads was found in a tomb from Chersonesos, demonstrating the possibility that certain superstitious beliefs, the evil eye in particular, were shared between far-flung areas (for an interpretation of the cowrie as evidence for belief in the evil eye, see Stolba 2009). Nevertheless, these shells may also be understood to have symbolic associations, as markers of belonging to a sacred sect, or as objects of personal pilgrimage to major sanctuaries beyond the region. *Monetaria annulus* examples with similar cut-away dorsum have been found at the Archaic sanctuary of Poseidon at Isthmia, where it has been suggested they are evidence of pilgrimage (Gebhard 1998). Similar cowrie shells have also been found at the 5th-c. BC altar of Aphrodite Ourania in Athens (Reese 1989, 68, pl. 16 d, e). Cowrie shells may also be an indicator of ethnicity. *Monetaria annulus* is somewhat atypical as a tomb offering since the majority of documented Cypraea finds from tombs in the region are Early Archaic with a strong indigenous character. This anomaly might be an indication of ethnicity, linking this burial and maybe some of the later inhabitants of Sant'Angelo Vecchio to an indigenous background. All is a matter of conjecture and attests to the dynamic nature that such a unique artifact brings to the site.

Cypraeidae

SAV Sh 01 *Monetaria annulus*
Common name: Ring cowrie or gold ring cowrie.
Provenance: Tomb 4; SA 79-965SS.
Condition: Complete, very worn.
Dimensions: l 2.3, w 1.7.
Comment: The domed back of the shell has been cut away to reveal the central whorl. The golden yellow band worn and intermittently visible. Found together with two *Chlamys glabra* valves (SAV Sh 04–SAV Sh 05) and one *Tapes decussates* valve (SAV Sh 19).
Date: 441–429 BC (weighted date range based on ceramics).
See also: T04-10.

1:2

Tonna galea

Tonna galea is one of the largest gastropods found in the Mediterranean. Although it may have been used as a food resource, it was perhaps known more for its aesthetic value. In archaeological contexts it is recovered on domestic sites, including those with ritual aspects, spanning the entire Greek period in southern Italy (in a domestic setting from the nearby farmhouse site at Sant'Angelo Grieco, SG81-256; Pantanello in a 6th-c. BC sanctuary context; Incoronata, IC78-459, and at Caulonia, Bernasconi 2007, 154; at Croton in a ritual deposit dating between the last quarter of the 5th and first quarter of the 4th c. BC, Ruga 2010, n. 42). As part of funerary assemblages, however, its frequency is rare, appearing only in Early-Roman-period tombs (for example, at Taranto in tomb contexts dating from the first half of the 2nd c. BC to the 1st c. AD, Colivicchi 2001, 96, no. 8.5). At Sant'Angelo Vecchio, it is found in Tomb 10 (SAV Sh 02) in association with a *Chlamys glabra* valve (SAV Sh 03) in a double burial consisting of a woman of advanced age and a child. The location of the shells near the head and on the chest is suggestive of their use as pendants. Taking into account that *Tonna galea* is often found as a singular recovery, in addition to its visually attractive aspect, these exceptional gastropods can be considered exotic objects and would have been commercially distributed or subject to some privileged form of distribution. The inclusion of this shell in the funerary assemblage marks a distinct differentiation of this older individual from the group, as these exotic items are rarely found as standard burial goods.

Tonna galea

SAV Sh 02 *Tonna galea*
Common name: Tun shell.
Provenance: Tomb 10, SAV 11.
Condition: Incomplete.
Dimensions: l 5.7, w 4.5.
Comment: Found together with a *Chlamys glabra* valve (SAV Sh 03).
Date: End of the 4th–first quarter of the 3rd c. BC.

1:2

Pectinidae

Generically referred to as "scallop shells," Pectinidae present a widespread distribution on Greek sites in southern Italy. Their function ranges from a source of food, both in domestic settings and as an element in ritual feasting (over 600 Pectinidea valves were recovered from various Greco-Roman contexts at Pantanello; *Pecten jacobeus* valves were also recovered from Caulonia, Bernasconi 2007, 154; in domestic ritual contexts from Fattoria Fabrizio, D'Annibale in *Fattoria Fabrizio*, 145–46, and from Croton, Ruga 2010, n. 42; and in ritual feasting contexts from the Sanctuary of Demeter at Siris Herakleia, Tschurtschenthaler and Gertl 2012), to utensils especially employed in cosmetic applications, a role that carries over to burials as well (tombs 48, 80, 136, 189, and 330 from the Pantanello and Saldone necropoleis, Carter and Hall in *Necropoleis*; from Incoronata tomb 58, Chiartano 1983, 165, fig. 87; Tursi-S. Maria D'Anglona in a 3rd-c. BC tomb, Bianco 1993b, 205; Taranto in a 2nd- to 1st-c. BC tomb, Hempel 2001, pl. 64, n. 16). With several species represented, Pectinidae shells constitute the most common mollusk at Sant'Angelo Vecchio. Twenty examples were recovered, from which three species could be identified: *Chlamys glabra*, *Aequipecten opercularis*, and *Pseudamussium clavatum*. As noted by numerous references in ancient literary texts, scallops were valued as a shellfish food source in the Greek world (Voultsiadou and Vafidis 2007, 112, 116). Nevertheless, the frequency of these shells at Sant'Angelo Vecchio is relatively too low to imply that they formed an integral part of the inhabitants' regular diet. The presence of scallop shells on site could relate instead to an occasional addition to a meal or as part of some specialized rite or feast. The appearance of some also suggests a fossil provenance, much like the case for the *Ostrea* recovered from the site.

Scallops achieve greater prominence in Magna Grecia, however, as an object of material culture in connection to mortuary practices. The choice of scallop shells over others is a matter of conjecture related to several aspects of social behavior and beliefs, among which personal collectibles rank as one of the more common explanations. The *Chlamys glabra* shell (SAV Sh 03) found in association with a *Tonna galea* shell (SAV Sh 02) in Tomb 10 is one example from which such a collection might be posited. This practice was expanded to the point where it became popular, transforming scallop shells into standardized objects of inclusion in mortuary assemblages—a custom especially noticeable in Hellenistic-period burials.

This popularity is due to its role as an essential toiletry item in daily life that continues to play a role in burial rites (as seen from a *pecten* shell with drilled holes on its wings for the attachment of rings to bind two valves from Tomb 193 in the Pantanello necropolis, Carter and Hall in *Necropoleis*. Similar shell cosmetic containers are also found at a 3rd- to 1st–c. BC tomb from Taranto, Colivicchi 2001, 171, no. 28.7, and from a female burial of the end of the 4th c. BC at Lilybaeum, see Bechtold 1999, 180–81, tav. LX, 2). Usually identified as cosmetic containers, many shells still retain traces of their original coloring substances on the valve's interior. The toiletry assemblage from Tomb 4 is enhanced by the presence of three shells, two of which presented remnants of coloring agents (SAV Sh 05, SAV Sh 19). Another functional role for these valves—a ladling utensil—can be detected from the smaller *Chlamys glabra* shell found in Tomb 4 (SAV Sh 05). Here, traces of a rosy-colored substance still adhered to both the interior and exterior of the valve. These features in association with the pyxides found alongside the burial suggest a direct link between these objects (for a similar association of shells with pyxides as part of a cosmetic set, see Prohászka in *Necropoleis*, 806. The scallop shell found within the metal pyxis displayed signs of grinding on the interior and was likely used as a grinding receptacle in association with a small grinding pebble. A similar combination of objects comes from a Hellenistic tomb from Eraclea: see Pianu 1990, 101, tav. XLVI, 2, where a shell was found within one of the compartments of a metal pyxis. This could be interpreted as a shell ladle for the handling of the toiletry substance probably found within): the ceramic vessel as the storage container for the substance, the smallest *Chlamys glabra* shell used as a ladle (SAV Sh 05), while the larger and more concave *Tapes decussates* shell served as a mixing container to produce the appropriate shade of color (SAV Sh 19). This toiletry assemblage can be regarded as standard goods included in gender-specific burials. These items, however, may have served a vital role in the burial event as a kit for the preparation or ornamentation of the body during the ceremonial rite of passage from the living to the afterlife.

Pectinidae

SAV Sh 03 *Chlamys glabra* (Linnaeus 1758)
Common name: Smooth scallop.
Provenance: Tomb 10; SAV 11.
Condition: Complete, very worn.
Dimensions: l 4.5, w 4.8.
Comment: Found together with a *Tonna galea* shell (SAV Sh 02).
Date: End of the 4th–first quarter of the 3rd c. BC.

1:2

SAV Sh 04 *Chlamys glabra* (Linnaeus 1758)
Common name: Smooth scallop.
Provenance: Tomb 4, Level 1, batt. 4; SA 79-968SS.
Condition: Complete.
Dimensions: l 4.1, w 4.3.
Comment: Found together with another *Chlamys glabra* valve (SAV Sh 05) and a *Tapes decussates* valve (SAV Sh 19).
Date: 441–429 BC (weighted date range).
See also: T04-11.

1:2

SAV Sh 05 *Chlamys glabra* (Linnaeus 1758)
Common name: Smooth scallop.
Provenance: Tomb 4, Level 1, batt. 4; SA 79-968SSb.
Condition: Complete.
Dimensions: l 3.1, w 3.4.
Comment: Traces of a rosy red substance adhering on interior and back of valve; found together with another *Chlamys glabra* valve (SAV Sh 04) and a *Tapes decussates* valve (SAV Sh 19).
Date: 441–429 BC (weighted date range).

1:2

SAV Sh 06 *Aequipecten opercularis* (Linnaeus 1758)
Common name: Queen scallop.
Provenance: Slope 3 W extension, Level 1, batt. 1; SA 79-63SS (see *Fig. 2.35*).
Condition: Near complete, left valve; missing umbo area.
Dimensions: l 6.6, w 6.5.
Date: Abandonment level above end-of-4th-c. BC layer.

1:2

SAV Sh 07 *Aequipecten opercularis* (Linnaeus 1758)
Common name: Queen scallop.
Provenance: Slope 3, W extension (continued N: D4), Level 1, batt. 5; SA 79-100SS (see *Fig. 2.35*).
Condition: Complete, right valve.
Dimensions: l 3.1, w 3.3.
Comment: Two other Pectinidae fragments found in association.
Date: Abandonment level above end-of-4th-c. BC layer.

1:2

Pectinidae cont.

SAV Sh 08 *Aequipecten opercularis* (Linnaeus 1758)
Common name: Queen scallop.
Provenance: Quadrant D5-C5, US 34; SAV 11.
Condition: Near complete, missing umbo area and auricle wings.
Dimensions: l 6.3, w 5.4.
Date: End of the 4th/first half of 3rd c. BC.

SAV Sh 09 *Pseudamussium clavatum*
Common name: Club scallop.
Provenance: Slope 3, Level 3, batt. 5; SA 79-898SS (Assemblage 13).
Condition: Complete.
Dimensions: l 4.4, w 4.3.
Date: End of the 4th/first half of the 3rd c. BC.

1:2

SAV Sh 10 *Pseudamussium clavatum*
Common name: Club scallop.
Provenance: Slope 3, D4 W, Level 3, batt. 5; SA 79-909SS (Assemblage 13).
Condition: Complete; damage on right edge.
Dimensions: l 4.8, w 4.6.
Date: End of the 4th/first half of the 3rd c. BC.

1:2

SAV Sh 11 Pectinidae
Common name: Scallop.
Provenance: F3, N Kiln interior, Level 2, batt. 1; SA 79-834SS (Assemblage 8).
Condition: Three fragments; found in association with one *Ostrea* sp. shell (SAV Sh 24).
Date: 2nd–1st c. BC.

SAV Sh 12 Pectinidae
Common name: Scallop.
Provenance: Slope 2, Sounding B, Level 3, batt. 2; SA 79-302SS (Assemblage 11).
Condition: Two fragments of one; ventral edge.
Date: End of the 4th/first half of the 3rd c. BC.

SAV Sh 13 Pectinidae
Common name: Scallop.
Provenance: Slope 3 W extension, Level 2, batt. 1; SA 79-391SS (Assemblage 13).
Condition: Two fragments.
Date: End of the 4th/first half of the 3rd c. BC.

SAV Sh 14 Pectinidae
Common name: Scallop.
Provenance: Slope 3, continued N (D4), Level 1, batt. 4; SA 79-87SS.
Condition: Fragment.
Date: End of the 4th/first half of the 3rd c. BC.

SAV Sh 15 Pectinidae
Common name: Scallop.
Provenance: Sounding C, Level 2, batt. 6; SA 79-204SS (see *Fig. 2.35*).
Condition: Fragment.
Date: 5th to 4th c. BC.

SAV Sh 16 Pectinidae
Common name: Scallop.
Provenance: Sounding C, Level 2, batt. 7; SA 79-207SS (see *Fig. 2.35*).
Condition: Fragment.
Date: 5th to 4th c. BC.

SAV Sh 17 Pectinidae
Common name: Bivalve.
Provenance: Slope 3, Quadrant C5, US 25; SAV 10.
Condition: Two fragments.
Date: End of the 4th/first half of the 3rd c. BC.

SAV Sh 18 Pectinidae
Common name: Scallop.
Provenance: US 35; SAV 11.
Condition: Fragment.
Date: End of the 4th/first half of the 3rd c. BC.

Tapes decussates

Tapes shells are quite common in prehistoric contexts, as demonstrated locally by their presence at the Neolithic site of Saldone and the Neolithic/Eneolithic site of Pantanello where they account for nearly a third of the mollusk assemblage (for Saldone, see Di Fraia 1970; the shell remains from the Pantanello Neolithic/Eneolithic site are in the process of analysis by the author). They are, however, virtually absent on later sites, occurring only once in a Late Roman context at Pantanello (shell recovered from a pit filled during the Late Roman period that also contained some Neolithic material to which the shell could pertain as well: Saggio 12 from 1974 University of Lecce excavations at Pantanello). Elsewhere, the Roman-to-Medieval components at Otranto are among the few documented occurrences of these bivalves in southern Italian archaeological contexts. The recovery of a *Tapes* shell at Sant'Angelo Vecchio would make it one of the few documented recoveries in the region for the Greek period (SAV Sh 19). Rarer still is its occurrence in a burial context in association with two other scallop shells (SAV Sh 04 and SAV Sh 05). Remnants of a red ochre-like substance on the interior of the shell would suggest its use as a cosmetic container (for use of shells as cosmetic containers with remnants of colored substances, see tomb 274 from Brindisi, Cocchiaro 1988, 79, no. 27; also from tombs 3, 15, and 54 at Taranto, Colivicchi 2001, 74, no. 3.14; 138, no. 15.16; 222, no. 54.8). Traces of a different colored substance were also visible on one of the scallop shells (SAV Sh 05), suggesting that these substances were possibly mixed to create various shades of color. These cosmetic containers could play a role in the funerary ritual whereby the individual was prepared for burial with facial make-up or coloring for the death repose.

Tapes decussates

SAV Sh 19 *Tapes decussates* (Linnaeus 1758)
Common name: (Crosscut) Carpet shell.
Provenance: H6, Tomb 4; Level 1, batt. 4; SA 79-968SS.
Condition: Near complete, two fragments; missing ventral edge.
Dimensions: l 2.5, w 3.5.
Comment: Faint traces of red substance on interior of valve. Found in association with two *Chlamys glabra* shells (SAV Sh 04 and SAV Sh 05).
Date: 441–429 BC (weighted date range).

1:2

Glycymeris sp.

Glycymeris shells appear consistently from prehistoric to Greco-Roman times on archaeological sites (*Glycymeris* species shell at the Neolithic/Eneolithic site at Pantanello as well as in the Greek and Roman components of the site. They are also found in low frequencies at Incoronata, Sant'Angelo Grieco, San Biagio, and Fattoria Fabrizio [D'Annibale in *Fattoria Fabrizio*]; for Incoronata, see also Dal Sasso 1995, 131. Other documented occurrences on southern Italian Greek sites include Caulonia-Casamatta, Bernasconi 2007, 154; and Croton, Ruga 2010, n. 42). Although their presence in the Metapontino is common, their frequency in the marine shell assemblage usually constitutes one of the minor components and in many cases is represented by a single example. On Greek sites, *Glycymeris* shells are found in a number of contexts that indicate a role beyond a food source. These shells are commonly found as objects related to personal adornment, and subsequently fulfill a role in the realm of ritual, especially connected to burials (*Glycymeris* finds in domestic ritual are to be seen at Croton, see n. 11 above. These shells are a common burial item as seen from tombs at Brindisi, Cocchiaro 1988, 79, no. 27; 174, no. 315; and Taranto, Colivicchi 2001, 74, no. 3.14; 109, no. 9.28; 138, no. 15.16; 151, no. 21.25; 222, no. 54.8, although in Late Hellenistic contexts). At Sant'Angelo Vecchio, only one fragment of *Glycymeris* of undetermined use was recovered, substantiating its repeated but limited appearance on archaeological sites (SAV Sh 20).

Glycymeris sp.

SAV Sh 20 *Glycymeris* sp.
Common name: (Dog) Cockle.
Provenance: Sounding D, Level 2, batt. 2; SA 80-50PL.
Condition: Fragment.

Ostrea edulis

The *Ostrea* assemblage at Sant'Angelo Vecchio consists of 11 specimens, of which seven are complete. Virtually all of the shells are encrusted by a layer of sandy concretion. Many of these were also fragmented, damaged, and worn, whereas another retained an *Ostrea* shell fragment adhering to the back of the valve. The appearance of the assemblage indicates that the shells are likely fossils. Fossilized shell banks are common occurrences as Pleistocene geologic strata in the Metapontine basin. Their presence in archaeological strata at Sant'Angelo Vecchio is likely a result of their displacement from one such sedimentary layer. The shells were re-deposited haphazardly throughout the site from a fossilized shell layer, which had been cut into perhaps during the Greek construction of the kilns, sheds, and tomb shafts (although the number of shells is low, several oyster and scallop shells were found in the kiln area that may hint at a possible role for these shells as flux agents during kiln firings). The use of shells in such circumstances is documented at Messana during the 5th c. BC, although in metallurgical contexts, Ingoglia et al. 2008. Still other instances for a high frequency of shells in kiln contexts are to be seen at the *kerameikos* dump deposit at Metaponto, D'Andria 1980a, 435, 444; and at the Roman kiln deposit at Pantanello, D'Annibale forthcoming). The spatial distribution of these fossilized *Ostrea* coincides with all building programs at the site: both sets of kilns, Slope 3 and Sounding C, and the farmhouse.

Ostrea edulis

SAV Sh 21 *Ostrea edulis*
Common name: European flat oyster.
Provenance: Slope 3, W extension. Level 2, batt. 2. SA 79-88SS (see *Fig. 2.35*).
Condition: Complete right valve; back encrusted by a layer of sandy concretion.
Dimensions: l 4.3, w 3.4.
Comment: One other fragment found in same lot.
Date: End of 4th/first half of the 3rd c. BC

SAV Sh 22 *Ostrea edulis*

Common name: European flat oyster.

Provenance: Sounding C, W of division, Level 2, batt. 3; SA 79-326SS (Assemblage 13).

Condition: Complete left valve; ventral edge damaged/chipped; back worn; interior encrusted by a layer of sandy concretion.

Dimensions: l 5.8, w 6.1.

Date: End of 4th/first half of the 3rd c. BC.

1:2

SAV Sh 23 *Ostrea edulis*

Common name: European flat oyster.

Provenance: House on the Hill, S half of Test Trench 1. Level 1, batt. 1; SA 79-562SS (see *Fig. 2.39*).

Condition: Complete left valve with some edge damage. Back encrusted by a layer of sandy concretion. Another Ostrea shell fragment adhering.

Dimensions: l 9.3, w 8.7.

Date: Second half of the 6th–5th c. BC.

1:2

SAV Sh 24 *Ostrea edulis*

Common name: European flat oyster.

Provenance: F3, N Kiln interior, Level 2, batt. 1; SA 79-834SS (Assemblage 8).

Condition: Complete; heavily encrusted with sandy concretion on interior of valve.

Dimensions: l 2.9, w 2.7.

Comment: Young specimen; found in association with three Pectinidae shell fragments (SAV Sh 11).

Date: Late Republican.

SAV Sh 25 *Ostrea edulis*

Common name: European flat oyster.

Provenance: Slope 3, W extension, continued N (D4). Level 1, batt. 5; SA 79-1008SS (see *Fig. 2.35*).

Condition: Complete right valve; heavily encrusted with sandy concretion on interior of valve.

Dimensions: l 9.3, w 7.8.

Comment: One other fragment found in same lot.

Date: End of 4th/first half of the 3rd c. BC.

1:2

Ostrea edulis cont.

SAV Sh 26 *Ostrea edulis*
Common name: European flat oyster.
Provenance: House on the Hill, S half of Test Trench 2, Level 1, batt.1; SA 79-558SS (see *Fig. 2.39*).
Condition: Complete right valve; heavily worn; encrusted with sandy concretion on interior of valve.
Dimensions: l 7.8, w 6.7.
Date: Second half of the 6th–5th c. BC.

SAV Sh 27 *Ostrea* sp.
Common name: Oyster.
Provenance: H3, Level 1, batt. 1; SA 79-680SS (see *Fig. 2.22*).
Condition: Ventral edge fragment.
Dimensions: l 3.8, w 5.4.
Date: End of 4th/first half of the 3rd c. BC.

SAV Sh 28 *Ostrea* sp.
Common name: Oyster.
Provenance: House on the Hill, S half of Test Trench 1. Level 1, batt. 2; SA 79-561SS (see *Fig. 2.39*).
Condition: Fragment, encrusted.
Dimensions: l 4.0, w 3.0.
Date: Second half of the 6th–5th c. BC.

Part III

Catalog Abbreviations

Classes

Amp	Transport Amphorae
AT	Architectural Terracottas
BG	Black-gloss Ware
BW	Banded Ware
CkW	Cooking Ware
Cor	Corinthian Pottery
GW	Grey Ware
Imp	Imperial Wares
KF	Kiln Furniture
LA	Lamps
Lag	Lagynoi
Li	Lithics
Lou	Louteria
LR	Late Roman Wares
LW	Loom Weights
M	Metal Objects
Min	Miniatures
Mor	Mortaria
OD	Opus Doliare
PCW	Plain and Coarse Wares
PLi	Prehistoric Lithics
PMP	Post-Medieval Pottery
RT	Roof Tiles
RW	Red Ware
Sh	Shells
U	Unguentaria

Quantification Indices

B	bases
H	handles
MNI	Minimum Number of Individuals
NFr	Number of Fragments
R	rims
S	body sherds
TNI	Typological Number of Individuals
US	Stratigraphic Unit

Measurements

All catalog measurements are given in cm unless otherwise indicated.

diam.	diameter
ext.	exterior
h	height
int.	interior
l	length
max.	maximum
min.	minimum
(rec.)	reconstructed
th	thickness
w	width

Other

b./batt.	battuta
HoH	House on the Hill
Inv. no.	Inventory Number (Museo Nazionale Archeologico di Metaponto)
L.	Level

13
Prehistoric Artifacts

Cesare D'Annibale

The recovery of prehistoric material at Sant'Angelo Vecchio is yet another instance where archaeological work aimed at Greco-Roman aspects of research has revealed unexpected older occupations of the locale.[1] Unlike substantial habitation sites, small special purpose or single activity prehistoric sites are inherently difficult to identify. After several seasons of field survey in the chora of Metaponto, prehistoric exploitation of the Metapontine landscape is extensively documented from the identification of well over 500 sites. The survey, although geared toward the understanding of the pattern of Greek settlement, provided an unexpected and unparalleled revelation of an unobtrusive prehistoric landscape. The fortuitous recovery of prehistoric material from these historic sites subjected to the finer scrutiny provided by excavation adds to the already overwhelming evidence that the Metapontino was intensively occupied especially during the latter stages of the Neolithic.

At Sant'Angelo Vecchio, the occupation of the Greek period and its components effectively removed any prior occupation strata that would allow us to contextualize the prehistoric material. With one exception, all of the prehistoric artifacts—five chipped stones and nine ceramics—are found in disturbed contexts dating to the Greek period or as surface finds. The sole exception is the recovery of an impasto ware base from strata below the Greek phase of occupation. Despite being found in a prehistoric level, its context is still unclear as this homogenous stratum is recognized across the site with no distinct features associated with it.

Chert

The typical prehistoric acquisition strategy for the procurement of tool stone within the Metapontine basin relied on the recovery of chert pebbles from eroded Pleistocene terrace slope bank deposits or riverbed sources. These locally available pebbles rarely reach 10 cm in size and seldom exceed 100 g in weight. The most common chert variety is distinguished by an olive gray color, from which two cores are represented on site (SAV PLi 01 and SAV PLi 02). Although not uncommon, the dark speckled bluish gray and black translucent varieties are among those that rank in the lower frequencies in site assemblages from prehistoric sites in the Metapontine basin (SAV PLi 03 and SAV PLi 04). Utilization of metaquartzite as a source of tool stone is also another infrequent occurrence in the local chipped stone industry (SAV PLi 05). The presence of this material on site is indicative of a diverse method of tool stone acquisition, one that focused specifically on riverbeds of large rivers such as the Basento and Bradano that transported a greater variety of raw materials from the interior such as this metaquartzite from the Calabrian Arc.

The five chipped stone artifacts recovered from Sant'Angelo Vecchio reflect typical reduction methodologies employed on pebbles.[2] Removal of flakes relies on direct percussion, the most effective manner to reduce the pebble utilizing natural ridges or angular platforms as starting points. Random reduction follows with minimal platform preparation as seen from the presence of flakes with a flat striking platform that still retains its cortical surface (SAV PLi 05). Some edge grinding is evident on platforms but not necessarily applied on a standard basis. Local pebble cherts were primarily reduced to provide a source of flakes destined to be utilized as ad hoc tools for a variety of functions. Most of these flakes were discarded after a single or few uses. In contrast, the bipolar flake is indicative of subsequent reduction of pebble cores after direct percussion becomes difficult due to their diminished size (SAV PLi 04). Bipolar technology maximizes the use of pebble cores as viable flake pro-

[1] Excavations conducted by the University of Texas at Austin at several Greco-Roman sites including Sant'Angelo Grieco, San Salvatore, Incoronata, Pantanello, the Pantanello necropolis, Fattoria Fabrizio (D'Annibale in *Fattoria Fabrizio*), and San Biagio (D'Annibale in *San Biagio*) have produced prehistoric material. This pattern has also been documented at other site excavations in the chora of Metaponto such as Fattoria Stefan, Pacciano, Musillo, Pascucci-Torraca, Bufalara, and Pagliarone Fattoria 1 (Amore 1992–1993, table 14) and Cozzo Presepe (Macnamara 1983).

[2] For a similar discussion on the chipped stone industry in the Metapontino, see D'Annibale in *San Biagio*, 61–62, on the material recovered from the Neolithic component of the San Biagio Late Roman farmhouse.

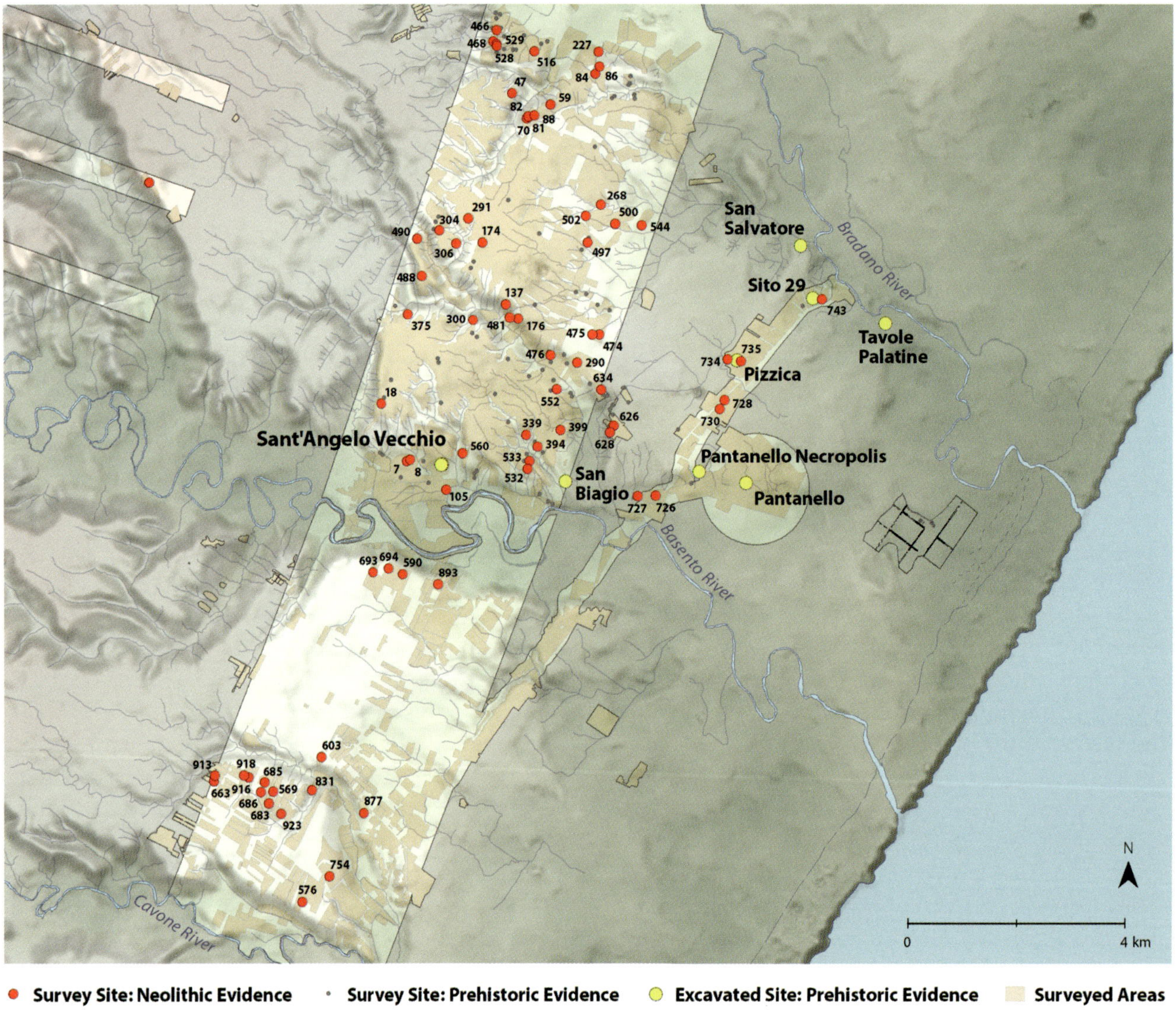

Figure 13.1 Map of prehistoric survey sites. (JT/ICA)

ducers. These splintered flakes would ultimately serve as wedges. In certain cases, bipolar flakes could result from the utilization of pebble cores as wedges, the battering of which would produce these splintered pieces. With the exception of this bipolar flake that may have been utilized no tools were found in the assemblage.

Ceramics

A total of nine prehistoric ceramics, representing the same number of vessels, were identified from the excavations at Sant'Angelo Vecchio. One is a fine ware or figulina sherd, while the remaining eight derive from various grades of depurated impasto coarse wares. The finds do not reveal a concentration of material in any particular area of the site, suggesting either a complete overhaul of the prehistoric landscape by historic building activities or that the dissociated finds are a reflection of sporadic or intermittent use of the area during the prehistoric period. The ceramics can be broadly allocated to the latter stages of the Neolithic, essentially the transition between Middle and Late Neolithic and possibly extending into the Eneolithic period (mid-5th to 3rd millennium BC).

The fine ware figulina fragment derives from a small bowl, the form of which links it to the extensive repertoire of small carinated bowls from the transition between Middle to Late Neolithic red-painted ceramics (SAV Figulina 01).[3] These are characteristically thin-walled and shallow, defined by a very angular juncture between the wall and base. The finely depurated impasto carinated bowl is somewhat atyp-

[3] For parallels from Cetrangolo, see Bianco 1983.

ical in terms of its fabric composition, however, the level of execution, form, and remnants of decoration closely parallels Serra D'Alto ceramic characteristics (SAV Impasto 01). The discernible chevron and meandering band motif is reminiscent of Serra D'Alto painted decoration that appears on bowls and cups.[4] The diffusion of this ware type is widespread throughout southern Italy with the finds from the regions of Basilicata and Apulia being particularly abundant. The fabric, however, would find a better connection to the Diana ceramic tradition, which is represented by both fine ware and impasto ceramics;[5] perhaps a blend or morphing of styles as these two cultural manifestations appear as contemporaneous entities especially during the mid-4th millennium.

For the most part the impasto ceramics are finely depurated with only one being a coarser variety. The fabric is well purified with occasional larger inclusions common but not dominating the fabric. In addition to the handled bowl/cup mentioned above, only two other forms can be identified from the impasto ceramic assemblage including a collared jug (SAV Impasto 03). Although not particularly diagnostic, a general allocation to the Neolithic can be proposed for this form as well. The collared jug is a typologically common form especially noted in Middle to Late Neolithic fine ware ceramics with the coarse ware form being equally represented. The base from an open container with a projected form pointing to a conical bowl (SAV Impasto 02) coupled with two sherds displaying classic technical features that are markers for Eneolithic ceramics, specifically Macchia a Mare wares, would tend to indicate a 3rd-millennium BC presence as well (SAV Impasto 05, SAV Impasto 06).[6]

One curious reworking of an impasto coarse ware sherd into a rounded form, most likely intended for use as a stopper, was also found (SAV Impasto 04). The practice of recycling broken or discarded items is chronologically widespread. As it is the product of an individual response to a unique situation, its cultural affiliation could represent either a contemporaneous reutilization during the Neolithic or a haphazard selection of a prehistoric sherd at some point during the Greek occupation of the site. The latter possibility finds many parallels in Greek-period contexts, as reworked ceramic sherd stoppers were an expedient method of sealing narrow mouthed containers.[7] However, several other uses for this object can be proposed, among which gaming pieces, tokens, or counters are but a few alternatives. At Sant'Angelo Vecchio, a case can be made for pieces such as this functioning as kiln spacers.[8]

The presence of impasto and figulina pottery and chipped stone likely derives from sporadic utilization of the spring at Sant'Angelo Vecchio during the prehistoric period, related to the latter stages of the Neolithic. Perennial springs have a long history of utilization in the Metapontino with prehistoric activity being recorded at many of these locations. Neolithic occupation has been documented adjacent to several spring localities such as Pantanello, San Biagio, and San Marco (*Fig. 13.1*).[9] With the identification of a Late Neolithic site some 400 m to the west, the spring at Sant'Angelo Vecchio is yet another locality that supported a major prehistoric settlement.[10] In addition to several other small sites in the surrounding area,[11] the occasional recovery of isolated prehistoric material suggests that Sant'Angelo Vecchio served as a subsidiary activity site in relation to the larger settlement. These outlying findspots are documented throughout the Metapontino indicating an extensive exploitation of the territory radiating from nodal centers.

The Basento River constitutes one of the major transportation corridors in the Metapontino from which prehistoric groups accessed the interior from the Ionian coast or vice versa. The site at Sant'Angelo Vecchio is strategically located at the entrance to a secondary access route into the Metapontine plateau from the Basento. This is one of many such localities from which prehistoric populations settling on fluvial terraces would insinuate the interior side valleys and uplands in order to support the economy of major sites such as that of site 7/8 nearby.

4 For general parallels in form and decorative motifs, see A. De Mitri 1978, pl. 3, pl. 5 no. 4; Cremonesi 1984, pl. 20, no. 3.

5 For Diana culture ceramic form parallels, see A. De Mitri 1978, pl. 6 nos. 1, 2, 7.

6 See Ingravallo 1980; Bianco 1983. The Pantanello Eneolithic material is currently under study.

7 For similar reworking or recycling of ceramic sherds into circular shapes, see references from catalog for SAV Impasto 04.

8 A similar suggestion was made for rounded ceramic sherds at Siris and at the *kerameikos* of Metaponto (Adamesteanu and Dilthey 1978, 523).

9 The Pantanello locality was the site of a major Middle/Late Neolithic to Eneolithic settlement (Ingravallo 1980; under study by the author). Near the San Biagio spring, Late Neolithic presence is documented from the excavation of a Late Roman farmhouse (D'Annibale in *San Biagio*); at the San Marco locality, another major Late Neolithic site was identified during the Metaponto field survey at site 227.

10 Metaponto field survey site 7/8.

11 Metaponto field survey sites 10, 103, 105, 554, 556, 558, and 560.

Catalog of Prehistoric Artifacts
Stone

SAV PLi 01 Core 1:2

Provenance: SAV 2011. US 32, pottery deposit pit fill.
Material: Chert.
Dimensions and technical features: l 5.0, w 3.5, th 1.8. Weight 38.7 g. Color: Munsell 5Y 4/2 to 5/3 olive gray to olive.
Condition: Two primary flakes removed.
Description: Tabular pebble core; two small broad primary flakes removed from one facet; 90% of cortical surface remains on one face.
Date: Prehistoric.

SAV PLi 02 Core 1:2

Provenance: SAV 2010. Square C4 US 18. Cleaning of the Retaining Wall.
Material: Chert.
Dimensions and technical features: l 4.6, w 3.2, th 1.1. Weight 18.5 g. Color: Munsell 5Y 4/2 to 5/3 olive gray to olive.
Condition: Two primary flakes removed.
Description: Tabular pebble core fragment; 100% of cortical surface remains on one face.

Comment: Same pebble as SAV PLi 01.
Date: Prehistoric.

SAV PLi 03 Core 1:2

Provenance: SAV 2010. Surface. Terrace slope east of farmhouse.
Material: Chert.
Dimensions and technical features: l 4.7, w 3.2, th 2.8. Weight 54.4 g. Color: Munsell gley 2 3/1 very dark bluish gray mottled and speckled with gley 2 5/1 bluish gray.
Condition: Eight flake removal scars evident.
Description: Randomly flaked pebble core; eight small broad and blade-like flakes removed on three platforms or facets; 35% of cortical surface remains; angular flaw lines within the pebble hinder flake removal producing snap ending flakes.
Date: Prehistoric.

SAV PLi 04 Bipolar Flake 1:2

Provenance: F7, Level 2, batt. 2. Lot: SA 80-5PL.
Material: Chert, translucent.
Dimensions and technical features: l 2.9, w 1.2, th 0.8. Weight 2.4 g. Color: Munsell 10YR 2/1 black to 10YR 3/1 very dark gray; quartz flaw lines.
Condition: Complete.
Description: Bipolar flake from pebble core fragment as several impact marks are visible on a striking platform; 5% of original cortical surface remaining.

Comment: Possible utilization as wedge.
Date: Prehistoric.

SAV PLi 05 Flake

1:2

Provenance: SAV 2010. Square C5, US 12.
Material: Metaquartzite.
Dimensions and technical features: l 2.1, w 2.8, th 0.8, weight 5.4 g. Color: Munsell 10R 3/3 black dusky red to 5YR 3/2 dark olive gray.
Condition: Proximal segment.
Description: Tertiary flake; cortical platform.
Comment: Likely a spall from percussion of nodule or cobble.
Date: Prehistoric.

Ceramics

SAV Figulina 01 Carinated Bowl

1:2

Provenance: Sounding D. Level 2, batt. 3. Lot: SA 80-51PL.
Material: Fine ware "figulina."
Dimensions and technical features: body th 0.60–0.85. Color: Munsell 10YR 8/2-8/3 very pale brown to 10YR 7/2 light gray; slip color 5YR 6/6 reddish yellow. Fabric type: grayish black core with very fine carbonate and sandy grain inclusions under 1 mm.
Condition: Carinated fragment; reddish yellow slip on exterior worn off but well conserved on interior.
Description: Very angular juncture at point of carination; slight turning/wiping grooves on interior. Reddish slip or paint on interior.
Comparison and comment: The porous appearance of the fabric and blackened core indicate the vessel was overfired. The form can be assumed under a generic red-painted ceramic repertoire of the Middle to Late Neolithic transition. Possible parallels can be made with forms from Cetrangolo (Bianco 1983, 7–8, figs. 3n, 4b, f) and from Grotta no. 3 at Latronico (Cremonesi 1984, pl. 20, no. 1).
Date: Neolithic.

SAV Impasto 01 Carinated Bowl

1:2

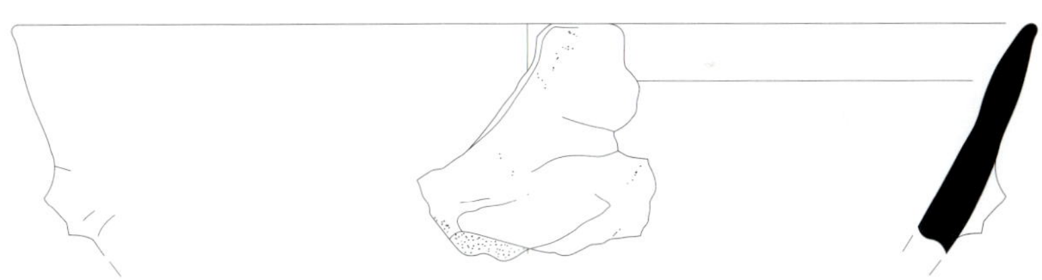

Provenance: SAV 2011 US 45.
Material: Fine impasto.
Dimensions and technical features: diam. 21.0, body th 0.6–0.8. Color: exterior Munsell 2.5YR 4/8-5/8 red; interior 2.5YR 4/8 red to 5R 5/6 yellowish red; black core. Fabric type: finely depurated clay with occasional sandy grain reaching 2 mm.
Condition: Rim to body carination with handle attachment.
Description: Lip thinned and rounded; rounded carination at widest part of vessel; strap-handle attachment at point of carination (3 cm wide, 0.8 thick); moderate burnishing on exterior and interior.
Decoration: Possible black painted chevron-like motif on interior; same type of line decoration on exterior possibly with a chevron/wavy line motif.
Comparison and comment: Although the fabric is not typical of Serra D'Alto ceramics, the form can be linked to Serra D'Alto type carinated bowl/cups especially if remnant of decorative motif is taken into account. This type of decoration is particularly evident on Serra D'Alto ceramics during the 4th millennium BC in southern Italy. For generic parallels to Serra D'Alto ceramics see Cremonesi 1984, pl. 20, no. 3 (from Latronico); A. De Mitri 1978, pl. 3, 1–2; pl. 5 no. 4 (from Grotta del Diavolo); from Saldone see Di Fraia 1970. The fabric is typical for Diana ceramics and the form as well also finds parallels in this tradition (A. De Mitri 1978, pl. 6, nos. 1, 2, and 7).
Date: Middle to Late Neolithic.

SAV Impasto 02 Open Container

1:2

Provenance: SAV 2010. W Section.
Material: Fine impasto.
Dimensions and technical features: diam. 9.5, body th 0.7–1.0. Color: surfaces and core Munsell gley 2.5/ black. Fabric type: very fine carbonate and sandy grain inclusions under 1 mm with the occasional carbonate grit reaching 3 mm.
Condition: Complete base with start of body.
Description: Flat base with some uneven concavity from pinching to shape wall; obtuse everted body; scarification still visible after smoothing on interior; exterior smoothed with some traces of scarification; two deep striations on underside of base; curious dimple depression on interior center of base from a finger or thumb imprint; from this central depression a series of radial traces of finger drags; low firing temperature.
Comparison and comment: This type of base and the angle at which the wall of the vessel deviates indicate a form that closely resembles flaring conical bowls usually associated with Eneolithic ceramics in particular Macchia a Mare wares. Similar bases were recovered from Pantanello (PZ75-584PL 1153).
Date: Late Neolithic or Eneolithic?

SAV Impasto 03 Closed Collared Jug

1:2

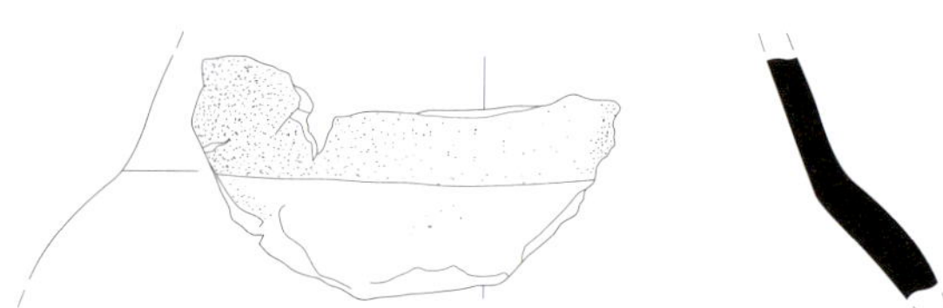

Provenance: Slope 4, N of Sounding A, found during clearing off of loose soil left by bulldozer. Lot: SA 79-52PL.
Material: Fine impasto.
Dimensions and technical features: neck diam. (base of neck) 11.0, body wall th 0.8, neck th 0.4. Color: exterior Munsell 5YR 5/4 reddish brown blotched with 7.5YR 3/1 very dark gray; interior 5YR 4/1 dark gray. Fabric: very fine sandy grain inclusions under 1 mm.
Condition: Neck and shoulder fragment.
Description: In-sloping neck with marked separation between neck and shoulder; moderate burnish on exterior; light to moderate burnish on interior; scarification still slightly visible on interior; etched/incised line demarcating juncture of neck and shoulder of vessel; line wavers.
Comparison and comment: See closed collared vessels from San Biagio (D'Annibale in *San Biagio*, 59, 2.1.1–2).
Date: Late Neolithic.

SAV Impasto 04 Body Sherd (reworked into Stopper)

1:2

Provenance: Slope 4. Level 1, batt. 2. Lot: SA 79-60PL (see *Fig. 2.35*).

Material: Fine impasto.

Dimensions and technical features: diam. 6.0–7.0, body th 0.6–0.9. Color: exterior Munsell 10R 6/6 light red to 10R 5/6 red; interior is mostly dark gray to black with blotches of 10R 5/6 red. Fabric: matrix includes a high percentage of carbonate grit with sandy grains, occasional carbonate grain can reach 3 mm.

Condition: Modified body sherd.

Description: Chipped all around; smoothed exterior and interior; low burnishing; low firing temperature.

Comparison and comment: The preferred functional use for this artifact type is as a stopper. The diameter of the reworked sherd would make it suitable for a lid for a Greek jug-type container. Ceramic stoppers are a common occurrence as they are an expedient method of sealing narrow mouthed vessels. They are widespread in Greek contexts some of which are documented at Siris (Adamesteanu and Dilthey 1978, 523) and at Sibari-Stombi (Guzzo et al. 1973, 255 no. 132, fig. 266). Nevertheless other uses for this object can be proposed amongst which gaming pieces, tokens or counters are but a few alternatives.

Date: Neolithic sherd possibly reutilized in Greek-period contexts.

SAV Impasto 05 Body

1:2

Provenance: Slope 3. Level 1, batt. 2. Surface cleaning. Lot: SA 79-59PL (see *Fig. 2.35*).

Material: Fine impasto.

Dimensions and technical features: body th 0.8–1.0. Color: exterior Munsell 10YR 6/2–6/3 light brownish gray to pale brown with 2.5Y 7/3–8/3 pale yellow and 2.5Y 4/1 dark gray blotches; interior 10YR 5/3-6/3 brown to pale brown. Fabric: black core; fine sandy grain inclusions under 1 mm.

Condition: Body wall fragment.

Description: Moderate to high burnishing on exterior; moderate burnishing on interior; some scarification slightly visible but burnished over; high firing temperature.

Comparison and comment: The technical aspects such as the high burnishing, fine impasto fabric, and color link this sherd to Eneolithic ceramics associated with Macchia a Mare wares.

Date: Eneolithic?

SAV Impasto 06 Body

1:2

Provenance: Slope 3, Level 1. batt. 2. Lot: SA 79-70PL (see *Fig. 2.35*).

Material: Semi-fine impasto.

Dimensions and technical features: body th 0.7–1.0. Color: exterior Munsell 10YR 6/2–6/3 light brownish gray to pale brown; interior 5YR 6/3–6/4 light brown with 10YR 4/1–3/1 blotches. Fabric: black core; very fine matrix with carbonate and sandy grain inclusions under 1 mm.

Condition: Body wall fragment.

Description: Moderate burnish on exterior and interior; broad scarification but smoothed over on interior.

Date: Neolithic.

SAV Impasto 07 Body

1:2

Provenance: Sounding D. Level 1, batt. 2. Lot: SA 80-43PL.
Material: Fine impasto.
Dimensions and technical features: body th 1.05–1.15. Color: exterior Munsell 10YR 4/1–3/1 dark gray; interior 10YR 5/2. Fabric: very fine sandy grain inclusions under 1 mm; black core.
Condition: Body wall fragment.
Description: Smoothed interior; moderate to high burnish exterior.
Comparison and comment: The technical aspects such as the high burnishing, fine impasto fabric, and color link this sherd to Eneolithic ceramics associated with Macchia a Mare wares.
Date: Neolithic.

SAV Impasto 08 Body

1:2

Provenance: Sounding D. Level 2, batt. 4. Lot: SA 80-52PL.
Material: Impasto.
Dimensions and technical features: body th 0.8–1.2. Color: exterior Munsell 10YR 7/4 very pale brown; black interior. Fabric: black core with occasional grain reaching 2–3 mm.
Condition: Body wall fragment (three fragments); worn especially on interior.
Description: Coarser fabric; low burnishing.
Date: Neolithic.

14

Corinthian Pottery

Francesca Silvestrelli

The excavations of the Archaic farmhouse yielded a body fragment of a Late Corinthian I or II convex pyxis (SAV Cor 01), one of the few Archaic imports identified at the site, where most of the pottery seems to be of local production. The form, popular at Taranto,[1] occurs also at Metaponto, where nonetheless only a few examples have been published.[2]

[1] Neeft 1995, 370–71.

[2] A fragmentary LC I convex pyxis was found in the Crucinia votive deposit: Lo Porto 1981, 295, no. 20, fig. 4.2. See also the convex pyxis with handles from tomb 89 at Torre di Mare: Nava 2004, 962, pl. LVII.4. See also Burn in *Necropoleis*, 602, CD-289-23 (pyxis or kothon, late 6th c. BC). Corinthian pottery from Metaponto and its territory is not extensively published. The evidence through the middle of the 6th c. BC is assembled in Lippolis 1995, 543, no. 32; 544, no. 41. The evidence cited by Lippolis is augmented by Corinthian vessels from sanctuaries (Favale votive deposit: M. Liseno 2004, 76–78; Crucinia votive deposit: Lo Porto 1981, 294–96, figs. 2–4) and tombs (Torre di Mare: Nava 2004, 962; San Salvatore, tomb 175: De Siena 2011, Carter 2012, 1110–11, fig. 3, ancient Corinthian alabastron possibly by the Dolphin Painter; Pantanello: Burn in *Necropoleis*, 593 and 601–2; Carter 2012, 1109–11). A Middle or Late Corinthian kothon is the only Corinthian import recovered in the Bradano-Basento survey: Tubelli in *Survey*, 150, 160, no. 7.

Catalog of Corinthian Pottery

SAV Cor 01 Convex Pyxis (with or without Handles)

Provenance: House on the Hill, Test Trench 1. Level 1, batt. 2. Lot: SA 79-595P (see *Fig. 2.39*).

Dimensions and technical features: l 6.8, w 2.8, th 0.4, body diam. 18.0. Clay Munsell 2.5Y 8/2 (pale yellow), hard and well-levigated.

Condition: Two joining fragments of the body.

Description: Black band, double dotted band, and two black bands on the shoulder.

Comparisons and comment: For the shape in general see *Corinth VII* (5), 43–44. For the decoration, *Corinth VII*, 5, 45, no. 51, pl. 4 (compared to vases from Asine of the first quarter of the 5th c. BC).

Date: Late Corinthian, middle or second half of the 6th c. BC.

15

Black-gloss Ware and Lamps

Emanuela Conoci

The excavations at Sant'Angelo Vecchio yielded 742 fragments of black-gloss pottery.[1] Of these, 169 fragments came from the House on the Hill, and the remaining 573 from the area at the foot of the hill (*Table 15.1*).[2] All of the material seems to have been produced locally, except for two Attic fragments (SAV BG 01 and SAV BG 02) and one other import (SAV BG 40).

The House on the Hill

Context and Chronology

The largest concentration of black-gloss pottery from the top of the hill was encountered in the area in and immediately around the House on the Hill. The recovered material dates mostly to the Late Archaic period (*Table 15.2*), the evidence for which far surpasses that available for all other periods.

Besides Late Archaic material, the excavation also yielded two lekythoi (SAV BG 34 and SAV BG 35), both datable to the 5th century BC and related to the later occupation of the area. The chronological range of the materials recovered within the structure[3] spans the second half of the 6th (B3b cup, SAV BG 07) through the first quarter of the 5th century BC, while the remaining materials date mostly to the first half of the 5th century. The single example that does not fit into this general context is an Early Hellenistic A-type skyphos found in the topsoil of Test Trench 2. The frequency of material dating between the Late Archaic period and the 5th century BC creates a solid chronology for the House on the Hill; the Early Hellenistic skyphos is probably an intrusion introduced by later occupation of the hill.

Shapes and Functions

The identified forms are primarily drinking vessels (*Fig. 15.1*). In the Late Archaic period, the forms are almost exclusively cups, especially Ionic-type B2 cups and skyphoi with offset rims. In terms of closed forms, a single olpe was identified, with a possible function of storing and pouring liquids. In the first half of the 5th century BC, cup-skyphoi, kylikes, and A-type skyphoi are also attested. Two perfume containers, a trefoil and a squat lekythos, complement the repertoire of attested forms.

The Pottery Deposit (Square D5) and the Retaining Wall (Squares C4–5)

Context and Chronology

Practically the entire corpus of black-gloss pottery collected along Slope 4 and in the subsequent excavation of Sounding C dates to the Late Archaic and Classical periods (*Table 15.5*). The Pottery Deposit[4] represents the most important assemblage at Sant'Angelo Vecchio by virtue of the frequency of black-gloss pottery (*Table 15.3*). The apparently homogeneous context is distinguished by a high degree of fragmentation, noted at the time of excavation, and by a correspondingly high number of reconstructable vessels (see SAV BG 08, SAV BG 11, and SAV BG 22), a situation not encountered in the other areas. In fact, the finds were preserved in part by their location inside the deposit, which offered protection from abrasive actions typically resulting in the "removal of material on a sherd's surface by a mechanical contact."[5]

The material recovered from the Pottery Deposit is mostly Late Archaic and Classical in date (*Table 15.4*). By the last quarter of the 5th century BC, the frequency of black-gloss pottery dropped noticeably, and the few subsequent attestations do not go beyond the third quarter of the 4th century BC.

Shapes and Functions

Remarkably, the Late Archaic and Classical vessels from the Pottery Deposit, comprising the most concentrated period of activity, are overwhelmingly associated with one particular function: drinking (*Fig. 15.2*). The Late Archaic period is characterized by the

---

[1] A preliminary study of the materials presented here was undertaken by E. Vittoria in 2008.

[2] Quantification has been organized according to three levels, based on the condition of the sherds: the total number of fragments (NFr), the minimum number of individuals (MNI), and the typological number of individuals (TNI), a minimum number weighted on the count of all body sherds that may be from the same vessel (sherd family) (Raux 1998, 13).

[3] See Ch. 2, "Site Phasing, Stratigraphy, and Site Assemblage."

[4] See Ch. 2, "Site Phasing, Stratigraphy, and Site Assemblage."

[5] Taylor 2000, 19.

Provenance	NFr	MNI	TNI
House on the Hill	169	30	77
Hillside	573	91	266
Total	**742**	**121**	**343**

Table 15.1 Black-gloss from Sant'Angelo Vecchio: general quantification.

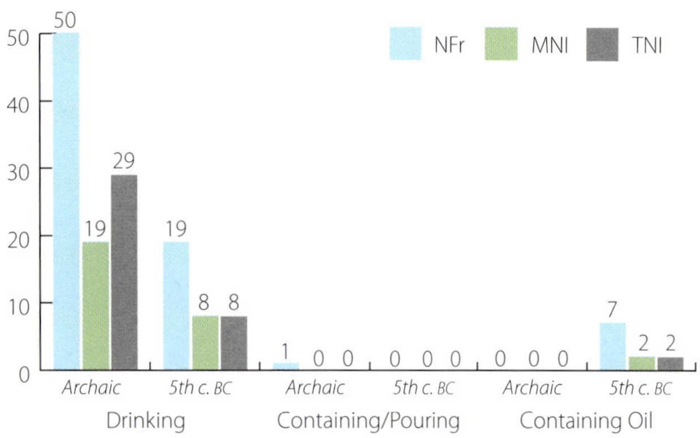

Figure 15.1 Black-gloss from the House on the Hill, organized by function. (LMJ/EC/ICA)

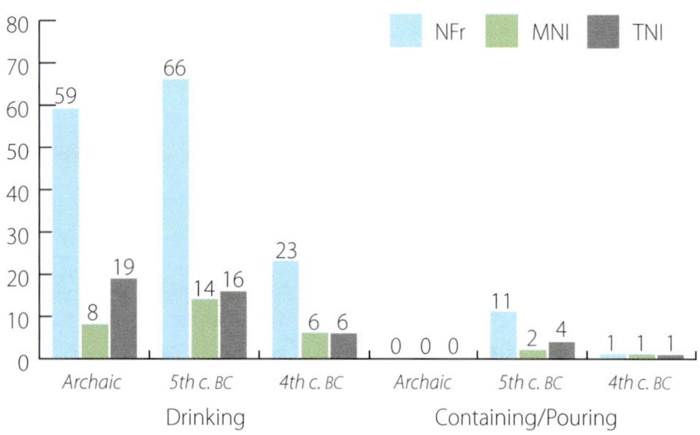

Figure 15.2 Black-gloss from the Pottery Deposit, organized by function. (LMJ/EC/ICA)

nearly exclusive presence of drinking cups, with Ionic-type B2 cups predominating over the skyphoi with offset rims. During the 5th century BC, the range of forms becomes more articulated, but nevertheless the assemblage is almost entirely composed of drinking vessels. The skyphos with concave rim is much more common than the other types of cup (cup-skyphos, kylix, skyphos with offset rim, A- and C-type skyphos), accompanied by an amphora, jug, and oinochoe as closed forms. In the 4th century BC, however, forms associated with drinking are limited to A- and C-type

Provenance	Form and Type	Date	NFr	MNI	TNI
Surface finds	Ionic-type B2 cup	Late Archaic	2	2	2
	Skyphos with offset rim	Late Archaic	5		1
	Open		3		2
	Closed		1		1
Test Trench 1	Ionic-type B2 cup	Late Archaic	3		2
	Open	Late Archaic	1		
Test Trench 2	Ionic-type B2 cup	520–480 BC	3		
	A-type skyphos	500–450 BC	1	1	1
		325–300 BC	1	1	1
	Open	Late Archaic	4		3
			2		1
Test Trench 3	B3b cup	550–500 BC	1	1	1
	Ionic-type B2 cup	Late Archaic	31	16	22
	Skyphos with offset rim	Late Archaic	8	2	5
	Open	Late Archaic	21		9
	Olpe	Late Archaic/ Classical	1		1
	Cup-skyphos	530–480 BC	1		
	C-type kylix	470–440 BC	9	1	1
	Kylix	Classical	2		
	Cup-skyphos	500–450 BC	1	1	1
	A-type skyphos	500–450 BC	3	2	2
		475–450 BC	1	1	1
		450–400 BC	1	1	1
	Trefoil lekythos	480–425 BC	2	1	1
	Squat lekythos	480–450 BC	5	1	1
	Skyphos		1		1
	Open		48	1	14
	Closed		10		8

Table 15.2 Black-gloss from the House on the Hill: quantification by form and date.

Provenance	NFr	MNI	TNI
House on the Hill	169	30	77
Hillside	356	60	186
Pottery Deposit (on hillside)	217	31	80
Total	**742**	**121**	**343**

Table 15.3 Black-gloss from Sant'Angelo Vecchio: quantification of the Pottery Deposit.

Phase	Form and Type	Date	NFr	MNI	TNI
Late Archaic	Ionic-type B2 cup		30	6	16
	Skyphos with offset rim		29	2	3
	Open		3		3
Classical (5th c. BC)	Skyphos with offset rim	500–450 BC	10	3	4
		5th c. BC	1	1	1
	Skyphos with concave rim	470–440 BC	33	4	5
			3	1	1
		Mid-5th c. BC	2	2	2
	Kylix	475–450 BC	2	1	1
			2		
	A-type skyphos	470–450 BC	5		
		430–400 BC	1		
	Cup-skyphos	430–400 BC	2	1	1
	C-type skyphos	450–425 BC	5	1	1
	Amphora		5	1	2
	Oinochoe	430–400 BC	5		1
	Jug	End of 5th c. BC	1	1	1
Classical (4th c. BC)	A-type skyphos	460–370 BC	15	1	1
	C-type skyphos	400–370 BC	4	1	1
		400–375 BC	2	2	2
		370–325 BC	2	2	2
	Jug	420–350 BC	1	1	1
Unident.	Skyphos		5		2
	Open		37		21
	Closed		12		8

Table 15.4 Black-gloss from the Pottery Deposit: quantification by form and date.

skyphoi. Another jug and a lamp (see below, p. 225, "Lamps") date to the same period. The forms of the small assemblage of Archaic and Late Archaic ceramics from the Retaining Wall (Slope 4) are also largely associated with drinking and containing liquids. Alongside the usual Ionic-type B2 cups appear two column kraters (SAV BG 02 and SAV BG 16), forms that are otherwise rare on the hill of Sant'Angelo. The majority of the materials are concentrated in the 5th century BC, when the drinking cups (skyphos with offset and concave rim, kylix) predominate, while a single jug signifies storage and pouring of liquids. Finally, three drinking cups (C-type skyphoi) are the only identified forms in the assemblage, with their latest date in the third quarter of the 4th century BC.

Provenance	Form and Type	Date	NFr	MNI	TNI
Surface finds	Ionic-type B2 cup	Late Archaic	9	3	4
	Skyphos with offset rim	Late Archaic	4	2	3
		5th c. BC	1		
	C-type skyphos	470–450 BC	1	1	1
	Open		12		7
	Closed		1		1
	Unidentified		1		
Bulldozed Stipe	Ionic-type B2 cup	Late Archaic	1		1
	Skyphos with offset rim	5th c. BC	1		1
	Mug	470–430 BC	5	1	1
	Dish	320–250 BC	3	1	1
	Dish	Second half of the 3rd c. BC	4	1	1
Slope 1	Cup-skyphos	430–400 BC	1	1	1
Slope 2	C-type skyphos	400–360 BC	2	1	1
Sounding B	Ionic cup B2	Late Archaic	2		
	Skyphos with offset rim		7		1
	Cup		3		1
	Squat lekythos		1		1
	Open		2		1
Slope 3, surface finds	Ionic-type B2 cup	Late Archaic	1	1	1
	Salt cellar	Late 4th–early 3rd c. BC	1	1	1
	Open		3		3
Slope 3, surface finds, and Slope 4	Bolsal	425–400 BC	5	1	1
Slope 3	Ionic-type B2 cup	Late Archaic	5	2	4
	Skyphos with offset rim	Late Archaic	26	1	2
		5th c. BC	1	1	1
	Skyphos with concave rim	5th c. BC	1		
	Olpe	525–450 BC	3	1	1
	Open	Late Archaic	4		2
	C-type kylix	5th c. BC	1	1	1
	Cup-skyphos	480–440 BC	2		
	A-type skyphos	450–400 BC	1	1	1
		350–300 BC	1	1	1
	C-type skyphos	470–450 BC	1	1	1
		470–430 BC	1		
		420–380 BC	1		
	Dish	325–275 BC	2		

Table 15.5 Black-gloss from the slopes: quantification by form and date.

Provenance	Form and Type	Date	NFr	MNI	TNI
Slope 3	Stemmed cup	320–250 BC	2	1	1
	Salt cellar	ca. 320–250 BC	1	1	1
	Bowl	275–225 BC	4	1	1
	Skyphos	Hellenistic	1	0	1
	Squat lekythos	4th c. BC	2	0	1
	Open		24	0	14
	Closed		6	0	6
Slope 4, surface finds	Ionic-type B2 cup	Late Archaic	1	0	1
	C-type skyphos	460–430 BC	1	0	0
	Open		4	0	4
	Closed		2	0	2
Slope 4	Krater	550–500 BC	1	1	1
		525–480 BC	1	0	0
	Ionic-type B2 cup	Late Archaic	7	2	3
	Skyphos with offset rim	5th c. BC	6	1	1
	Skyphos with concave rim	5th c. BC	3	2	2
	C-type kylix	5th c. BC	1	1	1
	C-type skyphos	First half of the 4th c. BC	1	1	1
	Jug	5th c. BC	1	0	1
	C-type skyphos	370–325 BC	2	2	2
	Amphora		3	0	1
	Open		41	0	16
	Closed		7	0	6
Sounding C	Ionic-type B2 cup	Late Archaic	4	0	1
	A-type skyphos	475–320 BC	1	1	1
	Open	5th c. BC	1	0	0
	Jug	480–470 BC	1	0	1
	Squat lekythos	ca. 450–325 BC?	1	0	1
	C-type skyphos	Early 4th c. BC	1	1	1
	Open		1	0	1
	Closed		3	0	3
Sounding D	Ionic-type B2 cup	Late Archaic	5	5	5
	Open		1	0	0
	Closed		2	0	2

Table 15.5 cont. Black-gloss from the slopes: quantification by form and date.

The Slopes (Squares D2–4)
Contexts and Chronology

The black-gloss pottery collected along the slopes is fragmentary (*Table 15.5*). The majority of the fragments date between the Late Archaic period and the 5th century BC. Early Hellenistic material was encountered at the depth of the concentration of terracotta objects corresponding to Slope 3 (SAV BG 41) and in the vicinity of Kiln 1 in square D4 (SAV BG 36 and SAV BG 39). These items are nevertheless associated with Late Archaic and Classical materials, which attests to the disturbance of the contexts and complicates a chronological reconstruction. The clear prevalence of materials associated with the early occupation phases (see SAV BG 25 and SAV BG 28) should be considered residual on the slopes, where the only identified context is the discard deposit of terracottas dating to the end of the 4th/beginning of the 3rd century BC.

The Workshop Area
(Squares H4–5, G4–7, F6–7)
Contexts and Chronology

The highest density of black-gloss ceramics in the area of the workshop shed, encountered in squares G6–7 and F7 (*Table 15.6*), should be analyzed in relation to the wall structures and identified tile falls, as well as to the nearby tombs that date mostly to the Early Classical period.[6]

The distribution of the materials is not chronologically homogeneous. Materials dating between the Late Archaic period and the 5th century BC were found mostly in the Republican upper tile fall 1 (squares G5–7), where a lekythos with a distinct shoulder, similar to the examples found in the graves, is attested. Materials dating to the 4th century BC and the Early Hellenistic period (SAV BG 38) were encountered only in square G7.

Tile Fall 2 (square F7) yielded a kantharos (SAV BG 40) datable to the second half of the 4th century BC. Other black-gloss materials were found in underlying levels, where evidence dated to the Archaic period (SAV BG 32) is associated with a C-type skyphos of the 4th century BC.

[6] See Ch. 4, "The Tombs at Sant'Angelo Vecchio"; Carter and Hall in *Necropoleis*, 443–47.

Provenance	Form and Type	Date	NFr	MNI	TNI
Surface finds	Ionic-type B2 cup	520–480 BC	1		
	Open		8		4
Square H3	Open		3		2
Square H5	C-type skyphos	Early 4th c. BC	1	1	1
	Open		1		
Square H6	Closed		1		1
Square H7	Ionic-type B2 cup	Late Archaic	1		1
Square G3	Open		1		1
Square G5	Skyphos with offset rim	500–450 BC	1	1	1
	Lekythos	Mid-5th c. BC	1	1	
	Open		1		1
Square G6	Skyphos with concave rim	First half of the 5th c. BC	1	1	1
	Pelike	5th c. BC	1	1	1
	One-handler	425–350 BC	1	1	
	Open		19		11
	Closed		5		4
Square G7	Skyphos with offset rim	500–450 BC	1	1	1
	Skyphos with concave rim	470–440 BC	1		
	C-type skyphos	370–325 BC	3	3	3
	Cup	350–275 BC	1	1	1
	Open		4		3
Square F6	Kylix	Mid-5th c. BC	1		
Square F7	Skyphos with offset rim	Late Archaic	2		
	Amphora	Central decades of the 6th c. BC	6	1	1
	A-type skyphos	450–400 BC	1	1	1
	C-type skyphos	370–325 BC	1	1	1
	Kantharos	350–300 BC	4		1
	Open		7		5
	Closed		1		1

Table 15.6 Black-gloss from the area of the Late Republican workshop shed: quantification by form and date.

The Kiln Complex (Squares E1–3, F2–3)

Contexts and Chronology

The Kiln Complex identified in squares E1–3 and F2–3 yielded very little material (*Table 15.7*). A single C-type skyphos found in the level below the pavement in square E2 and dated 370–325 BC can be considered a terminus post quem for the context.

Provenance	Form and Type	Date	NFr	MNI	TNI
Square F3, inside N Kiln	Ring-handled cup	225–150 BC	1	1	1
	Open		1		1
Sounding A	Ionic-type B2 cup	Late Archaic	1		1
	C-type skyphos	370–325 BC	1	1	1
	Open		5		2
Square E3	Bowl	Late 3rd–2nd c. BC	1	1	1
	Open		1		1
	Closed		1		1

Table 15.7 Black-gloss from the Kiln Complex (squares E1–E3, F2–F3): quantification by form and date.

Provenance	Form and Type	Date	NFr	MNI	TNI
Sounding C, Pottery Deposit	Circular	480–325 BC	3	1	1
Square G7	Globular	330–250 BC	1	1	1

Table 15.8 Lamps from Sant'Angelo Vecchio: quantification by form and date.

Other significant examples can be associated with the Late Republican occupation phase of the area. The bowl found in square E3 (SAV BG 43) is of a form that fits well with the production of the period, attested both in black-gloss and grey wares. The ring-handled cup in the debris level inside Kiln 5 (SAV BG 42) probably represents reuse of the vessel in the construction of the facility, which later collapsed over the space inside the combustion chamber. The ring-handled cup can therefore be considered a terminus post quem for the final construction phase of the kiln, probably in the second half of the 2nd century BC.[7]

Lamps

A few fragments belonging to two black-gloss lamps were found in the Pottery Deposit and in square G7 (*Table 15.8*). SAV LA 44 belongs to a circular lamp derived from models created in mainland Greece and related to variants of Howland's type 21. The low and wide reservoir forming a single curve from the rim to the direct and rather concave base, in particular, recalls variant 21 A from the Athenian Agora,[8] characterized by a single short nozzle with an oval hole and a hori-

[7] See also Ch. 17, "Red Ware."

[8] *Agora IV*, 45.

zontal handle opposite the nozzle. This type of lamp is dated between 480 and the beginning of the 3rd century BC, but the ceramic material from the Pottery Deposit appears to be no later than the third quarter of the 4th century BC.

Catalog of Black-Gloss Ware

Attic Black-glaze Pottery

Stemless kylix. SAV BG 01 belongs to a type of stemless kylix with concave lip, inset on the interior and exterior, known in the literature as a "Castulo cup." This type circulated widely throughout the Mediterranean basin, in particular in southern Italy and Sicily (see Shefton 1996, 89–94 for a synthesis), and there mostly in the second quarter of the 5th c. BC. The attestations indicate that this form was produced primarily in Attica and very likely intended for export rather than distribution within mainland Greece (*Agora XII*, 298, no. 471, fig. 5; Boulter 1953, 114, fig. 2, no. 199), where they are not documented before 480 BC, later than at several sites in Sicily (Nigro 2006b, 95, pl. 23.5). These kylikes are further documented among the imported black-glaze vessels at Gravisca, in Faliscan territory, at Populonia, Aleria, and Fano (*Gravisca 9*, 24–25, pls. 4–5, type 10). Local imitations datable to the middle of the 5th c. BC have recently been recognized at Velia (Ferrara 2010, 454) and Naples (Castel Capuano necropolis, Pontrandolfo 2000, 123, fig. 1). The type has so far not appeared in the morphological repertoire of the workshops of Metaponto (F. Silvestrelli, pers. comm., 2010).

Stemless Kylix

SAV BG 01 Attic Stemless Kylix

1:2

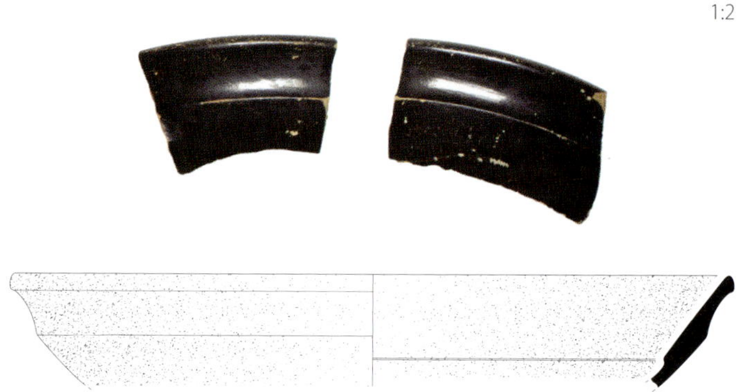

Provenance: Sounding C, Pottery Deposit. Lot: SA 79-763P A–B, Assemblage 15.

Dimensions and technical features: h 2.8, rim 18.3. Clay Munsell 5YR 7/6 (reddish yellow), very hard, well-fired, and levigated. Black gloss thick and lustrous, dipped.

Condition: Two rim fragments, not joining.

Description: Concave lip, inset on the interior and exterior.

Comparisons and comment: de la Genière and Greco 2010, 174, pl. 62, no. 535 (Poseidonia-Paestum, Heraion at mouth of Sele, 550–530 BC); *Greci, Enotri e Lucani*, 166, no. 2.30.25 (Chiaromonte-San Pasquale, T 227, 5th c. BC); *Poseidonia-Paestum II*, 151, fig. 70, no. 382 (Poseidonia-Paestum, plateia AN, 500–475 BC); *Kaulonia I*, 59, fig. 37, 133 (Caulonia, 475–450 BC); Hayes 1984, 24–25, fig. 10, no. 36 (475–450 BC or slightly later); Baldoni 2009, 225, pl. 25, no. 1015 (Marzabotto, 475–450 BC); Kustermann Graf 2002, 158, 88/O 369, pls. XLV, CXXII (Selinous, Manicalunga necropolis, T 8, 475–450 BC); *Agora XII*, 268, fig. 5, no. 471 (Athens, 470–450 BC); Boulter 1953, 114, fig. 2, no. 199 (Athens, mid-5th c. BC).

Date: 475–450 BC.

Krater. SAV BG 02 is a krater derived from Laconian models produced especially during the 6th c. BC (see most recently Coudin 2009, 25–26). They were common in Sicily and Etruria, and they are also attested in southern Italy (*Monte Sannace*, 41, pl. 221.2; Otranto: Semeraro 1992, 17, fig. 2:1, 9; *Sibari II*, 257, no. 141bis [25857], figs. 285, 385).

In Attica, the column krater, along with the bell krater, was the only type to have an "economy" black-glaze version in addition to the main figured version. The black-glaze production was not extensive (*Agora XII*, 54) but is attested at Athens until the 3rd c. BC, even in non-Attic fabric (Boulter 1953, 91). Fully glazed examples are in general rare among the imported ceramics found at Italian sites. The Attic black-glaze column kraters have a simple ring base or a truncated conical base with double "step." Both versions date to the first half of the 5th c. BC.

Krater

SAV BG 02 Column Krater
Provenance: Square C5, US 40, 1st cut.
Dimensions and technical features: h 7.6, rim 26.2. Clay Munsell 5YR 6/6 (reddish yellow), hard, with white and red rounded inclusions. Black gloss dilute and matte, painted.
Condition: One rim fragment.
Description: Collared rim, flat on the upper surface. Reserved upper surface of the rim, decorated with a band (0.5 cm).
Comparisons and comment: Agora XII, 54, 240, pl. 3, no. 58 (Athens, 500–480 BC); Boulter 1953, 91, pl. 33, nos. 98–99 (Athens, mid-5th c. BC); *Gravisca 9*, 32, pl. 10, no. 79 (Gravisca, Greek sanctuary, imported black-gloss); Lejsgaard Christensen 2010, 367, nos. C89–90 (Francavilla Marittima, Timpone della Motta).
Date: 550–500 BC.

1:2

Black-gloss of the Late Archaic and Classical Periods

Ionic-type B2 cup. Following the convention established for the farmhouses excavated by ICA in the chora of Metaponto, the Ionic-type cups are presented together with the black-gloss ceramics, although technically they do not form part of the latter class (Lanza Catti in *Fattoria Fabrizio*, 181). The majority of the identified examples (104 NFr, 35 MNI, 60 TNI) belong to type B2 (Villard and Vallet 1955). The attested types have a shallow body with a rigid and slightly convex or truncated conical profile, a feature long associated with the later colonial products and more common in Magna Grecia (Ciancio 1985, 98–99, pl. 43, nos. 5–6). The cups found at Sant'Angelo Vecchio can be ascribed to two distinct groups: the more common type, identified previously at Fattoria Fabrizio (Lanza Catti in *Fattoria Fabrizio*, 182, FF BG 02), has a shallow body and straight flaring rim (SAV BG 03), while a small number of fragments belong to the type with extremely low body and strongly flaring rim that appears to follow the orientation of the body (SAV BG 04). The examples from the two groups are linked by the presence of an accentuated carination, an important discriminating factor for the typological seriation of these vessels, which distinguishes the products made after 530 BC at Velia (Tubelli 2006, 47–48).

The base with truncated conical profile, fairly common in the Metapontine sphere (Van Compernolle 2000, fig. 2; Lo Porto 1981, 297, fig. 5; Lanza Catti in *Fattoria Fabrizio*, 182, FF BG 03), is very low on SAV BG 06, while on SAV BG 05, it is very sharply angled and marked by a shallow groove on the exterior. The gloss is applied with a brush. The upper exterior of the body is left reserved, but nevertheless decorated with a narrow band of gloss at the offset; the interior is completely painted, except for a narrow reserved line immediately below the rim. (The internal bottom is reserved on a single uncataloged example.) The handles are painted.

This type is attested throughout the Mediterranean basin and all along the Black Sea coast (Cook and Dupont 1998, 129), and it was also produced at several Italiote and Siceliote colonies (see Van Compernolle 2000 for a synthesis). Production of Ionic-type B2 cups in the urban center of Metaponto was recognized long ago (D'Andria 1980a, Archaic kiln) and is now confirmed also in the Metapontine chora (Carter 2006, 150; most recently, Lanza Catti et al. in

Survey, 152, nos. 8–26). Metapontine products have been identified at sites in the interior of Lucania, along the Agri valley and in the areas around Matera, Potenza, and Melfi (G. Greco 1991, 79; *Museo Taranto II.1*, 159).

The discovery of these cups in association with Attic kylikes (*Forentum I*, 155–56; Long/Miro/Volpe 1992, 203–4, fig. 8) has prompted moving their chronology to the first decades of the 5th c. BC. In terms of both parallels and morphological features, the examples from Sant'Angelo Vecchio fit comfortably within this range.

Ionic-type B2 Cups

SAV BG 03 Ionic-type B2 Cup

1:2

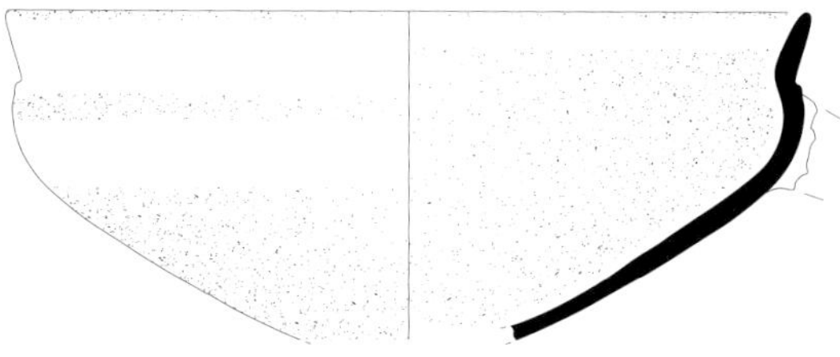

Provenance: House on the Hill. Lot: SA 79-1034P C (see *Fig. 2.39*).

Dimensions and technical features: h 5.5, rim 13.6. Clay Munsell 5YR 7/6 (reddish yellow), hard and well levigated. Black gloss thin and semi-lustrous, painted.

Condition: One rim fragment, missing the handle.

Description: Thick, offset lip with slightly out-turned rim; low body. Reserved line on the interior immediately below the rim; reserved upper exterior, decorated with a band on the upper exterior of the rim and at the offset.

Comparisons and comment: Da Zancle 1, 115, 117, Z/62 (Messina, via Torino, end of the 6th–beginning of the 5th c. BC); *Sibari II*, 92, figs. 68, 89, no. 45 (12515) (Sybaris, Stombi 1969); *Gravisca 4*, 165–66, pl. 10, no. 329 (Gravisca, Greek sanctuary, 580–520/500 BC); *Gravina II*, 66, fig. 27, no. 606 (Botromagno, Gravina di Puglia); Semeraro 1997, 40, fig. 2, no. 3 (Alezio, Monte d'Elia necropolis).

Date: 520–470 BC.

SAV BG 04 Ionic-type B2 Cup

1:2

Provenance: Sounding C, Pottery Deposit. Lot: SA 79-988PL, Assemblage 15.

Dimensions and technical features: h 4.15, rim 14.0. Clay Munsell 7.5YR 7/4 (pink), very hard, unevenly fired, and levigated, with luminous inclusions. Black gloss thick and lustrous, painted.

Condition: Two rim fragments and five body fragments, joining.

Description: Narrow, offset lip with sharply outward-turned rim; low body. Reserved line on the interior immediately below the rim; reserved upper exterior, decorated with a band on the upper exterior of the rim and at the offset.

Comparisons and comment: Olivero Ferrero 1989, 73–74, pl. XIX, no. 4 (Locri, 580–540 BC); *Forentum I*, 155–56, pl. 29, "kylix of the Ionic tradition" type 2.2 (Lavello, necropolis, second half of the 6th c. BC); La Torre and Mollo 2006, 179, pl. XLVIII.9, no. 37 (Blanda, end of the 6th c. BC, local fabric); G. Greco 1991, 79, fig. 194 (Serra di Vaglio, "casa dei pithoi," 6th–5th c. BC); *Gravisca 4*, 165, pl. 10, no. 328 (Gravisca, Greek sanctuary, 580–520/500 BC); Lanza Catti et al. in *Survey*, 162, no. 13 (Metapontine chora, site 622, 550–500 BC); Semeraro 1997, 302, fig. 251, no. 1138 (Vaste, Fondo S. Antonio).

Date: 500–470 BC.

SAV BG 05 Ionic-type B2 Cup
Provenance: Sounding C. Level 2, batt. 3. Lot: SA 79-324P, Assemblage 13.
Dimensions and technical features: h 2.1, base 5.0. Clay Munsell 7.5YR 7/4 (pink), very hard and levigated. Black gloss thin and semi-lustrous, painted.
Condition: Complete base.
Description: Conical base with slight groove on the exterior. Reserved underside, decorated with a band on the bottom inner side.
Comparisons and comment: Hayes 1984, 6–7, fig. 1, no. 5 (ca. 580–540 BC); *Gravisca 4,* 168, pls. 11, 19, no. 349, (Gravisca, Greek sanctuary, 580–520/500 BC); Semeraro 1997, 40, 42, fig. 2, no. 4 (Alezio, Monte d'Elia necropolis); *Tocra,* 120, fig. 56, pl. 87, no. 1222 (Tocra, Archaic Deposit); *Archeologia dei Messapi,* 214–15, no. 35 (Cavallino, T Asilo, Archaic).
Date: 540–500 BC.

1:2

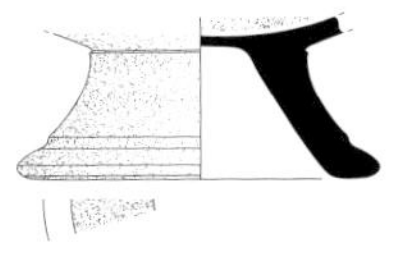

SAV BG 06 Ionic-type B2 Cup
Provenance: House on the Hill. Lot: SA 79-1062PL-F, Assemblage 1.
Dimensions and technical features: h 1.6, base 6.0. Clay Munsell 5YR 7/4 (pink), very hard, unevenly fired, and levigated, with sparse red and white inclusions. Black gloss thick and matte, painted.
Condition: Three base fragments, joining.
Description: Conical base. Reserved underside.
Comparisons and comment: D'Andria 1980a, 373, fig. 26, no. 38 (Metaponto, *kerameikos,* Archaic kiln); Gargano 2001, 69, no. 2 (Rutigliano, contrada Bigetti, T 3/1994, second half of the 6th–first decades of the 5th c. BC); *Gravisca 4,* 169, pl. 11, no. 352 (Gravisca, Greek sanctuary, 580–520/500 BC); Van Compernolle 2000, 93, figs. 2, 5 (Metaponto); Semeraro 1997, 40, fig. 2, no. 2 (Alezio, Monte d'Elia necropolis); *Poseidonia-Paestum II,* 87, fig. 59, no. K/1 (Poseidonia-Paestum, circular structure).
Date: 520–480 BC.

1:2

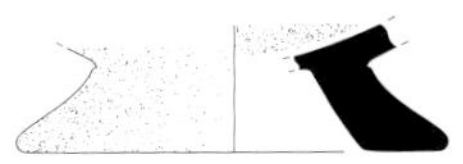

B3b cup. A single example (SAV BG 07) can be attributed to a type B3b cup (Villard and Vallet 1955; Van Compernolle 2000, fig. 1) or band cup (*Agora XII,* 262, nos. 389–391), characterized by a reserved band on the exterior at the level of the handles and another, narrower band below it. On the example from Sant'Angelo Vecchio the gloss is mostly lost, so that the typical narrow reserved band on the interior immediately below the rim could not be identified. Furthermore, without the truncated conical base and the handles, the identification is hypothetical and based essentially on the decorative scheme, which does not appear to include the red lines typical of the Archaic banded skyphos with canted handles of Attic inspiration (*Agora XII,* 258, nos. 330–31), to which the B3b cup can be compared in terms of the rim morphology. (Here it is worth noting the confusion evident in earlier publications, in which the profiles of B3b cups are sometimes presented alongside banded skyphoi with canted handles, as in *Cozzo Presepe,* 330.) The reconstructed diameter, 8.5 cm, is well outside of the norm for the type (ca. 17 cm), but production of smaller versions is confirmed by examples recovered in the necropolis of Taranto.

This type of cup was probably produced in the Metapontine sphere (*Cozzo Presepe,* 330). Circulation appears to have been widest in Campania and Calabria, while there are few attestations in Apulia (Taranto). It is also attested in several Siceliote contexts. Nevertheless, it is difficult to determine whether these were imports or local products.

B3b Cup

SAV BG 07 B3b Cup

1:2

Provenance: House on the Hill. Lot: SA 79-1062PL, Assemblage 1.

Dimensions and technical features: h 3.7, rim 8.5. Clay Munsell 5YR 7/6 (reddish yellow), powdery, with abundant luminous inclusions. Black gloss dilute and semi-lustrous, painted. Black gloss almost worn.

Condition: One rim fragment.

Description: Flared lip with direct tapering rim; convex deep body. Reserved handle zone and a narrow band below it.

Comparisons and comment: Sibari III, 103, figs. 86, 164, no. 168 (St. 34736) (Sybaris, Stombi 1971, ca. 550 BC); de la Genière and Greco 2010, 174, pl. 62, no. 535 (Poseidonia-Paestum, Heraion at mouth of Sele, 550–530 BC); de la Genière 1968, 301, 146, pl. 21.1, no. 3 (Sala Consilina, T A, second half of the 6th c. BC); Bacci 1977, 82, pl. 25, no. C. 12 (Locri, Centocamere, second half–end of the 6th c. BC); *Forentum I*, 71–72, 155, pl. 28 "skyphos of the Ionic tradition" type 1 (Lavello, necropolis, T 58, 550–500 BC); *Da Zancle 1*, 140–41, S/78 (Messina, via Industriale, 550–500 BC); *Cozzo Presepe*, 330, fig. 111, no. 178 (Cozzo Presepe, ca. 540–500 BC); *Museo Taranto I.3*, 230, 42.23 (Taranto, contrada Cortivecchie, prop. D'Acclavio, T 17.III.1914, ca. 500 BC); Maruggi 1982, 54, pl. XIV.3, no. 6 (Taranto, viale Magna Grecia T 10, ca. 470 BC); *Sibari IV*, 54, figs. 42, 146, no. 37 (St. 8446) (Sybaris, Stombi 1972).

Date: 550–500 BC.

Skyphos with offset rim. The skyphos with offset rim, together with the Ionic-type B2 cup, is the most common form in the Late Archaic contexts of Sant'Angelo Vecchio. The largest number of examples (81 NFr, 7 MNI, 15 TNI) belongs to the earliest type (SAV BG 08), featuring a straight and flaring rim separated from the body by a marked carination, a deep body tapering at the bottom, and bell-shaped handles. The ring base is sometimes decorated with narrow lines in shallow relief. The slip, very dilute and applied by brush, is of poor quality, as often encountered on Late Archaic products. The exterior has a reserved band at the level of the handles, the interior a narrow band below the rim; the underside of the base is reserved. This type of skyphos was probably produced alongside the Ionic-type B2 cups (see most recently Lanza Catti et al. in *Survey*, 153) with which it shares identical technical features, and from which it diverges in the deeper body and narrower rim (Lanza Catti in *Fattoria Fabrizio*, 177).

The skyphos with offset rim is currently considered a typically Metapontine product (see D'Andria 1980a, 372, fig. 26, no. 39), widely attested in the chora as well (Lanza Catti et al. in *Survey*, 165–67, nos. 27–40). It circulated extensively in southern Italy at both colonial (Adamesteanu 1974, 82; *Museo Taranto II.1*, 142–43, 1.7, 1.8) and indigenous (de la Genière 1968, pl. 16.3.2 and 6) sites, in funerary and domestic contexts. At Metaponto the type is documented mostly in Archaic contexts and appears to have been produced through the first decades of the 5th c. BC (M. Liseno 2004, 74). The completely black version with a disc base, datable to the 5th c. BC (Lanza Catti in *Fattoria Fabrizio*, 178), is less common than the Late Archaic banded version (22 NFr, 8 MNI, 10 TNI) and is accordingly represented in the catalog by two examples with a somewhat flaring offset rim and horseshoe handles which document the evolution of the type. SAV BG 09 has the deep body common to all of the examples published to date, while SAV BG 10 has a lower body that recalls small skyphoi with offset rim found recently at Metaponto.

The skyphos with offset rim is attested in mostly funerary contexts of the Late Archaic and Early Classical periods in Basilicata (see most recently Lanza Catti et al. in *Survey*, 215, nos. 5–6; Lanza Catti in *Fattoria Fabrizio*, 178) and Apulia, while at Sant'Angelo Vecchio the same vessel is, along with the skyphos with concave rim, typical of the first half of the 5th c. BC.

Skyphoi with Offset Rim

SAV BG 08 Skyphos with Offset Rim

Provenance: Sounding C, Pottery Deposit. Lot: SA 79-755P. Inv. no. 320002, Assemblage 15.

Dimensions and technical features: h 12.4, rim 13.8. Clay Munsell 7.5YR 7/4 (pink), very hard, unevenly fired, and levigated. Black gloss thin and matte, painted.

Condition: Nearly complete, missing the base.

Description: Offset, nearly vertical rim; convex body; canted, bell-shaped handles. Reserved external handle zone and handles; on the interior, reserved line immediately below the rim.

Comparisons and comment: Principi Imperatori Vescovi, 153, no. 4 (Canosa, pit grave in Largo Costantinopoli, 550–500 BC); *Cozzo Presepe*, 323, 327–29, fig. 111, no. 173 (Cozzo Presepe, ca. 540–500 BC); *Antichi Peucezi*, 51, 53, fig. 27 (Bitonto, via Traiana necropolis, T 9/1989, Archaic-Classical); M. Liseno 2004, 74, 7, pl. 36a–b (Metaponto, Favale votive deposit, second quarter of the 6th c.–480 BC); Carter and Hall in *Necropoleis*, 442, SS 28.3 (Metaponto, Saldone necropolis, T 28, 470–450 BC); *Sibari II*, 105, figs. 65, 86, no. 88 (12482) (Sybaris, Stombi 1969).

Date: 500–480 BC.

1:2

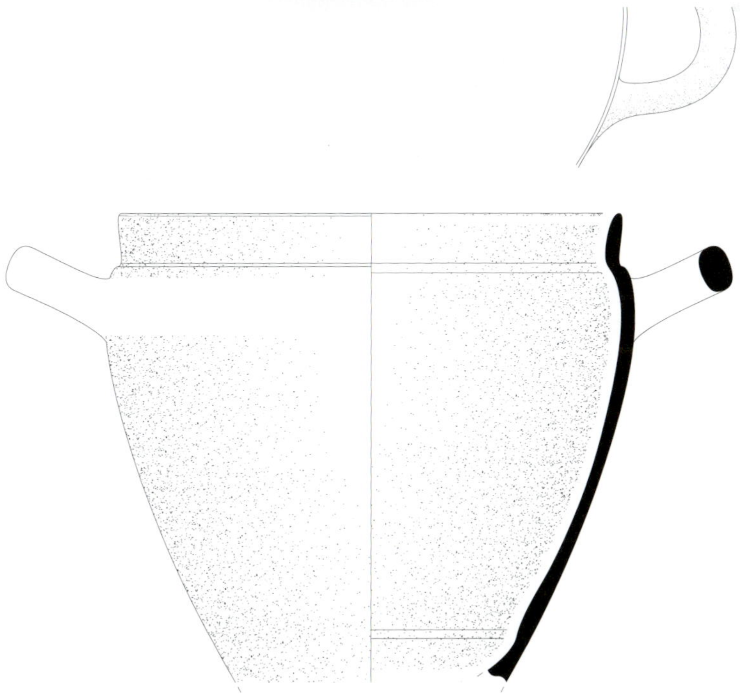

Skyphoi with offset rim cont.

SAV BG 09 Skyphos with Offset Rim

1:2

Provenance: Sounding C, Pottery Deposit. Lot: SA 79-892P A, Assemblage 15.

Dimensions and technical features: h 3.7, rim 10.0. Clay Munsell 7.5YR 7/4 (pink), very hard, unevenly fired, and levigated. Black gloss thick and matte, painted.

Condition: One rim fragment, two body fragments, and one handle fragment, joining.

Description: Narrow, offset, slightly outward-turned rim; slightly convex body; horseshoe-shaped handle.

Comparisons and comment: Lo Porto 1966, 210, fig. 64.3 (Metaponto, contrada Ricotta-Crucinia-Casa Teresa, T 29, ca. mid-6th c. BC); Lo Porto 1973, 171, pl. XIX.2, no. 16 (Pisticci, T 4, 550–500 BC); 225, pl. LXX.1.3, no. 4 (sporadic find, contrada Rifeccia necropolis, nearby Matera, 550–500 BC); Lo Porto 1981, 364–65, fig. 85.2 (Metaponto, Casa Ricotta necropolis, T 18 zone B, end of the 6th c. BC); Lo Porto 1996, 31–32, fig. 30, no. 17 (Ginosa, T Sangiorgio, 500–475 BC); *Museo Taranto II.1*, 50–51, 19.8 (Ginosa, T 25, 16.I.1967, 500–475 BC); Lo Porto 1988–1989, 332, figs. 32.10 and 35.1 (Montescaglioso, T 1, 475–450 BC); *Museo Taranto I.3*, 338–39, 118.22 (Ginosa, T via Vittorio Emanuele II, 13.I.1935, mid-5th c. BC); Semeraro 1997, 228, fig. 197, no. 783 (Oria, Monte Papalucio).

Date: 500–450 BC.

SAV BG 10 Skyphos with Offset Rim

1:2

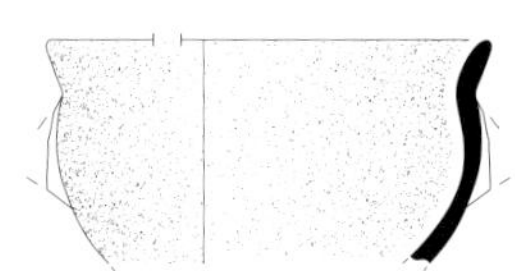

Provenance: Sounding C, Pottery Deposit. Lot: SA 79-988PL, Assemblage 15.

Dimensions and technical features: h 3.0. Clay Munsell 5YR 7/4 (pink), very hard and levigated. Black gloss thin and semi-lustrous, painted.

Condition: One rim fragment, missing the handle.

Description: Low, slightly outward-turned rim; convex body; possibly canted handle.

Comparisons and comment: Silvestrelli 2014, 100 (Metaponto, necropolis along the coast, T 697).

Date: 5th c. BC. (The vessel cited above as parallel from the Metaponto necropolis is associated with a Pisticci Painter hydria dated 440–420 BC.)

Skyphos with concave rim. At Sant'Angelo Vecchio, the skyphos with concave rim is as common as the completely black skyphos with offset rim (44 NFr, 10 MNI, 11 TNI). The preserved examples all belong to the type with deep body tapering toward the bottom, concave rim (SAV BG 11), and typical disc base with concave interior and convex exterior (SAV BG 12). The handles are canted and horseshoe-shaped. The dipped and consistent slip is absent only on the underside of the base, the bottom inner side of which is decorated with a band.

Recent studies interpret the type as an evolution of the completely black skyphos with offset rim, but derivation from Attic skyphoi with canted handles is possible judging from the strong similarities in the body profiles (*Agora XII*, 83–84, 258, no. 332, fig. 4; Lanza Catti in *Fattoria Fabrizio*, 178–89).

The type is attested in the Metapontino (see most recently Lanza Catti et al. in *Survey*, 215, no. 7; Lanza Catti in *Fattoria Fabrizio*, 187, FF BG 15) and in Apulia (*Museo Taranto II.2*, 491–92, pl. 38e), where it has been erroneously considered an Attic import; there are no significant attestations outside of these areas. The examples recovered at Sant'Angelo Vecchio are local products. On the basis of the comparanda and possible relationships with Attic prototypes dated ca. 500 BC and the completely black skyphos with offset rim, the examples from Sant'Angelo Vecchio can be dated between 470 and 440 BC.

Skyphoi with Concave Rim

SAV BG 11 Skyphos with Concave Rim

Provenance: Sounding C, Pottery Deposit. Lot: SA 79-760P + 988PL B, Assemblage 15.

Dimensions and technical features: h 7.4, rim 10.5. Clay Munsell 7.5YR 7/4 (pink), very hard, unevenly fired, and levigated. Black gloss thick and semi-lustrous, dipped.

Condition: Six rim fragments, eight body fragments, and one handle, joining.

Description: Concave rim; slightly swelling convex body; canted horseshoe-shaped handle.

Comparisons and comment: Lo Porto 1973, 225, pl. LXX, 1.6, no. 6, (sporadic find, contrada Rifeccia necropolis, nearby Matera, 550–500 BC); 159, pl. VI, 2.4, no. 4, (Pisticci, 525–500 BC) and 221, pl. LXV, 2.1 (Matera, Ponte San Giuliano, T 4, ca. 530–500 BC); *Museo Taranto II.2*, 491–92, pl. 38e (Rutigliano, Purgatorio necropolis, T 49, beginning of the 5th c. BC); Elliott in *Necropoleis*, 680, S23 (Pantanello necropolis, T 151, 460–440 BC).

Date: 470–440 BC.

1:2

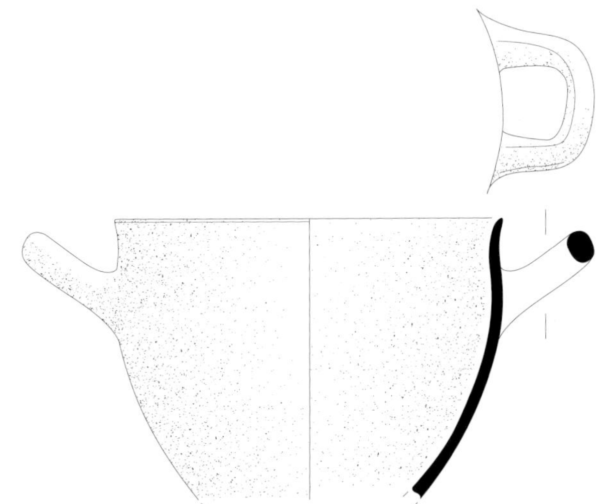

SAV BG 12 Skyphos with Concave Rim

Provenance: Sounding C, Pottery Deposit. Lot: SA 79-886P A, Assemblage 15.

Dimensions and technical features: h 3.3, base 6.7. Clay Munsell 5YR 6/6 (reddish yellow), very hard and unevenly fired, with sparse white and abundant luminous inclusions. Black gloss thick and lustrous, dipped/painted.

Condition: One base fragment and one body fragment, joining.

Description: Disc base with concave exterior and convex interior; convex body. Reserved underside of the base decorated with a band on the bottom inner side; bottom lower side filled with miltos.

Comparisons and comment: Lo Porto 1973, 221, no. 1, pl. LXV, 2.1 (Matera, Ponte San Giuliano, T 4, ca. 530–500 BC); *Museo Taranto II.2*, 491–92, pl. 38e (Rutigliano, Purgatorio necropolis, T 49, beginning 5th c. BC); Lo Porto 1988–1989, 332, fig. 32.9 (Montescaglioso, T 1, 475–450 BC); *Sibari V*, 220, figs. 200, 217, no. 172 (St. 2855) (Sybaris, Stombi 1974).

Date: 470–440 BC.

1:2

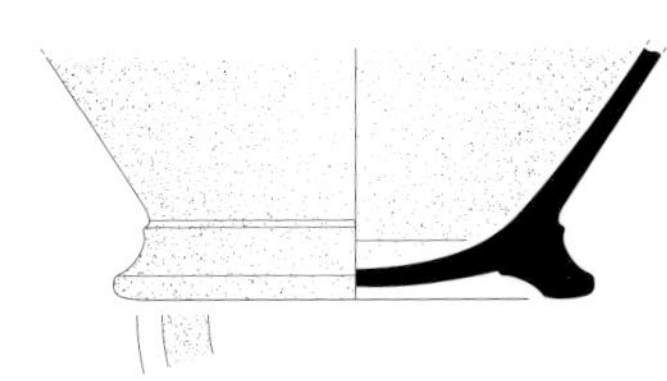

Cup-skyphos. The cup-skyphos is attested less frequently than the other Early Classical forms (7 NFr, 3 MNI, 3 TNI). Two types were identified. The first (SAV BG 13) has a shallow body and flaring rim slightly distinguished from the body, and recalls Attic forms of the 6th/beginning of the 5th c. BC (*Agora XII*, 109, 279, no. 569, fig. 6). The second (SAV BG 14) has a deep body with a single curve from rim to base, and therefore compares closely with the Attic forms with "light wall" which appeared around 440–430 BC (*Agora XII*, 110, 278, no. 595, fig. 6).

The earliest type is documented in the Pantanello necropolis and in other areas of the Metapontine chora in the first half of the 5th c. BC (Elliott in *Necropoleis*, 679–80, 682, CS1-3; Lanza Catti et al. in *Survey*, 216, nos. 9–11), while the thin-walled type appears at Pantanello in contexts dated to the third quarter of the same century (Elliott in *Necropoleis*, 679–80, 682, CS4). Both types are attested at other colonial sites (*Kaulonia I*, 55, 59, fig. 36) and indigenous centers of southern Italy, where they are often interpreted as Attic imports, in the second half of the 5th c. BC.

The small piece SAV BG 15 is quite unusual among the Metapontine productions: the wide disc base, quite high and thickened, is divided from the body by a reserved collar. The underside of the base is reserved as well and decorated by a painted band on the bottom inner side. Though the condition of the piece makes the typological attribution uncertain, the cup-skyphoi with concave lip (also called kotylai or simply skyphoi) give closer parallels. Mostly attested in the Attic black-figure and black-gloss production between the second half of the 6th and first quarter of the 5th c. BC, they were found in southern Italian and Etruscan contexts. For the Lucanian area, see most recently Bianco 2011, 77–78 (Guardia Perticara, Potenza, San Vito necropolis, T 210, 5th c. BC), displayed in the Museo Nazionale Archeologico di Metaponto (inv. no. 214622).

Cup-skyphoi

SAV BG 13 Cup-skyphos

Provenance: House on the Hill. Lot. SA 79-1062PL-D, Assemblage 1.

Dimensions and technical features: h 2.6, rim 15.0. Clay Munsell 7.5YR N6/ (gray), very hard, unevenly fired, and levigated. Black gloss thin and semi-lustrous, painted.

Condition: One rim fragment.

Description: Outward-turned rim; convex body.

Comparisons and comment: Lo Porto 1973, 188, no. 1, pl. XXXVI, 4, 1 (Montescaglioso, contrada Cappuccini, T 8, 500–480 BC, Attic fabric); Semeraro 1997, 202, no. 601, fig. 182 (Oria, Monte Papalucio, 500–480 BC, Attic fabric); *Forentum I*, 123, 195, pl. 30, type 3 (Lavello, necropolis, T 266, ca. 450 BC); *Antichi Peucezi*, 117, no. 15 (Bitonto, via Santoro necropolis, T 4/1982, 450–400 BC); *Museo Taranto II.2*, 498, pl. 40d (Rutigliano, Purgatorio necropolis, 450–400 BC); Ciancio 1997, 220, no. 278 (Botromagno, Gravina di Puglia, T 3 30/03/1994, 440–400 BC).

Date: 500–450 BC.

1:2

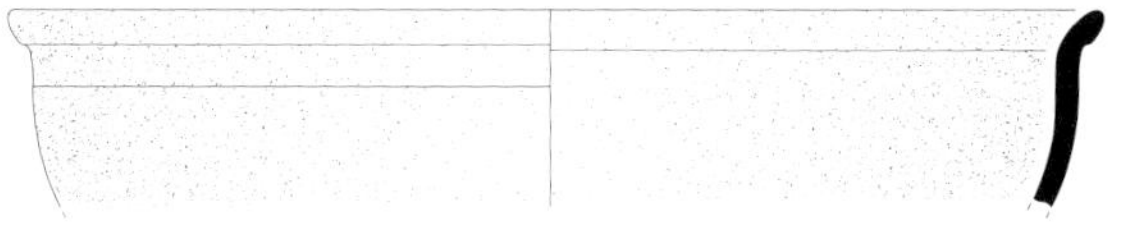

SAV BG 14 Cup-skyphos

Provenance: Sounding C, Pottery Deposit. Lot: SA 79-868P B, Assemblage 15.

Dimensions and technical features: h 4.4. Clay Munsell 10R 6/1 (reddish gray), very hard, unevenly fired, and levigated. Black gloss thick and semi-lustrous, dipped.

Condition: One rim fragment and one body fragment, joining.

Description: Direct, outward-sloping rim; convex, tapering body.

Comparisons and comment: Semeraro 1997, 219, fig. 191, no. 722 (Oria, Monte Papalucio, end of the 6th–beginning of the 5th c. BC, Tarentine fabric?); Rubinich 1992, 99, pl. XLV, no. 32 (Locri, shrine, 500–480 BC); Ciancio 1997, 222–23, no. 286 (Botromagno, Gravina di Puglia, T 3 30/03/1994, 440–400 BC).

Date: 430–400 BC.

1:2

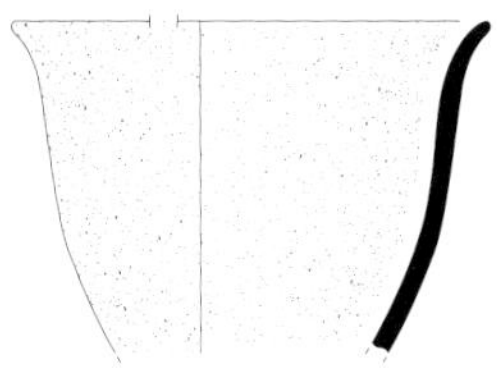

SAV BG 15 Cup-skyphos

Provenance: House on the Hill. Lot: SA 79-1062PL, Assemblage 1.

Dimensions and technical features: h 1.9, base 10.0. Clay Munsell 5YR 7/3 (pink), very hard, unevenly fired, and levigated. Black gloss dilute and matte, painted.

Condition: One base fragment.

Description: Disc base, low collar at the junction between base and body. Reserved collar and the underside of the base, decorated with a painted band on the bottom inner side.

Comparisons and comment: Museo Taranto I.3, 148, no. 2.16 (Taranto, via G. Mazzini angolo F. Cavallotti, tomb 8.VII.1931, 540–530 BC), 222, no. 41.12 (Taranto, Contrada Cortivecchie, viale Virgilio, tomb 2.V.1933, 520–500 BC); Hayes 1981, 25, 920.68.22, pl. 29, 7–9 (ca. 530–520 BC); Spadea 1999, 74–75 (Croton, Heraion at Capo Lacinio, edificio B, end of the 6th–beginning of the 5th c. BC); Baldoni 2009, 67, pl. 4, no. 31 (Marzabotto, 520–480 BC); Campus 1981, 58–61, nos. 37–39, tavv. LX–LXII, 95–96, no. 65, tav. LXIII (Tarquinia, ca. 490–480 BC).

Date: 530–480 BC.

1:2

Krater. The few fragments attributable to kraters found at Sant'Angelo Vecchio all come from the area of the Retaining Wall (2 NFr, 1 MNI, 1 TNI). One example is Attic (SAV BG 02), while the rest are Metapontine products. The base ring marked by a thin ridge at the junction with the body (SAV BG 16) represents a further attestation of the krater derived from Laconian models (see above, p. 227, "Krater"). This type of base occurs on completely black kraters with stirrup handles, dated to the second half of the 6th c. BC and found at various colonial and indigenous sites in southern Italy, generally considered Laconian products. An example believed to be a colonial product (Ferrara 2010, 449–50, fig. 129, VN 1) is dated to the first decades of the 5th c. BC.

These kraters were initially distributed in the eastern Mediterranean (Samos), Black Sea, Magna Grecia, Sicily, and Etruria as far as Massalia (Nafissi 1989, 71–72). In the final phase of production, the third quarter of the 6th c. BC, the traffic in these vessels appears to have been concentrated only in southern Italy and Sicily. The Laconian stirrup kraters were widely imitated in mainland Greece, Magna Grecia, and Sicily (Stibbe 1989, 54–57), where local production of *kelebai* is attested in Etruria (Caere), Campania (Capua, Palinuro), Peucetia (Bari, Gravina), and Messapia (Vaste: *Archeologia dei Messapi*, 62–63, no. 40). Examples are datable to the second half of the 6th c. BC. Both Laconian (Lanza Catti in *Fattoria Fabrizio*, 190–91, FF BG 21; Pisticci: Lo Porto 1973, 163, pl. X.2, no. 23) and local (Lanza Catti et al. in *Survey*, 218, no. 22) examples are known in the Metapontino.

Krater

SAV BG 16 Column Krater
Provenance: Square D5, US 39.
Dimensions and technical features: h 5.5, base 17.4. Clay Munsell 7.5YR 6/4 (light brown), hard, with white and rounded red inclusions. Black gloss thick and matte, almost worn.
Condition: One base fragment.
Description: Base-ring marked by a thin ridge at the junction with the body; reserved underside of the base.
Comparisons and comment: Kaulonia I, 69, fig. 46, no. 301 (Caulonia, second half of the 6th c. BC); *Sibari IV*, 145, fig. 132, no. 415 (St. 546) (Sybaris, Stombi 1972, second half of the 6th c. BC); Lattanzi 2007, 96–97 (Reggio Calabria, località San Gregorio, 6th–5th c. BC); Bonghi Jovino 1982, 65, no. 1, pls. 36.1 and 108.1 (Vico Equense, sporadic find); *Archeologia dei Messapi*, 62–63, no. 40 (Vaste, Fondo Melliche, third quarter of the 6th c. BC); *Leuca*, 100, pl. 54, A65 (Leuca, Grotta Porcinara).
Date: 525–480 BC.

1:2

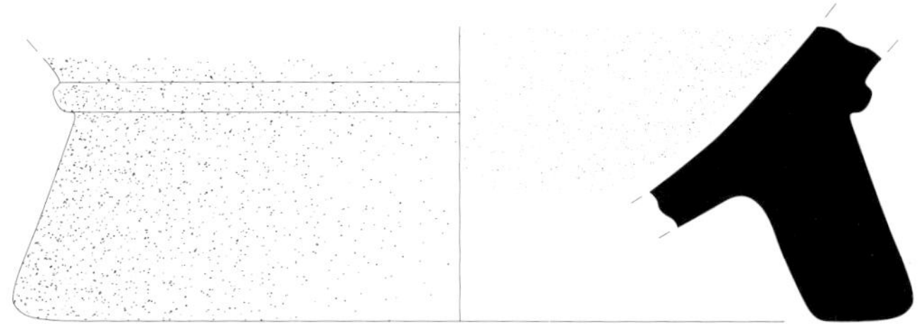

Attic-type skyphos. The Attic-type skyphos is attested by examples dating mostly to the 5th c. BC (14 NFr, 7 MNI, 7 TNI), featuring squat proportions, a straight rim, a more or less truncated conical profile, thick walls, and short and massive handles, similar to the *tronconiques trapus* types from Civita di Tricarico (*Civita di Tricarico*, 402). The types with wide convex body (SAV BG 17 and SAV BG 19) or clean truncated conical profile (SAV BG 18), with somewhat heavily proportioned base (SAV BG 20) and typically decorated with a circle and dot on the bottom lower side, date to the first half of the 5th c. BC. A single example (SAV BG 21) with more tapered body and *miltos* on the bottom lower side likely dates within the final decades of the 5th c. BC. These types, attested in Italy in Attic and colonial versions, were also produced in the Metapontine chora (Lanza Catti et al. in *Survey*, 220, nos. 27–28; Sant'Angelo Grieco, SG190.2).

SAV BG 22, with its barely inward-projecting rim and slightly tapering body, recalls Attic versions produced from the central decades of the 5th c. BC onward (*Agora XII*, 84). The type is attested among the waste from the *kerameikos* of Metaponto (Silvestrelli forthcoming, Deposit 12) and the materials from the chora (Sant'Angelo Grieco, SG319.1), circulating widely in southern Italy in the first half of the 4th c. BC.

Attic-type Skyphoi

SAV BG 17 Attic-type Skyphos
Provenance: House on the Hill. Lot: SA 79-1062PL, Assemblage 1.
Dimensions and technical features: h 4.0, rim 14.5. Clay Munsell 5YR 7/2 (pinkish gray), very hard and levigated. Black gloss dilute and semi-lustrous, possibly painted.
Condition: One rim fragment, missing the handle.
Description: Direct rim; hemispherical body.
Comparisons and comment: Sibari IV, 55, figs. 38, 136, no. 31 (St. 2979) (Sybaris, Stombi 1972) and 47, figs. 35, 136, no. 8 (St. 3415) (Sybaris, Stombi 1972, 600–550 BC, Corinthian fabric); *Kaulonia I*, 59, fig. 38, no. 159 (Caulonia, 5th c. BC); Baldoni 2009, 202, pl. 20, no. 446 (Marzabotto, 500–475 BC, Attic fabric); Ruga et al. 2005, 152, pl. LXI, fig. 14 (chora of Croton, Isola di Capo Rizzuto, 500–450 BC); Preacco Ancona 1992, 132, pl. XLIX, no. 87 (Locri, "Casa dei leoni," 350–300 BC).
Date: 500–450 BC.

1:2

1:2

SAV BG 18 Attic-type Skyphos

Provenance: House on the Hill. Lot: SA 79-1062PL, Assemblage 1.

Dimensions and technical features: h 3.6, rim 14.6. Clay Munsell 7.5YR 7/4 (pink), powdery, unevenly fired, and levigated. Black gloss thick and semi-lustrous, possibly painted.

Condition: One rim fragment.

Description: Direct rim; conical body.

Comparisons and comment: Long/Miro/Volpe 1992, 217–18, fig. 33.11 (Pointe Lequin shipwreck 1A, end of the 6th–beginning of the 5th c. BC, Attic fabric); Baldoni 2009, 202, pl. 20, no. 445 (Marzabotto, 500–475 BC, Attic fabric); *Mozia XII*, 229, pl. LIX, no. MD.03.1056/42 (Motya, Zone D, 5th c. BC); Ruga et al. 2005, 200–201, pl. LXXXIX, fig. 103.41, no. 39 (chora of Croton, Isola di Capo Rizzuto necropolis, site 17, ca. 450 BC, colonial fabric); *Sibari IV*, 107, figs. 99, 135, no. 248 (St. 6194) (Sybaris, Stombi 1972).

Date: 475–450 BC.

1:2

SAV BG 19 Attic-type Skyphos

Provenance: House on the Hill. Lot: SA 79-1062PL, Assemblage 1.

Dimensions and technical features: h 1.8, rim 14.0. Clay Munsell 5YR 7/4 (pink), hard and levigated, with sparse luminous and black inclusions. Black gloss thick and semi-lustrous, painted?

Condition: One rim fragment preserving the handle; slip partially worn.

Description: Direct rim; conical body; rounded handle.

Comparisons and comment: Semeraro 1997, 204, fig. 184, no. 635 (Oria, Monte Papalucio, 525–475 BC, Attic fabric); Dell'Aglio 1996b, 329, 331, no. 306 (Taranto, via Leonida 52, end of the 5th–beginning of the 4th c. BC); Baldoni 2009, 129, pl. 13, no. 178 (Marzabotto, end of the 5th–beginning of the 4th c. BC, Attic red-figure); Barbieri and Virtuoso 2002, 238, pl. 62, no. 37 (Ascoli Satriano, ca. 440–350 BC).

Date: 450–400 BC.

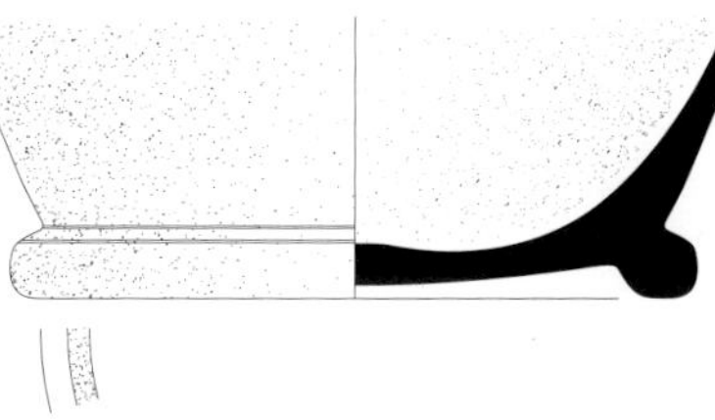

1:2

SAV BG 20 Attic-type Skyphos

Provenance: Sounding C, Pottery Deposit. Lot: SA 79-1022P B, Assemblage 15.

Dimensions and technical features: h 3.8, base 9.7. Clay Munsell 5YR 6/6 (reddish yellow), very hard and levigated. Slip oxidized.

Condition: One base fragment, gloss almost worn.

Description: Base-ring; convex lower body. Reserved underside; bottom inner side decorated with a band; circle and dot somewhat worn on the bottom lower side.

Comparisons and comment: Agora XII, 259, pl. 16, fig. 4, no. 342 (Athens, 470–460 BC); *Kaulonia I*, 59, fig. 38, no. 158 (Caulonia, mid-5th c. BC); Ruga et al. 2005, 161, pl. LXIV, fig. 26.6, no. 6 (chora of Croton, site 3 farmhouse, 450 BC, colonial fabric); *Sibari V*, 214, figs. 194, 214, no. 135 (St. 2438) (Sybaris, Stombi 1974).

Date: 470–450 BC.

Attic-type skyphoi cont.

SAV BG 21 Attic-type Skyphos

1:2

Provenance: Sounding C, Pottery Deposit. Lot: SA 79-887P B, Assemblage 15.

Dimensions and technical features: h 3.3, base 7.5. Clay Munsell 5YR 6/6 (reddish yellow), very hard, very well fired, and levigated. Black gloss thick and semi-lustrous, dipped/painted.

Condition: One base fragment.

Description: Base-ring; slightly convex lower body. Reserved bottom lower side, filled with *miltos*.

Comparisons and comment: Ruga et al. 2005, 152, pl. LXII, fig. 16 (chora of Croton, Isola di Capo Rizzuto, 500–450 BC, colonial fabric); Baldoni 2009, 135, pl. 14, no. 206 (Marzabotto, Attic owl-skyphos, 450–400 BC); Olivero Ferrero 1989, 97, pl. XXI, no. 48 (Locri, 425–400 BC); *Gravina II*, 116, fig. 37, no. 871 (Botromagno, Gravina di Puglia, ca. 450 BC).

Date: 430–400 BC.

SAV BG 22 Attic-type Skyphos

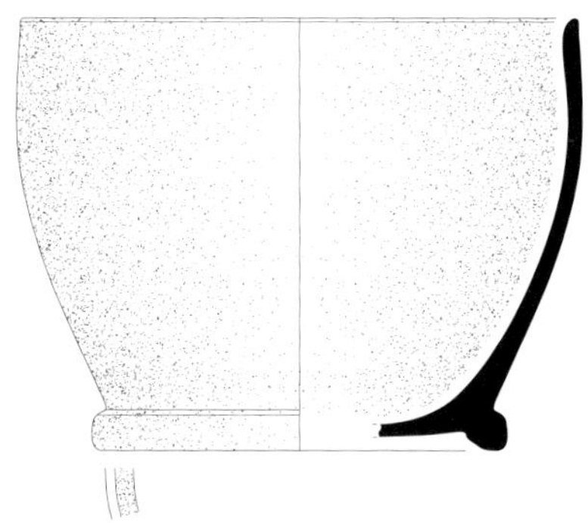

1:2

Provenance: Sounding C, Pottery Deposit. Lot: SA 79-762P + 867P B + 988PL B, Assemblage 15.

Dimensions and technical features: h 7.6, rim 10.0, base 7.5. Clay Munsell 5YR 6/3 (light reddish brown), very hard, and levigated. Black gloss thick and semi-lustrous, dipped/painted.

Condition: Three rim fragments, six base fragments, and six body fragments, nearly joining.

Description: Direct rim; slightly inward-sloping concave body; base-ring. Reserved underside; bottom inner side decorated with a band; bottom lower side filled with *miltos*.

Comparisons and comment: Silvestrelli forthcoming, fig. 5.11 (Metaponto, *kerameikos*, deposit 12, ca. 500–450 BC); Ruga et al. 2005, 201, pl. LXXXIX, fig. 103.42, no. 42 (chora of Croton, Isola di Capo Rizzuto necropolis, site 17, 470–460 BC, colonial fabric); *Forentum I*, 192–93, skyphos type 1.2, pl. 28 (Lavello, necropolis, ca. 450–350 BC); *Gravina II*, 116, fig. 36, no. 870 (Botromagno, Gravina di Puglia, 450–375 BC, perhaps later); *Armi*, 192–93, no. 7 (Montescaglioso, Contrada Sterpinia, T found in 1953, 400–350 BC); Buzzi Auxilia 1989, 131, pl. XXII, no. 86 (Locri, 400–350 BC); Dell'Aglio 1996b, 329, 331, no. 311 (Taranto, via Leonida, 52, 400–350 BC); *Museo Taranto III.1*, 248, fig. 185, phase A1 (Taranto, necropolis, 375–350 BC).

Date: 460–370 BC.

Corinthian-type skyphos. The Classical C-type skyphos is more common at Sant'Angelo Vecchio than the A-type. The 5th-c. BC examples are few (9 NFr, 3 MNI, 3 TNI), but nevertheless a sequence can be reconstructed, at the beginning of which are the examples with wide and not excessively tapering bodies and maximum diameter located at the rim (SAV BG 23), attested in Attica from the 6th through the first half of the 5th c. BC (*Agora XII*, 82–83). Two bases (SAV BG 24 and SAV BG 25) can be ascribed to the skyphoi with more ovoid bodies and rounded or horseshoe-shaped handles, attested at Athens in the all-black version in the second half of the 5th c. BC (*Agora XII*, 83, 258, pl. 15, no. 321). The sequence ends with examples that have a strongly ovoid profile (SAV BG 26) or a cylindrical upper body (SAV BG 27) and triangular handles (21 NFr, 16 MNI, 16 TNI). These skyphoi, dated through the late 4th c. BC, represent an evolution toward the Hellenistic types, on which the body is defined by a double curve. No base was found in association with this type.

The Early Classical types, documented in the Metapontino and in mostly Italiote contexts datable to the second half of the 5th c. BC, generally have a reserved band on the lower body and dot-and-circle decoration on the bottom lower side. The 4th-c. BC versions have numerous parallels at indigenous sites in southern Italy and Sicily.

Corinthian-type Skyphoi

SAV BG 23 Corinthian-type Skyphos

1:2

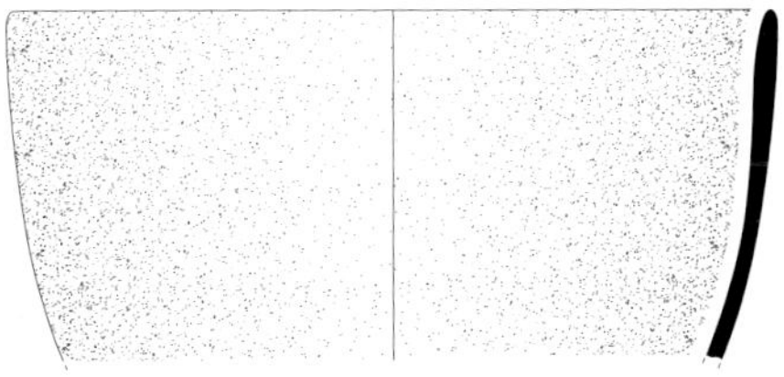

Provenance: Sounding C, Pottery Deposit. Lot: SA 79-867P A, Assemblage 15.
Dimensions and technical features: h 5.0, rim 11.5. Clay Munsell 5YR 7/6 (reddish yellow), very hard and levigated. Black gloss thick and semi-lustrous, dipped.
Condition: Two rim fragments and three body fragments, joining.
Description: Direct tapering rim; convex body.
Comparisons and comment: Oakley 1988, no. 41 (440–430 BC); Ruga et al. 2005, 161, pl. LXIV, fig. 26.5, no. 5 (chora of Croton, site 3 farmhouse, second half of the 5th c. BC); *Sibari IV*, 111, figs. 101, 135, no. 277 (St. 15918) (Sybaris, Stombi 1972, 5th c. BC); Ferrara 2010, 456, VN 15, pl. 52, no. 203 (Poseidonia-Paestum, Heraion at mouth of Sele, Thesauros).
Date: 450–425 BC.

SAV BG 24 Corinthian-type Skyphos

1:2

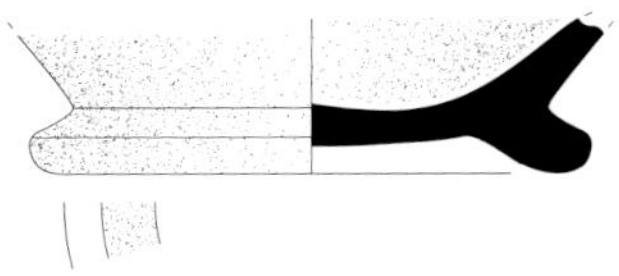

Provenance: Squares C4–5, US 18.
Dimensions and technical features: h 2.1, base 7.7. Clay Munsell 5YR 7/4 (pink), very hard and levigated. Black gloss thick and semi-lustrous, dipped/painted.
Condition: One base fragment, preserved to one-quarter.
Description: Disc base. Reserved underside; bottom inner side decorated with a band.
Comparisons and comment: Boulter 1953, 72, fig. 2, pl. 29, no. 25 (Athens, mid-5th c. BC); *Kaulonia I*, 59, fig. 38, no. 154 (Caulonia, not before mid-5th c. BC); Elliott in *Necropoleis*, 681, S56 (Pantanello necropolis, T 336, 480–440 BC); *Museo di Taranto*, 561, pl. CXXII, no. 40.2c (Taranto, contrada Vaccarella, T 5 25.I.1952, second/third quarter of the 5th c. BC); Ruga et al. 2005, 161, pl. LXIV, fig. 26.4, no. 4 (chora of Croton, site 3 farmhouse, ca. 450–430 BC, colonial fabric); *Museo Taranto II.1*, 65–66, no. 30.23 (Ginosa, T 30, 20.I.1967, end of the 5th–beginning of the 4th c. BC).
Date: 460–430 BC.

SAV BG 25 Corinthian-type Skyphos

1:2

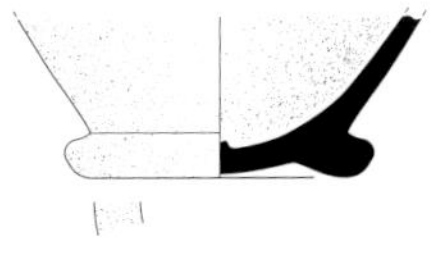

Provenance: Square C5. Level 1, batt. 2. Lot: SA 79-1063PL, Assemblage 13.
Dimensions and technical features: h 2.1, base 4.3. Clay Munsell 2.5YR 6/8 (light red), very hard and levigated. Black gloss thick and semi-lustrous, dipped/painted.
Condition: One base fragment, preserved to half.
Description: Disc base; slightly tapering body. Reserved bottom lower side painted with *miltos*.
Comparisons and comment: Oakley 1988, 181, fig. 1, no. 33 (470–460 BC); *Museo Taranto I.3*, 358, no. 125.52 (Taranto, corner of via Dante and via Minniti, T August 1934, end of the 5th–beginning of the 4th c. BC); *Museo Taranto II.2*, 171, 173, pl. 37c, no. 57.20 (Rutigliano, Purgatorio necropolis, T 57, ca. 350–340 BC); *Sibari IV*, 418, figs. 365, 382, no. 439 (PS 13014) (Sybaris, Prolungamento Strada 1972); *Forentum II*, 35, pl. CXXXVIII, no. 6 (Lavello, acropolis, T 598).
Date: 420–380 BC.

Corinthian-type skyphoi cont.

SAV BG 26 Corinthian-type Skyphos

1:2

Provenance: Sounding C, Pottery Deposit. Lot: SA 79-867P C, Assemblage 15.

Dimensions and technical features: h 6.4, rim 10.1. Clay Munsell 7.5YR 7/4 (reddish yellow), very hard and levigated, with very fine, sand-sized lithic inclusions. Black gloss thick and semi-lustrous, dipped.

Condition: Two rim fragments and two body fragments, joining; triangular handle missing.

Description: Direct tapering rim; ovoid body.

Comparisons and comment: Ruga et al. 2005, 152, pl. LXI, fig. 15 (chora of Croton, Isola di Capo Rizzuto, 500–450 BC, colonial fabric); *Museo Taranto I.3*, 339, no. 118.23 (Ginosa, via Vittorio Emanuele II, T 13.I.1935, beginning of the 5th c. BC); Gagliardi 2010, 73, fig. 5.2, no. 19 (Caulonia, necropolis, second half of the 5th c. BC); Dell'Aglio 1996b, 330–31, no. 314 (Taranto, via Leonida 52, first half of the 4th c. BC); Bruscella and Virtuoso 2005, 270, pl. XXXVII, no. 208 (Torre di Satriano, Lucanian sanctuary, last quarter of the 4th c. BC).

Date: 400–370 BC.

SAV BG 27 Corinthian-type Skyphos

1:2

Provenance: Square G7. Level 2, batt. 1. Lot: SA 79-793PL, (see *Fig. 2.20*).

Dimensions and technical features: h 3.7, rim 9.4. Clay Munsell 5YR 7/4 (pink), very hard and levigated. Black gloss thick and semi-lustrous, dipped.

Condition: One rim fragment.

Description: Direct, slightly outward-sloping rim; cylindrical body.

Comparisons and comment: Surra and Preacco Ancona 1997, 77, pl. 36, no. 85 (Pomarico Vecchio, 400–325 BC); *Ordona VII/2*, 508, fig. 280.8 (Ordona, T 123, ca. 350 BC); Elia 2006, 173, 193, fig. 12, FI203 (Terravecchia di Grammichele, second half of the 4th–3rd c. BC); *Civita di Tricarico*, 408, fig. 295, no. 267 (Civita di Tricarico).

Date: 370–325 BC.

Bolsal. A single example can be ascribed to the two-handled cup widely known as the bolsal (SAV BG 28). The convex profile of the body recalls the completely black Attic types with a disc base similar to that of the C-type skyphoi and a groove at the joint between the lower and the upper parts of the body. In Attica, the type is produced from the first quarter of the 5th c. BC onward (*Agora XII*, 107).

The bolsal with convex body and lacking decoration on the internal bottom is attested in unpublished Metapontine funerary contexts of the second half of the 5th c. BC on display in the National Archeological Museum at Metaponto (Torremare, T 76). Produced in the *kerameikos* of Metaponto in the second half of the 5th c. BC (Silvestrelli forthcoming, fig. 7.3, deposit 16), the type is also documented in the chora (Lanza Catti et al. in *Survey*, 234, no. 90) and the Pantanello necropolis, where examples with the same reserved bottom lower side have been dated between 420 and 370/360 BC.

Bolsals, both Attic and locally-produced, circulated in southern Italy and Sicily between the end of the 5th and the first half of the 4th c. BC. At Locri, the earliest examples with completely black external bottom or with reserved central zone are dated around the middle of the 4th c. BC (Buzzi Auxilia 1989, 110).

Bolsal

1:2

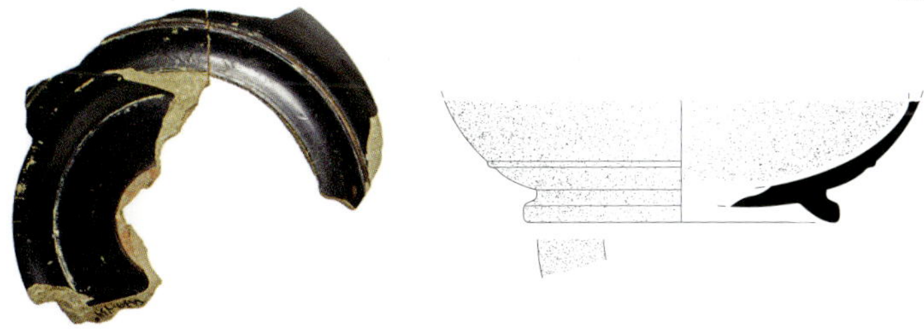

SAV BG 28 Bolsal

Provenance: Slope 3, surface cleaning, and Slope 4. Lot: SA 79-434P+1063PL, Assemblage 13.

Dimensions and technical features: h 2.4, base 6.5. Clay Munsell 5YR 6/4 (light reddish brown), very hard, unevenly fired, and levigated, with common luminous inclusions. Black gloss thick and lustrous, dipped/painted.

Condition: Two base fragments and two body fragments, joining (one body from square C5).

Description: Disc base; convex lower body with two incised lines on the exterior. The center of the bottom lower side is reserved and filled with *miltos*.

Comparisons and comment: Museo Taranto II.2, 500, pl. 41c (Rutigliano, Purgatorio necropolis, T 29, end of the 5th c. BC, Attic fabric); *Agora XII*, 273, pl. 24, fig. 6, no. 534 (Athens, ca. 430 BC); Kustermann Graf 2002, 234, 192/O 402, pls. XCIII, CXXIV (Selinous, Manicalunga necropolis, T 192, 425–400 BC, Attic fabric); Elliott in *Necropoleis*, 650, B1 (Pantanello necropolis, T 106, 420–370/360 BC); Orsi 1990, 131, pl. XC, T 1329 no. 1 (Camarina, necropolis, T 1329, 400–375 BC); *Sibari V*, 510, figs. 473, 503, no. 567 (PS 24929) (Sybaris, Prolungamento Strada 1974).

Date: 425–400 BC.

Mug. The rim SAV BG 29 is from a mug with offset lip, concave profile, and relatively deep body, datable between the late 6th and the beginning of the 5th c. BC (*Agora XII*, 71). The wide mouth (rim 12.0 cm) and the shape of the lip distinguish it from the more common types of one-handled mug with different proportions.

The technical features of the example, produced with reddish yellow (Munsell 5YR 6/6), levigated clay and coated with thick, semi-lustrous, and high-quality slip, are compatible with products widely attested in the Metapontine chora. The best parallel, in morphological terms, is an example from the Pantanello sanctuary (K. Swift, pers. comm., 2010), while the closest published parallels are one-handled *boccali* from funerary contexts in southern Italy and Sicily datable to the 5th c. BC.

Mug

SAV BG 29 Mug

Provenance: Bulldozed Stipe. Lot: SA 79-603PL, Assemblage 9.

Dimensions and technical features: h 5.1, rim 12.0. Clay Munsell 5YR 6/6 (reddish yellow), very hard, unevenly fired, and levigated. Black gloss thick and semi-lustrous, dipped.

Condition: Three rim fragments and two body fragments, partially joining.

Description: Offset lip with direct rim; slightly convex body.

Comparisons and comment: Greci, Enotri e Lucani, 166, no. 2.30.15 (Chiaromonte-San Pasquale, T 227, 5th c. BC); *Agora XII*, 249, pl. 11, no. 194 (Athens, 500–480 BC); Kustermann Graf 2002, 101–2, 13/O 494, pls. XV, CXXIX (Selinous, Manicalunga necropolis, T 13, mid-5th c. BC, Attic fabric).

Date: 470–430 BC.

1:2

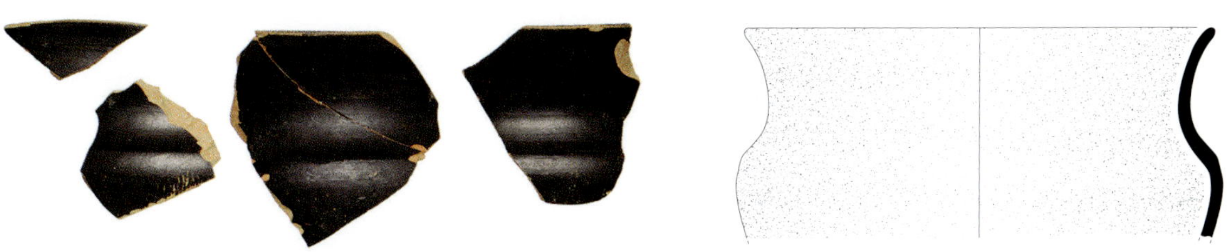

Jug. SAV BG 30 is a typical example of this form, known in the literature as "olpe," characterized by an S-shaped body profile, flaring and direct rim, and more or less pronounced ring base; it is attested at Sant'Angelo by a very small number of examples (2 NFr, 2 MNI, 2 TNI).

An origin for the type in Magna Grecia, possibly at Metaponto, has recently been proposed because of its concentrated circulation in the Achaian colony and in areas under Metapontine influence (*Museo Taranto II.2*, 483). Produced in the *kerameikos* of Metaponto between the end of the 5th c. and 350 BC (D'Andria 1980a, 408, fig. 55, no. 205, Deposit 1), it is attested in the chora by local examples (Elliott in *Necropoleis*, 688, J1; Lanza Catti in *Fattoria Fabrizio*, 215–16, FF BG 67; Sant'Angelo Grieco, SG285.11) and circulated widely throughout southern Italy in the course of the 4th c. BC, reaching both the Greek colonies and the indigenous centers of Lucania and Apulia (*Civita di Tricarico*, 397).

Jug

SAV BG 30 Jug

Provenance: Sounding C, Pottery Deposit. SA 79-1021P A, Assemblage 15.

Dimensions and technical features: h 4.4, base 3.6. Clay Munsell 5YR 7/4 (pink), very hard, unevenly fired, and levigated. Black gloss thick and semi-lustrous, dipped/painted.

Condition: One base fragment.

Description: Base-ring; slightly convex body. Reserved underside of the base, decorated with a band on the bottom inner side.

Comparisons and comment: Museo Taranto II.2, 483, pl. 36b (Rutigliano, Purgatorio necropolis, 475–beginning of the 4th c. BC); *Archeologia di una città*, 332, fig. 462, no. 715 (Bari, middle of the 5th c.–350 BC); Dell'Aglio 1996b, 331–32, no. 328 (Taranto, via Leonida 52, 4th c. BC); *Civita di Tricarico*, 397, fig. 293, no. 174 (Civita di Tricarico, 4th c. BC); Lo Porto 1988–1989, 350, fig. 46.3 and 56.2 (Montescaglioso, T 5, 400–350 BC); *Archeologia dei Messapi*, 141, no. 216 (Vaste, Fondo Melliche, T 597, ca. 400–350 BC); *Principi Imperatori Vescovi*, 357, no. 33 (Canosa, Ipogeo dei Vimini, 400–350 BC); *Ordona VII/2*, 371, fig. 196.48 (Ordona, T 92, ca. 370 BC); Interdonato 2002, 335, pl. 90, no. 19 (Ascoli Satriano, T 3); *Forentum II*, 50, type 1A, pl. CXXXII, no. 23 (Lavello, acropolis, T 669).

Date: 420–350 BC.

1:2

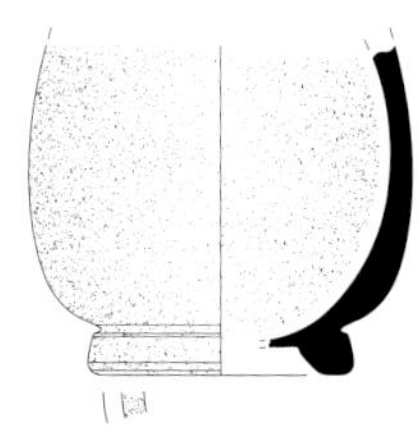

Olpe. SAV BG 31 belongs to a type of olpe derived from Attic models, most common between the 5th and 4th c. BC, which were typically completely black and had a sharply angled body, ribbon handle, and round mouth with thickened rim (*Agora XII*, 78–79). The example from Sant'Angelo Vecchio has a dilute and matte slip with reserved lower body and base, following a practice documented on vessels found in the Pantanello necropolis and datable to the first quarter of the 4th c. BC (Elliott in *Necropoleis*, 689, OL1).

The type is attested in Archaic contexts of the *kerameikos* of Metaponto and was produced at other colonial sites (Olivero Ferrero 1989, 103, pl. XXI, no. 59); examples that were probably imported have also been found in Archaic contexts of indigenous sites in southern Italy.

Olpe

SAV BG 31 Olpe

1:2

Provenance: Slope 3. Level 1, batt. 2. Lot: SA 79-90P (see *Fig. 2.35*).

Dimensions and technical features: h 10.0, base 3.4. Clay Munsell 7.5YR 7/4 (pink), powdery, unevenly fired, and levigated. Black gloss dilute, barely visible in patches, painted.

Condition: Two base fragments and one body fragment, joining; nearly complete vessel, missing only the rim and handle.

Description: Flat base; convex, elongated body. Reserved lower exterior; lower part of a band (0.4 cm) on the upper interior.

Comparisons and comment: Kaulonia I, 69, fig. 48, no. 328 (Caulonia, Archaic levels); Silvestrelli forthcoming, fig. 4.8 (Metaponto, *kerameikos*, Archaic deposit, ca. 550–500 BC); *Agora XII*, 254, pl. 13, no. 265 (Athens, ca. 500 BC); *Museo Taranto II.2*, 483, pl. 36a (Rutigliano, Purgatorio necropolis, T 35, 86, 500–450 BC); *Sibari V*, 542, figs. 528, 565, nos. 112–13 (CB 855, CB 856) (Sybaris, Casa Bianca 1974, ca. 500 BC); Semeraro 1997, 231, fig. 198, no. 800 (Oria, Monte Papalucio, 500–475 BC).

Date: 525–450 BC.

Amphora. The amphora is encountered infrequently at Sant'Angelo Vecchio (15 NFr, 2 MNI, 4 TNI). SAV BG 32 has a rim with rectangular section that recalls the rims of the Attic type B amphorae widely produced in figured versions (i.e., *Museo Taranto I.3*, 159–60, no. 12.5; 169, 171, nos. 13.8 and 13.9).

The closer parallels for this type of amphora come from the Tarentine Archaic tombs, where they are attested in black-gloss with overpainted narrow bands or reserved neck. In the Metapontine chora, the amphorae do not seem to be popular during the Archaic period. They become frequently included in funerary assemblages in the first half of the 5th c. BC, but a function as a wine container might also be proposed for the examples found in the farmhouses (Lanza Catti et al. in *Survey*, 202).

Amphora

SAV BG 32 Amphora

Provenance: Square F7. Level 3, batt. 4. Lot: SA 80-26PL A.

Dimensions and technical features: h 10.2, rim 17.4. Clay Munsell 7.5YR 7/4 (pink), very hard, well fired, and well levigated. Black gloss thin and semi-lustrous, painted.

Condition: One rim fragment and five body fragments, joining.

Description: Direct rim with concave exterior and offset neck. Reserved upper surface of the rim; reserved band (w 2.0) below the offset of the rim; reserved lower part of the interior (h 1.8 cm).

Comparisons and comment: Museo Taranto I.3, 175, 177, 14.5 (Taranto, contrada Lupoli, T 4, 550–530 BC); Esposito 1996, 311, nos. 299.1–2 (Taranto, contrada Tre Carrare, T 4 del 28 marzo 1958, 560–550 BC).

Date: Central decades of the 6th c. BC.

Oinochoe. The Pottery Deposit yielded several fragments of a "beaked" oinochoe (SAV BG 33) corresponding to Beazley's Shape X. The form derived from the so-called *prochoai* in bronze (Hill 1962, 57, 60), which are only rarely attested in southern Italy: an imported example datable to the beginning of the 5th c. BC was found in Messapia (Lo Porto 1994, 56, no. 3, figs. 7.4, 10.1–2).

The clay examples are generally small, with a distinct foot and ovoid or narrow body sharply articulated at the junction of the neck and shoulder, characterized by a pair of rotelles projecting above the rim (Green 1972, 8–9). The form is not attested before the late 6th c. BC and spread mostly in the second half of the following century, when it was produced in both Attic and Italiote red-figure, as well as black-gloss. In southern Italy, it is documented starting in the second half of the 5th c. BC, gaining a certain popularity in the second half of the following century, with examples in black-gloss, red-figure, and Gnathia.

Oinochoe

SAV BG 33 Oinochoe

Provenance: Sounding C, Pottery Deposit. Lot: SA 79-761P, Assemblage 15.

Dimensions and technical features: h 7.7, max. diam. 8.4. Clay Munsell 5YR 7/4 (pink), very hard, unevenly fired, and well levigated. Black gloss thick and semi-lustrous, dipped.

Condition: One handle and four body fragments, joining.

Description: Short and narrow neck; high handle; a pair of rotelles on the top of the neck, where the handle is attached.

Comparisons and comment: Museo Taranto I.3, 366, 368, 138.1 (Taranto, contrada Vaccarella, via Argentina, T 53, ca. 430 BC, Attic red-figure by the Washing Painter); *Museo Taranto II.2*, 140, 142, no. 48.14, and 534, pl. 49b, fig. 110 (Rutigliano, Purgatorio necropolis, T 48, 350–340 BC); Trendall 1989, 201, fig. 361 (red-figure by Asteas, Poseidonia-Paestum, T in contrada Gaudo, mid-4th c. BC); Hayes 1984, 114, no. 193 (T probably from Ordona, first half of the 4th c. BC); *Principi Imperatori Vescovi*, 259–60, no. 8 (Canosa di Puglia, Ipogeo Varrese, 4th c. BC); *Forentum I*, 187, pl. 112, fig. 200 type 2 epichysis (Lavello, necropolis, T 312, third quarter of the 4th c. BC).

Date: 430–400 BC.

Lekythoi/Squat lekythoi. Sant'Angelo Vecchio yielded few fragments attributable to perfume containers (12 NFr, 3 MNI, 5 TNI). The earliest fragment is the uncataloged disc base of a lekythos with a distinct shoulder datable to the middle of the 5th c. BC (see Elliott in *Necropoleis*, 665, L25 with tongue pattern).

Example SAV BG 34, on the other hand, belongs to the type with a high and narrow neck, trefoil rim, and globular body, attested in Deposits 19 and 22 of the *kerameikos* at Metaponto (Silvestrelli forthcoming, fig. 6.15) and datable between the end of the 5th and the beginning of the 4th c. BC. The form is rarely encountered outside of Metaponto; the example found in Tomb 2 in via Salentina at Ugento, which Lo Porto considered a proto-Italiote product (Lo Porto 1970–1971, 139), could very well be Metapontine. The chronology indicated by the materials recovered in the *kerameikos* appears to be confirmed by an example from Ordona datable to the first quarter of the 4th c. BC.

The lone squat lekythos in the catalog is nearly intact (SAV BG 35), missing only the echinus rim and the handle. The type, documented in the chora of Metaponto in the first quarter of the 4th c. BC (Elliott in *Necropoleis*, 685), follows Attic models with a bulbous body that appeared at the end of the Archaic period but did not become common before the second half of the 5th c. BC. The vessels were not produced in black-gloss before 450 BC (*Agora XII*, 153) and were certainly still exported in the 4th c. BC (*Museo Taranto III.1*, 254). The type is attested in the Metapontine sphere also in a ribbed version, and in southern Italy in both black-gloss and red-figure, with examples dating no later than the second half of the 4th c. BC (*Greci, Enotri e Lucani*, 267–68, no. 3.42.1; Giannotta 1996, 61–62, fig. 27, no. 47; 1997, 165–66, no. 18; *Archeologia dei Messapi*, 120, no. 168). As in Attica, production of the lekythos in southern Italy appears to end in the Early Hellenistic period, coinciding with the spread of the unguentaria (*Museo Taranto III.1*, 254).

Lekythoi/Squat Lekythoi

SAV BG 34 Trefoil Lekythos

1:2

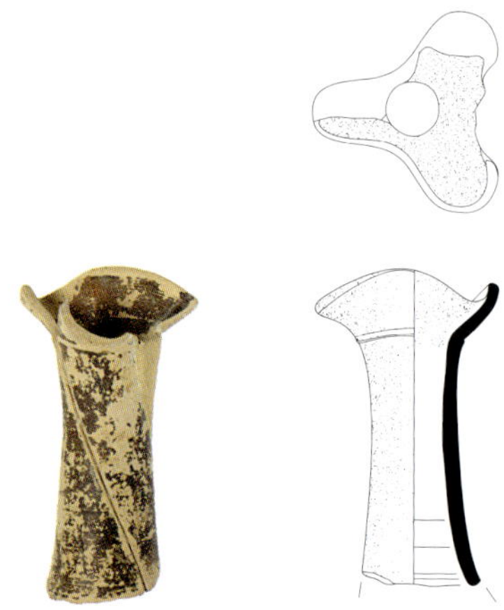

Provenance: House on the Hill. Level 4, batt. 1. Lot: SA 79-1057PL. Inv. no. 319956 (see *Fig. 2.39*).

Dimensions and technical features: h 6.0, rim 3.6. Clay Munsell 7.5YR 7/4 (pink), hard and levigated. Black gloss oxidized, thin and matte, painted?

Condition: Two rim fragments, joining.

Description: Trefoil mouth and narrow cylindrical neck. The junction of the neck and rim is marked by a shallow groove; the interior of the rim is partially slipped (h 1.5 cm).

Comparisons and comment: Museo Taranto I.3, 321, 323, no. 104.19 (Taranto, corner of via F. Di Palma and via F. Cavallotti, T X, 15.II.1937, 480–470 BC); Silvestrelli forthcoming, fig. 6.15 (Metaponto, *kerameikos*, deposit 19, ca. 450–400 BC); Lo Porto 1970–1971, 139–40, pl. LX.a, no. 29 (Ugento, T 2 via Salentina, end of the 5th c. BC); *Ordona III*, 50, figs. 20–21, pl. XXXIV, no. 15 (Ordona, T LX, ca. 375 BC).

Date: 480–425 BC.

SAV BG 35 Squat Lekythos

1:2

Provenance: House on the Hill, Test Trench 3. Level 1, batt. 1. Lot: SA 79-951P. Inv. no. 320025 (see *Fig. 2.39*).

Dimensions and technical features: h 7.4, base 5.0. Clay Munsell 7.5YR 7/4 (pink), very hard, unevenly fired, and levigated, with sparse luminous inclusions. Black gloss thin and semi-lustrous, painted.

Condition: Two base fragments and three body fragments, joining, missing only the rim and handle.

Description: Convergent, straight base-ring; bulbous body. The junction of the neck and body is marked by a shallow groove; reserved lower exterior decorated with a band (0.4 cm) on the external base.

Comparisons and comment: Archeologia di una città, 280, no. 628, figs. 364, 369 (Bari, T sarcophagus of via Amendola, 480–470 BC); Hayes 1984, 30, fig. 4, no. 48 (late 5th c. BC, Attic fabric); Carter et al. in *Survey*, 1095, site 737 (Metapontine chora, T 23-02, central decades of 5th c. BC); Lo Porto 1973, 161, no. 25, pl. VIII, 3, 6 (Pisticci, 5th–4th c. BC); Lo Porto 1988–1989, 321, fig. 20.3 (Casinello-San Teodoro, T 3, ca. 430–420 BC); *Museo Taranto III.1*, 253, fig. 189, phase A1 (Taranto, necropolis, 375–350 BC); Bitti 1989, 172, pl. XXVI, no. 162 (Locri, 350–300 BC).

Date: 480–450 BC.

Black-gloss of the Hellenistic Period

Dishes. Few pieces are from dishes datable to the Hellenistic period (9 NFr, 2 MNI, 2 TNI). The examples found at Sant'Angelo Vecchio belong to the type with narrow and projecting rim, diameter ranging between 13.0 and 14.0 cm, and slightly convex body with noticeably narrow base (SAV BG 36), or the type with body having an angular profile ending in a rounded, pendant rim (SAV BG 37). The slip, thick and semi-lustrous in earlier periods, suffers in the later examples, with a matte appearance and evident dribbling.

The origins of the dish are unclear. It seems to have appeared suddenly in many South Italian production centers (*Valesio I*, 147), where the earliest examples date to the last quarter of the 4th c. BC (*Civita di Tricarico*, 413). A consistent feature of grave assemblages, they are attested in tombs at Taranto only starting in the first half of the 2nd c. BC;

they are also common in domestic contexts. The examples found at Sant'Angelo Vecchio can be associated with the *à bord en z, bien marqué* and *à bord pendant arrondi* types from Civita di Tricarico (*Civita di Tricarico*, 412–13, 418–19).

SAV BG 36 is a variant of the Lamboglia 36 and can be interpreted as a forerunner of the dish with pendant rim. It appears in tombs dated between 410 and 275 BC in the Pantanello necropolis, is attested at Lavello, and is commonly found in Apulian tombs of the second half of the 4th c. BC (*Forentum I*, 199). Common at Valesio in contexts datable between 270 and 180 BC, the type is also frequently encountered at other sites in the vicinity of Brindisi and Taranto (*Valesio I*, 151). The *à bord pendant arrondi* type (SAV BG 37), on the other hand, corresponds to Lamboglia 36, attested in the Metapontino in both imported (Lo Porto 1981, fig. 69.3) and local versions (Lanza Catti et al. in *Survey*, 247, no. 163) and very common in Magna Grecia and Sicily between the end of the 4th and the 3rd c. BC (Pianu 2002, nos. 89–90; *Museo Taranto III.1*, fig. 204, phase C; *Tyndaris 1*, 137, pl. 2, VN/29). In the same period it is also documented at several indigenous centers of southern Italy (Morel 1970, fig. 24.6; Morel 1981, 1516b).

Dishes

SAV BG 36 Dish

1:2

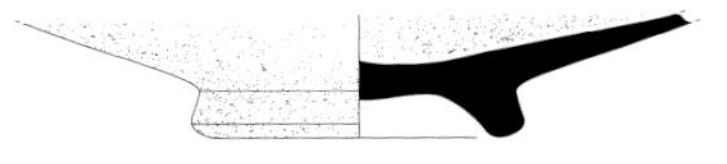

Provenance: Square D4 N, near baulk. Level 3, batt. 6. Lot: SA 79-913PL, Assemblage 13.
Dimensions and technical features: h 1.6, base 4.6. Clay Munsell 5YR 7/6 (reddish yellow), very hard and levigated. Black gloss thick and matte, dipped.
Condition: Two base fragments, joining.
Description: Base-ring; slightly convex body. Reserved underside of the base.
Comparisons and comment: Forentum I, 199, pl. 33, dish type 5 (Lavello, necropolis, T 69, 350–300 BC); *Civita di Tricarico*, 414, fig. 297, no. 328 (Civita di Tricarico, end of the 4th c. BC); Elliott in *Necropoleis*, 653, D2 (Pantanello necropolis, T 9, 310–275 BC); *Valesio I*, 151, form K03c, no. 225 (Valesio, 270/250–200/180 BC).
Date: 325–275 BC.

SAV BG 37 Dish

1:2

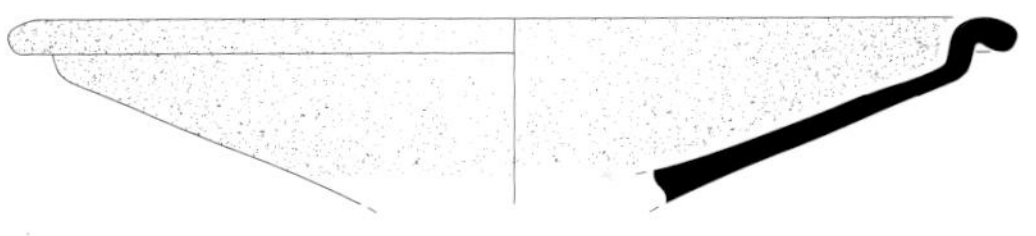

Provenance: Bulldozed Stipe. Lot: SA 79-603PL, Assemblage 9.
Dimensions and technical features: h 2.5, rim 13.9. Clay Munsell 10R 6/6 (light red), very hard and unevenly fired, with sparse luminous red inclusions. Black gloss thick and matte, dipped.
Condition: One rim fragment and two body fragments, joining.
Description: Angular lip with pendant projecting rim; straight wall. Reserved lower exterior.
Comparisons and comment: Surra and Preacco Ancona 1997, 67–68, pl. 31.11 (Pomarico Vecchio, last decades of the 4th c. BC); Preacco Ancona 1992, 123–24, 133, pl. XLIX, no. 97 (Locri, "Casa dei leoni," end of the 4th c.–275 BC); *Civita di Tricarico*, 422, fig. 299, no. 398 (Civita di Tricarico, end of the 4th c.–250 BC); Preacco Ancona 1989, 205, pl. XXVIII, no. 226 (Locri, end of the 4th c.–250 BC); *Valesio I*, 153, form K03d, no. 235 (Valesio, 325–250 BC); *Gravina II*, 77, fig. 29, no. 663 (Botromagno, Gravina di Puglia, 320–270 BC); *Monte Sannace*, 164, pl. 306.1, no. 2 (Monte Sannace, Hellenistic–Roman period).
Date: 320–250 BC.

Cup. A single fragment can be ascribed to a type of cup with a very deep body and wide, out-turned, slightly inward-sloping rim (SAV BG 38) attested mostly in southern Italy and Sicily. The rim recurs both on cup/one-handlers produced at Metaponto and on examples found in contexts at Valesio (Lanza Catti in *Fattoria Fabrizio*, 209, FF BG 53; *Valesio I*, 165, form K14b, no. 263). Other parallels exist with vessels produced locally in Magna Grecia and Sicily datable between the end of the 4th and the 3rd c. BC.

Cup

SAV BG 38 Cup
Provenance: Square G7. Level 2, cleaning around wall 4 to W. Lot: SA 79-1010PL.
Dimensions and technical features: h 4.4, rim 19.2. Clay Munsell 5YR 7/4 (pink), very hard, unevenly fired, and levigated, with sparse white inclusions. Black gloss thick and semi-lustrous, dipped.
Condition: One rim fragment.
Description: Flat, out-turned rim sloping inward and rounded on the upper surface; deep convex body.
Comparisons and comment: Preacco Ancona 1992, 128, pl. XLVI, no. 55 (Locri, "Casa dei leoni," 350–300 BC); Castiglione 1997, 311, fig. 3, no. 16 (Montagna dei Cavalli, end of the 4th c. BC); *Da Zancle 1*, 202, CST/101 (Messina, Via C. Battisti, Casa dello studente, 3rd c. BC); *Sibari V*, 467, figs. 459, 506, no. 450 (PS 32470) (Sybaris, Prolungamento Strada 1974) and 539, figs. 525, 564, no. 81 (CB 743) (Sybaris, Casa Bianca 1974).
Date: 350–275 BC.

1:2

Stemmed cup. The stemmed cup is attested by a single example, on which the hemispherical and thin (0.1 cm) body terminates in a strongly inward-projecting and barely thickened rim (SAV BG 39). The molded, truncated conical base normally associated with this form was not found.

The type is documented in the Metapontine sphere also by ribbed examples (Elliott in *Necropoleis*, 651, SC2–SC3). It was part of the red-figure repertoire (*Cozzo Presepe*, 356, fig. 126, no. 318) and was produced to a large extent in Gnathia ware (Delli Ponti 1996, 132, no. 72). It circulated in black-gloss in Lucania and Apulia between the end of the 4th and the 3rd c. BC (Lo Porto 1988–1989, 417, fig. 135.2; Barberis and Meirano 1997, 94, pl. 43, no. 157).

Stemmed Cup

SAV BG 39 Stemmed Cup
Provenance: Square D4. Level 3, batt. 4. Lot: SA 79-825PL, Assemblage 13.
Dimensions and technical features: h 2.7, rim 7.9. Clay Munsell 2.5YR 6/6 (light red), very hard and levigated. Black gloss thick and semi-lustrous, dipped.
Condition: Two rim fragments, joining.
Description: Direct rim; hemispherical body.
Comparisons and comment: *Antichi Peucezi*, 67, 70, fig. 75 (Bitonto, via Traiana, necropolis, T 3/1982, 4th c. BC, Gnathia ware); Lo Porto 1981, 362, fig. 83.2 (Metaponto, Casa Ricotta, necropolis, T 12 zone B, end of the 4th c. BC); Pianu 1990, 147, 228, pl. V skyphos type 1, no. 166.1 (Herakleia, necropolis, 330–280 BC); *Museo Taranto III.1*, 248, fig. 185, phase C (Taranto, necropolis, 275–225 BC).
Date: 320–250 BC.

1:2

Kantharos. The lone example that can be ascribed to a kantharos (SAV BG 40) is probably an imported product. The type appeared in Attica in the first half of the 4th c. BC with plain and ribbed versions (*Agora XII*, 286, fig. 7, pl. 29, nos. 700–704), often decorated with an X beneath the handle (*Panskoye I*, 128), and was immediately adopted in the black-gloss and Gnathia classes of southern Italy in both plain and ribbed versions. It circulated widely in the necropoleis of Messapia, where it was produced in local workshops between the end of the 4th and the late 3rd c. BC (Pagliara 1981, 217). It is attested in the chora of Metaponto by a single locally produced example (Lanza Catti et al. in *Survey*, 237, no. 109).

Kantharos

1:2

SAV BG 40 Ribbed Kantharos

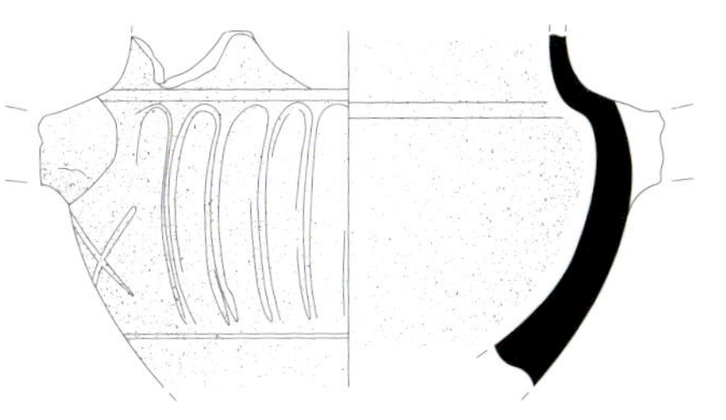

Provenance: Square F7. Level 2, batt. 6. Lot: SA 80-21PL A, Assemblage 4.

Dimensions and technical features: h 5.5, max. diam. 9.2. Clay Munsell 7.5YR 7/3 (pink), very hard and well fired, with sparse white, black, and red inclusions. Underslip reddish yellow 7.5YR 6/6. Black gloss thick and semi-lustrous, dipped.

Condition: Four body fragments, joining, preserving the attachment of the handle.

Description: Convex body, ribbed on the upper part, with scraped line at the end of the ribbing. X-shaped, pre-firing incision under the handle.

Comparisons and comment: Delli Ponti 1996, 172, no. 170 (350–300 BC); Giannotta 1997, 166–67, no. 20 (Vaste, T A Fondo Aia, 325–300 BC); *Panskoye I*, 133, no. B11, pls. 62, 70, 76 (Panskoye, Building U6, 320–310 BC); Giannotta 1996, 69, fig. 32, no. 62 (Rocavecchia, T 25, 325–300 BC, plain); Pagliara 1981, 217, pl. 56 (Vaste, 4th–3rd c. BC).

Date: 350–300 BC.

Bowl. A single base fragment from Sant'Angelo Vecchio can be ascribed to a bowl with straight or slightly inward-projecting rim and lacking handles (SAV BG 41). The type, not common in the Metapontino, is present in other parts of Lucania. It is attested in the sanctuary at Torre di Satriano by examples of varying sizes, with diameter ranging between 9.0 and 14.0 cm (Bruscella and Virtuoso 2005, 277), datable between the middle and end of the 4th c. BC. At Civita di Tricarico the *bol à bord peu rentrant* version (*Civita di Tricarico*, 435, fig. 305, no. 500) is encountered only in the wide and deep version datable between the second and third quarters of the 3rd c. BC.

In Campania, the mostly imported examples found in the necropolis of Pontecagnano (Serritella 1995, 19, 93) are probably central Italic products typical of the early 3rd c. BC (cf. Lamboglia 1952, Campana A, form 27). Furthermore, the type is well-attested in Apulia between the second half of the 3rd and the beginning of the 2nd c. BC (*Valesio I*, 172).

It is worth noting that the bowls "with incurved rim" are very commonly found in a large range of sizes in Hellenistic contexts at Athens (*Agora XXIX*, 161). They are believed to derive from the Classical "echinus bowls," of which they are a simplified version attested through the early 1st c. BC.

Bowl

SAV BG 41 Bowl

1:2

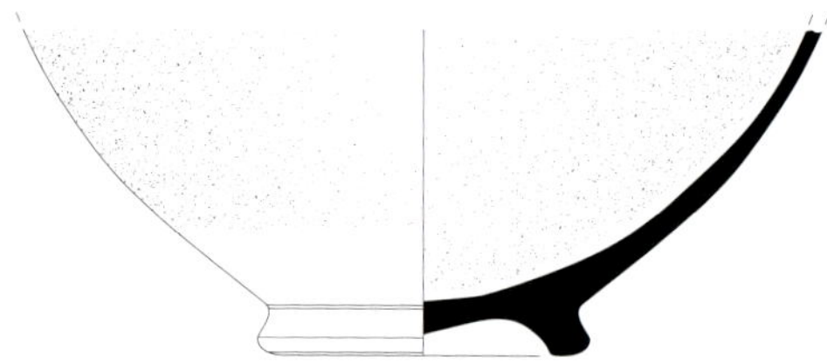

Provenance: Slope 3, W extension. Level 2, batt. 3. Lot: SA 79-127P. Inv. no. 319905 (see *Fig. 2.35*).

Dimensions and technical features: h 5.0, base 5.2. Clay Munsell 2.5YR 6/8 (light red), very hard and well fired, with luminous inclusions. Black gloss thick and semi-lustrous, dipped.

Condition: One base fragment and three body fragments, joining.

Description: Base-ring; deep convex body. Reserved lower exterior.

Comparisons and comment: *Gravina II*, 106–7, no. 831, fig. 34 (Botromagno, Gravina di Puglia, Late Republican); Cocchiaro and Andreassi 1988, 87, no. 49 (Brindisi, via Cappuccini, necropolis, T 268, end of the 3rd c.–175 BC); *Valesio I*, 173, no. 292, form K25 (Valesio, later 3rd/early 2nd c. BC).

Date: 275–225 BC.

Black-gloss of the Late Republican Period

Ring-handled cup. The rim fragment SAV BG 42 belongs to a cup with inward-sloping wall, two ring-handles immediately below the rim, and low, reserved base-ring. The slight carination of the body connects the example from Sant'Angelo Vecchio to the later evolution of the type (*Museo Taranto III.1*, 250). The earliest examples of the form with S-profile should date from the late 4th into the 3rd c. BC at Metaponto (most recently, Lanza Catti et al. in *Survey*, 191), but the relationship between the earlier examples and the carinated version presented here has yet to be firmly established (see black-gloss cup-kantharoi from the Pantanello sanctuary). Completely absent in northern Apulia (*Valesio I*, 191), it is documented instead in the Salento in contexts datable between the 3rd and the middle of the 2nd c. BC (Yntema 1990a, 179–80, pl. 82, form 7.4; D'Andria 2009, 167–68, fig. 37, no. 125), and in the Metapontino, where imported versions circulated alongside locally produced versions (Elliott in *Necropoleis*, 652; Lanza Catti et al. in *Survey*, 236, no. 100), attested until the first half of the 2nd c. BC.

Ring-handled Cup

SAV BG 42 Ring-handled Cup

1:2

Provenance: Square F3, N Kiln interior. Level 2, batt. 4. Lot: SA 79-883PL. Inv. no. 320050, Assemblage 8.

Dimensions and technical features: h 3.4, rim 8.2. Clay Munsell 10R 6/4 (pale red), very hard, unevenly fired, and levigated. Black gloss thick and matte, dipped.

Condition: One rim fragment.

Description: Direct and slightly outward-sloping rim; carinated body.

Comparisons and comment: Lo Porto 1981, 339, fig. 48.1 and 377, fig. 104.3 (Metaponto, Casa Ricotta, necropolis, T 2 zone A, 300–250 BC); *Museo Taranto III.1*, 249, fig. 186, phase C (Taranto, necropolis, 275–225 BC); Hempel 2001, 137, pl. 8, form 321/2 (Taranto, necropolis, ca. 225–160 BC); *Valesio I*, 193, form K44b, no. 338 (Valesio, end of the 3rd c.–150 BC).

Date: 225–150 BC.

Bowl. One fragment can be attributed to a bowl with pendant rim, basin concave in the upper part, and ring base (SAV BG 43). The type is part of the production of semi-glazed vessels with reserved lower exterior attested in southern Italy in black-gloss already at the end of the 3rd c. BC (Yntema 2005, 42). Examples of this type are widespread at various sites from Campania to Calabria, notably at Locri and Oppido Mamertina (see, most recently, Ferrara 2010, 465 n. 195). The type is analogous with the ones with grooved pendant rim identified in the Metapontino (Lanza Catti et al. in *Survey*, 249, no. 175) and dating to 300–200 BC. Similarities may also be drawn with the *bols à bord continu pendant* type from Civita di Tricarico in the second half of the 3rd c. BC (*Civita di Tricarico*, 433, fig. 304, no. 488). In Athens, examples of semi-glazed black-gloss bowls with concave-convex profile and ring base date between the late 2nd and early 1st c. BC (*Agora XXIX*, 159, 337–38, fig. 61, pl. 74, nos. 950–59), at which point they replace their entirely glazed antecedents. In southern Italy, the semi-glazed type begins to appear in the grey ware repertoire around 110 BC, with examples widely attested between Apulia and Basilicata during the 1st. c. BC. The examples appear to thin out between 30 and 10 BC.

Bowl

SAV BG 43 Bowl

1:2

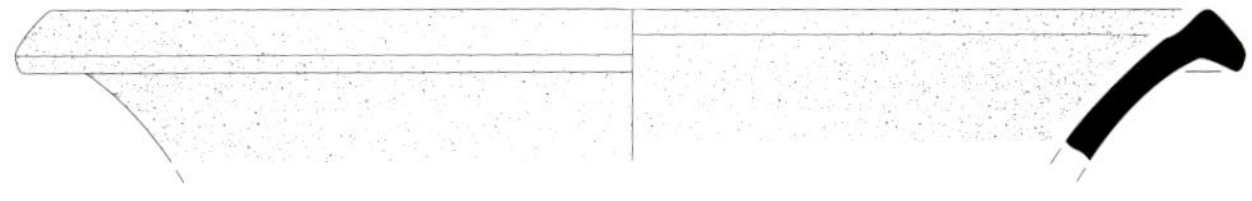

Provenance: Square E3, between Soundings A and B. Lot: SA 79-772PL, Assemblage 5.
Dimensions and technical features: h 2.0, rim 16.9. Clay Munsell 10YR 8/3 (very pale brown), very hard, with lithic and sparse rounded brown inclusions; abundant rounded voids. Reddish brown gloss Munsell 5YR 4/4, thin and matte, almost worn, dipped. The color of the piece seems to be altered by subsequent burning.
Condition: One rim fragment.
Description: Downturned rim, concave body; internal rim marked by a 0.6-cm slight groove.
Comparisons and comment: Ferrara 2010, 464–65, fig. 143, VN 30, tav. 48, no. 48 (Poseidonia-Paestum, Heraion at mouth of Sele, Sounding 5000, 3rd–2nd c. BC); *Gravina II*, 103, fig. 33, no. 811 (Botromagno, Gravina di Puglia, late 2nd–1st c. BC); Cotton 1979, 110, fig. 27, no. 14 (Francolise, San Rocco Villa, Late-Republican); Small et al. 1994, 211, fig. 7, no. 11 (grey ware, Botromagno, Gravina di Puglia, 80–70 BC).
Date: Late 3rd–2nd c. BC.

Lamps

SAV LA 44 Lamp

1:2

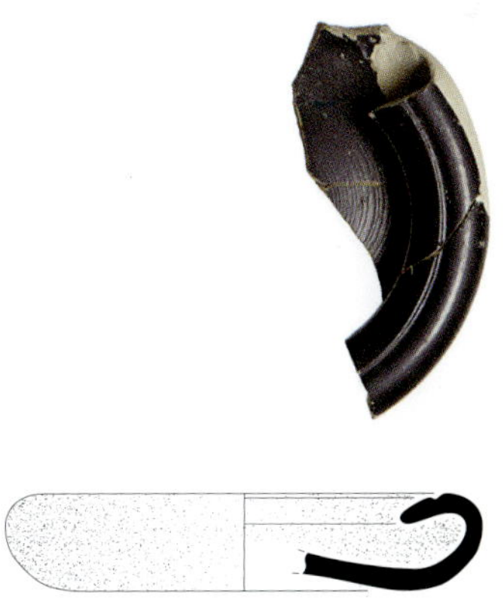

Provenance: Sounding C, Pottery Deposit. Lot: SA 79-865P, Assemblage 15.
Dimensions and technical features: h 1.7, rim 5.8, base 6.0. Clay Munsell 5YR 7/2 (pinkish gray), very hard, unevenly fired, and levigated. Black gloss thick and lustrous, dipped.
Condition: One rim fragment, one base fragment, and one body fragment, joining.
Description: Incurved and downward-sloping rim; concave base; circular reservoir. The upper surface of the shoulder is marked by a shallow groove; entirely slipped, except for the underside of the base.
Comparisons and comment: *Agora IV*, 45, no. 157, pls. 6, 34 (Athens, 500–480 BC); *Poseidonia-Paestum II*, 113, fig. 64, no. 153 (Poseidonia-Paestum, circular structure, 480 BC); Theseider Dupré 1992, 198, pl. LVI, no. 142 (Locri, "Casa dei leoni," end of the 5th c. BC); *Roccagloriosa I*, 291, fig. 193, no. 417 (Roccagloriosa, 5th c. BC); *Sibari V*, 226, figs. 203, 219, no. 205 (St. 3095) (Sybaris, Stombi 1974). For the groove, see *Archeologia dei Messapi*, 118, no. 163 (Vaste, Fondo Melliche, second half of the 4th c. BC).
Date: 480–325 BC.

16

Unguentaria

Francesca Silvestrelli

The Sant'Angelo Vecchio assemblages yielded only three unguentaria. Two complete or almost complete vessels (SAV U 01 and SAV U 02) were found in the large discard deposit on Slope 3, the formation of which is connected with the activity of the Hellenistic workshop.[1] The third vessel is from Sounding D. None of them is overfired or can be considered a discard. The unguentaria from Sant'Angelo can be referred to Type U3,[2] very common at Metaponto and its territory in contexts of the end of the 4th and the 3rd century BC. The local production of this type of unguentaria is confirmed by discarded specimens from the urban *kerameikos*.[3]

[1] See Ch. 2, "Site Phasing, Stratigraphy, and Site Assemblage."

[2] Vittoria in *Survey*, 351.

[3] D'Andria 1980a, 259 no. 18 fig. 4.

Catalog of Unguentaria

SAV U 01 Unguentarium

1:2

Provenance: Slope 3, W extension. Level 1, batt. 1. Lot: SA 79-73P. Inv. no. 319988 (see *Fig. 2.35*).

Dimensions and technical features: foot diam. 2.4, ext. rim diam. 1.6, max. h 7.5. Clay Munsell 7.5YR 7/6 (reddish yellow), compact, somewhat levigated, with few small inclusions. Black gloss (Munsell 7.5YR 3/0, black) opaque, uniform, present on the interior of the neck and over three-quarters of the exterior. Reserved foot.

Condition: Complete.

Description: Black-gloss unguentarium with flaring rim that has a nearly triangular section, cylindrical neck, ovoid body, and low foot with a disc profile.

Comparisons and comment: Lo Porto 1966, 144, fig. 3:1 (Metaponto-Castrum, 3rd–2nd c. BC); De Siena 1975, 39, fig. 46 and 48, fig. 54 (Metaponto, Temple of Apollo, 4th–3rd c. BC); M. Liseno 2004, 102, no. 2, pl. LXVII.c-d (Metaponto Favale votive deposit, last quarter of the 4th c. BC). Elliott in *Necropoleis*, 687, U46 (300–275 BC); Vittoria in *Survey*, 415, no. 268, Type U3b (also with parallels from Taranto and Herakleia).

Date: First half of the 3rd c. BC.

SAV U 02 Unguentarium

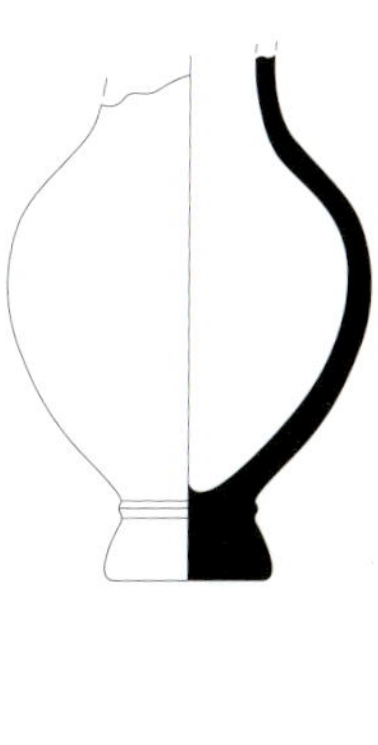

1:2

Provenance: Slope 3, square D4, outside stone circle to S. Level 3, batt. 5. Lot: SA 79-903P. Inv. no. 319928, Assemblage 13.

Dimensions and technical features: foot diam. 2.4, max. body diam. 4.3, max. h 7.0. Clay Munsell 2.5 YR 6/6 (red), very hard, levigated, with vughs.

Condition: Rim and upper part of the neck missing.

Description: Plain unguentarium with cylindrical neck, ovoid body, and low foot with disc base separated from the body by a groove.

Comparisons and comment: A similar type of unguentarium was found at Fattoria Stefan (inv. no. 316500).

Date: First half of the 3rd c. BC.

17
Red Ware

Mark Van Der Enden, Philip Bes, and Emanuela Conoci

Introduction[1]

Four fragments of red slipped pottery (representing four individual vessels) found at Sant'Angelo Vecchio may be ascribed, based on their technical features, to red ware.[2] The shape repertoire of red ware is often similar to that of black-gloss,[3] and for a long time the class was thought to be simply misfired black-gloss.[4] Recent studies, however, have recognized it as a deliberate and "autonomous group."[5] Despite close associations with black-gloss elsewhere, the red-ware fragments from Sant'Angelo Vecchio do not compare directly with the black-gloss repertoire common at sites in the Metapontino in either technical features or shapes.

Red ware seems to be a Late Hellenistic creation, becoming popular from the second half of the 2nd century BC onward. At the same time, the new shape known as the lagynos also begins to spread in eastern contexts, where it is attested in various classes (see Ch. 22, "Lagynoi").[6] Although concentrated in the East, imported examples of this shape also appeared in the western Mediterranean, for example, throughout Italy as well as in Spain.[7]

In southern Italy, the lagynos (attested in black-gloss, grey, banded, and plain wares)[8] is a new shape in the red-ware repertoire,[9] and recalls eastern Mediterranean models that were widely imitated in the West.[10] These came principally from the area of Pergamon and Knidos; Cyprus created its own version of the lagynos, and production centers of the shape have also been postulated for Argos, Samos and Rhodes, as well as in Israel and North Africa.[11] The lagynos is attested in southern Italy, mostly at Taranto and in the Metapontino,[12] in imported specimens and local productions,[13] dated between the Late Hellenistic and Early Imperial periods. The morphological features of the four red-ware examples from Sant'Angelo Vecchio appear to place them also in the Late Hellenistic to Early Imperial period (150 BC–AD 25).

Context and Chronology

All of the vessels discussed here (SAV RW 01–SAV RW 04) are associated with a Kiln Complex incorporating a courtyard and the remains of three kilns.[14] SAV RW 01 is attested in Level 2 of Sounding A and was found mixed with tile fragments belonging to the collapsed roof of the courtyard of the Kiln Complex. Two fragments of molds (SAV AT 08 and SAV AT 09) are also associated with this tile fall and may have been used in the construction of one of the kilns.[15] Other associated pottery finds include a louterion rim sherd with relief decoration (SAV Lou 01), a coarse-ware basin (SAV Imp 06), and a loom weight (SA 79-37T).

SAV RW 02, found inside Kiln 5 (square F3), was probably reused in the construction of the facility and was found associated with a black-gloss cup (SAV BG 42), an olla, a grey-ware boccaletto (SAV GW 01), and a pyramidal loom weight (SAV LW 30).[16] Fragments of coarse (SAV CkW 08 and SAV CkW 43) and grey

[1] Mark Van Der Enden (University of Leicester) and Philip Bes (independent archaeologist) warmly thank Francesca Silvestrelli for inviting us to study these finds. Unfortunately, we were unable to examine the finds in person and inspect their characteristics in the field. Valuable information in this regard was provided by Francesca Silvestrelli and Emanuela Conoci.

[2] The relationship of this ceramic class to the earliest terra sigillata forms has yet to be fully explored (Lippolis 1984, 33).

[3] Carrara Ronzani 1978, 219; most recently, Barbieri 2002, 263.

[4] In the context of this chapter, "black gloss" denotes vessels covered partially or fully with a black slip.

[5] Colangelo 2005b, 299.

[6] *Agora XXIX*, 296, fig. 37, pl. 49, nos. 494–498 (black gloss); *Paphos III*, no. 13, 20, fig. 11, pl. 5.4 (banded ware); *Agora XXIX*, 390–94, figs. 88–91, pls. 115–17, nos. 1503–1532 (lagynos ware); Vessberg and Westholm 1956, 59, fig. 22, no. 24 (various classes); see also below, SAV RW 04.

[7] Hempel 2001, 108.

[8] Yntema 2005, form 42; Alessio 1988, 330, pl. LXV, no. 30.6d, 350–51, pl. LXIX, no. 30.13; Hempel 2001, 107, pls. 26–27, 33, 35, 37, 39, 41–42, 46–47, 54–56, 59, 63–70.

[9] Lippolis 1984, 32.

[10] Yntema 2005, 80.

[11] *Agora XXIX*, 225; Hempel 2001, 108.

[12] *Museo Taranto III.1*, 250, fig. 187; Hempel 2001, pls. 4–5, form 114; Schojer 1988, 486–88, pl. CXII, nos. 37.11g–h; Lo Porto 1966, 189, fig. 42.1; *Da Leukania a Lucania*, 180, fig. 285, no. 3.

[13] Hempel 2001, 108.

[14] See Ch. 3, "The Structures at Sant'Angelo Vecchio"; Edlund 1982a, 91.

[15] See Ch. 2, "Site Phasing, Stratigraphy, and Site Assemblage"; Edlund 1982a, 92.

[16] See Ch. 2, "Site Phasing, Stratigraphy, and Site Assemblage"; Edlund 1982a, 110–11.

Provenance	Form and Type	Cat. No.	Date	NFr	MNI
Sounding A	Dish	SAV RW 01	Late Hellenistic–Early Imperial	1	1
North Kiln (square F3)	Juglet	SAV RW 02	Late Hellenistic–Early Imperial	2	1
Square E2	Juglet	SAV RW 03	Late Hellenistic–Early Imperial	1	1
South Kiln and Sounding A	Closed form	SAV RW 04	Late Hellenistic–Early Imperial	2	1

Table 17.1 Red ware from Sant'Angelo Vecchio: counts by form and date.

ware (SAV GW 01) are attested around and inside the North Kiln (Kiln 5).[17]

SAV RW 03 was found in the stratum beneath the pavement identified in square E2 (Sounding A, Level 3), where the kilns opened onto a small courtyard. Tile fragments were found in association.

SAV RW 04 consists of two joining fragments (SA 79-801PL and SA 81-7PL), the first of which, found inside the South Kiln (Kiln 4), was covered in mortar, and was therefore probably reused in the construction of the facility. The joining fragment (SA 81-7PL), completely burnt but bearing no traces of mortar, came from the tile fall (Level 2) identified next to the courtyard, at the end of the northern wall of the South Kiln.

The construction of the kilns dates to the 4th century BC, when they were used for the production of terracottas.[18] The latest phase of this complex, for which the production of tiles, coarse, and grey ware (finds found in association) is postulated, dates within the second half of the 2nd century BC. The red-ware fragments can be stratigraphically associated with this phase.

Discussion

SAV RW 01, a shallow dish with grooved rim, flat base, and convex wall, is a shape with no known parallels from sites in the Hellenistic and Roman eastern Mediterranean. The fragment from Sant'Angelo Vecchio has a red/orange slip that has survived poorly and is more thinly and less evenly applied. In shape, this fragment bears some similarity to the so-called flat-based pans *a orlo bifido* identified at a number of sites in the eastern Mediterranean, but probably of western origin. Examples from Ephesus and the Athenian Agora are cited here as morphological parallels.[19] Though the fabric, function, and finishing of SAV RW 01 differ completely from these pans, the morphological similarity perhaps provides a hint as to the chronology of the fragment from Sant'Angelo Vecchio. Most of the Athenian pans are dated, on the basis of their context, to the late 2nd/first half of the 1st century BC; two vessels illustrate that the form continued into the early 1st century AD.[20]

The shape displays similarities with these flat-bottomed vessels, attested in the Eastern Sigillata B repertoire produced in Asia Minor,[21] and this piece may well be a Late Hellenistic product influenced by the early Eastern Sigillatas or their predecessors. The fabric of SAV RW 01 is unusually micaceous, and may be directly related to eastern Mediterranean productions.

The absence of close parallels for SAV RW 01 tentatively suggests that this shape, with the characteristics described here, was not common in the Hellenistic and Roman eastern Mediterranean. A reddish slip suggests that this fragment may also be dated to the late 2nd century BC through the early 1st century AD. The presence of the other red-ware fragments, as well as the grey and coarse wares dated to the same time span, could support this. That the pans cited above display a similar dating might be indicative as well.

The juglet SAV RW 02 is a small closed shape with a flat base. The underside of the base was not completely slipped: the visible slip was the result of dipping. Distinctive to this fragment is its angular carination about halfway down the body, with the lower body being slightly convex. Above the carination, the body is more distinctively convex and narrows to (presumably) the neck (not preserved). The presence of a ribbon handle can be suggested, based on parallels with unpublished Late Hellenistic forms attested in the Metapontino.[22] This small juglet corresponds morphologically quite well to aryballoi from the Athenian Agora, contextually dated 225–150 BC; these are black-slipped.[23] The small size and diameter of the examples from the Athenian Agora also corre-

[17] See Ch. 2, "Site Phasing, Stratigraphy, and Site Assemblage"; Edlund 1982a, 101.

[18] See Ch. 2, "Site Phasing, Stratigraphy, and Site Assemblage," and Ch. 3, "The Structures at Sant'Angelo Vecchio"; Edlund 1982a, 126.

[19] *Agora XXXII*, fig. 89, cat. nos. 703–707; Steskal 2010, pl. 200, K547; Gassner 1997, 379, pl. 32 (large bowl, probably used for storage purposes); Lang-Auinger/Asamer/Müller-Kaspar 2003, K24, pl. 3.

[20] *Agora XXXII*, fig. 89, cat. nos. 704, 707, also for further bibliography.

[21] Hayes 1985, 56–57, forms 18–19, pl. XII, nos. 12–14.

[22] K. Swift, pers. comm. 2011.

[23] See *Agora XXIX*, 351–52, cat. nos. 1132, 1134 (both contextual dates).

spond to the fragment from Sant'Angelo Vecchio. An unguentarium from Ephesus[24] is also similar in the profile of its body and flat base.[25]

A type of biconical juglet identified at, for example, Athens and Knidos, also compares to SAV RW 02. The profile, in particular of the lower body, with its flat base and angular wall carination, is not dissimilar to the red-ware fragment from Sant'Angelo Vecchio,[26] but the upper walls of the Athenian and Knidian examples are not as convex as that of SAV RW 02. The flat base of the Knidian examples also differs in detail. More importantly, none of the Knidian or Athenian examples is fully red-slipped, although some of the Knidian vessels have an orange-brown/reddish-brown slip.[27] The examples from the Athenian Agora are all black-slipped, except one;[28] in profile, this vessel finds less correspondence with SAV RW 02, as it is more convex in shape but slipped red. This piece, of uncertain manufacture, has been dated 100–25 BC.[29] The other biconical juglets attested at the Athenian Agora are dated 140–130 BC,[30] and the Knidian biconical juglets referenced above are dated 225–150 and 125–50 BC. A biconical juglet from Pergamon,[31] which shows some affinity to SAV RW 02, is dated 100 BC–AD 125, while the biconical juglets catalogued by Robinson illustrate the continued existence of this form into the Early Imperial period.[32] Rotroff indicates that biconical juglets were rare at the Athenian Agora, citing only a few examples at other sites (Delos, Priene, Miletos, Paphos).[33] Kögler, however, identifies the biconical juglet as one of the most common locally produced shapes in the Late Hellenistic period at Knidos.[34] The red-slip finish of SAV RW 02 suggests a more recent date than that of the Athenian parallels, which are all black-slipped (with the one exception). A juglet from the Kestel excavations at Pergamon,[35] close in profile to SAV RW 02, has a contextual date of the beginning of the 2nd century BC (no details about the surface finish or measurements are given).

In shape, SAV RW 02 most closely resembles Athenian aryballoi, which at the Athenian Agora and Knidos have been dated to the 2nd and 1st centuries BC. In this respect, one fragment from Athens is of particular interest for its red slip.[36] Biconical juglets, with their flat base and carination of the lower body, also display certain similarities to SAV RW 02. The combination of a red slip and wall carination suggests, therefore, that SAV RW 02 ought to be dated within the 2nd century BC to early 1st century AD. The Athenian black-slipped aryballoi possibly indicate that this can be narrowed down to the latter part of the 2nd century BC to early 1st century AD, as indicated by the red-ware biconical juglet from the Athenian Agora and those from Knidos (some of which have a reddish-brown slip). Although few parallels for SAV RW 02 could be identified within the Hellenistic–Early Imperial eastern Mediterranean, a Late Hellenistic to Early Imperial dating for this fragment seems plausible.

SAV RW 03, like SAV RW 02, is a small, flat-based juglet. In size it is smaller than SAV RW 02, and the angular carination is missing. Instead, the profile of the body is double convex and squat, although a delicate delineation halfway down the body can still be discerned. The body slopes up to a thin and cylindrical neck, the transition from wall to neck is set off by a small ridge, and a vertical handle attachment is visible on one side. Athenian aryballoi can again be cited as potential parallels,[37] though none of these are red-slipped (two are brown-slipped). Two of these vessels have contextual dates encompassing the first half of the 2nd century BC: one is dated to the late 3rd–early 1st century BC, and another to the second half of the 2nd–early 1st century BC. Aryballoi from Knidos (form XXII) also show affinity with SAV RW 03, in their small size, somewhat squat globular body, and neck. The treatment of the base, however, is different in the case of all the Knidian examples of form XXII, and again none are red-slipped. This type has been dated by Kögler to ca. 150/125–75/50 BC. Other Knidian types that bear resemblance to SAV RW 03 are the small

[24] Mitsopoulos-Leon and Lang-Auinger 1991, pl. 218, no. O45.

[25] Mitsopoulos-Leon and Lang-Auinger 1991, 149. At Ephesus this vessel was dated to the 1st c. AD; probable Late Hellenistic predecessors are mentioned.

[26] Compare, e.g., Kögler 2010, fig. 26, cat. nos. E225, E227; *Agora XXIX*, fig. 38, nos. 512–513; *Agora V*, pl. 1, nos. F44, pl. 5, G88, pl. 18, M36, pl. 29, M269.

[27] E.g. Kögler 2010, fig. 26, cat. no. E225, fig. 27, cat. no. E233.

[28] *Agora XXIX*, fig. 38, cat. no. 514; cat. nos. 511–514 are not contextually dated.

[29] *Agora XXIX*, 297, cat. no. 514.

[30] *Agora XXIX*, 297, cat. nos. 511–513.

[31] Meyer-Schlichtmann 1988, pl. 25, kr. 2.

[32] *Agora V*, pl. 1, no. F44, pl. 5, G88, pl. 18, M36, pl. 29, M269 (the latter dated AD 375–400).

[33] *Agora XXIX*, 129, no. 47.

[34] Kögler 2010, 225.

[35] Özyiğit 2000, 101, fig. 5.

[36] *Agora XXIX*, fig. 38, no. 514.

[37] *Agora XXIX*, fig. 71, pl. 83, cat. nos. 1133, 1137, fig. 104, pl. 138, cat. nos. 1752–1753.

jug with shoulder handle (form XXIII),[38] a lekythos (form XXIV), and an amphoriskos (form XXXV), though none of these are fully red-slipped. The jug with shoulder handle identified at Knidos corresponds to SAV RW 03 in the presence of a vertical (strap?) handle, the transition between shoulder and neck, and its small size. All, however, have ring bases, and the body profile, though sometimes close,[39] is not an exact match. Two Knidian lekythoi are closer in overall shape, although both have a false ring base, and again none are red-slipped. The body profile, however, especially that of G108, is closely similar, as is the presence of a vertical handle and the shape of the neck. This shape has been dated to the Augustan-Tiberian period. Aryballoi or gutti from Pergamon also display affinity,[40] though these carry West Slope decoration, are black-slipped, and have ring bases. The body profiles of D73 and C26A compare to that of SAV RW 03, though they are squatter. Morel has also published a number of shapes not unlike SAV RW 03,[41] although no exact parallel could be identified. These are evidently slipped black and date to the Early Hellenistic period.

Two unguentaria from Ephesus can also be cited as potential parallels for SAV RW 03.[42] Both have a flat base and a body very similar in profile to that of SAV RW 03, a ridged transition from neck to shoulder, and a constricted neck. They have been dated to the 1st century AD, but could potentially have occurred already in the Late Hellenistic period.[43] A bulbous unguentarium from Corinth is also similar in profile, but a fully red-slipped surface and the ridged transition from shoulder to neck are lacking;[44] its context date is ca. AD 22–30. Characteristically with unguentaria, the examples cited here also do not have a handle attachment, visible nevertheless on SAV RW 03. The fabric(s) usually employed for the manufacture of unguentaria in the Late Hellenistic period also differ(s) from that of the fragment from Sant'Angelo Vecchio.

The profile of SAV RW 03, furthermore, is analogous to the perfume vessels documented in the Late Hellenistic necropoleis of Taranto and Herakleia, produced mostly in glass,[45] but also with red- to brown-slipped ceramic equivalents,[46] with which the fragment in question shares the squat, globular body and the slightly concave base. The fragment finds an additional parallel beyond southern Italy, in a "red-ware juglet" from the Hellenistic-Roman necropolis of Ancona.[47]

The presence of a handle attachment and the similarity in profile to the aryballoi cited above make it likely that, as in the case of SAV RW 02, SAV RW 03 is an aryballos. The fragment's small size and its close similarity in body profile to the unguentaria cited above remain noteworthy, and in this respect, it is important to mention that both aryballoi and unguentaria were oil/perfume containers.[48] If both SAV RW 02 and SAV RW 03 are indeed imports, we can hypothesize that they found their way to Sant'Angelo Vecchio for their contents. The parallels cited for SAV RW 03 favor a Late Hellenistic to Early Imperial date. The fully red-slipped nature of this fragment, a feature possessed by none of the cited parallels, is also in agreement. The bulbous unguentarium from Corinth cited above and the unguentaria from Ephesus illustrate that morphologically similar vessels apparently still occurred in the early 1st century AD.

The closed form SAV RW 04, a base (> 10 cm in diameter), is more difficult to place. As only the base and part of the lower wall have been preserved, the shape of the vessel cannot be confidently assessed. Traces of red slip are visible on the exterior. The vessel appears to have been closed, while the slip on the interior resulted from accidental dipping. The form of the base, which is undercut by a groove on the bottom lower side, is not distinct enough to allow a clear identification with a particular shape. The horizontal floor, the angle of the lower wall, and the slightly angular profile of the base, however, can be found in lagynoi.[49] Two lagynoi attested at the Athenian Agora[50] and dated 115–86/50 BC display a somewhat similar profile regarding the base and possibly the lower wall. Both lagynoi, however, were substantially larger than SAV RW 04 and carry the pink slip characteristic of

38 Kögler 2010, G109–110 (300–100 BC).

39 Kögler 2010, n. 123.

40 Schäfer 1968, pl. 5, nos. C26A, D73–75.

41 Morel 1981, pl. 171, 5443a1, 5453a1, pl. 172, 5454a1, pl. 173, 5471b1, 5472a1. For the carination, see Morel 1981, pl. 215, 8423a1.

42 Mitsopoulos-Leon and Lang-Auinger 1991, pl. 218, O47–O48.

43 Mitsopoulos-Leon and Lang-Auinger 1991, 149.

44 Slane Wright and Jones 1980, 156, 159, fig. 6, no. 100.

45 Maruggi 1988, 207, pl. XLVIII, no. 17.10b; D'Amicis 1988, 133, pl. XXVIII, N, no. 13.2a; *Da Leukania a Lucania*, 185, fig. 302, nos. 1–2.

46 Alessio 1988, 349, pl. LXIII, no. 30.13e; *Da Leukania a Lucania*, 186, fig. 302, no. 3.

47 Colivicchi 2002, 121–22, no. 14.3.

48 *Agora XXIX*, 171, 175–78.

49 See also Ch. 22, "Lagynoi."

50 *Agora XXIX*, fig. 91, cat. nos. 1530–1531.

lagynos ware.[51] The base of a lagynos from Sardis[52] is somewhat similar, although it is much higher and more sharply cut off on the interior. The convex shape of the lower wall, however, better matches SAV RW 04 than the angular lagynoi from Athens. The vessel is dated to the second half of the 1st century BC and carries a dull, pale yellow slip on the exterior and within the upper part of the internal neck.[53] A lagynos from Paphos,[54] with a much straighter wall than SAV RW 04, also shows similarity in the profile of its base. This vessel is dated to the late 2nd and early 1st century BC. Two lagynoi from Eretria can also be cited as examples,[55] even though they differ in detail. One of these lagynoi, dated ca. 100 BC,[56] has a base diameter of 9.9 cm; besides painted decoration, it has an orange-brown slip. A lagynos which is possibly closer in shape to SAV RW 04 comes from Patras.[57]

The relatively generic shape of SAV RW 04 makes a secure identification with a particular shape difficult, if not impossible. A general similarity with lagynoi, however, can be discerned. The parallels cited above are all dated to the late 2nd–first half of the 1st century BC. The general resemblance in the profile of the base and the presence of a reddish slip support such a date, as do associated finds and the other red-ware fragments attested in or around the kilns that are similarly dated. The presence of mortar makes it possible, however, that SAV RW 04 was used in the construction of the kiln. As such, it is possibly older than red-ware fragments SAV RW 01 and SAV RW 03. As stated above, however, the nature of the fragment is such that no definitive conclusions can be drawn.

Conclusions and Wider Context

On the basis of the archaeological context (including associated finds) and cited parallels, the four red-ware fragments presented above appear to date to the Late Hellenistic–Early Imperial period (ca. 150 BC–early 1st century AD). The fabric of all four fragments does not match the local repertoire, and the combination of shape and finish is not well attested within Italy in general and southern Italy in particular. The fragmented nature of the pieces precludes any certain identification, and the proposed dating of the fragments must be approached with some caution. All available indicators, however, point to a date between ca. 150 BC and the early 1st century AD. It is during this particular timespan that red-ware pottery came to characterize the (eastern) Mediterranean world and a number of classes of (eastern) terra sigillata emerged.

The most well-known and widespread of these is Eastern Sigillata A (ESA). Although no production infrastructure has yet been located, this ware is presumed to have originated somewhere in eastern Cilicia-Hatay, more specifically in the vicinity of Antioch.[58] ESA began to be manufactured shortly after 150 BC,[59] and production continued until the mid-3rd century AD.[60] The core zone of the distribution of ESA was the Levant (in particular its coastal zone) and south-central Turkey, as well as Cyprus.[61] It has been found throughout the eastern and central Mediterranean and beyond more occasionally, achieving its greatest distribution after ca. 50 BC, when it is found in large numbers primarily in the coastal areas of the eastern Mediterranean.[62] The peace and stability brought about by Augustus, together with changes in the organization of production, have been cited as important game-changers, which allowed ESA to achieve its wide, extra-regional distribution.[63]

Other eastern sigillatae of regional and supra-regional importance[64] are Eastern Sigillata B (ESB), manufactured at Ephesus and Tralleis and in the Maeander valley; Eastern Sigillata C (ESC, also known as Pergamene Sigillata), manufactured in the area of Pergamon; Eastern Sigillata D (ESD, also known as Cypriot Sigillata), most probably manufactured on Cyprus; and Sagalassos Red Slip Ware (SRSW). ESB began to be manufactured in the later 1st century BC, and displays a strong morphological influence from Italian sigillata. It is possible that related forerunners of this ware occurred already at the end of the 2nd century BC.[65] ESC began to be produced around the same time as ESA, the mid-2nd century BC,[66] although it did not become fully red-slipped until

[51] For a discussion of lagynos ware, see *Agora XXIX*, 225–28.
[52] Rotroff et al. 2003, pl. 55, cat. no. 334.
[53] Rotroff et al. 2003, 84.
[54] *Paphos III*, fig. XI, no. 13.
[55] Psalti 2004, pl. 246, no. 18877; Metzger 1994, pl. 36, no. 5.
[56] Metzger 1994, 74, 79.
[57] Kyriakou 1994, pl. 135, no. 6741.
[58] Lund 2005, 235–38; *Agora XXXII*, 13.
[59] Slane (1997, 269–74) identified a so-called Black Slipped Predecessor (BSP) of ESA; see also Berlin 2006, 13–14; *Agora XXXII*, 13–14.
[60] Reynolds 2010, 90; *Agora XXXII*, 19–20; Lund 2005, 234–37.
[61] Bes 2015.
[62] *Agora XXXII*, 18–19; Bes 2015.
[63] Poblome and Zelle 2002, 276–77.
[64] For an overview, see Bes 2015.
[65] Gassner 1997, 126.
[66] Rotroff et al. 2003, 84.

sometime around the mid-1st century BC. The production of SRSW began toward the later 1st century BC,[67] although the volume of production and the action radius of its distribution were smaller than that of the other eastern sigillatae. These examples serve to illustrate that during the second half of the 2nd century BC, production of terrae sigillatae sprang up throughout the Hellenistic East and achieved a greater distribution than most table wares that had circulated previously in that area. In addition to these major classes, terrae sigillatae—with a more restricted production and distribution framework—catered equally to the demand for table wares, and are gradually being recognized as such.[68]

All of these productions were characterized by the use of a red slip as finishing, covering the entire vessel's surface. The widespread distribution of these wares throughout the eastern Mediterranean signifies a real popularity and marked break from the earlier part of the Hellenistic period, in which black slip had dominated in both the East and the West.[69] This perceived dominance of black-slip pottery, however, must be somewhat nuanced. Although black slip is indeed prevalent, many table ware vessels produced in Hellenistic Asia Minor display a wide variety of slip colors (e.g., *feinegebrauchskeramik* and *firniskeramik mit schlickerdekor* at Ephesus).[70] Black dominated, but brown, gray, and various combinations were also common,[71] and red is also among the colors attested. Red slips dominate among the partially slipped pottery encountered at Sardis.[72] Although basically the result of varying conditions in the kiln, the widespread occurrence of non-black slips illustrates that the shift from black to red was perhaps not so pronounced as previously imagined (though not exclusively explained in association with chronological phases of transition). The different starting dates of the major manufacturers must also be considered, as they signify a more gradual transition from black to red. Slips other than a shiny black appear to have been acceptable to many consumers of table wares throughout the Hellenistic East. At Ephesus and Sardis, for example, fully black-slipped pottery was rare during the Hellenistic period.[73] At the Athenian Agora, about 60 percent of all vessels were fired an even black before ca. 150 BC. Thereafter, however, more vessels were fired red and brown,[74] illustrating that before red wares reached their apogee the color red was certainly not uncommon. The four red-ware fragments from Sant'Angelo Vecchio form part of this tradition of manufacture and are therefore part of the increasing popularity of red-ware table wares that eventually crystallized during the Augustan period.[75]

The four fragments discussed here find few morphological parallels in the repertoires of the (major) categories of red-ware pottery, though aryballoi are not particularly uncommon during the Late Hellenistic period. Also, all parallels known to us are primarily not red-slipped. It seems likely that these fragments are part of a transitional period, before terra sigillata products (Italian or Eastern) came to dominate the markets, in which potters reacted to the new taste for red-ware pottery. In fact, Italy had not yet embraced the tradition of terra sigillata, which came about only around 40/30 BC.[76] SAV RW 02 and SAV RW 03, juglets most closely resembling aryballoi, can be cited as examples of shapes whose parallels are mostly black-slipped. The lagynoi cited as parallels for SAV RW 04 also have a different finish from the red slip attested for this fragment. The latter are dated ca. late 2nd–first half of the 1st century BC, which would fit with the idea of a pre-Augustan period of experimentation in which the use of red slips became more and more common. The quality and appearance of the slip on our fragments, particularly the juglets/aryballoi, suggest that they are closer to terra sigillata products than to the majority of red-ware products of the later Hellenistic period, which generally do not achieve full coverage and/or the same appearance of the slip.[77]

The eastern context of the parallels cited above reinforces the idea that these four fragments are part of the Late Hellenistic–Early Imperial tradition of red-ware table wares. The origin of these fragments is to be sought tentatively somewhere in the eastern

67 Poblome 2006, 348.

68 For "Pink Ware" from Tel Anafa in northern Israel (to name but one), see Slane 1997, 371–75.

69 Hayes 1997, 37–38.

70 Ladstätter 2010, 90; Mitsopoulos-Leon and Lang-Auinger 1991, 17.

71 For Athenian evidence, see *Agora XXIX*, 11.

72 Rotroff et al. 2003, 24.

73 For Ephesus, see Mitsopoulos-Leon and Lang-Auinger 1991; Ladstätter 2003; Gassner 1997. For Sardis, see Rotroff et al. 2003, 19.

74 *Agora XXIX*, 11.

75 Poblome and Zelle 2002; Poblome 2004; Poblome et al. 2001.

76 Ettlinger 1990.

77 See, for instance, Kögler's phase II for an example of such later Hellenistic pottery (Kögler 2010, 25–26); see also the semi-slipped pottery of Sardis (Rotroff et al. 2003).

Mediterranean, although, given the chronology of most parallels cited, we may surmise that these vessels did not originate in Greece. In fact, local production of fully red-slipped table wares began to emerge by the mid-2nd century BC. Contacts between Italy and the East are well attested and provide a context in which red-ware imports could have arrived at Sant'Angelo Vecchio.[78] Since the late 3rd century BC, Roman traders, businessmen, and elites had gradually become important players, political and otherwise, in the eastern Mediterranean landscape, establishing themselves in a number of locations in the Greek East. In the framework that emerged, there would have been ample opportunity for the fragments found at Sant'Angelo Vecchio to have reached southern Italy, if indeed these originated somewhere in the Greek East. The distribution of ESA and Late Hellenistic Ephesian mold-made bowls provides a glimpse into the trade mechanisms responsible for the circulation of goods between the Greek East and Late Republican Italy.[79]

78 Wallace-Hadrill 2008.

79 Malfitana/Poblome/Lund 2005; Bilde 1993.

Catalog of Red Ware

SAV RW 01 Dish

1:2

Provenance: Sounding A. Level 2, batt. 2. Lot: SA 79-26PL. Assemblage 5.

Dimensions and technical features: h 2.3, rim 24.0. Clay Munsell 5YR 6/6 (reddish yellow), hard, evenly fired, and somewhat micaceous. Light red slip, Munsell 2.5YR 6, thin and matte, dipped?

Condition: Rim and part of the base.

Description: Flat-based dish with faint convex-concave profile of the wall. The rounded rim is grooved on top.

Comparisons and comment: Steskal 2010, K547, pl. 200 (Ephesus, Late Hellenistic–Roman, 1st-c. BC context); Gassner 1997, 379, pl. 32 (Ephesus, Late Hellenistic); Lang-Auinger/Asamer/Müller-Kaspar 2003, K24, pl. 3 (Ephesus, context of the third quarter of the 1st c. BC); *Agora XXIX*, no. 703 (Athens, context of 110–50 BC), 704 (Athens, context of 110 BC–AD 20), 705 (Athens, context of 115–86 BC), 706 (Athens, context of 115–75 BC), 707 (Athens, context of 20 BC–AD 10).

Date: Late Hellenistic–Early Imperial (late 2nd c. BC–early 1st c. AD).

SAV RW 02 Juglet

1:2

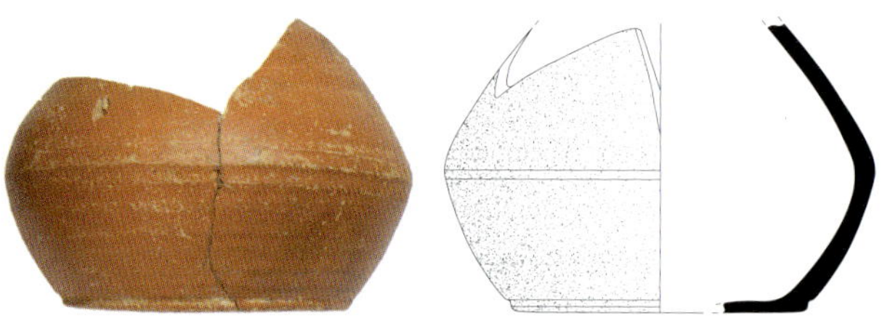

Provenance: Square F3, N Kiln interior. Level 2, batt. 2. Lot: SA 79-877P.

Dimensions and technical features: h 5.3, base 5.4. Clay Munsell 2.5YR 6/6 (light red), very hard and very finely granular although containing few macroscopically visible inclusions, except for very rare, extremely fine micaceous flakes on the surfaces of the fabric and also in the slip. Common voids, well-rounded, most < 0.5 mm, with a single anomalous, large irregular void, ca. 5 mm. Red slip, Munsell 2.5YR 5/8, thin and lustrous, slightly unevenly fired, dipped.

Condition: Two fragments of the base, joining.

Description: Biconical body, flat base marked by a slight groove.

Comparisons and comment: Hempel 2001, 135, pl. 5, Type 114/21 (Taranto necropolis, T E3-12, E3-16, E3-17, E3-40, third quarter of 2nd c. BC); Schojer 1988, 486–88, pl. CXII, no. 37.11g–h (Taranto, Contrada Corti Vecchie, Via Oberdan, *urna in terracotta*, 21.VI.1933, 1st c. BC); Montagna Pasquinucci 1972, 450–51, fig. 13, Form 148, no. 187 (Volterra necropolis, grey ware); Kögler 2010, cat. nos. E233, fig. 26 (Knidos, late 2nd–early 1st c. BC), E225, fig. 26 (Knidos, late 3rd c. BC–Augustan period), E227, fig. 26 (Knidos, late 3rd c. BC–Augustan period); Meyer-Schlichtmann 1988, kr. 2 (Pergamon, 100 BC–AD 125); Mitsopoulos-Leon and Lang-Auinger 1991, O45, pl. 218 (Ephesus, Late Hellenistic, 1st c. AD); Özyiğit 2000, 101, fig. 5 (Pergamon, early 2nd c. BC); *Agora V*, nos. F44, pl. 1 (Athens, AD 1–20), G88, pl. 5 (Athens, AD 1–25), M36, pl. 18 (Athens, AD 75–150), M269, pl. 29 (Athens, AD 375–400); *Agora XXIX*, nos. 512–513, fig. 38 (Athens, 140–130 BC), 514, fig. 38 (Athens, 100–25 BC), 1132, fig. 71 (Athens, 225–175 BC), 1134, fig. 71 (Athens, 240–190 BC).

Date: Late Hellenistic–Early Imperial (late 2nd c. BC–early 1st c. AD).

SAV RW 03 Juglet

1:2

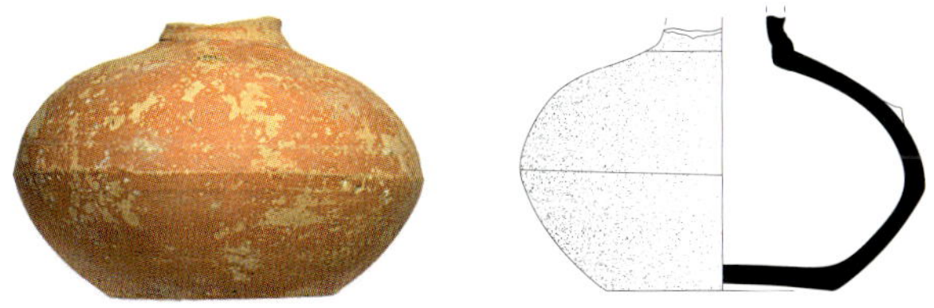

Provenance: Square E2, baulk at S Kiln. Level 3, batt. 1. Lot: SA 81-24P. Assemblage 5.

Dimensions and technical features: h 4.6, base 3.9. Clay Munsell 7.5YR 6/6 (reddish yellow), very hard and finely granular texture with frequent fine micaceous inclusions, sparse larger calcareous inclusions, most < 1 mm; the fabric is relatively porous with numerous irregular voids, < 2 mm, and occasional larger voids, with examples up to ca. 5 mm. Red slip, Munsell 2.5YR 5/8, dipped; matte and poorly adhered, lost in patches from the surface of the relatively porous and granular matrix of the fabric.

Condition: Complete base and body; ribbon handle, neck, and rim missing.

Description: Squat biconical body; slightly concave base.

Comparisons and comment: Agora XXIX, no. 1751, 421, fig. 104, pl. 138 (Athens, context of 325–200 BC), no. 1132, 351, fig. 71, pl. 83 (Athens, context of 225–175 BC), no. 1137, 352, fig. 71, pl. 83 (Athens, context of 150–86 BC); Alessio 1988, 349, pl. LXIII, no. 30.13e (Taranto, Piazza d'Armi, T I, 27.IV.1911, 2nd–1st c. BC); Colivicchi 2002, 121–22, no. 14.3 (Ancona, T 9 Corso G. Amendola 7.IV.1961, first half of the 2nd c. BC); *Da Leukania a Lucania*, 186, fig. 302, no. 3 (Herakleia, southern necropolis, T 608 prop. De Giorgio, third quarter of the 1st c. BC); *Museo Taranto III.1*, 362, 374, fig. 326 (Taranto, Contrada Vaccarella, T 2, 25.III.1922, 50–25 BC); Kögler 2010, form XXII, fig. 1 (Knidos, 150/125–75/50 BC), form XXIII, fig. 1 (Knidos, ca. 300–100 BC?), G108, fig. 41 form XXIV-D (Knidos, 25 BC–AD 75), G109, form XXIV-D, pl. 35, fig. 41 (Knidos, 25 BC–AD 75), G110, form XXIV-D, fig. 41 (Knidos, 25 BC–AD 75), E240, form XXXV, fig. 27 (Knidos, 125–50 BC), kn 123, form XXIII-C, fig. 58 (Knidos, ca. 250 BC); Mitsopoulos-Leon and Lang-Auinger 1991, O47–48, pl. 218 (Ephesus, Late Hellenistic, 1st c. AD); Schäfer 1968, no. C26A, D73-5, pl. 5 (Pergamon); Slane Wright and Jones 1980, no. 100, fig. 6, 156, 159 (Corinth, AD 22–30).

Date: Late Hellenistic–Early Imperial (late 2nd c. BC–early 1st c. AD).

SAV RW 04 Closed Form

1:2

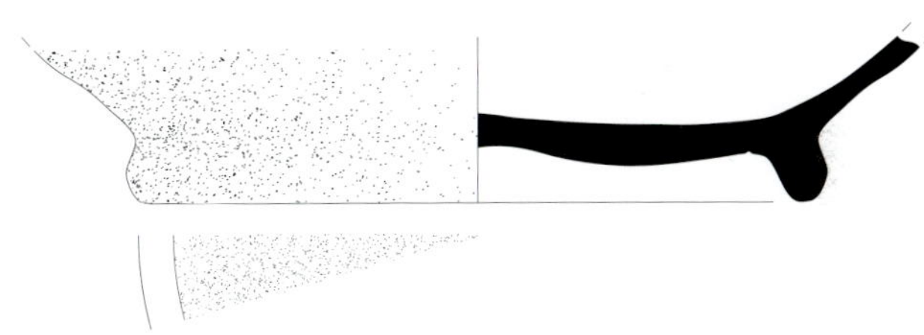

Provenance: Cut in baulk between N and S Kilns, Level 2, batt. 1–2, and S Kiln interior, Level 2, batt. 1. Lot: SA 79-801PL + SA 81-7PL (burnt). Assemblage 5.

Dimensions and technical features: h 2.9, base 12.8. Clay Munsell 5YR 6/6 (reddish yellow), very hard, evenly fired. Frequent fine, rounded voids particularly visible on unslipped surfaces, most < 1 mm. Red slip, Munsell 2.5YR 5/6-8, thin and matte, dipped.

Condition: Two bases and one body, joining.

Description: Base-ring; bottom lower side undercut by a narrow groove at the junction with the base. Dipped slip overall, including the underside except for the resting surface, possibly not well preserved; dribbles on the interior.

Comparisons and comment: Hempel 2001, pl. 4, Form 114/2 (Taranto necropolis, Phase E1, second quarter of the 2nd c. BC); *Agora XXIX*, no. 1503, 225, 390, fig. 88, pl. 115 (Athens, context of 150–110 BC), no. 1530, 393, fig. 91, pl. 117 (Athens, context of 115–86 BC), no. 1531, 393–94, fig. 91, pl. 117 (Athens, context of 115–50 BC); *Paphos III*, no. 13, 20, fig. 11, pl. 5.4 (Paphos, late 2nd–early 1st c. BC); *Da Leukania a Lucania*, 180, fig. 285, no. 3 (Herakleia necropolis, T 92 Via Umbria, beginning of 1st c. BC?); Kyriakou 1994, no. 6741, pl. 135 (Patras); Metzger 1994, no. 5, pl. 36 (Eretria, ca. 100 BC); Rotroff et al. 2003, no. 334, pl. 55 (Sardis, second half of 1st c. BC); Psalti 2004, no. 18877, pl. 246 (Eretria).

Date: Late Hellenistic–Early Imperial (late 2nd c. BC–early 1st c. AD).

18
Grey Ware

Eloisa Vittoria

The final occupation phase of the Sant'Angelo Vecchio site is attested by the presence, albeit scarce, of grey-ware pottery, which suggests use of the site through the Late Republican period.[1]

A total of 15 individual vessels were identified during the study of the 60 recovered grey-ware fragments. The majority of these fragments come from the area of the Kiln Complex and from the area immediately to the east. They are for the most part dishes (SAV GW 03; SAV GW 07; SA 79-438PL), as well as a pair of boccaletti (SAV GW 01 and SAV GW 02) and several wall fragments from open and closed forms. According to the excavation logs, these items were found in the occupation levels, rather than in the surface levels, with the result that they are stratigraphically reliable.

The patently misfired base of an open form (SAV GW 07) is particularly interesting in that blistering marks the entire exterior surface. As it comes from the lowest battute of Sounding B, it provides proof of the production of grey ware at the site. Additional proof is supplied by a small wall fragment of a probably closed form found in the first level of Sounding D (SA 80-42PL), likewise misfired, with cracking and blistering of the fabric from wall to wall. The production of grey ware at Sant'Angelo Vecchio is further confirmed by the recovery of other fragments inside the kilns, especially the North Kiln (Kiln 5): the nearly complete boccaletti (SAV GW 01 and SAV GW 02), and a fragment of a vertical handle (SAV GW 15). A dish was recovered between the North Kiln and the northern wall of the complex (SAV GW 06). Another dish fragment (SA 79-438PL) was found in square F2, corresponding to the South Kiln (Kiln 4). Sounding D, east of the two larger kilns, yielded a small cup fragment (SAV GW 10), the base of a closed form (SAV GW 14), and the base of an open form with roulette decoration (SAV GW 12). Another base with roulette decoration (SAV GW 11) comes from the area of the structures northwest of the Kiln Complex, while other dishes (SAV GW 04 and SAV GW 05) and a cup (SAV GW 08) were recovered in the surface levels between the Kiln Complex and the abovementioned structures. Finally, a fragmentary base of an open form, probably a dish (SAV GW 13), was discovered in square H7, in the western area, during the widening of the excavation zone in 1980.

Typology

The typology utilized in this study is based on the typology developed by the author for the study of the materials from the Bradano-Basento survey.[2] The entire class of Metapontine grey ware awaits a more systematic study, taking into account the large quantities of materials recovered in stratified contexts such as the small kiln at Pantanello and the farmhouse at Sant'Angelo Grieco.

Fabric and Technical Features

The clay employed in Metapontine grey-ware production is well levigated, generally gray in color (Munsell N5/) with occasional darker tones (Munsell N4/) and rare lighter tones (Munsell N6/). Under the microscope, the structure of the fabric is characterized by the presence of rare small vughs, which occur in quantities less than or equal to 1 percent on the Munsell scale.

The gloss coating, obtained by dipping, is generally homogeneous but thin and opaque, with a low level of reflection. On two dishes (SA 79-438PL and SAV GW 06) and several wall fragments, however, it is notably shiny. The coating technique often leaves telltale trickles along the lower exterior of a piece, an area which is normally reserved (SAV GW 13, SAV GW 03, SAV GW 01 and SAV GW 02, SAV GW 06). Minor decorative motifs were applied on both the interior and exterior surfaces with a roulette (SAV GW 11, SAV GW 01, SAV GW 12).

[1] For details on the morphology of the site and the chronological sequence of the structures, see Ch. 2, "Site Phasing, Stratigraphy, and Site Assemblage," and Ch. 3, "The Structures at Sant'Angelo Vecchio." Preliminary research on the grey ware from the Metapontine chora was done by Keys (1983); the evidence from Sant'Angelo Vecchio was first presented in Edlund 1982a and Burgers 1991. For quantification by TNI, see Ch. 15, note 2.

[2] Vittoria in *Survey*. Munzi 1999, 95, no. 28. Bats 1988, 55.

Form	Type	Date	Provenance	NFr	MNI	TNI
Boccaletto	Bo1 - Form 21	50–1 BC	Square F3, Level 2, SE end of N Kiln near baulk Square F3, N Kiln interior, Level 2, batt. 2	18	2	2
Cup	C2 - Form 18	150–100 BC	Square F6, Level 1, batt. 1	1	1	1
	C3 - Form 20	150–100 BC	Sounding D, Level 1, batt. 1	1	1	1
Dish	D1 - Forms 4-5	150–80 BC	Square F4, Level 1, batt. 1	4	1	1
	D2 - Forms 4-5	100–1 BC	Platform at junction between N Kiln + N wall under tile layer, Level 3, batt. 1 Square F2, Level 2, batt. 1	5	1	2
	Form 2	150–80 BC	Square F5, Level 1, batt. 2	1	0	1
	D4 - Form 7	100–50 BC	Sounding A, S area of Room 1, Level 2, batt. 3	2	1	1
	Misfired	Unknown	Sounding B, Level 3, batt. 10	6	1	1
Closed form	Unknown	Unknown	Sounding D, Level 1, batt. 2 (foot) Sounding D, Level 1, batt. 1 (closed bodies)	4	1	1
Open form	roulette dec.	Unknown	Square G4, Level 2, batt. 1 Sounding D, Level 2, batt. 7	2	2	2
	Unknown	Unknown	Square E1, cleaning, Level 2 (rim) Square H7, Level 1, batt. 2 (foot) Surface finds (SA 79 7PL) Sounding A, S area of Room 1, Level 2, batt. 2 Square G6, Level 2, batt. 3, E of wall Square H3, Level 1, batt. 1 (2 frags.) (SA 79 677PL and 671PL) Sounding D, Level 1, batt. 1–3 (7 frags.) (SA 80 45PL) Square F7, Level 2, batt. 2	15	2	2
Unknown form	Unknown	Unknown	Square F3, N Kiln interior, Level 2, batt. 2	1	0	0
Totals				**60**	**13**	**15**

Table 18.1 Quantification of grey ware from Sant'Angelo Vecchio.

Shapes and Functions

Despite the relatively small size of the sample of recovered material, it is nevertheless noteworthy that 13 of the 15 identified vessels are open forms associated with table service; only the base SAV GW 14 can be ascribed to a closed form, and possibly also the vertical ribbed handle SAV GW 15 which might belong to a lagynos. Previous studies have demonstrated, furthermore, that in the Hellenistic period, closed forms for holding liquids were distinguished by a preference for plain and banded wares, since these styles improved the refrigeration of the contents.

Catalog of Grey Ware

The choice of starting the catalog with the boccaletti was a deliberate one, primarily because this is the only form for which a complete profile is available, but also because, as will be discussed in detail, the form is strongly associated with the Metapontine chora. They are followed by the dishes, the most commonly attested form in this study, then by the cups and the bases of open and closed forms. For the detailed history of each form and its associations with other ceramic productions, see Vittoria in *Survey*.

Boccaletto

The recovery of this type of drinking vessel inside the North Kiln (Kiln 5) confirms the hypothesis of local production of a form that appears to be peculiar to the Metapontino. To date, the form has no recognizable comparanda beyond this area, but it is well attested, as the catalog demonstrates, at Sant'Angelo Vecchio, Sant'Angelo Grieco, and Pantanello, as well as in the castrum of Metaponto. Three additional fragments of grey-ware boccaletti were recovered in the course of the Bradano-Basento survey (*Survey*); two of these fragments come from sites located on the northern side of the Avinella stream (sites 372 and 502), the third from the Basento valley not far from Sant'Angelo (site 105). For a general introduction to the form, which was influenced by metal vessels, and its relationships with other forms and classes (such as black-gloss ring-handled cups, and several jugs), see Vittoria in *Survey*. The two examples from Sant'Angelo Vecchio are of average size, which corresponds to a rim diameter range of 8.0–16.0 cm.

Boccaletti

SAV GW 01 Boccaletto

1:2

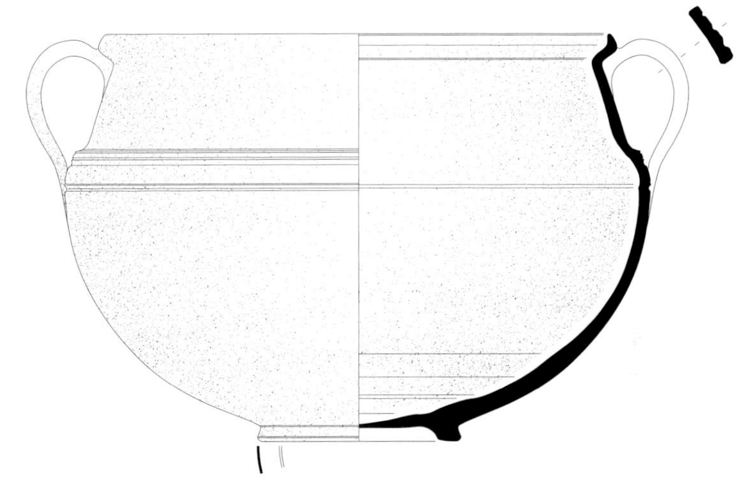

Provenance: Square F3, N Kiln interior. Level 2, batt. 2. Lot: SA 79-928P. Inv. no. 319912. Assemblage 5.

Dimensions and technical features: h 10.8, rim 14.2, foot 5.63, th 0.25–0.4, handle 1.75 x 0.45. Clay Munsell N5/ (gray), hard and smooth to the touch. Gloss Munsell N3/ and N4/ (dark gray), matte, not very homogenous, and dipped.

Condition: Composed of 12 fragments; the complete profile is preserved along with ¼ of the rim, the entire foot, and one handle.

Description: Vertical rim with triangular section and concave upper surface, attached to a straight and oblique neck, which is in turn attached to a globular body; ring foot with parallelepiped section; molded vertical handle with sub-rectangular section attached below the rim and to the maximum diameter of the body; on the exterior surface of the body, a roulette decoration consisting of impressed lozenges, each measuring ca. 0.55 x 0.15 cm, arranged in seven/eight parallel rows. Exterior and interior painted except for the lower exterior body, where trickles of gloss are visible. On the interior and exterior surfaces grooves left by the potter's wheel are visible.

Comparisons and comment: Survey Type GW Bo1. Giardino 1980, pl. 84, no. 50 (Metaponto, castrum, "Boccale di Tipo *b*"). Burgers 1991, Form 21 (PZ81.W.217.1, second half of 1st c. BC, which corresponds to Carter 1983a, 475, fig. 50, no. 154, called kantharos). Lissi Caronna 1998–1999, 194, fig. 16, no. 3 (Metaponto, farmhouse in *contrada* Casa Teresa, decorated with "false baccellature"). Sant'Angelo Grieco SG17.2, SG17.4, SG46.1, and SG235.16 have the same double grooving on the shoulder and an identically ribbed handle (study in progress).

Date: Second half of the 1st c. BC.

Boccaletti cont.

SAV GW 02 Boccaletto

1:2

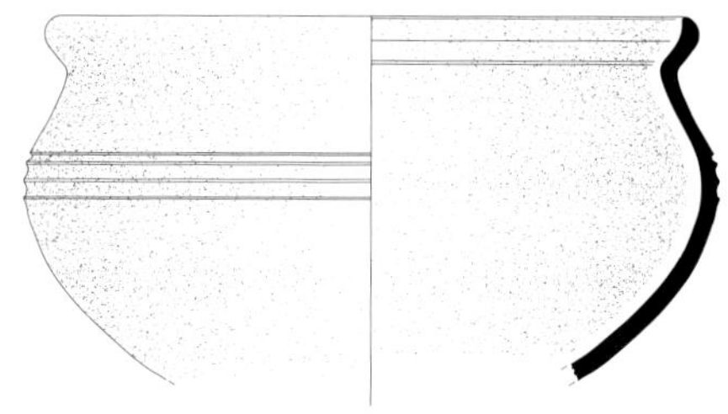

Provenance: Square F3, SE end of N Kiln near baulk. Level 2. Lot: SA 79-880PL. Assemblage 5.

Dimensions and technical features: h 5.6, rim 10.0, th 0.2–0.25. Clay Munsell gray N5/, hard and smooth to the touch. Gloss N3/ (dark gray), matte, homogenous, and dipped.

Condition: Composed of 6 fragments; the complete profile is preserved along with 1/20 of the rim and body.

Description: Vertical rim with triangular section and concave upper surface, attached to a concave neck, which is in turn attached to a globular body emphasized at the widest point by three parallel impressed grooves. Exterior and interior painted except for the lower exterior body, where trickles of gloss are visible. On the exterior surfaces grooves left by the potter's wheel are visible.

Comparisons and comment: See SAV GW 01.

Dish with Out-turned Rim

The form, represented by a single example, appears to be strongly associated with the black-gloss morphological repertory and is rarely attested in the South Italian grey-ware series that have received the most attention to date. This dish is absent from the rich typological assemblage documented in the kiln waste-deposit at Pantanello, but examples of it have been found at Sant'Angelo Grieco (Burgers 1991, Form 7, SG35.11 and SG320.1). Only three examples of the form were recovered in the castrum (Giardino 1980, 260). It is, finally, completely absent at Crucinia (De Siena 1993) and Casa Teresa (Lissi Caronna 1998–1999).

Dish with Out-turned Rim

SAV GW 03 Dish

1:2

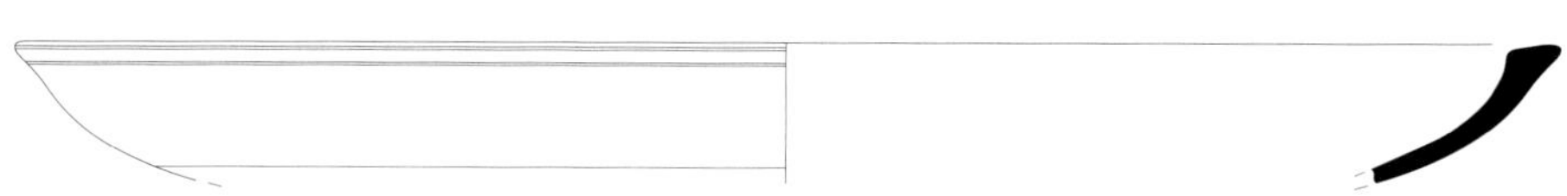

Provenance: Sounding A, S area of Room 1. Level 2, batt. 3. Lot: SA 79-29PL (see *Fig. 2.30*).

Dimensions and technical features: h 2.16, rim 24.0, th 0.3–0.4. Clay Munsell dark gray N4/, hard and smooth to the touch, with very rare and tiny vughs and tiny inclusions. Very dark gray gloss N3/, matte, homogenous, thin, and dipped.

Condition: Composed of two fragments; 1/14 of the rim and body is preserved.

Description: Barely out-turned rim with elliptical section; concave body. Painted exterior and interior, but in all likelihood the lower body was reserved.

Comparisons and comment: Survey GW-D4. Giardino 1980, 260, pl. 82, no. 16. Burgers 1991, Form 7 (SG35.11 and SG320.1, from Sant'Angelo Grieco, dated first half of 1st c. BC on the basis of the stratigraphy only). For the black-gloss version see Prag 1992a, 286, fig. 28, no. 648 (Gravina, 300–270 BC). Morel 2237a1 (Cuma, early 3rd c. BC). See also Vittoria 2011c, 19, fig. 29.

Date: The dating cannot be precisely determined with the information currently available; on the basis of the comparanda the first half of the 1st c. BC is a possibility, but the supposed association with morphological features peculiar to black-gloss production might push the dating back to the 2nd c. BC.

Dish with Wide Rim

The dish with wide, offset rim often distinguished by a ledge at the joint between the rim and the body wall seems to represent a transition in the history of this table vessel. Initially, in fact, the dish is still linked to the black-gloss morphological tradition, as is apparently demonstrated by the example SAV GW 03, but then it evolves into the type with wide rim, until the diffusion of the type with vertical rim which, however, is strongly influenced by sigillata production (see Ch. 17, "Red Ware").

In general, the dish with wide rim—whether emphasized with a ledge or bearing a flat or slightly concave upper surface—does not seem to be among the most commonly attested forms in Metapontine grey-ware production, as is clear from the materials recovered at Pantanello (Burgers 1991, Form 2) and Metaponto (Giardino 1980, 259–60); in both cases, in fact, the quantity of dishes with wide rim is manifestly reduced and appears to be complemented by the dishes with simple offset rim, which are likewise evidence of the late arrival of a form originating in black-gloss production. A similar situation has been documented at Valesio (Yntema 2001, 216, Form L02) and Taranto (Hempel 2001, 244, form 622.1-3; Hempel 1996a, 336, 338, form 2), while a different picture has emerged from the studies of materials from other sites such as Gravina (Prag 1992b, 136, 298, pl. 40), where on the contrary the dish with wide rim was appreciated and circulated widely.

Dish with Wide Rim

SAV GW 04 Dish

1:2

Provenance: Square F5. Level 1, batt. 2. Lot: SA 79-464PL.

Dimensions and technical features: h 5.4, w 2.2, th 0.35–0.45. Clay Munsell N5/ (gray), hard and smooth to the touch, with very rare and tiny vughs. Gloss Munsell 5Y 4/1 (greenish-gray), matte, homogenous, thin, and dipped.

Condition: One fragment of the body near the rim is preserved.

Description: Slightly convex rim with interior ridge near the body wall. Painted exterior and interior. Near the ridge the diameter is ca. 23 cm.

Comparisons and comment: Burgers 1991, Form 2 (second half of 2nd c. BC); Yntema 2005, 23, Form 2b (approximately mid-2nd to early 1st c. BC). Prag 1992b, 298, pl. 40, no. 970 (Gravina). For the types with wide rim attested in black gloss, see Lanza Catti et al. 2011, 246, no. 161 (site 416-09). Gravina (Prag 1992a, 286, pl. 28, no. 638, 3rd or possibly 2nd c. BC). Morel 1314–1315. Valesio (Yntema 2001, 158, Form K04). De Siena 1993, 129, 132, nos. 177 and 193 (Metaponto-Crucinia, T324 and T317, second half of 2nd–early 1st c. BC). Lo Porto 1981, 351, fig. 64, no. 2 (Metaponto, *contrada* Casa Ricotta, T5, second half of 3rd and perhaps part of 2nd c. BC).

Date: 150–80 BC.

Dish with Vertical Rim

Both types of this form appear in this study: the version characterized by a gentle transition from the rim to the body, and the version rendered distinct by a neat carination and a nearly horizontal body. The form is very common in grey-ware production not only at Metaponto (Giardino 1980, 253, 282; Burgers 1991, Form 4/5), but also at Herakleia (Giardino 1990, pl. XXIV, type A), Taranto (Hempel 2001, 114), and Gravina (Prag 1992b, 141–43).

The fragment SAV GW 05 can be ascribed to the series of so-called lid/plates, containers with a peculiar, low rim which allowed them also to serve as lids (see Prag 1992b, 141; Vittoria in *Survey*, 280). For this type, which has a more oblique body wall allowing for greater depth than the following type, Yntema has proposed a derivation from bowls of the South Italian and Tyrrhenian black-gloss traditions (Yntema 2005, 26); Prag, too, indicates a similar link with bowls for the materials from Gravina (Prag 1992b, 143, 300, fig. 42.1006, dated 150–90 BC; 303, fig. 45.1061, dated to the end of the 3rd or perhaps the beginning of the 2nd c. BC).

The type with more pronounced carination and shallower body appears at first to overlap chronologically with the preceding type, only to take its place in a reciprocal exchange of morphological influences that originate in black gloss, permeate grey ware, and terminate in sigillata. For the details of the form's evolution, see Vittoria in *Survey*, 280–81. The dimensions of the examples from Sant'Angelo Vecchio are average among the comparanda considered: similar dishes, in fact, generally measure between 18.0 and 37.0 cm, although even larger versions are known.

Dish with Vertical Rim

SAV GW 05 Dish

Provenance: Square F4. Level 1, batt. 1. Lot: SA 79-456PL. Inv. no. 320044.

Dimensions and technical features: h 4.7, rim 30.5, th 0.5–0.6. Clay Munsell N5/ (gray) and 2.5Y 6/1 (yellowish-gray) on the surface, hard and smooth to the touch, with rare and tiny vughs. Gloss Munsell N3/ (very dark gray), matte, homogenous, and dipped. Slightly misfired.

Condition: Composed of four fragments; ⅕ of the rim and body is preserved.

Description: Low, indistinct rim with elliptical section curving into the straight body wall with a gentle carination. Painted exterior and interior. On the exterior surface traces left by the potter's wheel are visible.

Comparisons and comment: Survey GW-D1. De Siena 1993, 129, no. 176 (Metaponto-Crucinia, T 324, second half of 2nd–early 1st c. BC). Hempel 2001, 244, pl. 13, Form 622, no. 4 (Taranto, phase E2-E3, second half of 2nd–first half of 1st c. BC).

Date: Middle of the 2nd–beginning of the 1st c. BC.

1:2

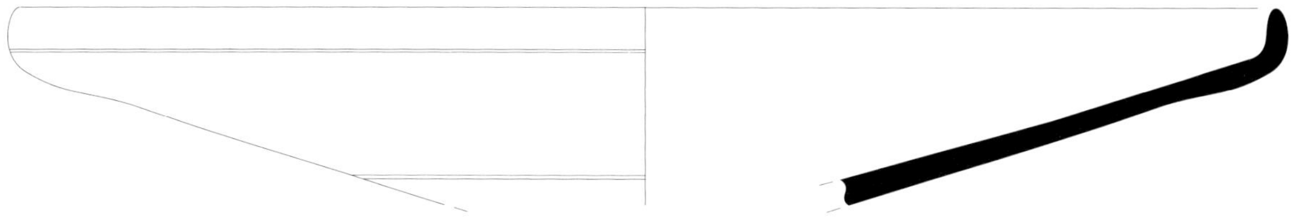

SAV GW 06 Dish

Provenance: Platform at junction between N Kiln + N Retaining Wall under tile layer. Level 3, batt. 1. Lot: SA 81-23PL B-C-D.

Dimensions and technical features: h 2.25, rim 30.0, th 0.5–0.6. Clay Munsell N5/ (gray) with gray core N6/, hard and smooth to the touch, with rare and tiny vughs. Gloss Munsell 10YR 3/1 (dark brownish-gray), bright, homogenous, dipped, and gritty to the touch; on the exterior surface the gloss is partly iridescent.

Condition: Composed of four fragments; 1⁄20 of the rim and body is preserved.

Description: Slightly oblique rim with sub-rectangular section; on the interior surface the body wall tends to curve inward slightly. Painted exterior and interior except for lower exterior of the body, where trickles of gloss are visible. On the interior and exterior surface traces left by the potter's wheel are visible.

Comparisons and comment: Survey GW-D2b. Giardino 1980, pl. 81, no. 4 (1st c. BC). Burgers 1991, Form 4, SG35.15 (Metaponto, Sant'Angelo Grieco, 1st c. BC); PZ81.W.327P (dated second half of 2nd c. BC on the basis of comparison with a dish from a Metapontine tomb). De Siena 1993, 129, no. 178 (Metaponto-Crucinia, tomb 324, second half of 2nd–early 1st c. BC). Prag 1992b, 300, fig. 42, no. 994 (Gravina, end of 2nd–1st c. BC). Small et al. 1994, fig. 8, no. 29 (Gravina, trench F202, dated 75±15 BC). Yntema 2001, 220, no. 383, form L05, dated generally from 90/70 BC to the Augustan period (from Valesio).

Date: Second half of the 2nd–1st c. BC.

1:2

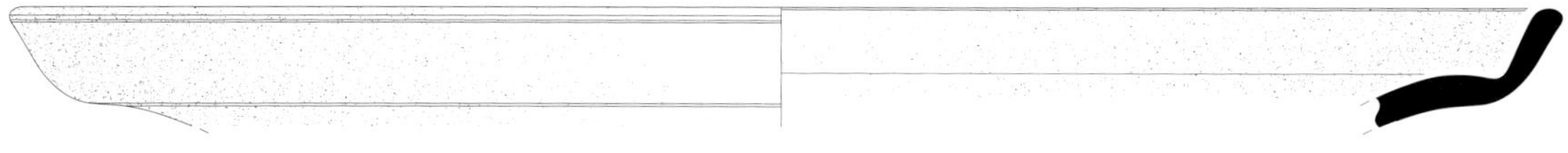

SAV GW 07 Dish

Provenance: Sounding B. Level 3, batt. 10. Lot: SA 79-433P (see *Fig. 2.35*).

Dimensions and technical features: h 2.1, foot 8.2, th 0.25–0.5. Clay Munsell 5YR 3/3 (dark brown) and N2.5/ (black), hard and gritty to the touch. Gloss N4/ (dark gray), matte, homogenous, and dipped.

Condition: Composed of six fragments; ⅓ of the foot is preserved. Misfired, with bubbling on the exterior surface.

Description: Ring foot with trapezoidal section and slightly oblique resting surface. Painted interior surface.

1:2

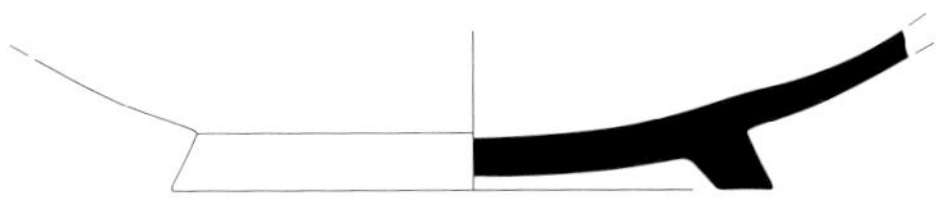

Comparisons and comment: De Siena 1993, 134, no. 215 (Metaponto-Crucinia, T317, second half of 2nd–early 1st c. BC).

Date: Second half of the 2nd–1st c. BC.

Cup with Offset Rim

Unlike the dish with vertical rim, the cup with offset rim is not commonly attested in the Metapontino. The most convincing comparanda come from Pantanello and two sites discovered by the field survey in the Basento valley (*Survey*), sites 105 (one fragment lacking the groove beneath the rim) and 556 (two fragments bearing the grooves beneath the rim); given the proximity of these two sites to the production facility at Sant'Angelo Vecchio, it is possible that the vessels represented by the three fragments were produced at the latter site.

The shape of the rim, offset and mounted on a deep body with oblique or nearly vertical walls, is very similar to that of the bowls (Burgers 1991, Form 12A, "bowl with convex-concave wall and outcurving rim;" Giardino 1980, 267–69, *coppa a parete espansa*, tipo a) which in the next phase definitively supplant the cups, evolving into the bowl characterized by an oblique wall and a rim with distinct hanging end (Burgers 1991, Form 13, "bowl with convex-concave wall and drooping rim;" Giardino 1980, 267–69, *coppa a parete espansa*, tipo c).

Cup with Offset Rim

SAV GW 08 Cup

Provenance: Square F6. Level 1, batt. 1. Lot: SA 79-1053PL (see *Fig. 2.23*).

Dimensions and technical features: h 3.65, rim int. 18.0, th 0.35–0.4. Clay Munsell N5/ (gray) with core N4/ (dark gray), hard and smooth to the touch, with rare tiny vughs. Gloss N5/ (gray) on the exterior surface with slightly gritty feel, gloss 5GY 5/1 (greenish-gray) on the interior surface, matte, not very homogenous, thin, and dipped.

Condition: ¹⁄₂₀ of the rim and body is preserved, but the rim is chipped. The gloss, almost completely worn, is better preserved on the rim and on the groove below the rim. The workmanship is not very careful, and a small amount of excess clay is visible on the interior surface.

Description: Outward-turned rim with triangular section, differentiated on the exterior by a groove; the body wall is oblique and straight. Painted exterior and interior. On the exterior surface faint traces of the potter's wheel are visible.

1:2

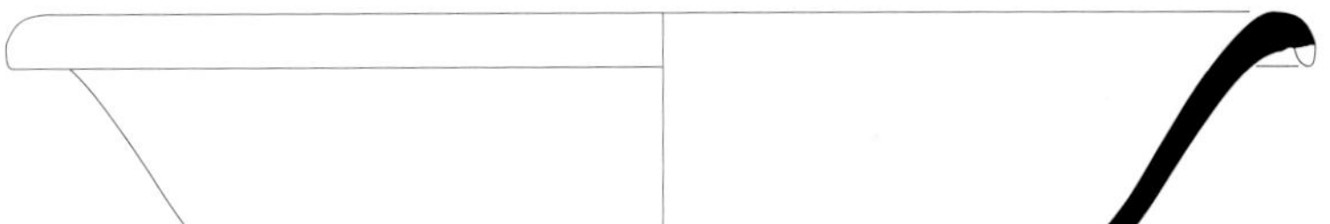

Comparisons and comment: Survey GW-C2b. Burgers 1991, Form 18, PZ81.NW.791.5 (second half of 2nd c. BC). Yntema 2001, 228, no. 401, form L31 (from an unstratified context: the form is dated generally to mid–late 2nd c. BC). Yntema 1993, 123, fig. 46, no. 320 (Oria, middle–end of 2nd c. BC). Costamagna 1983, pl. 49, no. 17 (Taranto). Prag 1992b, 302, fig. 44, no. 1047 (Gravina, 2nd c. BC, probably late). Mazzei and Pietropaolo 1995, 246, no. 159 (Arpi). Giardino 1978, pl. 68, B14 (Leuca).

Date: Second half of the 2nd c. BC.

Cups with offset rim cont.

SAV GW 09 Cup

1:2

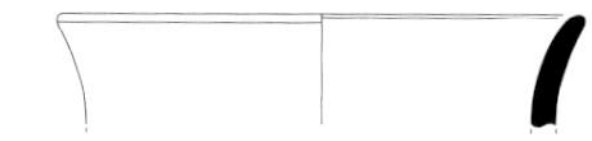

Provenance: Square E1, cleaning. Level 2. Lot: SA 79-735PL (see *Fig. 2.30*).

Dimensions and technical features: h 1.5, rim 7.0, th 0.4. Clay Munsell N5/ (gray), hard and smooth to the touch, with rare and tiny vughs. Gloss N5/ (gray), matte, not very homogenous, thin, and dipped.

Condition: 1/10 of the rim and body is preserved. The gloss is almost completely worn.

Description: Slightly outward-turned rim with rectangular section; the body wall is slightly concave. Painted exterior and interior.

Comparisons and comment: Survey GW-C1? The fragment is too small to be attributed to a specific form; the hint of the body wall suggests that the fragment may belong to a cup with offset rim, for which see Burgers 1991, Form 18, PZ81.NW.751.21 (second half of 2nd c. BC) and Lo Porto 1981, 351, fig. 64, no. 1 (Metaponto, *contrada* Casa Ricotta, T5, second half of 3rd and perhaps part of 2nd c. BC).

Date: Second half of the 2nd c. BC.

Hemispherical Cup

The small fragment that attests the production of this drinking vessel at Sant'Angelo Vecchio is almost completely lacking the characteristic groove(s) on the exterior of the body below the rim—in this case, the groove is very roughly indicated with a shallow and barely perceptible depression. The fragment's modest dimensions would also suit the bowls, a form with a similar shape of rim but shallower body (see Vittoria in *Survey*, 292, Type GW-B2). Nevertheless, the diameter of the rim (10.0 cm) is more suggestive of the hemispherical cups, which generally measure 9.5–13.5 cm, while bowls are typically slightly larger (11.0–20.0 cm).

Hemispherical Cup

SAV GW 10 Cup

1:2

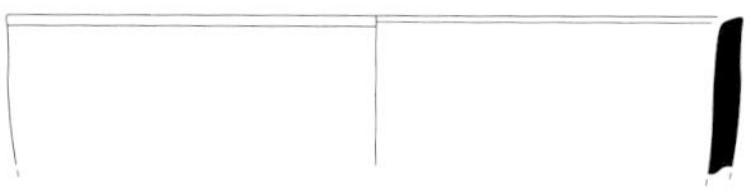

Provenance: Sounding D. Level 1, batt. 1. Lot: SA 80-42PL.

Dimensions and technical features: h 2.0, rim 10.0, th 0.35. Clay Munsell N4/ (dark gray), hard and smooth to the touch. Gloss Munsell N4/ (dark gray) on the exterior surface, gloss N5/ (gray) on the interior surface, matte, homogeneous, and dipped.

Condition: 1/20 of the rim and body is preserved.

Description: Indistinct, slightly outward-turned rim with triangular section; on the preserved fragment the body wall is straight. Painted exterior and interior.

Comparisons and comment: Survey GW-C3. D'Andria 1976, pl. XXXVI A, no. 31957 (Metaponto, castrum). Burgers 1991, Form 20, PZ81.W.377.1 (second half of 2nd c. BC). Prag 1992b, 299, fig. 41, no. 988 (Gravina, late 3rd–late 1st c. BC).

Date: Second half of the 2nd c. BC.

Unknown Open Form

The three base fragments from open forms recovered during the excavations preserve too small a portion of the body wall for secure attributions to specific forms, but nevertheless in all three cases the section of preserved wall is clearly straight and has a slightly oblique shape, features which might suggest a dish (see, e.g., Burgers 1991, forms 1, 2, 4, 5, and 14. Yntema 2005, forms 1–5) or a bowl (Burgers 1991, forms 12 and 13; Yntema 2005, form 16). No help in identifying the form is provided, on the other hand, even by the roulette decoration visible on the interior floor of fragments SA 79-460PL A and SA 80-58PL A. This type of decoration is in fact commonly encountered in both forms (see catalog). The decorative motifs on these fragments are applied in concentric circles on the interior floor. On SA 79-460PL A there are two clearly distinct lines of lozenges, while on SA 80-58PL A there are four lines of sub-triangular elements that are slightly larger, and less clearly defined, than the lozenges (cf. GW dec.1b from the field survey).

As has been emphasized elsewhere (Vittoria in *Survey*, 296), the chronological range of the rouletting cannot be defined, although this type of decoration, despite its modest presence on black-gloss products (Morel 1542a1, from Bolsena, dated end of the 3rd c. BC, 2653e1, from Volterra, first half of 2nd c. BC; Prag 1992a, 288, fig. 30, no. 711, dish from Gravina dated 250–200 BC), appears to have been infrequent in the earliest grey-ware production (see also Yntema 2005, 6). The unpublished kiln waste-deposit at Pantanello offers a guide to the types of open forms which bore this decorative scheme: it appears to be entirely absent on the dishes with offset rim, which clearly refer to black-gloss types (Burgers 1991, form 1; Yntema 2005, 21, form 1), while it becomes more common on the dishes with molded rim, where it is present especially on the rim itself as well as on the exterior body (Vittoria 2001, 284, type GW-D5; Burgers 1991, form 3; Giardino 1980, pl. 81, no. 4; Yntema 2005, form 3), and on the dishes with vertical rim (Vittoria in *Survey*, 280–81, type GW-D2; Burgers 1991, forms 4-5; Giardino 1980, pl. 81, no. 10; Yntema 2005, forms 4-5). These latter types gradually break free of the influence of black-gloss production and absorb ideas from sigillata production, where this type of decoration is common.

Unknown Open Form

SAV GW 11 Open Form

1:2

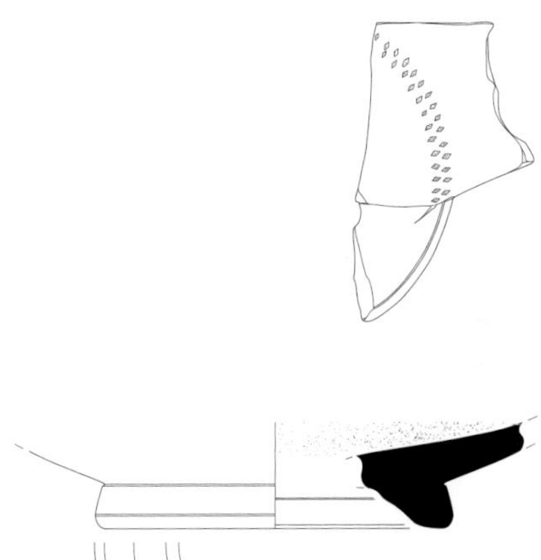

Provenance: Square G4. Level 2, batt. 1. Lot: SA 79-460PL A. Assemblage 3.

Dimensions and technical features: h 2.15, foot 7.5, th 0.55–0.75. Clay Munsell N5/ (gray), hard and smooth to the touch, with rare vughs. Gloss N5/ (gray), matte, homogenous, and dipped.

Condition: ⅓ of the foot is preserved.

Description: Ring foot with trapezoidal section and slightly oblique resting surface. Painted interior surface bearing a roulette decoration consisting of impressed lozenges, each measuring 0.2 x 0.1 cm, arranged in two concentric rows. On the interior surface faint traces left by the potter's wheel are visible.

Comparisons and comment: For the foot, see Giardino 1980, pl. 81, no. 6 (dish from the castrum of Metaponto). Burgers 1991, Form 2 (dish), Form 4 (dish, PZ81. W327P), Form 13 (bowl, PZ81.NW.769.1) and Form 14 (dish, PZ81.NW.855.41), from Metaponto-Pantanello. De Siena 1993, 129, no. 178 (Metaponto-Crucinia, T 324, second half of 2nd–early 1st c. BC). Giardino 1990, pl. 24, B (Herakleia, via Umbria, Tomb 108). Prag 1992b, 298, pl. 40, no. 971 (Gravina). For the rouletting see Burgers 1991, Form 12A (bowl, PZ81.W.358P) and Form 12B (bowl, PZ81.W.558P). Prag 1992b, 298, pl. 40, no. 984 (Gravina). Yntema 2005, Form 4 (Crucinia, Pantanello, and Taranto).

Date: Second half of the 2nd–1st c. BC.

Unknown open forms cont.

SAV GW 12 Open Form

1:2

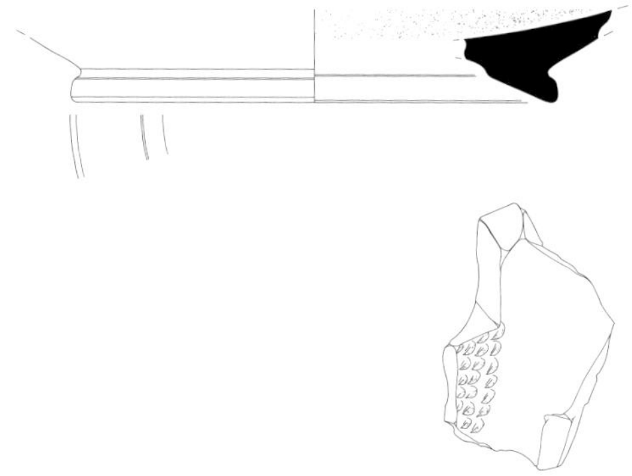

Provenance: Sounding D. Level 2, batt. 7. Lot: SA 80-58PL A (see *Fig. 2.21*).

Dimensions and technical features: h 1.85, foot 10.0, th 0.5–0.6. Clay Munsell N6/ (light gray), hard and smooth to the touch, with rare and tiny vughs and two very tiny inclusions. Gloss N5/ (gray), matte, homogenous, and dipped.

Condition: 1⁄7 of the foot is preserved.

Description: Ring foot with trapezoidal section and oblique resting surface. Painted interior surface bearing a roulette decoration consisting of sub-triangular impressions, each measuring ca. 0.3 x 0.15 cm, arranged in four concentric rows.

Comparisons and comment: For the foot, see Burgers 1991, Form 12A (bowl, PZ81.W.359.4). Giardino 1980, pl. 82, no. 21 (bowl). Prag 1992b, 300, pl. 42, no. 996 (dish from Gravina). For the rouletting, see Survey GW dec.1b. Giardino 1980, pl. 86, no. 65 (open shape). Prag 1992b, 303, pl. 45, no. 1054 (bowl from Gravina).

Date: Second half of the 2nd–1st c. BC.

SAV GW 13 Open Form

1:2

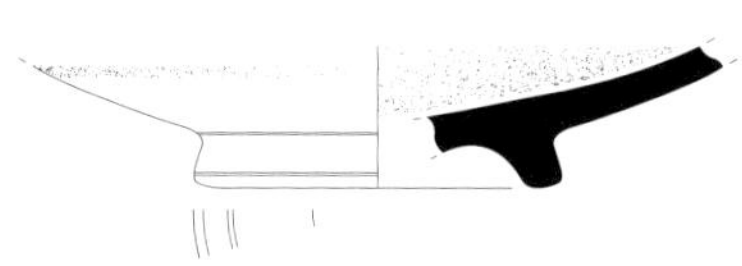

Provenance: Square H7. Level 1, batt. 2. Lot: SA 80-37PL A (see *Fig. 2.21*).

Dimensions and technical features: h 1.9, foot 4.4, th 0.35–0.5. Clay Munsell N6/ (gray), hard and smooth to the touch, with rare and tiny vughs. Gloss Munsell N3/ (very dark gray), matte, homogenous, and dipped.

Condition: 1⁄7 of the foot is preserved.

Description: Ring foot with trapezoidal section and flat resting surface. Painted interior and exterior surfaces except for reserved lower body, foot, and underside. On the interior of the body and foot faint traces left by the potter's wheel are visible.

Comparisons and comment: D'Andria 1976, pl. XXXVI A, no. 31975 (ribbed bowl from the castrum of Metaponto). Giardino 1980, pl. 82, no. 23 (bowl from the castrum of Metaponto). Burgers 1991, Form 12A (dish from Metaponto-Pantanello). De Siena 1993, 133, no. 200 ("coppetta" from Metaponto-Crucinia, T 317, second half of 2nd–early 1st c. BC). Yntema 2001, 220, Form L16 (bowl from Valesio).

Date: Second half of the 2nd–1st c. BC.

Unknown Closed Form

From the deeper exploration of Sounding D in 1980 comes the only diagnostic fragment belonging to a closed form. The amount of the body wall preserved is not sufficient to identify the specific form of the vessel; nonetheless the catalog compares it to a container with a similar type of base.

Unknown Closed Form

SAV GW 14 Closed Form

Provenance: Sounding D. Level 1, batt. 2. Lot: SA 80-43PL A.

Dimensions and technical features: h 1.65, foot 7.6, th 0.4–0.6. Clay Munsell N5/ (gray), hard and smooth to the touch, with very rare and tiny vughs.

Condition: ⅓ of the foot is preserved.

Description: Ring foot with trapezoidal section and slightly oblique resting surface; irregular workmanship of the foot. No traces of glaze are visible.

1:2

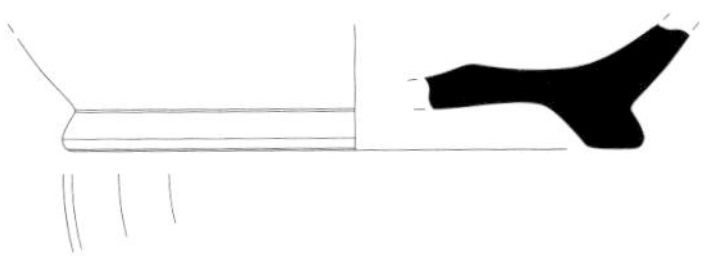

Comparisons and comment: Survey GW Cl1. Burgers 1991, Form 30 (jug/oinochoe PZ81.W.393P).

Date: Second half of the 2nd–1st c. BC.

Other

A ribbed vertical handle was recovered inside one of the kilns. The dimensions suggest a vessel of moderate size, very probably closed, such as a jug or lagynos.

Other

SAV GW 15 Unknown Form

Provenance: Square F3, N Kiln interior. Level 2, batt. 2. Lot: SA 79-878P. Assemblage 5.

Dimensions and technical features: h 10.35, l 2.4, w 1.0. Clay Munsell N5/ (gray), hard and smooth to the touch, with very rare and tiny vughs. Glaze Munsell N4/ (dark gray), matte and not very homogeneous.

Condition: Only the molded handle is preserved.

Description: Vertical handle with sub-rectangular section and two ribs. The entire fragment is painted. Fingerprints are visible in several places.

Comparisons and comment: De Siena 1993, 130, 133, nos. 187 and 212 (lagynoi from Metaponto-Crucinia, tombs 324 and 317, dated second half of 2nd–early 1st c. BC and cited in Yntema 2005, 80, Form 42 b).

Date: Second half of the 2nd–1st c. BC.

1:2

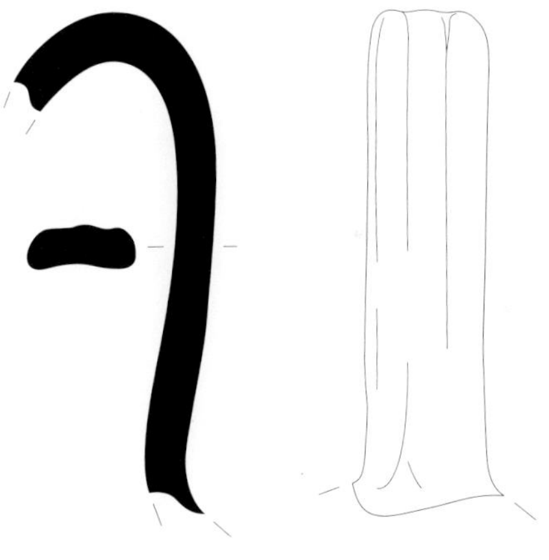

19
Miniatures

Anna Cavallo

Four small vessels among the materials recovered at Sant'Angelo Vecchio can be unequivocally identified as votive miniatures by their morphological features and the frequent presence of identical vessels in religious and funerary contexts.[1] Both of the shapes belong to the well-known repertoire of Metapontine miniature vessels, the morphological range of which was limited and associated, albeit very vaguely, with the full-sized vessels that inspired them.[2] One of the vessels (SAV Min 01) comes from Sounding C, while two others (only SAV Min 02 is in the catalog) were found in the Pottery Deposit discovered in the vicinity.[3] A third fragment was found in the Bulldozed Stipe.

Form	Date	NFr	MNI
Closed shape		1	1
		1	1
Krateriskos	520–300 BC	1	1
Lebes	520–400 BC	1	1
Total		**4**	**4**

Table 19.1 Quantification of miniatures.

[1] Following the model established in the study of the miniatures from the farmhouses excavated by ICA and the Pantanello sanctuary, small vessels which could not be ascribed with certainty to the votive class are not included in this study; they appear instead in the chapters dedicated to the individual ceramic classes to which they belong; see Cavallo in *Fattoria Fabrizio*.

[2] This study does not employ function as the discriminating factor for the identification of votive miniatures: the unique association of these vessels with the sacred world divorced them from the everyday functionality of the full-sized versions.

[3] The majority of Metapontine votives come from sacred areas or necropoleis: see, e.g., Lo Porto 1966; 1981; Morel 1970; Adamesteanu 1975; De Siena 1980; Dilthey 1980; Carter 1982; 1983a. The rare publications of miniatures from domestic contexts include Lo Porto 1966; Cotton 1977. On the votives from the Metapontine farmhouses excavated by ICA, see Cavallo in *Fattoria Fabrizio*.

Catalog of Miniatures

Krateriskos

Krateriskoi are among the best-documented forms encountered in the Metapontino. Their profile and size are in general highly variable, giving rise to a rather elaborate typology. They are typically found at cult sites, with few examples known from other contexts (Cavallo in *Fattoria Fabrizio*, with bibliography on Metapontine contexts). The example from Sant'Angelo Vecchio is very small, with a flaring rim, carinated body wall, small horizontal handles, and foot with cylindrical section. The two tiny handles are attached horizontally to the belly. The vessel was recovered in Sounding C. The identified parallels suggest a chronology between the end of the 6th c. and 300 BC.

Krateriskos

SAV Min 01 Krateriskos

1:2

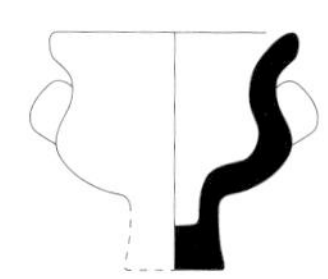

Provenance: Sounding C, Level 2, batt. 4. Lot: SA 79-317P. Inv. no. 320043. SAV 207 (see *Fig. 2.35*).

Dimensions and technical features: h 3.2, rim 3.4. Munsell 5YR 6/2. Poorly fired.

Condition: Complete profile.

Description: Flaring rim with continuous profile; low, squat body; small horizontal handles; foot with cylindrical section.

Comparisons and comment: Cotton 1977, no. 396 (Cozzo Presepe, 300 BC); Lo Porto 1981, 314, no. 231, fig. 23.14 (Metaponto, votive deposit in contrada Crucinia); Lo Porto 1991, 167, no. 240, pl. LXXIX (Timmari, votive deposit, 520–480 BC).

Date: 520–300 BC.

Lebes

The shape, very close to that of the full-size version, is characterized by a vertical collar-neck and compressed globular body. Small versions of the lebes, otherwise known in the literature as stamnoid or globular pyxides, are attested in Lucania and in the chora of Sybaris in domestic, sacred, and funerary contexts. The best-preserved examples have handles attached at the shoulder and a small cylindrical foot. Similar vessels have been recovered in the Metapontino: one from the votive deposit in contrada Crucinia (Lo Porto 1981, 311–12, nos. 184–186, fig. 23.4), one from the Pantanello necropolis (Carter in *Necropoleis*, 691, MIN LG6), and 14 examples from sites documented by the field survey between the Bradano and Basento (Vittoria in *Survey*, Min2, pl. 9.25). The example from Sant'Angelo Vecchio comes from the Pottery Deposit.

Lebes

SAV Min 02 Lebes

1:2

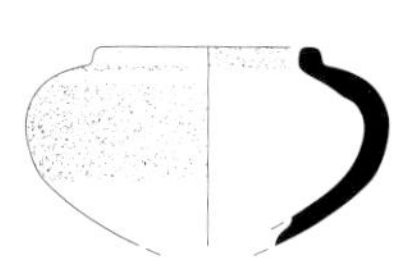

Provenance: Sounding C, Pottery Deposit. Lot: SA 79-918P. SAV 213. Assemblage 15.

Dimensions and technical features: h 2.6, rim int. 2.5. Clay type 4, Munsell 2.5YR 8/3. Very hard, compact, medium-grained, well-levigated, with tiny light and dark stone inclusions. Black and brown glaze.

Condition: Fragment of the rim and shoulder

Description: Thickened, vertical rim; globular body. On the interior, band corresponding to the rim; on the exterior, surface coated from the rim to midway down the body wall. Handle attached at the shoulder.

Comparisons and comment: *Sibari II*, 323, no. 459, fig. 320 (Stombi); *Sibari IV*, 14, no. 419, figs. 131, 147 (Stombi); Lo Porto 1981, 311–12, nos. 184–186, fig. 23.4 (Metaponto, votive deposit in contrada Crucinia, 520–480 BC); Lo Porto 1991, 167, no. 242, pl. LXXIX (Timmari, votive deposit, 520–480 BC); Carter in *Necropoleis*, 691, MIN LG6 (Metaponto, Pantanello necropolis, 450–400 BC); Vittoria in *Survey*, Min2, pl. 9.25.

Date: 520–400 BC.

20
Banded Ware

Anna Cavallo

This chapter is dedicated to the ceramic class characterized by well-levigated fabric with banded decoration or, more rarely, a surface partially glossed by dipping. The class is found from the Late Archaic period onward in the Bradano watershed, where it circulated widely, as in Magna Grecia generally.[1]

As repeatedly emphasized in recent publications about other contexts,[2] banded pottery has an undeniable affinity with plain wares: they generally share technical features of production, a utilitarian character,[3] and, at least in part, a repertoire of attested forms. Parallel analysis by context of the plain- and banded-ware fragments from the excavations at Sant'Angelo Vecchio demonstrates both the range of forms shared by the two groups and their specific morphological features. Nevertheless the two groups are treated as independent classes,[4] and are presented separately.[5]

Materials: Quantitative and Distributive Analysis

The excavations at Sant'Angelo Vecchio yielded a total of 290 fragments of banded pottery, including 126 rims, 32 feet, 2 lids, 13 handles, and 117 bodies. A note of caution about the tabulation of the non-diagnostic fragments: often fragments of banded vessels are recovered without traces of the decoration, which was generally minimal, simplified, and limited to small sections of various parts of the surface, and consequently they are frequently counted among plain wares. Indeed, the number of undecorated body fragments recovered at the site is decidedly disproportionate to the quantity of diagnostic fragments ascribed to the plain wares (see Ch. 21, "Plain and Coarse Wares"). The same situation applies to the feet and bases, which are often found without any traces of decoration.[6]

A rather modest number of fragments was recovered in the excavations of the House on the Hill: 58 fragments belonging to 14 individuals found in the topsoil. Among the open forms, the most common is the one-handler (10 examples, see SAV BW 05 and SAV BW 08). Less frequent are the lekanai (three examples, see SAV BW 16), while the closed forms are rarer, with sporadic attestations of the olpe with circular rim (SAV BW 43) and the pelike (SAV BW 32). One rim fragment probably comes from a banded hydria documented elsewhere in Magna Grecia in the second half of the 6th century BC (SAV BW 40). Despite the limitations imposed by the size of the sample, the analysis of the materials from the House on the Hill appears to indicate that banded ware was relatively rare at the site, with preference given to plain wares (see Ch. 21, "Plain and Coarse Wares"). Comparison of the repertoires of banded and plain wares, moreover, demonstrates a significant functional distinction in the attested forms. Among the plain wares, containers for the preparation of foodstuffs and closed forms for the storage of liquids predominate, while small open forms are almost completely absent. This picture is enriched by the study of the black-gloss pottery, which demonstrates a predominance of drinking vessels such as the skyphos and kylix.[7] Very few banded fragments were found in Sounding C, while from the Pottery Deposit came 93 fragments, among which at least 29 individual containers could be identified. Analysis of the banded morphological repertoire in this case reveals trends similar to those documented for the other attested

[1] That banded wares remain lesser understood is probably to due to the uncertainties associated with their very definition, which has been rendered especially difficult by problems in identifying the contributions of eastern, Italo-Geometric, and indigenous ceramics. For a thorough review of the studies on banded ware in Magna Grecia, and more specifically in Basilicata, see Cavallo in *Fattoria Fabrizio*.

[2] Bagnasco Gianni 1997, 99; Cuozzo 2006, 88–89; Caravelli 2008, 451–52.

[3] On the inclusion of non-figured fine ware in the broader category of plain wares, see Cerchiai 2007, 1.

[4] For a thoughtful definition of the concept of ceramic class, see Panella 1996.

[5] On the basis of the results obtained from the analysis of the identified ceramic materials, this option was also adopted for the study of the pottery from the other Metapontine farmhouses excavated by ICA. Other contributions are noteworthy for their contextual presentation of the two ceramic classes: see the study of the fragments from the intensive field survey of the Metapontine chora between the Bradano and Basento (Vittoria in *Survey*) and the recent publication on Civita di Tricarico (Caravelli 2008). For the Etruscan sphere, see Threipland and Torelli 1970; Donati 1994; Bagnasco Gianni 1997.

[6] The hypothesis that some bases ascribed to plain ware originally belonged to banded-ware vessels is supported also by the observations on the recovered feet of plain jugs, attested in a significantly larger quantity compared to the known rims (see Ch. 21, "Plain and Coarse Wares").

[7] See Ch. 15, "Black-gloss Ware and Lamps."

Fabric	Group	Description	Munsell	Observations	Forms
SAV 1	Group A—marine clays	Hard, compact, and fine-grained, with tiny white and black stone inclusions. Powdery surface.	5YR 7/6 5YR 6/6 5YR 7/8	Fine fabric of marine origin, with numerous tiny rounded vacuoles and occasional minute inclusions, including micro-crystals which, together with small carbonate elements, are part of the matrix.	
SAV 2	Group B—mixed clays	Hard, porous, and fine-grained, with tiny white and black stone and mica inclusions. Powdery surface.	5YR 7/6 7.5YR 7/4 5YR 7/4 5YR 7/6	Fine-grained fabric containing microfossils of marine origin and abundant carbonate inclusions. The concentration of iron-rich clays, characterized by a darker color after firing, suggests a mixed fabric, as do the fine-grained crystalline inclusions.	Large containers, basins, jugs, deep basins.
SAV 3	Group A—marine clays	Hard, compact, fine-grained, and well-levigated.	5YR 7/3 5YR 7/6 7.5YR 3/1	Fine-grained fabric with clean reddish fracture.	One-handlers, basins, plates, dinoi, bowls.
SAV 4	Group B—mixed clays	Very hard, compact, fine-grained, and levigated, with tiny light and dark stone inclusions.	5YR 6/3 5YR 6/4 2.5YR 6/6 2.5YR 6/6	Granular fabric with concentrations of iron-rich clay. Several microfossils noted.	Basins, large containers, tubs, mortaria.
SAV 5	Group C—alluvial and fluvial clays	Very hard and compact, with occasional vacuoles in fracture and numerous tiny white stone inclusions.	2.5YR 6/6 5YR 7/8	Fine, clean, and reddish matrix with gray interior indicating the presence of organic material in the original clay. Thin crystalline and sandy inclusions of uniform size. The sand could be of marine origin, from the beaches created by the fracturing of the marine terraces.	Basins, wide bowls, bowls.

Table 20.1 Fabric types attested in the plain and banded wares at Sant'Angelo Vecchio. For a summary of the groups, see *Fattoria Fabrizio*, table 11.1; for a concordance of fabric types across the Metapontine farmhouses excavated by ICA, see *Fattoria Fabrizio*, table 11.4. (KPS/ICA).

ceramic classes: as in the black gloss, the open forms predominate, including numerous one-handlers (SAV BW 02, SAV BW 04, SAV BW 06, SAV BW 07, SAV BW 09, SAV BW 11, SAV BW 13) and there are a significant number of carinated lekanai (SAV BW 17, SAV BW 21, SAV BW 22). Noteworthy, too, are the remains of four banded lebetes (SAV BW 45 and SAV BW 46).

The Workshop Area (squares F7, G5, G6, and G7) yielded a total of 41 banded fragments, including materials dating between the second half of the 4th and the early 3rd centuries BC, such as footless plates with inverted body wall (SAV BW 28), a skyphoid cup (SAV BW 15), and large bowls (SAV BW 24 and SAV BW 25), as well as containers with comparanda suggesting broader date ranges. In this context there is a marked decrease in the presence of one-handlers, which had been prevalent at the site in the preceding occupation phases, and a strong increase in the number of large open forms. The area of the slopes yielded very rare examples of banded containers.

Materials: Technical-Morphological Analysis

The clays employed in the production of the banded wares found at Sant'Angelo Vecchio are compact and well-levigated, very similar to the clays employed in the production of the plain wares, except that the banded wares do not use the coarsest fabrics. On the whole, the plain and banded wares found at Sant'Angelo Vecchio use five fabric types, which can be associated with the three macro-groups of Metapontine clays documented by microscopic and archaeometric analyses (*Table 20.1*).[8]

Among the banded wares from Sant'Angelo Vecchio, the most common fabric is SAV 3, while the rarest is SAV 1. The relative homogeneity of the materials suggests a common production context and excludes the presence of imports. The quality of the workmanship and the attention to detail are on the whole skillful: the surfaces are generally smoothed and often covered with a thin layer of dilute clay serving as a compact and homogeneous base for the application of the decoration.

The decoration is minimal—typical of this ceramic class—and generally limited to opaque black, brown-black, and (more rarely) brown bands. The bands were made with a clay solution, sometimes very dilute, that was applied mostly by brush while the dried vessel was

[8] The information derived from macroscopic and archaeometric analysis of the clays was obtained by K. Swift; initial macroscopic analysis of the fabrics was performed by the author, followed by the subdivision into groups.

Form	Type	Date	NFr	MNI
Small bowl/ one-handler	1	5th–beginning of the 3rd c. BC	13	8
	2	5th–4th c. BC	33	13
	3	End of the 6th–first half of the 4th c. BC	18	9
	Unknown		22	5
Skyphoid cup		Second half of the 3rd–first half of the 2nd c. BC	1	1
Lekane	1	5th–first half of the 4th c. BC.	23	8
	2	Second half of the 6th–5th c. BC	5	5
	3	5th c. BC	2	1
	Unknown		7	3
Basin		5th c. BC	1	1
Dish			2	1
Plate		4th–3rd c. BC	3	3
Large bowl		4th–3rd c. BC	11	2
Deep bowl			4	2
Jug	1	5th c. BC	6	5
	2	4th–beginning of the 3rd c. BC	5	4
	3	5th–4th c. BC	4	3
	4	End of the 6th–5th c. BC	2	2
	Unknown		19	4
Olpe		6th–5th c. BC	3	1
Lebes	1	End of the 6th–5th c. BC.	2	2
	2	5th–4th c. BC	1	1
	Unknown		1	
Lid			2	2
Rim	FA		1	1
	FC		1	1
Base	FA		6	
	FC		3	
Wall	FA		12	
	FC		55	
Unknown			22	
Totals			**290**	

Table 20.2 Banded ware from Sant'Angelo Vecchio.

turned on the potter's wheel. In several cases, the decoration is quite accurate and precise, but often it is the product of a rather summary movement, resulting in a band with imprecise boundaries and evident overlap of the strokes. Noteworthy among the examples decorated with this technique are the one-handlers with completely glossed interiors, which likely can be ascribed to the earliest phase of occupation of the site at Sant'Angelo Vecchio. More rarely, the decorative technique employed partial immersion in the diluted clay solution, especially for the rims of the closed forms. Although this approach facilitated the finishing of the vessel, it also frequently resulted in obvious defects such as long dribbles. The bands are typically positioned on the rim, the attachment of the foot, and the handles, as well as occasionally on the body wall.

The repertoire of forms attested at Sant'Angelo Vecchio enriches those documented at the other sites in the chora investigated by ICA, resulting in a rather high total number of forms. This considerable array does not, however, represent all of the known forms in the banded-ware corpus. The range of forms documented at the other farmhouses is rather limited, the majority identified as the most common and standardized types such as the one-handlers and, to a lesser extent, the jugs.

Criteria for Classification

The methodological approach adopted for the study of the banded pottery from Sant'Angelo Vecchio follows a strictly morphological criterion, identifying the original forms by analyzing the structural attributes of the fragments. This choice was dictated especially by the attested repertoire, in which many forms (in particular open forms like the one-handler and bowl) lent themselves to multiple uses by virtue of their features, therefore seriously complicating a morphological-functional approach.

The finds were subjected to macroscopic analysis and attributed to the relevant reference forms on the basis of their specific characteristics. They were then articulated, where possible, into types corresponding to associations of morphological features regularly repeated in the corpus of materials studied.[9] Despite the frequently very poor state of preservation, nearly all of the identified fragments were able to be ascribed to a specific form, leaving a minimal number of unidentified rims and feet. The variants were identified in a subsequent phase: these are considered variations in the shape or the decoration that do not contribute to the definition of the type.

[9] On the concept of type and typology applied to prehistoric materials, see Peroni 1967, 156.

Catalog of Banded Ware

The attribution of the fragments to their original forms reveals a clear predominance of open forms, which account for 73% of the total (calculations based on MNI; for a definition of MNI, see n. 2 on p. 221). Among these, the one-handlers are the most common, represented by 33 examples. The range of closed forms represented includes pelikai, hydriai, table amphorae, and jugs, the latter of which accounts for 27% of the total. Alongside these forms, there is a small corpus of olpai and containers that performed various functions, including personal use.

Small Bowl/One-handler

Basilicata provides an excellent example of the diffusion typical of the one-handlers in Magna Grecia (for a thorough review of the history of the form and its circulation in Greece and Magna Grecia, see Cavallo in *Fattoria Fabrizio*). At a very early point the Lucanian populations acquired the form from the nearby Greek colonies and before the middle of the 6th c. BC made it one of the most common containers in their repertoire of open forms (Yntema 1990a, 173; *Valesio I*, 63). The production of one-handlers in the Metapontine area is attested starting in the second half of the 6th c. BC (Silvestrelli forthcoming, figs. 4, 6, 7), and continued in the Bradano watershed until at least the 3rd c. BC, with the occasional example documented also in subsequent periods (Macnamara 1983a, 337; Saunders and du Plat Taylor 1992, 15; Marchegiani 1997, 165; Caravelli 2008, 472). Local production is still documented at Taranto in the early 4th c. BC (*Arte e artigianato*, 329–30), while in Apulia it endured through the 2nd c. BC (*Valesio I*, 67–68).

In the course of the excavations at Sant'Angelo Vecchio, 86 fragments of one-handlers were identified (51 rims, 15 feet, 17 bodies and 3 handles), corresponding to at least 35 individuals and constituting the most common container at the site. The one-handlers are characterized by a hemispherical body, indistinct rim, foot with flat or slightly concave resting surface, and small horizontal or slightly raised handle attached immediately beneath the lip. Taken together with the finds from the other excavated farmhouses in the chora, the Pantanello necropolis, and the field survey between the Bradano and Basento Rivers, they confirm the existence of a single typological reference system for the Metapontino, which corresponds to the most common model in southern Italy beginning in the Late Archaic period. The types equipped with a ring foot, however—well documented in mainland Greece (*Agora XII*, 125, fig. 8, pl. 30) and in several South Italian contexts (see, for example, Caravelli 2008, 472–73; Cuozzo and D'Andrea 1991, 84, type 38 A2; *Palinuro II*, 112, T XXIII, pl. 15.1, T XXV, pl. 16.1; *Cairano*, 81–82, type A1.a; Cuozzo 2006, no. TTA261, pl. 21.6)—are not attested. The footless one-handler is documented at Sant'Angelo Vecchio also in plain ware (SAV PCW 01), attested in the same chronological phase but less common (see Ch. 21, "Plain and Coarse Wares"), while completely glossed footless examples are unattested.

The decorative syntax of the one-handlers is highly varied but always divided between interior and exterior. Examples decorated by dipping part of the vessel, as documented at other sites in southern Italy, especially in the earliest phases of circulation (see, among others, Macnamara 1983a, 336–37, nos. 193–96; Lissi Caronna 1990–1991, 333, fig. 161.11; *Fratte*, 220, t. 44/1963, no. 3, fig. 362; La Torre 2002, 202, 204, nos. 233, 234, 242, fig. 36) were not found at Sant'Angelo Vecchio. On all of the fragments examined, the decoration was applied by brush. The most common variant has a band halfway down the body on the exterior, with a matching band on the interior of the body or at the level of the floor. The rim of this variant has a narrow, generally rather careless brushstroke, and the handle, too, can be partially or fully glossed. There is nevertheless a substantial number of individuals with fully glossed interiors, using generally rather thick and dark black gloss. The majority of one-handlers from the excavated Metapontine farmhouses having this kind of decoration come from Sant'Angelo Vecchio, where it is attested among the finds from the House on the Hill and occurs frequently among the finds of the Pottery Deposit, suggesting a possible terminal chronology by the middle of the 5th century BC.

The variable diameter of the rim, between 6.4 and 14.0 cm, indicates the availability of different sizes, probably to be used for different purposes. The full profile could be reconstructed for 10 of the preserved containers; these are rather limited in height, between ca. 3.0 and 5.5 cm. Among the fragmentary examples, however, even larger versions are attested.

Type 1—indistinct and slightly inward curving rim. The first type is characterized by an indistinct, rounded or thinned, and barely inward-curving rim; hemispherical body; flat or slightly raised resting surface; and a handle with sub-circular section or ribbon handle, mounted beneath the rim and arched slightly upward. Eight individuals, represented by 13 fragments, recovered at Sant'Angelo Vecchio belong to this type with rim diameter between 6.4 and 10.2 cm. One case, lacking the handle, has a fully reconstructable profile (SAV BW 01).

The majority of the fragments have a decorative band on both the interior and the exterior, halfway down the body, and the rim bears a thin brushstroke of gloss (variant 1.a). Occasionally, the decoration covers all of the interior instead, with an exterior band (variant 1.b). The handle, where preserved, is a partially glossed ribbon.

Type 2—indistinct rim with vertical orientation. The second type of banded one-handler is characterized by an indistinct, rounded or slightly tapering rim with vertical orientation, hemispherical body, and a flat or slightly concave resting surface. The handle, where preserved, is a ribbon or has a sub-circular section, is located immediately beneath the rim, and is horizontal or arched slightly upward. To this type belong 13 individuals found at Sant'Angelo Vecchio (33 fragments). The rim diameters range from 8.4 to 14.0 cm, suggesting that the vessels were of medium size. Almost all of the examples come from the Pottery Deposit in Sounding C and the House on the Hill, and can be dated to the 5th c. BC.

Two distinct variants of the type were identified based on the decoration. In variant 2.a, the interior surface is decorated with two bands, one on the rim and the other midway down the body wall; the exterior surface has a strip on the rim and a band midway down the body wall. In variant 2.b, the interior surface is completely glossed or the tondo is left reserved (SAV BW 10), while the exterior surface has a band midway down the body wall.

Several examples yielded a complete profile, but in no case was the handle preserved (SAV BW 06, SAV BW 07, SAV BW 08, and SAV BW 09). Among these, SAV BW 06 and SAV BW 07 are noteworthy for their greater than average size compared to the other examples in the group. Both vessels have a band set high on the interior at the level of the rim and small added decorative elements. SAV BW 07 has a dotted exterior decoration immediately beneath the rim, while SAV BW 06 has a circular band on the interior floor and a line on the diameter of the exterior floor. Both were found in the Pottery Deposit.

Type 3—indistinct rim and truncated conical body. The third type of banded one-handler is characterized by an indistinct, rounded, and slightly tapering rim; oblique, nearly straight walls; truncated conical body; flat or slightly raised resting surface; and a handle with semicircular section or ribbon handle attached immediately beneath the rim. Within the type there are several decorative schemes. The first variant (variant 3.a) has bands on the interior and the exterior of the body and on the rim, with the rare addition of a circular strip on the interior floor. In the second variant (variant 3.b), the interior is completely glossed, while the exterior has a band, occasionally very wide, halfway down the body.

Among the finds from Sant'Angelo Vecchio, nine individuals (18 fragments) can be ascribed to this type. The attested vessels have highly variable diameters, with rims that range from 7.7 to 10.6 cm. Reconstructed profiles are between 3.0 and 5.0 cm in height. The comparanda suggest a chronology between the end of the 6th and the first half of the 4th c. BC.

Small Bowl/One-handlers

SAV BW 01 Small Bowl/One-handler Type 1.a

Provenance: Sounding C, Pottery Deposit. Lot: SA 79-757P. SAV 210. Inv. no. 319960. Assemblage 15.

Dimensions and technical features: h 3.0, rim 6.4, base 2.7. Fabric SAV 3, Munsell 5YR 7/3, exterior Munsell 7.5YR 3/1–5YR 3/1, hard, compact, fine-grained, and well-levigated. Dilute, opaque, and black gloss applied by brush in an irregular manner, with evident overlapping.

Condition: Reconstructed from two fragments.

Description: Indistinct and slightly inward-curving rim with tapering termination; hemispherical body wall. Small horizontal handle, arched upward, attached beneath the rim; the termination has a decorative strip with irregular borders. On the interior, low band halfway down the body; on the exterior, band with irregular borders halfway down the body wall.

1:2

Comparisons and comment: Kaulonia I, 53, nos. 89 and 91, fig. 34 (Caulonia, 5th c. BC); Marchegiani 1997, 164, n. 11, pl. 57 (Pomarico Vecchio, 320–280 BC); Carter and Parmly Toxey in *Necropoleis*, 703, T 345-2 (Pantanello necropolis, 410–360 BC); de la Genière and Greco 2010, 203, tav. 67, n. 689 (Heraion di foce Sele, second half of the 6th–first half of the 5th c. BC).

Date: 5th–4th c. BC.

Small bowl/one-handlers cont.

SAV BW 02 Small Bowl/One-handler Type 1.a

1:2

Provenance: Bulldozed votive deposit. Lot: SA 79-603PL. SAV 138. Assemblage 9.

Dimensions and technical features: h 2.9, rim 10.2. Fabric SAV 3, Munsell 5YR 7/6, hard, compact, fine-grained, and well-levigated. Brown-black, thin, and opaque gloss on the rim; traces of a band on the body.

Condition: One fragment of the rim and body.

Description: Indistinct rim with oblique termination; hemispherical body. Horizontal handle attachment immediately below the rim preserved; surface imprecisely smoothed. On the interior, low band at the level of the rim; on the exterior, band on the lower part of the body.

Comparisons and comment: Ruga et al. 2005, 194, n. 13, tav. LXXXV, fig. 88 (Croton, 5th c. BC); Colangelo 2001, 72, no. 91 (Torre di Satriano, end of the 4th–beginning of the 3rd c. BC); Colangelo 2005a, 254–55, no. 181 (Torre di Satriano, sanctuary; second half of the 4th–beginning of the 3rd c. BC).

Date: 4th–beginning of the 3rd c. BC.

SAV BW 03 Small Bowl/One-handler Type 1.a

1:2

Provenance: Square G6, E of wall. Level 2, batt. 2. Lot: SA 79-739PL. SAV 139. Assemblage 3.

Dimensions and technical features: h 2.0, rim 9.6. Fabric SAV 3, Munsell 5YR 7/4, hard, compact, fine-grained, and well-levigated. Brown-black, thin, and opaque gloss.

Condition: One fragment of the rim, body, and handle.

Description: Indistinct rim with oblique termination; hemispherical body; horizontal handle with sub-circular section arched upward. On the interior, narrow band at the level of the rim; traces of gloss on the handle.

Comparisons and comment: See above, SAV BW 02.

Date: 4th–beginning of the 3rd c. BC.

SAV BW 04 Small Bowl/One-handler Type 1.b

1:2

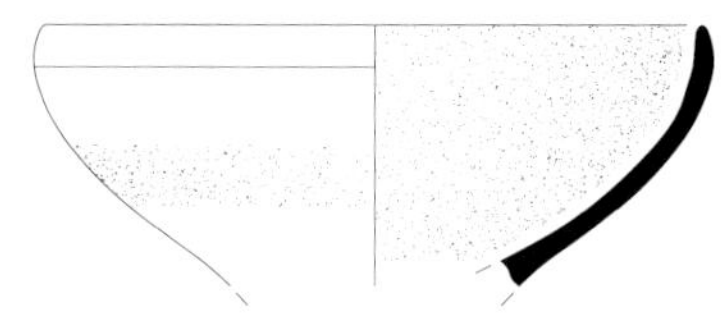

Provenance: Sounding C, Pottery Deposit. Lot: SA 79-988PL. SAV 83. Assemblage 15.

Dimensions and technical features: h 3.4, rim 8.8. Fabric SAV 3, Munsell 5YR 6/4, hard, compact, fine-grained, and well-levigated. Black, thin, opaque, and iridescent gloss on the interior; dilute and opaque gloss on the exterior.

Condition: Three fragments of the rim and body.

Description: Indistinct rim with rounded termination; hemispherical body. Completely glossed interior; on the exterior, band halfway down the body.

Comparisons and comment: *Testimonianze*, 131, T2; inv. no. 60279 (Tolve, second half of the 5th c. BC); Silvestrelli forthcoming, fig. 7.12 (Metaponto, *kerameikos*, deposit 16, 425–350 BC); de la Genière and Greco 2010, 203, tav. 67, n. 655 (Heraion di foce Sele, second half of 6th–first half of 5th c. BC).

Date: Second half of the 5th–4th c. BC.

SAV BW 05 Small Bowl/One-handler Type 2.a

1:2

Provenance: House on the Hill. Lot: SA 79-1062PL. SAV 56. Assemblage 1.

Dimensions and technical features: h 3.0, rim 10.0. Fabric SAV 1, Munsell 5YR 7/6, hard, compact, and fine-grained, with tiny white and black stone inclusions. Powdery surface. Black, dense, and opaque gloss.

Condition: One fragment of the rim and body.

Description: Indistinct rim with flat termination; hemispherical body. On the interior, wide band at the level of the rim; on the exterior, band halfway down the body.

Date: 5th–4th c. BC.

SAV BW 06 Small Bowl/One-handler Type 2.a

Provenance: Sounding C, Pottery Deposit. Level 2, batt. 1. Lot: SA 79-758P. SAV 211. Assemblage 15.

Dimensions and technical features: h 5.4, rim 14.0. Fabric SAV 3, Munsell 5YR 6/4, hard, compact, fine-grained, and well-levigated. Thick, opaque, and misfired gloss, overfired to black. On the exterior, narrow band and horizontal line on the floor; on the interior, wide band on the rim, narrow band halfway down the body, and small circle on the floor.

Condition: Partially reconstructed from six fragments.

Description: Indistinct rim with rounded termination; hemispherical body; flat and indistinct resting surface. On the interior, band at the level of the rim and halfway down the body, narrow strip on the floor; on the exterior, band halfway down the body and linear stroke along the diameter of the floor. Variations in the color and consistency of the clay likely due to exposure to fire.

1:2

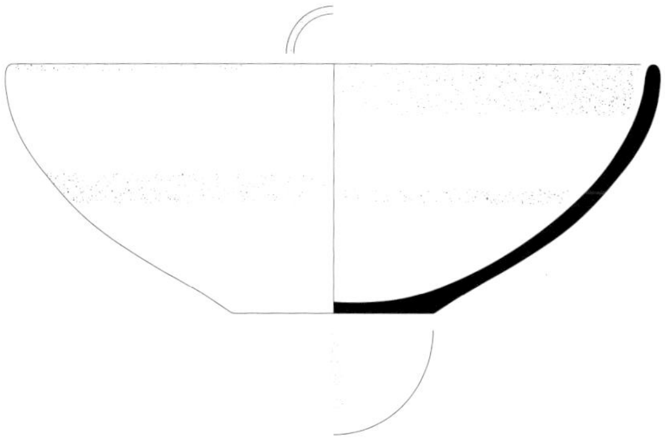

Comparisons and comment: Saunders and du Plat Taylor 1992, 15, no. 84, fig. 4 (Gravina di Puglia, 5th–4th c. BC); *Kaulonia I*, 53, no. 94, fig. 34 (Caulonia, 5th c. BC); Silvestrelli forthcoming, fig. 6, no. 19 (Metaponto, second half of the 5th c. BC).

Date: 5th–4th c. BC.

SAV BW 07 Small Bowl/One-handler Type 2.a

Provenance: Sounding C, Pottery Deposit. Lot: SA 79-759P. SAV 205. Inv. no. 320045 Assemblage 15.

Dimensions and technical features: h 4.8, rim 14.0. Fabric SAV 3, Munsell 5YR 6/4, hard, compact, fine-grained, and well-levigated. Brown-black, thin, and opaque gloss.

Condition: Profile reconstructed from two fragments.

Description: Indistinct rim with rounded termination; hemispherical body; indistinct resting surface with slightly concave lower profile. On the exterior, strip at the termination of the rim and band on the lower body; on the interior, strip at the level of the rim and halfway down the body, narrow band on the floor.

1:2

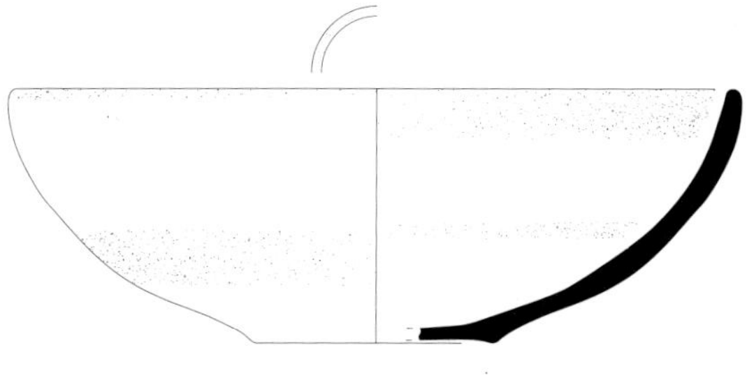

Comparisons and comment: See above, SAV BW 06.

Date: 5th–4th c. BC.

SAV BW 08 Small Bowl/One-handler Type 2.b

Provenance: House on the Hill. Level 1, batt. 1. Lot: SA 79-1038P. SAV 200 (see *Fig. 2.39*).

Dimensions and technical features: h 3.8, rim 8.8. Fabric SAV 1, Munsell 5YR 7/4, hard, compact, and fine-grained, with tiny white and black stone inclusions. Powdery surface. Brown-black, thin, and opaque gloss, misfired and flaking.

Condition: One fragment of the rim, body, and handle attachment.

Description: Indistinct rim with oblique termination; hemispherical body; horizontal attachment for handle with sub-circular section arched upward; small and barely distinct resting surface. Interior completely glossed; on the exterior, wide band with irregular boundaries halfway down the body.

1:2

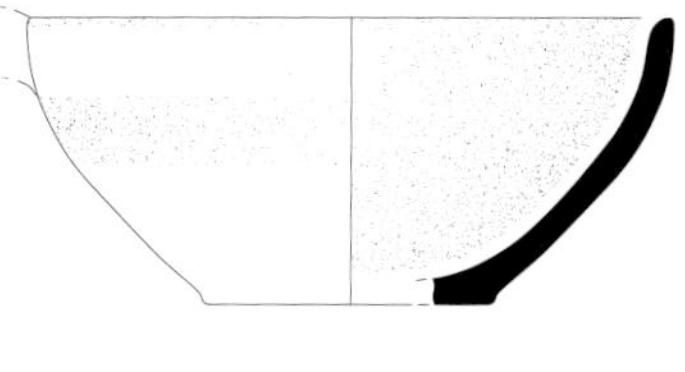

Comparisons and comment: See below, SAV BW 09.

Date: 5th–4th c. BC.

Small bowl/one-handlers cont.

SAV BW 09 Small Bowl/One-handler Type 2.b

1:2

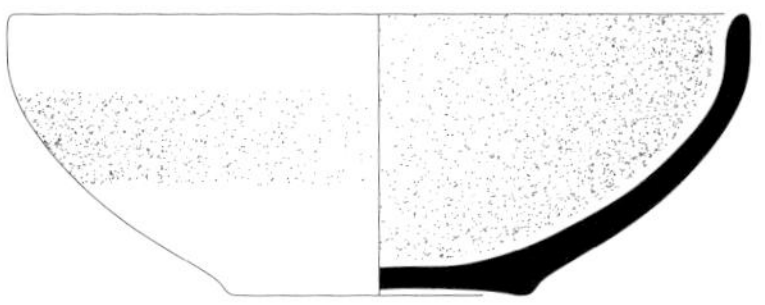

Provenance: Sounding C, Pottery Deposit. Lot: SA 79-756P. SAV 209. Inv. no. 319959. Assemblage 15.

Dimensions and technical features: h 3.8, rim 9.9, base 3.8. Fabric SAV 3, Munsell 5YR 6/6, hard, compact, fine-grained, and well-levigated. Black gloss.

Condition: Partially reconstructed from six fragments.

Description: Indistinct rim with oblique termination; hemispherical body; barely distinct resting surface with slightly concave lower profile. Completely glossed interior; on the exterior, wide band halfway down the body.

Comparisons and comment: Carter and Parmly Toxey in *Necropoleis*, 702, T 128-11 (Pantanello necropolis, 400–360 BC); Silvestrelli forthcoming, fig. 5, no. 14 (Metaponto, first half of the 5th c. BC); M. Liseno 2004, 97, no. 3, pl. LIV. g (Metaponto, end of the 5th–beginning of the 4th c. BC); Saunders and du Plat Taylor 1992, 15, no. 80, fig. 4 (Gravina di Puglia, 5th c. BC).

Date: 5th–4th c. BC.

SAV BW 10 Small Bowl/One-handler Type 2.b

1:2

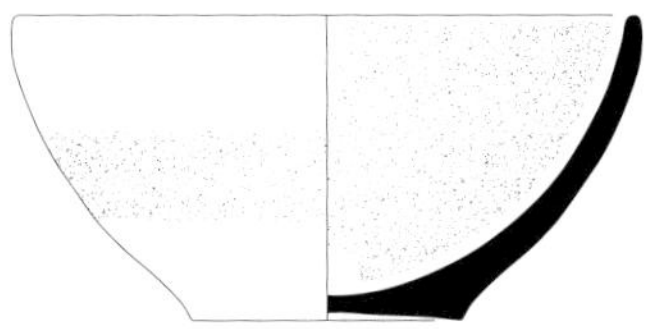

Provenance: Sounding C, Pottery Deposit. Lot: SA 79-988PL. Assemblage 15.

Dimensions and technical features: h 4.0, rim 8.4. Fabric SAV 3, Munsell 5YR 7/6, hard, compact, fine-grained, and well-levigated. Brown-black, thin, and opaque gloss.

Condition: Profile reconstructed from seven fragments.

Description: Indistinct rim with flat termination, hemispherical body, barely distinct resting surface with slightly concave lower profile. Interior completely glossed except for reserved circle in center; external wide band halfway down the body.

Comparisons and comment: See above, SAV BW 09.

Date: 5th–4th c. BC.

SAV BW 11 Small Bowl/One-handler Type 2.b

1:2

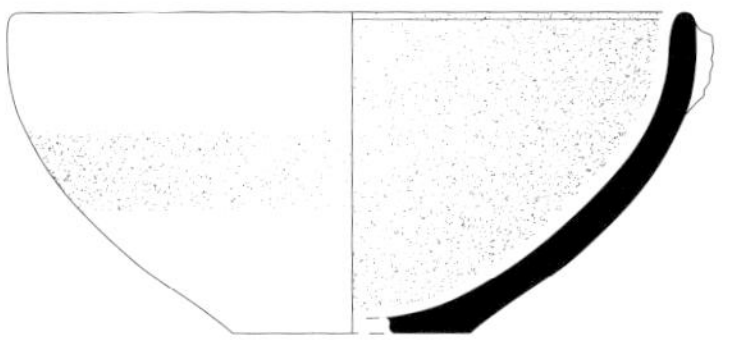

Provenance: Sounding C, Pottery Deposit. Lot: SA 79-988PL. SAV 84. Assemblage 15.

Dimensions and technical features: h 4.3, rim 9.1. Misfired. Thick, opaque, and black gloss; irregular line on the exterior.

Condition: Profile reconstructed from three fragments.

Description: Indistinct rim with flat termination; hemispherical body; beneath the rim, attachment for a horizontal ribbon handle, arched upward. Completely glossed interior; on the exterior, band with irregular borders halfway down the body.

Comparisons and comment: See above, SAV BW 09.

Date: 5th–4th c. BC.

SAV BW 12 Small Bowl/One-handler Type 3.a

1:2

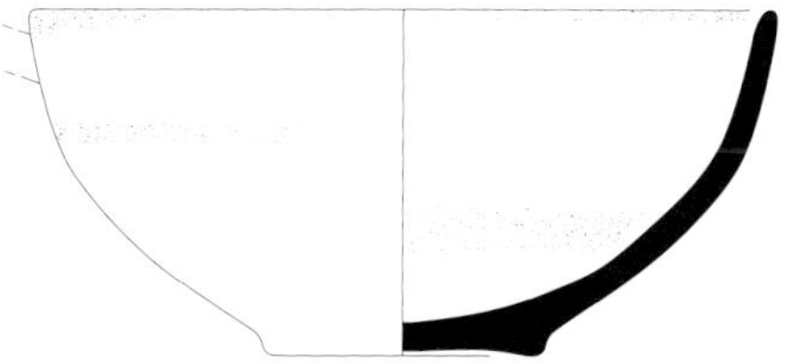

Provenance: Square D4. Level 3, batt. 4. Lot: SA 79-934P. SAV 199. Assemblage 13.

Dimensions and technical features: h 5.0, rim 10.6. Fabric SAV 3, Munsell 5YR 6/8, hard, compact, fine-grained, and well-levigated. Black, thin, and opaque gloss.

Condition: Profile reconstructed from five fragments.

Description: Indistinct rim with rounded termination; truncated conical body; small and barely distinct resting surface with concave lower profile. Strip with irregular profile on the interior and exterior termination of the rim. On the interior, low band on the floor; on the exterior, band with irregular borders halfway down the body.

Comparisons and comment: Marchegiani 1997, 164, no. 13, pl. 57 (Pomarico Vecchio); Carter and Parmly Toxey in *Necropoleis*, 703, T 198-4 (Pantanello necropolis, 410–360 BC); Gassner 2003, 306, Ic11, pl. 10; 317, IIa113, pl. 18 (Elea, 500–475 BC).

Date: 5th–4th c. BC.

SAV BW 13 Small Bowl/One-handler Type 3.b

1:2

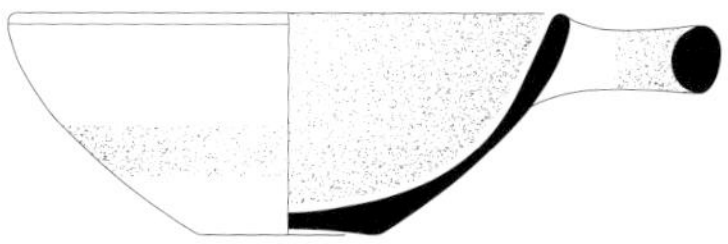

Provenance: Sounding C, Pottery Deposit. Lot: SA 79-993P. SAV 67. Assemblage 15.

Dimensions and technical features: h 3.0, rim 7.7, base 2.3. Fabric SAV 3, Munsell 2.5YR 6/6, hard, compact, fine-grained, and well-levigated. Black, thick, and lustrous gloss applied by brush.

Condition: Profile reconstructed from two fragments.

Description: Indistinct rim with rounded termination; truncated conical body; flat resting surface; small horizontal handle with sub-circular section arched upward and attached immediately beneath the rim. Completely glossed interior; on the exterior, strip halfway down the wall; handle partially glossed by dipping.

Comparisons and comment: Carter and Parmly Toxey in *Necropoleis*, 702, T 95-10 (Pantanello necropolis, 440–400 BC); D'Andria 1980b, 123, pl. XI (Metaponto, 5th c. BC).

Date: 5th–4th c. BC.

SAV BW 14 Small Bowl/One-handler Type 3.b

1:2

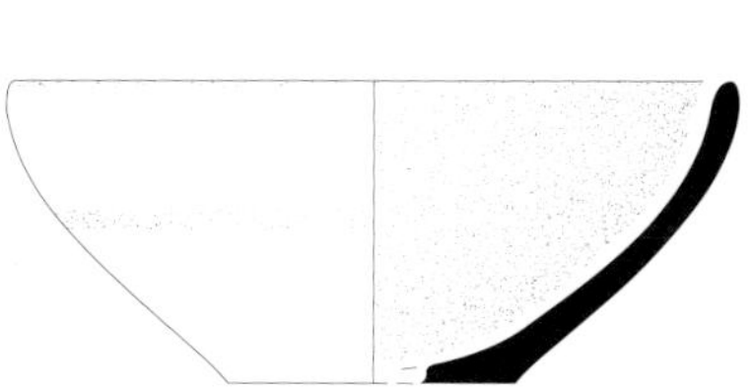

Provenance: Sounding C, Pottery Deposit. Lot: SA 79-988PL-5. SAV 91. Assemblage 15.

Dimensions and technical features: h 4.1, rim 10.0. Fabric SAV 3, Munsell 5YR 6/8, hard, compact, fine-grained, and well-levigated. Black, thin, and opaque gloss.

Condition: Profile reconstructed from three fragments.

Description: Indistinct rim with rounded termination; truncated conical body; flat resting surface. Interior completely glossed; on the exterior, traces of a band.

Comparisons and comment: Carter and Parmly Toxey in *Necropoleis*, 703, T 251-2 (Pantanello necropolis, 460–380 BC); M. Liseno 2004, 92, no. 6, pl. LIV. f, (Metaponto, end of the 5th–beginning of the 4th c. BC); Casalicchio 2012, 334, no. 18, fig. 21 (Rotondella, 520–450 BC); Macnamara 1983a, 337, no. 199, fig. 115 (Cozzo Presepe, 550–500 BC); Silvestrelli forthcoming, fig. 4.7 (Metaponto, 550–500 BC); Russo Tagliente 1995, 90, no. 218, fig. 89 (Armento, first half of the 4th c. BC).

Date: Second half of the 6th–first half of the 4th c. BC.

Skyphoid Cup

Among the open forms with banded decoration attested at Sant'Angelo Vecchio is a series of two-handled cups featuring an indistinct rim, deep body, truncated conical foot, and ribbon-shaped ring-handle or handle with squashed elliptical section attached immediately below the rim. The form is widely documented in southeastern Italy and is common in the black-gloss, Gnathia, and grey-ware repertoires, from which the less common banded version probably derives. Local production in the Metapontino is attested by the end of the 4th c. BC in the *kerameikos* (Silvestrelli forthcoming), while the distribution appears to have been more widespread within the first half of the 3rd c. BC, beyond which point there are no published examples, as demonstrated by numerous tombs at Pantanello and in the urban necropolis (Lo Porto 1966, 197; 1981, 341, 378; Scarano 1992, 19, fig. 5; Elliott in *Necropoleis*, 651; De Siena 2007a, pls. XXII.b, T626; XXIII, T558).

In the same period, the skyphoid cup circulated at Cozzo Presepe (Morel 1970, 102, nos. 4–5, fig. 23; Prag 1983, 352, no. 290, fig. 125), Monte Irsi (*Monte Irsi*, nos. 78–79, fig. 22), and Pomarico (Marchegiani 1997, no. 21, pl. 58). The form is present, furthermore, at Herakleia (Pianu 1990, 228, pl. 5, skyphos type 2; Casagrande 2002, 159, no. 151, fig. 52), Taranto (*Museo Taranto III.1*, 250, fig. 186), and Ginosa (*Museo Taranto II.1*, 168, no. 43.64). At Taranto there are examples of the skyphoid cup until the 2nd c. BC (*Museo Taranto III.1*, 250, fig. 186), while at Gravina the examples continue to the middle of the 1st c. BC (Saunders and du Plat Taylor 1992, no. 221, fig. 10).

The form is also widely documented, with slight variations limited to the handles, in Apulia (Giannotta 1995, 56, fig. 24, cat. no. 25; *Valesio I*, form C35, 85–87; Mastronuzzi 2005, fig. 16.2; a plain example, dated generically by comparanda to the first half of the 4th c. BC, is published in Masiello 2006, 330, pl. 7.a), where a derivation from the black-gloss kantharos of the Apulian tradition has been proposed (*Valesio I*, 85; for a thorough review of the debate on the origin of the form, see Lanza Catti in *Fattoria Fabrizio*, 206).

Skyphoid Cup

SAV BW 15 Skyphoid Cup

Provenance: Square G5, E of wall. Level 2, batt. 1. Lot: SA 79-777PL. Assemblage 3.

Dimensions and technical features: h 2.3, rim 12.4. Fabric SAV 3, Munsell 5YR 6/4, hard, compact, fine-grained, and well-levigated. Black, thin, and opaque gloss.

Condition: One fragment of the rim and body.

Description: Out-turned rim, flat upper surface. Black gloss only on the interior and on the upper surface of the rim.

Comparisons and comment: Carter and Parmly Toxey in *Necropoleis*, 696, T 125-7 (Pantanello necropolis, 325–300 BC); *Valesio I*, 87, n. 122, form C35 (Valesio, late 3rd–first half of the 2nd c. BC).

Date: Second half of the 3rd–first half of the 2nd c. BC.

1:2

Carinated Lekane

The so-called carinated lekanai—characterized by a projecting rim, rather deep and carinated body, and (where preserved) ring or raised disc foot—are among the most common banded forms in all of the Metapontine farmhouses excavated by ICA. These containers can be identified first in the eastern Greek tradition, starting at the end of the 7th c. BC with a wave decoration and becoming common in the 6th c. BC, when they feature the typical banded decoration (Py 1993, 438, GREC-OR-Cp5a; banded carinated cups are present, for example, on Samos: *Samos III*, 95–96, fig. 19, nos. I/11, I/12, fig. 20; 120, no. I/14, fig. 23; *Samos IV*, 158–59, nos. 589–596, pl. 19; on the Ionic examples from Gravisca, see *Gravisca 4*, 240, nos. 499–506, pls. 23–24). Containers with similar morphological features are also found in the Italic Geometric tradition, from which the numerous Italiote examples probably derived (d'Agostino 1968, 105, fig. 20; *Kaulonia 1*, 68–69).

In the Bradano watershed, the lekane appeared in the first half of the 6th c. BC at Cozzo Presepe (du Plat Taylor 1983, 338–39), while in the Metapontino the carinated lekane circulated especially in the 5th and 4th c. BC. Examples dated 420–325 BC were found in the Pantanello necropolis (Carter and Parmly Toxey in *Necropoleis*, 709–10), and numerous fragments were recovered during the ICA field survey between the Bradano and Basento Rivers (Vittoria in *Survey*, 363–65) and among the finds in the excavated farmhouses. Local production is securely attested throughout the 5th c. BC by the finds from the *kerameikos* of Metaponto (Silvestrelli forthcoming, figs. 5.12, 6.20).

The decoration of the carinated lekanai is fairly standardized, consisting generally of wide bands of gloss covering the upper surface of the rim, the exterior surface immediately below the lip, and the upper section of the interior. There is, furthermore, another decorative scheme that has multiple overlapping strips on the interior. Completely undecorated examples are notably rarer in the excavated farmhouses (see Ch. 21, "Plain and Coarse Wares," SAV PCW 05). The varying sizes of the lekanai recovered in the farmhouses indicate their probable use for different functions.

At Sant'Angelo Vecchio, 37 lekanai fragments (29 rim, 1 foot, 7 body) were found that can be attributed to at least 17 individuals. The comparanda suggest a chronology between the second half of the 6th and the first half of the 4th c. BC. The carinated lekanai were classified in three types based on the shape of the rim:

Type 1—flaring, pendant rim;
Type 2—flaring, horizontal rim; and
Type 3—projecting rim.

Type 1—flaring, pendant rim. The first type is characterized by a downturned and slightly pendant projecting rim, flaring, with flat or slightly convex or concave upper surface. The carinated body wall has a straight profile; the foot, where preserved, is a ring or disc with raised floor. At Sant'Angelo Vecchio, 23 fragments (8 individuals) of lekanai of the first type were found. These can be dated between the 5th and the first half of the 4th c. BC. The rim diameter ranges between 23 and 50 cm.

Type 2—flaring, horizontal rim. The second type is characterized by a projecting and horizontal rim with flat and slightly convex or concave upper surface and carinated body. The type is rarely attested in the Metapontine farmhouses: at Sant'Angelo Vecchio, only five examples were found. The examples have diameters ranging from 28 to 40 cm and are dated to the second half of the 6th–5th c. BC.

Type 3—projecting rim. The third type is characterized by an externally projecting rim with flat upper surface and sub-triangular section and a carinated body. The containers ascribed to this type are generally small and encountered rarely in the Metapontine farmhouses. A single individual found at Sant'Angelo Vecchio belongs to this type.

Carinated Lekanai

SAV BW 16 Carinated Lekane Type 1

Provenance: House on the Hill, Trench 3. Level 1, batt. 2 + surface finds 10 m from southwest corner of abandoned farmhouse, north of Sounding A. Lot: SA 79-1035P-B + SAV79-39PL (see *Fig. 2.39*).

Dimensions and technical features: h 5.3, rim 23.0. Fabric SAV 1, Munsell 2/5YR 6/8, hard, compact, and fine-grained, with tiny black and white stone inclusions. Powdery surface. Black and opaque gloss.

Condition: Eleven fragments of the rim and body.

Description: Flaring and pendant rim with rectangular section and flat upper profile; pronounced carination of the body, beneath which the wall becomes oblique. Band on upper surface of rim and internal narrow band at level of the rim.

Comparisons and comment: Mastronuzzi 2013, 91, no. 131, fig. 56 (Oria, Monte Papalucio sanctuary, Votive Deposit no. 3, 480–470 BC).

Date: First half of the 5th c. BC.

1:2

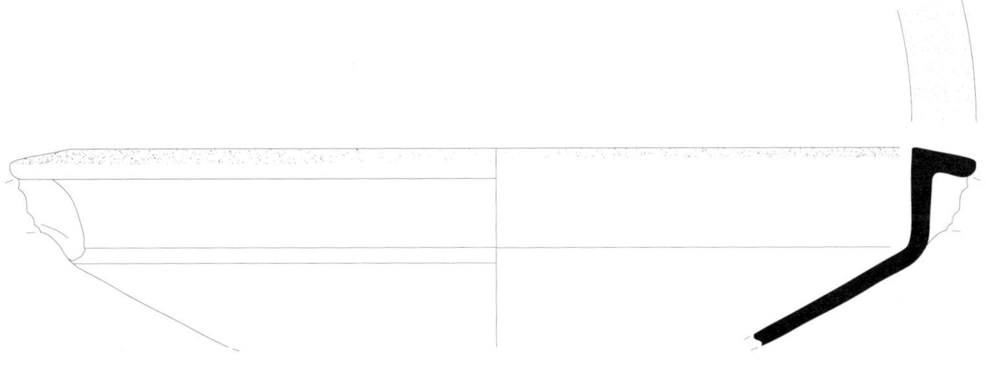

Carinated lekanai cont.

1:3

SAV BW 17 Carinated Lekane Type 1
Provenance: Sounding C, Pottery Deposit. Lot: SA 79-988PL. SAV 80. Assemblage 15.
Dimensions and technical features: h 4.5, rim ext. 24.5. Fabric SAV 3, Munsell 5YR 6/4, hard, compact, fine-grained, and well-levigated. Brown-black, thin, and opaque gloss, dilute on the interior.
Condition: Two fragments of the rim and body.
Description: Flaring and pendant rim with rectangular section and slightly convex upper profile; pronounced carination of the body, beneath which the wall becomes oblique. Band on the rim and on the interior band corresponding to the rim.
Comparisons and comment: Kaulonia I, 68–69, no. 274, fig. 45 (Caulonia); Silvestrelli forthcoming, fig. 5, no. 12 (Metaponto, from deposits 19 and 22, 500–450 BC); Vittoria in *Survey*, 365, CLne6c, no. 40; Ruga et al. 2005, 162, no. 8, tav. LXIV, fig. 26/8 (Croton, first half of the 5th c. BC).
Date: 5th–first half of the 4th c. BC.

SAV BW 18 Carinated Lekane Type 1
Provenance: Sounding C, Pottery Deposit. Lot: SA 79-988PL-2. SAV 79. Assemblage 15.
Dimensions and technical features: h 4.2, rim ext. 51.0. Fabric SAV 3, Munsell 5YR 7/6, hard, compact, fine-grained, and well-levigated. Brown-black, dilute, and opaque gloss.
Condition: One fragment of the rim and body.
Description: Flaring and pendant rim with rectangular section, slight interior protrusion, and slightly convex upper profile; pronounced carination of the body, beneath which the wall becomes oblique. Band on the upper surface of the rim; on the interior, strip beneath the rim.
Comparisons and comment: See above, SAV BW 17.
Date: 5th–first half of the 4th c. BC.

1:3

SAV BW 19 Carinated Lekane Type 1

Provenance: Square D4 W. Level 3, batt. 5. Lot: SA 79-907PL. SAV 176. Assemblage 13.

Dimensions and technical features: h 5.6, rim int. 27.4. Fabric SAV 3, Munsell 5YR 7/4–5YR 3/1, hard, compact, fine-grained, and well-levigated. Brown, dilute, and opaque gloss on the exterior; brown, thin, and opaque gloss on the interior.

Condition: One fragment of the rim and body.

Description: Flaring and pendant rim with rectangular section and flat upper profile; pronounced carination of the body, beneath which the wall becomes oblique; horizontal handle attachment. Band on upper surface of rim and internal band reaching the carination. Band on upper surface of rim and internal band at level of the rim.

Comparisons and comment: Carter and Parmly Toxey in *Necropoleis*, 711, no. T343-1 (Pantanello necropolis, 420–360 BC); Vittoria in *Survey*, 365, CLne7, no. 41; M. Liseno 2004, 94, no. 25, pl. LVIII, c (Metaponto, 420–360 BC); similar to Silvestrelli forthcoming, fig. 6, n. 20 (Metaponto, from deposits 19 and 22, 450–400 BC).

Date: Second half of the 5th–first half of the 4th c. BC.

SAV BW 20 Carinated Lekane Type 2

Provenance: Sounding B. Level 3, batt. 8. Lot: SA 79-366P. SAV 154 (see *Fig. 2.35*).

Dimensions and technical features: h 3.6, rim ext. 40.2. Fabric SAV 3, Munsell 5YR 7/6, hard, compact, fine-grained, and well-levigated. Black, thick, and opaque gloss on the exterior; black, dilute, and opaque gloss on the interior.

Condition: One fragment of the rim and body.

Description: Flaring and horizontal rim with rectangular section and slightly inclined upper profile; pronounced carination of the body, beneath which the wall becomes oblique. Band on upper surface of rim and internal wide band reaching the carination.

Comparisons and comment: Vittoria in *Survey*, 365, CLne6a, no. 38; similar in morphology, but different in size and decorative syntax to an example from the Metapontine chora, Lecce 2010–2011, 24, type 1a, fig. 6.3 (Metaponto, second half of the 6th–first half of the 5th c. BC).

Date: Second half of the 6th–5th c. BC.

Carinated lekanai cont.

SAV BW 21 Carinated Lekane Type 2
Provenance: Sounding C, Pottery Deposit. Lot: SA 79-988PL. SAV 81. Assemblage 15.
Dimensions and technical features: h 3.1, rim ext. 28.6. Fabric SAV 1, Munsell 5YR 7/6, hard, compact, and fine-grained, with tiny black and white stone inclusions. Powdery surface. Brown-black, thin, and opaque gloss.
Condition: One fragment of the rim and body.
Description: Flaring, horizontal rim with rectangular section and flat upper profile; pronounced carination of the body, beneath which the wall becomes oblique. Band on upper surface of rim and internal wide band reaching the carination.
Comparisons and comment: See above, SAV BW 20.
Date: Second half of the 6th–5th c. BC.

1:2

SAV BW 22 Carinated Lekane Type 3
Provenance: Sounding C, Pottery Deposit. Lot: SA 79-988PL. SAV 82. Assemblage 15.
Dimensions and technical features: h 3.6, rim 15.7. Fabric SAV 3, Munsell 5YR 7/3, hard, compact, fine-grained, and well-levigated. Brown-black, thin and opaque gloss.
Condition: Two fragments of the rim and body.
Description: Flaring, pendant rim with triangular section and flat upper profile; pronounced carination of the body, beneath which the wall becomes oblique. Poorly defined bands made with very dilute gloss on the brim, on the interior immediately beneath the lip, and near the bottom of the body wall. Traces of exposure to fire.

1:2

Comparisons and comment: Kaulonia 1, 68–69, no. 272, fig. 45 (Caulonia). Mastronuzzi 2013, 69, no. 18, fig. 36 (Oria, Monte Papalucio sanctuary, votive deposit no. 1, 480–470 BC).
Date: 5th c. BC.

SAV BW 23 Carinated Lekane Type 3

Provenance: Sounding C, Pottery Deposit. Lot: SA 79-988PL.

Dimensions and technical features: h 6.4, base 11.7. Fabric SAV 3, Munsell 5YR 7/6, hard, compact, fine-grained, and well-levigated. Brown, thin, and opaque gloss.

Condition: One fragment of the foot and body.

Description: Slightly raised disc foot attached to an oblique wall. On the interior, narrow brown-red band; on the interior floor, traces of a thin circle of gloss.

Comparisons and comment: Carter and Parmly Toxey in *Necropoleis*, 709, T 7-2 and T 15-4 1 (Pantanello necropolis, 420–360 BC).

Date: 5th–4th c. BC.

1:2

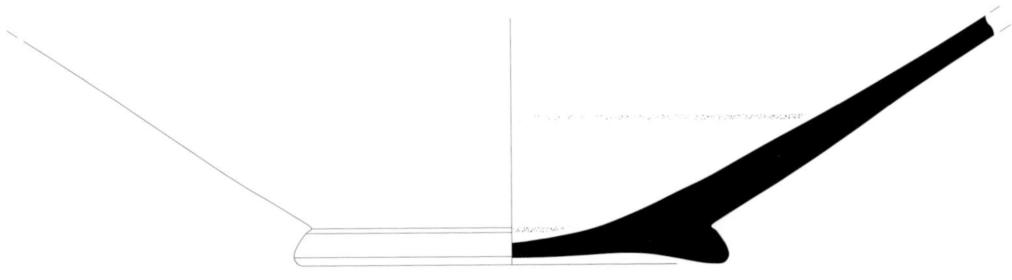

Large Bowl

The banded ceramic assemblages recovered at Sant'Angelo Vecchio include two individuals of a wide, open container featuring a flaring and horizontal lip with slightly pendant edge (SAV BW 24 and SAV BW 25). The identified parallels are rather scant and occasionally shared with plain examples, and a function related to serving foodstuffs is possible. SAV BW 24 is made with a fabric rich in inclusions that is not typical of banded ware.

Large Bowls

SAV BW 24 Large Bowl Type 1

Provenance: Square G7 W of wall 4, Level 2, batt. 1. Square G6 W of wall 1, Level 2, batt. 2. Lot: SAV 79-793PL + SAV 79-796PL + SAV 79-749PL. SAV 39 (see *Fig. 2.20*).

Dimensions and technical features: h 3.7, rim int. 28.4. Fabric SAV 5, Munsell 2.5YR 6/6, very hard and compact, with occasional vacuoles in fracture and many tiny white stone inclusions. Traces of red-brown, thin, and opaque gloss.

Condition: Ten fragments of the rim and body.

Description: Thin and flaring rim with convex upper profile; curving wall. On the wall, groove incised with a stick; on the rim, traces of gloss.

Comparisons and comment: Precise parallels could not be identified; a generic parallel can be found in Caravelli 2008, 476, no. 756, fig. 317 (Civita di Tricarico, 4th–3rd c. BC).

Date: 4th–3rd c. BC.

1:2

Large bowls cont.

SAV BW 25 Large Bowl Type 1

1:2

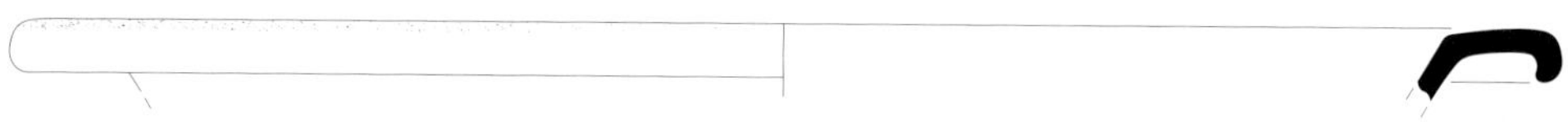

Provenance: Square G7, E of wall. Level 2, batt. 2. Lot: SA 79-797PL. SAV 41. Assemblage 17.

Dimensions and technical features: h 1.2, rim 25.8. Fabric SAV 3, Munsell 2.5YR 6/6, hard, compact, fine-grained, and well-levigated. Brown, thin, and opaque gloss.

Condition: One fragment of the rim.

Description: Flaring and horizontal rim with convex upper surface and hanging, slightly inward-projecting edge. The entire rim surface is covered by a band, which spills into the interior.

Comparisons and comment: De Siena 1980, 86, pl. VIII.10 (Metaponto, Sanctuary of Apollo Lykeios, end of the 4th–3rd c. BC); Amore 1992–1993, 110, tav. V (Metaponto, Fattoria Musillo, 4th–3rd c. BC); similar to Vittoria in *Survey*, 361, no. 26, Pl/Lne 4c.

Date: 4th–3rd c. BC.

Deep Bowl

Two samples of a container characterized by a very deep body and a high and flaring rim were found in Sounding C (SAV BW 26) and in Sounding B (uncataloged). No parallels could be found.

The type is characterized by a high, flaring rim with flat upper surface mounted via a clear separation on a deep body. The interior is completely glossed; the upper surface of the rim has a band.

Deep Bowl

SAV BW 26 Deep Bowl

Provenance: Sounding C. Level 2, batt. 7. Lot: SA 79-205PL. SAV 126 (see *Fig. 2.35*).

Dimensions and technical features: h 5.5, rim ext. 30.2. Fabric SAV 3, Munsell 5YR 7/3, hard, compact, fine-grained, and well-levigated. Black, dilute, and opaque gloss.

Condition: One fragment of the rim and body.

Description: Flaring lip and rim with flat upper surface; slightly convex wall. The interior is completely glossed except for the band on the lip between the rim and the start of the body wall. Band on the upper surface of the rim.

1:2

Dish

The dish SAV BW 27, found in squares G5 and H5, can be associated morphologically with the deep plates, although it is smaller in size (diam. 16 cm). This type of dish, for which close parallels could not be found in banded ware, is similar to black-gloss forms found in both the field survey of the Metapontine chora (Lanza Catti et al. in *Survey*, 241, no. 128) and the Pantanello sanctuary.

Dish

SAV BW 27 Dish

Provenance: Square G5, E of wall, Level 2, batt. 1. Square H5. Level 1, batt. 2. Lot: SA 79-777PL + SAV 79-546PL. SAV 145 + SAV 42. Assemblage 3.

Dimensions and technical features: rim 16.0. Fabric SAV 1, Munsell 5YR 7/6, hard, compact, and fine-grained, with tiny white and black stone inclusions. Powdery surface. Brown, dilute, and opaque gloss.

Condition: Two fragments of the rim and body, not joining.

1:2

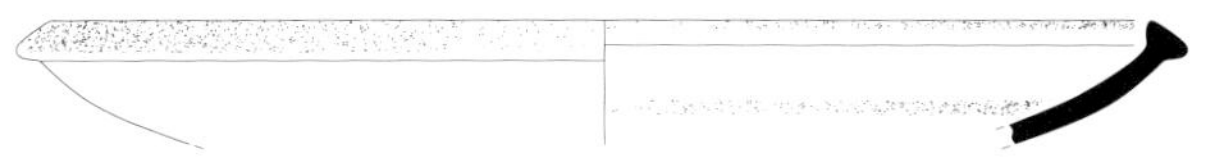

Description: Small, slightly thickened, and protruding rim with inclined upper surface; convex wall. Band on the rim.

Plate

The plate with a barely raised and footless floor, inverted wall, and rounded rim is documented in both banded and plain versions in the Metapontino. The decoration of the plate often consists of an interior band, which reaches the start of the body wall, and an exterior strip at the level of the rim. Other common decorative patterns are known, however, some of which have the entire interior surface covered in gloss or a combination of glossed interior floor and multiple interior bands.

Metapontine production of footless plates is documented, at least between the end of the 4th and the beginning of the 3rd c. BC, by the finds in the urban *kerameikos* (Silvestrelli forthcoming, fig. 14). Containers of this type are rare in Metapontine necropoleis (Scarano 1992, 20, tt. 29–30, fig. 6), but they are present, albeit in fairly meager quantities, in domestic and sacred contexts (Morel 1970, 99, fig. 24.14–15; Prag 1983, 358, no. 329, fig. 129; Postrioti 1996, fig. 10.c; Bianco and Deodato 1997, 83, pl. 75.122–23; M. Liseno 2004, 97, pl. LXI.d; Vittoria in *Survey*, 357–58). Similar plates have also been found in burials at Herakleia and Taranto, although they are rare (Giardino 1990, 114, pl. XVIII; Giardino 1998, 78, tomb Belluno 9B, pl. XVIII; Pianu 1990, 334, pl. LXXV, no. 278.1; *Museo Taranto III.1*, 272, fig. 204; Casagrande 2002, 144, no. 97, fig. 49; 173, no. 218, fig. 50).

The comparanda identified in Basilicata indicate a chronology between the second half of the 4th and the end of the 3rd c. BC, while examples dating as late as the 1st c. BC come from Taranto (*Museo Taranto III.2*, 197, fig. 38.8). Three fragments of dishes belonging to as many individual vessels were identified at Sant'Angelo Vecchio (two uncataloged).

Plate

SAV BW 28 Plate

Provenance: Square G6, W of wall. Level 2, batt. 2. Lot: SA 79-749PL. SAV 64 (see *Fig. 2.20*).

Dimensions and technical features: h 1.9, base 22.4. Fabric SAV 1, Munsell 5YR 7/6, hard, compact, and fine-grained, with tiny black and white stone inclusions. Powdery surface. Brown-black, thin, and opaque gloss.

Condition: Profile.

Description: Concave wall; rim distinguished by an indentation in the upper surface of the wall, with oblique exterior edge. On the interior, glossed wall; on the exterior, band on the exterior profile of the rim. Band on the floor.

Comparisons and comment: Morel 1970, 99, fig. 24.14 (Cozzo Presepe, end of the 4th c. BC); Lo Porto 1973, 224, no. 2.5, pl. LXIX (Lucignano, 4th–3rd c. BC); G. Greco 1991, 51, fig. 116, black-gloss version, (Serra di Vaglio, 320–250 BC); similar to Scarano 1992, 20, tt. 29–30, fig. 6 (Metaponto, Hellenistic necropolis, first half of the 3rd c. BC); *Museo Taranto III.1*, 272, fig. 204 (Taranto, 250–200 BC); Postrioti 1996, fig. 10.c (Metaponto, Temple E votive deposit, beginning of the 3rd c. BC); similar to Bianco and Deodato 1997, 83, pl. 75.122, plain-ware version, (Pomarico Vecchio, 4th–3rd c. BC); Casagrande 2002, 144, no. 97, fig. 49 (Herakleia, end of the 4th–3rd c. BC); M. Liseno 2004, 97, pl. LXI.d (Metaponto, Favale votive deposit, 4th c. BC); *Civita di Tricarico*, 425, nos. 411–13, fig. 300 (Civita di Tricarico, first half of the 3rd c. BC); Vittoria in *Survey*, 357, Pl1 (chora of Metaponto)

Date: 4th–3rd c. BC.

1:2

Jugs

The medium-sized and large containers for liquids, generally featuring a generous capacity and wide resting surface, are one of the most common banded forms in the excavated farmhouses, in particular one- and two-handled oinochoai with circular rims, pelikai, hydriai, and table amphorae. The number of rims, feet, and handles belonging to these forms recovered in the excavated farmhouses is very high, but no complete vessel or example with complete profile is preserved. In the majority of cases, therefore, the evolution of the form, the proportional relationships between the various parts, and the arrangement of the handles cannot be reconstructed. Because of this, the fragments potentially belonging to oinochoai with circular rims, pelikai, hydriai, and table amphorae are presented here as a group, with suggestions as to the specific identity of each individual, where relevant.

The large number of fragments ascribed to this group suggests widespread use of these vessels for containing liquids in all of the contexts investigated, where they doubtless responded to various needs. The rare small examples were perhaps used at the table, while the more common medium-sized and large examples were probably used for storing liquid foodstuffs. Thus the functional landscape of the banded vessels is very different from that of their plain counterparts, where the small and medium-sized jugs, better suited for regular use at the table, are more numerous (see Ch. 21, "Plain and Coarse Wares"). Further confirmation is found in the neat distinction in the morphological repertoire between the banded and plain classes, with few rare exceptions, giving the banded forms examined here unique structural features.

The excavations yielded 36 fragments of jugs, of which 21 were classified as rims, nine as handles, three as body sherds, and three as ring or truncated conical feet, belonging to 18 individuals. The decoration generally is found on the upper surface of the rim and its interior and exterior profiles; on the foot, however, the decoration consists of one or more narrow bands on the exterior profile or at the start of the body wall. The low fragment counts (see above, p. 279, "Materials: Quantitative and Distributive Analysis") are probably due to the inclusion of banded foot fragments lacking traces of decoration among the plain wares. Feet belonging to medium-sized and large jugs are notably absent from the group, although such vessels are well attested by rims with large diameters. The jugs found at Sant'Angelo Vecchio were divided into 4 types:

Type 1—thickened rim with sub-rectangular section;
Type 2—thickened rim with sub-triangular section;
Type 3—flaring rim; and
Type 4—flaring, thickened rim with elliptical section.

Type 1—thickened rim with sub-rectangular section. The first type is characterized by a thickened rim with sub-rectangular section mounted via a clear separation on a cylindrical neck which generally has a concave exterior profile. The exterior profile of the rim can be convex (variant 1.a) or straight and oblique (variant 1.b, not present at Sant'Angelo Vecchio). The handle, where preserved, is mounted either on the rim, indicating a one-handled pitcher, or on the neck, indicating a pelike or amphora for the table and pantry. The decoration is generally applied by brush on the interior and exterior surfaces of the rim and characterized by irregular borders, frequent dribbles, and evident overlapping of the brush-strokes. Examples of decoration by dipping are rarer (SAV BW 29); in these cases the glossed portion of the exterior is more extensive.

Similar containers, the profiles of which can be reconstructed through the complete examples recovered from the Pantanello necropolis (Carter and Parmly Toxey in *Necropoleis*, 711–14), occur at many Italiote sites starting in the Late Archaic period and are very common in Basilicata between the end of the 6th and the end of the 4th c. BC. At Sant'Angelo Vecchio, Type 1 is attested by five examples, none of which preserves a complete profile.

Type 2—thickened rim with sub-triangular section. The second type is characterized by a thickened rim with sub-triangular section and variable exterior profile: convex (variant 2.a), straight (variant 2.b, not present at Sant'Angelo Vecchio), or concave (variant 2.c). The rim is mounted via a clear separation on the body with cylindrical section. The handle is attached to either the rim or the neck, suggesting that the rims belonged to one-handled jugs or table amphorae/pelikai. The decoration is applied by brush over all of the interior and exterior of the rim but not on the neck.

Containers with these characteristics circulated very widely, and the type is one of the best attested in the Metapontine farmhouses. The parallels from the Bradano watershed suggest a chronology between the last quarter of the 4th and the beginning of the 3rd c. BC. Four individuals were recovered at Sant'Angelo Vecchio.

Type 3—flaring rim. The third type is characterized by a flaring rim mounted seamlessly on a cylindrical neck; the rim can be indistinct (variant 3.a) or slightly projecting (variant 3.b). The handle attachments, where (rarely) preserved, are located on both the rim and the neck, so that this type of rim, too, could be used to make one-handled jugs or table amphorae/pelikai. The fairly precise decoration, applied by brush or dipping (SAV BW 38), covers the interior and exterior surfaces of the rim.

This profile is the least common in the Metapontine farmhouses, and more generally it did not circulate widely in the known contexts of Basilicata. The comparanda come from contexts dated to the 4th and the beginning of the 2nd c. BC. Three individuals represented by four fragments among the materials from Sant'Angelo Vecchio can be ascribed to this type. The rims have diameters ranging from 6.6 to 13.2 cm, indicating small and medium-sized containers.

Type 4—flaring, thickened rim with elliptical section. The fourth type is characterized by a flaring, thickened rim with elliptical section mounted on a truncated conical neck. The upper surface can have a convex (variant 4.a) or concave profile (variant 4.b). The rim is attached to a generally truncated conical neck, mounted on the shoulder via a clear separation. The rims belonging to this type are on the whole not commonly encountered in the farmhouses. The decoration consists of one or more narrow bands especially on the upper surface, and occasionally on the exterior and interior profiles. None of the recovered fragments preserves the handle attachment, but the identified parallels are mostly hydriai with a chronology between the 6th and 4th c. BC. Two individuals from Sant'Angelo Vecchio can be ascribed to this type.

Jugs

SAV BW 29 Jug Type 1.a

1:2

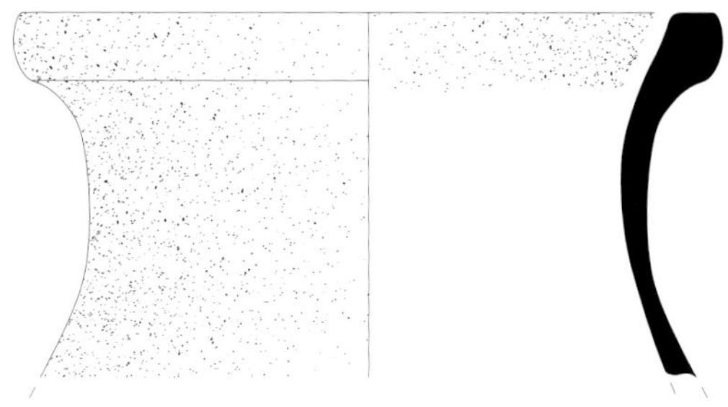

Provenance: Sounding C, Pottery Deposit. Lot: SA 79-1067P. SAV 257. Inv. no. 319958. Assemblage 15.

Dimensions and technical features: h 6.7, rim 13.0. Fabric SAV 3, Munsell 2.5 YR 7/6, hard, compact, fine-grained, and well-levigated. Black, thick, and opaque gloss.

Condition: Two fragments of the rim and neck.

Description: Massive rim with a thickened flange that has a convex exterior profile, mounted on a neck with concave walls. Extant exterior completely slipped to the upper interior.

Comparisons and comment: Swift forthcoming, PZ 77.0965.1 (Pantanello, from the area of the spring, a context dating to the 5th c. BC, perhaps Late Archaic, Phase 2), similar to Carter and Parmly Toxey in *Necropoleis*, 711, 300-2 (Pantanello necropoleis, pelike, 460–440 BC).

Date: 5th c. BC.

SAV BW 30 Jug Type 1.a

1:2

Provenance: Surface find. Lot: SA 79-7PL. SAV 181.

Dimensions and technical features: h 3.2, rim 12.5. Fabric SAV 3, Munsell 5YR 7/4, hard, compact, fine-grained, and well-levigated. Brown, thin, and opaque gloss.

Condition: One fragment of the rim and neck.

Description: Thickened rim with sub-rectangular section, rounded exterior profile, and upper surface inclined toward the interior; truncated conical neck. On the interior, band with irregular borders at the level of the rim; on the exterior, band with irregular profile covering most of the rim.

Comparisons and comment: Saunders and du Plat Taylor 1992, 36, no. 359, fig. 20 (Gravina di Puglia, 4th–3rd c. BC); for the general shape, see Vittoria in *Survey*, 386, J8b, no. 133 (chora of Metaponto).

Date: Perhaps 5th–3rd c. BC.

SAV BW 31 Jug Type 1.a

1:2

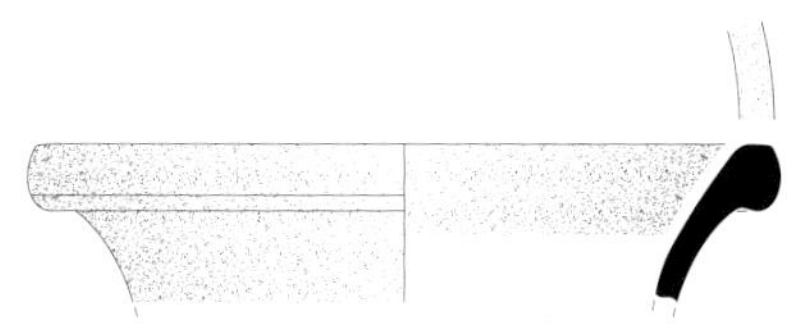

Provenance: Sounding C, Pottery Deposit. Lot: SA 79-988PL. SAV 86. Assemblage 15.

Dimensions and technical features: h 2.0, rim 9.5. Fabric SAV 1, Munsell 5YR 7/4, hard, compact, and fine-grained, with tiny white and black stone inclusions. Powdery surface. Brown-black, thin, and opaque gloss.

Condition: One fragment.

Description: Thickened rim with flat upper surface and vertical exterior edge. Band on the upper surface of the rim.

Comparisons and comment: Silvestrelli forthcoming, fig. 6.21 (Metaponto *kerameikos*, deposits 19 and 22, second half of the 5th c. BC).

Date: 5th c. BC.

SAV BW 32 Pelike (?) Type 1.a

1:2

Provenance: House on the Hill. Lot: SA 79-1062PL. SAV 55. Assemblage 1.

Dimensions and technical features: h 3.5, rim int. 15.5. Fabric SAV 1, Munsell 5YR 7/6, hard, compact, and fine-grained, with tiny white and black stone inclusions. Powdery surface. Brown, dilute, and opaque gloss.

Condition: One fragment of the rim and neck.

Description: Rim with thickened flange, upper surface slightly inclined toward the interior, concave interior profile, and convex exterior profile; neck with vertical wall. On the interior and exterior, wide band with irregular borders at the level of the rim. No traces of handles, but the form is compatible with the pelike based on comparanda.

Comparisons and comment: Silvestrelli forthcoming, fig. 5.13 (Metaponto *kerameikos*, from a context of the first half of the 5th c. BC); Carter and Parmly Toxey in *Necropoleis*, 711, T324-2 (Pantanello necropolis, pelike, 460–440 BC); Vittoria in *Survey*, 386, no. 130, type J7a2 (chora of Metaponto).

Date: 5th c. BC.

SAV BW 33 Jug Type 2.a

1:2

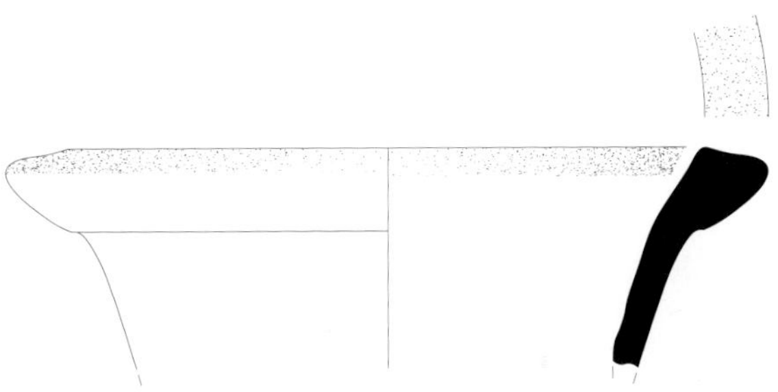

Provenance: Square G5. Level 1, batt. 2. Lot: SA 79-600PL. SAV 32. Assemblage 3.

Dimensions and technical features: h 4.3, rim 12.6. Fabric SAV 1, Munsell 10YR 7/4, hard, compact, and fine-grained, with tiny white and black stone inclusions. Powdery surface. Traces of thin and opaque gloss on the rim and interior surface.

Condition: One fragment of the rim and truncated conical neck.

Description: Flanged rim with concave upper surface and convex exterior profile; neck. On the interior, band with irregular profile at the level of the rim.

Comparisons and comment: Similar to Marchegiani 1997, 169–70, no. 44, pl. 62 (Pomarico Vecchio, 320–280 BC).

Date: Perhaps 4th–beginning of the 3rd c. BC.

SAV BW 34 Jug Type 2.a

1:2

Provenance: Square D4 within stone circle. Level 3, batt. 5. Lot: SA 79-896PL. SAV 186. Assemblage 13.

Dimensions and technical features: h 5.2, rim ext. 11.8. Fabric SAV 3, Munsell 7.5YR 7/4, hard, compact, fine-grained, and well-levigated. Brown, thin, and opaque gloss.

Condition: One fragment of the rim and neck.

Description: Flanged rim with concave upper surface and convex exterior profile; vertical neck. On the rim and interior, band with irregular profile at the level of the rim.

Comparisons and comment: Similar to du Plat Taylor 1983, 345, n. 265, fig. 121 (Cozzo Presepe, site A, topsoil).

Date: From levels of Phase 4.

Jugs cont.

SAV BW 35 Jug Type 2.c

1:2

Provenance: Slope 3, W extension topsoil. Level 1, batt. 4. Lot: SA 79-83PL. SAV 97 (see *Fig. 2.35*).

Dimensions and technical features: h 4.2, rim ext. 14.0. Fabric SAV 1, Munsell 7.5YR 7/4, hard, compact, and fine-grained, with tiny white and black stone inclusions. Powdery surface.

Condition: One fragment of the rim and neck.

Description: Rim with sub-triangular section, protruding edge, flaring exterior profile, and slightly concave upper surface; small overhang beneath the rim on the exterior, truncated conical neck.

Comparisons, comment, and date: See above, SAV BW 33.

SAV BW 36 Jug Type 2.c

1:2

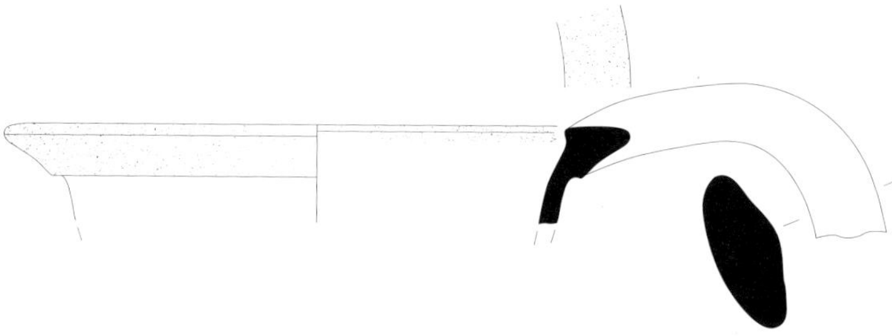

Provenance: Slope 3, W extension. Level 2, batt. 1. Lot: SA 79-132PL. SAV 96. Assemblage 11.

Dimensions and technical features: h 2.1, rim 13.5. Fabric SAV 1, Munsell 5YR 7/6, hard, compact, and fine-grained, with tiny white and black stone inclusions. Powdery surface. Traces of brown-red, thin, and opaque gloss on the handle and rim.

Condition: Two fragments of the rim, neck and handle.

Description: Rim with sub-triangular section, flaring exterior profile, and slightly concave upper surface; slight constriction on the interior; small overhang beneath the rim on the exterior; attachment of the vertical ribbon handle at the rim. Traces of gloss on the rim and handle.

Comparisons and comment: Similar to Vittoria in *Survey*, 390, no. 149, J17a (chora of Metaponto).

SAV BW 37 Jug Type 3.a

1:2

Provenance: House on the Hill. Lot: SA 79-1062PL. SAV 60. Assemblage 1.

Dimensions and technical features: h 3.25, rim int. 13.2. Fabric SAV 1, Munsell 5YR 7/6, hard, compact, and fine-grained, with tiny white and black stone inclusions. Powdery surface. Traces of black, thin, and opaque gloss.

Condition: Two fragments of the rim and attachment of the neck.

Description: Indistinct, flaring, and slightly thickened rim with rounded termination. Vertical ribbon handle, arched upward and attached at the rim. Traces of gloss on the handle and exterior of the rim.

Comparisons and comment: Swift forthcoming a, PZ 76.0020.5 (Pantanello, from the area of the sanctuary, context dating to the 4th-c. BC oikos, likely before ca. 320 BC); Lecce 2012, 36, fig. 10.43 (Metaponto, 5th c. BC).

Date: 5th–4th c. BC.

SAV BW 38 Jug Type 3.b

1:2

Provenance: Sounding C, Pottery Deposit. Lot: SA 79-988PL. SAV 85. Assemblage 15.

Dimensions and technical features: h 3.6, rim 6.6. Fabric SAV 3, Munsell 5YR 7/6, hard, compact, fine-grained, and well-levigated. Red, thick, and opaque gloss.

Condition: One fragment of the rim and neck.

Description: Thickened and slightly flaring rim with rounded termination turned outward above an overhang; concave neck. Internal band with irregular profile at the level of the rim; on the exterior, uniform coating of gloss.

Comparisons and comment: Similar to Carter and Parmly Toxey in *Necropoleis*, 708, T350-11 (Pantanello necropolis, 460–440 BC) is an example of a pitcher featuring a similar rim mounted on a slight overhang, but the closest parallel is Vittoria in *Survey*, 385, no. 127, J6 (chora of Metaponto).

Date: 5th c. BC.

SAV BW 39 Jug Type 3.b

1:2

Provenance: Slope 3, W extension topsoil. Level 1, batt. 4. Lot: SA 79-83PL. SAV 98 (see *Fig. 2.35*).

Dimensions and technical features: h 4.5, rim 13.6. Fabric SAV 1, Munsell 5YR 7/6, hard, compact, and fine-grained, with tiny white and black stone inclusions. Powdery surface. On the exterior, brown-black, thin, and opaque gloss; on the interior, black, thick, and opaque gloss.

Condition: One fragment of the rim and neck.

Description: Flaring and tapering flanged rim with concave upper profile. On the interior, narrow band at the level of the rim and another at the neck; on the exterior, narrow band on the lower edge of the rim and another, irregular band on the neck.

Comparisons and comment: None found, at present.

Date: From levels of Phase 4.

SAV BW 40 Hydria Type 4.a

1:2

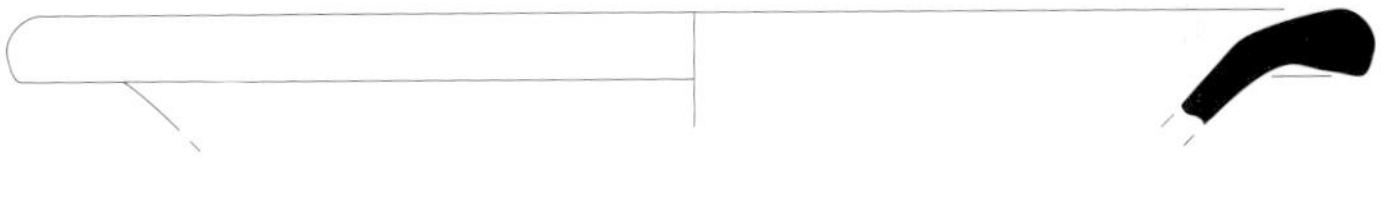

Provenance: House on the Hill. Lot: SA 79-1062PL. SAV 59. Assemblage 1.

Dimensions and technical features: h 1.5, rim 18.8. Fabric SAV 1, Munsell 7.5YR 8/4, hard, compact, and fine-grained, with tiny white and black stone inclusions. Powdery surface. Faint traces of brown-black, thin, and opaque gloss.

Condition: One fragment of the rim and attachment of the neck.

Description: Flaring, horizontal, and thickened rim with convex upper profile; oblique neck and walls. Traces of gloss on the rim.

Comparisons and comment: Swift forthcoming, PZ 76.0187.1, Pantanello sanctuary from the general area of the Late-Archaic sanctuary structure (from the first quarter of the 5th c. BC); Vittoria in *Survey*, 382, no. 112 (chora of Metaponto).

Date: Perhaps 5th c. BC.

Jugs cont.

SAV BW 41 Hydria Type 4.b

1:2

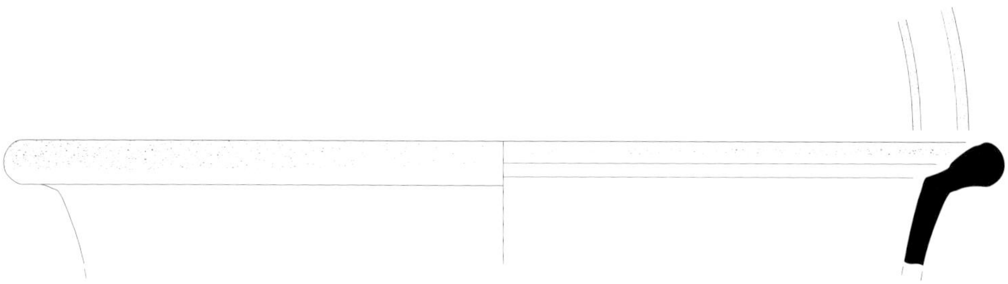

Provenance: Square G7, from sand below N-S wall level 4. From cleaning. Lot: SA 80-30PL. SAV 18.

Dimensions and technical features: h 3.1, rim 25.0. Fabric SAV 1, Munsell 5YR 7/3, hard, compact, and fine-grained, with tiny white and black stone inclusions. Powdery surface. Black, thin, and opaque gloss.

Condition: One fragment of the rim and neck.

Description: Flaring, thickened rim with rounded exterior edge and convex upper surface, narrowing at the attachment of the body wall; truncated conical neck. On the rim, two painted lines; on the exterior, band at the level of the rim.

Comparisons and comment: Similar examples, but with a different inclination of the rim: du Plat Taylor 1983, 343, nos. 235, 237, fig. 119 (Cozzo Presepe, first half of the 6th c. BC.); Vittoria in *Survey*, 380, H1b (chora of Metaponto).

Date: 600/575–480 BC, based on comparisons.

Foot

SAV BW 42 Jug

1:2

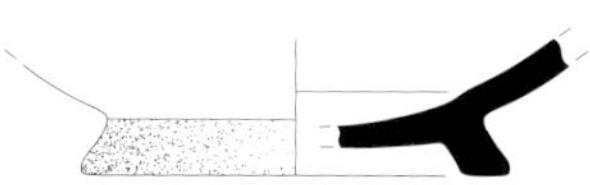

Provenance: Square G6, E of wall. Level 2, batt. 3. Lot: SA 79-986PL. SAV 183. Assemblage 3.

Dimensions and technical features: h 1.8, base 5.4. Fabric SAV 3, Munsell 5YR 6/4, hard, compact, fine-grained, and well-levigated. Brown-black, thin, and opaque gloss.

Condition: One fragment of the foot and attachment of the body wall.

Description: Ring foot with oblique exterior profile; convex walls. Band on the exterior of the foot.

Olpe

The excavations at Sant'Angelo Vecchio yielded fragments of three banded olpai featuring a circular rim, flat resting surface, and baggy body which can be confidently ascribed to the vessels of East Greek tradition that had the greatest success in southern Italy between the second half of the 6th and the 3rd c. BC. Metapontine production of small banded olpai is attested in the *kerameikos* of Metaponto beginning in the first half of the 5th c. BC (Silvestrelli forthcoming, fig. 4.8).

The form is rarely encountered in the excavated farmhouses, limited mostly to small fragments of the rim or base that do not offer enough evidence for a reliable typological seriation in either banded or plain ware. The results of the distribution analysis of the olpai are reinforced by the equivalent results for the banded and plain oinochoai, which are also very rarely attested among the finds from the farmhouses (see p. 336 in Ch. 21, "Oinochoe"). The same picture emerged from the field survey of the chora and the excavation of the Pantanello necropolis, therefore indicating a generalized local situation. Other containers for liquids, such as banded and plain jugs, were probably used at the table, or perhaps small containers with handles were used to draw liquids from larger containers (see Ch. 21, "Plain and Coarse Wares").

Olpe

1:2

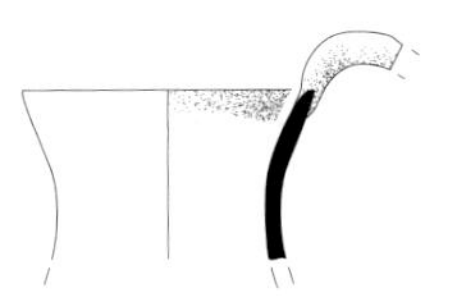

SAV BW 43 Olpe
Provenance: House on the Hill. Lot: SA 79-1062PL. SAV 57. Assemblage 1.
Dimensions and technical features: h 3.0, rim 4.0. Fabric SAV 1, Munsell 5YR 7/6, hard, compact, and fine-grained, with tiny black and white stone inclusions. Powdery surface. Black, thin, and opaque gloss.
Condition: One fragment of the rim and attachment of the handle.
Description: Indistinct, tapering, and slightly flaring rim. Small vertical ribbon handle attached at the rim. On the interior, band at the level of the rim; completely glossed handle.

Base

1:2

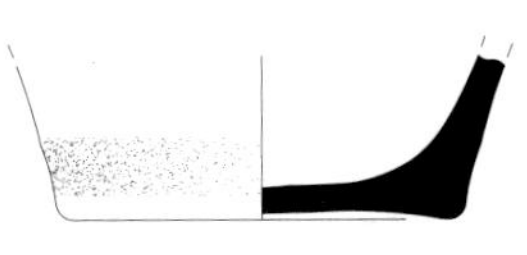

SAV BW 44 Olpe
Provenance: Bulldozed Stipe. Lot: SA 79-603PL. SAV 137. Assemblage 9.
Dimensions and technical features: h 2.3, rim 5.2. Fabric SAV 3, Munsell 5YR 6/4; hard, compact, fine-grained, and well-levigated. Sloppy band of brown-black, thin, and opaque gloss (upward skip of the brush).
Condition: One fragment of the base.
Description: Base of a small olpe with concave interior surface. On the exterior, band at the level of the floor.
Date: From levels of Phase 4.

Lebes (or Stamnos?)

Already present in various sizes at Samos, Miletus, Rhodes, and Clazomenae between the middle and the end of the 6th c. BC, the lebetes or stamnoi belong to the repertoire of East Greek forms borrowed and developed in Magna Grecia and Sicily (on the debate associated with the name given to these containers, see Amyx 1958, 190–93; for examples of stamnoi from Rhodes, see *ClRh III*, 244–45, t. CCXXVIII, fig. 243; 257, t. CCXLVIII, fig. 254; 255, t. CCXXXIV, fig. 249; *ClRh IV*, 58, t. VI, fig. 34; 134, t. XLIV, fig. 131; 198, t. C, fig. 213; 253, t. CXXI, fig. 275; 215, t. CIX, fig. 234; for examples from Miletus, see *IstMitt 32*, 1982, 43, 122–23, fig. 12, nos. 68–71; for examples from Clazomenae, see Ersoy 1993, nos. 530, 531).

Although several large individuals are known, the best-attested variety in the western Greek world is represented by the examples commonly called lebetes gamikoi, distinguished by a short collared lip with flat rim, rounded shoulder, tapering or compressed body, truncated conical foot, and horizontal handles with elliptical sections attached at the shoulder. These were produced especially in black-gloss, figured, or Gnathia-style wares, but numerous individuals featuring banded decoration on the lip, body, foot, and handle are known, with small tongues or dribbles of gloss applied to the shoulder with the brush.

The function of these containers is debated. Some specialists emphasize a close connection with marriage and the storage of purifying water for the female ablutions that preceded the *gamos* (see, among others, Daremberg and Saglio VII, "lebes"; Lippolis 1994b, 262). Others see in the lebetes substitutes for the krater (Kanowski 1984, 87) or, more cautiously, generic containers for liquids (Elliott in *Necropoleis*, 659).

Attested from the end of the 6th/beginning of the 5th c. BC and spreading notably from the second quarter of the 5th c. until the end of the 4th c. BC, the banded lebetes circulated very widely—even in the Metapontino, where local production is documented starting in the second quarter of the 5th c. BC with wasters from production sites both in the chora and in the urban *kerameikos* (Lecce 2012, 28; Silvestrelli forthcoming, fig. 5.6). Varied examples come from the Pantanello necropolis, moreover, where the form appears before the middle of the 5th c. BC (Carter and Parmly Toxey in *Necropoleis*, 659–63 and 699–700). An earlier example was found in the votive deposit in contrada Crucinia (Lo Porto 1981, 308, fig. 17.2). Three fragmentary lebetes were identified at Sant'Angelo Vecchio, belonging to two types distinguishable by the direction of the rim.

Type 1—vertical collared or slightly inclined rim. This type features a low vertical neck with continuous profile attached to the shoulder via a clear separation. Two examples were recovered at Sant'Angelo Vecchio, of which SAV BW 45, from the Pottery Deposit, preserves the remains of a fairly massive horizontal handle with elliptical section and an exterior decorative scheme consisting of horizontal bands with, between them, a rather smoothly rendered vertical *a filetti* motif.

Type 2—rim with thickened vertical collar. The second type is distinguished by a thickened collar rim mounted on the body wall via a clear separation. A single example was found at Sant'Angelo Vecchio, in the Pottery Deposit.

Lebetes (or Stamnoi?)

SAV BW 45 Lebes Type 1

Provenance: Sounding C, Pottery Deposit. Lot: SA 79-894P A. SAV 259. Assemblage 15.

Dimensions and technical features: h 4.2, rim int. 8.7. Fabric SAV 3, Munsell 2.5 YR 7/4, hard, compact, fine-grained, and well-levigated. Black and dilute gloss applied imprecisely.

Condition: One fragment of the rim, shoulder, and handle.

Description: Low vertical, collared rim; rounded shoulder and tapering globular body; horizontal ribbon handle, arched upward, attached at the shoulder. Strip of gloss at the level of the rim; on the exterior shoulder, strip of gloss with, beneath it, pattern of oblique ovoid dribbles; beneath the shoulder, occasionally dilute strip; fully glossed lower belly; nearly completely glossed handle.

1:2

Comparisons and comment: Lecce 2012, 28, stamnos tipo 2, fig. 8.21 (Metaponto, end of the 6th–first half of the 5th c. BC); Lo Porto 1981, 308, no. 158, fig. 17.2 (Metaponto, votive deposit in contrada Crucinia, 520–480 BC); Ruga et al. 2005, 194, no. 91, tav. LXXXV, 91 (Croton, 5th c. BC); Elliott in *Necropoleis*, 662, LG28 (460–440 BC).

Date: End of the 6th–5th c. BC.

SAV BW 46 Lebes Type 2

Provenance: Sounding C. Pottery Deposit. Lot: SA 79-988PL. SAV 87. Assemblage 15.

Dimensions and technical features: h 1.7, rim 8.2. Fabric SAV 3, Munsell 5YR 7/4, hard, compact, fine-grained, and well-levigated. Brown-black, thin, and opaque gloss.

Condition: One fragment of the rim and shoulder.

Description: Low vertical and slightly thickened rim, distinct from the rounded shoulder; globular body. On the interior and exterior, band at the level of the rim.

1:2

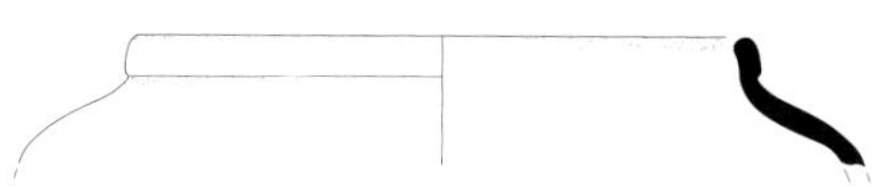

Comparisons and comment: For this type, no specific parallel was found. See generally Carter and Parmly Toxey in *Necropoleis*, 700, no. T 306-10/14 (Pantanello necropolis, 400–370 BC).

Date: From a context of Phase 3.

Feet from Unidentified Open Forms

Two fragments of feet belonging to banded open forms—most likely bowls, large bowls, or carinated lekanai—are presented separately in this section because they could not be attributed securely to any of the known forms. The decoration is concentrated on the exterior of the foot and on the interior floor of the body. The diameters range from 7 to 9 cm.

Feet from Unidentified Open Forms

SAV BW 47 Foot

1:2

Provenance: Slope 3, W extension. Level 2, batt. 3. Lot: SA 79-108PL. SAV 155 (see *Fig. 2.35*).

Dimensions and technical features: h 2.1, base 9.0. Fabric SAV 3, Munsell 2.5YR 5/6, hard, compact, fine-grained, and well-levigated. Black (Munsell 2.5YR 3/3), thin, and opaque gloss.

Condition: One fragment of the foot and attachment of the body.

Description: Foot of an open form with rounded exterior surface; oblique walls. On the interior, completely glossed circular area on the floor and band above the floor.

Date: From levels of Phase 4.

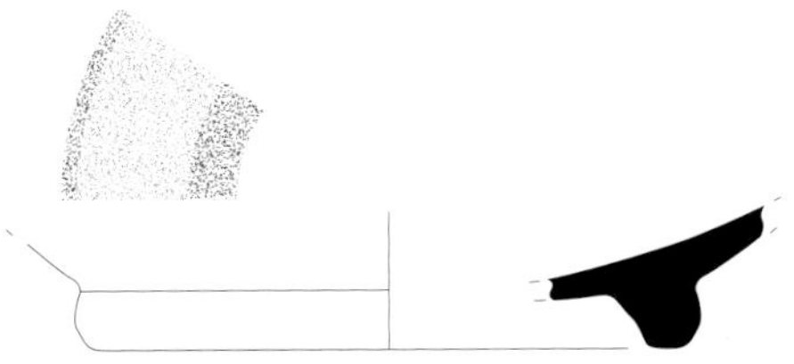

SAV BW 48 Foot

1:2

Provenance: Sounding C, Pottery Deposit. Lot: SAV 79-988PL-5. SAV 90. Assemblage 15.

Dimensions and technical features: h 1.4, base 7.0. Fabric SAV 3, Munsell 2.5YR 5/6, hard, compact, fine-grained, and well-levigated. Brown-black, thin, and opaque gloss.

Condition: One fragment of the foot and attachment of the body.

Description: Foot of an open form with rounded exterior profile. On the interior, completely glossed circular area on the floor of the body.

Date: From a context of Phase 3.

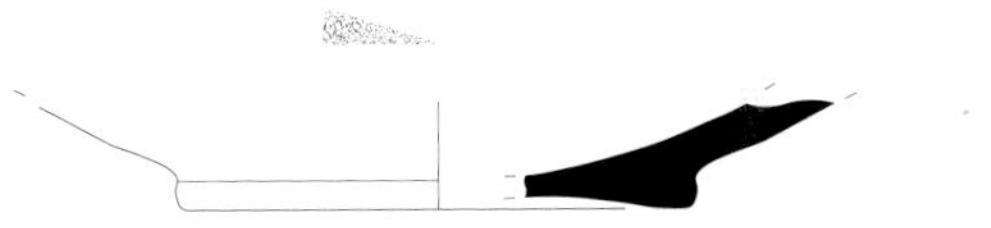

Lids

Among the finds from Sant'Angelo Vecchio is one banded lid fragment featuring oblique walls and a lid-resting surface. The rather remarkable diameter indicates that the lid was paired with a large open form, such as a large bowl, lekane, or flanged lekane. Similar lids in banded ware were found at Fattoria Fabrizio.

Lids

SAV BW 49 Lid

Provenance: Sounding C. Level 2, batt. 7. Lot: SA 79-205PL. SAV 125 (see *Fig. 2.35*).

Dimensions and technical features: h 4.5, rim 38.0. Fabric SAV 1, Munsell 5YR 7/6, hard, compact, and fine-grained, with tiny white and black stone inclusions. Powdery surface. Thin and opaque gloss.

Condition: One fragment of the rim and edge.

Description: Lid with oblique rim, tapered edge, and internal lid-resting surface with triangular section. Traces of gloss on the exterior.

Comparisons and comment: Two samples from Fattoria Fabrizio (Cavallo in *Fattoria Fabrizio*, FF BW 41 and FF BW 42)

Date: After ca. 320 BC.

1:2

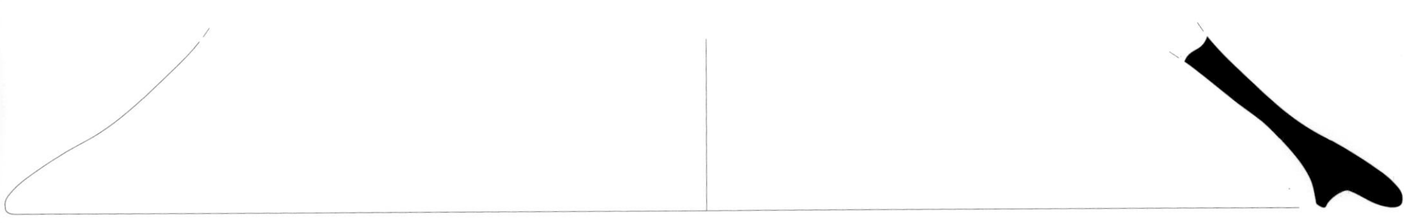

21

Plain and Coarse Wares

Anna Cavallo

The excavations at Sant'Angelo Vecchio yielded a sizeable quantity of undecorated pottery dated to the Late Archaic, Classical, and Hellenistic/Late Republican period.[1] This class is characterized by a modest variety of forms, which are in turn occasionally represented by a wide range of types. Although they often bear features suitable for various uses, the examples included are traditionally interpreted as wares used in the preparation of foodstuffs, the short-term storage of solid and liquid foods, and—to a lesser extent—the consumption of food at the table.

The study of the undecorated wares from the Metapontine farmhouses excavated by ICA was carried out in a contextual fashion,[2] examining the materials from the individual sites with the goal of creating a typological repertoire for all of them. The results of the study were compared to, and augmented by, the published materials,[3] beginning with the Metapontino[4] and proceeding to areas within or near the Bradano watershed such as Cozzo Presepe, Gravina di Puglia, Pomarico Vecchio, Oppido Lucano, Torre di Satriano, Civita di Tricarico, and Roccagloriosa.[5] Other colonial sites like Taranto, Herakleia, Sybaris, Caulonia, Laos, and Locri served as additional fixed points of reference.

Quantitative and Distribution Analysis

Of the ceramic materials recovered during the excavations at Sant'Angelo Vecchio, 1283 fragments were identified as belonging to undecorated wares. Some body fragments (238) among these could not be attributed securely to open or closed shapes because of their very poor state of preservation; 529 fragments were attributed to closed shapes and 20 to open shapes. Calculation of the plain and coarse body fragments is nevertheless complicated by the presence of fragments certainly belonging to undecorated parts of banded vessels. This situation is even more evident when the quantification of the plain and coarse wares is compared to that of the banded wares, where the number of body fragments is far lower as a proportion of the entire class (see Ch. 20, "Banded Ware").

The area investigated yielded 355 diagnostic fragments, including 175 rims, 6 complete profiles, 113 feet, and 67 handles, for 254 total individual vessels. The poor state of preservation of most of the materials limited the reconstruction of complete profiles, and the typology is proposed on the basis of parallels with similar examples from other contexts.

House on the Hill. The excavation of the Archaic farmhouse on the hilltop yielded 308 fragments of plain and coarse ceramics. Although the majority of the finds concentrated in the area of Test Trench 3 and in the topsoil are poorly preserved body fragments (236, of which 144 fragments belong to closed shapes and one to an open shape), a small number of diagnostic fragments was identified, including 28 rims, nine handles and 19 bases, for a minimum number of individuals of 37. Small open shapes are almost entirely absent among these, being attested by a single example of a small bowl/one-handler (SAV PCW 01); such shapes are more commonly encountered among the banded repertory documented at the site. The most common form is the lekane/basin, represented by at least seven examples of Types 1–4 and 8 (SAV PCW 06, SAV PCW 09, SAV PCW 14, SAV PCW 15, and SAV PCW 18).

[1] On the history and status of the study of plain and coarse ceramics, see Cavallo in *Fattoria Fabrizio*, 261–64.

[2] The contexts considered include the farmhouses Fattoria Fabrizio, Sant'Angelo Grieco, Sant'Angelo Vecchio, and Fattoria Stefan, all of which are part of the Metaponto Project initiated at the end of the 1970s with the ICA excavations.

[3] Although recent studies at several sites have revealed an unexpectedly high rate of importation of less precious products (*Gravisca 12*, 23–24; Camporeale 1997, 15–23; Gassner 2003, 73; Nigro 2006a, 79–80), plain and coarse pottery had a primarily local character associated with the production and selection of the containers most frequently used at the table, in the kitchen, and in the pantry. This study therefore sought comparanda for the Metapontine materials mostly among contexts belonging to the same cultural milieu.

[4] The fundamental background for the Metapontine finds are the ceramic studies for the ICA intensive field survey (Vittoria in *Survey*), the Pantanello necropolis (Carter and Parmly Toxey in *Necropoleis*), the urban *kerameikos* (D'Andria 1980a; Silvestrelli forthcoming), the Favale votive deposit (M. Liseno 2004), and several necropoleis (Lo Porto 1966; 1981; 1988–1989).

[5] Among others, Cotton 1977; Lissi Caronna 1980, 1983, 1990–1991; Fracchia and Keith 1990; Cotton 1992a; Bianco and Deodato 1997; Rinaldi 2005; Caravelli 2008.

Alongside these were encountered nine jugs, including SAV PCW 46, one of the rare containers recovered in the Metapontine farmhouses that preserved the complete profile except for the handle, and a small fragment possibly from an oinochoe (not in catalog).

Pottery Deposit. Among the 171 plain and coarse fragments identified in the Pottery Deposit, 42 body fragments belong to closed shapes and 13 to unidentified open shapes, with an MNI of 36. Represented especially by lekanai/basins (28 fragments belonging to 14 individuals) of Types 1, 2, and 8 (SAV PCW 07, SAV PCW 10, and SAV PCW 19) and by a bowl, open forms predominate. Jugs are rarer, numbering eight examples (SAV PCW 47). Also noteworthy is a form that is almost completely unknown in the plain and banded repertoires at Metaponto, a trefoil oinochoe (SAV PCW 73). The datable examples from this context indicate a chronology in the 5th century BC, but in all cases the number of parallels identified was rather limited. The dinos SAV PCW 45 is unique, its morphological features finding parallels dated at least to the end of the 4th century BC. Considering the features of the fragments recovered in the votive deposit and belonging to the other ceramic classes, it is probably an intrusion originating in the nearby waste deposit discovered during the excavation of Sounding C and in the vicinity of Slope 4, which yielded Early Hellenistic material.

Workshop Area. A group of 180 fragments belonging to at least 56 vessels comes from the Hellenistic workshop and Late Republican workshop shed, in addition to a much smaller number of banded fragments. Small open shapes are almost completely absent, while open shapes such as the basins predominate—especially the deep (three examples) and the large basins (nine examples), forms with a function that could also be related to the manufacturing activity of the area. This hypothesis is also supported by the absence of fragments belonging to vessels of this form in the contemporary deposits of the slopes. Among the closed forms are 14 jugs, two table amphorae dating to the beginning of the 3rd century BC, and two jars. The chronological context of most of the datable fragments (SAV PCW 02, SAV PCW 03, and SAV PCW 58) lies between the late 4th and the 3rd century BC.

Slope Area. Deposits excavated on the slopes which were dated to the phase of the Hellenistic workshop yielded 240 fragments comprising 79 vessels. Jugs (37), deep basins (6) and lekanai/basins (11) are the best attested shapes.

Kiln Complex. The 62 fragments from the Kiln Complex can be ascribed to 21 vessels. Four fragments belonging to the same number of individuals were found within the small kiln (Kiln 3) in E2–E3; their state of preservation precludes secure identification and date. The deposit in the North Kiln interior (Kiln 5) yielded SAV PCW 53, the Type 1 jar SAV PCW 75, and the basin SAV PCW 32, the latter reused in the construction of the kiln. Other identifiable forms documented among them are large basins and a single large jar (SAV PCW 76).[6]

Clays and Decorations

Nearly 65 percent of the clays identified are fine-grained and well levigated, displaying features very similar to those of more precious ceramics. These features suggest a common production context, probably local in scope. Early analyses of the fabrics, moreover, revealed an unmistakable affinity with those documented among the banded ware (see Ch. 20, "Banded Ware," and *Table 20.1*). In addition to the most commonly encountered fabrics, there is a small group of rather large vessels, such as large basins (SAV PCW 34–SAV PCW 36), made with a fabric that tends to be less well-levigated and fine (see fabrics SAV 2 and 4 in *Table 20.1*).

The containers analyzed are markedly homogeneous even in the technical features of their production: the workmanship is consistently good, the potter's wheel is used regularly, and the surfaces are generally smoothed by hand using fingers or a stick. Many fragments suggest a final application of a sponge to remove excess water left over from the working of the clay and to improve the smoothness of the surfaces. In some cases, furthermore, a fine layer of clay the same color as the fabric was applied to ensure a uniform finish. Several vessels identified as misfired due to excess or insufficient heat in the firing chamber were found distributed around the entire excavation area.

Decoration is rare, limited to motifs incised with a round-tipped instrument. The predominant mo-

[6] For other Late Republican pottery from the site, see Ch. 17, "Red Ware," Ch. 22, "Lagynoi," and Ch. 26, "Roman and Late Roman Wares."

tif consists of thin concentric circles on the rims of lekanai/basins (see, for example, SAV PCW 06–SAV PCW 08) and on the walls of deep basins (like in SAV PCW 40), but small stylized leaves are also attested (SAV PCW 39).

Catalog of Plain and Coarse Wares

Classification Criteria

The methodological approach adopted in the study of the plain and coarse pottery aims to identify the forms to which the fragments belong through analysis of their morphological characteristics, limiting hypotheses about relative functions to the strict description of the form (on the principles associated with this approach, also followed in the study of the materials from the other Metapontine farmhouses excavated by ICA because of their generally poor state of preservation and the multifunctional character typical of the most common containers, see Cavallo in *Fattoria Fabrizio*, 261–88; *Pontecagnano II.1*; *Gravisca 12*; Bagnasco Gianni 1997, 2007; Brizzi 1999a; Nigro 2006a). All of the recognizable forms were identified, and a typological seriation was developed for each of them where the group of fragments was sufficiently large. The second level of classification entailed the distinction of morphological or decorative variants that are not strictly necessary for the definition of the type. The typology developed for this study refers to the master typology developed for the materials recovered from all of the Metapontine farmhouses investigated by ICA.

Small Bowl/One-handler

Plain and coarse small bowl/one-handlers circulated in Magna Grecia at the same time as the banded examples, for which they served as the most common alternative (for a more detailed analysis of the form, see Ch. 20, "Banded Ware"). In the Metapontino, plain and coarse small bowls/one-handlers are attested in the urban center, in the chora, and in the Pantanello necropolis. Five individuals were found in the House on the Hill, eight were found in levels of Phase 4 (Early Hellenistic period) in the area of the slopes and in the Workshop Area. The reduced number of specimens reveals a clear preference for the banded version of the form (see Ch. 20, "Banded Ware").

The fragments of the form identified among the Metapontine farmhouses excavated by ICA, in both banded and plain ware, belong to three types (Cavallo in *Fattoria Fabrizio*, 265). SAV PCW 01 belongs to the only type attested in plain ware at Sant'Angelo Vecchio, Type 3, which is identified by a direct, flaring rim and truncated conical body.

Small Bowl/One-handler

SAV PCW 01 Small Bowl/One-handler Type 3
Provenance: House on the Hill. Lot: SA 79-1062PL. SAV 62. Assemblage 1.
Dimensions and technical features: h 3.5, rim 8.4. Fabric SAV 1, Munsell 5YR 7/6, hard, compact, and fine-grained, with tiny white and black stone inclusions. Powdery surface.
Condition: Two fragments of the rim and body.
Description: Indistinct rim with rounded termination; hemispherical body.

1:2

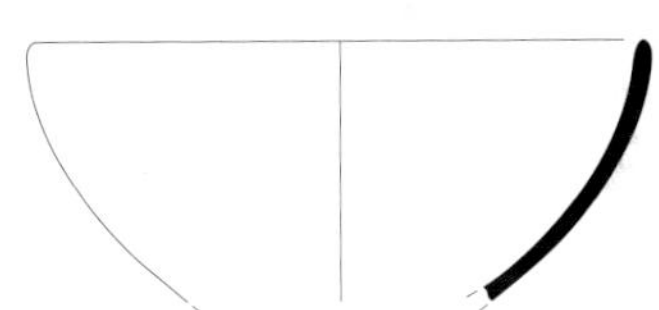

Comparisons and comment: Cotton 1983, 365, no. 368 (Cozzo Presepe, 550–500/480 BC).
Date: End of the 6th–beginning of the 5th c. BC.

Skyphoid Cup

Four rims can be attributed to two-handled cups with indistinct rims, deep bodies, truncated conical feet, and handles that either are ribbon-shaped rings or have squashed elliptical sections attached immediately beneath the rim, similar to the banded series of vessels (for the features and chronology of the form, see p. 288 in Ch. 20, "Skyphoid Cup"). The highly fragmentary state of the materials does not preclude the possibility that they come from partially decorated vessels preserving no trace of the decoration.

Skyphoid Cups

SAV PCW 02 Skyphoid Cup

1:2

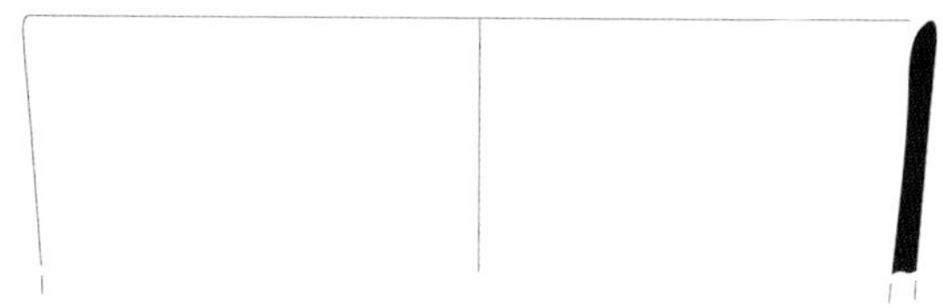

Provenance: Square G7, E of wall. Level 2, batt. 1. Lot: SA 79-802PL. SAV 48. Assemblage 17.

Dimensions and technical features: h 3.8, rim 14.2. Fabric SAV 3, Munsell 2.5YR 6/6, hard, compact, fine-grained, and well-levigated.

Condition: One fragment of the rim.

Description: Indistinct rim; straight body wall.

Comparisons and comment: Scarano 1992, 19, T22, inv. no. 297193, fig. 5 (Metaponto, 320–250 BC, partially glossed).

Date: 320–250 BC.

SAV PCW 03 Skyphoid Cup

1:2

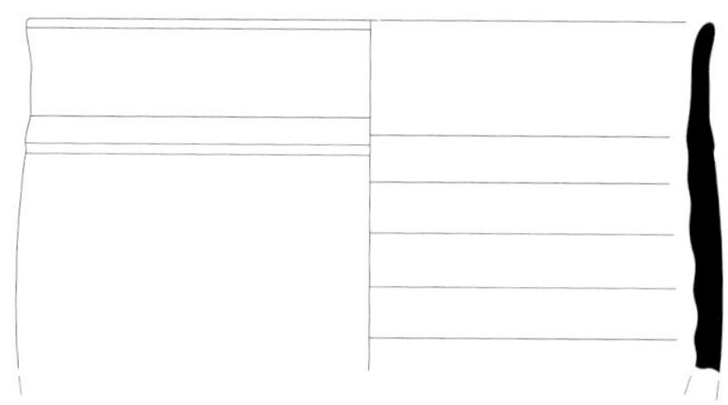

Provenance: Square F6. Level 1, batt. 1. Lot: SA 79-1053PL (topsoil) (see *Fig. 2.23*).

Dimensions and technical features: h 6.0, rim 12.0. Fabric SAV 3. Munsell 7.5YR 6/6, compact, slightly porous, with small and sparse white inclusions.

Condition: One fragment of the rim.

Description: Indistinct rim; straight body wall.

Plate

SAV PCW 04 Plate

1:2

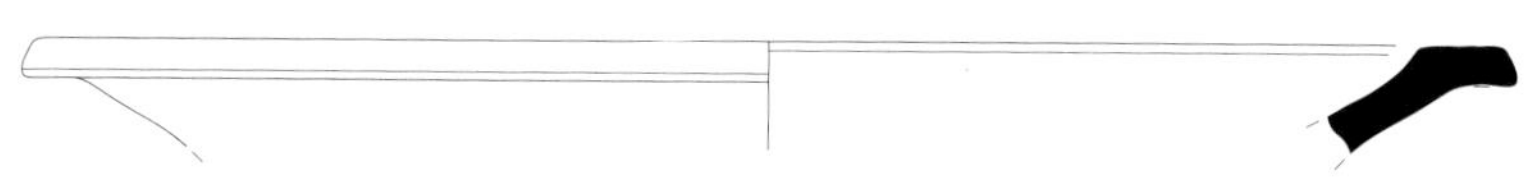

Provenance: Square G6, W of wall. Level 2, batt. 2. Lot: SA 79-1001PL. SAV 193 (see *Fig. 2.20*).

Dimensions and technical features: h 1.4, rim 20.3. Fabric SAV 4. Munsell 2.5YR 6/8, hard, compact, and fine-grained, with sparse and tiny white and black stone inclusions. Powdery surface.

Condition: Two fragments of the rim.

Description: Flaring rim with flat upper profile and oblique walls.

Comparisons and comment: No precise comparisons have been found so far. Nonetheless, the fragment bears some similarities to dishes from Gravina-Botromagno, Accurso necropolis: Andriani and Laricchia 2007, 9, tomb 4 (end of the 4th c. BC). See also the bowl in *Roccagloriosa I*, 270, no. 280, fig. 187 (Roccagloriosa, late 4th–early 3rd c. BC), and the plates in Vittoria in *Survey*, 360, nos. 21–22, Pl/Lne4b1 (Metaponto, chora).

Carinated Lekane

SAV PCW 05 is closest in shape to carinated lekanai of Type 1 in banded ware (on the form, see p. 288 in Ch. 20, "Carinated Lekane"). It features a pendant rim with triangular section, carinated body, and handle with elliptical section attached at the lip. That the vessel is a plain ware example is unusual. It also differs from the banded specimens in the handle type. Its fabric is very similar to that of cooking vessels, which further enhances its status as an anomaly in Metapontine lekane production.

Carinated Lekane

SAV PCW 05 Lekane

Provenance: Square D4, outside stone circle to N near baulk (from SA 79-904PL). Level 3, batt. 5. Lot: SA 79-905P. Inv. no. 319957. SAV 208. Assemblage 13.

Dimensions and technical features: h 4.0, rim 31.0. Fabric SAV 5, Munsell 2.5YR 5/6, very hard and compact with occasional vughs in fracture, with many tiny white stone inclusions. Heavily encrusted.

Condition: One fragment of the rim and body.

Description: Flaring, pendant rim with triangular section, flat upper profile, and rounded edge; pronounced carination of the body, beneath which the walls become oblique; raised horizontal handle with elliptical section attached at the rim and decorated with a finger impression. On the interior, a small ledge at the level of the carination.

Comparisons and comment: Conti 1989, 305, no. 361, pl. XLI (Locri, 4th–3rd c. BC).

Date: From Early Hellenistic levels (Phase 4).

1:2

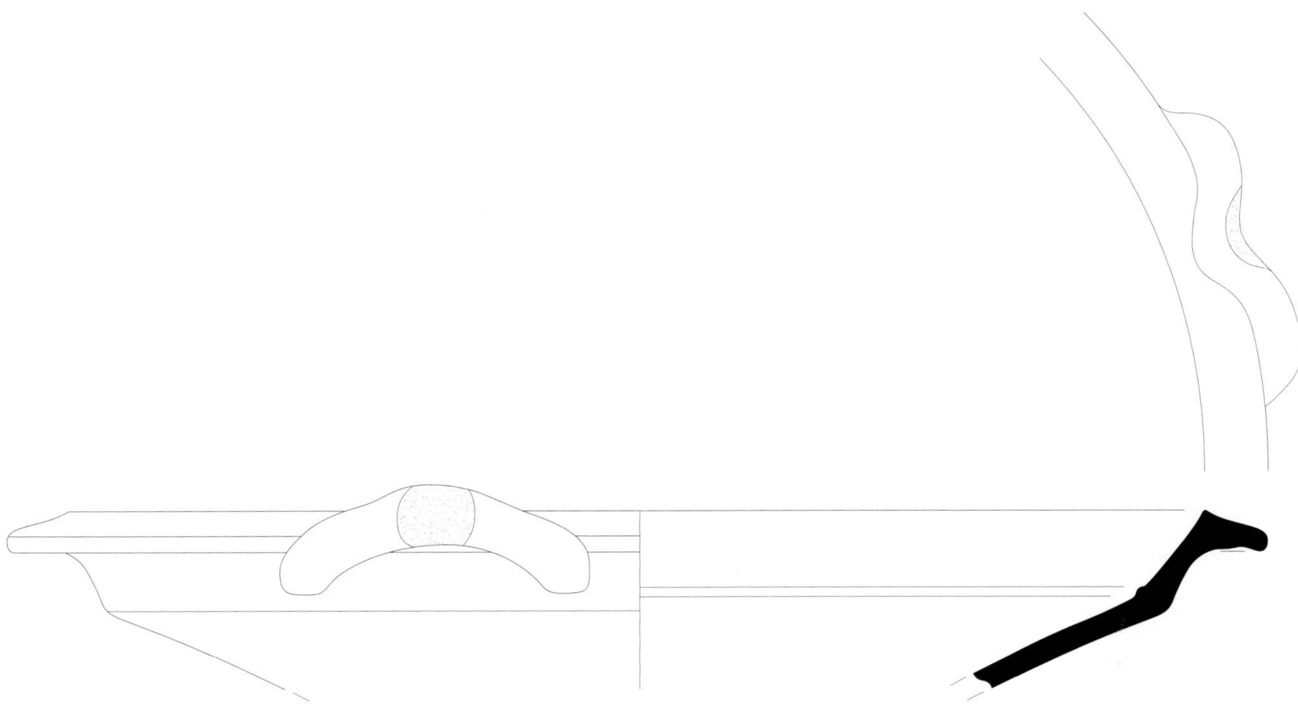

Lekane/Basin

The lekanai and basins represent one of the most common and diffuse shapes in the Metapontine chora (see Vittoria in *Survey*, 371–72); at Sant'Angelo Vecchio they constitute approximately 19% of the identified individuals. This group includes 116 fragments belonging to 48 open vessels distinguished generally by large size and a deep body with mostly oblique walls (on the choice of nomenclature see Cavallo in *Fattoria Fabrizio*, 267). The multifunctional nature of the form assured its success. Lekanai/basins were useful for various activities associated with the preparation of foodstuffs, the washing of small utensils, and the short-term storage of foods in the pantry (Amyx 1958, 204–5; Sparkes 1962, 128–29; *Agora XII*, 211). Lekanai/basins may also have been used, moreover, to carry prepared foods in large quantities to the table to be shared. The documented profiles find parallels dating between the 6th and the 1st c. BC, with the greatest concentration between the 5th and 3rd c. BC.

On the lekanai/basins, the rim is the only reliable morphologically diagnostic part. A remarkably varied typology has been documented by seven types among the Metapontine farmhouses excavated by ICA (Cavallo in *Fattoria Fabrizio*, 267), not all of which are attested among the materials from Sant'Angelo Vecchio. Only Types 1–4 were present, to which we may add an eighth type not previously recognized:

- Type 1—horizontal flange;
- Type 2—upturned flange;
- Type 3—pendant flange;
- Type 4—thickened flange with rectangular section; and
- Type 8—inverted rim with convex upper profile and pendant lip.

Type 1. This type is characterized by a horizontal brim with a flat or slightly concave upper surface, which in several cases is decorated with incised parallel lines. Two examples come from the House on the Hill, and three from the area of the slopes, from levels belonging to Phase 3.

Type 2. The second type features a flange with flat or slightly concave upper surface, inclined upward, and occasional decoration consisting of concentric incisions made with a sharp instrument. All of the examples have a pendant edge which is often vertically oriented and inclined toward the exterior and has a convex profile. The walls of the best-preserved examples are in general slightly curved, while examples with oblique walls are rarer. The lone basin preserving the complete profile has a ring foot. The handle with a large elliptical section, where preserved, attaches directly to the rim and stands over it. The rim often bears a decorative pattern of concentric grooves made using an instrument with a rounded tip. The comparanda indicate a chronology between the 5th and the 3rd c. BC, with rare attestations through the 1st c. BC.

At Sant'Angelo Vecchio Type 2 is the most common, with three vessels from the House on the Hill and 14 from the lower area. The vessels feature a rim with a diameter in the range of 36 to 44 cm.

Type 3. This type was attested at Sant'Angelo Vecchio by a single example found in the House on the Hill. The rim of the third type has a slightly pendant brim with flat or convex upper surface, and the body is characterized by rounded walls.

Type 4. The fourth type features a thickened flange with sub-rectangular section and concave or flat upper surface. The rim, which can be horizontal (variant 4a) or oblique (variant 4b), may have an occasionally pendant edge. The body is deep and sturdy, often more voluminous than the average basin recovered in the excavated farmhouses. Three examples recovered at Sant'Angelo Vecchio can be ascribed to this type.

Variant 4a appears to derive from the lekane; the examples often share the same articulation of the rim. The basins, however, have a much deeper body with flaring walls; the generally flat base can nevertheless also feature a small, atrophied ring-foot. Variant 4b has a brim rim and flat upper surface bearing a groove, and is documented beginning in the 4th c. BC (Lamboley 1999, 67–68, pl. 5, 74, 77). The attestations become more frequent in the course of the 3rd c. BC, with a pendant lip variant.

Type 8. Among the plain ceramics from Sant'Angelo Vecchio two fragments of a basin of a previously unidentified type (Type 8), characterized by an inverted rim with convex upper profile and pendant lip, were found. The type has a remarkably broad diameter, consistently thick rim and wall, and low body features, suggesting a use similar to mortaria in the preparation of foodstuffs. This type of basin is known in the Metapontino (Vittoria in *Survey*, 343, lekane/basin, variants Lne/Bas1 and Lne/Bas3, chora of Metaponto; Cotton 1983, 366, nos. 376, 378, fig. 136, Cozzo Presepe, where the form is attested from the end of the 6th c. BC and throughout the Hellenistic period). The three vessels from Sant'Angelo Vecchio are associated with Late Archaic and Classical levels (Phases 2–3).

Lekanai/Basins

SAV PCW 06 Lekane/basin Type 1

1:3

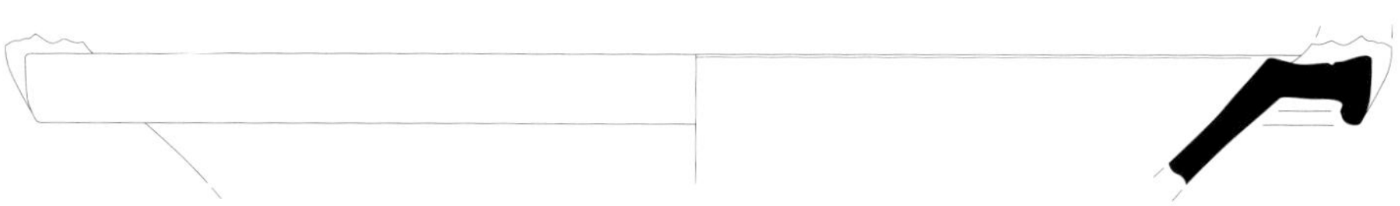

Provenance: House on the Hill, S half of Test Trench 2 on hilltop; Level 1, batt. 1. Lot: SA 79-557PL and SA 79-616PL A. SAV 131 (see *Fig. 2.39*).

Dimensions and technical features: h 3.2, rim 35.7. Fabric SAV 3, Munsell 2.5YR 6/6, hard, compact, fine-grained, and well-levigated.

Condition: Two fragments of the rim and body and the attachment of the handle.

Description: Flaring rim with oblique rim with a pendant vertical edge; oblique body wall; handle with elliptical section attached to the rim. On the rim, incision made with a stick.

Comparisons and comment: Hänsel 1973, 451, fig. 31, no. 2 (Herakleia, ca. 430–400 BC); Vittoria in *Survey*, 368, no. 55, Lne9a (Metaponto, chora).

Date: 5th c. BC.

SAV PCW 07 Lekane/basin Type 1

Provenance: Sounding C, Pottery Deposit. Lot: SA 79-988PL. Assemblage 15.

Dimensions and technical features: h 6.3, rim 34.0. Fabric SAV 1, Munsell 5YR 6/6, hard, medium-grained, and levigated, with sparse tiny light and dark gray stone inclusions.

Condition: Two fragments of the rim and handle.

Description: Flaring rim with horizontal flange and pendant vertical edge.

Comparisons and comment: Carter 1983a, 478, no. 89, fig. 53 (Pantanello, deposit over the clay pit).

Date: 5th c. BC.

1:3

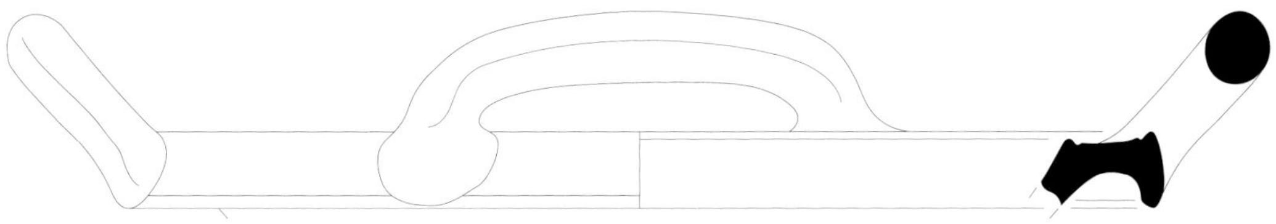

SAV PCW 08 Lekane/basin Type 1

Provenance: Square D4 W. Level 3, batt. 5. Lot: SA 79-907PL. SAV 177. Assemblage 13.

Dimensions and technical features: h 2.8, rim 38.4. Fabric SAV 4, Munsell 5YR 6/6, very hard, compact, medium-grained, and levigated, with tiny light and dark stone inclusions.

Condition: One fragment of the rim and body.

Description: Flaring horizontal rim with rectangular section, concave upper profile, and slightly tapering exterior profile; on the interior, at the level of the body, a small but pronounced ridge. On the rim, four incisions made with a stick.

Comparisons and comment: The rim bears similarities to *Valesio I*, 264, no. 464, decorated by a similar grooving on the upper surface (Valesio, 4th–3rd c. BC); the same type of decoration also occurs on basins from Gravina (Cotton 1992, 170, no. 1232, fig. 59, Gravina VIIIb); also similar to *Kaulonia I*, 63, 220, fig. 41 (Caulonia, 4th c. BC). For the decoration see Vittoria in *Survey*, 367, no. 49 (Lne5, Metaponto, chora) and Cavallo in *Fattoria Fabrizio*, FF PCW 15 (Metaponto, chora, Fattoria Fabrizio, 400–100 BC).

Date: Late 4th–3rd c. BC.

1:3

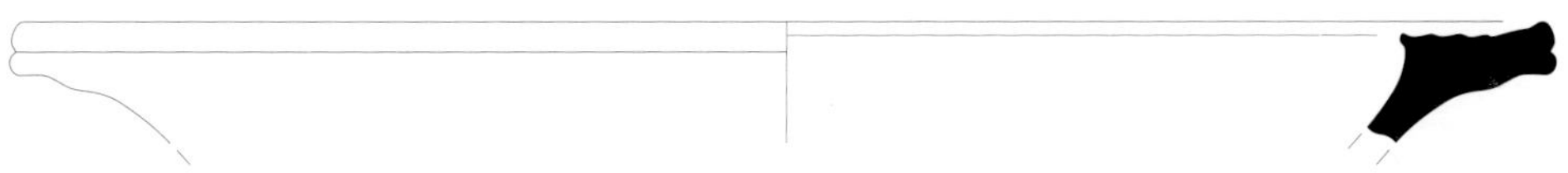

SAV PCW 09 Lekane/basin Type 2

Provenance: House on the Hill, Test Trench 1 and S half of Test Trench 2 on hilltop. Level 1, batt. 2. Lot: SA 79-558PL + SA 79-557PL. SAV 133 (see *Fig. 2.39*).

Dimensions and technical features: h 4.0, rim ext. 39.5. Fabric SAV 3, Munsell 5YR 6/4, hard, compact, fine-grained, and well-levigated.

Condition: Three fragments of the rim and body.

Description: Flaring rim with oblique rim which has a pendant vertical and slightly inward-projecting edge; the joint between the rim and the body is marked on the interior by a small but pronounced ridge; oblique body wall. Incision on the rim.

Date: From Late Archaic contexts (Phase 2).

1:3

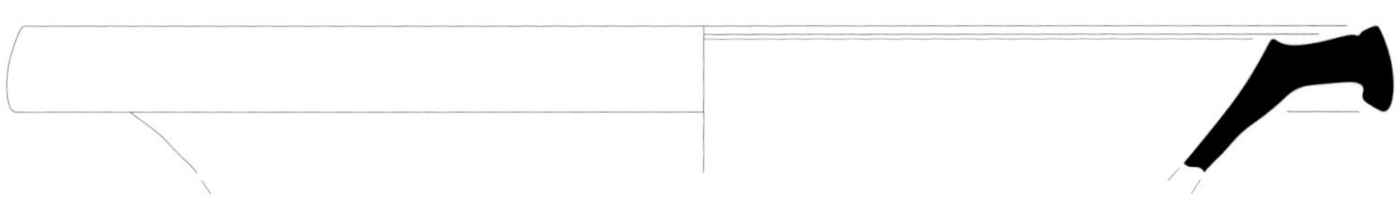

Lekanai/basins cont.

SAV PCW 10 Lekane/basin Type 2

Provenance: Sounding C, Pottery Deposit. Lot: SA 79-988PL. Assemblage 15.

Dimensions and technical features: h. 6.6, rim 36.0. Fabric SAV 3, Munsell 5YR 6/6, hard, compact, fine-grained, with rare tiny white and dark gray inclusions.

Description: Seven fragments of the rim and body.

Comparisons and comment: Very similar to specimens from Metaponto, *kerameikos*, deposit no. 12 (second quarter of the 5th c. BC): Silvestrelli, pers. comm.

Date: First half of the 5th c. BC.

1:3

SAV PCW 11 Lekane/basin Type 2

Provenance: Sounding C. Level 2, batt. 6. Lot: SA 79-201PL. SAV 107 (see *Fig. 2.35*).

Dimensions and technical features: h 4.0, rim 38.0. Fabric SAV 4, Munsell 5YR 6/4, very hard, compact, medium-grained, and levigated, with tiny light and dark stone inclusions.

Condition: One fragment of the rim and body.

Description: Flaring rim with oblique rim which has a vertical pendant edge; convex body wall. Incision on the rim.

Date: From Early Hellenistic levels (Phase 4).

1:3

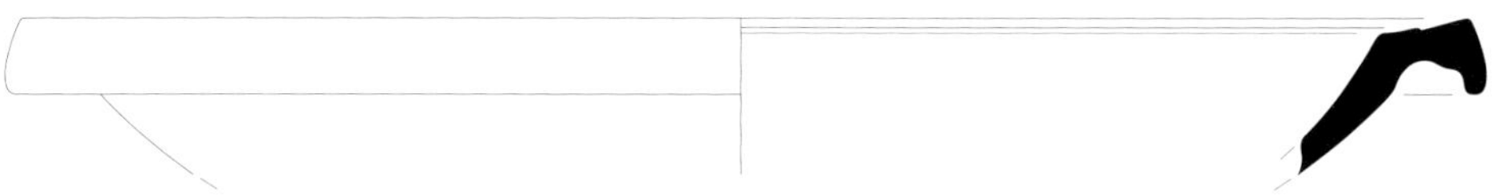

SAV PCW 12 Lekane/basin Type 2

Provenance: Slope 4, surface cleaning (Sounding C, E extension). Lot: SA 79-421PL. SAV 152.

Dimensions and technical features: h 2.7, rim 43.0. Fabric SAV 4, Munsell 5YR 6/3, very hard, compact, medium-grained, and levigated, with tiny light and dark stone inclusions.

Condition: One fragment of the rim and body.

Description: Flaring rim with oblique rim which has a barely pronounced vertical edge; oblique body wall. On the rim, firing bubble and two incisions.

Comparisons and comment: Bears similarities to Cotton 1992a, 170, nos. 1230, 1232, fig. 59 (Gravina di Puglia, 200–50 BC).

Date: 4th–3rd c. BC.

1:3

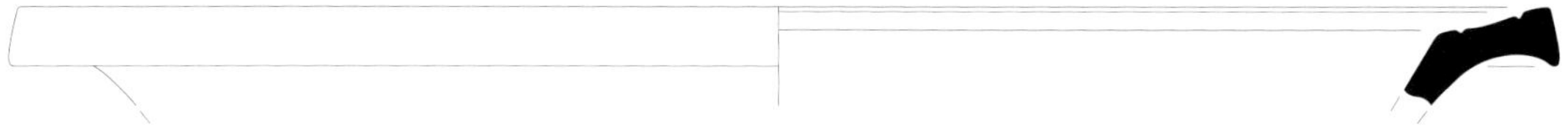

SAV PCW 13 Lekane/basin Type 2

Provenance: Sounding C. Level 2, batt. 6. Lot: SA 79-201PL. SAV 106 (see *Fig. 2.35*).

Dimensions and technical features: h 9.6, rim 44.5. Overfired.

Condition: One fragment of the rim and body.

Description: Flaring rim inclined upward; pendant edge with convex exterior profile; oblique body wall. Overfired.

Comparisons and comment: Lo Porto 1981, 368, t. 23, fig. 89.3 (Metaponto, 250 BC); Cotton 1983, 366, no. 379, fig. 136 (Cozzo Presepe, Site B, Phase V, 300 BC); Cotton 1992a, 168, no. 1205, fig. 57 (Gravina di Puglia, 2nd c. BC); Vittoria in *Survey*, 369, no. 57, Lne10 (Metaponto, chora).

Date: 3rd–2nd c. BC.

1:3

SAV PCW 14 Lekane/basin Type 3

Provenance: House on the Hill, Test Trench 2. Level 1, batt. 1. Lot: SA 79-557PLA. SAV 192 (see *Fig. 2.39*).

Dimensions and technical features: h 5.0, rim 40.4. Fabric SAV 3, Munsell 2.5YR 6/4, hard, compact, fine-grained, and well-levigated.

Condition: One fragment of the rim and body.

Description: Flaring rim with oblique rim which has a vertical pendant edge; the joint between the rim and the body is marked on the inside by a small but pronounced ridge; oblique body wall. On the rim, two grooves incised with a stick.

Comparisons and comment: Carter 1983a, 487, no. 163, fig. 53 (Pantanello); *Kaulonia I*, 63, no. 208, fig. 41 (Caulonia, 5th c. BC).

Date: 5th c. BC.

1:3

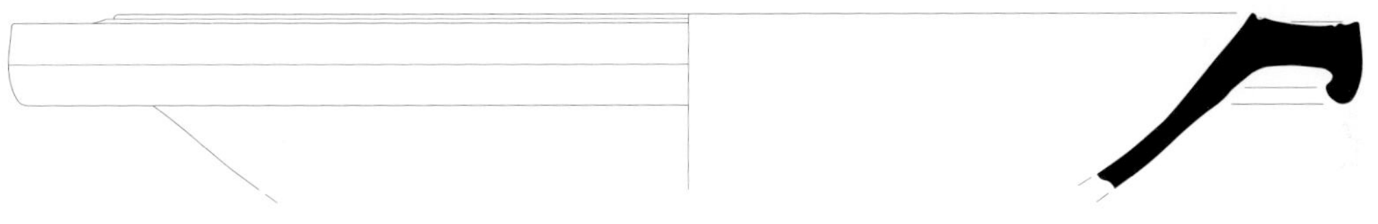

SAV PCW 15 Lekane/basin Type 4a

Provenance: House on the Hill, S half of Test Trench 2. Level 1, batt. 1. Lot: SA 79-557PL. SAV 130 (see *Fig. 2.39*).

Dimensions and technical features: h 4.1, rim int. 34.0. Fabric SAV 1, Munsell 5YR 7/6, hard, compact, and fine-grained, with tiny white and black stone inclusions. Powdery surface.

Condition: One fragment of the rim and body.

Description: Massive flaring, pendant rim with rectangular section and contracting exterior profile; oblique body wall.

Comparisons and comment: Bears similarities to the type 1 basins from Oria, Monte Papalucio sanctuary: Mastronuzzi 2013, 293–95, pl. 15 (from a context dated to 550–490 BC). See also Conti 1989, 300, no. 351, pl. XXXIX (Locri, from layers of the 5th and 4th c. BC); Carè and Cavallo 2006, 253, CC14, fig. 3 (from Terravecchia di Grammichele, Catania, disturbed votive deposit yielding pottery from the 6th to the 4th c. BC; the fragment is dated to the 4th–3rd c. BC).

Date: 6th–5th c. BC.

1:3

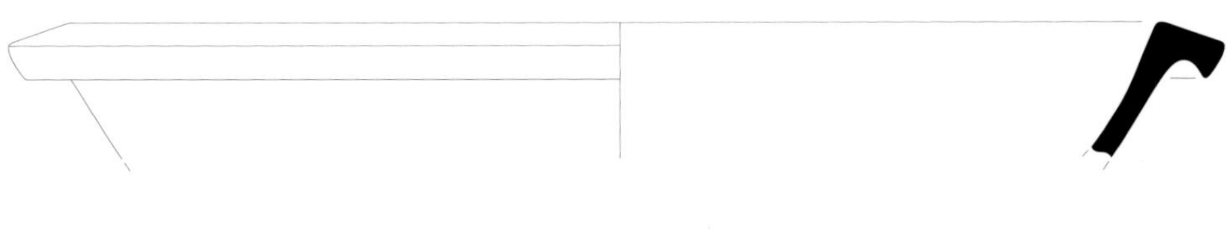

Lekanai/basins cont.

SAV PCW 16 Lekane/basin Type 4b

Provenance: Sounding D. Level 1, batt. 1. Lot: SA 80-42PL. SAV 2.

Dimensions and technical features: h 10.2, rim ext. 34.2. Fabric SAV 1, Munsell 5YR 7/3, hard, compact, and fine-grained, with tiny white and black stone inclusions. Powdery surface.

Condition: One fragment of the rim and body.

Description: Flaring rim with thick horizontal flange, oblique upper profile, and hanging edge; convex body wall. Incision on the rim.

Comparisons and comment: Carter 1983a, 485, no. 132, fig. 53 (Pantanello); Yntema 2001, form M55 (Valesio, 4th–3rd c. BC); Hänsel 1973, fig. 31.3 (Herakleia, 4th–3rd c. BC); *Metaponto II*, 367, fig. 136, nos. 280, 282 (Cozzo Presepe); *Pomarico Vecchio I*, 178, tav. 69, n. 89; *Kaulonia I*, 63, fig. 41, n. 205 (Caulonia, 4th c. BC); bears similarities to Vittoria in *Survey*, 366, no. 43, Lne2a-b (Metaponto, chora); Cotton 1992a, 166–68, fig. 56, no. 1196; fig. 57, no. 1206 (Gravina, late 2nd–1st c. BC) and fig. 59, no. 1230 (4th–3rd c. BC).

Date: End of the 4th–2nd c. BC.

1:3

SAV PCW 17 Lekane/basin Type 4b

Provenance: Slope 4, surface cleaning (Sounding C, E extension). Lot: SA 79-421PL. SAV 151.

Dimensions and technical features: h 4.8, rim ext. 34.1. Fabric SAV 3, Munsell 5YR 7/4, hard, compact, fine-grained, and well-levigated.

Condition: One fragment of the rim and body.

Description: Flaring rim with thick horizontal flange and pendant edge; oblique body wall; horizontal handle attached to the rim. On the rim, incision made with a stick.

Comparisons and comment: See SAV PCW 16.

Date: From Early Hellenistic levels (Phase 4).

1:3

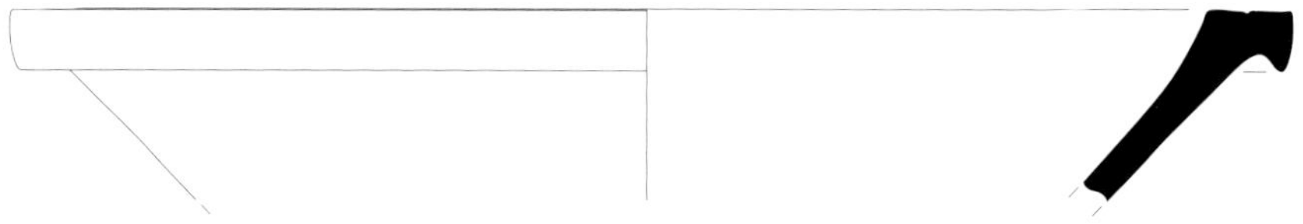

SAV PCW 18 Lekane/basin Type 8

Provenance: House on the Hill, Test Trench 3. Level 1, batt. 1. Lot: SA 79-992P. SAV 201 (see *Fig. 2.39*).

Dimensions and technical features: h 5.1, rim int. 24.0. Fabric SAV 3, Munsell 5YR 7/6, hard, compact, fine-grained, and purified.

Condition: Four fragments of the rim and body.

Description: Inverted rim with convex upper profile and pendant lip. Hemispherical body-wall.

Comparisons and comment: Cotton 1983, 366, no. 376, fig. 136 (Cozzo Presepe, Site A, Phase IV, Archaic period); Gassner 2003, no. I a 41, pl. 3 (Elea, 520 BC); Vittoria in *Survey*, 371, n. 67 (Metaponto, chora).

Date: From Late Archaic levels (Phase 2).

1:3

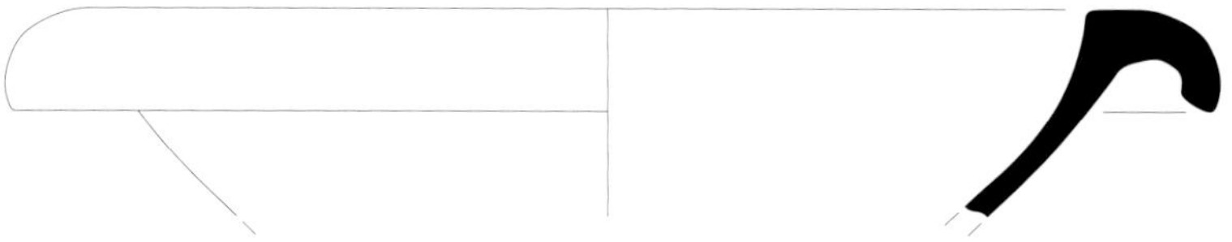

SAV PCW 19 Lekane/basin Type 8

1:3

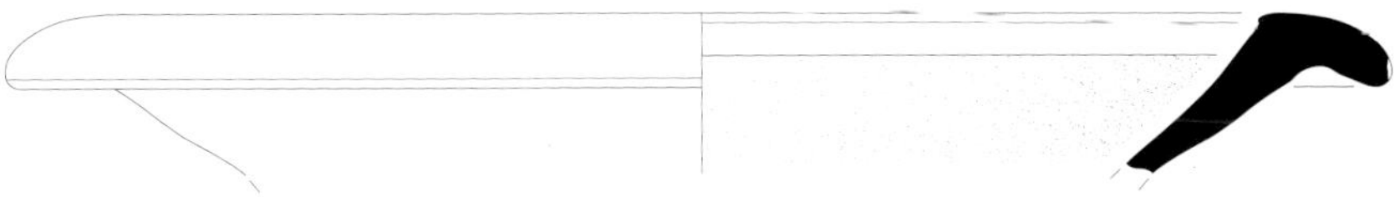

Provenance: Sounding C, Pottery Deposit. Lot: SA 79-988PL. SAV 76. Assemblage 15.
Dimensions and technical features: h 3.8, rim 34.0. Fabric SAV 3, Munsell 10Y 6/6, hard, compact, fine-grained, and well-levigated. Red-brown, opaque gloss.
Condition: One fragment of the rim and body. Interior surface glossed.
Description: Out-turned rounded rim, hemispherical body.
Comparisons and comment: Lo Porto 1973, no. 1.1, pl. XXV (Pisticci, via Negrelli, beginning of the 5th c. BC); Bottini and Lecce 2013, bacile tipo 1 (Pisticci, Santa Maria del Casale, tomb 1/1986, first quarter of the 4th c. BC). Vittoria in *Survey*, 371 no. 64 (Metaponto, chora) with rim more slightly oblique.
Date: 5th c. BC.

Plate/Lid

Sant'Angelo Vecchio yielded four basins featuring a simple flaring rim—either horizontal and elongated (variant 8a) or slightly pendant (variant 8b)—attached to a body with truncated conical section. The foot, where preserved, is low, barely distinguished, and slightly raised. These examples can be counted among the plate/lids, which were particularly handy due to the small size of their feet and their smaller overall size compared to the other basins. The type, attested in similarly small quantities also at Fattoria Stefan, has no parallels.

Plate/Lids

SAV PCW 20 Plate/lid Type 8a

1:3

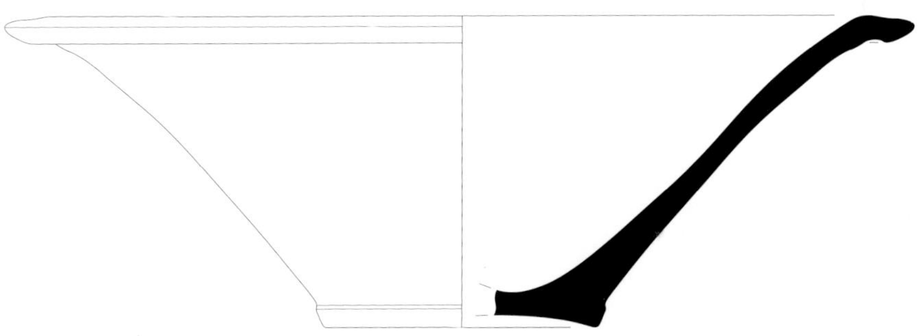

Provenance: Slope 3, W extension, grey votive deposit; Level 2, batt. 1. Slope 3, W extension, continued N (square D4); Level 2, batt. 1. Slope 3, square D4, N stone circle near baulk; Level 3, batt. 5. Lot: SA 79-591P, SA 79-424PL, SA 79-92PL, + SA 79-942TA. Inv. no. 319995. SAV 92 and SAV 204. Assemblage 14.
Dimensions and technical features: h 11.5, rim 34.8, base 10.1. Fabric SAV 4, Munsell 2.5YR 7/4, very hard, compact, medium-grained, and levigated, with tiny light and dark stone inclusions.
Condition: Profile reconstructed from 15 fragments.
Description: Horizontal flaring rim with oblique exterior face; body wall with truncated conical section; barely distinguished foot with flaring exterior profile.
Date: From Early Hellenistic levels (Phase 4).

SAV PCW 21 Plate/lid Type 8a

1:3

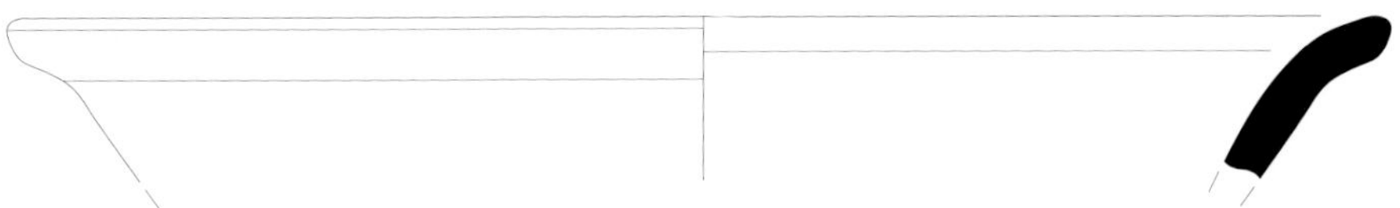

Provenance: Sounding D. Level 2, batt. 6. Lot: SA 80-56PL. SAV 6.
Dimensions and technical features: h 4.1, rim 36.0. Fabric SAV 1, Munsell 5YR 7/4, hard, compact, and fine-grained, with tiny white and black stone inclusions. Powdery surface.
Condition: One fragment of the rim and edge.
Description: Basin with oblique rim and flaring edge with vertical exterior profile.
Comparisons and comment: Bears some similarities to the basin (of larger dimensions) from Locri: Conti 1989, no. 363, pl. XLI, XLIX (4th–first half of the 3rd c. BC).
Date: 4th–3rd c. BC?

Lekanai/basins cont.

SAV PCW 22 Plate/lid Type 8b

1:3

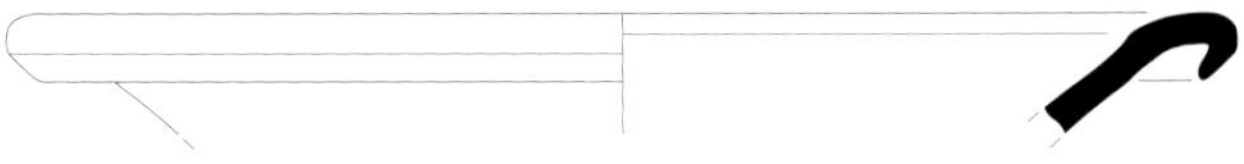

Provenance: Square H5. Level 1, batt. 2. Lot: SA 79-546PL (see *Fig. 2.21*).

Dimensions and technical features: h 2.4, rim 25.0. Fabric SAV 4, Munsell 2.5YR 6/8, nucleus Munsell 7.5YR 6/1, compact, slightly porous, fine and sparse black inclusions, small and fairly frequent white inclusions.

Condition: One fragment of rim and body.

Description: Hooked rim, straight body wall.

Comparisons and comment: Vittoria in *Survey*, 361, no. 25, Pl\Lne4b4 (Metaponto, chora).

Date: From Early Hellenistic levels (Phase 4).

Large Bowl

The materials recovered in the excavated farmhouses included a rim very similar to that found on several black-gloss bowls and plates, albeit on a much larger scale. The identified parallels are fairly scant and often shared with partially glossed vessels. Similar profiles were observed among the banded materials. A single type (Type 2) is attested at Sant'Angelo Vecchio among those identified in the excavated farmhouses.

Type 2. The second type is characterized by a thickened rim projecting slightly outward, attached to a moderately deep hemispherical body. Type 2 is rarely encountered in the excavated farmhouses, but two examples were found at Sant'Angelo Vecchio. The parallels indicate a chronology between the 4th and 3rd c. BC.

Large Bowl

SAV PCW 23 Large Bowl Type 2

1:3

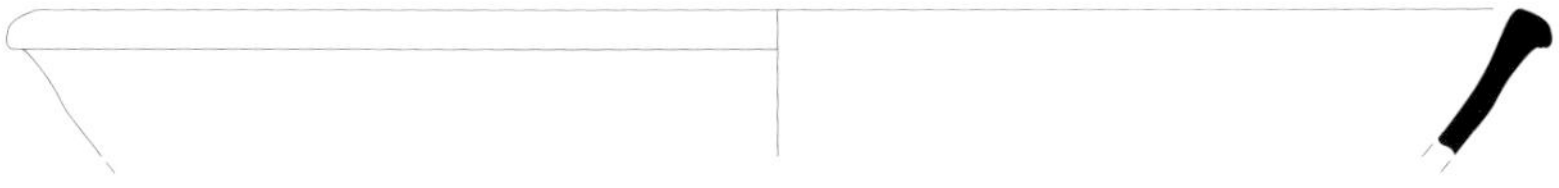

Provenance: Square G7. Level 1, batt. 2. Lot: SA 79-796PL. SAV 53. Assemblage 16.

Dimensions and technical features: h 2.9, rim 31.6. Fabric SAV 2, Munsell 7.5YR 7/4, hard, porous, and fine-grained, with tiny white, black, and micaceous inclusions. Powdery surface.

Condition: One fragment of the rim and body.

Description: Thickened rim projecting toward the exterior with convex and sloping exterior profile; oblique body wall with irregular profile.

Comparisons and comment: The fragment bears similarities with large bowls from Pomarico Vecchio: Bianco and Deodato 1997, 180, nos. 106–7, pl. 72 (320–280 BC). See also the bowl (of smaller diameter) from Locri: Conti 1989, 317–18, no. 378, pl. XLIII (4th–3rd c. BC).

Date: 4th–3rd c. BC.

Foot of Open Vessel

The following base can be ascribed to an open vessel featuring a wide and deep body, possibly identifiable as a lekane/basin by the raised disc foot with oblique exterior profile. The body walls are sturdy and moderately thick.

Foot of Open Vessel

SAV PCW 24 Base of an Open Vessel, possibly a basin

1:3

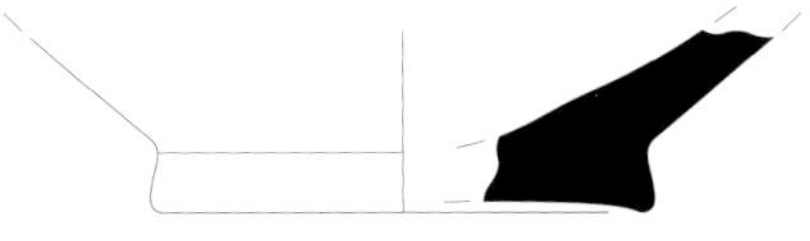

Provenance: Surface find. Lot: SA 79-7PL. SAV 182.

Dimensions and technical features: h 3.7, base diam. 10.25. Fabric SAV 3, Munsell 5YR 6/4, hard, compact, and fine-grained, with tiny white and black stone inclusions. Powdery surface.

Condition: Fragment of the foot and attachment of the body wall.

Description: Foot of an open vessel with oblique and rounded exterior profile.

Foot of Open Form: Large Bowl/Lekane

The following foot fragments may be ascribed to the large bowls and lekanai/basins attested at Sant'Angelo Vecchio. These vessels have a ring foot attached to a very capacious body. Despite the fragments' lack of decoration, which classes them with the plain and coarse wares, originally they could have been attached to partially glossed bodies.

Feet of Open Forms: Large Bowl/Lekane

SAV PCW 25 Base of Lekane/basin

1:3

Provenance: House on the Hill, Test Trench 1; Level 1, batt. 2. Test Trench 2; Level 1, batt. 1; Level 1, batt. 2. Lot: SA 79-558PL, SA 79-557PL, + SA 79-681P. SAV 132 (see *Fig. 2.39*).
Dimensions and technical features: h 9.6, base 18.0. Fabric SAV 3, Munsell 5YR 7/4, hard, compact, fine-grained, and well-levigated, with rare small gray inclusions.
Condition: Three fragments of the foot and attachment of the wall.
Description: Foot of an open form with curving exterior profile; convex wall.
Date: From Late Archaic levels (Phase 2).

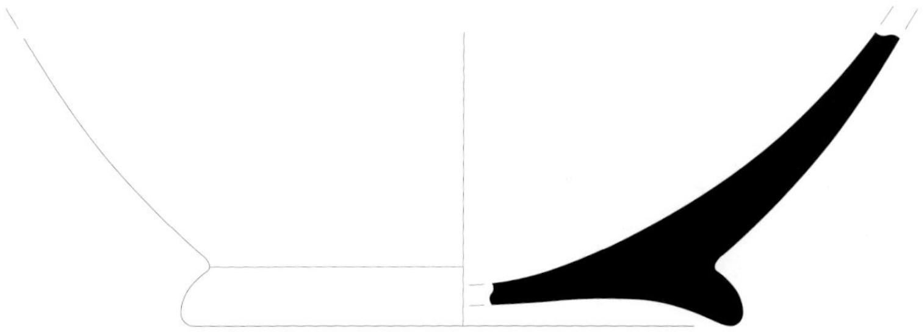

SAV PCW 26 Base of Lekane/basin

1:3

Provenance: Sounding C, Pottery Deposit. Lot: SA 79-988PL. Assemblage 15.
Dimensions and technical features: h 3.0, base 15.8. Fabric SAV 1, Munsell 2.5YR 6/8, hard, compact, and fine-grained, with rare tiny white inclusions.
Condition: Three fragments of the foot and attachment of the body wall.

Description: Foot of an open form with curving exterior profile and slightly convex bottom surface.
Date: 5th–4th c. BC.

SAV PCW 27 Base of an Open Vessel

1:3

Provenance: Square G6, W of wall. Level 2, batt. 1. Lot: SA 79-733PL. SAV 122 (see *Fig. 2.20*).
Dimensions and technical features: h 2.5, base 11.3. Fabric SAV 1, Munsell 5YR 7/4, hard, compact, and fine-grained, with tiny white and black stone inclusions. Powdery surface.
Condition: One fragment of the foot and attachment of the body.

Description: Foot of an open form with oblique and contracting exterior profile.
Date: From levels of Phase 4 (Early Hellenistic period).

Large Basin

A group of basins is distinguished by their remarkable size: wide rim, massive walls, and deep body. Their size probably allowed these vessels to serve a variety of functions not exclusively associated with food. No precise parallels were found. (SAV PCW 34–SAV PCW 36) are classed in a single group. Considering their remarkable volume, they could have been likely used for various functions in domestic and manufacturing contexts. The type is represented at Sant'Angelo Vecchio by four examples, all of which were found, significantly, inside the workshop and in the vicinity of the kiln area. All of the fragments studied have a massive flaring rim with sub-rectangular section and oblique or slightly convex wall.

Large Basins

SAV PCW 28 Large Basin

Provenance: House on the Hill, Test Trench 2, W side. Level 1, batt. 2. Lot: SA 79-569T. SAV 157 (see *Fig. 2.39*).

Dimensions and technical features: h 6.9, rim int. 34.0. Clay Munsell 5YR 6/6, hard, compact, and medium-grained, with numerous micaceous inclusions.

Condition: One fragment of the rim and body wall.

Description: Massive flaring, inclined rim with rounded exterior profile; oblique wall. On the rim, shallow groove.

Date: From Late Archaic levels (Phase 2).

1:3

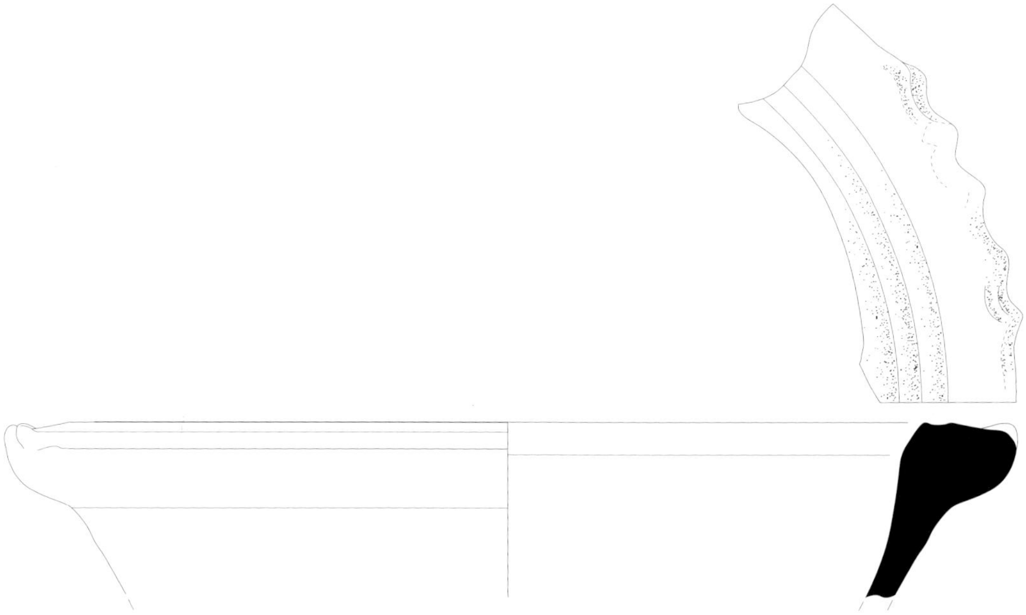

SAV PCW 29 Large Basin

Provenance: Square G6, E of wall. Level 2, batt. 2. Lot: SA 79-739PL. SAV 134. Assemblage 3.

Dimensions and technical features: h 7.8, rim 29.3. Clay Munsell 5YR 7/6, greenish. Evident smoothing and wrinkling of the surface.

Condition: One fragment of the rim and body.

Description: Thickened rim with sub-circular section, with slightly downward-sloping upper surface.

Date: Early Hellenistic (Phase 4).

1:3

SAV PCW 30 Large Basin
Provenance: Square G6 (from SA 79-737T). Level 2, batt. 2. Lot: SA 79-746PL. SAV 159. Assemblage 3.
Dimensions and technical features: h 11.4, rim ext. 34.0. Clay Munsell 5YR 7/6, hard and medium-grained, with numerous small dark stone inclusions.
Condition: One fragment of the rim and body wall.
Description: Massive flaring, inclined rim with vertical exterior profile and oblique lower surface; slightly oblique wall. On the rim, shallow groove.
Date: From Early Hellenistic levels (Phase 4).

1:3

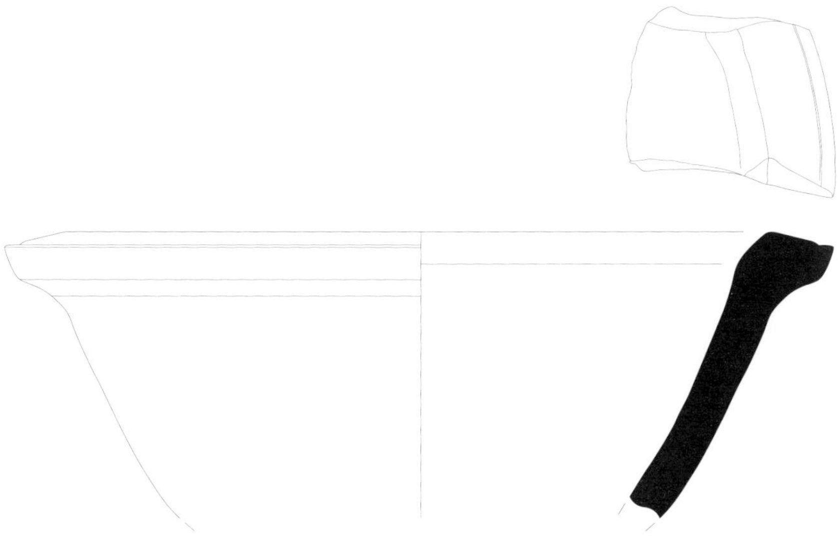

SAV PCW 31 Large Basin
Provenance: Square G7, cleaning around wall 4 to W. Lot: SA 79-1010PL.
Dimensions and technical features: h. 10.2, rim 34.0. Fabric SAV 4, Munsell 2.5YR 6/8 hard, low porosity, rare small white inclusions, rare small grog, rare silver mica.
Description: One fragment of the rim and body. The fragment preserves the attachment of the handles.
Comparisons and comment: P. Bottini 1988, 195, fig. 33, no. 270220 (S. Evraso, second half of the 4th–3rd c. BC).
Date: 4th–3rd c. BC.

1:3

Large basins cont.

SAV PCW 32 Large Basin

Provenance: Square F3, inside N Kiln. Level 2, batt. 1 and 4. Lot: SA 79-873PL + SA 79-883PL. Inv. no. 320050. SAV 190. Assemblage 8.

Dimensions and technical features: h 9.4, rim ext. 26.3. Fabric SAV 4, Munsell 2.5YR 5/6, very hard, compact, medium-grained, and levigated, with tiny light and dark stone inclusions.

Condition: Profile reconstructed from two fragments.

Description: Massive, thickened rim protruding toward the exterior; oblique wall; flat base.

Date: From Late Republican levels (Phase 5).

1:3

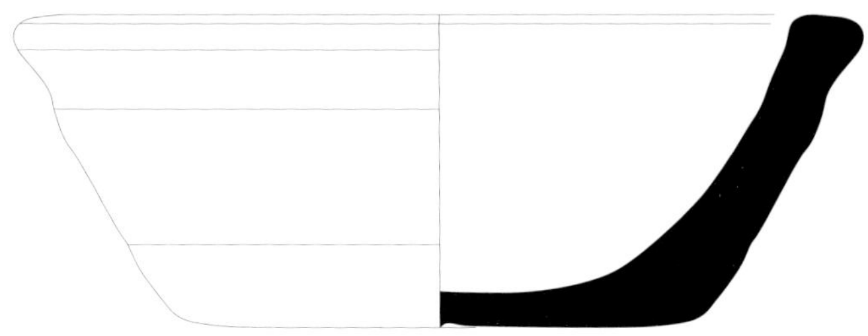

SAV PCW 33 Large Basin

Provenance: Square H3, cleaning of wall. Level 1, batt. 3. Lot: SA 79-1024PL (topsoil). SAV 173. Assemblage 2.

Dimensions and technical features: h 7.3, rim 23.0. Fabric SAV 4, Munsell 2.5YR 6/6, very hard, compact, medium-grained, with tiny light and dark stone inclusions.

Condition: Two fragments of the rim and base.

Description: Flaring and barely distinct rim and flat base.

1:3

SAV PCW 34 Large Basin

Provenance: Square H3. Level 1, batt. 1. Lot: SA 79-560PL (topsoil). SAV 36 (see *Fig. 2.22*).

Dimensions and technical features: h 9.6, rim 29.6. Fabric SAV 4, Munsell 2.5YR6/6, very hard, compact, medium-grained, and levigated, with tiny light and dark stone inclusions.

Condition: Six fragments of the rim and body wall.

Description: Massive flaring, elliptical rim with rectangular section, joined to the body on the interior by an oblique surface.

1:3

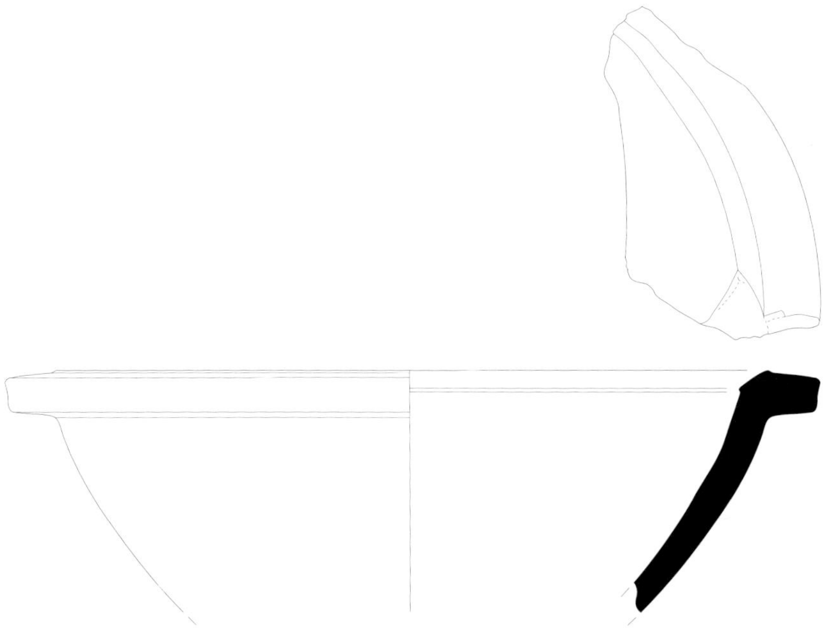

SAV PCW 35 Large Basin

Provenance: Square H3. Level 1, batt. 1. Lot: SA 79-677PL (topsoil). SAV 29 (see *Fig. 2.22*).

Dimensions and technical features: h 8.6, rim 28.0. Fabric SAV 4, Munsell 5YR 6/4, very hard, compact, medium-grained, and levigated, with tiny light and dark stone inclusions.

Condition: One fragment of the rim and body wall.

Description: Massive, flaring, and flat rim with vertical exterior profile and oblique lower surface; oblique wall.

Comparisons and comment: For the profile, see Cotton 1983, 372, no. 423, fig. 142 (Cozzo Presepe, 250 BC).

Date: From Late Republican levels (Phase 5).

1:3

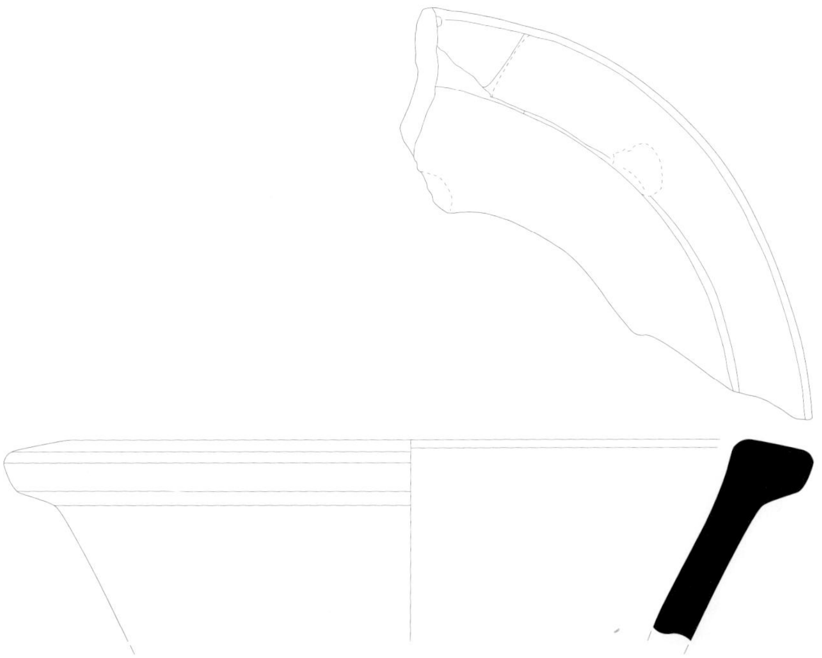

SAV PCW 36 Large Basin

1:3

Provenance: Square H3. Level 1, batt. 1. Lot: SA 79-677PL. SAV 30 (see *Fig. 2.22*).

Dimensions and technical features: h 4.9, rim unknown. Fabric SAV 4, Munsell 2.5YR 6/6, very hard, compact, medium-grained, and levigated, with tiny light and dark stone inclusions.

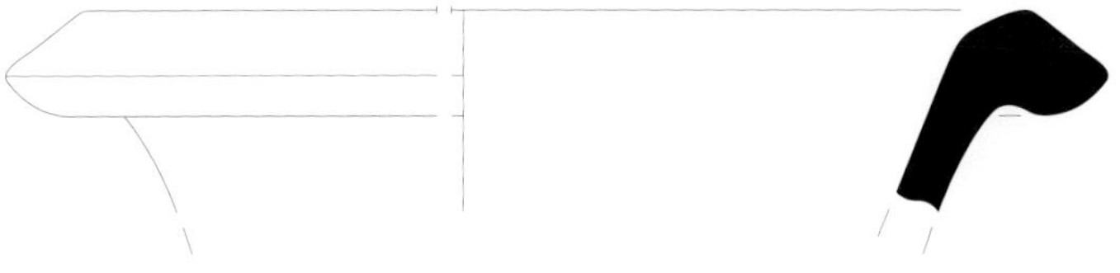

Condition: One fragment of the rim and body wall.

Description: Flaring rim, protruding toward the exterior, with parallelepiped section, downward-sloping exterior surface, and molded lower edge; oblique body with elliptical continuation.

Date: From Late Republican levels (Phase 5).

Feet of Open Vessels, Possibly of Large Basins

SAV PCW 37 Large Basin

1:3

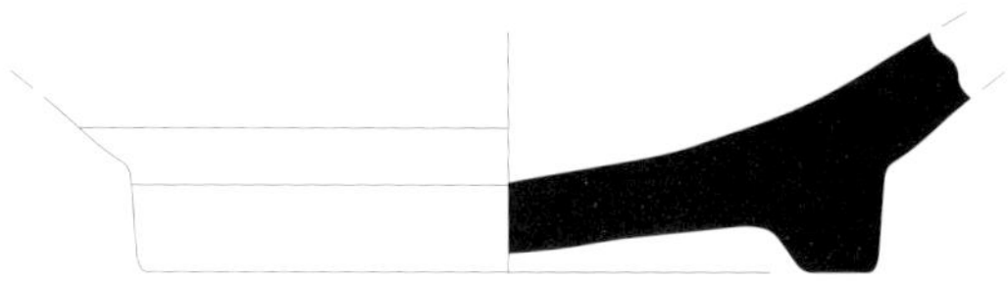

Provenance: House on the Hill, Test Trench 2. Level 1, batt. 1. Lot: SA 79-616PL. SAV 140 (see *Fig. 2.39*).
Dimensions and technical features: h 4.9, base 15.5. Fabric SAV 1, Munsell 5YR 7/8, hard, compact, and fine-grained, with tiny white and black stone inclusions. Powdery surface.
Condition: One fragment of the foot and attachment of the body wall.
Description: Massive foot of an open form with vertical exterior surface; convex walls.
Date: From Late Archaic levels (Phase 2).

SAV PCW 38 Large Basin

1:3

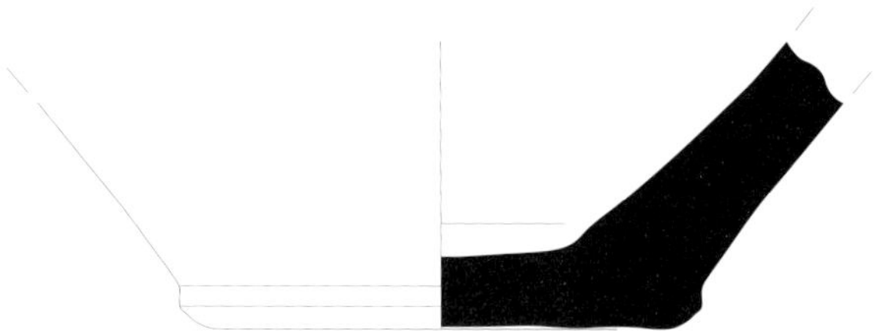

Provenance: Cut in baulk between N and S Kilns. Level 2, batt. 1–2. Lot: SA 81-7PL. Assemblage 5.
Dimensions and technical features: h 7.4, base 13.85. Fabric SAV 4, Munsell 2.5YR 5/4, very hard, compact, medium-grained, and levigated, with tiny light and dark stone inclusions.
Condition: One fragment of the foot.
Description: Low, distinct foot of a large and deep vessel with inclined exterior profile and flat resting surface. Possibly reused in the construction of one of the kilns.
Date: From a Late Republican context (Phase 5).

Deep Basin

Ten examples of deep basins, featuring a thickened rim and vertical and slightly convex walls, were recovered at Sant'Angelo Vecchio. These vessels, though commonly encountered in the other excavated Metapontine farmhouses, have few parallels in other contexts: a similar example was found in the Pantanello necropolis (Carter and Parmly Toxey in *Necropoleis*, 725, T64-1), while others were found at Herakleia (Pianu 1990, 32, pl. 13.2) and in Apulia (Giannotta 1997, 173, fig. 4), with a chronological range between the last quarter of the 4th and the beginning of the 2nd c. BC. The form's peculiar structural properties lent it to various uses in a domestic context, including the storage of foodstuffs for short and medium periods. The examples found in the excavated Metapontine farmhouses were classified in three types based on the variations in the shape of the rim; Types 1 and 3 are attested at Sant'Angelo Vecchio.

Type 1. The first type features a thickened and rounded rim attached to a slightly oblique wall. The foot, when present, is narrow and barely distinct. The half-moon handles are pressed against the wall immediately beneath the rim. Very often the wall bears a decoration of one or more lines incised with a stick. SAV PCW 39 has three grooves separating two parallel rows of tiny oblique leaf-like markings oriented in a mirror pattern on either side of the central groove, while SAV PCW 40 bears an incised braid motif beneath the groove.

The type is also attested at the Pantanello necropolis (Carter and Parmly Toxey in *Necropoleis*, 725, T64-1) and in the Metapontine chora (Vittoria in *Survey*, 404, DBas2a). Three individual vessels, with rim diameters in the range of 38.0 to 40.0 cm, belong to this type.

Type 3. The first variant of the third type (3a) is characterized by an outward-projecting rim with triangular section, attached to a thin and oblique wall. The rim is wide, and the body is very deep. The second variant (3b) features a thin and elongated outward-projecting rim. The type, infrequently encountered among the excavated Metapontine farmhouses, is dated to the 4th c. BC by rare parallels.

Deep Basins

SAV PCW 39 Deep Basin Type 1

Provenance: Slope 3, W extension, E of rock feature, votive deposit S. Level 3, batt. 1. Lot: SA 79-1060P. Inv. no. 320047. SAV 170. Assemblage 13.

Dimensions and technical features: h 18.8, rim ext. 37.0. Fabric SAV 3, Munsell 5YR 7/6, hard, compact, fine-grained, and well-levigated.

Condition: Two fragments of the rim and body.

Description: Thickened rim projecting toward the exterior with convex upper profile; oblique wall contracting toward the bottom. On the interior, numerous grooves left by the potter's wheel; on the wall, three deep incisions made with a stick, between which run two oblique leaf-like motifs arranged in a mirror pattern on either side of the central incision.

Comparisons and comment: Pianu 1990, 32, pl. 13.2 (Herakleia, 350–325 BC); Vittoria in *Survey*, 404, DBa-s2a (Metaponto, chora).

Date: 4th–3rd c. BC.

1:3

SAV PCW 40 Deep Basin Type 1

Provenance: Slope 4, W extension; Level 1, batt. 2. Sounding C, E of and including division; Level 2, batt. 3. Lot: SA 79-125P (from SA 79-119PL) + SA 79-322PL. SAV 197.

Dimensions and technical features: h 8.6, rim 37.0. Fabric SAV 2, Munsell 5YR 7/6, hard, porous, and fine-grained, with tiny white, black, and micaceous inclusions. Powdery surface.

Condition: Two fragments of the rim and body.

Description: Thickened rim projecting toward the exterior with convex upper profile; oblique walls; horizontal grip attached beneath the rim. Two deep grooves made with a stick at the level of the rim; one groove on the wall, beneath which runs a braid pattern made with a rounded instrument.

Comparisons and comment: See SAV PCW 39.

Date: 4th–3rd c. BC.

1:3

Deep basins cont.

SAV PCW 41 Deep Basin Type 1

1:3

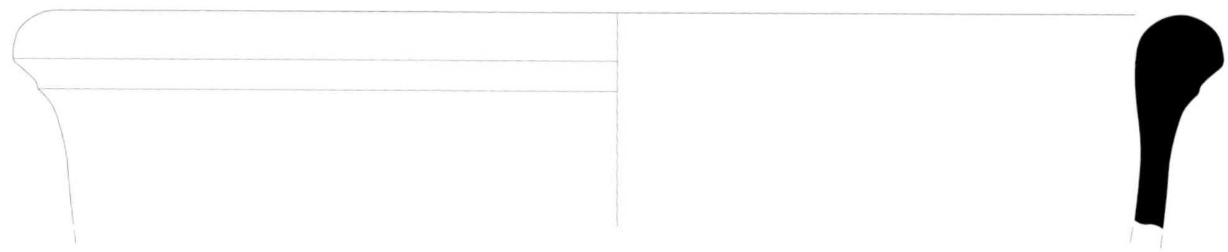

Provenance: Slope 4, W extension. Level 1, batt. 1. Lot: SA 79-119PL. SAV 180 (see *Fig. 2.35*).

Dimensions and technical features: h 6.8, rim 40.0. Fabric SAV 2, Munsell 7.5YR 7/4, hard, porous, and fine-grained, with tiny white, black, and micaceous stone inclusions. Powdery surface.

Condition: One fragment of the rim.

Description: Thickened rim projecting toward the exterior with convex upper profile; vertical body wall.

Comparisons and comment: See SAV PCW 39.

Date: 4th–3rd c. BC.

SAV PCW 42 Deep Basin Type 3a

1:3

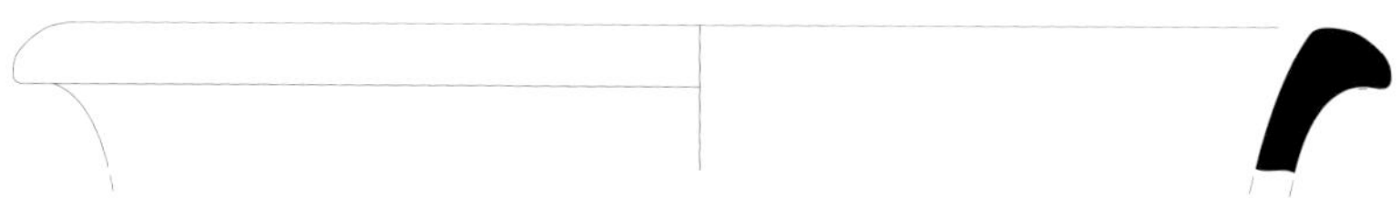

Provenance: Square H5. Level 2, batt. 2. Lot: SAV79-806PL. SAV 50 (see *Fig. 2.21*).

Dimensions and technical features: h 3.2, rim 31.0. Fabric SAV 1, Munsell 5YR 7/6, hard, compact, and fine-grained, with tiny white and black stone inclusions. Powdery surface.

Condition: One fragment of the rim and body.

Description: Thickened rim projecting toward the exterior with rounded upper profile.

Date: From Early Hellenistic levels (Phase 3).

SAV PCW 43 Deep Basin Type 3a

1:3

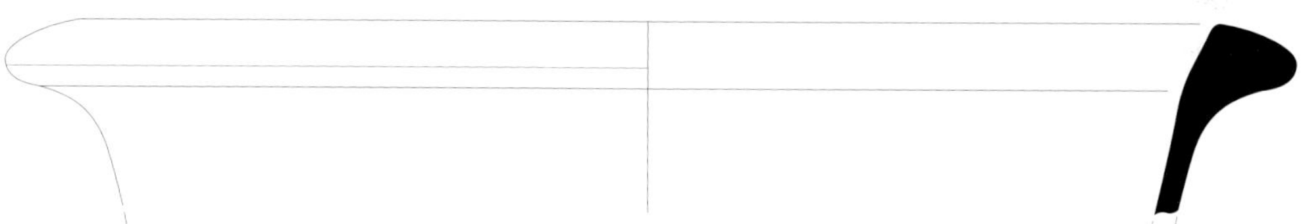

Provenance: Sounding C. Level 2, batt. 7. Lot: SA 79-205PL. SAV 128 (see *Fig. 2.35*).

Dimensions and technical features: h 5.8, rim ext. 35.8. Fabric SAV 4, Munsell 5YR 7/6, very hard, compact, medium-grained, and levigated, with tiny light and dark stone inclusions.

Condition: Two fragments of the rim and body.

Description: Thickened rim projecting toward the exterior with oblique upper profile; vertical body wall.

Comparisons and comment: Bears similarities to Cotton 1992a, 176, no. 1320, fig. 65; *Con il fuso*, 217, no. 99 (Montemurro, 4th c. BC).

Date: 4th c. BC.

SAV PCW 44 Deep Basin Type 3b

1:3

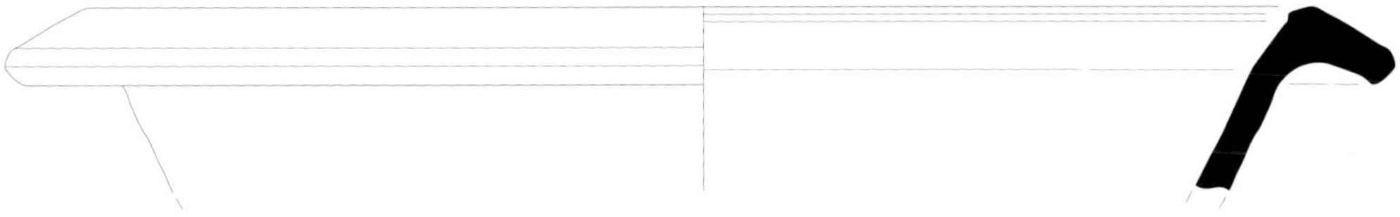

Provenance: Square G6, W of wall; Level 2, batt. 2. Slope 3, W extension, stipe N. Level 2, batt. 1. Lot: SA 79-749PL + SA 79-385PL. SAV 65 and SAV 110. Assemblage 13.

Dimensions and technical features: h 5.2, rim int. 35.8. Fabric SAV 1, Munsell 10YR 7/3, hard, compact, and fine-grained, with tiny white and black stone inclusions. Powdery surface. Misfired.

Condition: Two fragments of the rim and body.

Description: Flaring oblique rim with rounded exterior profile, oblique walls. A deep offset on the interior marks the transition from the rim to the body wall.

Comparisons and comment: Similar to Bianco and Deodato 1997, 178, no. 88, pl. 69, which presents a more inclined rim (Pomarico, second half of the 3rd c. BC).

Date: From Early Hellenistic levels (Phase 4).

Dinos

The excavated Metapontine farmhouses yielded an abundant group of open forms equipped with wide mouths and deep bodies, identified as dinoi. All of the examples have a thickened and slightly inward-curved rim that protrudes on the exterior, mounted on a nearly vertical wall. The form is also frequently attested in the Metapontine chora, although it is generally more common in the banded version (Cavallo in *Fattoria Fabrizio*, 275). Because of its resemblance to the better-known Archaic vessels, in the study of the materials from the chora it was named *dinos* (Dn) (Vittoria in *Survey*, 376–79). It is not attested in the Pantanello necropolis.

Several hypothetical functions have been proposed for the Metapontine dinos, most notably storage of foodstuffs in the pantry and the mixing of beverages, as is known for the kraters (containers with analogous profile are found among the "storage vessels" in Saunders and du Plat Taylor 1992, 30–31, and in *Valesio I*, 97, 271).

None of the examples recovered in the Metapontine farmhouses or in the Bradano-Basento field survey preserves the complete profile; the shape of the base is therefore unknown. The handles, when present, can be ribbon-shaped and attached immediately beneath the rim with a circular section (SAV PCW 39) or, as on several banded examples, have a sub-circular section and be pressed against the body wall (Cavallo in *Fattoria Fabrizio*, 247, FF BW 22).

The rim is the most significant diagnostic element for the typological seriation. Four types were identified among the materials recovered in the excavated Metapontine farmhouses, only one of which is attested at Sant'Angelo Vecchio with a single individual vessel.

Type 1. The rim of Type 1 has the form of a "T" and is slightly inward-projecting; the exterior edge is thickened and protruding. The mouth is wide, and the wall is slightly oblique, suggesting a notable depth.

Dinos

SAV PCW 45 Dinos Type 1

1:3

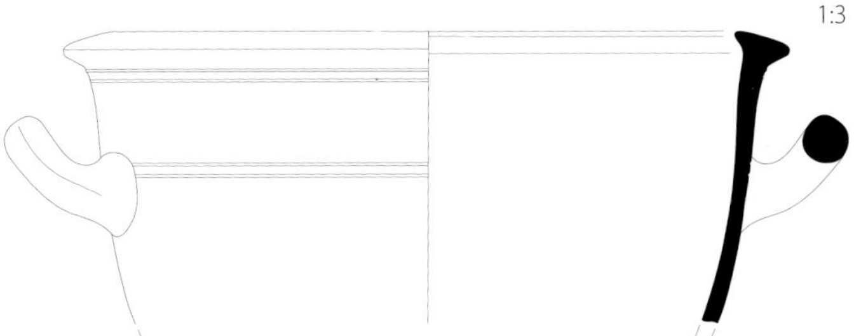

Provenance: Sounding C, Pottery Deposit. Lot: SA 79-988PL 2. SAV 78. sAssemblage 15.

Dimensions and technical features: h 9.2, rim 23.6. Fabric SAV 3, Munsell 5YR 7/6, hard, compact, fine-grained, and well-levigated.

Condition: Two fragments of the rim, handle, and body wall.

Description: Thickened, inward-projecting rim, protruding on both the interior and the exterior; ovoid or globular body. On the body, two horizontal incisions made with a stick.

Comparisons and comment: Bianco and Deodato 1997, 179, pl. 71, no. 98 (Pomarico Vecchio, late 4th–early 3rd c. BC, with further comparisons). Vittoria in *Survey*, 377, no. 93, Dn5a (Metaponto, chora). Cavallo in *Fattoria Fabrizio*, 276, FF PCW 24 (Metaponto, Fattoria Fabrizio, 350–300 BC).

Date: 4th c. BC?

Closed Shapes (Jug, Pelike, Hydria, and Table Amphora)

Among the undecorated materials at Sant'Angelo Vecchio were found 210 fragments comprising 87 individuals, generally in a very fragmentary state, that belonged to closed forms with a narrow neck. This broad identification cannot be specified beyond a larger group including jugs, pelikai, hydriai, and table amphorae.

Noteworthy among the materials is a fragment of a jug rim (SAV PCW 62) made with a fabric very similar to that used in cooking ware. The quantity and types of inclusions, alongside the typical reddish coloring, are highly unusual among the undecorated ceramics documented in the Metapontino, which are nearly always characterized by well-levigated fabrics. The attested presence of this fabric at all of the excavated farmhouses excludes an origin among the cooking wares, giving the issue greater complexity than is at first apparent. In fact, although they are indisputably close to the undecorated jug types, the morphological features of the rims in question are often similar to those found on small pots. The same fabric is attested sporadically in other undecorated forms, such as some oinochoai and basins, as well as one lekane (SAV PCW 05). The use of this fabric beyond cooking vessels makes the presence of jugs in unlevigated fabric less of an anomaly. The use of refractory fabrics in the production of jugs and other vessels is confirmed also at Torre di Satriano and Valesio (Rinaldi 2005, nos. 94–96, pl. XXVI; *Valesio I*, 279; for the other Metapontine farmhouses, see Cavallo in *Fattoria Fabrizio*, 264).

The range of morphological variation in the jugs documented at the Metapontine farmhouses is extraordinarily broad, with seven types identified (Cavallo in *Fattoria Fabrizio*, 277). Examples belonging to Types 1–5 were found at Sant'Angelo Vecchio. The rims have diameters in the range of 10 to 18 cm, indicating vessels of medium and large size.

Type 1. The first type is characterized by a flaring, slightly protruding, and thickened rim; the neck can have a cylindrical or truncated conical profile, separated from the body by a carination. The rim diameters are in the range of 10.0 to 18.0 cm. SAV PCW 46 and SAV PCW 47, probably one-handled jugs, are paralleled by a banded example of the same form with the handle attached to the rim (SAV BW 38). Type 1 is attested by 93 fragments belonging to 14 individuals, mainly from levels connected to the Archaic, Late Archaic, and Early Classical occupation.

Type 2. The second type features a flaring, rectangular, flat-brimmed rim and a narrow, truncated conical neck. The type is documented by three individual vessels.

Type 3. The third type is characterized by a rim mounted on a short, vertical, cylindrical neck. Three fragments were found at Sant'Angelo Vecchio.

Type 4. The fourth type is characterized by a rim with listel with triangular section mounted on a cylindrical neck. The type is also attested—and much more commonly—in banded ware (see p. 296 in Ch. 20, "Jugs"). The parallels from the Bradano watershed date between the first half of the 5th and the end of the 4th c. BC, with attestations through the beginning of the 1st c. BC at Gravina di Puglia. Twelve fragments were found at Sant'Angelo Vecchio, representing 10 vessels.

Type 5. Type 5 features a flaring and slightly thickened rim with indented exterior profile. The type is attested at all of the excavated Metapontine farmhouses with the exception of Sant'Angelo Grieco. In most of the cases examined, the fabric is fairly similar to that used in cooking ware. The morphological features are repetitive and standardized: the handle, when preserved, is always pressed against the rim, suggesting a one-handled jug. A single example was recovered at Sant'Angelo Vecchio. The comparanda indicate a chronology between the end of the 4th and the 3rd c. BC.

Closed Shapes (Jugs, Pelikai, Hydriai, and Table Amphorae)

SAV PCW 46 Jug Type 1

1:3

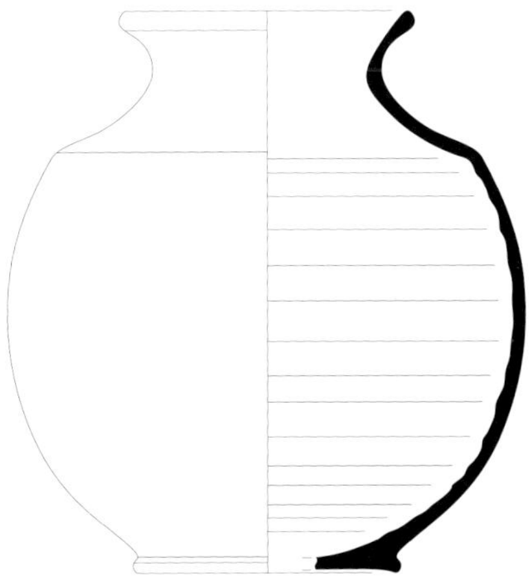

Provenance: House on the Hill. Level 1, batt. 2. Lot: SA 79-1061P. SAV 174 (see *Fig. 2.39*).

Dimensions and technical features: rim h 11.7, base h 11.3, rim 9.6, base 8.5. Fabric SAV 3, Munsell 5YR 7/8, hard, compact, fine-grained, and well-levigated.

Condition: Reconstructed from 20 fragments.

Description: Flaring, thickened rim with rounded exterior profile mounted on a splayed neck connected to the shoulder by a marked edge; globular body; low, distinct ring foot.

Comparisons and comment: Corchia/Pancrazzi/Tagliente 1982, 26–27, pl. 20,3 and 23,1 (Cavallino, Fondo "Aiera Vecchia," second half of the 6th–first quarter of the 5th c. BC). Mastronuzzi 2013, 277 (Oria, Monte Papalucio sanctuary, banded ware, jug type 1, 6th–5th c. BC).

Date: End of the 6th–5th c. BC.

SAV PCW 47 Jug Type 1

1:3

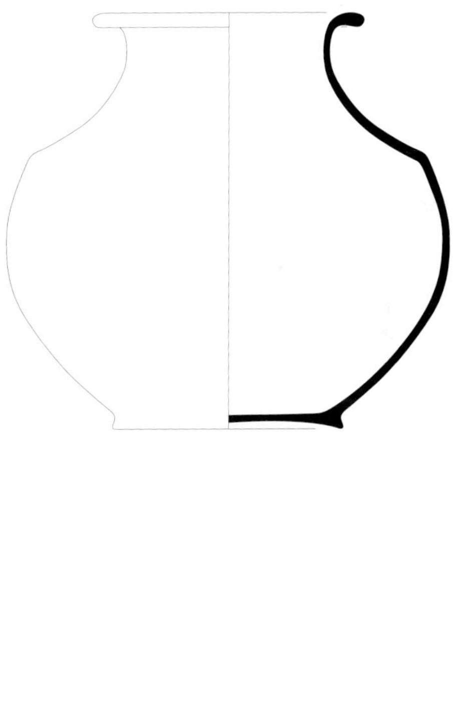

Provenance: Sounding C, Pottery Deposit. Inv. no. 319911. Lot: SA 79-926P. Assemblage 15.

Dimensions and technical features: h 16.7, rim 11.0. Fabric SAV 1, Munsell 5YR 6/6 with core Munsell 5YR 5/3, hard and compact, with very rare small white inclusions. Rare elongated vacuoles.

Condition: Entire profile reconstructed from 40 fragments.

Description: Everted rim, wide neck, carinated shoulder, and ring foot with low convex surface.

Comparisons and comment: Corchia/Pancrazzi/Tagliente 1982, 26–27, pl. 20,5 and 23,2 (Cavallino, Fondo "Aiera Vecchia," second half of the 6th–first quarter of the 5th c. BC). Bears similarities to the banded ware jug in Mastronuzzi 2013, 87, no. 113, fig. 52 (Oria, Monte Papalucio sanctuary, votive deposit no. 3, 480–470 BC).

Date: 5th c. BC.

Closed shapes cont.

SAV PCW 48 Jug Type 1

1:3

Provenance: Sounding C; Level 2, batt. 6. Sounding C; Level 2, batt. 7. Lot: SA 79-201PL + SAV79-205PL. SAV 102 (see *Fig. 2.35*).

Dimensions and technical features: h 6.5, rim 11.5. Fabric SAV 1, Munsell 5YR 7/8, hard, compact, and fine-grained, with tiny white and black stone inclusions. Powdery surface.

Condition: Four fragments of the rim and neck.

Description: Flaring and slightly thickened rim with rounded upper profile, mounted on a concave neck.

Comparisons and comment: P. Bottini 1988, no. 270067; *Kaulonia I*, 69, no. 326, fig. 48 (Caulonia).

Date: 5th c. BC.

SAV PCW 49 Jug Type 1

1:3

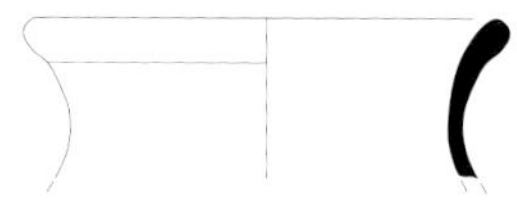

Provenance: Sounding C, Pottery Deposit. Lot: SA 79-988PL. SAV 89. Assemblage 15.

Dimensions and technical features: h 3.2, rim 10.0. Fabric SAV 3, Munsell 5YR 7/4, hard, compact, fine-grained, and well-levigated.

Condition: One fragment of the rim and neck.

Description: Indistinct, flaring, and slightly thickened rim with rounded termination; concave neck.

Date: From Classical levels (Phase 3).

SAV PCW 50 Jug Type 1

1:3

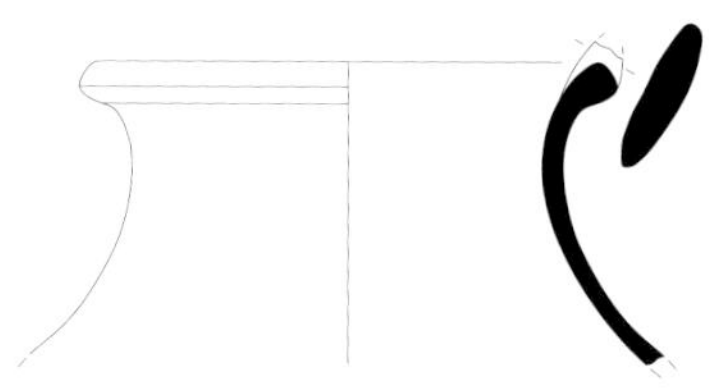

Provenance: Sounding C, Pottery Deposit. Lot: SA 79-888P. SAV 206. Assemblage 15.

Dimensions and technical features: h 6.2, rim 11.0. Fabric SAV 3, Munsell 7.5YR 7/4, hard, compact, fine-grained, and well-levigated.

Condition: Two fragments of the rim and neck.

Description: Flaring and slightly thickened rim with rounded exterior edge and sharply angled upper profile, mounted on a flaring neck; vertical handle attached to the rim.

Comparisons and comment: Carter 1983a, 485, fig. 53 (Pantanello); P. Bottini 1988, inv. no. 270067, fig. 29 (S. Evraso, 320–280 BC); *Kaulonia I*, 69, no. 323, fig. 48 (Caulonia, chronology not clearly indicated).

Date: From Classical levels (Phase 3).

SAV PCW 51 Jug Type 1

1:3

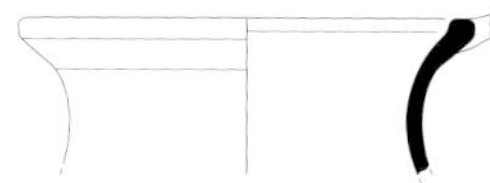

Provenance: House on the Hill. Lot: SA 79-1062PL. SAV 58. Assemblage 1.

Dimensions and technical features: h 3.0, rim 9.0. Fabric SAV 3, Munsell 5YR 7/6, hard, compact, fine-grained, and well-levigated.

Condition: One fragment of the rim and neck.

Description: Thickened, vertical rim with sub-rectangular section, convex exterior surface, and straight interior profile; concave neck.

Date: From Late Archaic levels (Phase 2).

SAV PCW 52 Jug Type 1

1:3

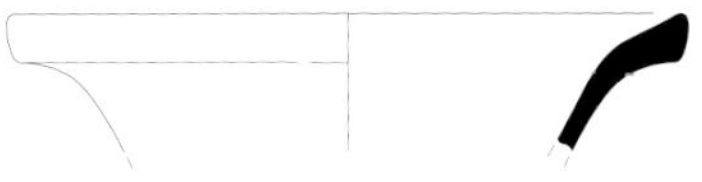

Provenance: Sounding D, E of Sounding A. Level 2, batt. 5a. Lot: SA 80-53PL. SAV 14.

Dimensions and technical features: h 2.8, rim 13.8. Fabric SAV 1, Munsell 5YR 7/8, hard, compact, and fine-grained, with tiny white and black stone inclusions. Powdery surface.

Condition: One fragment of the rim and neck.

Description: Flaring and thickened rim with vertical exterior edge; oblique neck.

Comparisons and comment: Bears similarities to Cotton 1983, 369, no. 407 (Cozzo Presepe, 300 BC).

Date: 4th c. BC?

SAV PCW 53 Rim of a Closed Shape, possibly a Jug Type 1

1:3

Provenance: Square F3, N Kiln interior. Level 2, batt. 1. Lot: SA 79-873PL. SAV 188. Assemblage 8.

Dimensions and technical features: h 10.3, rim 14.5.

Condition: Four joining fragments of rim, neck, and body.

Description: Flaring rim with slightly convex interior surface; short neck to which the handles are attached; globular body.

Comparisons and comment: Pantanello, kiln deposit (PZ 80.0222.101, 2nd–1st c. BC).

Date: From Late Republican levels (Phase 5).

SAV PCW 54 Jug Type 2

1:3

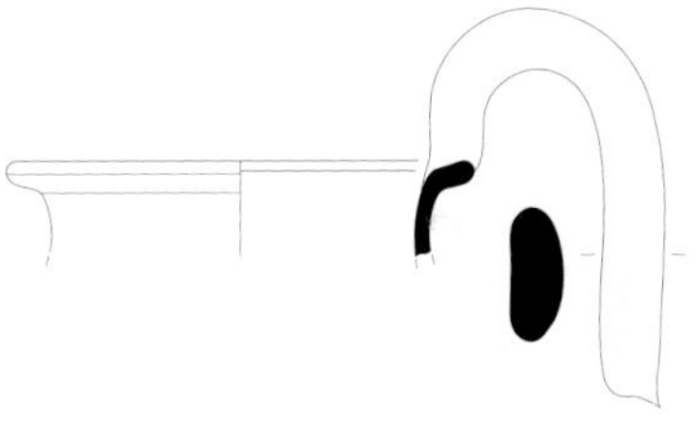

Provenance: Sounding C, Pottery Deposit. Lot: SA 79-988PL. SAV 88. Assemblage 15.

Dimensions and technical features: h 8.2, rim 10.1. Fabric SAV 3, Munsell 5YR 7/6, hard, compact, fine-grained, and well-levigated.

Condition: One fragment of the rim and neck.

Description: Flaring rim with flat upper profile and externally rounded edge; oblique neck.

Date: From Classical levels (Phase 3).

SAV PCW 55 Hydria or Table Amphora Type 2

1:3

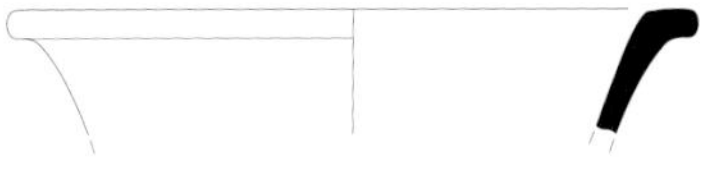

Provenance: House on the Hill. Lot: SA 79-1062PL. SAV 72. Assemblage 1.

Dimensions and technical features: h 2.5, rim 14.4. Fabric SAV 3, Munsell 5YR 7/4, hard, compact, fine-grained, and well-levigated.

Condition: One fragment of the rim and neck.

Description: Flaring, horizontal rim with flat upper surface; oblique neck.

Comparisons and comment: Bears similarities to Vittoria in *Survey*, 383, no. 116, J1e (Metaponto, chora).

Date: From Late Archaic levels (Phase 2).

Closed shapes cont.

SAV PCW 56 Closed Shape Type 2

Provenance: F7. Level 1, batt. 3. Lot: SA 80-3PL. SAV 16 (see *Fig. 2.20*).

Dimensions and technical features: h 4.3, rim 18.2. Clay Munsell 7.5YR 6/6, hard, porous, and fairly well-levigated, with rare small white and gray and very rare large black inclusions.

Condition: One fragment of rim.

Description: Flaring, horizontal rim with flat upper surface; oblique neck.

Date: From Early Hellenistic levels (Phase 4).

1:3

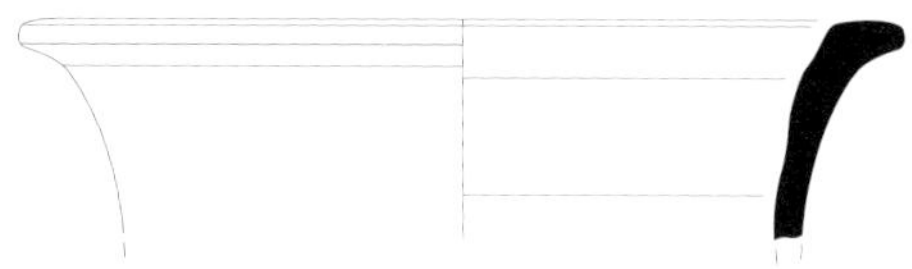

SAV PCW 57 Jug Type 3

Provenance: Square G7. Level 1, batt. 2. Lot: SA 79-796PL. SAV 52. Assemblage 16.

Dimensions and technical features: h 3.9, rim ext. 11.9. Fabric SAV 1, Munsell 5YR 7/8, hard, compact, and fine-grained, with tiny white and black stone inclusions. Powdery surface.

Condition: One fragment of the rim and neck.

Description: Rim with listel, nearly vertical exterior profile, concave interior profile, and flat upper surface; vertical neck.

1:3

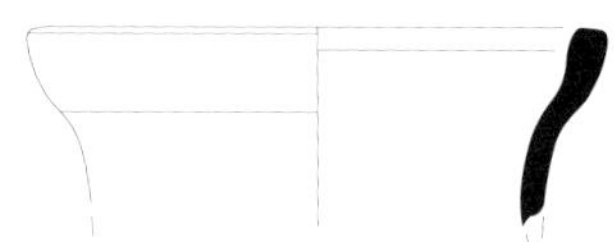

Comparisons and comment: Casagrande 2002, 161, no. 164, fig. 56 (Eraclea, jug type 4, 375–270 BC).

Date: 4th–3rd c. BC.

SAV PCW 58 Jug or Amphora Type 3

Provenance: Square G6, W of wall. Level 2, batt. 1. Lot: SA 79-733PL. SAV 120 (see *Fig. 2.20*).

Dimensions and technical features: h 2.4, rim 11.2. Fabric SAV 1, Munsell 5YR 7/4, hard, compact, and fine-grained, with tiny white and black stone inclusions. Powdery surface.

Condition: One fragment of the rim and neck.

Description: Flaring rim with short neck, convex exterior profile, and oblique upper edge; ovoid or globular body.

1:3

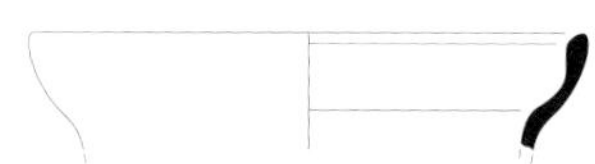

Comparisons and comment: P. Bottini 1988, fig. 31, inv. no. 270084 (S. Evraso, 320–280 BC); Postrioti 1996, 123, D25, fig. 8 (Metaponto, votive deposit from Temple E, 3rd c. BC).

Date: 3rd c. BC.

SAV PCW 59 Jug Type 4

Provenance: Square D4 W. Level 3, batt. 5. Lot: SA 79-907PL. SAV 179. Assemblage 13.

Dimensions and technical features: h 4.8, rim 13.8. Fabric SAV 3, Munsell 7.5YR 8/4, hard, compact, fine-grained, and well-levigated.

Condition: One fragment of the rim and neck.

Description: Rim with listel with flat upper surface and flaring exterior profile; vertical neck.

Comparisons and comment: Similar profile in banded ware in Carter and Parmly Toxey in *Necropoleis*, 713, T260-2 (Pantanello, 460–440 BC).

Date: From Early Hellenistic levels (Phase 4).

1:3

SAV PCW 60 Jug Type 4

Provenance: Slope 3, W extension Level 2, batt. 1. Lot: SA 79-132PL. SAV 95. Assemblage 11.

Dimensions and technical features: h 3.2, rim 13.2. Clay Munsell 2.5YR 5/8, quite hard, compact, few small white inclusions, few medium-sized black inclusions, very rare large red-brown inclusions, rare silver mica.

Condition: Three fragments of the rim preserving the attachment of the handle.

1:3

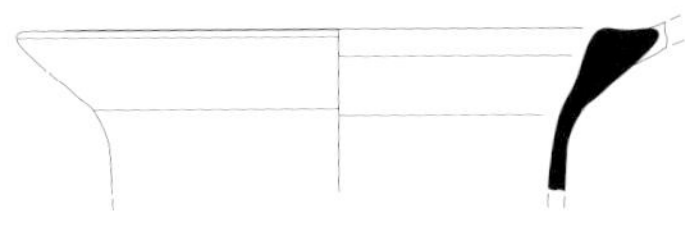

Description: Rim with triangular section and slightly convex upper surface.

Date: Late 4th–3rd c. BC.

SAV PCW 61 Jug Type 4

Provenance: Slope 3, W extension. Level 2, batt. 1. Lot: SA 79-132PL. Assemblage 11.

Dimensions and technical features: h 2.8, rim 15.0. Clay Munsell 2.5YR 5/8, hard, low porosity, few small white and black inclusions.

Condition: One fragment of the rim.

Description: Rim with triangular section, separated from the cylindrical neck by a cyma, and a slightly convex interior surface.

1:3

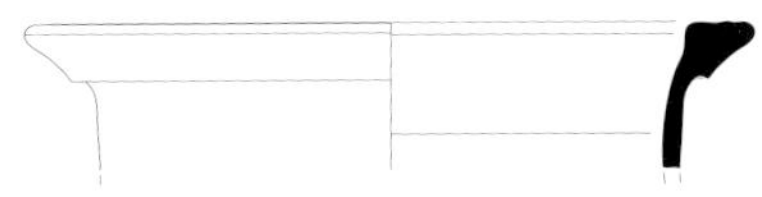

Comparisons and comment: Cotton 1983, 370, no. 412, fig. 140 (table amphora, Cozzo Presepe, site B, phase I, before the end of the 4th c. BC); see also SAV BW 37.

Date: Late 4th–3rd c. BC.

SAV PCW 62 Jug Type 5

Provenance: Square D4 W. Level 3, batt. 5. Lot: SA 79-907PL. SAV 250. Assemblage 13.

Dimensions and technical features: h 3.9, rim 12.0. Fabric SAV 5, Munsell 5YR 7/8, very hard and compact with occasional vughs in fracture, with numerous tiny white stone inclusions.

Condition: One fragment of the rim and neck.

Description: Flaring and slightly thickened neck; the exterior profile is grooved and slightly contracted at the center; vertical neck.

Comparisons and comment: Similar to Cotton 1983, 369, no. 402, fig. 138 (Cozzo Presepe, area B, phase III, end of the 4th–3rd c. BC); Bianco and Deodato 1997, 183, no. 141, pl. 77 (Pomarico, 320–280 BC); Brizzi 1999b, 320, no. 666, fig. 317 (Oppido Mamertina, 3rd–2nd c. BC); Rinaldi 2005, 229, no. 96, fig. XXVI (Torre di Satriano, 3rd c. BC).

1:3

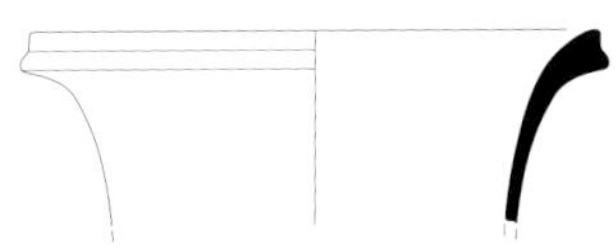

Date: 4th–3rd c. BC.

Foot of Closed Shape

The feet identified at Sant'Angelo Vecchio include 78 fragments from closed forms, distinguished by curving walls, which on the interior generally bear clear signs of having been thrown on the potter's wheel. The fragments belong to jugs, jars, table amphorae, and hydriai, which can be difficult to distinguish from the feet alone. Among the many identified examples, the most numerically significant have a distinct ring foot with truncated conical or rounded profile. Footless vessels are also attested (SAV PCW 70–SAV PCW 72). The diameter of the fragments is in the range of 6.6 to 16.5 cm, indicating a wide range of vessel sizes.

Feet of Closed Shapes

SAV PCW 63 Jug

1:3

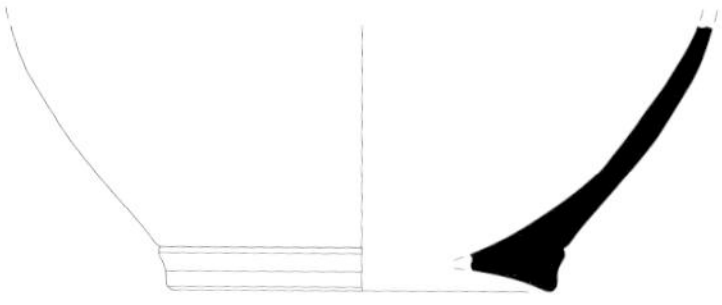

Provenance: Square H4. Level 1, batt. 1. Lot: SA 79-664PL (topsoil). SAV 147 (see *Fig. 2.21*).

Dimensions and technical features: h 5.3, base 8.0. Fabric SAV 5, Munsell 2.5YR 5/8, very hard and compact with occasional vughs in fracture, with numerous tiny white stone inclusions (see Ch. 25, "Cooking Ware").

Condition: Two fragments of the foot and attachment of the body wall.

Description: Distinct foot of a closed form with vertical exterior surface marked by mid-level indentation; oblique wall.

SAV PCW 64 Jug

1:3

Provenance: Sounding D. Level 2, batt. 1. Lot: SA 80-48PL. SAV 28.

Dimensions and technical features: h 1.8, base 10.0. Fabric SAV 5, Munsell 2.5YR 5/6 with gray core, very hard and compact with occasional vughs in fracture, with numerous tiny white stone inclusions.

Condition: One fragment of the foot.

Description: Ring foot of a closed form with oblique and then inward-projecting exterior profile.

SAV PCW 65 Jug

1:3

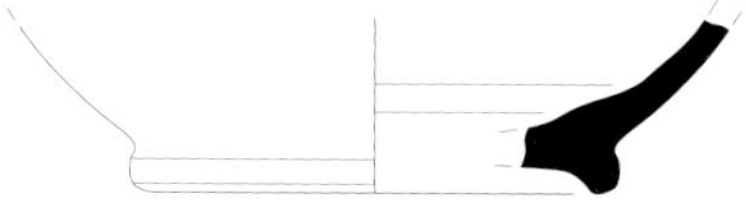

Provenance: Square F5. Level 1, batt. 1. Lot: SA 79-455PL (topsoil). SAV 37 (see *Fig. 2.22*).

Dimensions and technical features: h 3.5, base 10.0. Fabric SAV 3, Munsell 2.5YR 6/6, hard, compact, fine-grained, and well-levigated.

Condition: One fragment of the foot.

Description: Distinct foot of a closed form with vertical exterior face and convex resting surface.

SAV PCW 66 Jug

1:3

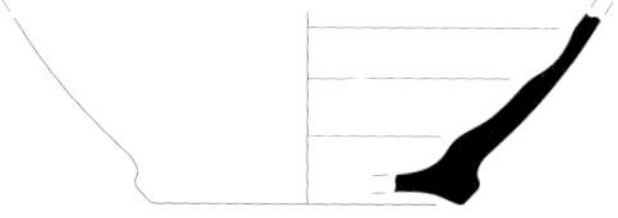

Provenance: Square G7, W of wall. Level 2, batt. 1. Lot: SA 79-793PL. SAV 40 (see *Fig. 2.20*).

Dimensions and technical features: h 3.7, base 6.6. Fabric SAV 5, Munsell 2.5YR 5/6, very hard and compact with occasional vughs in fracture, with numerous tiny white stone inclusions.

Condition: One fragment of the foot.

Description: Distinct ring foot of a closed form with inward-projecting exterior profile and concave lower surface.

Date: From Early Hellenistic levels (Phase 4).

SAV PCW 67 Jug

1:3

Provenance: Slope 3 W extension. Level 3, batt. 1. Lot: SA 79-131PL. SAV 101. Assemblage 13.
Dimensions and technical features: h 1.6, base 12.0. Fabric SAV 3, Munsell 2.5YR 6/6, hard, compact, fine-grained, and well-levigated.
Condition: One fragment of the foot and attachment of the body wall.
Description: Distinct foot of a closed form with rounded exterior profile; thin, convex walls; very thick base.
Date: From Early Hellenistic levels (Phase 4).

SAV PCW 68 Jug

1:3

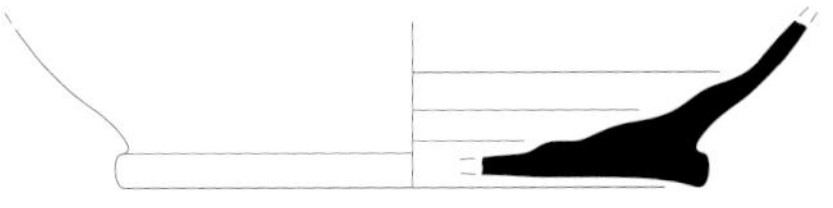

Provenance: Slope 3, votive deposit SW extension. Level 3, batt. 1. Lot: SA 79-376PL. SAV 108. Assemblage 11.
Dimensions and technical features: h 3.4, rim 12.4. Fabric SAV 3, Munsell 5YR 7/6, hard, compact, fine-grained, and well-levigated.
Condition: One fragment of the foot and attachment of the body wall.
Description: Distinct foot of a closed form with rounded exterior edge and concave lower profile; convex wall.
Date: From Early Hellenistic levels (Phase 4).

SAV PCW 69 Jug

1:3

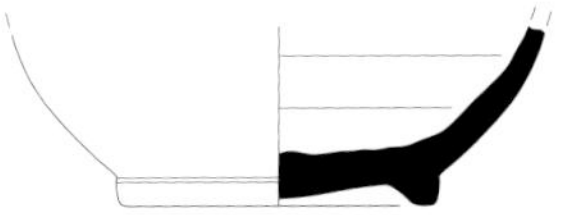

Provenance: Square G6, W of wall. Level 2, batt. 1. Lot: SA 79-733PL. SAV 123 (see *Fig. 2.20*).
Dimensions and technical features: h 3.4, base 6.7. Fabric SAV 1, Munsell 5YR 7/6, hard, compact, and fine-grained, with tiny white and black stone inclusions. Powdery surface.
Condition: Two fragments of the foot and attachment of the body wall.
Description: Ring foot of a closed form with vertical exterior profile; convex wall.
Date: From Early Hellenistic levels (Phase 4).

SAV PCW 70 Jug

1:3

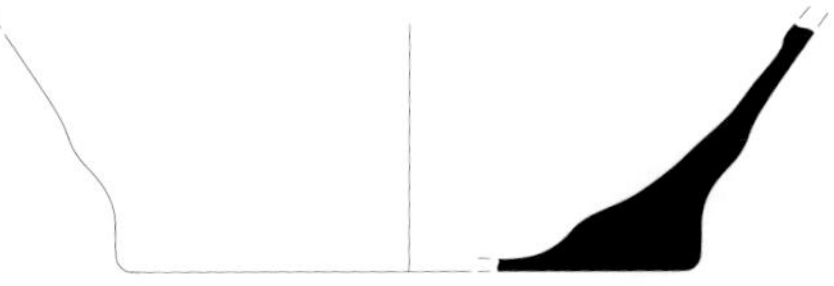

Provenance: House on the Hill. Lot: SA 79-1062PL. SAV 68. Assemblage 1.
Dimensions and technical features: h 4.9, base 12.0. Fabric SAV 3, Munsell 5YR 7/6, hard, compact, fine-grained, and well-levigated.
Condition: One fragment of the base.
Description: Indistinct foot of a closed form with flat resting surface; oblique walls with firing bubbles.
Date: From Late Archaic levels (Phase 2).

Feet of closed shapes cont.

SAV PCW 71 Jug

1:3

Provenance: House on the Hill. Lot: SA 79-1062PL. SAV 71. Assemblage 1.
Dimensions and technical features: h 1.4, base 4.5. Fabric SAV 1, Munsell 5YR 7/6, hard, compact, and fine-grained, with tiny white and black stone inclusions. Powdery surface.
Condition: One fragment of the base.
Description: Footless base of a closed form with concave resting surface.
Date: From Late Archaic levels (Phase 2).

SAV PCW 72 Jug

1:3

Provenance: Sounding B. Level 3, batt. 7. Lot: SA 79-358PL. SAV 112 (see *Fig. 2.35*).
Dimensions and technical features: h 3.4, base 8.5. Fabric SAV 1, Munsell 7.5YR 7/4, hard, compact, and fine-grained, with tiny white and black stone inclusions. Powdery surface.
Condition: One fragment of the base.
Description: Footless base of a closed form with flat resting surface.
Date: From Early Hellenistic levels (Phase 4).

Oinochoe

A single undecorated fragment from the House on the Hill can be identified as belonging to a trefoil oinochoe. The rarity of the form is further confirmed by the studies of the materials found in the other Metapontine farmhouses excavated by ICA, Sant'Angelo Grieco, Fattoria Fabrizio, and, especially, Fattoria Stefan, where the number of vessels recovered is larger and therefore statistically more significant. Other shapes were probably used to pour liquids—see above, p. 328, "Closed Shapes (Jug, Pelike, Hydria, and Table Amphora)"—although not the olpai, which were found in the farmhouses in truly scarce quantities (see Ch. 20, "Banded Ware").

Oinochoe

SAV PCW 73 Trefoil Oinochoe

1:3

Provenance: Pottery Deposit. Lot: SA 79-988PL. Assemblage 15.
Dimensions and technical features: h 4.9. Clay Munsell 7.5YR 6/4, soft and barely porous, with tiny white inclusions; rare elongated vacuoles.
Condition: One fragment of the rim.
Description: Upper part of the rim of a trefoil oinochoe.
Date: From Classical levels (Phase 3).

Jar

Four closed vessels at Sant'Angelo Vecchio are identifiable as jars, featuring a globular or ovoid body. The examples recovered in the Metapontine farmhouses demonstrate a wide typological range articulated in six types (see Cavallo in *Fattoria Fabrizio*, 280), of which only Type 1 is attested at Sant'Angelo Vecchio. None of the recovered examples preserved a complete profile.

Type 1. The first type features a flaring rim, marked with a clear separation on the interior, mounted seamlessly on a curving body wall.

Jars

SAV PCW 74 Jar Type 1

Provenance: Square C5. Level 1, batt. 2. Lot: SA 79-1063PL. SAV 119. Assemblage 13.

Dimensions and technical features: h 2.5, rim 17.8. Clay Munsell 7.5YR 7/4, low porosity, few small black inclusions, rare mica, and very small rounded vacuoles.

Condition: One fragment of the rim and attachment of the shoulder.

Description: Flaring rim inclined upward with rounded exterior edge, mounted on a low and flaring neck and globular body.

1:3

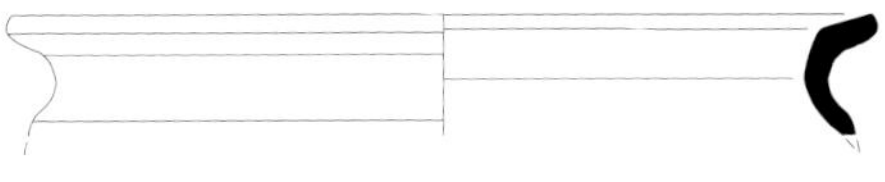

Comparisons and comment: Vittoria in *Survey*, 374, no. 81 (Ja3a, Metaponto, chora), with further comparisons.

Date: End of the 4th–3rd c. BC.

SAV PCW 75 Jar Type 1

Provenance: Square F3, N Kiln interior. Level 2, batt. 4. Lot: SA 79-883PL. Assemblage 8.

Dimensions and technical features: h 5.3, rim 22.5. Clay Munsell 5YR 6/8 friable, slightly porous, few very small white inclusions.

Condition: One fragment of the rim and body.

Description: Flaring rim separated from the rounded body by a convexity.

1:3

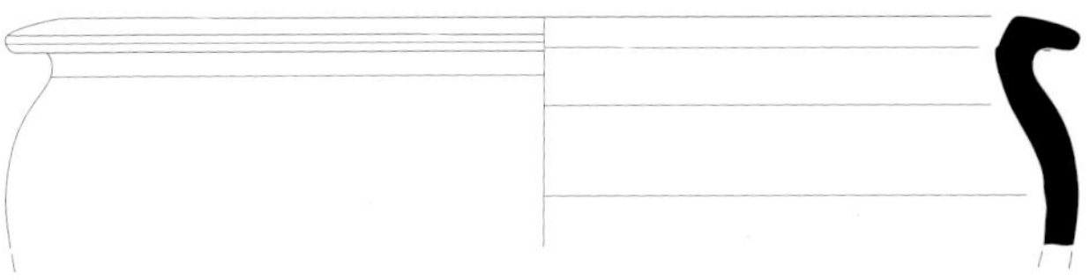

Comparisons and comment: Although lacking precise parallels, the fragment bears similarities to Caravelli 2008, 460, no. 630, fig. 310 (Civita di Tricarico, 320–280 BC), and to vessels from Gravina (Cotton 1992a, fig. 58, nos. 1218, 1220, and 1221, end of the 2nd c. BC). See also jars from Giancola (Br): (Biffino 2012, 232, tav. XXXII, end of the 2nd c. BC).

Date: From Late Republican levels (Phase 5).

Large Jar

Among the materials from the Metapontine farmhouses excavated by ICA, several large containers featuring a deep body and ovoid or globular body were identified as large jars. The dimensions are consistently large, but the thickness of the body wall is variable, suggesting that the form was used for various functions, probably and especially the storage of foodstuffs in the medium and long term. The variable morphology of the rim results in three types, two of which (Types 1 and 3) are represented at Sant'Angelo Vecchio.

Type 1. Characterized by a low rim with short, vertical neck mounted on a wide globular or ovoid body with thick walls, the first type is known from vessels documented over a wide chronological range, from the Archaic period through the 1st c. BC. The type is attested at Sant'Angelo Vecchio by a single individual.

Type 3. The third type features a protruding rim with flat or rounded upper profile mounted on a very deep ovoid body. In Lucania, the form is attested mostly between the 4th and 3rd c. BC, while in Apulia it is encountered until the middle of the 1st c. BC. A single example of the type was found at Sant'Angelo Vecchio.

Jars

SAV PCW 76 Large Jar-like Container Type 1
Provenance: Cut in baulk between N and S Kilns. Level 2, batt. 1–2. Lot: SA 81-7PL. SAV 19. Assemblage 5.
Dimensions and technical features: h 13.1, rim 36.0. Fabric SAV 4, Munsell 5YR 7/4, very hard, compact, medium-grained, and levigated, with tiny light and dark stone inclusions. Heavily encrusted and corroded surface. The fragment was reused for the construction of one of the kilns.
Condition: One fragment of the rim and shoulder.
Description: Vertical rim with rounded upper surface mounted with a clear separation on a globular or ovoid body.
Comparisons and comment: Bears similarities to the pithos in Cotton 1992c, 201, no. 1567, fig. 88 (Gravina di Puglia).
Date: From Late Republican levels (Phase 5).

1:3

SAV PCW 77 Large Jar-like Container Type 3
Provenance: Slope 3 W extension. Level 2, batt. 2. Lot: SA 79-84PL. SAV 94 (see *Fig. 2.35*).
Dimensions and technical features: h 4.4, rim ext. 32.7. Fabric SAV 1, Munsell 5YR 7/4, hard, compact, and fine-grained, with tiny white and black stone inclusions. Powdery surface.
Condition: One fragment of the rim and body.
Description: Thickened rim, protruding toward the exterior, with rounded upper surface; curving body wall.
Comparisons and comment: P. Bottini 1988, fig. 34, inv. no. 270269 (S. Evraso, 320–280 BC).
Date: From Early Hellenistic levels (Phase 4).

1:3

Lid

SAV PCW 78 Lid
Provenance: Square D4 W. Level 3, batt. 5. Lot: SA 79-907PL. SAV 178. Assemblage 13.
Dimensions and technical features: h 2.6, rim 21.8. Fabric SAV 3, Munsell 7.5YR 7/4, hard, compact, fine-grained, and well-levigated.
Condition: One fragment of the rim and edge.
Description: Lid with oblique rim, which has an indistinct edge and vertical exterior surface.
Date: From Early Hellenistic levels (Phase 4).

1:3

22

Lagynoi

Carlo De Mitri

The earliest attestations of the *lagynos*, a form used to contain and transport various liquids, date to the middle of the 3rd century BC,[1] and in the course of the 2nd century BC, it achieved a greater diffusion in the Aegean basin, where it is well documented until the second quarter/middle of the 1st century BC.[2]

The recent study of the material from the excavations in the Athenian Agora,[3] considered together with the finds from the Levant,[4] has led to the identification of at least six sub-classes of lagynoi, all produced in the larger Aegean/eastern Mediterranean area and distinguished by their technical features. White-ground lagynoi are covered in a white wash with brown vegetal or (in many cases) figural decorations painted primarily on the shoulder. On banded ware lagynoi, the exterior surface is covered with a hazelnut-colored wash and decoration consisting mostly of bands painted on the body wall at the joint between the neck and the shoulder and on the rim. This subclass also includes orange-brown vegetal or (rarely) figural motifs on the shoulder. Red-slip lagynoi, easily distinguished by the gloss that covers the entire body, appears to originate on Cyprus and was subsequently adopted at Pergamon and, later still, in Eastern Sigillata A. Coarse- (or household-) ware lagynoi,[5] which lack decoration entirely, were used most probably for wine transport. The existence of multiple workshops producing these vessels is attested by the different fabrics that have been identified: from granular and rich in inclusions, to well-levigated with a wash on the exterior wall (the latter were probably used at the table). Hellenistic mold-made relief lagynoi were produced in both the Aegean and Italic areas, and lagynoi in faïence, attributed to an Alexandrian production, had a rather limited circulation.[6]

The success of the form is clearly reflected in the speed with which it was introduced into the morphological repertory of many productions. At Athens, for example, it appeared in both black-gloss and West Slope ware.[7] The first appearance of these containers in the western Mediterranean is documented at the beginning of the 2nd century BC, particularly in funerary contexts. The imported examples are accompanied by models made in various locally-produced wares, perhaps even to the point of being replaced.[8]

The lagynos can be distinguished by the features of the body, especially the wall, in four principal types with several variants:[9]

Type I has a truncated conical body with straight walls; the shoulder is generally convex, but on several examples it is straight and oblique and (rarely) nearly horizontal; the high cylindrical neck, tapering near the top, terminates mainly in a collar rim, rarely in an echinus rim. Ribbon handle; base with ring foot.

Type II has a lenticular body with convex walls; the body is biconical, often "squat"; the neck terminates in a flaring echinus

[1] The lagynos is generally identified as a container for wine, the use of which was associated with the symposium and several religious festivals (Pierobon 1979). Its production in different ceramic classes may be evidence of different functions: both a container for the table and the pantry, and a transport vessel for not only wine but also scented oils. On the form, and for a synthesis of the state of scholarship on it, see Sciarma 2005, with abundant bibliography.

[2] On the presence of lagynoi in the Aegean/eastern Mediterranean area and their distribution in the Mediterranean in general, see Hubner 2000; for mid-1st-c. BC attestations, see the data from the Antikythera shipwreck in Vivliodetis 2012.

[3] *Agora XXIX*, 225–31; *Agora XXXIII*, 82–84.

[4] For the excavations on Cyprus, see Vessberg and Westholm 1956 and *Paphos III*.

[5] *Agora XXXIII*, 82.

[6] For a definition of this ceramic group, see Seif El-Din and Nenna 1993. On its presence in the western Mediterranean, see Colivicchi 2002, 260, no. 41.1.

[7] *Agora XXIX*, 127.

[8] In southern Italy, the production of lagynoi has been localized at Egnatia (Berloco and Schiavariello 2012) and Brindisi (Manacorda and Pallecchi 2012) on the Adriatic coast and in the main settlements of the Ionian coast, Taranto, Metaponto, and Herakleia. The argument for local production at these sites is nevertheless based on the technical features and macroscopic analyses of the clays; no kiln wasters have been identified.

[9] Various typologies have been proposed for the study of lagynoi. Besides those of Leroux (1913) and Vessberg and Westholm (1956) pertaining to the eastern productions, see the proposals for the productions in Italy: Giardino 1990 on Herakleia and Sciarma 2005 on Etruria, where the preceding seriations are assembled. The attempt to rationalize the data in order to create a common typology for the entire category of lagynoi is in progress.

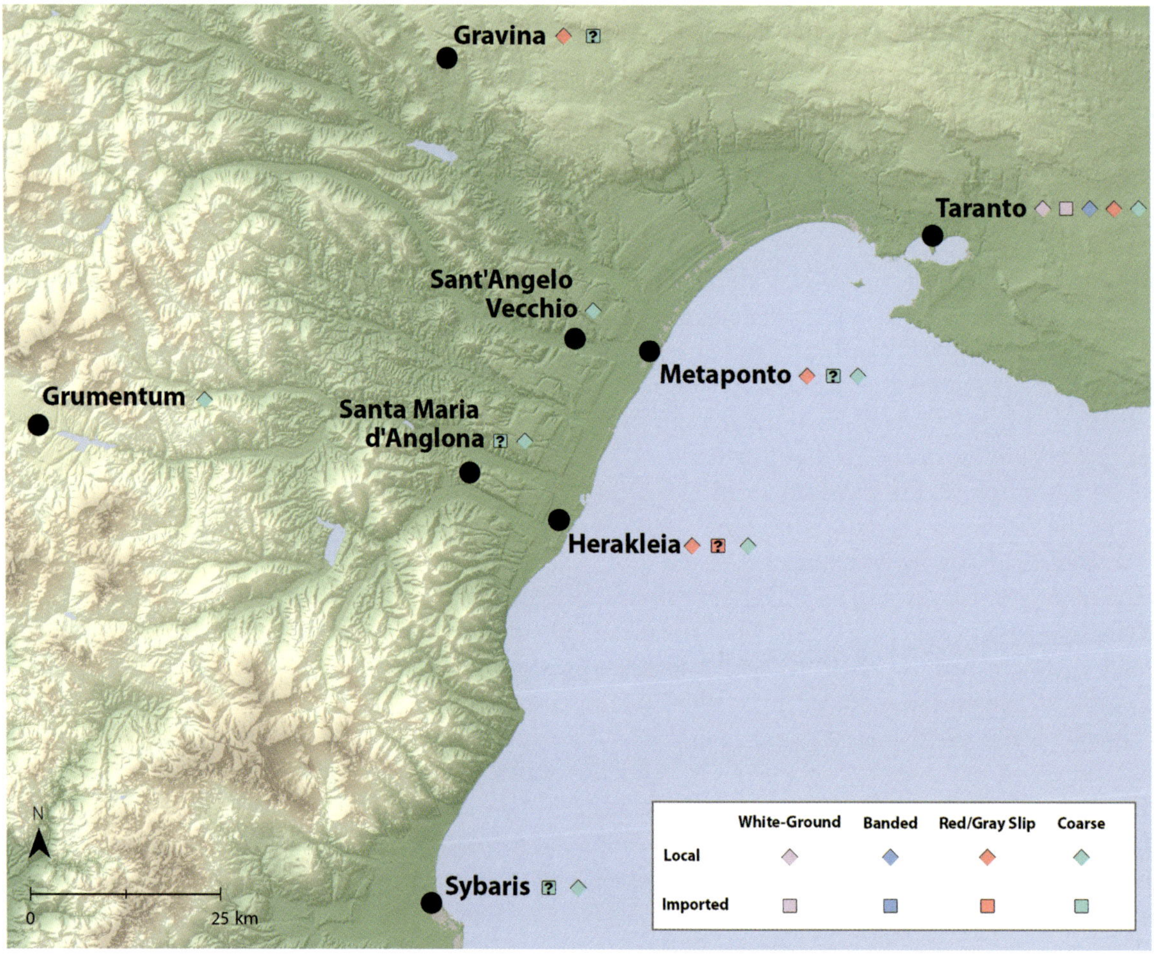

Figure 22.1 Distribution map of lagynoi in the central Ionian arc. (JT/ICA)

rim. Ribbon handle; base with ring or flat, slightly concave foot.

Type III has a high body with a slightly rounded carination at the junction with the shoulder, cylindrical neck with collar rim, ribbed ribbon handle, and ring foot. It may be considered a transitional type between I–II and IV.

Type IV has a globular body, ovoidal or piriform. High neck tapering toward the top with rounded or triangular collar rim. Ribbon handle; base with ring foot. Lagoena (Latin word for lagynoi) and olpai produced in the Early Imperial period derive from this type.

Lagynoi are attested at various settlements along the entire Ionian arc, especially in funerary contexts, by examples that can be ascribed mainly to local productions (*Fig. 22.1*). Many examples have been recovered at Taranto, mostly as grave goods in necropoleis dating to the 2nd and 1st centuries BC, and continuing into the 1st century AD. The latest examples, however, are limited to Type IV.[10] The documentation for the first half of the 2nd century BC includes mostly coarse, black-gloss, and grey-ware lagynoi, as well as one imported example in white-ground ware. In the second half of the 2nd century BC, local imitations of these productions appeared,[11] accompanied by coarse, banded, and especially red-slip examples.

The attestations of the form along the central Ionian arc are concentrated mostly in the two main coastal sites, Metaponto and Herakleia. At Metaponto, the lagynos is well documented in funerary assemblages of the 2nd and 1st centuries BC. Type II is practically the only type attested, in both coarse and red-slip versions.

[10] For the typological sequence of the lagynoi at Taranto, see Hempel 2001, 106–9.

[11] For a discussion of this aspect, see Lippolis 2005, 256, pl. XV.2.

The majority of the grey-ware lagynoi, along with the horizontal shoulder variant, can be attributed to Type I. All of the examples can be ascribed to local productions, with the lone exception of an example found in an early 2nd-century BC tomb.[12]

At Herakleia, lagynoi are a recurring feature of funerary assemblages as early as the beginning of the 2nd century BC, but the attestations increase notably in the second half of the same century.[13] Here, too, the predominant version is Type II, produced in coarse ware or red-slip.[14] Types III and IV are also common, produced in red-slip and, especially in the 1st century BC, in coarse ware.[15] The form is also attested in the chora of Herakleia, at Santa Maria d'Anglona, in Tombs G1[16] and G3.[17]

Lagynoi are attested in the Lucanian-Apulian hinterland at Gravina, where red-slip examples belonging to regional productions have been found, along with coarse models that could be imports based on the characteristics of the clays.[18] A molded gray slip lagynos was also found at Monte Irsi.[19] A coarse lagynos was also found in a 2nd/1st-century BC burial in the necropolis of Grumentum, in the upper Agri valley.[20]

The form is also documented in the plain of Sybaris, in the southern Ionian arc. In fact, a re-examination of the published data has led to the identification of at least two examples from the settlement, probably imports.[21] Other attestations with similar features to those found in the urban center have been observed at the surrounding rural sites.[22]

Several fragments of lagynoi have been identified at Sant'Angelo Vecchio. Most of them come from the Kiln Complex, active until the beginning of the Imperial period, while SAV Lag 01 comes from Sounding D, a concentration of chronologically heterogeneous material washed down from the plateau above. The absence of elements belonging to the bodies of these containers precludes their precise typological identification, and therefore they are assigned the generic chronology of 2nd–1st century BC.

12 According to the published description (Lo Porto 1966, 139, tomb 2), this example could be an import.

13 Giardino 1990.

14 In one case there is also an example in black gloss, irregularly diluted to red (Giardino 1993, 180, tomb 92, fig. 285).

15 On the continuity of use of the coarse lagynoi, see C. Greco 2009, 807, fig. 13. Judging from the description, a globular red-slip lagynos found in a grave assemblage dated to the second half of the 1st c. BC may be an import (Bianco 1993a, 184, tomb 119, no. 295).

16 Rüdiger 1969, 193.

17 The lagynos from Tomb G3 is an intermediate form between Types I and II, produced in coarse ware, and may be an import (Bianco 1993b, 199, no. 328).

18 On the finds at Gravina, see Cotton 1992a, fig. 50, no. 1123, fig. 62, nos. 1253–1256.

19 Jentel 1977, 164–65.

20 P. Bottini 1997, 268, no. 2.

21 *Sibari I*, 54, fig. 48, nos. 15, 58, fig. 53, no. 24.

22 Roncoroni 2004, 63, nos. 8–9.

Catalog of Lagynoi

SAV Lag 01 Lagynos

1:2

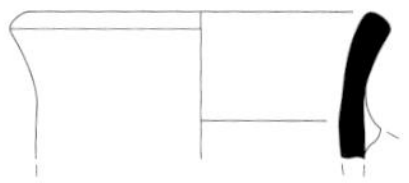

Provenance: Sounding D. Level 1, batt. 2. Lot: SA 80-43PL.

Dimensions and technical features: h 1.9, rim 4.6. Clay Munsell 7.5YR 6/4, friable and barely porous, with rare tiny white and black inclusions and rounded vacuoles.

Condition: One fragment of the rim and body.

Description: Flaring rim with rounded lip; cylindrical neck preserving the attachment for the handle.

Comparisons and comment: The fragment has a fairly close parallel in an example with biconical body found in a tomb at Metaponto (De Siena 1993, tomb 317, no. 9, fig. 209; Metaponto, second half of the 2nd–beginning of the 1st c. BC).

Date: End of the 2nd–1st c. BC.

SAV Lag 02 Lagynos

1:2

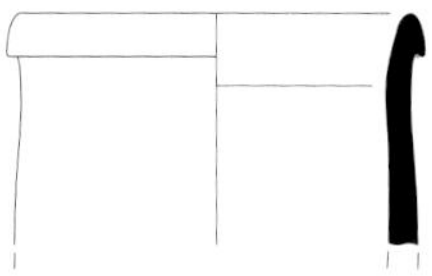

Provenance: Square F3, SE end of N Kiln near baulk. Level 2. Lot: SA 79-880PL. Assemblage 8.

Dimensions and technical features: h 3.0, rim 5.3. Clay Munsell 5YR 6/8, with rare tiny white inclusions.

Condition: One fragment of the rim and body.

Description: Collar neck with triangular section and slightly pendant lip; cylindrical neck.

Comparisons and comment: No close parallels were found, but several plain lagynoi that have a collar neck with triangular section have been found at Taranto (Hempel 2001, 196, tomb E3-2, no. 15; Taranto, beginning of the 1st c. BC) and Egnatia (Berloco and Schiavariello 2012, 389; Egnatia, dated generically to the end of the 2nd–beginning of the 1st c. BC). The prototype appears at Athens in Late Hellenistic contexts (*Agora XXXIII*, 256, no. 113; Athens, beginning of the 2nd c. BC).

Date: End of the 2nd–1st c. BC.

SAV Lag 03 Lagynos

1:2

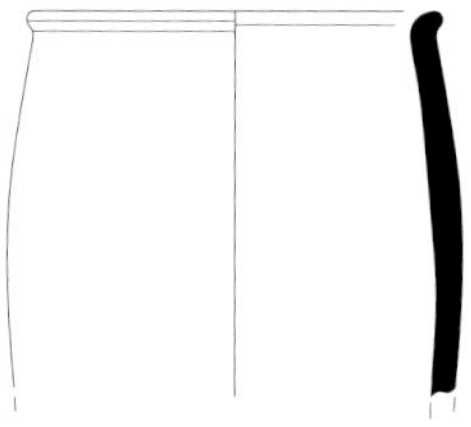

Provenance: Cut in balk between N and S kilns. Level 2, batt. 1–2. Lot: SA 81-7PL. Assemblage 5.

Dimensions and technical features: h 5.0, rim 5.5. Clay Munsell 2.5YR 5/6, friable and moderately porous, with rare tiny micaceous and black inclusions.

Condition: Two fragments of the rim and body.

Description: Collar rim distinguished from the high, swollen neck by an offset.

Comparisons and comment: The example is morphologically similar to the containers with globular, nearly spherical body typical of the funerary assemblages of the F and G phases of the necropolis of Taranto (Hempel 2001, 135, form 114, no. 28; Taranto, end of the 2nd–beginning of the 1st c. BC; Lippolis 1994, 251, fig. 187, 1st c. BC–1st c. AD).

Date: End of the 2nd–1st c. BC.

SAV Lag 04 Lagynos

Provenance: Square F3, N Kiln interior. Level 2, batt. 4. Lot: SA 79-883PL. SAV 189. Assemblage 8.

Dimensions and technical features: h 4.0, rim 4.9. Fabric SAV 3, Munsell 5YR 6/6, hard, compact, fine-grained, and well-levigated, with rare tiny white inclusions.

Condition: One fragment of the rim and neck.

Description: Rounded and slightly flaring collar rim; high, swollen neck.

Comparisons and comment: Despite the difference in the rim, the swollen neck groups this example with the preceding one.

Date: End of the 2nd–1st c. BC.

1:2

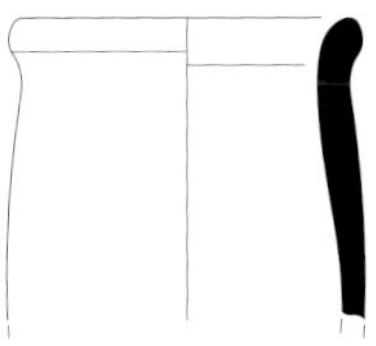

23

Louteria

Annalisa Concilio

Louteria—vessels with a wide and shallow body, rim that varies in shape, and supporting base—were used as containers for water for domestic and religious needs.[1] They circulated widely at Metaponto and in its chora, where they occur principally in domestic contexts[2] and, to an apparently lesser extent, in sanctuaries.[3] Nevertheless, these objects have never been the subject of systematic study.

The excavations at Sant'Angelo Vecchio yielded a total of eight louterion fragments belonging to at least five individuals, all made of clay. Due to the highly fragmentary state of the material it cannot be determined whether any body and stand joined to form a single vessel.

[1] For an exhaustive evaluation of the bibliography on the study and functions of louteria, see Iozzo 1981; Pimpl 1997, 7–8. For louteria in sanctuaries, see Krauskopf 2005.

[2] For the materials recovered in the Metapontine chora during the Bradano-Basento field survey, see Vittoria in *Survey*, 352, 418–19; most of the examples were found in sites interpreted as farmhouses. The use of louteria in the domestic sphere is further confirmed by the findings in the excavated farmhouses: Fattoria Pacciano in contrada Saldone (4th–3rd c. BC; Uggeri 1969, 62; Adamesteanu 1974, 85; Amore 1992–1993, 107); Fattoria Arezzo in the Lago del Lupo locality, which yielded material of the 4th c. BC (Adamesteanu 1973, 59, fig. 4; Amore 1992–1993, 131, pl. XXVII; the louterion is compared to Iozzo 1981, 189, type D75, dated to the beginning of the 4th c. BC); and Fattoria Fabrizio (Cavallo in *Fattoria Fabrizio*, which also mentions examples found at Fattoria Stefan and Sant'Angelo Grieco).

[3] A louterion fragment was found in the sanctuary identified at site 397, Avinella (Vittoria in *Survey*, 352, Lou2a). Additional examples from Pantanello are being studied. For a marble louterion from the Sanctuary of Apollo Lykeios at Metaponto, see Adamesteanu 1975, 231, fig. 242c.

SAV Lou 01 and SAV Lou 02, both featuring a relatively deep body, can be compared to Iozzo's Group A of louteria found in Calabria,[4] which includes examples from the Orientalizing, Archaic, and Classical periods. SAV Lou 03, on the other hand, with a flat bottom and a markedly distinct rim, belongs to a type considered by Iozzo an evolution of Group A,[5] attested especially from the 5th century BC through the Hellenistic period. This type of body was probably associated with stands that had a splayed base, such as SAV Lou 06.

The fragments were all found in secondary deposition. SAV Lou 01 was reassembled from three fragments, two of which came from the collapse of the roof of the building datable to the 4th century BC (square G7; see Ch. 2, "Site Phasing, Stratigraphy, and Site Assemblage"), while the third fragment was found in the collapse of the roof of the Kiln Complex courtyard (Sounding A). The entire surface of the latter bears so thick a layer of concretion as to indicate its probable reuse as building material in one of the kilns located immediately to the west. Alteration of the clay color and consistency in SAV Lou 02 was probably caused by repeated exposure to a heat source, corroborating the documentation of its findspot inside the North Kiln (Kiln 5) of the production facility.

[4] Iozzo 1981, 147–63.

[5] Iozzo 1981, 147, 178.

Cat. No.	Inv. No.	Provenance	Description	Date	NFr	MNI
SAV Lou 01	319906	SA 79-26PL + SA 79-997PL + SA 79-813T Sounding A, S area of Room 1, Level 2. batt. 2. Square G7, Level 2, batt. 2–4. Square G7, Level 2, batt. 5.	rim	5th–4th c. BC	3	1
SAV Lou 02		SA 79-873PL Square F3, N Kiln interior, Level 2, batt. 1.	rim	End of the 5th–early 4th c. BC	1	1
SAV Lou 03		SA 79-454P Square F2, kilns, Level 2, batt. 1.	rim	Second half of the 4th c. BC	1	1
SAV Lou 04		SA 79-600PL Square G5, Level 1, batt. 2.	base	End of the 5th–4th c. BC	1	
SAV Lou 05		SA 79-746PL Square G6, wall and tombs, Level 2, batt. 2.	base	End of the 5th–4th c. BC	1	1
SAV Lou 06		SA 79-322PL Slope 4, Sounding C, E of & including division, Level 2, batt. 1.	base	2nd–1st c. BC	1	1

Table 23.1 Quantification of the louteria recovered at Sant'Angelo Vecchio.

Catalog of Louteria

SAV Lou 01 Louterion

Provenance: Sounding A, S area of Room 1, Level 2, batt. 2. Square G7, Level 2, batt. 2–4. Square G7, Level 2, batt. 5. Lots: SA 79-28P (from SA 79-26PL) + SA 79-813T + SA 79-997PL. Assemblage 5.

Dimensions and technical features: h 7.0, rim 56.0. Clay Munsell 5YR 5/8 (yellowish red), hard and compact, with frequent elongated vacuoles, numerous small calcite and black clastic inclusions, and rare large grog fragments.

Condition: Three fragments of the rim and body.

Description: Lip with continuous profile, molded rim, and flattened top surface. Low body with rounded interior profile and slightly irregular exterior profile. The top surface of the rim bears an impressed wave motif running to the left, not well preserved in the two fragments from square G7. On the exterior, immediately beneath the rim, are two parallel grooves, the upper one pronounced, the lower one slighter.

Comparisons and comment: Allegro 1982, 125, fig. 24, no. 53 (Himera, 5th–4th c. BC); bears similarities to Iozzo 1981, 158, pl. XXXVIII.1, fig. 2, no. 17, Group A (Reggio Calabria, dated by the decoration to the end of the 6th–5th c. BC) and to La Torre 2002, fig. 44, K 6 (Campora S. Giovanni, end of the 6th c. BC); compares closely to Pimpl 1997, fig. 12, no. 3 (Delos, Late Archaic). The example from Delos is in marble, and the decoration on the rim is replaced with a mutilated inscription that provides the basis for the chronology. The wave motif occurs frequently on louteria of the 4th c. BC: see, for example, Group B in Iozzo 1981, fig. 3, no. 31, pl. XLI.4 (Caulonia); fig. 4, no. 33, pl. XLII.2 (Reggio Calabria); fig. 3, no. 34, pl. XLII.3 (Caulonia); fig. 4, no. 36, pl. XLIII.1 (Locri). Compare also the basin in *Monte Sannace*, pl. 354, no. 1 (Monte Sannace, 4th–3rd c. BC).

Date: 5th–4th c. BC.

Louteria cont.

SAV Lou 02 Louterion

Provenance: Square F3, N Kiln interior. Level 2, batt. 1. Lot: SA 79-873PL. Assemblage 8.

Dimensions and technical features: h 4.2, rim 48.0. Clay Munsell 2.5Y 6/2 (light brownish gray)–7.5YR 7/3 (pink), hard, with frequent large and small red-brown and black clastic inclusions, rare quartzite and gold mica.

Condition: One fragment of the rim.

Description: Rim with rectangular section slightly inclined toward the exterior; body with taut profile.

Comparisons and comment: Different in inclination but morphologically similar to Cotton 1977, 372, fig. 142, no. 424 (Cozzo Presepe, 420–390 BC). The exact provenance of the example from Cozzo Presepe is unknown, and the chronology is based on comparison with an example from Athens (*Agora XII*, fig. 16, 1865); see also *Sibari IV*, fig. 143, inv. no. 9563; similar to Radić 1991, 156, fig. 1 (Lošinj, 3rd–2nd c. BC).

Date: End of the 5th–beginning of the 4th c. BC.

1:3

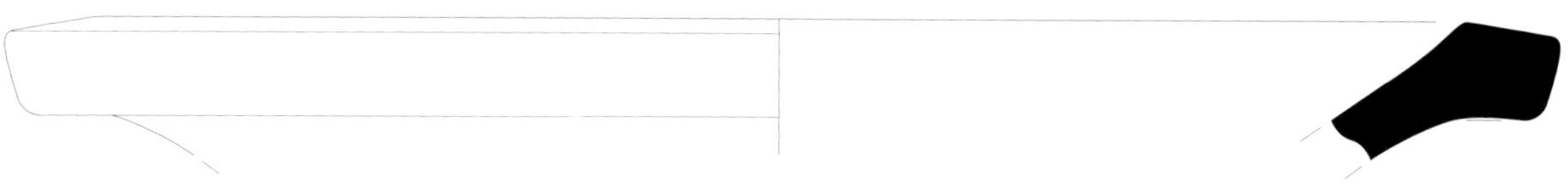

SAV Lou 03 Louterion

Provenance: Square F2, kilns. Level 2, batt. 1. Lot: SA 79-454P (see *Fig. 2.30*).

Dimensions and technical features: Clay Munsell 2.5YR 5/8, very hard and compact, with numerous red clastic inclusions, frequent small quartzite and calcite inclusions, and abundant mica.

Condition: One fragment of the rim.

Description: Very low body with thickened rim that is flattened on the top.

Comparisons and comment: The example is similar in shape and size to louteria with highly flattened body known both in Greece and at other sites in Magna Grecia. It lacks decoration and the ledge on the inside of the body that often marks the transition from the rim to the floor on this type of louteria (Iozzo 1981, 163). Similar to Iozzo 1981, 189, fig. 3, Group B, no. 29 (Reggio Calabria, second half of the 4th c. BC); similar to Scafuro 2009, 154, fig. 67 (Fratte, second half of the 4th–mid-3rd c. BC).

Date: Second half of the 4th c. BC.

1:3

SAV Lou 04 Stand

Provenance: Square G5. Level 1, batt. 2. Lot: SA 79-600PL. Assemblage 3.

Dimensions and technical features: h 6.2, base 30.0. Clay Munsell 5YR 5/8 (yellowish red), hard and compact, with frequent rounded vacuoles, numerous small calcite inclusions, rare black clastic inclusions and gold mica.

Condition: One fragment of the base.

Description: Hollow bell-shaped support base with distinct shoulder. Immediately below the shoulder there is a groove made with a stick.

Comparisons and comment: Similar to Iozzo 1981, 178, fig. 9, Group D, no. 76 (Croton-Capo Colonna, early 4th c. BC), 150, Group A, fig. 1, no. 7 (Sybaris, mid-6th c. BC); Berti and De Santis 2005, 441, pl. III, nos. 16–17 (Spina, end of the 5th–4th c. BC). It bears similarities in the composition and color of the clay to both SAV Lou 01 and SAV Lou 03. The most convincing typological association involves the first type of basin, but it is known that, if they did not consist of a single unit, these vessels could be made by changing their component parts, so that a given type of base could support different types of basins.

Date: End of the 5th–4th c. BC.

1:3

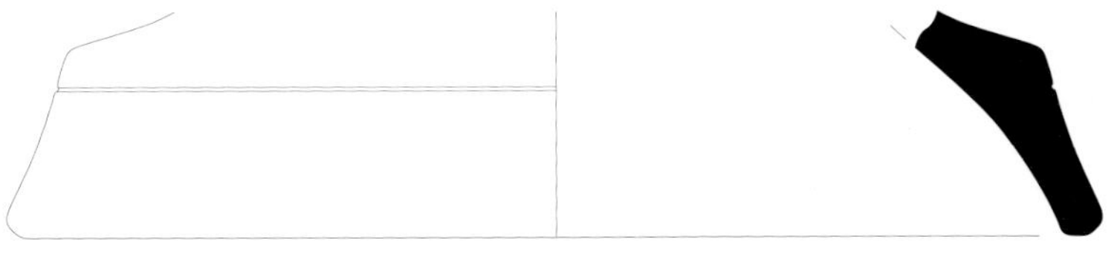

SAV Lou 05 Stand

Provenance: Square G6, wall and tombs. Lot: SA 79-746PL (from SA 79-737T). Assemblage 3.

Dimensions and technical features: h 6.2, base 34.0. Clay Munsell 7.5YR 7/4 (pink), hard, porous, and powdery to the touch, with frequent irregular vacuoles, numerous medium-sized black, gray, and red clastic inclusions, and rare mica.

Condition: One fragment of the base.

Description: Hollow bell-shaped support base with distinct shoulder, flat upper surface, and two parallel grooves made with a stick.

Comparisons and comment: Similar to Iozzo 1981, fig. 9, Group D, no. 76 (Croton-Capo Colonna, early 4th c. BC) and 150, Group A, fig. 1, no. 7 (Sybaris, mid-6th c. BC); bears similarities to de Cazanove et al. 2008, fig. 344, no. 1252 (Civita di Tricarico, 250–200 BC).

Date: End of the 5th–4th c. BC.

1:3

Louteria cont.

SAV Lou 06 Stand
Provenance: Sounding A, kilns, subdivision of Room 1. Level 2, batt. 1. Lot: SA 79-16PL. Assemblage 5.
Dimensions and technical features: h 13.7, base 38.6. Clay Munsell 7.5YR 5/4 (brown), very hard and compact, with numerous small and medium-sized elongated and rounded vacuoles, numerous small black, gray, and (rare) red clastic inclusions, and very rare quartzite.
Condition: One fragment of the base.
Description: High molded, horn-shaped foot.
Comparisons and comment: Iozzo 1981, 189, fig. 9, Group D, no. 75 (Torre del Mordillo, early 4th c. BC), G. Greco 1991, 62, fig. 127 (Serra di Vaglio, end of the 4th c. BC), P. Bottini 1992, 78, no. 68064 (Tolve, 3rd c. BC), *Gravina*, 319, fig. 61, no. 1252 (Gravina di Puglia, 5th/4th to 1st c. BC), Maturo 1997, pl. 100, no. 23 (Pomarico Vecchio, surface find), Radić 1991, 156, fig. 1 (Lošinj, 3rd–2nd c. BC), 157, fig. 6 (Lastovo, found in the cargo of a shipwreck in association with amphorae produced in southern Italy at the beginning of the 2nd c. BC).
Date: 2nd–1st c. BC.

1:3

24

Mortaria

Annalisa Concilio

Cat. No.	Provenance	Description	Date	NFr	MNI
SAV Mor 01	SA 80-21PL + SA 79-796PL Square F7, Level 2, batt. 6. Square G7, Level 1, batt. 2	rim	4th c. BC	2	1
SAV Mor 02	SA 79-786PL Square G7, W of wall, Level 2, batt. 2	rim	Late 4th c. BC	1	1
SAV Mor 03	SA 79-739PL Square G6, E of wall: Level 2, batt. 2	rim	4th–second half of the 3rd c. BC	1	1
SAV Mor 04	SA 80-45PL Slopes, Sounding D	rim	Second half of the 6th–4th c. BC	1	1
SAV Mor 05	SA 79-636PL Square G5	base	4th c. BC	1	

Table 24.1 Quantification of the mortaria from Sant'Angelo Vecchio.

Mortaria made of stone and ceramic were commonly used for mixing in domestic, sacred, and industrial contexts. The vessels generally consisted of a wide and shallow body with thick walls, a massive rim, and a flat base. These features guaranteed the vessel's solidity and stability during smashing and grinding of materials with the mortarium's accompanying pestle.[1] For further reinforcement, mortaria were made using coarse fabrics, with the occasional addition of stone inclusions on the interior floor. They were also often equipped with a spout for pouring off liquid or semi-liquid mixtures after preparation. Produced in Greece starting in the Archaic period, mortaria quickly spread throughout the Greek West.[2]

The excavations at Sant'Angelo Vecchio yielded eight fragments belonging to six ceramic mortaria. All of the fragments were found in secondary depositions, and almost all of them came from the Workshop Area (squares G5–G7 and F7). According to their comparanda, they date between the 4th and 3rd centuries BC. SAV Mor 01 belongs to the Corinthian type frequently attested in Lucania in numerous variants.[3] SAV Mor 04 is the only example from the area east of the Kiln Complex (Sounding D; see Ch. 2, "Site Phasing, Stratigraphy, and Site Assemblage"), and compares closely to examples dating between the middle of the 6th and the middle of the 4th century BC.

[1] On the definition and function of mortaria in the Greek world, see Villing and Pemberton 2010, 557–59, with preceding bibliography. For a pestle found at Sant'Angelo Vecchio, see SAV KF 14.

[2] Corinth is indicated in the literature as among the main production centers of the form, and Corinthian mortaria are the most common and widely imitated types at many colonial sites. The morphological evolution of Corinthian mortaria can be traced from the middle of the 7th c. BC to the Hellenistic period. An exhaustive and detailed analysis of the subject can be found in Villing and Pemberton 2010; see also Bianco and Deodato 1997, 176.

[3] The field survey of the Metapontino yielded 176 fragments of mortaria, subdivided into 11 types, of which 71 fragments can be ascribed to the Corinthian type (Vittoria in *Survey*, 347). Moreover, the majority of the large number of mortarium fragments (ca. 50) found in the sanctuary and tile factory at Pantanello also belong to this type.

Catalog of Mortaria

SAV Mor 01 Mortarium

1:3

Provenance: Square G7. Level 1, batt. 2. Square F7. Level 2, batt. 6. Assemblage 4.
Lot: SA 79-796PL + SA 80-21PL.
Dimensions and technical features: h 5.0, rim 36.8. Clay Munsell 5YR 6/6 (reddish yellow), hard and compact, with frequent small and medium-sized red-brown inclusions, small black inclusions, rare medium-sized quartzite and calcite inclusions, and mica.
Condition: Two fragments of the rim and body. The attachment of the base is preserved.
Description: Thickened lip with rounded upper surface and nearly elliptical section; low wall with rounded interior profile and straight exterior profile transitioning into the flat base.
Comparisons and comment: Similar to Vittoria in *Survey*, 396, type Mor5c no. 173; Bianco and Deodato 1997, no. 71, pl. 66 (Pomarico Vecchio, end of the 4th–second half of the 3rd c. BC); *Kaulonia I*, no. 237, fig. 42 (Caulonia).
Date: 4th c. BC.

SAV Mor 02 Mortarium

1:3

Provenance: Square G7, W of wall. Level 2, batt. 2. Lot: SA 79-786PL (see *Fig. 2.20*).
Dimensions and technical features: h 5.5, rim 31.5. Clay Munsell 7.5YR 6/6 (reddish yellow), hard and compact, with frequent brown clasts, small calcite inclusions, and rare gold and silver mica. Light-colored slip over the entire surface.
Condition: One fragment of the rim and body.
Description: Enlarged lip with inclined and rounded upper surface; distinct, slightly tapering, and rounded rim; low body with concave interior profile and straight exterior profile. A shallow groove on the interior marks the transition from the lip to the body.
Comparisons and comment: P. Bottini 1992, 103, inv. no. 69841 (Tolve, late 4th–first half of the 3rd c. BC); Vittoria in *Survey*, 396, type Mor7b no. 176.
Date: Late 4th c. BC.

SAV Mor 03 Mortarium

1:3

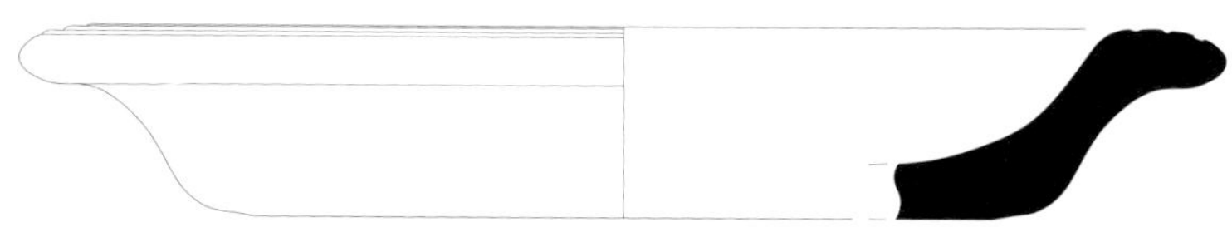

Provenance: Square G6, E of wall. Level 2, batt. 2. Lot: SA 79-739PL. Assemblage 3.
Dimensions and technical features: h 4.6, rim 30.0. Clay Munsell 5YR 5/8 (yellowish red), hard and compact, with frequent small and medium-sized red and black stone inclusions, as well as rare medium-sized quartzite inclusions.
Condition: Complete profile.
Description: Thick lip with brim and slightly rounded upper surface featuring three parallel and equally spaced incised grooves; rounded and slightly tapering rim; low body with oblique walls transitioning continuously into the flat base.
Comparisons and comment: Bianco and Deodato 1997, 75, pl. 67, no. 74 (Pomarico Vecchio, end of the 4th–second half of the 3rd c. BC), Carter 1983a, 453, fig. 28, no. 26 (Pantanello, kiln complex). Similar to Yntema 2001, 263, M56, no. 468 (Valesio, later 3rd–early 2nd c. BC).
Date: 4th–second half of the 3rd c. BC.

SAV Mor 04 Mortarium

1:3

Provenance: Slopes, Sounding D. Level 1, batt. 2. Lot: SA 80-45PL.

Dimensions and technical features: h 7.7, rim 24.4. Clay Munsell 5YR 7/6 (reddish yellow), hard, with irregularly shaped vacuoles, abundant medium-sized and large stone inclusions present also on the interior surface, frequent small black inclusions, medium-sized quartzite inclusions, and rare mica and grog.

Condition: One fragment of the rim and body.

Description: Almond-shaped lip with rounded and slightly tapering rim; truncated conical body of medium depth with straight wall.

Comparisons and comment: Cotton 1977, 368, fig. 136, no. 387 (Cozzo Presepe, site A, phase IIIb, 600/575–550 BC); Vittoria in *Survey*, 394, n. 166; *Kaulonia I*, 64, fig. 42, no. 228 (Caulonia), Rinaldi 2005, 238, pl. 27, no. 103 (Torre di Satriano, mid-4th c. BC); Blasi 2001, 64, fig. 49, no. 65 (Torre di Satriano, mid-4th c. BC). Similar to Cotton 1992b, 194, fig. 83, no. 1507 (Gravina di Puglia, mid-5th c. BC).

Date: Mid-6th–4th c. BC.

SAV Mor 05 Mortarium

1:3

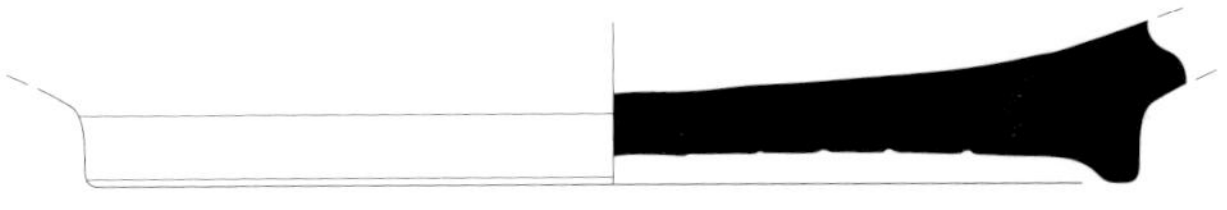

Provenance: Square G5, wall and tombs. Level 2, batt. 1. Lot: SA 79-636PL. Assemblage 3.

Dimensions and technical features: h 3.1, base 21.9. Clay Munsell 5YR 6/6 (reddish yellow), hard and compact, with frequent medium-sized and small dark inclusions and abundant small and medium-sized calcite inclusions.

Condition: One fragment of the base.

Description: Barely distinct foot with small resting surface; flat and slightly recessed base featuring five irregular concentric grooves.

Comparisons and comment: Precise parallels were not found. The composition and color of the fabric is similar to the one of SAV Mor 01.

Date: 4th c. BC.

25
Cooking Ware

Antonietta Di Tursi

The cooking ware is attested at Sant'Angelo Vecchio by a small number of fragments recovered from different parts of the site used in different periods (*Fig. 25.1*), indicating different functions of these areas over the course of time.[1]

Classification

The classification of the materials is based on the identification of functional forms distinguished by types, which are in turn subdivided into variants.[2] The catalog, organized by form (indicated by letter), type, and variant (indicated by Arabic numerals),[3] briefly describes the history, morphological evolution, geographic circulation, and chronology of each. It also proposes a calibrated dating for the most chronologically sensitive forms. The discussion of the comparanda does not always refer to similar items, but instead aims to document the attestation of the forms and types and describe their geographic circulation and chronological development.

The typology for cooking ware, flexible for the addition of new types and variants, expands upon that established for Fattoria Fabrizio,[4] from which it also borrows the nomenclature of the forms indicating vessel functions: pan, casserole, pot, and lid cover, joined at Sant'Angelo Vecchio by the *klibanos*. Unlike at Fattoria Fabrizio, at Sant'Angelo Vecchio the pots are presented with an articulated morphology that distinguishes types and variants traceable to the Greek tradition (chytra and *kakkabè*) or the Italic tradition.

Technical Characteristics and Fabrics

In the absence of chemical and petrographic analyses, the materials were examined macroscopically. The majority of the fragments are characterized by a homogeneous fabric resistant to thermal and mechanical stresses and reddish or reddish-orange in color, with a few brown or blackish-gray examples (e.g., SAV CkW 30). The hackly fracture is rich in filler materials of various origins and sizes, in particular black and white stone inclusions and a small quantity of mica. These materials give the fabric a surface that is hard, porous, and rough to the touch.[5] Visible signs were left by the potter's wheel and a cursory smoothing with a stick, which failed to entirely eliminate the roughness of the surfaces. A few examples are coated on the interior surface (SAV CkW 05 and SAV CkW 23). Most of the fragments have more or less evident discoloration caused by both exposure to fire and fumes and steam (likely from cooking). The pieces with blackening from fire exposure, which is concentrated on the floors and walls but also appears on the rims, lid-resting ledges, and handles, are found exclusively in Late Archaic contexts (SAV CkW 31, House on the Hill; see also SAV CkW 10, Pottery Deposit; SAV CkW 33, Sounding D). The color variations caused by fumes and steam instead are found on fragments from the area of the Hellenistic kilns,[6] and thus represent possible evidence for the production of cooking ware at Sant'Angelo Vecchio. Local production cannot be confirmed, however, due to the absence of cooking-ware wasters.

Decoration is very rare, found mostly on the *klibanoi* (SAV CkW 26 and SAV CkW 27) in the form of patterns of small marks and incised undulating lines made while the clay was still soft (*Fig. 25.2*). One pot rim bears a braid motif (SAV CkW 37).

Contexts

House on the Hill. The cooking ware from the House on the Hill (*Table 25.1*) came mostly from the area

[1] For a synthesis of the problems inherent in the class, see Blotti in *Fattoria Fabrizio*, 297, n. 1.

[2] A detailed description of the methodology adopted for this class is available in Blotti in *Fattoria Fabrizio*.

[3] Lowercase letters are sometimes utilized to distinguish between specific examples. On TNI, see n. 2 on p. 220.

[4] Blotti in *Fattoria Fabrizio*.

[5] The classification of the fabrics was accomplished using the method proposed at the Settefinestre villa site (Carandini and Ricci 1985, 14).

[6] SAV CkW 04 (Sounding B); SAV CkW 05 (Slope 1, surface cleaning); SAV CkW 19 (Slope 1, surface cleaning); SAV CkW 20 (Sounding C, E of and including division); SAV CkW 23 (Slope 3, N extension); SAV CkW 26 (D4, outside stone circle to N near baulk); SAV CkW 27 (Slope 3, votive deposit, SW extension); SAV CkW 28 (D4, within stone circle); SAV CkW 34 (Sounding B); SAV CkW 35 (D4, within stone circle); SAV CkW 39 (Sounding C); SAV CkW 44 (D4 W); SAV CkW 45 (H5); SAV CkW 46 (Slope 3, votive deposit, SW extension).

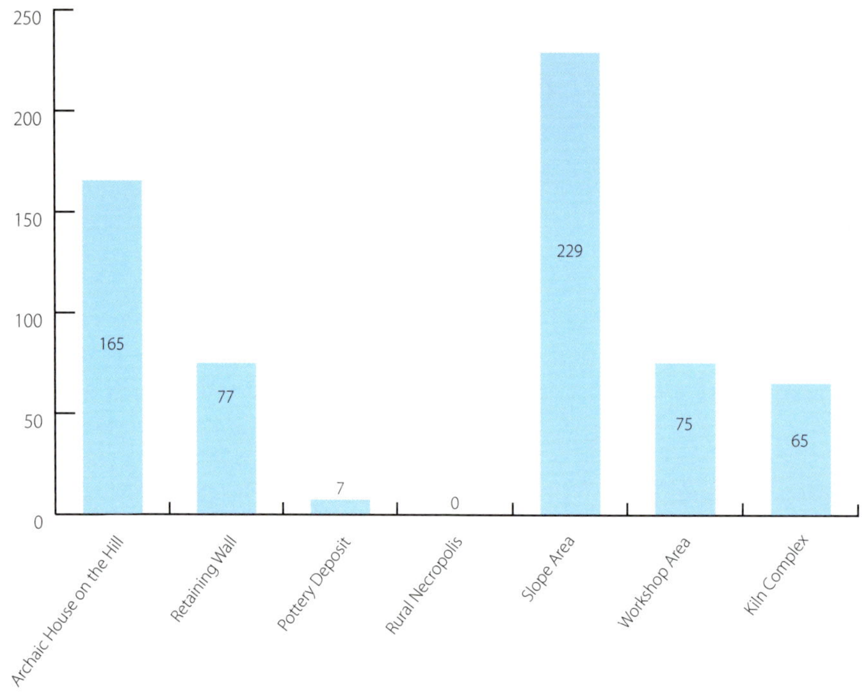

Figure 25.1 Distribution of cooking ware at Sant'Angelo Vecchio by number of fragments. (LMJ/ICA)

Form	Type	Date	Provenance	MNI	TNI	NFr
Pot	P1.2	End of the 6th c. BC	House on the Hill	1	1	3
	P1.5	End of the 6th c. BC	Surface finds from area around abandoned farmhouse on hill	14	1	18
	Unknown	Unknown	Surface finds from area around abandoned farmhouse on hill; Test Trench 1 on top of hill	0	0	106
Unknown form	Unknown	Unknown	S half of Test Trench 2 on top of hill Test Trench 2, W side	0	0	38
Totals				**15**	**2**	**165**

Table 25.1 Quantification of the cooking ware from the House on the Hill at Sant'Angelo Vecchio.

Form	Type	Date	Provenance	MNI	TNI	NFr
Casserole	C3.1	End of the 6th–first half of the 5th c. BC	Sounding C, Pottery Deposit	1	1	3
Pot	P1.5	End of the 6th–first half of the 5th c. BC	Sounding C, Pottery Deposit	2	1	3
	P3.1	End of the 6th–first half of the 5th c. BC	Sounding C, Pottery Deposit	1	1	1
Totals				**4**	**3**	**7**

Table 25.2 Quantification of the cooking ware from the Pottery Deposit at Sant'Angelo Vecchio.

Form	Type	Date	Provenance	MNI	TNI	NFr
Pan	Unknown	Unknown	C4, US 18	0	0	1
Casserole	C4	First half of the 5th c. BC	D5, US 39	1	1	1
	C6.2	End of the 4th–first half of the 3rd c. BC	C5, US 35 (extension 2)	1	1	1
Pot	P1.2	5th–4th c. BC	C5, US 40, 1 cut	1	1	3
	P1.5	6th–5th c. BC	C5, US 40, 1 cut C5, US 35 (extension 2) C4, US 9, 18	4	1	15
	P3.1	6th–5th c. BC	C4, US 9, 18	1	1	3
	Unknown	Unknown	C5, US 40, 1 cut C4/C5, US 41 C5, Level 1, batt. 5 US 45, sporadic find	0	0	31
Lid cover	Li2	End of the 4th–first half of the 3rd c. BC	C4, cleaning	1	1	1
	Unknown	Unknown	C5, Level 1 batt. 2	0	0	1
Open form	Unknown	Unknown	C4, cleaning of N section and Retaining Wall US 40, 1 cut US 45 sporadic find C4/C5, US 41 D5, US 39	0	0	13
Unknown form	Unknown	Unknown	C4, cleaning of Retaining Wall US 40, 1 cut D5, US 39, D5/D4 cleaning of Retaining Wall	0	0	7
Totals				**9**	**6**	**77**

Table 25.3 Quantification of the cooking ware from the area of the Retaining Wall at Sant'Angelo Vecchio.

inside the only identified room. The only form attested in this area is the chytra of the Greek tradition (P1). At least fifteen examples can be ascribed to two variants (P1.2 and P1.5), documented from the Archaic through the Hellenistic period without obvious morphological variation.[7] Nevertheless, the absence of forms such as casseroles and pans and the presence of black-gloss pottery datable to the Late Archaic period suggest a chronology between the end of the 6th and the first half of the 5th centuries BC for the examples from this area. Many of the rims found in the House on the Hill have traces of blackening due to exposure to fire, confirming that these vessels were part of the structure's kitchen equipment. The use of deep containers with narrow mouth, curving floor, and three-ribbed vertical ribbon handles attached to the rim reflects the cooking and eating habits of the time, in which foods boiled in chytrai were preferred.[8]

Figure 25.2 Decoration on the klibanoi SAV CkW 26 (*left*) and SAV CkW 27 (*right*); scale 1:3. (CW/ICA)

Pottery Deposit. The Pottery Deposit (*Table 25.2*) yielded a very small quantity of cooking-ware fragments, in particular a partly reconstructable casserole (SAV CkW 10) bearing traces of exposure to fire. Soundings carried out in 2010–2011 for cleaning and verification in squares C5 and D5 at the Retaining Wall revealed levels containing materials similar to those found in the Pottery Deposit. Among these were found a very small number of cooking-ware frag-

[7] For discussion of the chronology and bibliography, see the catalog entries for SAV CkW 31 and SAV CkW 34.

[8] Bats 1988, 45.

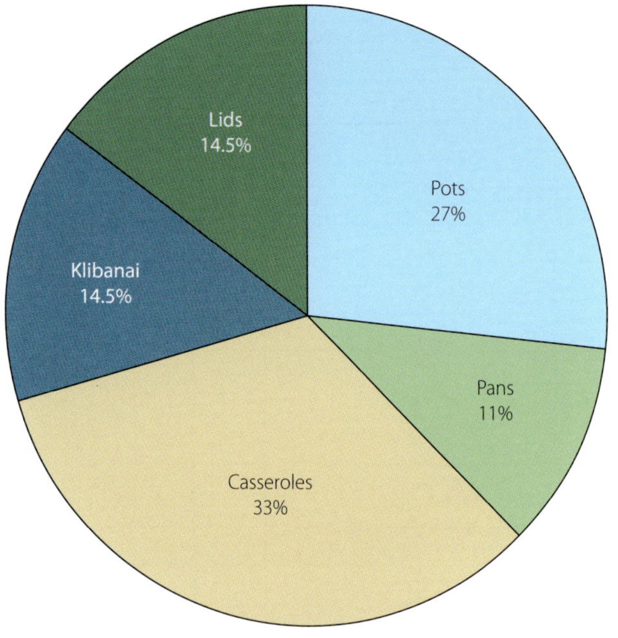

Figure 25.3 Quantification by form of the fragments from the Slope Area at Sant'Angelo Vecchio. (LMJ/ICA)

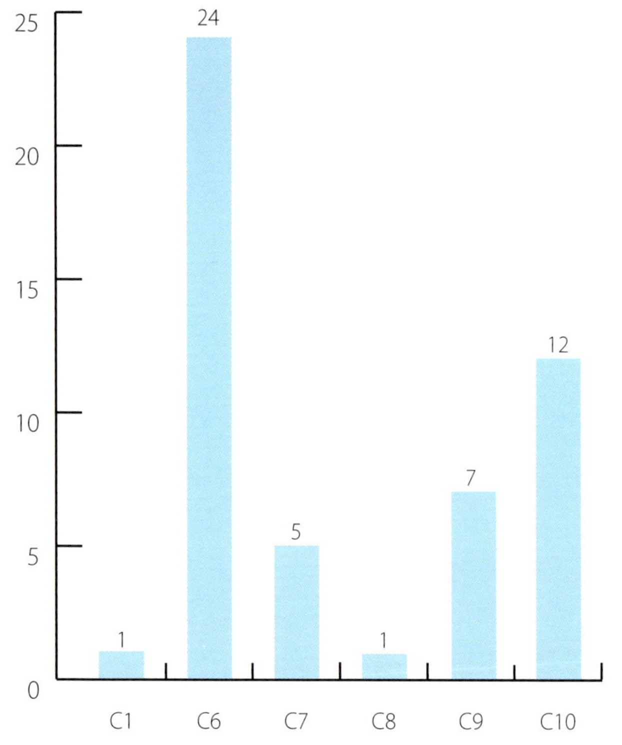

Figure 25.4 Quantification by type of the casserole fragments from the Slope Area at Sant'Angelo Vecchio. (LMJ/ICA)

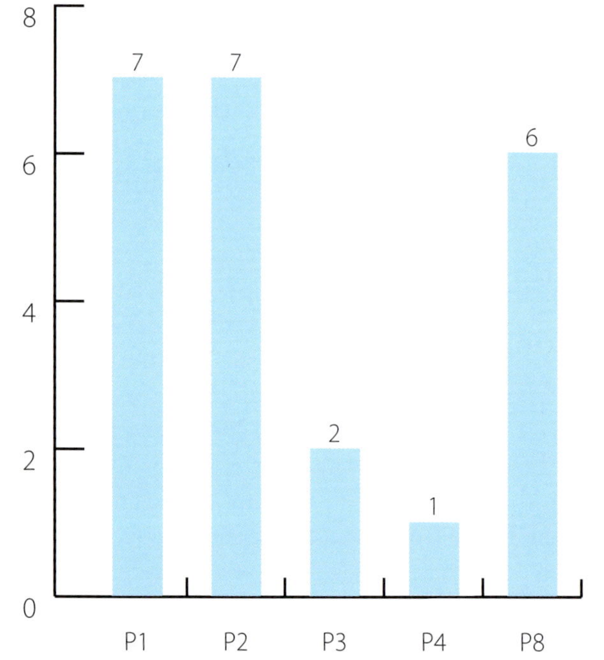

Figure 25.5 Quantification by type of the pot fragments from the Slope Area at Sant'Angelo Vecchio. (LMJ/ICA)

ments belonging to chytrai of type P1 and a casserole of type C4, datable between the end of the 6th and the first half of the 5th century BC (*Table 25.3*).

Slope Area. The majority of the cooking-ware fragments come from the levels associated with the workshop active at Sant'Angelo Vecchio in the late 4th and 3rd centuries BC, and in particular from the waste deposits found along the slope of the hill in the area where kilns were discovered (squares D3–4; *Table 25.4–Table 25.6*). The uniformity of these deposits, excavated in levels and battute despite having been recognized immediately as part of a single feature, is demonstrated by the numerous joining fragments found among both the cooking wares[9] and the terracotta plaques. The cooking ware, datable between the end of the 4th and the first half of the 3rd century BC, belongs to a wide range of forms (*Fig. 25.3*).

The most common form is the casserole, which has a remarkable variety of types and variants. The best attested is type C6, found especially in square D4; it is characterized by a carinated, oblique, and shallow body, which circulated in the Metapontino between the 4th and the first half of the 3rd century BC.[10] Types

[9] E.g., the casserole SAV CkW 23, the fragments of which (SA 79-381PL; SA 79-297PL; SA 79-325PL; SA 79-205PL; SA 79-300PL) come from Slope 3, N extension, Level 3, batt. 1; Sounding C, Level 2, batt. 2 and 7; Sounding B, Level 3, batt. 2.

[10] For discussion of the chronology and bibliography, see the catalog en-

Form	Type	Date	Provenance	MNI	TNI	NFr
Pan	PD2	End of the 4th–3rd c. BC	Slope 1, Surface cleaning	1	1	1
	PD3.3	End of the 4th–3rd c. BC	Slope 1, Surface cleaning	1	1	1
Casserole	C9.1	End of the 4th–3rd c. BC	Slope 1, Surface cleaning	1	1	1
	Unknown	Unknown	Slope 1, Surface cleaning	0	0	1
Pot	P1.5	End of the 4th–3rd c. BC	Sounding D, Level 1, batt. 1; Level 2, batt. 6	2	1	5
	Unknown	Unknown	Sounding D, Level 1, batt. 3; Level 2, batt. 1	0	0	4
Klibanos	Kl2	End of the 4th–3rd c. BC	Slope 1, Surface cleaning	0	0	1
Lid	Li1/L2	End of the 4th–3rd c. BC	Slope 1, Surface cleaning	0	0	2
Unknown	Unknown	Unknown	Sounding D, Level 1, batt. 2–3; Level 2, batt. 1	0	0	4
Totals				**5**	**4**	**20**

Table 25.4 Quantification of the cooking ware from Slope 1 (square E3) and Sounding D at Sant'Angelo Vecchio.

with rounded walls and thickened rim (C7, C8, and C9) were frequently encountered in Sounding B (*Fig. 25.4*). Also attested is type C1 (SAV CkW 09) with rounded walls and curving floor—its prototype may have been a Metapontine creation dating to the first half of the 4th century BC.[11] Along with type C1, type C10,[12] which occurs especially in contexts dated to the first half of the 3rd century BC, defines the chronological limits of the cooking ware attested in this phase.

Figure 25.6 Pan handle SA 80-3PL (F7, Level 1, batt. 3); scale 1:1. (CW/ICA).

The pot is the second most common form, in particular the chytrai of types P1 and P4 and, to a lesser extent, the *kakkabai* of type P8 (*Fig. 25.5*), all datable between the end of the 4th and the beginning of the 3rd centuries BC. The chytrai were distributed homogeneously throughout the waste deposit with the lone exception of type P4, featuring a brim rim, which circulated especially in the Metapontine chora in the same period. Type P8, a *kakkabè* with interior lid-resting ledge, is attested by two individual vessels recovered in Sounding B.

The lids are attested by types featuring an oblique dome with indistinct and rounded edge, distinguished by a cylindrical (Li1) or button (Li2) knob. Excluding the three examples with reconstructable profiles, the majority of the fragments preserve only the rim and belong to types Li1 and Li2. Both of these types recall a third that circulated widely in the Metapontino and in Magna Grecia generally between the middle of the 4th and the 3rd centuries BC.[13]

The pan, another form that circulated widely in the same period, is attested by 23 fragments from 15 individual vessels. The examples belong to types PD2 and PD3, dating between the end of the 4th and the first half of the 3rd centuries BC. There are also numerous fragments of *klibanoi* of type Kl2, variants *a* and *b*, attested by at least five individuals and datable to the same period as the pans.

Finally, eight fragments of cooking ware—all walls, except for a pan handle (*Fig. 25.6*) and type C10 casserole rim (SAV CkW 24)—were found in the area northwest of the kilns (squares F7 and G7). Here the remains of a foundation wall and a collapsed roofing system using tiles stamped *Herakas* are dated to the Hellenistic period.[14] Production of cooking ware in

tries for SAV CkW 12 and SAV CkW 14.

[11] Quercia 2003, 195.

[12] SAV CkW 21–SAV CkW 23.

[13] For discussion of the chronology and bibliography, see the catalog entries for SAV CkW 28 and SAV CkW 29.

[14] See Ch. 2, "Site Phasing, Stratigraphy, and Site Assemblage," and Ch. 3, "The Structures at Sant'Angelo Vecchio."

Form	Type	Date	Provenance	MNI	TNI	NFr
Pan	PD2	End of the 4th–3rd c. BC	Slope 3, W extension. Level 2, batt. 3 Slope 3, votive deposit S. Level 3, batt. 2 Sounding B. Level 2, batt. 5	4	1	5
	PD3.2	End of the 4th–3rd c. BC	Slope 3, votive deposit S. Level 3, batt. 2	3	1	3
Casserole	C1	End of the 4th–3rd c. BC	Slope 3, votive deposit S. Level 3, batt. 2	1	1	1
	C6.2	End of the 4th–3rd c. BC	Slope 3, W extension. Level 2, batt. 1, 3 Slope 3 NW extension. Level 1, batt. 5 Slope 3, votive deposit S. Level 3, batt. 2 Slope 3, W extension. Level 1, batt. 4	3	1	4
	C6.3	End of the 4th–3rd c. BC	Slope 3, W extension. Level 2, batt. 1, 2 Slope 3, W extension. Level 3, batt. 1	2	1	4
	C7	End of the 4th–3rd c. BC	Slope 3 NW extension. Level 1, batt. 5	1	1	3
	C8.2	End of the 4th–3rd c. BC	Slope 3, W extension. Level 3, batt. 1	1	1	1
	C9.1	End of the 4th–3rd c. BC	Slope 3, W extension. Level 2, batt. 3	1	1	1
	C9.2	End of the 4th–3rd c. BC	Slope 3, W extension. Level 3, batt. 1	1	1	2
	C9.3	End of the 4th–3rd c. BC	Slope 3, W extension. Level 2, batt. 1	1	1	1
	C10.1a	End of the 4th–3rd c. BC	Slope 3, W extension. Level 2, batt. 1 Slope 3, W extension, votive deposit N. Level 2, batt. 2	1	1	3
	C10.1b	End of the 4th–3rd c. BC	Slope 3, W extension. Level 2, batt. 1	1	1	1
	C10.1c	End of the 4th–3rd c. BC	Slope 3 N extension. Level 3, batt. 1 Sounding C. Level 3, batt. 2, 7 Sounding B. Level 3, batt. 2	1	1	6
	Unknown	Unknown	Slope 3, votive deposit S. Level 3, batt. 2 Slope 3, W extension. Level 2, batt. 1, 2/Level 3, batt. 1 Slope 3 NW extension. Level 1, batt. 5 Sounding B. Level 3, batt. 2	0	0	8
Klibanos	Kl2	End of the 4th–3rd c. BC	Slope 3, W extension. Level 3, batt. 1 Slope 3, votive deposit S. Level 3, batt. 2 Sounding B. Level 2, batt. 1	1	1	9
	Unknown	Unknown	Slope 3 NW extension. Level 1, batt. 5 Slope 3, W extension. Level 2, batt. 1 Slope 3, cleaning of grass and topsoil. batt. 1	0	0	3
Pot	P1.8	5th–4th c. BC	Sounding B. Level 3, batt. 1	1	1	1
	P2.3	2nd c. BC	Slope 3, W extension. Level 3, batt. 1	1	1	1
	P4	End of the 4th–3rd c. BC	Slope 3, W extension. Level 2, batt. 1	1	1	1
	P8.3	End of the 4th–3rd c. BC	Slope 3, W extension. Level 2, batt. 1	1	1	4
	Unknown	Unknown	Slope 3 surface cleaning Slope 3, W extension. Level 2, batt. 1–2; Level 3, batt. 1 Slope 3, W extension votive deposit. N Level 2, batt. 2 Sounding B. Level 2, batt. 2; Level 3, batt. 3	0	0	13
Lid	Li1.1	End of the 4th–3rd c. BC	Slope 3, W extension. Level 2, batt. 1	5	1	9
	Li1/L2	End of the 4th–3rd c. BC	Slope 3, cleaning of grass and topsoil. batt. 1 Slope 3, W extension. Level 1, batt. 4; Level 2, batt. 2 Slope 3, W extension. Level 3, batt. 1 Slope 3, W extension votive deposit. N Level 2, batt. 2	3	1	4
	Unknown	Unknown	Surface finds, grassy slope above Slope 3 Slope 3, W extension. Level 2, batt. 1 Slope 3, votive deposit S. Level 3, batt. 2	0	0	3
Open form	Unknown	Unknown	Slope 3, W extension. Level 2, batt. 1, 3 Sounding B. Level 2, batt. 5	0	0	10
Unknown	Unknown	Unknown	Slope 3; Sounding B, S extension. Level 2, batt. 1	0	0	1
Totals				**34**	**20**	**102**

Table 25.5 Quantification of the cooking ware from Slope 3 (squares D3–4) at Sant'Angelo Vecchio.

Form	Type	Date	Provenance	MNI	TNI	NFr
Pan	PD2	End of the 4th–3rd c. BC	D4, within stone circle. Level 3, batt. 5	2	1	3
	PD3.2	End of the 4th–3rd c. BC	D4 W. Level 3, batt. 5 D4, within stone circle. Level 3, batt. 5 Sounding C. Level 2, batt. 7	4	1	5
	Unknown	Unknown	Sounding C. Level 2, batt. 7	0	0	5
Casserole	C6.1	End of the 4th–3rd c. BC	Sounding C, E of and including division. Level 2, batt. 5 Sounding C. Level 2, batt. 6, 7	3	1	4
	C6.2	End of the 4th–3rd c. BC	Slope 4, W extension. Level 1, batt. 1 D4 W. Level 3, batt. 5 Slope 4. Level 1	3	1	8
	C6.3	End of the 4th–3rd c. BC	D4, within stone circle. Level 3, batt. 5 Sounding C. Level 2, batt. 6, 9	2	1	4
	C7	End of the 4th–3rd c. BC	Sounding C. Level 2, batt. 3 D4 W and Slope 3, W extension. Level 3, batt. 1, 4	2	1	6
	C9.3	End of the 4th–3rd c. BC	Sounding C, E of and including division. Level 2, batt. 5	1	1	1
	C10.1b	End of the 4th–3rd c. BC	Sounding C, E of and including division. Level 2, batt. 5	1	1	2
	Unknown	Unknown	Sounding C. Level 2, batt. 2, 6–7 D4 W. Level 3, batt. 5	2	1	8
Klibanos	Kl2	End of the 4th–3rd c. BC	D4. Level 3, batt. 4 D4, within stone circle. Level 3, batt. 5 D4, N near baulk. Level 3, batt. 6 Sounding C. Level 2, batt. 2, 3 Sounding C. Level 2, batt. 6 Sounding C, E of and including division. Level 2, batt. 5	4	1	17
	Unknown	Unknown	Sounding C. Level 2, batt. 3	0	0	1
Pot	P1.5	End of the 4th–3rd c. BC	Slope 4, N of Sounding A found during clearing of loose soil left by bulldozer	1	1	1
	P2.2	End of the 4th–3rd c. BC	D4, within stone circle. Level 3, batt. 5	3	1	5
	P2.3	End of the 4th–3rd c. BC	D4, within stone circle. Level 3, batt. 5	1	1	1
	P3.1	End of the 4th–3rd c. BC	Sounding C. Level 2, batt. 5 Slope Area 3 cleaning of grass and topsoil	1	1	2
	P8.1	End of the 4th–3rd c. BC	D4 W. Level 3, batt. 5	1	1	1
	P8.2	End of the 4th–3rd c. BC	Sounding C, E of and including division. Level 2, batt. 5	1	1	1
	Unknown	Unknown	Sounding C. Level 2, batt. 2, 4–7 Sounding C, E of and including division. Level 2, batt. 5 Slope 4. Level 2, batt. 5 D4, N near baulk. Level 3, batt. 6	1	1	18
Lid	Li2	End of the 4th–3rd c. BC	Sounding C. Level 1, batt. 3 Slope 4, N of Sounding A found during clearing of loose soil left by bulldozer	1	1	1
	Li1/L2	End of the 4th–3rd c. BC	D4, within stone circle. Level 3, batt. 5 D4, N near baulk. Level 3, batt. 6 Sounding C, E of and including division. Level 2, batt. 5 Sounding C. Level 2, batt. 6	8	1	12
Unknown	Unknown	Unknown	Slope 4, N of Sounding A found during clearing of loose soil left by bulldozer	0	0	1
Totals				**42**	**19**	**107**

Table 25.6 Quantification of the cooking ware from Sounding C and Slope 4 (square D4) at Sant'Angelo Vecchio.

Form	Type	Date	Provenance	MNI	TNI	NFr
Pan	PD4	2nd–1st c. BC	Sounding A, S sub-section of Room 1 down to wall. Level 2, batt. 2	1	1	8
	PD5	2nd–1st c. BC	F3, inside kiln. Level 2, batt. 2	1	1	1
	PD6	2nd–1st c. BC	F3, SE end of N Kiln near baulk. Level 2	1	1	3
Casserole	C10.1b	2nd–1st c. BC	F3, SE end of N Kiln near baulk. Level 1, batt. 1, 4	1	1	2
Klibanos	KL1	2nd–1st c. BC	E3, between Soundings A and B. Level 2, batt. 1 F3, inside kiln Level 2, batt. 2	2	1	8
	Unknown	Unknown	F3, SE end of N Kiln near baulk. Level 2, batt. 4	0	0	2
Pot	P1.5	6th–3rd c. BC	Sounding A, S end of Room 1. Level 3, batt. 1	1	1	1
	P5	2nd c. BC	F3, SE end of N Kiln near baulk. Level 2	1	1	4
	P6.1	2nd c. BC	F3, N Kiln interior. Level 2, batt. 1	1	1	4
	P6.2	2nd c. BC	F3, N Kiln interior. Level 2, batt. 2	1	1	6
	Unknown	Unknown	E1 cleaning. Level 2. F1. Level 2, batt. 1. F2. Level 2, batt. 1 Cut in baulk between N and S Kilns. Level 2, batt. 1, 2 Sounding A, S sub-section of Room 1 down to wall. Level 2, batt. 1 Platform at the junction between the N Kiln and the N wall. Level 3, batt. 1	0	0	7
Lid	Li4	2nd–1st c. BC	F1. Level 2, batt. 1 F3, N Kiln interior. Level 2, batt. 2 Platform at the junction between the N Kiln and the N wall. Level 3, batt. 2	4	1	14
	Unknown	Unknown	Sounding A, S sub-section of Room 1. Level 2, batt. 3 F3, SE end of N Kiln near baulk. Level 2, batt. 1, 4 Cut in baulk between N and S Kilns. Level 2, batt. 1, 2	3	1	4
Open form	Unknown	Unknown	E1, cleaning. Level 2	0	0	1
Totals				**17**	**11**	**65**

Table 25.7 Quantification of the cooking ware from the Kiln Complex at Sant'Angelo Vecchio.

this phase cannot be confirmed due to the absence of materials that can be securely identified as wasters. Despite the presence of color variations due to fumes and steam, the high degree of preservation of most of the fragments make identification possible. The morphological repertoire of the cooking ware recovered in the deposit in the area of the slopes dates between the end of the 4th and the first half of the 3rd century BC, and perfectly reflects that documented in the Metapontino in the same period.[15] There was an apparent preference for open forms (*Table 25.5* and *Table 25.6*) with shallow bodies, such as the casserole and pan, which seem to have been in greater demand and used more frequently, probably in connection with changes in cooking habits. Prolonged cooking of foods (e.g., boiled in chytrai) was replaced by a faster method based on roasting and frying and performed under a lid, *sub testu*,[16] for meats, vegetables, flatbreads, and cakes cooked over a low and diffuse fire. The *kakkabai*, the deep pots with lid-resting ledge, are scarcely attested despite the form's wide circulation in the Metapontine chora during the 4th century BC.[17] An analogous situation has been documented in deposit 1 of the *kerameikos* of Metaponto, where the form had a limited production in the first quarter of the same century.[18]

Kiln Complex. Around the middle of the 3rd century BC, the small kilns along the slope were abandoned. Production resumed in the 2nd and 1st centuries in the Kiln Complex.[19] The cooking ware from this context is represented by some fragments (*Table 25.7*) recovered in the North Kiln and in the courtyard in front of it. A few fragments of Hellenistic date found in the courtyard suggest that this area had also been used in the preceding period.

The most noteworthy result of this study is the identification of new types belonging to the Italic morpho-

[15] See, in particular, the intensive field survey between the Bradano and the Basento (Gabrieli in *Survey*), Fattoria Stefan, Fattoria Fabrizio (*Fattoria Fabrizio*), Metaponto (M. Liseno 2004; Quercia 2003), Cozzo Presepe (Cotton 1977), and Pomarico Vecchio (Bianco and Deodato 1997).

[16] On *klibanoi* in general, see Cubberley/Lloyd/Roberts 1988, 98–119.

[17] For discussion of the chronology and bibliography, see the catalog entry for SAV CkW 47.

[18] Quercia 2003, 195.

[19] See Ch. 3, "The Structures at Sant'Angelo Vecchio."

Form	Type	Date	Provenance	MNI	TNI	NFr
Casserole	C8.a	End of the 4th–3rd c. BC	G5. Level 2, batt. 1 H5. Level 1, batt. 2	1	1	3
	C6.3	End of the 4th–3rd c. BC	H3, cleaning of wall	1	1	1
	Unknown	Unknown	F6. Level 1, batt. 2 G6. Level 1, batt. 4 G6, side of wall. Level 2, batt. 3 H3. Level 1, batt. 1	0	0	7
Pot	P1.3	6th–1st c. BC	H3, cleaning of wall. Level 1, batt. 3	1	1	2
	P2.4	2nd c. BC	G6. Level 1, batt. 5 G6. Level 3, batt. 1	1	1	2
	P3.3	2nd c. BC	G6, E of wall. Level 2, batt. 3–4 G5, wall. Level 2, batt. 4-1	3	1	19
	P8.3	End of the 4th–3rd c. BC	H5. Level 1, batt. 2	1	1	4
	Unknown	Unknown	G5. Level 1, batt. 2 G6. Level 1, batt. 4 G6. Level 1, batt. 5 H3. Level 1, batt. 1 H5. Level 2, batt. 1	0	0	6
Lid	Li1/Li2	End of the 4th–3rd c. BC	G6, E of wall. Level 2, batt. 2 G6, side of wall. Level 2, batt. 3 H4. Level 1, batt. 1	4	1	6
	Unknown	Unknown	G6, W of wall. Level 1, batt. 2	1	0	2
Open form	Unknown	Unknown	G6. Level 1, batt. 5 H5. Level 2, batt. 1	0	0	3
Unknown form	Unknown	Unknown	G6, E of wall. Level 3, batt. 1	0	0	1
Totals				**13**	**7**	**56**

Table 25.8 Quantification of the cooking ware from the Workshop Area (squares G5–6, F6, and H3–5) at Sant'Angelo Vecchio.

logical repertoire. The nearly fully preserved examples, mostly from the interior of the North Kiln, form half of an olla (SAV CkW 43), a pan (SAV CkW 07), a *klibanos* (SAV CkW 25), and two mugs (SAV CkW 48 and SAV CkW 49), datable between the 2nd and 1st centuries BC. A lid (SAV CkW 30) found on the platform located between the entrance of the North Kiln (Kiln 5) and the northern wall of the Kiln Complex and beneath the collapse of the kiln, as well as a pan (SAV CkW 06) found in the courtyard preceding the North Kiln can be dated to the same period.

Some of the examples, such as the *klibanos* (SAV CkW 25) and the pan (SAV CkW 07), both reused in the construction of the kiln, have deformities due to misfiring that suggest local production. There is not enough evidence, however, to suggest hypotheses about the method of production employed.

Workshop Area. Several cooking-ware fragments were also found in the area of the Late Republican workshop shed, datable between the end of the 2nd and the beginning of the 1st century BC: four examples of a globular olla (SAV CkW 37, SAV CkW 39, and SAV CkW 40) with flat bottom and pendant rim and lacking handles, typical products of the Italic tradition.[20] The presence of fragments dating between the end of the 4th and the middle of the 3rd century BC from surface levels documents an earlier occupation as well as the agrarian use of the area in recent times, which has almost completely obliterated the Late Republican remains of the site (*Table 25.8*).

Conclusions

The cooking-ware fragments recovered at Sant'Angelo Vecchio, a site occupied almost continuously from the Late Archaic through the Early Imperial period, document the changes in the Greek morphological repertoire, probably indicative of dietary changes and the transition to the Roman phase. The presence of vessel types belonging to the Roman tradition is a symptom of an evident adaptation, in which Roman influence took root even in the most practical aspects of daily life.

[20] See the type P3 pots from square G6.

Catalog of Cooking Ware

Open Forms

Pans. The pan group includes various vessels with a flat or curving base, shallow body, and optional handle(s), probably used for frying and roasting. The shape appeared in Athens in the middle of the 5th c. BC (*Agora XII*, 228) and spread in the 4th and 3rd c. BC, arriving in southern Italy during the Hellenistic period (Conti 1992, Guzzo 1978).

The pans from the Metapontine chora constitute five types, subdivided into variants according to the shape of the rim. The examples recovered at Sant'Angelo Vecchio—53 fragments from at least 24 individual vessels—belong to four of these types (PD2, 3, 4, and 5) and a newly recognized sixth (PD6). A single handle fragment with elliptical section and decorative lugs appears to imitate a pan in bronze (*Fig. 25.6*).

Types PD2 and PD3 are datable to the 4th and 3rd c. BC. Types PD4, PD5, and PD6 are morphologically similar to the Roman *patinae* that circulated in southern Italy beginning in the 3rd c. BC. They also differ from types PD2 and PD3 in their fabric, which is rich in medium and large inclusions, as well as in the better finishing of their surfaces, which are carefully smoothed with a stick.

PD2. Pan type PD2 is characterized by an indistinct rim, carinated and oblique body, and convex base. At Sant'Angelo Vecchio there are nine fragments belonging to seven vessels, all found in the Slope Area. The type is attested along the Ionian coast in contexts dating between the end of the 4th and the beginning of the 3rd c. BC, with parallels from the Metapontino, the agora of Herakleia, and Calabria.

PD3. Pan type PD3 is characterized by a brim rim. The type, attested at Sant'Angelo Vecchio by 25 fragments from 13 vessels, is found in the chorai of Metaponto (SAV CkW 50) and Siris (Hänsel 1973) as early as the end of the 5th c. BC, with the curving base superseded by the flat base. This version is present in South Italian contexts of the 4th and 3rd c. BC, in particular in the Metapontino, as demonstrated by the numerous examples recovered during the intensive field survey between the Bradano and Basento (Gabrieli in *Survey*). Three variants were identified by the details of the rim shape.

The first variant of PD3 (PD3.1) is characterized by a brim rim inclined toward the exterior. This is the most common variant, with 22 fragments belonging to ten vessels, widely circulated in the 4th and 3rd c. BC. The second variant of PD3 (PD3.2) is characterized by a brim rim inclined toward the interior. The variant is attested by two fragments of the same date found in the Slope Area. These pans are widely documented at contemporary indigenous sites and Metapontine farmhouses, as well as in Calabria and Apulia.

The third variant of PD3 (PD3.3) is characterized by a deep body with a diameter of approximately 32 cm and very thick, oblique walls made impermeable by a poorly applied coating. The lone example recovered at Sant'Angelo Vecchio has parallels with materials from the Metapontine chora, Pomarico Vecchio, and Locri in contexts dated to the 4th and 3rd c. BC.

PD4. Pan type PD4 is characterized by an indistinct rim and low, oblique body. The form is attested at Sant'Angelo Vecchio by a single example from the Kiln Complex. Parallels exist at Herakleia, Minturnae, Valesio, Gravina, and Torre S. Sabina (Brindisi) in contexts dating between the 3rd c. BC and (especially) the Late Republican period.

PD5. Pan type PD5 is characterized by a rim modeled to accommodate a lid and base with concave exterior profile. The type, attested at Sant'Angelo Vecchio by a single fragment, finds parallels in 2nd–1st-c. BC contexts at Sutri near Viterbo and in Gaul.

PD6. Pan type PD6 is characterized by a body with a rounded profile and a rim with small flange to accommodate a lid. A single fragment was recovered at Sant'Angelo Vecchio in the Kiln Complex. The type is comparable to pans found at nearby Cozzo Presepe and at Vaste (Salento) in contexts dating to the 4th and 3rd c. BC. It is also attested at Valesio (Salento) in the 2nd c. BC.

Pans

SAV CkW 01 Pan Type PD2

Provenance: Sounding B. Level 2, batt. 5. Lot: SA 79-312P (see *Fig. 2.35*).

Dimensions and technical features: h 3.1, rim 18.0, wall th 0.4, w 10.3. Clay Munsell 5YR 6/8 reddish yellow, hackly fracture, hard and rough to the touch, with moderate quantities of small white and black stone inclusions distributed uniformly. Surface mica and vughs. Local fabric.

Condition: One fragment.

Description: Indistinct rim rounded at the top; shallow and oblique body; convex base.

1:2

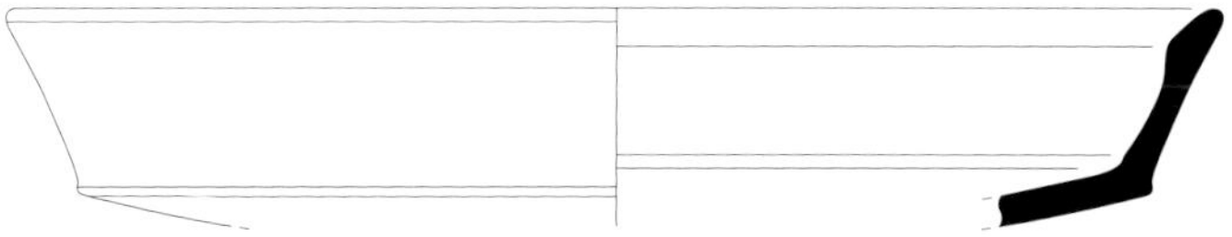

Comparisons and comment: Guzzo 1978, 470, fig. 5, no. 28 (from Acquappesa); Casagrande 2002, 401, fig. 50, type 6a (from the agora of Herakleia, 320–270 BC); Gabrieli in *Survey*, 444, PD2-4 Ck 260-02 (Metaponto field survey).

Date: End of the 4th–first half of the 3rd c. BC.

SAV CkW 02 Pan Type PD3.1

Provenance: Sounding C. Level 2, batt. 7. Lot: SA 79-205PL (see *Fig. 2.35*).

Dimensions and technical features: h 2.4, rim 28.0, wall th 0.6, w 17.2. Clay Munsell 2.5YR 4/4 reddish brown, hackly fracture, hard and smooth to the touch, with very few small white stone inclusions distributed uniformly. Surface vughs. Local fabric.

Condition: Fragmentary.

Description: Brim rim with concave profile inclined toward the exterior; shallow and oblique body; flat base.

Comparisons and comment: Conti 1983, 37, pl. X, no. 14 (from Locri); Rotroff and Oakley 1992, 26, T 112-2 (from Athens, 310–275 BC); Rossi 1989, 190, pl. 335, no. 4 (from Monte Sannace); Campagna 1995, 251, fig. 17, no. 1 (from Fondo S. Antonio-Vaste, end of the 4th–beginning of the 3rd c. BC); Carter and Parmly Toxey in *Necropoleis*, 728, T 111-2 casserole (310–275 BC); Bianco and Deodato 1997, 192, pl. 82, no. 182 (from Pomarico Vecchio).

Date: End of the 4th–first half of the 3rd c. BC.

1:2

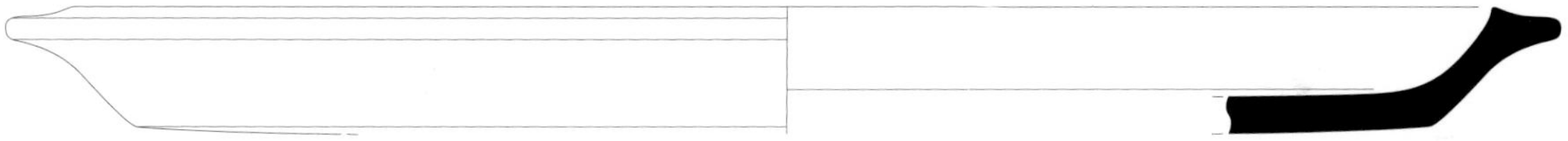

SAV CkW 03 Pan Type PD3.2a

Provenance: Square D4, within stone circle. Level 3, batt. 5. Lot: SA 79-896PL-1. Assemblage 13.

Dimensions and technical features: h 2.1, rim 26.0, wall th 0.5, w 8.5. Clay Munsell 5YR 5/8 yellowish red, hackly fracture, hard and rough to the touch, with moderate quantities of small white and black stone inclusions distributed uniformly. Surface mica and vughs. Local fabric.

Condition: One fragment.

Description: Brim rim with concave profile and small rise; low and oblique body; slightly convex base. On the interior, the transition from the wall to the base is marked by a distinct recess.

Comparisons and comment: Brizzi 1999b, 312, fig. 315, no. 619 (from Oppido Mamertina).

Date: End of the 4th–first half of the 3rd c. BC.

1:2

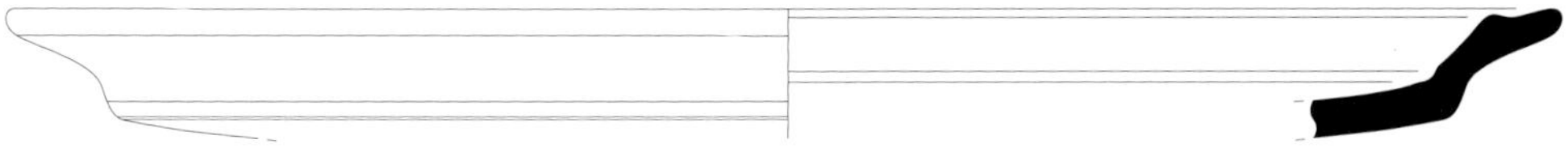

Pans cont.

SAV CkW 04 Pan Type PD3.2b
Provenance: Sounding B. Level 3, batt. 2. Lot: SA 79-301P. Assemblage 11.
Dimensions and technical features: h 1.9, rim 24.2, wall th 0.5, w 7.4. Clay Munsell 2.5YR 4/1 dark reddish gray, hackly fracture, hard and rough to the touch, with large quantities of small and medium-sized white stone inclusions distributed uniformly. Surface mica and vughs. Local fabric.
Condition: One fragment, misfired.
Description: Brim rim slightly inclined toward the interior; shallow and oblique body; slightly concave base.
Comparisons and comment: Guzzo 1978, 470, fig. 4, no. 43 (from Acquappesa).
Date: End of the 4th–first half of the 3rd c. BC.

1:2

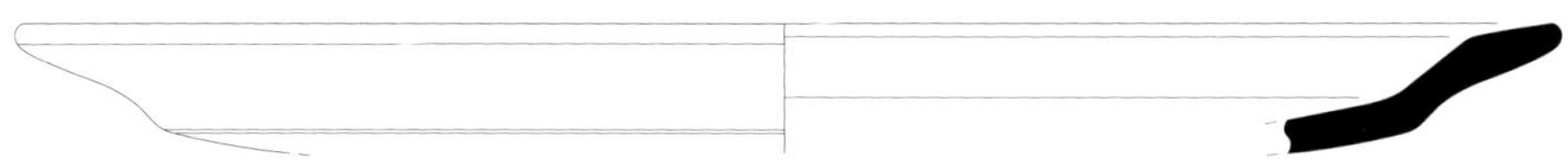

SAV CkW 05 Pan Type PD3.3
Provenance: Slope 1, surface cleaning. Lot: SA 79-359PL-2.
Dimensions and technical features: h 4.7, rim 32.0, wall th 1.3, w 9.3. Clay Munsell 2.5YR 4/6 red, shaly fracture, hard, smoothed, and rough to the touch, with moderate quantities of small white stone inclusions distributed uniformly. Surface vughs. Traces of coating on the interior surface. Local fabric.
Condition: One fragment, misfired.
Description: Brim rim inclined toward the interior; deep and oblique body.
Comparisons and comment: Conti 1989, 284, pl. XXXVIII, no. 325 (from Locri, 4th–3rd c. BC); Bianco and Deodato, 192, pl. 82, no. 183 (from Pomarico Vecchio); Pietropaolo 1997, 264, fig. 14, no. 2 (from Torre S. Sabina-Brindisi); Gabrieli in *Survey*, 443, PD1-1 Ck 341-02 (Metaponto field survey).
Date: End of the 4th–first half of the 3rd c. BC.

1:2

SAV CkW 06 Pan Type PD4

Provenance: Sounding A, S area of Room 1. Level 2, batt. 2. Lot: SA 81-26PL. Assemblage 5.

Dimensions and technical features: h 3.1, rim 24.0, wall th 0.9. Clay Munsell 2.5YR 5/6 red, hackly fracture, hard and rough to the touch, with large quantities of white and black stone inclusions of varying size distributed uniformly. Surface vughs. Local fabric.

Condition: Reconstructed.

Description: Indistinct rim, rounded near the top; low, oblique body. On the exterior, there is a distinct bulge marking the transition from the wall to the flat base.

Comparisons and comment: Kirsopp Lake 1934–1935 106, pl. XIX, no. 16, (mid-3rd c. BC from Minturnae); Vegas 1973, 41, fig. 14, no. 3 (1st c. BC–1st c. AD); Boersma and Yntema 1987, 79, fig. 32, no. 4 (from Valesio); Cotton 1992b, 184, fig. 74, no. 1042 (from Gravina, 2nd–1st c. BC); Pietropaolo 1997, 265, fig. 15, no. 4 (from Torre S. Sabina-Brindisi, end of the 2nd–first half of the 1st c. BC); Yntema 2001, 320, Form N51 no. 587 (from Valesio); Casagrande 2002, 172, fig. 50, type 7 (from agora of Herakleia, 375–270 BC).

Date: 2nd–1st c. BC.

1:2

Pans cont.

SAV CkW 07 Pan Type PD5

Provenance: Square F3, N Kiln interior. Level 2, batt. 2. Lot: SA 79-924P. Assemblage 8.

Dimensions and technical features: h 3.7, rim 31.1, wall th 0.8. Clay Munsell 5YR 6/6 reddish yellow, hackly fracture, hard and smooth to the touch, with small quantities of medium-sized and large stone inclusions distributed uniformly. Exterior surface vughs. Local fabric.

Condition: Full profile.

Description: Modeled rim and low body with oblique profile; base with concave exterior profile.

Comparisons and comment: Duncan 1964, 68, fig. 16, no. 196 (from Sutri); Bats 1993, 360, COM IT 6a (200–25 BC).

Date: 2nd–1st c. BC.

1:2

SAV CkW 08 Pan Type PD6
Provenance: Square F3, SE end of N Kiln near baulk. Level 2, batt. 4. Lot: SA 79-880PL. Assemblage 8.
Dimensions and technical features: h 3.3, rim 30.0, wall th 0.9, w 11.1. Clay Munsell 10R 5/4 weak red, hackly fracture, hard and smooth to the touch, with large quantities of small and medium-sized black and white stone inclusions distributed uniformly. Surface vughs. Local fabric.
Condition: Fragmentary.
Description: Forked rim; low body with rounded wall.
Comparisons and comment: Cotton 1977, 378, fig. 148, no. 483 (from Cozzo Presepe); Campagna 1995, 251, fig. 9, no. 2 (from Fondo S. Antonio-Vaste, 3rd c. BC); Yntema 2001, 283 Form N01a no. 497 (from Valesio, first half of the 2nd c. BC).
Date: 2nd–1st c. BC.

1:2

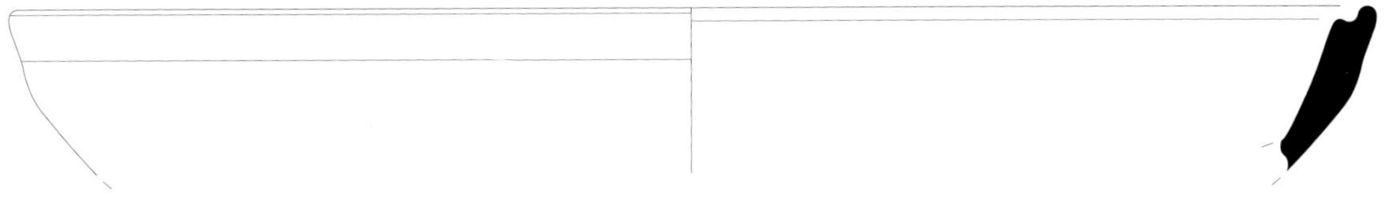

Casseroles. The casserole, also known as the *lopas*, appears in Greece in the third quarter of the 5th c. BC. It was utilized especially for cooking fish and was very popular in the Hellenistic period (*Agora XXXIII*, 178–86). The form is distinguished by a wide mouth, interior lid-resting ledge, rounded or carinated body of variable depth, and curving base that becomes flat in the latest versions. It is normally equipped with handles with elliptical section adhering to the body. The shallow depth of the body made the casserole particularly useful for steaming. Casseroles represent the most commonly attested forms at Sant'Angelo Vecchio, with 97 fragments belonging to 36 vessels. The varied and well-articulated typology created from these fragments is based on the shape of the body, with numerous variants determined by details of the rim. Eight types were found at Sant'Angelo Vecchio.

C1. Casserole type C1 is characterized by a sizable, curved body and lid-resting ledge. The type is derived from the earlier lopas, which circulated in Greece starting in the first half of the 5th c. BC (Bats 1988, 48). In the first phase of production, the type accompanied the traditional chytra, and to a certain extent even replaced it, especially for foods requiring long cooking times, where use of the casserole was found to be more practical.

The lone example recovered at Sant'Angelo Vecchio has a thin rim and pronounced shoulder. After making its appearance in Greece at the end of the 5th c. BC (*Agora XII*, 227), the form spread widely throughout Magna Grecia in the course of the following century.

C3. On type C3, the body, with concave exterior profile, is joined to a base with convex profile via a pronounced carination. It is attested at Sant'Angelo Vecchio by a single example from the Pottery Deposit, which yielded material datable to the first half of the 5th c. BC. The type appeared at Sybaris at the end of the 6th c. BC and spread in the Hellenistic period, in particular at Metaponto and the farmhouses in its chora, as well as in Lucania and Apulia generally.

C4. Casserole type C4 is characterized by a flaring rim with a very pronounced, vertical lid-resting ledge and lug handle attached at the rim. This relatively uncommon type dates to the Late Archaic period. Close parallels are currently not available, with the sole exception of an example with similar rim from Caulonia found in a context dated generically prior to the Hellenistic period.

C6. Casserole type C6 is characterized by a low, straight-walled body inclined on the exterior and handle with elliptical section attached at the rim. The diameter ranges between 18 and 24 cm. The type, attested at Sant'Angelo Vecchio by 26 fragments belonging to at least 15 vessels, circulated widely between the end of the 4th and the middle of the 3rd c. BC in the farmhouses of the Metapontine chora, Apulia (Gravina, Monte Sannace), Campania (Roccagloriosa), and Calabria (Locri). Three variants were identified by the inclination of the body wall and the rim shape. The first variant of casserole type C6 (C6.1) is characterized by an indistinct rim with a flat upper edge. It is attested by three individual vessels. The second variant of casserole type C6 (C6.2) is characterized by a rim with rounded upper edge. Seven individual vessels were identified as belonging to this variant. The third variant of casserole type C6 (C6.3) is characterized by a flaring rim with a rounded upper edge. The variant is attested by five individual vessels.

C7. Casserole type C7 is characterized by a flaring rim with pronounced and oblique lid-resting ledge, body wall with rounded exterior profile, and curving base. The type, attested by nine fragments belonging to three individual vessels, all from contexts of the late 4th and first half of the 3rd c. BC, is encountered in Metapontine farmhouses, at Herakleia, and in Apulia (Otranto, Monte Sannace, Gravina).

C8. Casserole type C8 is characterized by a flaring rim with a triangular section and a rounded body. Two vessels of this type were recovered at Sant'Angelo Vecchio from contexts dating between the end of the 4th and the first half of the 3rd c. BC. The type appeared at Metaponto in the 5th c. BC (M. Liseno 2004, 98) and circulated widely, especially between the end of the 4th and the beginning of the 3rd c. BC in Apulia (Monte Sannace) and the Metapontino (Cozzo Presepe and Difesa S. Biagio-Montescaglioso). It occurs frequently at Fattoria Stefan (in course of study), where a large number of variants are distinguished by the rim shape.

C9. Casserole type C9 is characterized by a flaring rim and a deep body with a rounded profile. The type is attested at Sant'Angelo Vecchio by six fragments belonging to five individual vessels from the Slope Area. None of the fragments preserves a handle. It circulated widely at the same time in the chorai of Metaponto and Siris, and in Apulia, Lucania, and Calabria generally. Two variants were identified on the basis of the rim shape.

The first variant of casserole type C9 (C9.1) is characterized by a flaring rim with a flat upper edge. This variant is attested at Sant'Angelo Vecchio by two vessels from contexts dating between the end of the 4th and the first half of the 3rd c. BC. It has close parallels with contemporary examples in Apulia, Lucania, and Calabria. The second variant of casserole type C9 (C9.2) is characterized by a flaring rim inclined toward the interior. A single fragment of the type was recovered at Sant'Angelo Vecchio. It has contemporary parallels in the chorai of Metaponto and Siris.

C10. Casserole type C10 is characterized by a large, carinated body with an oblique profile, thickened and flaring rim, flat base, and solid handles adhering to the body wall. The diameter ranges between 30 and 38 cm. The type, attested at Sant'Angelo Vecchio by 15 fragments from six vessels, is subdivided into two variants by the shape of the rim. Comparanda dating from the end of the 4th through the 3rd c. BC come from Pomarico, Cozzo Presepe, and Herakleia, as well as Apulia and Calabria. The first variant of casserole type C10 (C10.1) is characterized by a thickened and flaring rim. This variant, attested by five individuals, has contemporary parallels in the chorai of Metaponto and Siris. Fragment C10.1*c* (SAV CkW 23) is noteworthy for its unusual size (rim diam. ca. 38 cm) and its solid handles with elliptical section decorated with finger impressions and adhering to the body wall. This casserole from Hellenistic contexts recalls a smaller but analogous example from Berenice (Riley 1977), where many casseroles were found with finger decorations on the handles. The second variant of casserole type C10 (C10.2) is characterized by a thickened and flaring rim with concave upper edge. This variant, documented at Sant'Angelo Vecchio by a single example, is present also at Valesio.

Casseroles

SAV CkW 09 Casserole Type C1

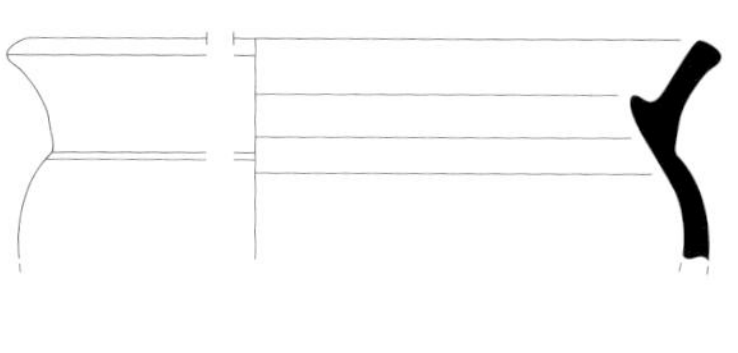

1:2

Provenance: Slope 3, votive deposit S. Level 3, batt. 2. Lot: SA 79-382PL-3. Assemblage 11.

Dimensions and technical features: h 2.9, rim unknown, wall th 0.3, w 2.8. Clay Munsell 2.5YR 4/8 red, hackly fracture, hard and rough to the touch, with large quantities of small and medium-sized stone inclusions distributed uniformly. Surface mica and vughs. Local fabric.

Condition: One fragment.

Description: Flaring rim with pronounced, oblique lid-resting ledge. On the exterior, the point of transition between the rim and the convex shoulder is marked by a distinct bulge.

Comparisons and comment: Agora XII, fig. 18, no. 1965 (see also pl. 95, 350–325 BC); *Kaulonia I*, fig. 57, no. 387; Fracchia and Keith 1990, 265, fig. 185, no. 239 (from Roccagloriosa); Casagrande 2002, 173, fig. 48, type 11, sheet 211 (from agora of Herakleia); Garaffa 2011, 45, fig. 4.1 (from Garaguso-deposito Autera, 6th–5th c. BC); Blotti in *Fattoria Fabrizio* (Fattoria Fabrizio).

Date: End of the 4th–first half of the 3rd c. BC.

SAV CkW 10 Casserole Type C3

Provenance: Sounding C, Pottery Deposit. Lot: SA 79-919P A–C. Assemblage 15.

Dimensions and technical features: h 3.7, rim 25.0, wall th 0.5, w 18.4. Clay Munsell 5YR 7/6 reddish yellow, shaly fracture, hard and smooth to the touch, with large quantities of small white and black stone inclusions distributed uniformly. Surface mica and vughs. Traces of exposure to fire, particularly on the rim. Local fabric.

Condition: Fragmentary.

Description: Flaring rim with flat upper surface and lid-resting ledge; handle with elliptical section attached obliquely so as to stand over the rim and adhering to the low body; curving base.

Comparisons and comment: Sibari V, 205, fig. 215, no. 88 (end of the 6th–beginning of the 5th c. BC), 486, fig. 500, no. 503 (end of the 5th c. BC); Rinaldi 2005, 226, pl. XXIV, no. 69 (from Torre di Satriano, 4th–3rd c. BC); Quercia 2003, 187, fig. 5, C3b (from *kerameikos* of Metaponto, first quarter of the 4th c. BC); Campagna 1995, 259, fig. 20, no. 7 (from Fondo S. Antonio-Vaste); *Gravina III*, 131, fig. 54, no. 303 (520–450 BC); Blotti in *Fattoria Fabrizio* (Fattoria Fabrizio); Di Tursi, in course of study (Fattoria Stefan).

Date: End of the 6th–beginning of the 4th c. BC.

1:2

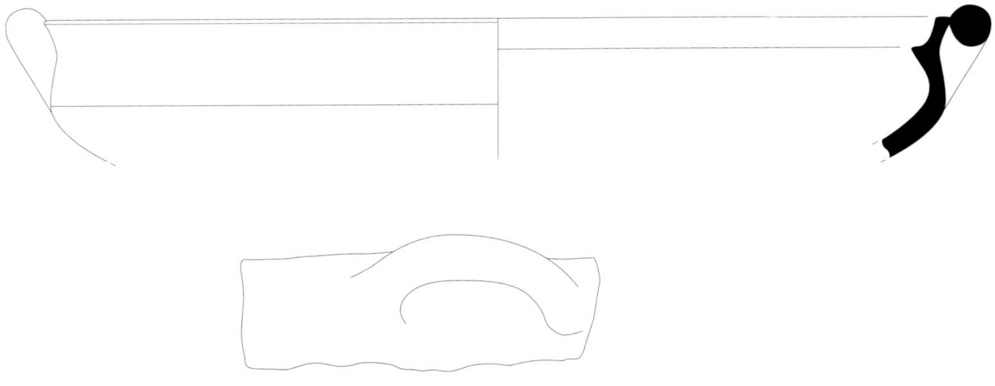

SAV CkW 11 Casserole Type C4

Provenance: Square D5. US 39.

Dimensions and technical features: h 3.8, rim 24.5, wall th 0.6, w 6.8. Clay Munsell 7.5YR 5/4 brown, hackly fracture, hard, with moderate quantities of small white stone inclusions distributed uniformly. Surface mica and vughs. Interior smoothed. Local fabric.

Condition: One fragment.

Description: Flaring rim with thinned upper surface and pronounced vertical lid-resting surface; low, carinated body with oblique walls; irregular handle attached at the rim; convex base.

Comparisons and comment: Kaulonia I, 90, fig. 60, no. 437.

Date: First half of the 5th c. BC.

1:2

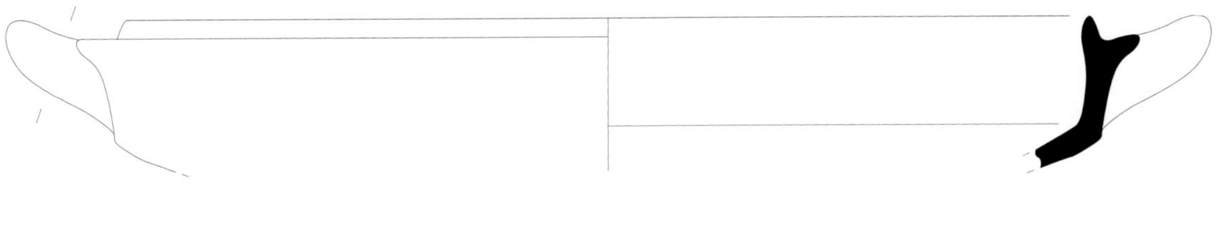

Casseroles cont.

1:2

SAV CkW 12 Casserole Type C6.1

Provenance: Sounding C. Level 2, batt. 6. Lot: SA 79-201PL (see *Fig. 2.35*).

Dimensions and technical features: h 3.9, rim 18.0–20.0, wall th 0.4, w 6.6. Clay Munsell 2.5YR 4/6 red, hackly fracture, hard and rough to the touch, with large quantities of small and medium-sized white and black inclusions distributed uniformly. Surface vughs. Local fabric.

Condition: One fragment.

Description: Indistinct rim with flat upper surface and barely pronounced lid-resting surface; carinated body with oblique walls on the exterior; convex base.

Comparisons and comment: Cotton 1977, 377, fig. 147, no. 465 (from Cozzo Presepe 325–300 BC); Cotton 1992b, 179, fig. 70, no. 1358 (3rd–1st c. BC, from Gravina); *Kaulonia I*, 85–86, fig. 59, nos. 419–420 (5th–4th c. BC); Vecchio 2002, 205, pl. 1, no. 1 (from Motya, 4th–3rd c. BC); Quercia 2003, 184, fig. 4, C2 (Metaponto, *kerameikos*, deposit no. 1, first quarter of the 4th c. BC); Gabrieli in *Survey*, 448, CC1c-19 Ck 122-01 (Metaponto field survey).

Date: End of the 4th–first half of the 3rd c. BC.

SAV CkW 13 Casserole Type C6.2

Provenance: Slope 3, NW extension. Level 1, batt. 5. Lot: SA 79-907PL, SA 79-83PL.

Dimensions and technical features: h 3.9, rim 18.0-20.0, wall th 0.5, w 11.6. Clay Munsell 2.5YR 4/6 red, hackly fracture, hard and smooth to the touch, with large quantities of small white and black stone inclusions distributed uniformly. Large vughs on the interior of the body. Local fabric.

Condition: Fragmentary.

Description: Indistinct rim rounded at the top with barely pronounced lid-resting surface; horizontal handle with elliptical section attached beneath the rim and adhering to the carinated body wall with oblique exterior profile.

Comparisons and comment: Kirsopp Lake 1934–1935, 105, pl. XVII, no. 2c (from Minturnae, middle of the 3rd c. BC); Cotton 1977, 377, fig. 147, no. 468 (from Cozzo Presepe, ca. 325–300 BC); Cotton 1992b, 181, fig. 71, no. 1368 (from Gravina); de Cazanove 1996, 925, fig. 12 (from Civita di Tricarico, end of the 4th–3rd c. BC); Caravelli 2008, 619, fig. 330, 960 (from Civita Tricarico); Di Tursi, in course of study (Fattoria Stefan).

Date: End of the 4th–first half of the 3rd c. BC.

1:2

SAV CkW 14 Casserole Type C6.3

Provenance: Square D4, outside stone circle to N near baulk. Level 3, batt. 5. Lot: SA 79-904PL. Assemblage 13.

Dimensions and technical features: h 3.1, rim 22.0–24.0, wall th 0.4, w 4.4. Clay Munsell 2.5YR 5/8 red, hackly fracture, hard and rough to the touch, with large quantities of small white and black stone inclusions distributed uniformly. Surface mica and vughs. Local fabric.

Condition: Fragmentary.

Description: Flaring rim rounded at the top; horizontal handle with elliptical section attached beneath the rim and adhering to the carinated body wall with oblique exterior profile.

Comparisons and comment: Kirsopp Lake 1934–1935, 105, pl. XVII, no. 2a (from Minturnae, middle of the 3rd c. BC); Morel 1970, 105, fig. 28, no. 1 (from Cozzo Presepe, end of the 4th–first half of the 3rd c. BC); Rossi 1989, 189, pl. 332, no. 1 (from Monte Sannace, middle of the 2nd c. BC); Fracchia and Keith 1990, 267, fig. 185, no. 247 (from Roccagloriosa, end of the 4th–beginning of the 3rd c. BC); Conti 1992, 250, pl. LXXVI, no. 235 (from Locri, mid-4th–mid-3rd c. BC); Casagrande 2002, 173, fig. 54, no. 222 (from agora of Herakleia, 4th–3rd c. BC); Caravelli 2008, 619, fig. 330, 963 (from Civita Tricarico); Gabrieli in *Survey*, 448, CC1a-18 Ck 157-01 (Metaponto field survey).

Date: End of the 4th–first half of the 3rd c. BC.

1:2

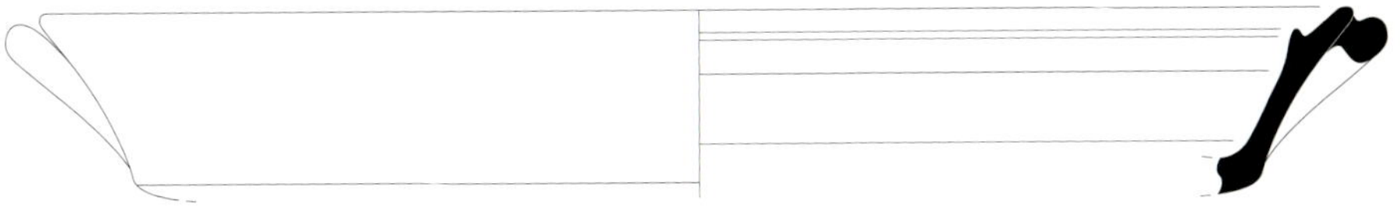

SAV CkW 15 Casserole Type C7*a*

Provenance: Square D4. Level 3, batt. 4. Lot: SA 79-826PL + SA 79-131PL. Assemblage 13.

Dimensions and technical features: h 5.8, rim 18.0, wall th 0.4. Clay Munsell 2.5YR 4/8 red, hackly fracture, hard and smooth to the touch, with moderate quantities of small and medium-sized white and black stone inclusions distributed uniformly. Local fabric.

Condition: Partially reconstructed.

Description: Flaring rim with rounded top and pronounced, oblique lid-resting ledge; horizontal handle with elliptical section attached beneath the rim and adhering to the body wall with rounded exterior profile; curving base.

Comparisons and comment: Cotton 1977, 377, fig. 147, no. 477 (from Cozzo Presepe, ca. 300 BC); Semeraro 1983, 191, pl. 93, no. 251 (from Otranto); Rossi 1989, 189, pl. 330, no. 19 (from Monte Sannace); Cotton 1992b, 180, fig. 70, no. 1363 (from Gravina); Casagrande 2002, 171, fig. 48, no. 210 (from agora of Herakleia); Gabrieli in *Survey*, 446, CG2b-12 Ck 296-01 (Metaponto field survey); Di Tursi, in course of study (Fattoria Stefan).

Date: End of the 4th–first half of the 3rd c. BC.

1:2

Casseroles cont.

SAV CkW 16 Casserole Type C7*b*

1:2

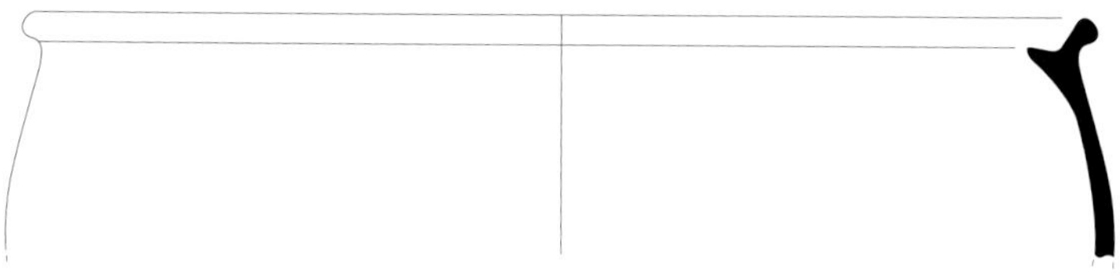

Provenance: Slope 3, NW extension. Level 1, batt. 5. Lot: SA 79-92PL (see *Fig. 2.35*).

Dimensions and technical features: h 4.0, rim 18.2, wall th 0.4, w 6.7. Clay Munsell 2.5YR 4/8 red, hackly fracture, hard and smoothed, with large quantities of small and medium-sized white and black stone inclusions distributed uniformly. Surface mica and vughs. Local fabric.

Condition: One fragment.

Description: Flaring and rounded rim with pronounced, oblique lid-resting ledge; deep body with rounded exterior profile. The transition between the rim and body is marked by a distinct bulge.

Comparisons and comment: Cotton 1977, 379, fig. 147, no. 473 (from Cozzo Presepe, ca. 325–300 BC); Cotton 1992b, 180, fig. 70, no. 1361 (from Gravina); Campagna 1995, 252, fig. 17, no. 7 (from Fondo S. Antonio-Vaste, end of the 4th–beginning of the 3rd c. BC). Fiorani 2011, 125, fig. 2,c (Timmari).

Date: End of the 4th–first half of the 3rd c. BC.

SAV CkW 17 Casserole Type C8*a*

1:2

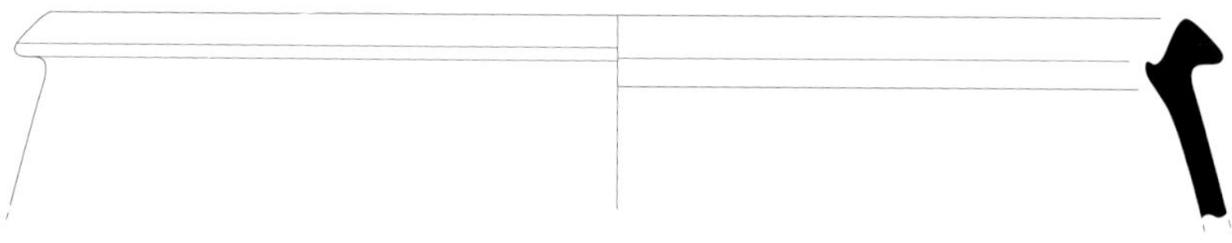

Provenance: Square H5. Level 1, batt. 2. Lot: SA 79-546PL, SA 79-636PL. Assemblage 3.

Dimensions and technical features: h 3.4, rim 20.0, wall th 0.4, w 9.6. Clay Munsell 5YR 4/1 dark gray, hackly fracture, hard and rough to the touch, with large quantities of small stone inclusions distributed uniformly. Surface mica and vughs. Local fabric.

Condition: Fragmentary.

Description: Thickened rim with triangular section and barely pronounced, oblique lid-resting ledge; rounded body.

Comparisons and comment: Cotton 1977, 377, fig. 147, no. 472 (from Cozzo Presepe, 300 BC); Rossi 1989, 190, pl. 332, no. 6; pl. 331, no. 12 (from Monte Sannace); Roubis 1996, 249, fig. 9, no. 10 (from Difesa S. Biagio-Montescaglioso, first half of the 3rd c. BC); Yntema 2001, 289, Form N03 no. 5178 (from Valesio); M. Liseno 2004, 98, pl. LXIV b (from Metaponto, 460–350 BC); Di Tursi, in course of study (Fattoria Stefan).

Date: 4th–3rd c. BC.

SAV CkW 18 Casserole Type C8*b*

1:2

Provenance: Slope 3 W extension. Level 3, batt. 1. Lot: SA 79-131PL. Assemblage 13.

Dimensions and technical features: h 3.0, rim unknown, wall th 0.4, w 4.0. Clay Munsell 2.5YR 5/8 red, shaly fracture, hard and rough to the touch, with moderate quantities of small and medium-sized white and black stone inclusions distributed uniformly. Surface mica and vughs. Traces of coating on the exterior surface. Local fabric.

Condition: One fragment.

Description: Thickened and flaring rim turned toward the exterior with a pronounced and oblique lid-resting ledge; shoulder with rounded exterior profile.

Comparisons and comment: Caravelli 2008, 619, fig. 330, 964 (from Civita Tricarico); Di Tursi, in course of study (Fattoria Stefan).

Date: End of the 4th–first half of the 3rd c. BC.

SAV CkW 19 Casserole Type C9.1

1:2

Provenance: Slope 1, surface cleaning. Lot: SA 79-361P.
Dimensions and technical features: h 5.7, rim 24.0, wall th 0.5, w 10.4. Clay Munsell 2.5YR 4/8 red, shaly fracture, hard, compact, and smooth to the touch, with moderate quantities of small stone inclusions distributed uniformly. Surface mica and vughs. Local fabric.
Condition: Fragmentary, misfired.
Description: Flaring rim flattened at the top with barely pronounced lid-resting ledge; body with convex exterior profile.
Comparisons and comment: Conti 1983, 34, pl. X, no. 5; Conti 1989, 277, pl. XXXVII, no. 317 (from Locri); *Kaulonia I*, 85, fig. 58, no. 410 (Hellenistic period); Cotton 1992b, 182, fig. 71, no. 1381 (from Gravina); Lissi Caronna 1990–1991, 271, fig. 96, no. 61 (from Oppido Lucano, mid-4th–beginning of the 3rd c. BC); Brizzi 1999b, 320, fig. 317, no. 661 (from Oppido Mamertina); Fracchia 2001, 41, fig. 27, AN56 (from Roccagloriosa); Yntema 2001, 289, Form N03 no. 521 (from Valesio).
Date: End of the 4th–first half of the 3rd c. BC.

SAV CkW 20 Casserole Type C9.2

1:2

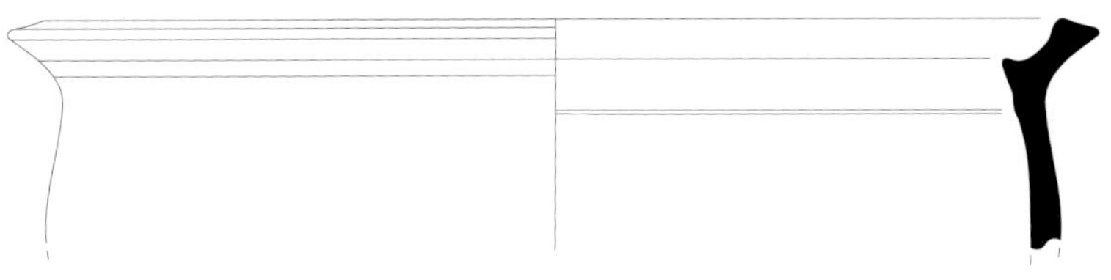

Provenance: Sounding C, E of and including division. Level 2, batt. 5. Lot: SA 79 322PL. Assemblage 20.
Dimensions and technical features: h 4.2, rim 18.0–20.0, wall th 0.5, w 5.4. Clay Munsell 2.5YR 4/6 red and 2.5YR 4/2 weak red, hackly fracture, hard and rough to the touch, with large quantities of small stone inclusions distributed uniformly. Surface vughs. Local fabric.
Condition: One fragment, misfired.
Description: Flaring lip and rim inclined toward the exterior with oblique lid-resting ledge; shoulder with straight exterior profile.
Comparisons and comment: Rossi 1989, 189, pl. 330, no. 8 (Monte Sannace, 2nd–1st c. BC); Caravelli 2008, 617, fig. 328, 936 (from Civita di Tricarico, 250–200 BC).
Date: End of the 4th–first half of the 3rd c. BC.

SAV CkW 21 Casserole Type C10.1*a*
Provenance: Slope 3, W extension. Level 2, batt. 1. Lot: SA 79-135PL, SA 79-423PL. Assemblage 13.
Dimensions and technical features: h 3.8, rim 30.0–32.0, wall th 0.2, w 13.3. Clay Munsell 2.5YR 5/8 red, hackly fracture, hard and rough to the touch, with large quantities of small and medium-sized white and black inclusions distributed uniformly. Surface mica and vughs. Local fabric.
Condition: One fragment.
Description: Thickened and flaring rim with pronounced oblique lid-resting ledge; handle with elliptical section attached beneath the rim and adhering to the body wall; body with oblique exterior profile.
Comparisons and comment: Cotton 1977, 377, fig. 147, no. 467 (from Cozzo Presepe, ca. 325–300 BC); Cotton 1992b, 180, fig. 71, no. 1366 (from Gravina); Bianco and Deodato 1997, 190, pl. 81, no. 174 (from Pomarico Vecchio); Casagrande 2002, 174, fig. 54, no. 223 (from agora of Herakleia, 375–270 BC); Caravelli 2008, 616, fig. 327, 927 (from Civita di Tricarico)
Date: End of the 4th–first half of the 3rd c. BC.

1:2

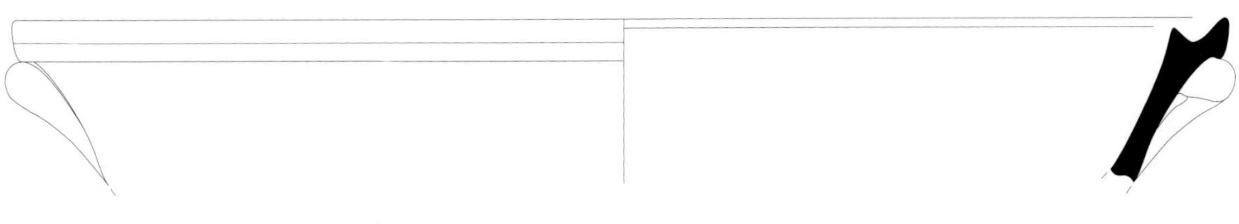

Casseroles cont.

SAV CkW 22 Casserole Type C10.1*b*

1:3

Provenance: Sounding C, E of and including division. Level 2, batt. 5. Lot: SA 79-322PL.

Dimensions and technical features: h 4.0, rim 32.0, wall th 0.5, w 7.5. Clay Munsell 2.5YR 5/8 red, hackly fracture, hard and rough to the touch, with large quantities of small and medium-sized white and black stone inclusions distributed uniformly. Surface mica and vughs. Local fabric.

Condition: One fragment.

Description: Flaring and thickened rim with pronounced oblique lid-resting ledge; body with oblique exterior profile.

Comparisons and comment: Semeraro 1992, fig. 4.1, no. 291 (from Otranto, Hellenistic period); *Kaulonia I*, 90, fig. 62, no. 452. Yntema 2001, 283, form N01a no. 496 (from Valesio).

Date: End of the 4th–first half of the 3rd c. BC.

SAV CkW 23 Casserole Type C10.1*c*

Provenance: Slope 3, N extension. Level 3, batt. 1. Lot: SA 79-381P. Assemblage 11.

Dimensions and technical features: h 6.5, rim 38.6, wall th 0.7, max. w 14.9. Clay Munsell 2.5YR 5/8 red and 2.5YR 4/3 reddish brown, hackly fracture, hard and rough to the touch, with large quantities of small and medium-sized white and black inclusions distributed uniformly. Surface vughs. Traces of coating on the interior of the body. Local fabric.

Condition: Fragmentary, misfired.

Description: Flaring and thickened rim with pronounced oblique lid-resting ledge; horizontal handle with elliptical section with finger impressions attached beneath the rim and adhering to the body wall; deep body with oblique exterior profile; slightly convex base.

Comparisons and comment: Riley 1977, 244, fig. 98, no. 426 (Hellenistic period); Kirsopp Lake 1934–1935, 105, pl. XVII, no. 2f (from Minturnae, mid-3rd c. BC); Casagrande 2002, 174, fig. 55, no. 225 (from Herakleia).

Date: End of the 4th–first half of the 3rd c. BC.

1:3

SAV CkW 24 Casserole Type C10.2

Provenance: Square G7. Level 1, batt. 2. Lot: SA 79-796PL. Assemblage 16 (see *Fig. 2.20*).

Dimensions and technical features: h 4.75, rim unknown, wall th 0.2, w 4.5. Clay Munsell 2.5YR 5/8 red, hackly fracture, hard and powdery to the touch, with moderate quantities of small and medium-sized white stone inclusions distributed uniformly. Surface mica and vughs. Local fabric.

Condition: One fragment.

Description: Flaring and thickened rim with concave upper edge and pronounced interior lip-resting ledge; deep body with oblique walls; horizontal handle with elliptical section attached beneath the rim.

1:3

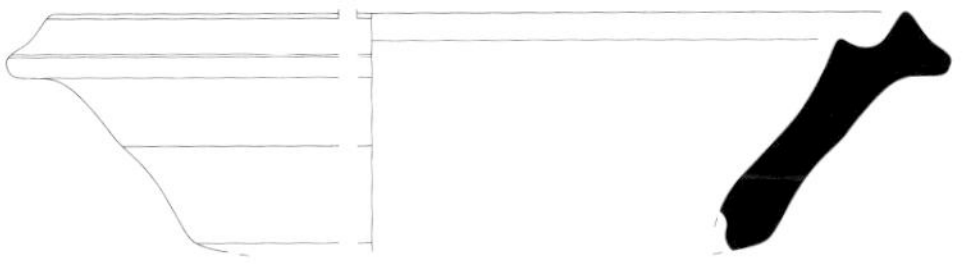

Comparisons and comment: Yntema 2001, 283, form N01b no. 501 (from Valesio, end of the 3rd–2nd c. BC). Caravelli 2008, 619, fig. 330, 964 (from Civita di Tricarico).

Date: First half of the 3rd c. BC.

Klibanos. Cooking under a lid, or *sub testu*, documented as early as the Bronze Age, was very common in Greek and Roman times and continued into the Middle Ages. It was used especially for baking cakes and flatbreads, as well as meats and vegetables, though to a lesser extent. The food was placed over a low and well-ventilated fire on tiles or in low-walled vessels, such as pans or dishes, and then covered with the *klibanos*, which was itself covered with embers or burning coals (Cubberley/Lloyd/Roberts 1988). At Sant'Angelo Vecchio, *sub testu* cooking is attested by 41 fragments of *klibanoi* from seven vessels of two distinct types datable between the 4th and 2nd c. BC.

Kl1. Type Kl1 can be reconstructed from a nearly complete example. It has a high, curved profile and a solid handle with elliptical section attached to the center of the dome. The diameter is ca. 30 cm. Two vessels were found in contexts of the 2nd c. BC. The type was common in southern Italy beginning in the middle of the 3rd c. and is documented through the 2nd and 1st c. BC.

Kl2. Klibanos type Kl2 is characterized by a low dome with a convex profile. The edge is defined by a high and modeled listel. The diameter ranges between 32 and 34 cm. A ring handle (not preserved in the recovered examples) was probably attached at the center of the dome. The exterior surface of the dome is decorated with a wave motif together with fingernail marks (SAV CkW 27) and/or concentric circles (SAV CkW 28) created in the drying phase before firing. All of the examples from Sant'Angelo Vecchio are made with a reddish fabric rich in medium-sized and large inclusions, which was hastily smoothed so that the surface roughness was not entirely removed.

At Sant'Angelo Vecchio the type is attested by 27 fragments from five vessels recovered in the Slope Area, in contexts dating between the end of the 4th and the middle of the 3rd c. BC. In this period, the type circulated widely in southern Italy in conjunction with low-walled containers such as baking and frying pans, which formed ensembles used for rapid cooking of bread, flatbreads, cakes, and roasted or fried foods.

Klibanoi

SAV CkW 25 Klibanos Type Kl1

Provenance: Square F3, N Kiln interior. Level 2, batt. 2. Lot: SA 79-923P–SA 79-1048PL. Inv. no. 319997. Assemblage 8.

Dimensions and technical features: h 15.0, rim 29.6, wall th 0.9. Clay Munsell 10R 4/6 red, hackly fracture, hard and rough to the touch, with large quantities of white and black stone inclusions of varying size distributed uniformly. Vughs. Smoothed. Local fabric.

Condition: Partially reconstructed.

Description: Flat vertical rim thickened at the bottom with flaring listel at the middle of the dome; handle with elliptical section attached horizontally to the central section of the dome, where the profile is convex.

Comparisons and comment: Kirsopp Lake 1934–1935, 106, pl. XIX, no. 12 (from Minturnae, middle of the 3rd c. BC); Small 1992, 191, fig. 81, no. 1481 (from Gravina); Rinaldi 2005, 227, pl. XXV, no. 80 (from Torre di Satriano, 3rd–1st c. BC); von Mehren 2007, 112, fig. 104, AA 50 (from Pontecagnano, 2nd–1st c. BC). Caravelli 2008, 621, fig. 332, 981 (from Civita di Tricarico); Laurenzana 2011, 268, fig. 1.e (from Rossano di Vaglio).

Date: 2nd–1st c. BC (based on context).

1:3

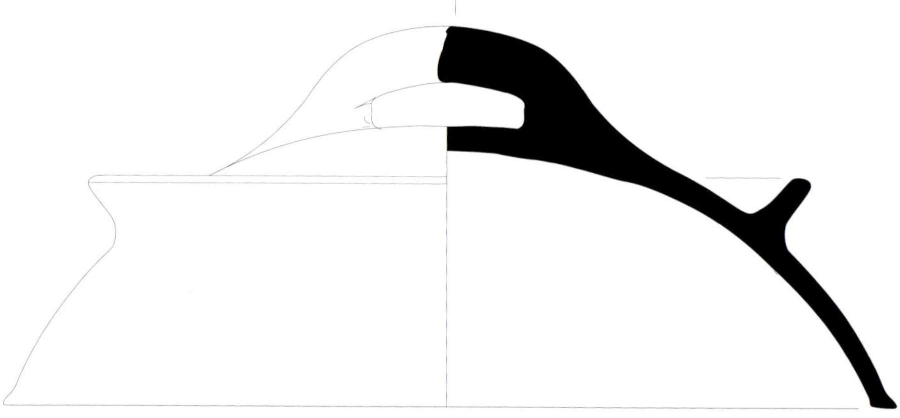

SAV CkW 26 Klibanos Type Kl2*a*

1:3

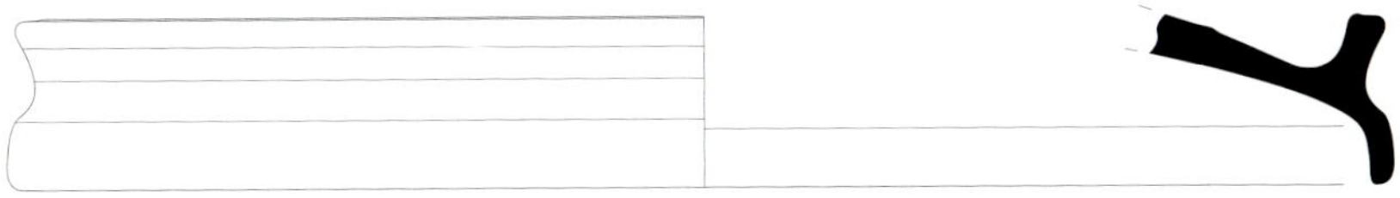

Provenance: Square D4, outside stone circle to N near baulk. Level 3, batt. 5. Lot: SA 79-904PL. Assemblage 13.

Dimensions and technical features: h 4.0, rim 32.0–34.0, wall th 0.9, w 10.9. Clay Munsell 2.5YR 4/6 red, hackly fracture, hard and rough to the touch, with large quantities of small and medium-sized stone inclusions distributed uniformly. Surface vughs. Smoothed. Local fabric.

Condition: Fragmentary, misfired.

Description: High vertical listel rounded at the bottom; brim with flaring profile at the top; dome with convex profile. On the dome there is a double series of incised markings, with fingernail marks over concentric circles interspersed with a wave motif.

Comparisons and comment: Conti 1983, 37, pl. X, no. 9 (from Locri); Conti 1989, 292, pl. XXXVIII, no. 338 (from Locri, 4th–first half of the 3rd c. BC); Small 1992, 192, fig. 81, no. 1482 (from Gravina, end of the 2nd–1st c. BC); de Cazanove 1996, 926, fig. 13 (from Civita di Tricarico); Caravelli 2008, 621, fig. 332, 991 (from Civita di Tricarico); Yntema 2001, 305, form N06d no. 567 (from Valesio).

Date: End of the 4th–first half of the 3rd c. BC.

SAV CkW 27 Klibanos Type Kl2*b*

1:3

Provenance: Slope 3, votive deposit SW extension. Level 3, batt. 1. Lot: SA 79-376PL. Assemblage 11.

Dimensions and technical features: h 2.9, w 9.75, max. wall th 1.0. Clay Munsell 2.5YR 5/8 red, hackly fracture, hard and rough to the touch, with large quantities of small and medium-sized white and black stone inclusions distributed uniformly. Local fabric.

Condition: One fragment, misfired.

Description: Dome with oblique profile and incised concentric circles interspersed with a wave motif.

Comparisons and comment: Rossi 1989, 190, pl. 333, no. 6 (from Monte Sannace, 3rd–2nd c. BC).

Date: End of the 4th–first half of the 3rd c. BC.

Lid/Cover

The frequency of vessels bearing a lid-resting ledge, such as casseroles and *kakkabai*, explains the large number of fragments ascribed to lids. Lids were also used on vessels that did not have an internal resting surface, such as chytrai and pans. A large majority (51 fragments) of the 68 lid fragments found at Sant'Angelo Vecchio preserve only the rim. Among these, 36 individual vessels were recognized, dating between the middle of the 4th and the 2nd c. BC. The form has a dome with oblique profile and central knob, which serves to distinguish the various types. The poor state of preservation of the materials placed severe limitations on the study of the form, as associations among knobs, bodies, and rims could not be made with certainty. Three distinct types were identified by the shape of the knob: cylindrical (Li1), hollow button (Li2), and shapeless (Li4). Twenty-six fragments with indistinct and rounded rims, attributable to 17 individual vessels, could not be assigned with certainty to type Li1 or Li2 due to the absence of the knob.

Li1. Lid type Li1 is characterized by a fully cylindrical knob, dome with oblique profile and rounded edge; the diameter is ca. 14.0 cm. This type is suitable for small vessels with or without a resting surface, such as, for example, chytrai of types P1 or P3.1. The fabric is reddish and rich in small inclusions, with surface vughs; many fragments are discolored through exposure to fire. The type is attested at Sant'Angelo Vecchio by nine fragments from five vessels, perhaps produced between the end of the 4th and the first half of the 3rd c. BC. It circulated widely at Metaponto and in its territory, and in Magna Grecia generally, in the 4th and 3rd c. BC.

Li2. Lid type Li2 is characterized by a knob with hollow button. The fabric is reddish and rich in small inclusions, with surface vughs; many fragments are discolored through exposure to fire. The diameter, ca. 17.0 cm, makes this type suitable for medium-sized vessels equipped with a lid-resting ledge, such as casseroles of types C7 and C10. The type, attested at Sant'Angelo Vecchio by a single individual, was fairly common in southern Italy and Sicily in the Hellenistic period.

Li4. Lid type Li4 is characterized by a shapeless cylindrical knob and dome with truncated conical profile and diameter of ca. 17.0 cm. The type is distinguished from the others also by its fabric, which is rich in large stone inclusions and has surface vughs. Attested at Sant'Angelo Vecchio by 15 fragments belonging to five vessels from 2nd-c. BC contexts, the type has parallels in Apulia and Calabria.

Lids/Covers

SAV CkW 28 Lid Type Li1

1:2

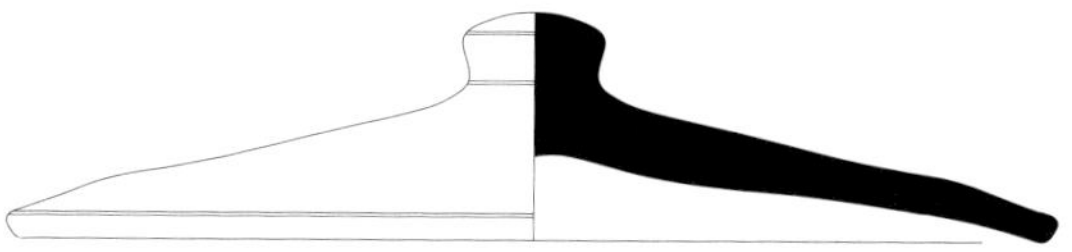

Provenance: Square D4, within stone circle. Level 3, batt. 5. Lot: SA 79-896PL. Assemblage 13.

Dimensions and technical features: h 3.0, rim 14.3, wall th 0.5, w 10.9. Clay Munsell 2.5 YR 5/8 red, hackly fracture, hard and rough to the touch, with large quantities of small white and black stone inclusions distributed uniformly. Surface vughs. Local fabric.

Condition: Missing part of the dome, misfired.

Description: Dome with oblique profile and indistinct, rounded edge; slightly undulating body wall raised above the supporting surface; cylindrical knob with subtle recess at the base.

Comparisons and comment: Cotton 1977, 378, fig. 148, no. 478 (from Cozzo Presepe, ca. 325–300 BC); *Kaulonia I*, 90, fig. 62, no. 543; Cotton 1992b, 192, fig. 82, no. 1486 (from Gravina, end of the 2nd–1st c. BC); de Cazanove 1996, 926, fig. 3 (from Civita di Tricarico, end of the 4th–3rd c. BC); Quercia 2003, 192, fig. 7, E3 (from *kerameikos* of Metaponto, first quarter of the 4th c. BC); *Agora XXXIII*, 198, fig. 91, no. 734 (from Athens, 2nd–1st c. BC).

Date: End of the 4th–first half of the 3rd c. BC.

SAV CkW 29 Lid Type Li2

1:2

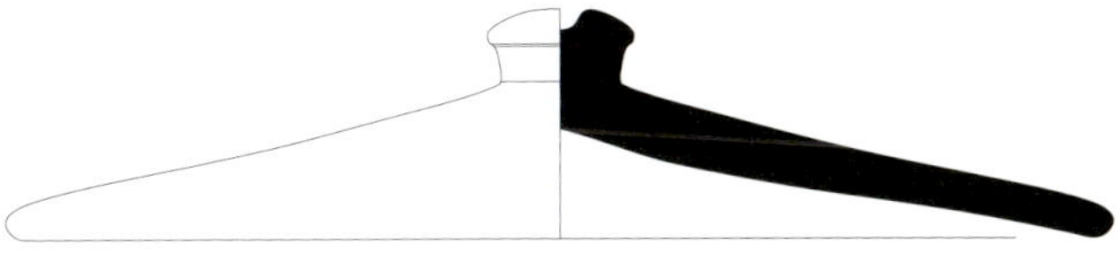

Provenance: Slope 3, W extension. Level 3, batt. 1. Lot: SA 79-131PL. Assemblage 13.

Dimensions and technical features: h 3.4, rim 17.0, wall th 0.7, w 13.9. Clay Munsell 2.5YR 6/8 light red, shaly fracture, hard and powdery to the touch, with moderate quantities of small stone and micaceous inclusions distributed uniformly. Surface vughs. Local fabric.

Condition: Partially reconstructed.

Description: Dome with oblique profile and indistinct, rounded edge; body wall raised above the supporting surface; button knob with rim inclined toward the exterior and central hollow with circular section.

Comparisons and comment: Cotton 1992b, 192, fig. 82, no. 1483 (from Gravina, end of the 2nd–1st c. BC); de Canzanove 1996, 926, fig. 13 (from Civita di Tricarico, end of the 4th–3rd c. BC); Conti 1989, 293, pl. XXXVIII, no. 341 (from Locri, 5th–3rd c. BC); Caravelli 2008, 624, fig. 334, 1044 (from Civita di Tricarico); Di Tursi, in course of study (Fattoria Stefan).

Date: End of the 4th–first half of the 3rd c. BC.

SAV CkW 30 Lid Type Li4

1:2

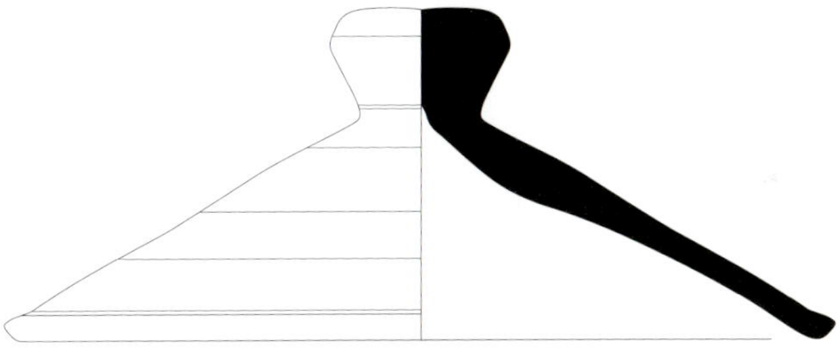

Provenance: Platform between the junction of the N Kiln and N wall. Level 3, batt. 1. Lot: SA 81-23PL.

Dimensions and technical features: h 6.5, rim 17.1, wall th 0.6. Clay Munsell 5Y 5/1 dark gray, hackly fracture, hard and rough to the touch, with large quantities of small and medium-sized white and black stone inclusions distributed uniformly. Surface vughs. Local fabric.

Condition: Reconstructed.

Description: Dome with truncated conical profile and rounded edge; body wall raised above the supporting surface; shapeless cylindrical knob.

Comparisons and comment: Brizzi 1999b, 324, fig. 318, no. 677 (from Oppido Mamertina); Yntema 2001, 301, form N06a no. 557 (from Valesio, 3rd–2nd c. BC); Laurenzana 2011, 268, fig. 1.f (from Rossano di Vaglio).

Date: 2nd–1st c. BC (based on context).

Closed Forms

Pots. The pots include deep vessels used principally for foods requiring long cooking times, such as soups or very dense vegetable stews. The typology is divided into chytrai, *kakkabai* of the Greek tradition, and pots of the Italic tradition.

Chytrai P1. The type P1 chytra is a form documented from the 8th c. BC (Bats 1988, 45) through the Hellenistic period. Detailed analyses of the type examples associated with their contexts (which would enable an accurate seriation) are lacking. The vessel is deep with a globular profile and the flaring rim transitioning seamlessly to the cylindrical neck (with or without handles arching slightly over the rim). It lacks a lid-resting ledge in order to facilitate pouring of the contents. The base is continuous and curving. Due to the poor state of preservation of the examples from Sant'Angelo Vecchio, which are mostly fragments of rims, handles, and necks, the exact shape of the body cannot be reconstructed. Type P1, characterized by the absence of a lid-resting ledge and its cylindrical neck, is derived from the earlier chytrai that circulated widely in Greece and Magna Grecia between the 6th and 3rd c. BC. Attested by 65 fragments belonging to 22 vessels, it is subdivided into four variants, three of which were identified by the shape of the rim.

One variant of chytra type P1 (P1.2) is characterized by a flaring rim rounded at the top. This small vessel, attested by two individuals, circulated in Greece and Magna Grecia as early as the 6th c. BC, achieving wide distribution in the 4th and 3rd c. BC and continuing through the Late Hellenistic period. Another variant of pot type P1 (P1.3) is characterized by a slightly flaring rim rounded at the top. The variant, attested at Sant'Angelo Vecchio by a single example, is associated with Late Hellenistic contexts but widely documented from the Archaic period onward. Variant P1.5 is characterized by a flaring rim with a flat top. This variant is well attested at Sant'Angelo Vecchio, with 56 fragments from 18 vessels. Like the other chytrai, it was in use for a very long time, from the middle of the 6th c. BC through the Hellenistic period, all the while retaining the same morphological features. This chronology is confirmed in full by the contexts at Sant'Angelo Vecchio, and the examples recovered here were distributed equally between the Late Archaic House on the Hill and the Hellenistic kilns. Variant P1.8 is characterized by a pendant rim with a flat upper surface and a short neck with concave exterior profile. This type of chytra is represented by a single example found in a late 4th/early 3rd-c. BC context. In the same period it circulated widely, especially in the Metapontino, although it is attested in the urban center of Metaponto as early as the end of the 5th c. BC.

Chytrai P2. Chytra type P2 is characterized by a pendant rim and short, cylindrical neck. The type can be identified as the transitional element between the Greek chytra and the Roman olla, due to its affinity with Hellenistic prototypes and also with the chytrai with oblique rims and distinct shoulders (P5) that are widely attested in the Republican and Imperial periods (Caravelli 2008, 498). The type is attested at Sant'Angelo Vecchio by nine fragments from five vessels and is subdivided into three variants.

Variant P2.2 is characterized by a pendant brim rim and short, cylindrical neck on offset shoulder. The variant is represented by two vessels recovered with materials dated end of the 4th–first half of the 3rd c. BC. In the same period it circulated widely in the Metapontine chora and in southern Italy, where it is encountered in Late Republican contexts. Variant P2.3, with a rounded brim rim, is documented at Sant'Angelo Vecchio, as well as in the Metapontino, in contexts dating between the end of the 4th and the beginning of the 3rd c. BC. In the same period, it circulated widely in southern Italy, with even greater diffusion in the Late Republican period. Variant P2.4, attested at Sant'Angelo Vecchio by a single individual vessel, has a pendant rim with rounded upper surface decorated with an impressed wave motif. Late Republican in date, it has exact parallels in the area of Viterbo.

Chytra P3. Type P3 includes vessels of medium capacity that have a flaring rim with a small ledge to accommodate a lid and an ovoid body without handles. The rim diameter varies between 14.0 and 16.0 cm. Considered a variant of the chytra and described as "wide-mouthed and lidded chytra" (*Agora XII*, 225, 620–490 BC) or "lidded chytra" (Rotroff and Oakley 1992, 121), it appeared at Athens at the beginning of the 5th c. BC and spread widely starting in the second quarter of the 4th c. BC. In the Hellenistic period, the morphological features of the rim changed, and the small lid-resting surface on the interior was lost, but the type remained very popular (P3.3–P3.4). This type is attested at Sant'Angelo Vecchio by 25 fragments from five vessels.

The first variant of this type, P3.1, is characterized by a thin, flaring and rounded rim inclined toward the interior with a barely pronounced lid-resting surface. The variant, attested by a single example, is encountered throughout the Metapontine chora in the Hellenistic period. Variant P3.2 is characterized by a flaring and rounded rim with concave interior. This variant is attested by a single misfired example with deformed rim. It is dated between the end of the 4th and the first half of the 3rd c. BC and circulated among the farmhouses of the Metapontine chora in the same period. Variant P3.3 is characterized by a flaring rim rounded at the top and lacking an interior lid-resting surface. This variant, represented by 18 fragments from three individual vessels, attests the passage from the Greek chytrai to the pots of Italic tradition. The distinctive features are the absence of the handles and the flat base, which is useful for cooking without a support and serving directly at the table (Bats 1988, 65).

P4. This type has a brim rim with large diameter (22 cm). It appeared initially at Athens as a storage container with the name of hydria between the end of the 6th and the beginning of the 5th c. BC and soon became part of the typical Greek kitchen furnishings (*Agora XII*, 200–201), with the body becoming more globular over time. The type, attested at Sant'Angelo Vecchio by a single example, circulated widely in the 4th and 3rd c. BC in the Metapontine chora, as indicated by the materials recovered during the intensive field survey between the Bradano and Basento, at Pomarico Vecchio, and in the farmhouses excavated by ICA in the chora.

P5. Chytra type P5 is characterized by an almond-shaped rim with diameter of ca. 14 cm, ovoid body lacking handles, and flat base. The type, attested by a single example from the Kiln Complex, belongs to the Roman morphological repertoire and has parallels among the materials found in the production facility at Marcianella-Chiusi, at Cosa, and at Athens.

Kakkabai, P6. Type P6 includes the *kakkabè*, distinguished from the chytra by its shallower body, greater width, and especially the presence of the lid-resting ledge on the interior. Documented at Athens from the end of the 6th c. BC alongside the chytra (*Agora XII*, 225), it spread in both Greece and Magna Grecia, in particular in the Classical and Hellenistic periods, when the rim experienced greater variability in morphology and typology. The type is widely attested in Magna Grecia beginning at the end of the 4th c. BC and continuing through the 3rd c. BC, especially in the farmhouses of the Metapontine chora. No relevant fragments were found, on the other hand, in waste deposit 1 of the *kerameikos* of Metaponto, dated to the first quarter of the 4th c. BC. The type features a flaring lip, rim inclined toward the interior, and a lid-resting ledge, subdivided into three variants distinguished by the shape of the rim. At Sant'Angelo Vecchio, it is represented by ten fragments from four individual vessels.

The first variant of *kakkabè*, P6.1, is characterized by a low, flaring lip on a rim slightly inclined toward the interior with pronounced vertical lid-resting ledge. This variant, attested by a single example, circulated widely in the Metapontino between the middle of the 4th and the 3rd c. BC. Variant P6.2 is characterized by a high, flaring lip on a rim inclined toward the interior with pronounced oblique lid-resting ledge. This variant, attested by two vessels, has parallels with contemporary materials found in Magna Grecia, and from there it was imported into Athens and Corinth starting in the middle of the 3rd c. BC. Variant P6.3 is characterized by a high, flaring lip, rim inclined toward the interior, and oblique and barely pronounced lid-resting ledge. A single fragment of the type was recovered at Sant'Angelo Vecchio. It has contemporary parallels in the chorai of Metaponto and Siris. Variant P6.4 is characterized by a low, flaring lip on rim inclined toward the interior with pronounced oblique lid-resting ledge. Represented by two vessels, this variant has parallels with contemporary materials found in Magna Grecia.

Olla P8. Type P8, the olla, includes small pots (rim 8.0–10.0 cm) with indistinct rims (P8.1) or slightly flaring rims (P8.2) and ovoid bodies probably lacking both foot and handles. The unusual reddish fabric, purified to a greater degree and polished with a stick, and the thickness of the walls suggest similarity with—or probable imitation of—thin-walled ware, especially cups, although the examples from Sant'Angelo Vecchio diverge from thin-walled ware in the irregularity and relatively hasty treatment of their surfaces and in the thickness of the walls. It is possible that these vessels cannot be strictly considered as cooking ware, but they could have been used to warm up liquids near the fire. Similar technical idiosyncrasies have been encountered on mugs from Torre S. Sabina near Brindisi (Pietropaolo 1997, 262) and fragments from the castrum of Metaponto (L. Giardino, pers. comm.). The type, attested at Sant'Angelo Vecchio by two examples from the Kiln Complex, has parallels with contemporary materials from Sutri.

Pots

SAV CkW 31 Pot Type P1.2

1:2

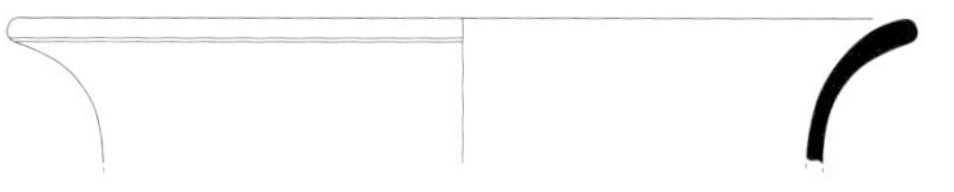

Provenance: House on the Hill. Lot: SA 79-1062PL. Assemblage 1.

Dimensions and technical features: h 1.9, rim 12.0, wall th 0.2, w 3.7. Clay Munsell 2.5YR 4/1 dark reddish gray, hackly fracture, hard and rough to the touch, with large quantities of small and medium-sized white and black stone inclusions distributed uniformly. Surface mica and vughs. Traces of exposure to fire on the exterior of the rim. Local fabric.

Condition: One fragment.

Description: Flaring rim rounded at the top. Short cylindrical neck.

Comparisons and comment: Hänsel 1973, 407, fig. 30, no. 1 (from Herakleia, 7th–6th c. BC); Guzzo 1978, 472, fig. 5, no. 8 (from Acquappesa, 4th–first half of the 3rd c. BC); *Kaulonia I*, 81, fig. 56, no. 372; Conti 1992, 252, pl. LXXVII, no. 247 (from Locri, mid-4th–mid-3rd c. BC); Cotton 1992b, 184, fig. 75, no. 1407 (from Gravina, 4th–3rd c. BC); Pietropaolo 1997, 361, fig. 10, no. 6 (from Torre S. Sabina-Brindisi, 2nd c. BC); Quercia 2003, 180, fig. 2, A1c (from the *kerameikos* of Metaponto, first quarter of the 4th c. BC); Caravelli 2008, 612, fig. 323, 840 (from Civita di Tricarico); Blotti in *Fattoria Fabrizio* (Fattoria Fabrizio); Di Tursi, in course of study (Fattoria Stefan).

Date: End of the 6th c. BC.

SAV CkW 32 Pot Type P1.3

1:2

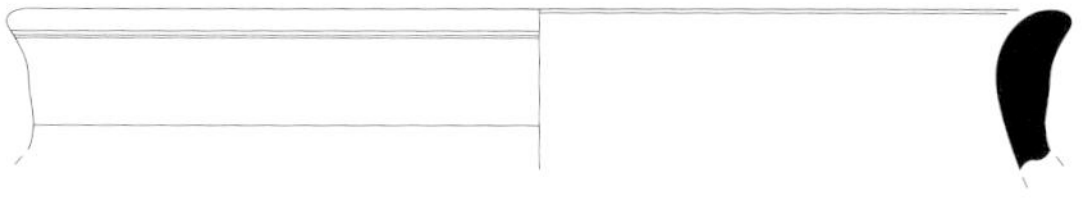

Provenance: Square H3, cleaning. Level 1, batt. 3. Lot: SA 79-1024PL. Assemblage 2 (see *Fig. 2.22*).

Dimensions and technical features: h 2.1, rim 14.0, wall th 0.5, w 3.5. Clay Munsell 2.5YR 5/8 red, hackly fracture, hard and rough to the touch, with large quantities of small and medium-sized stone inclusions distributed uniformly. Surface mica and vughs. Local fabric.

Condition: One fragment.

Description: Slightly flaring rim rounded at the top. A small recess marks the transition from the rim to the neck.

Comparisons and comment: *Agora XII*, 224, fig. 18, no. 1932 (from Athens, 575–560 BC); Boersma and Yntema 1987, 79, fig. 32, no. 2 (from Valesio, end of the 4th–beginning of the 2nd c. BC); Conti 1989, 265, pl. XXXVI, no. 299 (from Locri, 5th–mid-3rd c. BC); *Kaulonia I*, 81, fig. 56, no. 371; Cotton 1992b, 184, fig. 75, no. 1411 (from Gravina); Pietropaolo 1997, 262, fig. 11, no. 8 (from Torre S. Sabina-Brindisi); Brizzi 1999b, 642, fig. 317, no. 651 (from Oppido Mamertina, 4th–1st c. BC); Yntema 2001, 293, Form N04 no. 532 (from Valesio); Caravelli 2008, 612, fig. 323, 843 (from Civita di Tricarico).

Date: 6th–1st c. BC.

SAV CkW 33 Pot Type P1.5

1:2

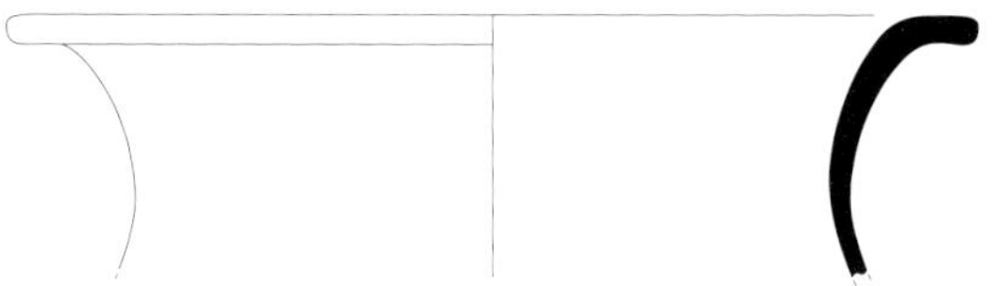

Provenance: Sounding D. Level 1, batt. 1. Lot: SA 80-42PL.

Dimensions and technical features: h 3.7, rim 14.0, wall th 0.2. Clay Munsell 5YR 6/8 yellowish red, hackly fracture, hard and rough to the touch, with large quantities of small and medium-sized white and black stone inclusions distributed uniformly. Surface mica and vughs. Evident traces of exposure to fire on the exterior of the rim. Local fabric.

Condition: One fragment of the rim.

Description: Flaring rim with flat top; cylindrical neck.

Comparisons and comment: *Sibari IV*, 66, fig. 66, no. 91, pl. 142, no. 9542 (from Sybaris, 6th c. BC); Cotton 1977, 375, fig. 145, no. 440 (from Cozzo Presepe, 550; 500/480–325; 300 BC); Cotton 1992b, 184, fig. 75, no. 140 (from Gravina, end of the 2nd–1st c. BC); Quercia 2003, 180, fig. 2, A1b (from *kerameikos* of Metaponto, first quarter of the 4th c. BC); M. Liseno 2004, 98, pl. LXIV d (from Metaponto); Distasi and Perretti 2006, 241, no. 120 (from Montemurro, mid-4th–beginning of the 3rd c. BC); Caravelli 2008, 613, fig. 324, 862 (from Civita di Tricarico); Blotti in *Fattoria Fabrizio* (Fattoria Fabrizio); Di Tursi, in course of study (Fattoria Stefan).

Date: End of the 6th–first half of the 3rd c. BC.

SAV CkW 34 Pot Type P1.8

1:2

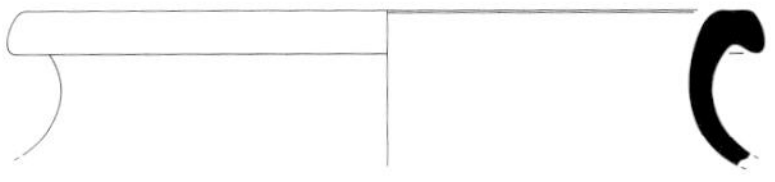

Provenance: Sounding B. Level 3, batt. 1. Lot: SA 79-304PL (see *Fig. 2.35*).

Dimensions and technical features: h 2.0, rim 10.0, wall th 0.3, w 5.4. Clay: Munsell 2.5YR 4/16 red, hackly fracture, hard and rough to the touch, with large quantities of small and medium-sized white and black stone inclusions distributed uniformly. Surface mica and vughs. Local fabric.

Condition: One fragment, misfired.

Description: Flaring rim with flat top and lacking lid-resting ledge; short neck with concave exterior profile.

Comparisons and comment: Guzzo 1978, 472, fig. 5, no. 71 (from Acquappesa, 4th–first half 3rd c. BC); Conti 1983, 34, pl. X, no. 2 (from Locri); Cotton 1992b, 190, fig. 79, no. 1462 (from Gravina, 4th–3rd c. BC); Bianco and Deodato 1997, 188, pl. 79, no. 161 (from Pomarico Vecchio, middle of the 3rd c. BC); Quercia 2003, 181, fig. 2, A1e (from *kerameikos* of Metaponto, first quarter of the 4th c. BC); Silvestrelli forthcoming, 6, no. 13 (Metaponto, last quarter of the 5th–end of the 4th c. BC); Caravelli 2008, 613, fig. 324, 857 (from Civita Tricarico); Gabrieli in *Survey*, 450, PJ2-29 Ck 260-03 (Metaponto field survey); Blotti in *Fattoria Fabrizio* (Fattoria Fabrizio); Di Tursi, in course of study (Fattoria Stefan).

Date: End of the 4th–first half of the 3rd c. BC.

SAV CkW 35 Pot Type P2.2

1:2

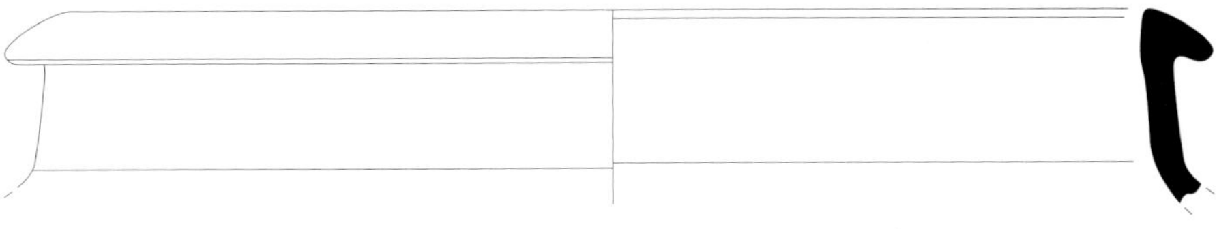

Provenance: Square D4, within stone circle. Level 3, batt. 5. Lot: SA 79-896PL-3. Assemblage 13.

Dimensions and technical features: h 3.8, rim 22.0, wall th 0.2, w 9.1. Clay: Munsell 2.5YR 5/6 red, hackly fracture, hard and rough to the touch, with moderate quantities of small and medium-sized stone inclusions distributed uniformly. Surface vughs. Local fabric.

Condition: One fragment, misfired.

Description: Pendant brim rim; short cylindrical neck; offset shoulder.

Comparisons and comment: Duncan 1964, 60, fig. 11, no. 92 (from Sutri, 2nd–1st c. BC); Cotton 1977, 375, fig. 145, no. 443 (from Cozzo Presepe, 300–250 BC); Fracchia and Keith 1990, 87, fig. 9, no. PC2 (from Roccagloriosa, end of the 3rd–1st c. BC); Cotton 1992b, 189, fig. 79, no. 1458 (from Gravina, 3rd–1st c. BC); Carter and Parmly Toxey in *Necropoleis*, 726 T 12-6 (300–275 BC); Caravelli 2008, 615, fig. 326, 909 (from Civita di Tricarico).

Date: End of the 4th–first half of the 3rd c. BC.

SAV CkW 36 Pot Type P2.3

1:2

Provenance: Slope 3, W extension. Level 3, batt. 1. Lot: SA 79-131PL. Assemblage 13.

Dimensions and technical features: h 2.7, rim unknown, wall th 0.3, w 3.7. Clay Munsell 2.5YR 5/8 red, hackly fracture, hard and rough to the touch, with large quantities of small and medium-sized white and black stone inclusions distributed uniformly. Surface mica and vughs. Local fabric.

Condition: One fragment.

Description: Offset rim with brim that becomes thinner and rounded at the top; straight neck.

Comparisons and comment: Cotton 1992b, 190, fig. 79, no. 1460 (from Gravina, end of the 2nd–1st c. BC); Aprosio and Pizzo 2003, 239, pl. LIX, RT VIII.8.1 (from production facility at Marcianella-Chiusi, end of the 2nd–beginning of the 1st c. BC); Dyson 1976, 23, fig. 3, CF 38 (from Cosa, 2nd–1st c. BC); Gabrieli in *Survey*, 450, PJ1-28 Ck 113-02 (Metaponto field survey); Caravelli 2008, 615, fig. 326, 908 (from Civita di Tricarico); Di Tursi, in course of study (Fattoria Stefan).

Date: End of the 4th–first half of the 3rd c. BC.

Pots cont.

SAV CkW 37 Pot Type P2.4

1:2

Provenance: Square G6. Level 1, batt. 5. Lot: SA 79-684PL (see *Fig. 2.22*).

Dimensions and technical features: h 3.4, rim 16.0–18.0, wall th 0.6, w 5.2. Clay Munsell 2.5YR 4/6 red, hackly fracture, hard and rough to the touch, with large quantities of small and medium-sized white and black stone inclusions distributed uniformly. Surface vughs. Local fabric.

Condition: One fragment.

Description: Pendant rim rounded at the top and bearing impressed wave motif; short cylindrical neck.

Comparisons and comment: Duncan 1964, 60, fig. 11, no. 93 (from Sutri, 2nd–1st c. BC).

Date: 2nd–1st c. BC.

SAV CkW 38 Pot Type P3.1

1:2

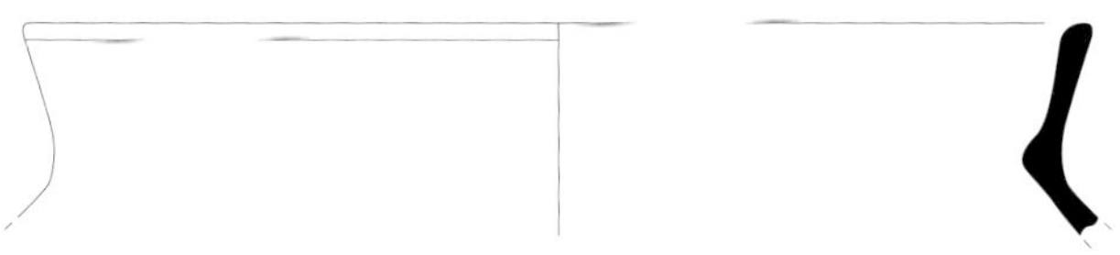

Provenance: Square D5. US 9, 18.

Dimensions and technical features: h 3.1, rim 16.0, wall th 0.4, w 6.4. Clay Munsell 2.5YR 6/6 light red, hackly fracture, hard and rough to the touch, with large quantities of small and medium-sized white and black stone inclusions distributed uniformly. Surface mica and vughs. Local fabric.

Condition: One fragment.

Description: Thin, flaring rim rounded at the top and inclined toward the interior with oblique, barely pronounced lid-resting surface.

Comparisons and comment: Agora XII, 225, 372, fig. 18, no. 1947 (520–490 BC); Rotroff and Oakley 1992, 29, 121, fig. 25, no. 325, pl. 58 (from Athens, end of the 5th c. BC); *Kaulonia I*, 90, fig. 61, no. 444 (Hellenistic period); Cotton 1977, 375, fig. 146, no. 456 (from Cozzo Presepe, ca. 300 BC); Di Tursi, in course of study (Fattoria Stefan).

Date: 5th–4th c. BC.

SAV CkW 39 Pot Type P3.2

1:2

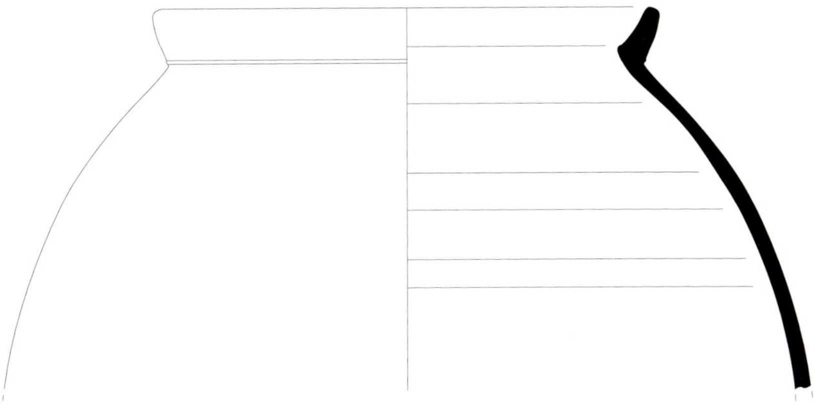

Provenance: Sounding C. Level 2, batt. 7. Lot: SA 79-205PL (see *Fig. 2.35*).

Dimensions and technical features: h 10.1, rim 14.0, wall th 0.4, w 16.3. Clay Munsell 2.5YR 5/6 reddish brown, hackly fracture, hard and smoothed, with moderate quantities of small white and black stone inclusions distributed uniformly. Surface vughs. Local fabric.

Condition: Partially reconstructed, deformed, and misfired.

Description: Flaring rim rounded at the top with concave interior surface and small interior lid-resting surface. A small channel beneath the rim marks the transition between the shoulder with continuous profile and the globular body.

Comparisons and comment: Brizzi 1999b, 317, fig. 316, no. 645 (from Oppido Mamertina, 4th–1st c. BC); Blotti in *Fattoria Fabrizio* (Fattoria Fabrizio); Di Tursi, in course of study (Fattoria Stefan).

Date: End of the 4th–first half of the 3rd c. BC.

SAV CkW 40 Pot Type P3.3*a*

1:2

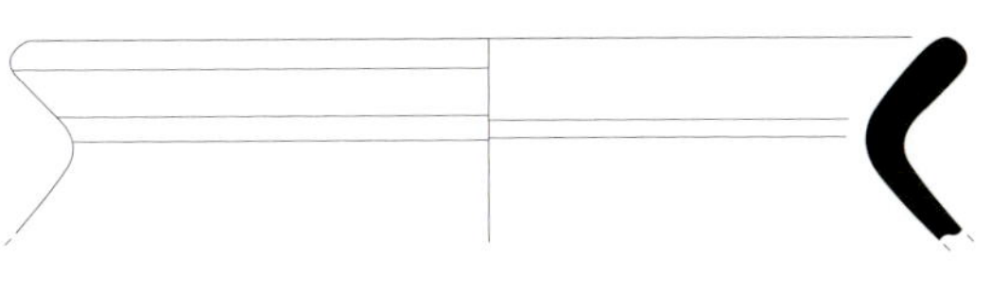

Provenance: Square G6, E of wall (from SA 79-1015PL). Level 2, batt. 4. Lot: SA 79-1016PL (see *Fig. 2.22*).

Dimensions and technical features: h 2.9, rim 14.0, wall th 0.4, w 8.2. Clay Munsell 2.5YR 4/6 red, hackly fracture, hard and rough to the touch, with large quantities of small and medium-sized white and black stone inclusions distributed uniformly. Surface mica and vughs. Local fabric.

Condition: One fragment.

Description: Flaring rim rounded at the top.

Comparisons and comment: Cotton 1977, 375, fig. 146, no. 450 (from Cozzo Presepe, ca. 300 BC); *Kaulonia I*, 85, fig. 56, no. 273 (Brizzi 1999b, 318, fig. 317, no. 655, from Oppido Mamertina); Caravelli 2008, 612, fig. 323, 832 (from Civita di Tricarico).

Date: 2nd–1st c. BC.

Pots cont.

SAV CkW 41 Pot Type P3.3*b*

1:2

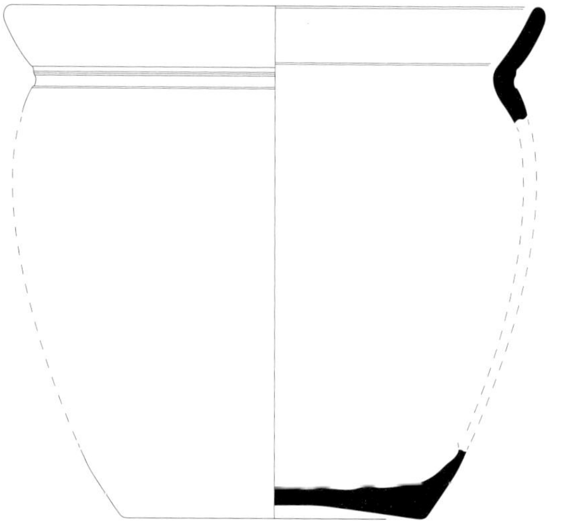

Provenance: Square G6, E of wall. Level 2, batt. 3. Lot: SA 79-986PL-1 and 2. Assemblage 3.

Dimensions and technical features: h rim 2.9, h base 1.75, rim 14.0, base 8.0, wall th 0.4, max. rim w 5.8; max. base w 8.0. Clay Munsell 2.5YR 5/8 red, hackly fracture, hard and rough to the touch, with moderate quantities of small and medium-sized white and black stone inclusions distributed uniformly. Surface mica and vughs. Local fabric.

Condition: Fragmentary.

Description: Flaring rim rounded at the top; flat base. The point of transition between the rim and the shoulder is marked by a pair of moldings.

Comparisons and comment: Corinth VII.3, 122, pl. 27, no. 651 (146 BC); Pietropaolo 1997, 261, fig. 10, no. 5 (from S. Sabina-Brindisi, mid-2nd–1st c. BC); Aprosio and Pizzo 2003, 228, pl. LIII, RT VIII 4.3 (from production facility at Marcianella-Chiusi, end of the 3rd–mid-2nd c. BC).

Date: 2nd–1st c. BC.

SAV CkW 42 Pot Type P4

1:2

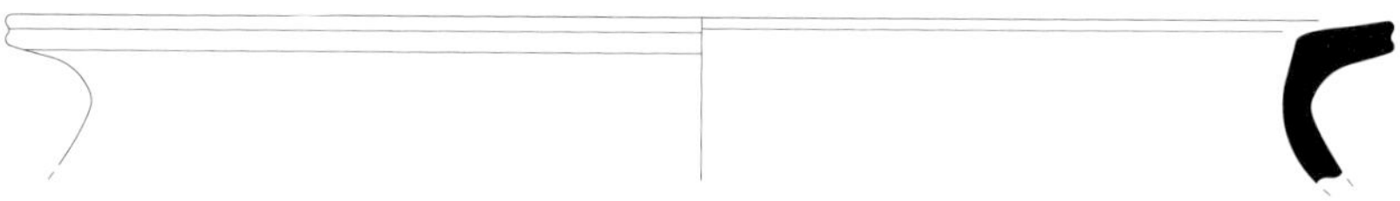

Provenance: Slope 3, W extension, votive deposit N. Level 2, batt. 1. Lot: SA 79-385PL. Assemblage 13.

Dimensions and technical features: h 2.5, rim 21.9, wall th 0.4, w 6.1. Clay Munsell 2.5YR 5/6 red, hackly fracture, hard and rough to the touch, with large quantities of small and medium-sized white and black stone inclusions distributed uniformly. Surface vughs. Local fabric.

Condition: One fragment.

Description: Rim with triangular brim slightly inclined toward the interior and modeled on the exterior.

Comparisons and comment: Morel 1970, 104, fig. 28, no. 6 (from Cozzo Presepe, end of the 4th–beginning of the 3rd c. BC); Cotton 1977, 375, fig. 145, no. 449 (from Cozzo Presepe, ca. 300 BC); Cotton 1992b, 86, fig. 76, no. 1422 (from Gravina, 2nd–1st c. BC); Bianco and Deodato 1997, 188, pl. 79, no. 162 (from Pomarico Vecchio); Gabrieli in *Survey*, 451, P2-12 Ck 113-04 (Metaponto field survey); Caravelli 2008, 616, fig. 327, 922 (from Civita di Tricarico); Di Tursi, in course of study (Fattoria Stefan).

Date: End of the 4th–first half of the 3rd c. BC.

SAV CkW 43 Pot Type P5

Provenance: Square F3, inside kiln. Level 2, batt. 2. Lot: SA 79-876PL+ SA 79-880PL. Assemblage 8.

Dimensions and technical features: h 14.0, rim 13.0–14.0, wall th 0.5, w 12.6. Clay Munsell 2.5YR 5/6 reddish brown, hackly fracture, hard and smooth to the touch, with moderate quantities of small and medium-sized white and black stone inclusions distributed uniformly. Surface vughs. Local fabric.

Condition: Partially reconstructed.

Description: Almond-shaped rim rounded at the top; ovoid body. A small recess marks the point of transition between the shoulder and the neck with convex exterior profile.

Comparisons and comment: Aprosio and Pizzo 2003, 237, pl. LVIII, RT VIII 6.2 (from production facility at Marcianella-Chiusi, end of the 2nd–beginning of the 1st c. BC); Dyson 1976, 25, fig. 2, CF22 (from Cosa, 2nd c. BC); *Agora XXXIII*, 178, fig. 81, nos. 628–629 (from Athens).

Date: 2nd–1st c. BC.

1:2

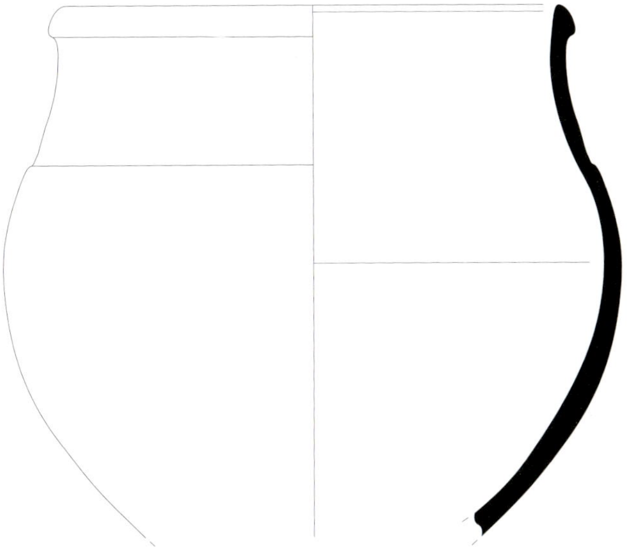

SAV CkW 44 Pot Type P6.1

Provenance: Square D4 W. Level 3, batt. 5. Lot: SA 79-907PL. Assemblage 13.

Dimensions and technical features: h 4.0, rim 24.0–30.0, w 2.0. Clay Munsell 2.5YR 5/1 gray, hackly fracture, very hard and rough to the touch, with large quantities of medium-sized and large white and black stone inclusions distributed uniformly. Surface vughs. Local fabric.

Condition: One fragment, misfired.

Description: Low, flaring lip on rim inclined slightly toward the interior with pronounced vertical lid-resting ledge. The transition between the rim and the shoulder is marked by a distinct channel.

Comparisons and comment: Cotton 1977, 376, fig. 146, no. 459 (from Cozzo Presepe, 325–250 BC); Cotton 1992b, 187, fig. 77, no. 1431 (from Gravina, 3rd–1st c. BC); M. Liseno 2004, 98, pl. LXIV c (460–350 BC); Gabrieli in *Survey*, 452, P1d-35 Ck 167-02 (Metaponto field survey).

Date: End of the 4th–first half of the 3rd c. BC.

1:2

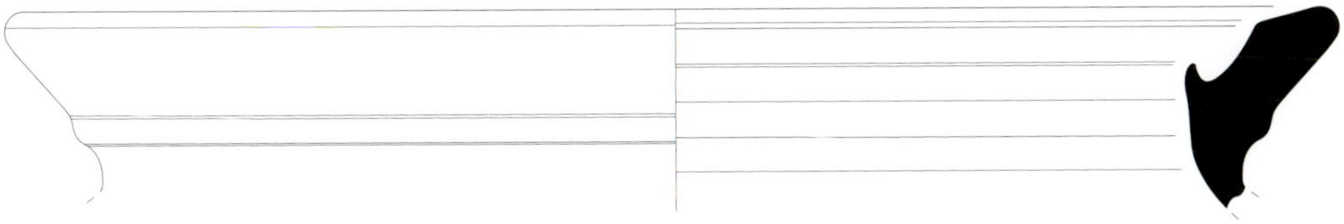

Pots cont.

SAV CkW 45 Pot Type P6.2

1:2

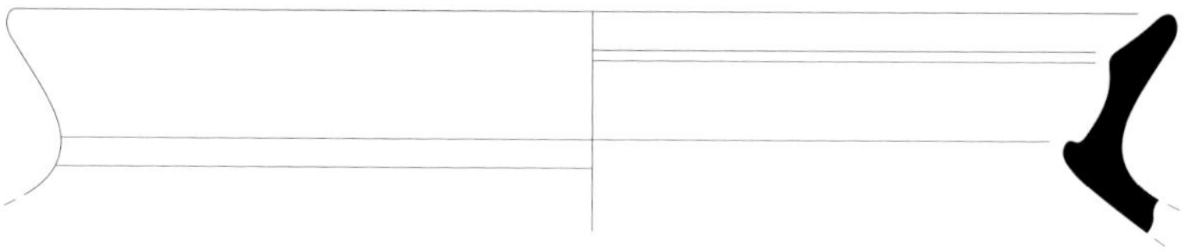

Provenance: Square H5. Level 1, batt. 2. Lot: SA 79-322PL (see *Fig. 2.21*).

Dimensions and technical features: h 3.7, rim 20.0, wall th 0.6, w 8.8. Clay Munsell 10R 5/8 yellowish brown, hackly fracture, hard and rough to the touch, with large quantities of small and medium-sized white and black stone inclusions distributed uniformly. Surface vughs. Local fabric.

Condition: One fragment, misfired.

Description: High flaring lip on rim rounded at the top and inclined toward the interior with oblique and barely pronounced lid-resting ledge.

Comparisons and comment: Corinth VII.3, 122–23, pl. 27, no. 656 (146 BC); Cotton 1977, 375, fig. 146, no. 451 (from Cozzo Presepe, ca. 325–300 BC); Rossi 1989, 189, pl. 330, no. 7 (from Monte Sannace, 2nd–1st c. BC); *Kaulonia I*, 85, fig. 57, no. 392 (Cotton 1992b, 188, fig. 78, no. 1445; from Gravina, end of the 2nd–1st c. BC); Semeraro 1992, 67, fig. 4.1, no. 293 (from Otranto); Bianco and Deodato 1997, 189, fig. 78, no. 167 (from Pomarico Vecchio, 320–second half of the 3rd c. BC); Casagrande 2002, 170, fig. 47, 7 (from agora of Herakleia, 4th–3rd c. BC); *Agora XXXIII*, fig. 78, no. 614 (from Athens, 250–210 BC); *Mozia XII*, 180, pl. XXXV, MD.03.1009/8 (from Motya, end of the 5th–beginning of the 4th c. BC); Gabrieli in *Survey*, 451, P1b-33 Ck 487-01 (Metaponto field survey); Di Tursi, in course of study (Fattoria Stefan).

Date: End of the 4th–first half of the 3rd c. BC.

SAV CkW 46 Pot Type P6.3

1:2

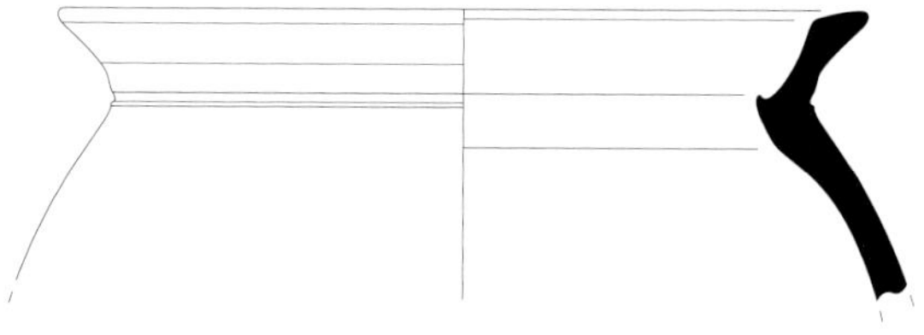

Provenance: Slope 3, votive deposit SW extension. Level 3, batt. 1. Lot: SA 79-376PL-1.

Dimensions and technical features: h 4.6, rim 12.8, wall th 0.5, w 6.9. Clay Munsell 2.5YR 5/6 red, hackly fracture, hard and rough to the touch, with large quantities of small white and black stone inclusions distributed uniformly. Surface mica and vughs. Local fabric.

Condition: One fragment, misfired.

Description: High, flaring lip, rim rounded on the top and slightly inclined toward the interior, and oblique and barely pronounced lid-resting ledge. The transition between the rim and shoulder with convex exterior profile is marked by a small bulge.

Comparisons and comment: Morel 1970, 105, fig 28, no. 5 (from Cozzo Presepe); Casagrande 2002, 170, fig. 47, no. 206 (from agora of Herakleia).

Date: End of the 4th–first half of the 3rd c. BC.

SAV CkW 47 Pot Type P6.4

1:2

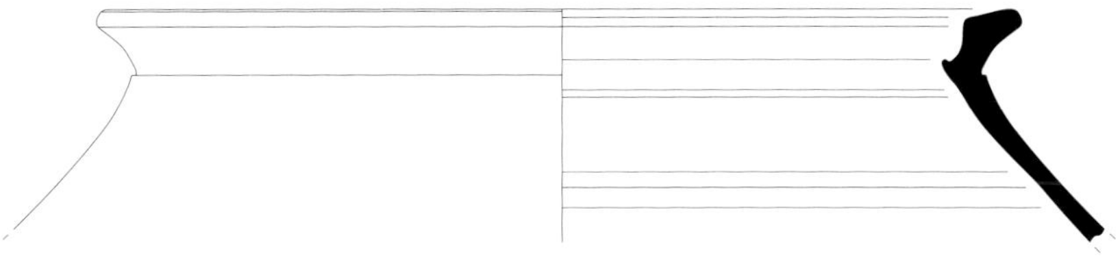

Provenance: Square H5. Level 1, batt. 2. Lot: SA 79-546PL, SA 79-542PL (see *Fig. 2.21*).

Dimensions and technical features: h 4.5, rim int. 16.0, wall th 0.4, w 5.4. Clay Munsell 2.5YR 6/8 light red, hackly fracture, hard and rough to the touch, with large quantities of small white and black stone inclusions distributed uniformly. Surface mica and vughs. Local fabric.

Condition: Fragmentary.

Description: Low, flaring lip on rim rounded at the top and inclined toward the interior, with oblique and barely pronounced lid-resting ledge. A shallow channel at the base of the rim marks the transition to the shoulder with oblique exterior profile.

Comparisons and comment: Morel 1970, 104, fig. 28, no. 5 (from Cozzo Presepe, end of the 4th–beginning of the 3rd c. BC); *Kaulonia I*, 85, fig. 57, no. 392; Distasi and Perretti 2006, 242, no. 126 (from Montemurro, mid-4th–beginning of the 3rd c. BC); Caravelli 2008, 611, fig. 322, 816 (from Civita di Tricarico); Di Tursi, in course of study (Fattoria Stefan).

Date: End of the 4th–first half of the 3rd c. BC.

SAV CkW 48 Pot Type P8.1

1:2

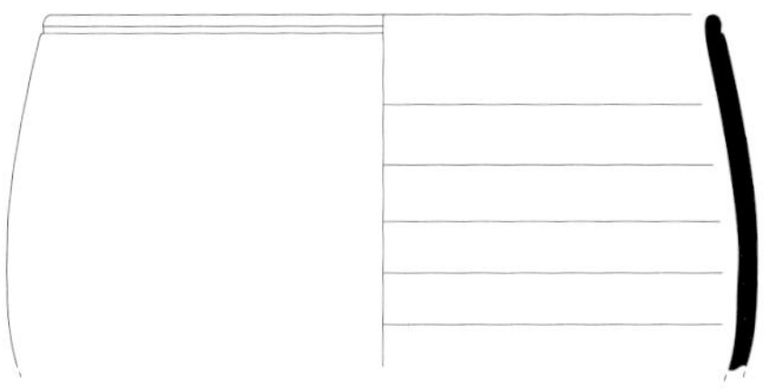

Provenance: Square F3, N Kiln interior. Level 2, batt. 2. Lot: SA 79-875PL. Assemblage 8.

Dimensions and technical features: h 5.1, rim 10.0, w 5.2. Clay Munsell 2.5YR 4/8 red, hackly fracture, hard and smooth to the touch, with large quantities of very small white stone inclusions distributed uniformly. Local fabric.

Condition: Fragmentary.

Description: Thin, indistinct rim rounded at the top and emphasized by a small channel; body with ovoid profile.

Comparisons and comment: No comparanda could be identified.

Date: 2nd–1st c. BC, based on context.

SAV CkW 49 Pot Type P8.2

1:2

Provenance: Square F3, N Kiln interior. Level 2, batt. 1. Lot: SA 79-874P. Assemblage 8.

Dimensions and technical features: h 5.9, rim 8.0, wall th 0.9, w 4.6. Clay Munsell 2.5YR 5/6 red, hackly fracture, hard and smooth to the touch, with moderate quantities of small and medium-sized white and black stone inclusions distributed uniformly. Surface vughs. Local fabric.

Condition: Fragmentary.

Description: Thin, indistinct, and slightly flaring rim rounded at the top; body with ovoid profile.

Comparisons and comment: Duncan 1964, 51, fig. 5, no. 13 (from Sutri, 2nd c. BC).

Date: 2nd–1st c. BC.

Base

Due to the poor state of preservation of the cooking wares from Sant'Angelo Vecchio, only in very few cases was it possible to attribute specific types of bases to specific forms with any certainty. This problem is particularly prevalent among the closed forms, where the normally curving base was recovered in a highly fragmentary state. Type F2 belongs to deep containers used for prolonged cooking directly over embers or above them by means of a support.

Base

SAV CkW 50 Base of Closed Form Type F2

Provenance: Square D4, N near baulk. Level 3, batt. 6. Lot: SA 79-913PL. Assemblage 13.

Dimensions and technical features: h 4.5, wall th 0.3, w 12.5. Clay Munsell 2.5YR 4/8 red, hackly fracture, hard and rough to the touch, with large quantities of small white and black stone inclusions distributed uniformly. Local fabric.

Condition: Fragmentary.

Description: Indistinct and curving base.

Comparisons and comment: Quercia 2003, 190, fig. 7, F1 (from *kerameikos* of Metaponto, first quarter of the 4th c. BC); Blotti in *Fattoria Fabrizio* (Fattoria Fabrizio).

Date: End of the 4th–first half of the 3rd c. BC.

1:2

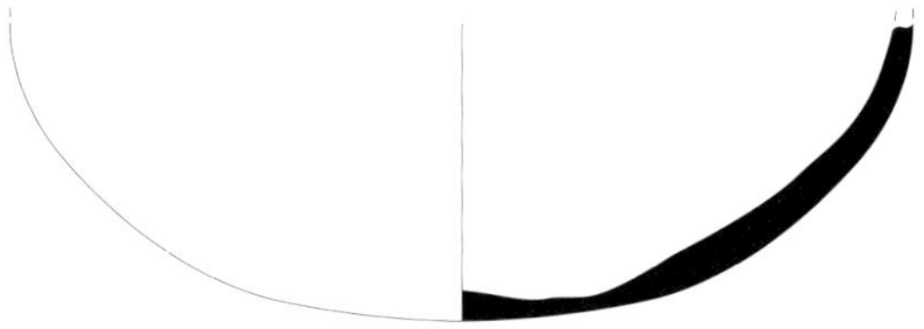

26

Roman and Late Roman Wares

Erminia Lapadula

Only a small number of ceramic fragments from Sant'Angelo Vecchio date to the Imperial and Late Roman periods. This scarcity of information precludes comprehensive hypotheses and analyses, and makes it impossible to formulate general or specific considerations about the activities of settlement, production, and commerce that took place at the site during these periods. The studies of the chora in the Roman period are similarly inconclusive, offering a preliminary and still incomplete picture of the Roman Metapontino.[1] The goal of this brief study is to present the information about the Roman period that can be gleaned from the stratigraphy of Sant'Angelo Vecchio, and thereby to fill in some of the gaps in our understanding of the ancient countryside during the Imperial and Late Roman periods.

Roman Imperial Ware

The Roman ceramic fragments from Sant'Angelo Vecchio can be subdivided into two distinct categories on the basis of their chrono-morphological elements: a more numerous group of Early Imperial fragments and a small number of Late Roman examples. Fragments of table and storage vessels (like an Italian terra sigillata plate rim) and a more numerous group of plain ware characterized by open (cups, basins, plate-lid, and lid) and closed forms (jugs, juglets, bottle, and ollae) date to the Early Imperial period.

The Italian terra sigillata plate SAV Imp 01 belongs to productions of the Late Augustan period and first half of the 1st century AD that have been documented in various forms in the Metapontino. Among Metapontine sites, the castrum of Metaponto has yielded the greatest number of Italian terra sigillata fragments, while the numbers recovered at rural sites have been quite low—with the exception of Pantanello.[2] Even the recently published results of the field survey between the Bradano and Basento Rivers only offer meager evidence for Italian terra sigillata: a total of 21 fragments, of which five come from Pantanello (site 428) and six from Sant'Angelo Vecchio (site 105).[3] These ceramics evidently reached mainly urban sites (Metaponto) and the rural settlements with the greatest economic and social significance (Pantanello, Sant'Angelo Vecchio, site 726). Conversely, the number of examples becomes very small in the farmhouses and small rural facilities (Masseria Durante; sites 1, 95, 395, and 735).[4] The situation in the Metapontino corresponds to a wider regional pattern. The urban centers at Herakleia, Grumentum, and Venusia and the large villas of Termitito and San Giovanni di Ruoti produced large numbers of Italian terra sigillata, in contrast with many small and medium-sized rural settlements that yielded decidedly smaller numbers.[5]

The current absence of complete studies of the Italian terra sigillata that circulated in the Metapontino complicates any potential hypotheses about commercial traffic with the areas where these items were produced (northern and central Italy, Campania, and Latium). It may be generally said that Metaponto and its chora were certainly fully active in the commercial traffic of the Adriatic coast. Though our picture of the distribution of the Italian terra sigillata in the Metapontino is still incomplete and imprecise, the presence of these products along the coast and in the interior demonstrates that proximity to communication routes was critical for a settlement. If, in fact, the presence of these ceramics is closely correlated with communication routes and commercial traffic, the quantity is instead proportional to the purchasing power of the society that inhabited the settlement.

[1] See, most recently, Giardino in *San Biagio*.

[2] The material from the castrum of Metaponto is unpublished. For a synthesis see, most recently, Giardino in *San Biagio* with relevant bibliography. The sanctuary at Pantanello will be the subject of an upcoming volume in *The Chora of Metaponto* series.

[3] For details of the Italian terra sigillata fragments from the field survey, see Lapadula in *Survey*.

[4] Masseria Durante: Lissi Caronna 1998–1999. For the field survey sites, see Lapadula in *Survey*.

[5] Herakleia: De Siena and Giardino 1994; Grumentum: P. Bottini 1997, Mastrocinque 2009; Venosa: Salvatore 1991. The ceramic material from Termitito is currently unpublished. For a synthesis of the site, see De Siena and Giardino 2001; De Siena 2005. For the Italian terra sigillata from San Giovanni di Ruoti, see the recent update with relevant bibliography in *Felicitas Temporum*, 572–74.

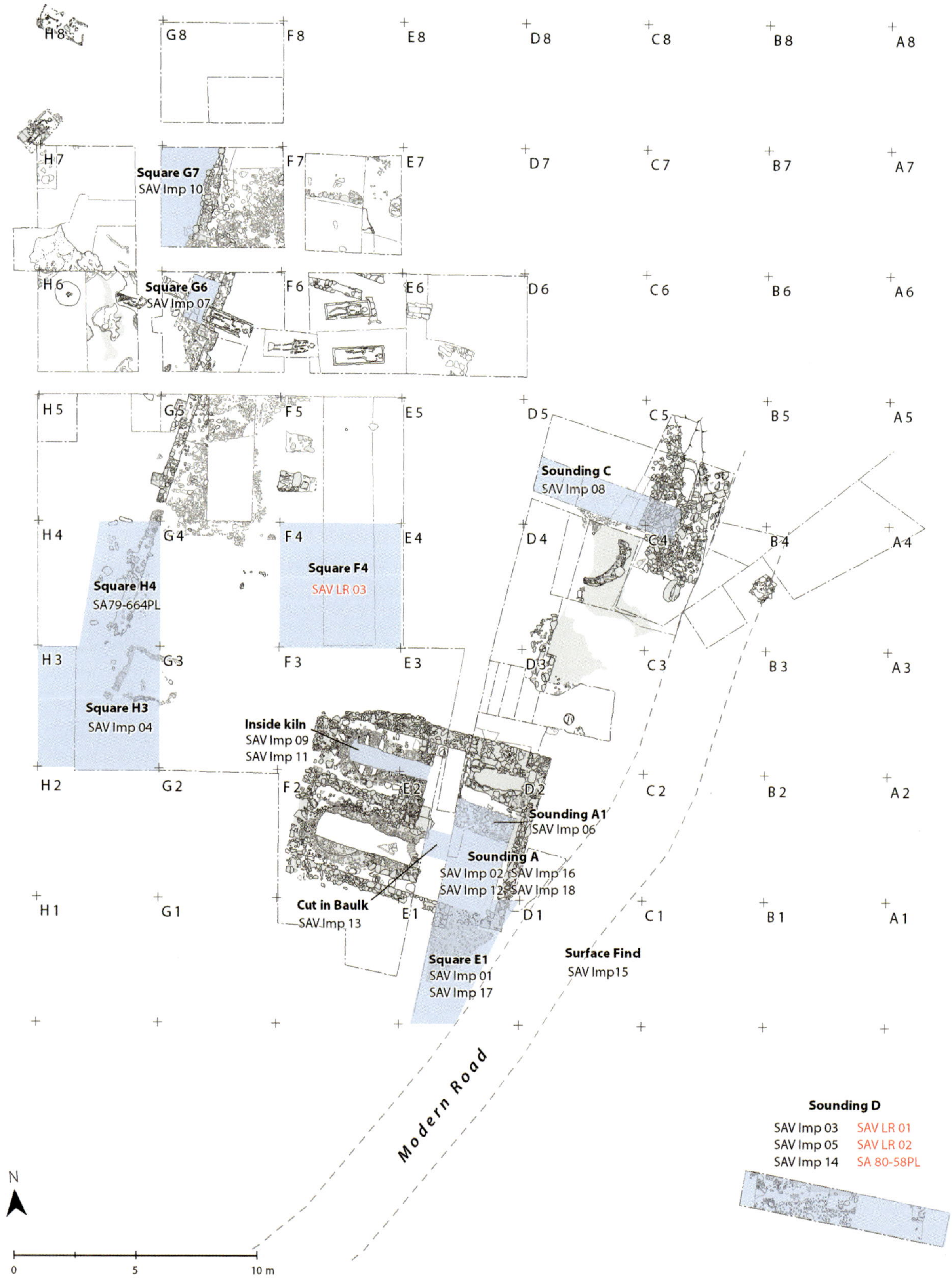

Figure 26.1 Distribution of the Imperial (*black*) and Late Roman (*red*) wares from Sant'Angelo Vecchio; areas highlighted in purple indicate general locations of listed finds. (JT/ICA).

Class	Cat. No.	Form	Chronology	Provenance	MNI
ITS	SAV Imp 01	Plate (Goud. 34)	Early 1st c. AD	E 1	1
Plain ware	SAV Imp 02	Bowl	1st c. AD	Sounding A	1
	SAV Imp 03		Early 1st c. AD	Sounding D	1
	SAV Imp 04		Late 1st c. BC–1st c. AD	H 3	1
	SAV Imp 05		Early Imperial	Sounding D	1
	SAV Imp 06	Basin	Late 1st c. BC–1st c. AD	E 2/Sounding A1	2
	SAV Imp 07	Plate-lid	1st–2nd c. AD	G 6	1
	SAV Imp 08	Lid	1st c. AD	Sounding C	1
	SAV Imp 09	Jug	1st c. AD	F 3	1
	SAV Imp 10		Early Imperial	G 7	
	SAV Imp 11		Early Imperial	E 2/E 3	1
	SAV Imp 12	Bottle	1st c. BC–1st c. AD	Sounding A	1
	SAV Imp 13	Olla	1st c. BC–1st c. AD	Kiln area	1
	SAV Imp 14		Early Imperial	Surface	1
	SAV Imp 15		Early Imperial	Sounding D	1
	SAV Imp 16		Late Republican–Early Imperial	Sounding A	1
	SAV79-772PL	Open form	Late 1st c. BC–1st c. AD	E 3; L. 2, b. 1	1
	SAV Imp 17	Closed form	Late 1st c. BC–1st c. AD	E 1	1
	SAV Imp 18		1st c. AD	Sounding A	1
	SAV79-664PL		Late 1st c. BC–1st c. AD	H 4; L. 1, b. 1	1

Table 26.1 Quantification of the Imperial wares from Sant'Angelo Vecchio.

Roman plain ware is documented at Sant'Angelo Vecchio by equal proportions of open and closed forms. The morphological and technological features of the vessels reflect the general situation of undecorated pottery in Italy between the end of the Roman Republic and the first century of the Empire. Although plain table wares were generally produced with locally available clays (unlike the fine wares), the fabrics and shapes of these objects identify them as belonging to a Roman cultural environment. Fabrics are distinguished by their color, which ranges from light nuances of beige-hazelnut all the way to orange. The surface is often powdery to the touch and in several cases has a reddish slip. Among the new shapes that appear beginning in the Late Republic, closed forms find success in varied articulations of the jug and bottle. At Sant'Angelo Vecchio, the profile of the bowl SAV Imp 02 and the two lids SAV Imp 07 and SAV Imp 08 imitate the morphological features of African products in red slip and plain ware. Likewise, the Roman ceramic culture is attested by a rope-like bowl and other applied decorative elements, such as on the bowl rim SAV Imp 04.

Without thorough studies on Roman plain ware, scholarly understanding of the local distribution of forms in the Metapontino is poor, but material is ripe for this crucial work to be done. The unpublished contexts of the castrum of Metaponto have yielded the largest and most authoritative collection of both plain ware and Italian terra sigillata.[6] More complete, specific, and recently published studies that might provide comparanda come from several contexts in Apulia, in particular the urban site of Ordona (FG) and the production facility at Giancola (Brindisi).[7]

The Late Republican and Early Imperial ceramics from Sant'Angelo Vecchio came mainly from a well-defined and limited area, corresponding to the southern zone of the site (*Fig. 26.1*) where Kilns 3, 4, and 5 are located (mainly from the surface contexts of Soundings A, C, and D, and squares E1–3, F3, and H3). The finds appear to belong to the last phase, when the kilns were active. Several technical details of a few items (e.g., SAV Imp 06) suggest that some of these objects may have been produced on the site or may have belonged to the skilled craftsmen who oversaw production (for details of this phase of the settlement, see Ch. 5, "Ceramic Production at Sant'Angelo Vecchio and in the Metapontine Chora").

Late Roman Ware

Eight fragments of table and storage vessels and transport amphorae can be ascribed to the Late Roman period. Among these are a body fragment of a slipped common olla, a very small body fragment of an open form (too small to be included in the catalog), and a rather wide, plain ribbed ribbon handle

[6] The author wishes to thank L. Giardino for the information.

[7] Ordona: De Stefano 2008; Giancola: Biffino 2012, 227–60.

Class	Cat. No.	Form	Chronology	Provenance	MNI
Slipped common ware	SAV LR 01	Olla	4th–5th c. AD	Sounding D	1
	SA 80-58PL	Closed form	Late Roman	Sounding D; L. 2, b. 7	1
Plain ware	SAV LR 02	Amphora	Late Roman	Sounding D	1
	SA 80-58PL	Unknown	Late Roman	Sounding D; L. 2, b. 7	3
Transport amphorae	SA 80-58PL	LRA 2	5th–7th c. AD	Sounding D; L. 2, b. 7	1
	SAV LR 03	Eastern production	5th–6th c. AD	F4	1

Table 26.2 Quantification of the Late Roman wares from Sant'Angelo Vecchio.

that indicates a Late Roman or Early Medieval date. There is similar ribbing on the bodies of the transport vessels, both those of possible local production imitating eastern forms and those imported from the eastern Mediterranean.

A modest number of published studies has filled gaps in the knowledge of regional and local Late Roman ceramics more fully than what is known of Early Imperial ceramics. The northeastern part of the region and the Metapontino represent the best-known and -studied territories.[8] Slipped common ware is the "fossil guide" of Late Roman contexts. The features of this class of material—including the prevalence of the olla, basin, and amphora shapes; decorative lines incised prior to firing; and colored slips—have been documented by advanced and broad study.[9] The olla attested at Sant'Angelo Vecchio fits squarely with all these typical elements (SAV LR 01). Slipped common wares circulated at all of the Late Roman settlements in the Metapontino in generally large quantities, regardless of their type.[10]

Scholarship on the transport amphorae, also well advanced and constantly updated, includes the countryside and settlements of Basilicata and the Metapontino in the Late Roman period.[11] The amphora body fragment SAV LR 03 found at Sant'Angelo Vecchio can be ascribed either to eastern production or to a local, regional, or sub-regional production that imitated the most common eastern forms. The incised parallel lines on its exterior surface suggest amphora models of the LRA 1 and 2 families. Although the vessel's type cannot be securely identified, its chronology is nevertheless Late Roman, possibly extending into the Early Middle Ages. This dating is corroborated by the presence of a small fragment of a Late Roman 2 amphora.

The few Late Roman fragments from Sant'Angelo Vecchio were mainly found in Sounding D (*Fig. 26.1*). Their presence here could be associated with the abandonment phase of the site, or, as an alternative, with a frequentation of the spring in connection to a later occupation of the nearby area. Data from the field survey suggest the possible presence of a Roman rural settlement at sites 384, 387, and 388, all located a short distance from the kilns at Sant'Angelo Vecchio (*Fig. 1.9*).[12] The items recovered on the surface of these three sites indicate occupation between the Imperial and Late Roman periods.

[8] For northeastern Basilicata, see Di Giuseppe 1998, Di Giuseppe and Capelli 2005, and most recently *Felicitas Temporum* with relevant bibliography. For the Metapontino, see Giardino in *San Biagio*, which synthesizes her long-running studies. For the Late Roman period, in particular, the materials from the castrum were augmented by the important stratigraphy of the Late Antique port in *località* Mele.

[9] Di Giuseppe 1998, Di Giuseppe and Capelli 2005, Turchiano 2000a, Leone 2000, Leone 2002, Greco and Lapadula 2004, Lapadula 2007, Aprosio 2008, 286–90. See also Lapadula in *San Biagio*, 115–17.

[10] On the spread of painted ceramics in the Metapontino and in Basilicata, see Lapadula in *San Biagio*, 115–16.

[11] See the discussion in Lapadula in *Survey*, 1148–53.

[12] For sites 382, 387, and 388, see Lapadula in *Survey*.

Catalog of Roman and Late Roman Wares

Imperial Ware

Italian Terra Sigillata

SAV Imp 01 Plate

1:2

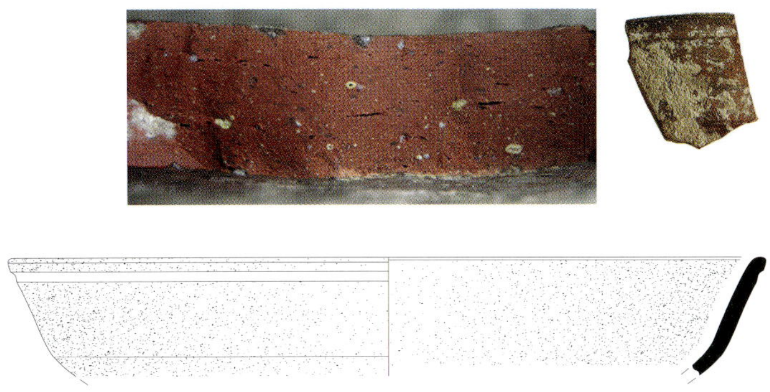

Provenance: Square E1, cleaning. batt. 2. Lot: SA 79-735PL.

Dimensions and technical features: h 3.0, diam. (rec.) 18.4, th 0.4. Clay Munsell 2.5YR 5/6 (red) in fracture, non-homogeneous dark red to reddish brown on the surface due to exposure to flame, with tiny white uniformly distributed inclusions, rare vacuoles, and clean fracture.

Condition: Fragment of the rim.

Description: Plate with wall inclined toward the exterior and distinct rim with a groove on the exterior wall.

Comparisons and comment: Goudineau 34/*Atlante II*, Form XIII, pl. CXXIII, no. 3. The exact start of production of this plate is not known, but it circulated mostly during the reign of Tiberius. The fragment from Sant'Angelo Vecchio has been notably altered by exposure to flame, which has changed its technical properties. The area of production of this item is difficult to identify, although it might have come from Campanian ateliers.

Date: Early 1st c. AD.

Plain Ware

SAV Imp 02 Bowl

Provenance: Sounding A, S area of Room 1. Level 2, batt. 2. Lot: SA 79-26PL.

Dimensions and technical features: rim diam. 26.7, h 4.3, th 0.8. Clay Munsell 7.5YR 7/4 (pink) on the surface and in fracture, with very tiny white inclusions, rare large dark inclusions, and rare tiny shiny inclusions (mica).

Condition: Fragment of the rim with part of the wall.

Description: Rim with triangular section; wall with curved profile.

Comparisons and comment: Bowls with similar articulations of the rim suggest examples produced in Italian terra sigillata with several variants in morphology and size (e.g., *Monte Irsi*, cup no. 294, fig. 35). The form generically suggests Pucci's type XXXV (Pucci 1985, 395, XXV), which had great success and circulated widely in Campania beginning in the reign of Tiberius (Soricelli 1994, 112). Parallels in plain ware come from Early Imperial contexts at Ordona (Annese 2000, 257, pl. I, 4.1) and from late 1st-c. BC contexts, which in turn have parallels in examples from Early Imperial contexts at Benevento (De Stefano 2008, 72, pl. III, 1.2).

Date: 1st c. AD.

1:2

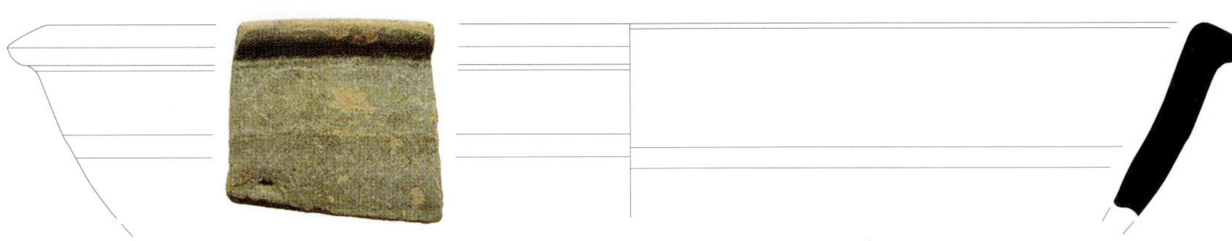

Plain ware cont.

SAV Imp 03 Bowl
Provenance: Sounding D. Level 1, batt. 1. Lot: SA 80-42PL.
Dimensions and technical features: rim diam. ca. 18.0, h 2.9, th 0.9. Clay Munsell 5YR 6/6 (reddish yellow) on the surface and in fracture, with rare tiny white inclusions, medium-sized red inclusions, and tiny dark uniformly distributed inclusions.
Condition: Fragment of the rim with part of the wall.
Description: Rim with flaring brim with rounded extremity; wall with curved profile.
Comparisons and comment: An example of a bowl with similar articulation of the rim comes from contexts dating to the late 1st c. BC/early 1st c. AD at Ordona (De Stefano 2008, 72, type 2, pl. III, 2.1).
Date: Early 1st c. AD.

1:3

SAV Imp 04 Bowl
Provenance: Square H3. Level 1, batt. 1. Lot: SA 79-677PL.
Dimensions and technical features: diam. int. 21.0, th 1.1. Clay Munsell 7.5YR 7/6 (reddish yellow) on the surface and in fracture, with rare tiny white and dark inclusions.
Condition: Fragment of the rim with short segment of the wall.
Description: Brimmed rim with applied rope-like decoration serving as handle.
Comparisons and comment: A fairly precise parallel is a bowl from a context dated late 1st c. BC–late 1st c. AD at Ordona (De Stefano 2008, 73, type 8, pl. IV, 8.1). It is worth noting the presence of the same morphological features in the incense-burners that were generally produced in coarser and refractory fabrics and circulated widely in the Roman world between the Late Republic and the first two centuries of the Empire (Olcese 2003, 91–92, pl. XXI). Vessels with rope-like appliqués were attested in various Roman-period contexts. A mortarium with characteristics similar to the example from Sant'Angelo Vecchio was found at Brindisi (Aprosio 2008, 290, pl. VI, n. VI.9).
Date: Late 1st c. BC–1st c. AD.

1:3

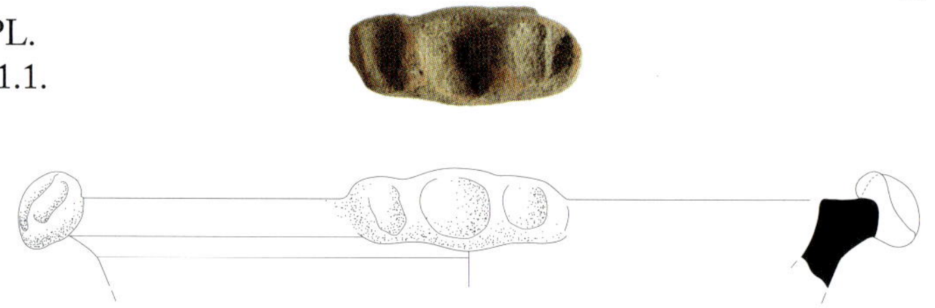

SAV Imp 05 Bowl
Provenance: Sounding D. Level 1, batt. 2. Lot: SA 80-45PL.
Dimensions and technical features: diam. 8.2, h 2.5, th 1.0. Clay Munsell 5YR 6/6 (reddish yellow) on the surface and in fracture, with tiny white and dark uniformly distributed inclusions and rare large dark inclusions.
Condition: Fragment of the base.
Description: Ring foot with flaring extremity with squared profile.
Comparisons and comment: The articulation of the base suggests bowls of the Roman tradition produced in both undecorated and sigillata wares in the Early Imperial period.
Date: Early Imperial.

1:3

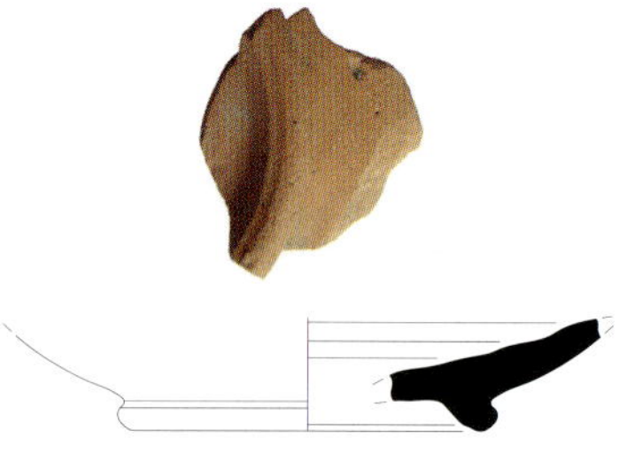

SAV Imp 06 Basin

Provenance: Square E2, Sounding A-1. Level 2, batt. 1. Lot: SA 79-933P and SA 79-984P.

Dimensions and technical features: rim 38.0, h 18.6, foot 10.0, max. th 1.1. Clay Munsell 7.5YR 6/2 (pinkish gray), Munsell 7.5YR 6/3 (light brown), and Munsell 2.5YR 6/3 (light reddish brown) in fracture, with rare large white and dark inclusions and tiny white and dark non-uniformly distributed inclusions. The variations in the color of the fabric suggest poor firing of the vessel; the surface is completely covered in encrustations.

Condition: Partially reconstructed.

Description: Rim with short flaring brim with squared profile and slightly hanging lower extremity; curving wall; flat base.

Comparisons and comment: The flat base SA 81-23PL likely belongs to the same type. A precise parallel comes from a context dated end of the 1st c. BC–end of the 1st c. AD at Ordona (De Stefano 2008, 73, type 9, pl. IV, 9.1). This type of basin is widely attested in central and southern Italy (Cosa, Ostia, Capua) between the Augustan period and the first centuries of the Empire (De Stefano 2008, 73).

Date: Late 1st c. BC–1st c. AD.

1:3

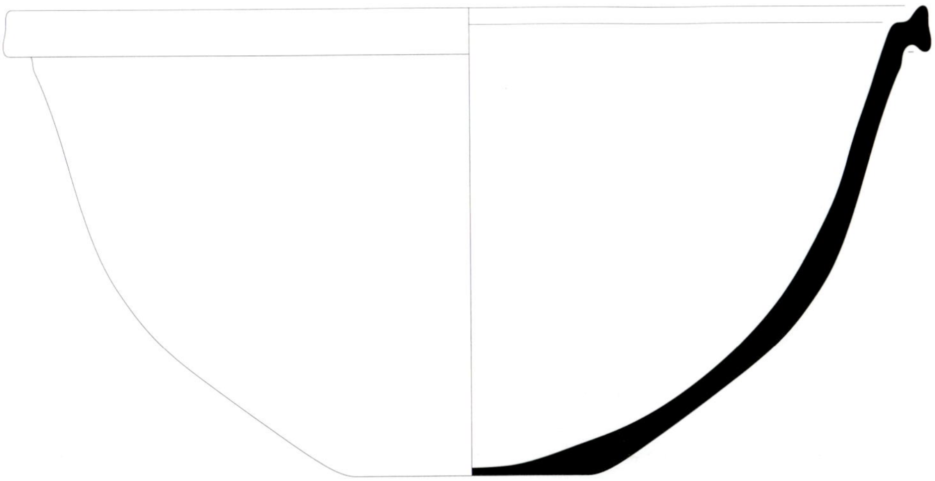

SAV Imp 07 Plate-lid

Provenance: Square G6, W of wall. Level 2, batt. 2. Lot: SA 79-1001PL.

Dimensions and technical features: diam. 21.5, h 4.7, th 1.0. Clay Munsell 5YR 6/6–6/8 (reddish yellow) on the surface and in fracture, with very tiny white uniformly distributed inclusions and rare very tiny dark inclusions; a large fragment of rock is visible in fracture.

Condition: Fragment of the rim and part of the wall.

Description: Indistinct rim with rounded extremity and straight wall.

Comparisons and comment: Lids with similar features occur among African Red Slip Ware of the Early and Middle Empire (e.g., Hayes 196A, common between the 2nd and 3rd c. AD but attested through the 5th c. AD).

Date: 1st–2nd c. AD.

1:3

Plain ware cont.

SAV Imp 08 Lid

Provenance: Sounding C. Level 2, batt. 6. Lot: SA 79-201PL.

Dimensions and technical features: diam. 22.0, h 3.0, th 0.6. Clay Munsell 10YR 8/2 (very pale brown) on the surface, Munsell 5YR 6/6 (reddish yellow) in fracture, with tiny white randomly distributed inclusions and rare tiny shiny inclusions (mica). A thin and light-colored slip on the surface.

Condition: Fragment of the rim.

Description: Raised rim and flaring wall with slight sharp rise near the rim.

Comparisons and comment: Lids with similar features are known among the Late Republican and Imperial cooking wares from various Italian sites. A generic parallel comes from contexts of the 1st c. AD at Albintimilium (Olcese 1993, 246–47, fig. 55, no. 179). Lids with raised rim and rounded or squared profile are attested in Roman territory in the Late Republic and Early Empire (3rd–1st c. BC; Olcese 2003, 89–90, pl. XIX, type 2). The type furthermore generically suggests lids in African plain ware (Pellegrino 2009, 186, fig. 14a, no. 12) and red slip ware (Hayes form 182) common in the Mediterranean basin in the first centuries of the Empire (Biffino 2012, 230, pl. XXXI, 60).

Date: 1st c. AD.

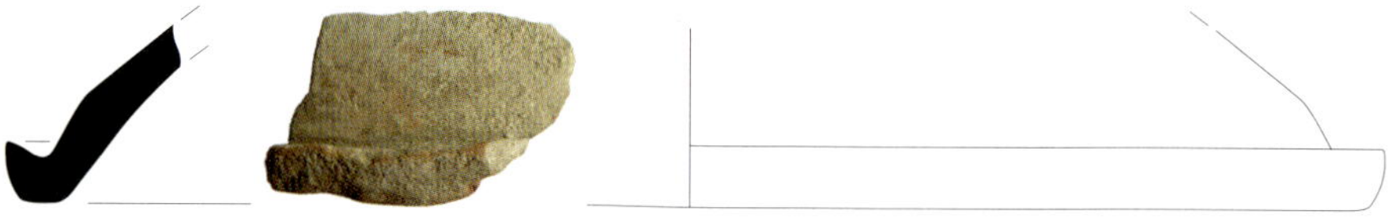

1:2

SAV Imp 09 Jug

Provenance: Square F3, N Kiln interior. Level 2, batt. 2. Lot: SA 79-927P. Inv. no. 319910.

Dimensions and technical features: h 17.5, th 0.4. Clay Munsell 7.5YR 5/8 (strong brown) on the surface and in fracture, granular and rough to the touch, with tiny and medium-sized randomly distributed inclusions (visible also on the surface) and rare tiny shiny inclusions (mica).

Condition: Jug missing the rim. Partially reconstructed.

Description: Body with elliptical profile, high neck flaring toward the exterior, and flat base.

Comparisons and comment: Jugs with similar features have been documented in Latium, in particular in the area of Rome, in contexts of the 1st and 2nd c. AD, predominantly in the 1st c. (Olcese 2003, 93, pl. XXV, type 2).

Date: 1st c. AD.

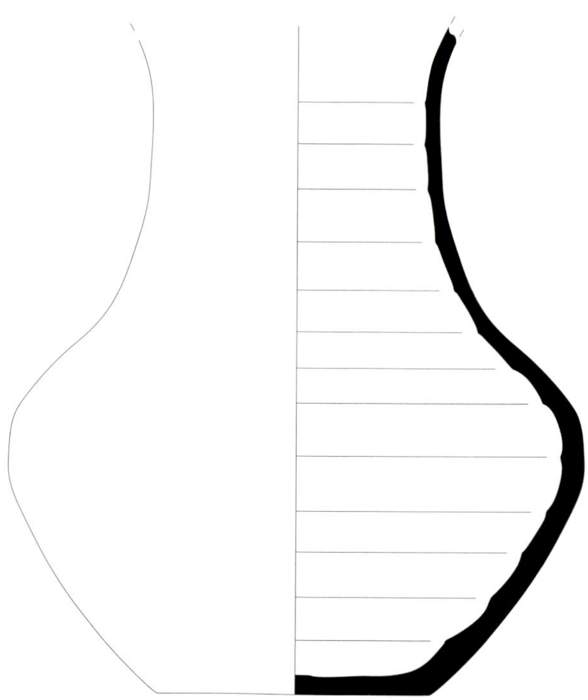

1:2

SAV Imp 10 Jug

Provenance: Square G7, area of wall and tombs, cleaning around wall to W. Level 2. Lot: SA 79-1011P.

Dimensions and technical features: rim diam. 7.75, h 11.2, th 0.35. Clay Munsell 7.5YR 7/4 (pink) on the surface and in fracture, with rare very tiny white inclusions and rare dark inclusions.

Condition: Partially reconstructed. Missing the base.

Description: Jug with ovoid body and band-rim with pronounced interior recess perhaps used to support the lid.

Comparisons and comment: The shape, attested among the jugs at Albintimilium, circulated in several variants in numerous contexts of central and southern Italy in the Augustan and Early Imperial periods (Olcese 1993, 283–86). Generic parallels involving mainly the articulation of the rim come from Pompeii and Naples (Carminiello ai Mannesi) in contexts of the 1st c. AD (Fracchia 1994, 175, fig. 77).

Date: Early Imperial.

SAV Imp 11 Jug

Provenance: Square E2/E3, kiln interior. Level 2, batt. 2. Lot: SA 79-768PL.

Dimensions and technical features: h 4.7, w 3.3, th 1.0. Clay Munsell 7.5YR 6/4 (light brown) in fracture, with very tiny white randomly distributed inclusions and rare large white inclusions (calcite) visible in fracture.

Condition: Fragment of the handle, covered in calcareous encrustations.

Description: Ribbon handle featuring two deep lateral grooves.

Comparisons and comment: This type of handle, with several variants in size and minor details of morphology, is typical of the closed forms (jugs, mugs, and juglets) of the Early Imperial period.

Date: Early Imperial.

Plain ware cont.

SAV Imp 12 Bottle

Provenance: Sounding A. Level 2, batt. 2. Lot: SA 79-26PL.

Dimensions and technical features: h 7.2, th 0.9. Clay Munsell 5YR 6/8 (reddish yellow) on the surface and in fracture, with tiny white and dark uniformly distributed inclusions and rare tiny shiny inclusions (mica). A microfossil is visible in fracture.

Condition: Fragment of the neck.

Description: Presumably hourglass-shaped neck; the attachment of the shoulder is emphasized on the exterior by a ledge and a groove, on the interior by a ridge projecting from the rounded profile.

Comparisons and comment: Bottles with similar features circulated in numerous Augustan contexts. A generic parallel comes from the production site at Giancola (Biffino 2012, 230, pl. XXXI, 60).

Date: 1st c. BC–1st c. AD.

1:2

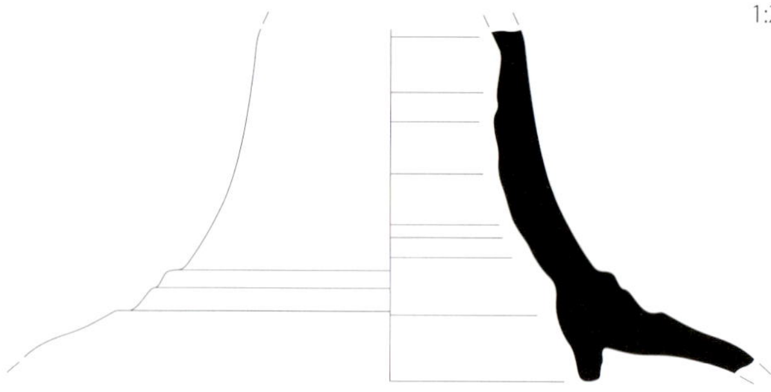

SAV Imp 13 Olla

Provenance: Cut in baulk between N and S Kilns. Level 2, batt. 1–2. Lot: SA 81-7PL.

Dimensions and technical features: rim diam. 18.0, h 6.5, th 1.0. Clay Munsell 2.5YR 6/6 (light red) on the surface and in fracture, with tiny white randomly distributed inclusions, very tiny dark uniformly distributed inclusions, and rare large dark inclusions.

Condition: Two joining fragments of the rim, with encrustations on the surface.

Description: Flaring rim with rounded extremity.

Comparisons and comment: The type is attested at Ordona between the Late Republic and the Early Empire (De Stefano 2008, 81, type 19, pl. VIII, 19.1–19.2). The form is attested at other sites in Italy in the same chronological span (De Stefano 2008, 81).

Date: 1st c. BC–1st c. AD.

1:2

SAV Imp 14 Olla

Provenance: Sounding D. Level 2, batt. 5a. Lot: SA 80-53PL.

Dimensions and technical features: rim diam. 8.65, h 2.5, th 0.4. Clay Munsell 5YR 7/6–6/6 (reddish yellow) on the surface, Munsell 7.5YR 7/1 (light gray) in fracture, with tiny white and dark uniformly distributed inclusions.

Condition: Fragment of the rim.

Description: Rim with vertical band and pointed extremity.

Comparisons and comment: The type is common in plain and cooking wares. The form circulated widely between the 1st c. BC and the 2nd c. AD and is attested in southern Italy between Campania (Pompeii, Herculaneum) and Apulia (Ordona) (Turchiano 2000b, 259–60).

Date: Early Imperial.

1:2

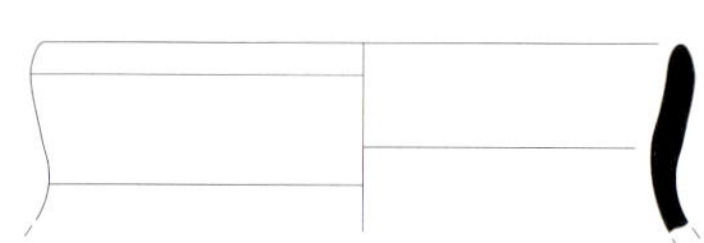

SAV Imp 15 Olla

Provenance: Surface finds, along modern road to east, beyond the Slope Areas. Lot: SA 79-462PL.

Dimensions and technical features: rim diam. 22.5, h 5.0, th 1.1. Clay Munsell 2.5YR 8/3 (pale brown) on the surface, Munsell 7.5YR 7/4 (pink) in fracture, with very tiny and rare dark inclusions.

Condition: Fragment of the rim.

Description: Indistinct vertical rim with squared section and flattened upper extremity.

Comparisons and comment: Generic parallels for the type come from several sites in southern Italy (De Stefano 2008, 83, type 24) in contexts dating between the 2nd c. BC and the 2nd c. AD, as well as in southern France (Pellegrino 2009, 178, fig. 10, nos. 7–8) in contexts of the 1st c. AD.

Date: Late Republican–Early Imperial.

1:2

SAV Imp 16 Olla

Provenance: Sounding A, S sub-section of Room 1 down to wall. Level 2, batt. 1. Lot: SA 79-18PL.

Dimensions and technical features: rim diam. 10.8, h 2.5, th 0.3. Clay Munsell 7.5YR 7/4–7/6 (reddish yellow) on the surface and in fracture, with very tiny and rare white and dark inclusions and rare tiny shiny inclusions.

Condition: Fragment of the rim.

Description: Vertical rim with short brim and flattened upper extremity.

1:2

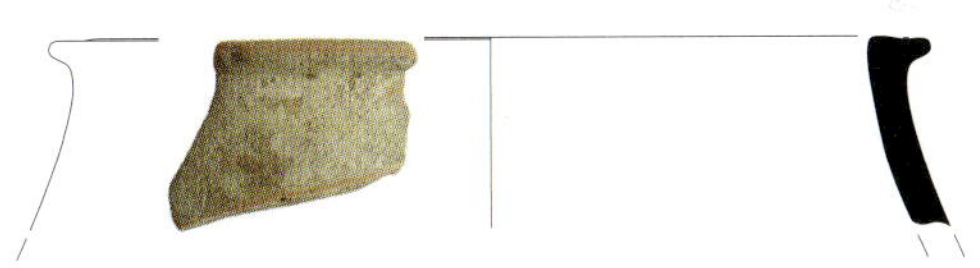

Comparisons and comment: See SAV Imp 15.

Date: Late Republican–Early Imperial.

Plain ware cont.

SAV Imp 17 Closed Form
Provenance: Square E1, cleaning. Lot: SA 79-735PL.
Dimensions and technical features: h 2.7, diam. (rec.) 10.9, th 0.7. Clay Munsell 5YR 6/8 (reddish yellow) on the surface and in fracture, with numerous very tiny white and dark uniformly distributed inclusions.
Condition: Fragment of the base.
Description: Base with flat resting surface; the attachment with the flaring wall is marked by a small and irregular ledge.
Comparisons and comment: The closed shape and the features of the fabric suggest the jug neck SA 79-26PL, but the different find context precludes an origin of the two fragments in the same vessel. An imprecise but very similar parallel comes from the late 1st-c. AD materials at Carminiello ai Mannesi in Naples (Fracchia 1994, 178–79, fig. 79, no. 48).
Date: 1st c. BC–1st c. AD.

1:2

SAV Imp 18 Closed Form
Provenance: Sounding A, S sub-section of Room 1 down to wall. Level 2, batt. 1. Lot: SA 79-18PL.
Dimensions and technical features: h 3.3, diam. (rec.) 15.75, th 1.3. Clay Munsell 5YR 6/8 on the surface, Munsell 2.5YR 5/8 (red) in fracture, with very tiny white and dark uniformly distributed inclusions. The color of the core has been altered (Munsell 10YR 5/1, gray) due to the elevated firing temperature.
Condition: Fragment of the base.
Description: Ring foot with rounded exterior and flat resting surface.

1:2

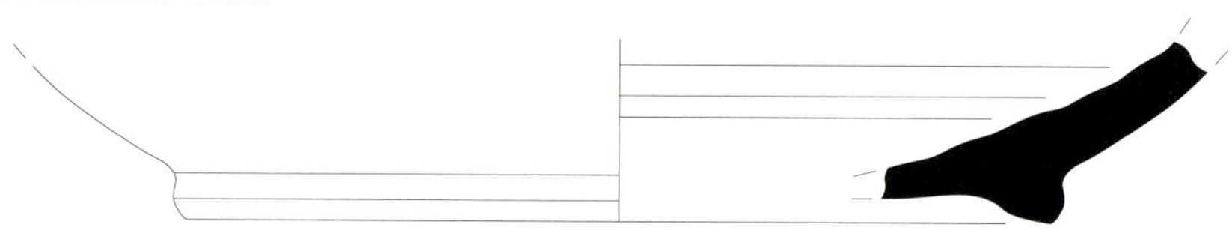

Comparisons and comment: A generic and rather imprecise parallel is the base of a closed form attested in the abandonment phase of the production site at Giancola (Biffino 2012, 261, pl. XLIV, 189).
Date: 1st c. AD.

Late Roman Ware and Early Medieval Ware

Slipped Common Ware

1:2

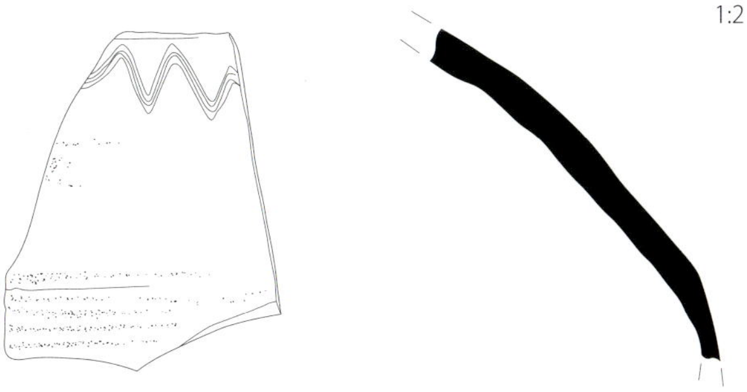

SAV LR 01 Olla

Provenance: Sounding D, Level 2, batt. 7. Lot: SA 80-58PL.

Dimensions and technical features: h 7.0, w 5.8, th 0.7. Clay Munsell 7.5YR 8/4 (pink) on the surface and in fracture, with rare tiny white and dark inclusions. Traces of red on the parallel incised lines and below the incised undulating line.

Condition: Fragment of the wall/shoulder.

Description: Wall/shoulder of an olla with wave motif incised near the neck and parallel incised lines at the juncture between the neck and body.

Comparisons and comment: Ollae of different types and sizes are widely attested in Late Roman slipped common wares in numerous South Italian contexts. In Basilicata, the main attestations come from the site of San Giovanni di Ruoti and the numerous rural settlements (villas and farmhouses) of the northeastern part of the region. In addition, there is the important and unpublished context of the castrum of Metaponto. For generic comparanda for the form, see Di Giuseppe 1998, Freed 1994, Di Giuseppe and Capelli 2005.

Date: 4th–5th c. AD.

Plain Ware

1:2

SAV LR 02 Amphora

Provenance: Sounding D. Level 1, batt. 1. Lot: SA 80-42PL.

Dimensions and technical features: h 6.4, w 6.0, th 1.0. Clay Munsell 7.5YR 7/6 (reddish yellow) on the surface, Munsell 7.5YR 6/4 (light brown) in fracture, with very tiny dark uniformly distributed inclusions, rare tiny shiny inclusions (mica), and rare large dark red and white inclusions (calcite).

Condition: Fragment of the handle.

Description: Ribbed ribbon handle.

Comparisons and comment: A characteristic feature of Medieval amphorae is the rather wide ribbon handle. The type nevertheless begins to circulate in the Late Roman/Early Medieval period.

Date: Late Roman.

Transport Amphorae

SAV LR 03 Amphora with Grooved Walls

Provenance: Square F4. Level 1, batt. 1. Lot: SA 79-459P.

Dimensions and technical features: h 5.0, w 5.2, th 1.2. Clay Munsell 5YR 6/8 (yellowish red) in fracture, 7.5YR 6/4 (light brown) on the surface, with rare tiny shiny inclusions and tiny white uniformly distributed inclusions.

Condition: Fragment of the wall.

Description: Wall with parallel incised lines on the exterior surface.

Comparisons and comment: Walls of unknown origin with varying concentrations of grooves are attested at several Late Roman and Early Medieval sites in Italy (e.g., Val Pescara: Verrocchio 1998, 695, fig. 25.10–13). The morphological features suggest production in the East (LRA 1–2 and similar), while the fabrics indicate local production. The wall fragment from Sant'Angelo Vecchio is also part of this rather vague picture, but it is nevertheless datable to the Late Roman–Early Medieval period.

Date: 5th–6th c. AD.

1:2

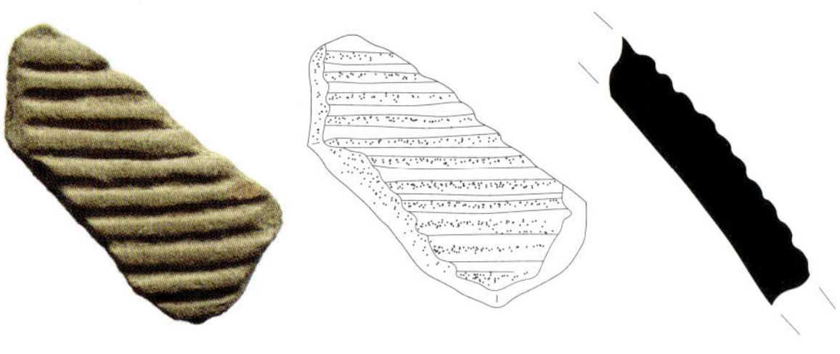

27

Transport Amphorae

Teresa Oda Calvaruso

In contemporary scholarship, the study of transport amphorae has provided valuable new documentation for the reconstruction of life at ancient sites.[1] The recent contribution by Swift[2] examines the circulation of transport amphorae in the Metapontino between the 7th and 1st centuries BC based on the corpus of the various amphora types documented by the ICA field survey between the Bradano and Basento Rivers. Analysis of the distribution of these amphorae across the territory demonstrates that in the Archaic period, products were transported in Corinthian vessels of Types A and B, as well as in vessels similar to the B type, such as the Western Greek amphorae produced in Calabria. In the Classical period, the dominant type appears to have been the Corinthian B amphora, which was also produced locally in several places, including the Metapontine chora.[3] In the Hellenistic period, the Greco-Italic amphora predominated over the other types in the Metapontino, just as at other Italiote and Siceliote sites. This container was produced at several centers in Magna Grecia and Sicily,[4] but by the end of the 4th century BC, the most widely circulated versions in the Metapontino were the local ones.

The number of amphora fragments recovered in each of the investigated sectors of Sant'Angelo Vecchio is rather meager compared to the other ceramic classes, but they nevertheless provide information useful for identifying the chronology of the various excavated contexts. The most serious limiting factor in the classification of the amphora fragments is their physical condition—apart from the absence of whole examples, often the fragments are so small that they cannot always be ascribed to a known type. This is especially true of the body fragments, which form the majority of the unidentified examples. Of the 264 total amphora fragments recovered at Sant'Angelo Vecchio,[5] only 53 are diagnostic (i.e., belonging to the rim, handles, or toe) and therefore useful for the identification of the type.[6] The catalog describes the fragments that represent identified types.

The physical distribution of the diagnostic fragments within the site is as follows: 1 on the surface, 3 from the Archaic farmhouse, 10 from the Workshop Area, 24 from the Kiln Complex and the Slope Area, 4 from the Pottery Deposit, and 11 from the cleaning of the Retaining Wall (*Fig. 27.1*). Due to the highly fragmentary state of the materials, this study by necessity takes body fragments into consideration alongside diagnostic ones for the more general information they can provide on production and chronology (*Fig. 27.2*).

The types of amphorae identified at Sant'Angelo Vecchio are those typically encountered at sites in Magna Grecia between the 6th and the 2nd century BC (*Fig. 27.3* and *Fig. 27.4*). Where possible, the typological classification and chronological identification of the materials presented in the catalog follow Swift's study of the amphorae from the ICA field survey between the Bradano and Basento, as well as parallels encountered in published Italiote and indigenous Italic contexts. The individual amphora types are designated with a code consisting of a letter indicating the form (A), an Arabic numeral indicating the type, another Arabic numeral indicating the variant, and a lower-case letter indicating the specific example.[7]

---

[1] The bibliography on specific types is large: see, e.g., Koehler 1979; Koehler 1981; Lyding Will 1982; Gras 1987; Vandermersch 1994; Olcese 2010. There are also numerous synthetic studies: e.g., Empereur and Garlan 1986; *Amphores romaines*; *Lattara 6*; Sciallano and Sibella 1994; Caravale and Toffoletti 1997; Auriemma 2004; Sourisseau 2011. More recent studies are characterized by a new attention to archaeometric analyses aimed at defining the production areas of the various containers: Olcese 2010, 2012; Swift in *Survey*.

[2] Swift in *Survey*.

[3] Swift in *Survey*, 461–62.

[4] *Locri IV*; Vandermersch 1994; Keay and Williams 2005; Olcese 2010.

[5] This study considers both the fragments recovered in the excavations of 1979–1981 and those collected in the cleaning of the Retaining Wall in 2010–2011.

[6] A total of 29 rim fragments was recovered, few of which preserve any portion of the neck or the handle attachment, as well as 21 handles and 3 toes.

[7] Type 1 indicates the Laconian amphora, Type 2 the amphora of extra-regional (Eastern Greek?) production, Type 3 the Archaic Type B series, Type 4 the Corinthian A series, Type 5 the Type B series, Type 6 the Greco-Italic series, and Type 7 the Metapontine amphora with collar rim.

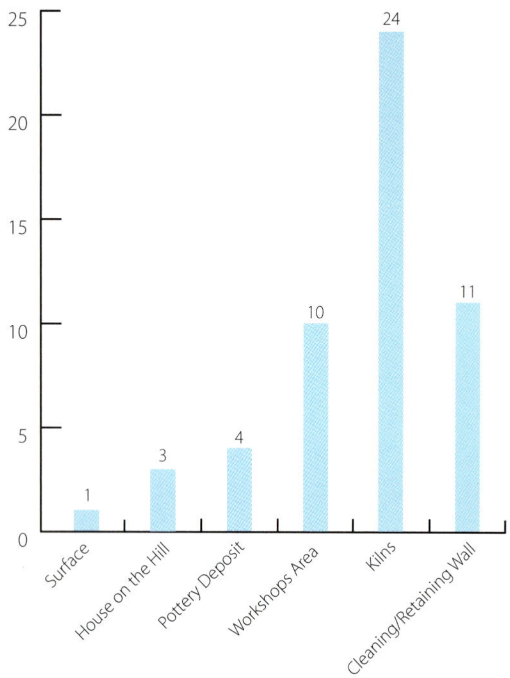

Figure 27.1 Quantification of the diagnostic amphora fragments at Sant'Angelo Vecchio.

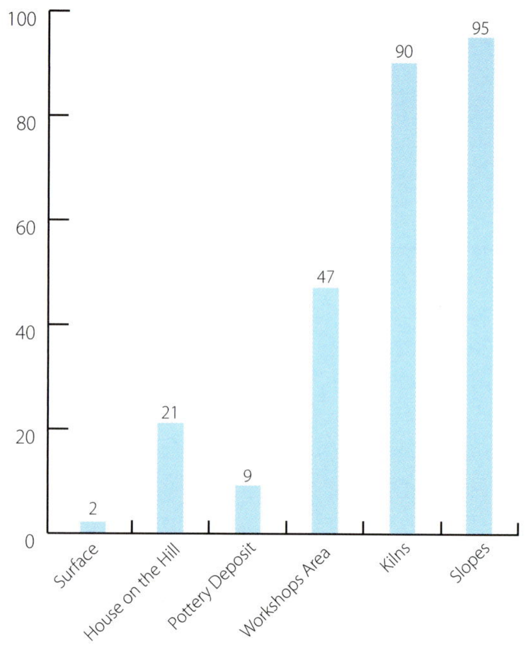

Figure 27.2 Quantification of the total amphora fragments at Sant'Angelo Vecchio.

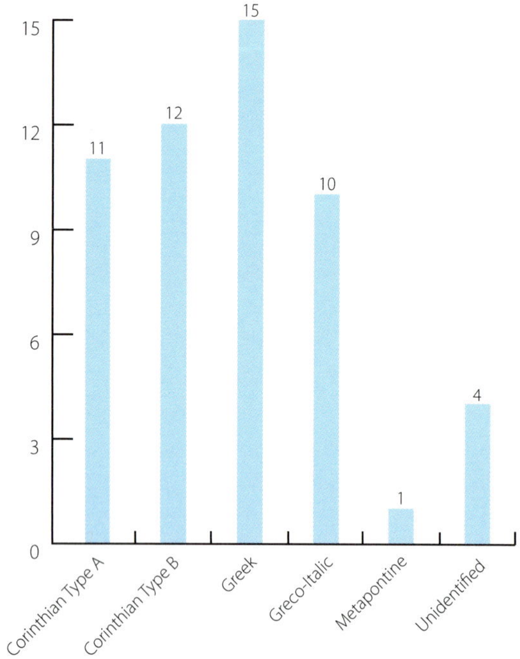

Figure 27.3 Quantification of the diagnostic amphora fragments by type at Sant'Angelo Vecchio.

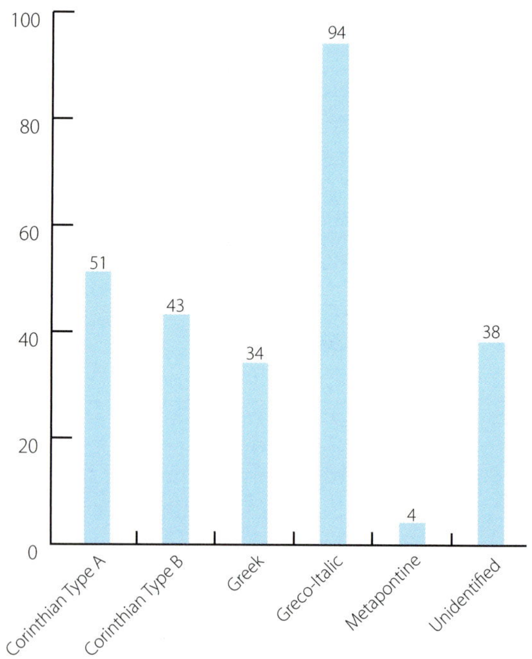

Figure 27.4 Quantification of the total amphora fragments by type at Sant'Angelo Vecchio.

Context	Amphora Type	NFr
Late Archaic	Laconian	1
	Archaic Type B	32
	Corinthian Type A	19
	Eastern Greek	1
	Total	**53**

Table 27.1 Quantification of the amphora fragments of the Late Archaic period at Sant'Angelo Vecchio.

The amphorae recovered in the House on the Hill belong mostly to the Late Archaic period. Several body fragments identifiable by the fabric[8] belong to a Corinthian Type A amphora of the end of the 6th/beginning of the 5th century BC.[9] This vessel, traditionally considered an oil container, is one of the oldest encountered on the site. Several body fragments of Corinthian Type B amphorae, probably used for transporting wine, also belong to the Late Archaic period.[10] A puffed cushion rim of an Archaic Type B amphora (SAV Amp 04) dates to the first half of the 5th century BC. A handle with oval section and inscribed cross/X stamp (SAV Amp 02), on the other hand, appears to belong to an extra-regional production, perhaps Eastern Greek. A rim of a Laconian amphora (SAV Amp 01), dating to the first half of the 6th century BC, and a toe with flat floor of Western Greek manufacture, dating to the second half of the 6th century BC, were residual in the area of the slopes. They confirm the occupation of the area around the spring in the second half of the 6th century BC. Their recovery in this area is probably due to having slid down the slope from the Archaic frequentation of the spring.

The Pottery Deposit in Sounding C yielded several fragments documenting continuity in the importation of the Corinthian Types A′ and B amphorae. The two examples of Corinthian Type B (SAV Amp 19 and

Context	Amphora Type	NFr
Pottery Deposit	Corinthian Type A	7
	Corinthian Type B	2
	Total	**9**
Cleaning of the Retaining Wall	Archaic Type B	13
	Corinthian Type A	30
	Corinthian Type B	12
	Greco-Italic	35
	Unidentified	5
	Total	**95**
Area of the kilns	Laconian	1
	Archaic Type B	11
	Corinthian Type A	6
	Corinthian Type B	14
	Greco-Italic	36
	Metapontine	3
	Unidentified	19
	Total	**90**
Area of the tombs and storerooms	Archaic Type B	4
	Corinthian Type A	3
	Corinthian Type B	10
	Greco-Italic	22
	Unidentified	8
	Total	**47**

Table 27.2 Quantification of the amphora fragments from the hillside at Sant'Angelo Vecchio.

SAV Amp 20), the rim SAV Amp 14, the handle SAV Amp 15, and body fragments of Corinthian Type A′ can be dated by comparanda to the second half of the 5th century BC.

The cleaning of the Retaining Wall performed in 2010–2011[11] yielded a Corinthian Type A brim rim (SAV Amp 13) of the end of the 6th/beginning of the 5th century BC, two Archaic Type B puffed cushion rims (SAV Amp 08 and SAV Amp 09) of the first half of the 5th century, and two handles of the second half of the 4th century BC (one Corinthian B and one Metapontine Greco-Italic).

The majority of the amphora fragments found in the Slope Area and the Kiln Complex date to the Early Hellenistic period. The most commonly attested type is Greco-Italic,[12] which is believed to have transported

[8] The five body fragments from Corinthian A amphorae are characterized by the typical clay of Archaic amphorae: rough, pinkish beige in color, with large and frequent inclusions of various materials.

[9] Beginning in the second half of the 8th c. BC, the Corinthian A amphora was the most common container in southern Italy and Sicily: Koehler 1979; Koehler 1981.

[10] Production of the Corinthian B amphora began in the middle of the 6th c. BC, particularly on Corcyra, but recent archaeometric analyses have demonstrated the existence of other production centers in Calabria, along the Adriatic coast of Apulia, and in the Metapontino (Swift in *Survey*, 461–62). The clay of the five recovered body fragments—pale yellow, soft, and purified—suggests production in Calabria.

[11] The operation was performed by F. Silvestrelli.

[12] In the last few decades publications of Hellenistic contexts have provided important new documentation for the study of these amphorae: Manacorda 1986; *Locri IV*; Termini 1997; Brizzi 1999a; Campagna 2000;

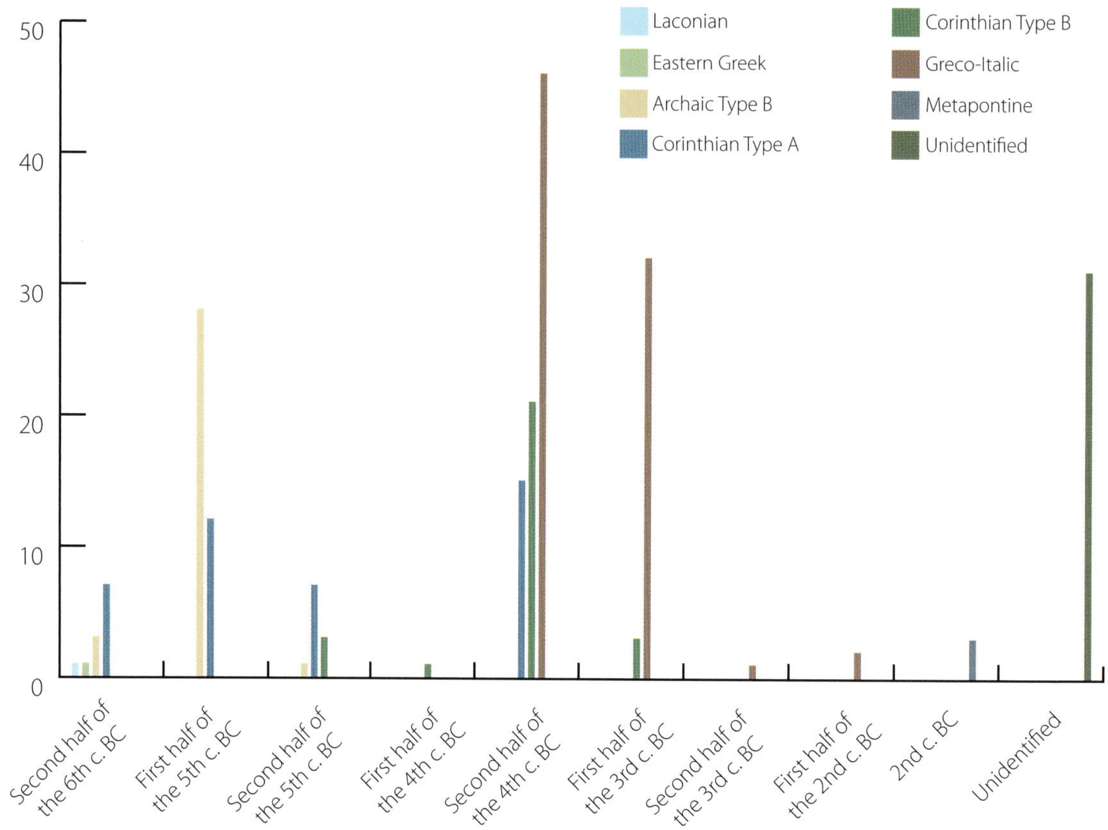

Figure 27.5 Chronological distribution of the diagnostic amphora fragments at Sant'Angelo Vecchio.

wine, documented in the end of the 4th–beginning of the 3rd century BC only by body fragments. One handle with a circular section and one body fragment can be ascribed to Corinthian Type A of the end of the 4th century BC, and three rims (SAV Amp 21–SAV Amp 23) and a toe (SAV Amp 27) to Type B. One Corinthian Type A′ rim, dated to the second half of the 6th century, and two Archaic Type B rims (SAV Amp 05, SAV Amp 07) and two handles (SAV Amp 11) dating to the 5th century BC are probably residual, having been found mostly in the area of Kilns 1 and 2. They could be materials from the earlier phase reused in the construction of the kilns. One handle (SAV Amp 31), two shoulder fragments (SAV Amp 32 and SAV Amp 33), and 20 body fragments belong to Greco-Italic amphorae of the end of the 3rd through the first half of the 2nd century BC.

The latest examples of amphorae, dating between the end of the 3rd and the middle of the 2nd century BC, are two rims of Greco-Italic amphorae (SAV Amp 28 and SAV Amp 29) found in the Kiln Complex and a rim of a Metapontine amphora from Sounding D (SAV Amp 34). No amphora fragments dating to the 1st century BC were found.

The amphora fragments found in Workshop Area—four Greco-Italic (SAV Amp 30) and four Corinthian Types A and B (SAV Amp 16, SAV Amp 18)[13]—likewise date to the beginning of the Hellenistic period. Here the latest examples, two Corinthian Type B handles (SAV Amp 24 and SAV Amp 25), belong to the first half of the 3rd century BC. An Archaic Type B rim (SAV Amp 06) is noteworthy, as are several body fragments datable to the first half of the 5th century BC with morphological features suggesting Western Greek production, but made with fabrics indicating an origin in Magna Grecia.[14] These materials were found in the vicinity of the concentration of tombs and therefore could be residual, introduced into the later contexts by

Corretti and Capelli 2003; Olcese 2004; Tramontana 2008; C. De Mitri 2009a, b; Olcese 2010. On the types and the problem of the production centers, see Vandermersch 1989, n. 2, 60, 86; Olcese 2010; Swift in *Survey*.

13 Five fragments (including one rim) were identified as belonging to a single vessel that, unfortunately, could not be recognized among the known types.

14 Macroscopic analysis of the fabrics appears to suggest production in Calabria: Vandermersch 1989; *Locri IV*, 208–11; Swift in *Survey*, 472.

the construction of the Hellenistic building. All of the remaining body fragments can be ascribed to the Greco-Italic group datable between the middle of the 4th and the middle of the 3rd century BC.

The 53 diagnostic amphora fragments recovered at Sant'Angelo Vecchio represent the types commonly encountered in South Italian contexts between the Archaic and Hellenistic periods (*Fig. 27.5*), and therefore offer important evidence about the possible acquisition paths that led to the deposition of the amphorae at the site during its lifetime. As containers used to transport specific products, amphorae reflect the agricultural and economic conditions of their production centers, but they were also used to store and preserve foodstuffs (wine, oil, cereals) in houses and storerooms.

In the Archaic period, goods seem to have traveled generally from east to west, despite the apparent lack of economic logic in importing large quantities of products, since foodstuffs could be produced locally and therefore short-range trade associated with daily needs was more likely.[15] Greek imports in the Archaic period, documented in the House on the Hill by the Corinthian Type A amphora, suggest that oil was the primary necessity. The type and the morphological features of the few identified fragments reflect continuity in importing Corinthian Type A′ amphorae through the end of the 4th century BC. After a minimal presence in the Archaic period, wine containers, on the other hand, are documented especially beginning in the 5th century BC, represented by Corinthian Type B amphorae that can be assigned to two groups on the basis of their clays: pale yellow in several examples, pinkish in others. At the same time, production of colonial Western Greek amphorae increased, indicating that the goods previously imported in Eastern Greek amphorae were being substituted with local products.

By the Hellenistic period, the products consumed at the site came from local and regional sources in southern Italy, a situation reflected in the predominance of the Greco-Italic amphorae over the other types. As in the chora,[16] examples of vessels made with a reddish fabric attributable to local Metapontine production have been found among the Hellenistic amphorae in this area of the site.[17]

The chronological seriation of the amphora types demonstrates continuity in the occupation of the site from the middle of the 6th through the end of the 5th century BC. The first half of the 4th century BC is attested by only two examples of Corinthian Type B amphora, but in the second half of the century, the number of amphorae returned to a high level. This was sustained through the end of the 3rd century BC, plunging again in the 2nd century BC, with only two examples of Greco-Italic amphora and a single example of a Metapontine amphora. These last fragments are the latest in the amphora corpus.

In terms of the circulation of these vessels, this study indicates a rather homogeneous and qualitatively continuous reconstruction in line with the broader context of southern Italy, including for each period the types most commonly encountered at practically all known contemporary Italiote and Siceliote sites, and in particular the larger Metapontino.[18] In terms of the production areas, finally, it should be emphasized that in the Archaic and Classical periods the largest number of examples comes from outside of the region (from Calabria in particular), while local products predominate over imports beginning in the 4th century BC.

[15] *Locri IV*, 227.

[16] Swift in *Survey*, 463–68.

[17] Beyond Pantanello, there are two sites identified by the field survey where amphorae were produced: 266 and 153 (Swift in *Survey*, 466–68).

[18] Swift in *Survey*; *Fattoria Fabrizio*.

Catalog of Transport Amphorae

Laconian Amphora

Although the Laconian amphora was identified only recently, its circulation appears to have been rather limited. Broad study reveals that it circulated essentially in Sicily and Etruria, beginning with the initial considerations of M. Martelli in 1979 and P. Pelagatti in 1980 and continuing through Pelagatti's comprehensive analysis of the Sicilian evidence in 1995, which clarified the type's features and created a precise typological seriation. In the Metapontino, the type is attested at Incoronata (Cavagnera 1995, 44–45).

From the third quarter of the 7th c. BC at least, the oil produced in Laconia, in Greece, was exported in vessels very similar to the Attic SOS and *à la brosse* amphorae. In fact, the Laconian amphorae belong to the family of completely or partially glossed Archaic commercial vessels. Particular features include a variably molded rim emphasized by a listel, short neck, and bulging body terminating in a ring foot. The earliest examples, through the second quarter of the 6th c. BC, are completely glossed, while their successors, from the second half of the 6th century until the last quarter of the 5th c. BC, have more reserved zones, especially on the neck and shoulder.

This rim is the only fragment of the type so far recovered at Sant'Angelo Vecchio. It has parallels with several Sicilian examples dated to the first half of the 6th c. BC.

Laconian Amphora

SAV Amp 01 Laconian Amphora, Type A1.1a

Provenance: Sounding D, Level 2, batt. 7. Lot: SA 80-58PL.

Dimensions and technical features: h 5.42, rim 14.5. Clay Munsell 2.5YR 5/6, brick red, with small white and black inclusions, rare medium-sized red inclusions, and vacuoles. Heavily cracked gloss.

Condition: Fragment of the rim and neck.

Description: Rim with rounded profile and flat upper surface with small underlying collar inserted onto a pronounced ledge on the neck. Black gloss on the flat part of the rim, the attachment of the small collar, and the small collar itself, as well as on the interior; reserved neck.

Comparisons and comment: Boitani 1990, 57, no. 98, fig. 135 (from Gravisca, first half of the 6th c. BC); Pelagatti 1990, fig. 49, no. 203 (from Monte San Mauro, second quarter of the 6th c. BC.).

Date: First half of the 6th c. BC.

1:2

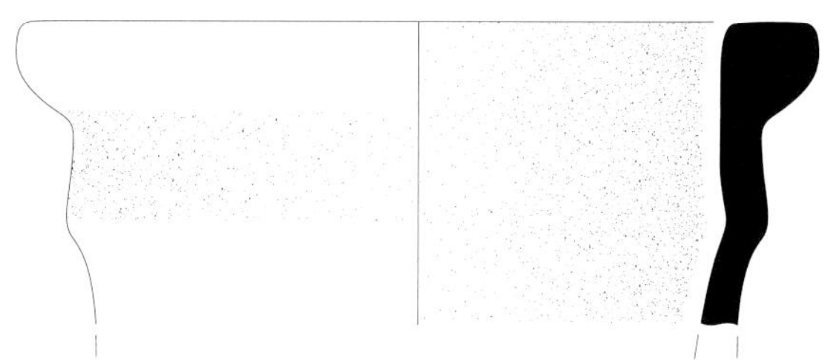

Extra-regional (Eastern Greek?) Amphora

Although this fragment could not be ascribed to a precise fabric and type, its clay suggests that it was produced in an extra-regional location, perhaps mainland Greece, implied also by the stamp type, which can be dated by comparanda to the 5th c. BC. This is the only example attested at Sant'Angelo Vecchio.

Extra-regional (Eastern Greek?) Amphora

SAV Amp 02 Eastern Greek Amphora, Type A2.1

Provenance: House on the Hill. Level 1, batt. 1. Lot: SA 79-1056P (see *Fig. 2.39*).

Dimensions and technical features: h 4.32, th 1.78. Clay Munsell 2.5YR 4/8 or 5/8, compact, dark orange fabric with occasional white and very rare black inclusions, and abundant extremely fine silver and gold mica inclusions.

Condition: Fragment of a handle.

Description: Handle with oval section; on the upper section there is a stamp with an inscribed cross or wheel.

Comparisons and comment: For the stamp, Grace 1956, 171–72, no. 220, pl. 76 (from Athens, end of the 5th–mid-4th c. BC); Cordano 2007, 17–18, fig. 3 (from the Camarina necropolis, 6th c. BC); Archaeological Museum of Pithekoussai (from Stipe dei Cavalli, 7th c. BC).

Date: Second half of the 6th c. BC.

1:2

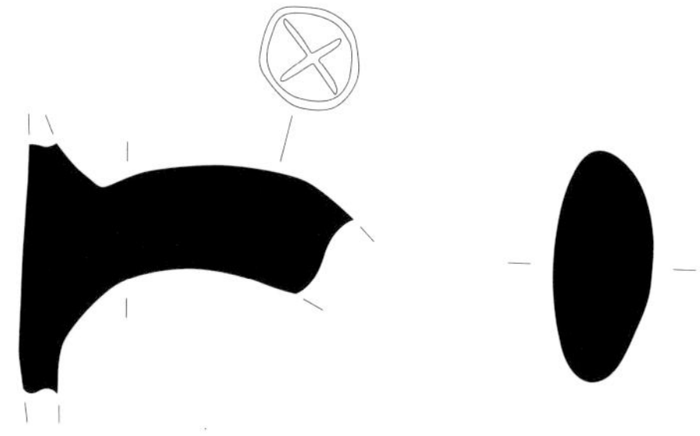

Archaic Type B Amphora Series

The "Archaic Type B" designation includes a type of amphora that is morphologically similar to the containers circulating in the Mediterranean beginning in the Archaic period and conventionally called "Ionic-Massaliote," "of the Ionic tradition" (Gassner 1996), "Corinthian B," or "pseudo-Chiote." In reality, the vessel was widely imitated in the western Greek world between the 6th and 5th c. BC. The production locations are varied: Velia (Gassner 1996), Locri (Barra Bagnasco 1992), and, it seems, Poseidonia-Paestum (Gassner 1996, 167 n. 9). The fabrics seem to indicate production also in Sicily, at Palermo and Himera (Albanese Procelli 1996, 113) and perhaps on Motya (Di Sandro 1990, 239).

At Sant'Angelo Vecchio, the Archaic Type B amphora is attested by 12 examples. Two types are present, with several variants in the shape of the rim. One type has a small, rounded, puffed cushion rim formed by folding the upper edge toward the exterior, thus occasionally creating an air pocket; the other type has a higher and thinner, almost almond-shaped rim. On both types there is a ledge of variable size at the juncture between the rim and neck, which is generally straight; the ribbon-shaped handles are attached under the rim. The only toe recovered has a flat and full floor imitating those of Corinthian Type A amphorae (Albanese Procelli 1996, 113). The color of the clay varies from roseate to beige, occasionally with a grayish core and a light cream-colored slip. The fabric is rather soft and doughy, with stone and augite inclusions, and rich in small flecks of gold mica. These characteristics suggest that the vessels were produced in Calabria.

Archaic Type B Amphora Series

SAV Amp 03 Archaic Type B Amphora, Type A3.1a

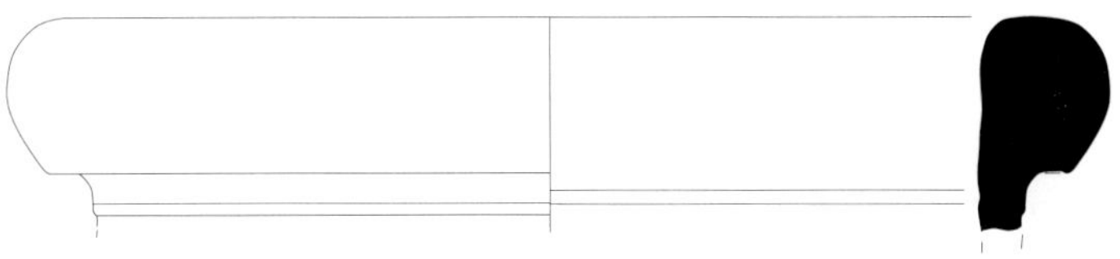

Provenance: Surface find. Lot: SA 79-979P.
Dimensions and technical features: h 3.5, rim 17.0. Clay Munsell 10YR 8/3, gritty and hard with frequent small stone inclusions. Slip Munsell 7.5YR 8/3.
Condition: Fragment of the rim and neck.
Description: Rim in the shape of a large puffed cushion with rounded upper section; beneath the rim there is a small ledge.
Comparisons and comment: Locri IV, 231, pl. LX, no. 182 (from Locri, first half of the 5th c. BC, amphora with puffy cushion rim); Swift in *Survey*, no. 8, 472 (from survey, late 6th–first half of the 5th c. BC, Archaic type B).
Date: First half of the 5th c. BC.

SAV Amp 04 Archaic Type B Amphora, Type A3.1b

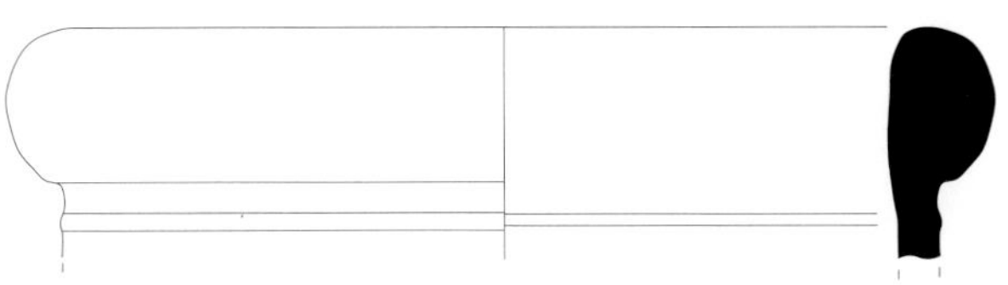

Provenance: House on the Hill. Lot: SA 79-1062PLC. Assemblage 1.
Dimensions and technical features: h 3.7, rim 14.5. Clay Munsell 5YR 6/6 or 7/6, gritty and hard with numerous small and medium-sized inclusions.
Condition: Fragment of the rim and neck.
Description: Rim in the shape of a puffed cushion with a very small ledge.
Comparisons and comment: Edlund 1982a, 30, fig. 13; *Locri IV*, pl. LX, no. 184 (from Locri, first half of the 5th c. BC, amphora with puffy cushion rim); Sourisseau 2011, fig. 9, no. 6, 180 (from Rifriscolaro necropolis, Camarina, last quarter of the 6th–first half of the 5th c. BC, Greco-Italic amphorae form 1a).
Date: First half of the 5th c. BC.

Archaic type B series cont.

SAV Amp 05 Archaic Type B Amphora, Type A3.1c

1:2

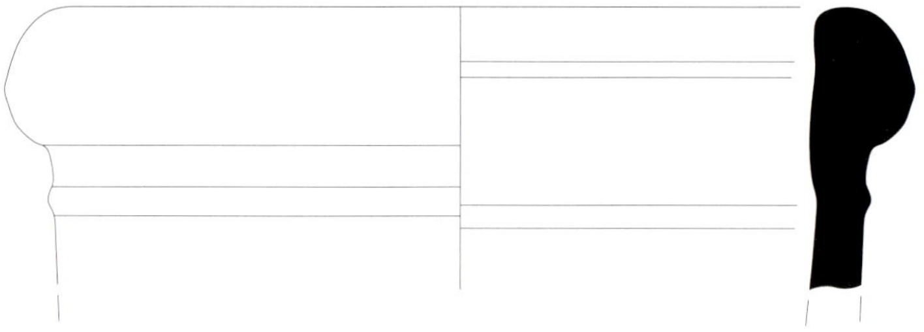

Provenance: Square D4, within stone circle. Level 3, batt. 5. Lot: SA 79-896PL. Assemblage 13.

Dimensions and technical features: h 5.7, rim 15.5. Clay Munsell 2.5YR 4/4, very gritty, with numerous white calcareous inclusions. Core Munsell 2.5YR 5/6.

Condition: Fragment of the rim and neck.

Description: Rim in the shape of a small cushion; on the neck there is a small ledge.

Comparisons and comment: Metapontine production. *Locri IV*, pl. LX, no. 184 (from Locri, first half of the 5th c. BC, amphora with puffy cushion rim).

Date: First half of the 5th c. BC.

SAV Amp 06 Archaic Type B Amphora, Type A3.2a

Provenance: Square E6. Level 2, batt 4. Lot: SA 80-62PL (see *Fig. 2.23*).

Dimensions and technical features: h 4.7, rim 15.8. Clay Munsell 7.5R 8/3, porous, with black and white inclusions and biotite flakes.

Condition: Fragment of the rim and neck.

Description: Rim in the shape of a small puffed cushion, folded over to form an air pocket; beneath the rim there is a small raised area between two grooves.

Comparisons and comment: Calabrian production. Di Sandro 1986, 39, pl. 8, no. sg 88 (from Pithekoussai, first half of the 5th c. BC, Massaliote amphorae); *Locri IV*, pl. LX, no. 187, B1 (from Locri, first half of the 5th c. BC, amphora with puffy cushion rim); Swift in *Survey*, no. 8, 472 (from survey, late 6th–first half of the 5th c. BC, Archaic type B).

Date: First half of the 5th c. BC.

1:2

SAV Amp 07 Archaic Type B Amphora, Type A3.3a

1:2

Provenance: Square F7. Level 3, batt. 3. Lot: SA 80-24PL (see *Fig. 2.20*).

Dimensions and technical features: h 4.9, rim 14.6. Clay Munsell 7.5YR 6/4, gritty and hard, with numerous white and black stone and mica inclusions.

Condition: Fragment of the rim and neck.

Description: Elongated rim with almond section and angular termination; beneath the rim there is a listel.

Comparisons and comment: Cavalier 1985, 37, fig. 6a (from Lipari, final decades of the 6th–first half of the 5th c. BC, Ionic amphora); Corretti and Capelli 2003, 291–92, pl. LIII, no. 8 (from Entella, second half of the 6th–beginning of the 5th c. BC, Western Greek amphora); Sourisseau 2011, fig. 12, no. 2, 183 (from Rifriscolaro necropolis, Camarina, last quarter of the 6th–first decades of the 5th c. BC, Western Greek amphora); Swift in *Survey*, no. 10, 473 (similar, from survey, end of the 6th–beginning of the 5th c. BC, Archaic type B).

Date: First half of the 5th c. BC.

SAV Amp 08 Archaic Type B Amphora, Type A3.3b
Provenance: Retaining Wall, US 9/18.
Dimensions and technical features: h 5.75, rim 13.5. Clay Munsell 7.5YR 6/6, hard and gritty, with numerous small stone inclusions.
Condition: Fragment of the rim and neck.
Description: Rim with almond section and small ledge on the underside.
Comparisons and comment: *Locri IV*, pl. LXI, no. 197 (from Locri, 5th–4th c. BC, amphora with almond-shaped rim).
Date: First half of the 5th c. BC.

1:2

SAV Amp 09 Archaic Type B Amphora, Type A3.4a
Provenance: Retaining Wall, US 39.
Dimensions and technical features: h 5.2, rim 11.5. Clay Munsell 5Y 7/6, gritty, with small stone inclusions.
Condition: Fragment of the rim and neck.
Description: Rim in the shape of a puffed cushion with rounded upper section and a small ledge on the neck.
Comparisons and comment: *Locri IV*, pl. LX, no. 182 (similar, from Locri, first half of the 5th c. BC, amphora with puffy cushion rim); Corretti and Capelli 2003, 292, no. 10, pl. LIII (from Entella, second half of the 6th–beginning of the 5th c. BC, Western Greek amphora); Sourisseau 2011, fig. 12, no. 3, 183 (from Rifriscolaro necropolis, Camarina, last quarter of the 6th–first decades of the 5th c. BC, Western Greek amphora).
Date: First half of the 5th c. BC.

1:2

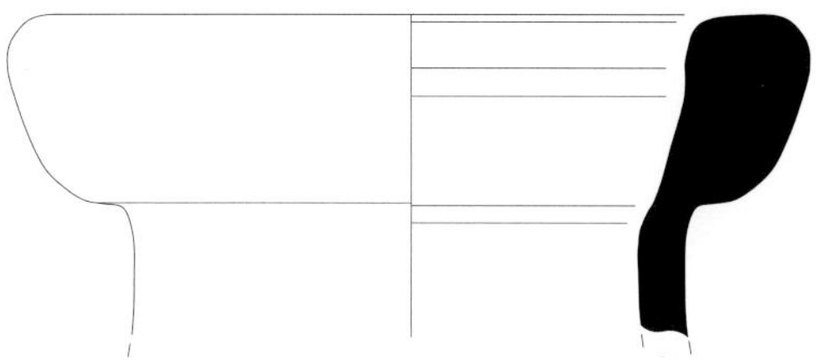

SAV Amp 10 Archaic Type B Amphora, Type A3
Provenance: Square F3, N Kiln interior. Level 2, batt. 1. Lot: SA 79-817P. Assemblage 8.
Dimensions and technical features: h 5.4, foot diam. 7.1. Clay Munsell 5YR 6/6, compact, with small white inclusions and a few black ones.
Condition: Fragment of the toe.
Description: Full toe with flat floor and slight carination on the exterior.
Comparisons and comment: For the form, Semeraro 1997, 91–94, no. 172b, fig. 47 (from Leuca, first half of the 6th c. BC, Ionic amphora); Cavalier 1985, 34–37, fig. 6a–c (from Lipari, second half of the 6th c. BC, Ionic amphora).
Date: Second half of the 6th c. BC.

1:2

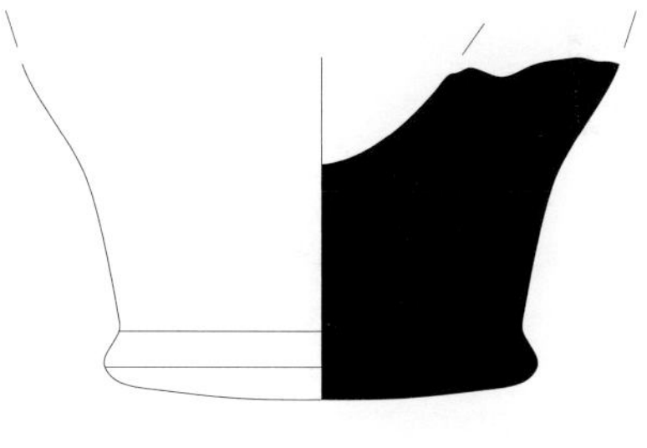

Archaic type B series cont.

SAV Amp 11 Archaic Type B Amphora, Type A3

Provenance: Slope 4, W extension, Level 1, batt. 1 (from SA 79-141PL). Lot: SA 79-983P (see *Fig. 2.35*).

Dimensions and technical features: h 6.2, th 1.9. Clay Munsell 10YR 7/4, gritty and hard, with numerous white and black stone and mica inclusions.

Condition: Fragment of a handle.

Description: Handle with oval section.

Comparisons and comment: Calabrian production. Edlund 1982a, 66; Cavalier 1985, 34–37, fig. 6a–c (from Lipari, end of the 6th–first half of the 5th c. BC); for the clay, Swift in *Survey*, no. 17, 475 (from survey, end of the 6th–beginning of the 5th c. BC, Archaic Type B).

Date: Second half (?) of the 5th c. BC.

1:2

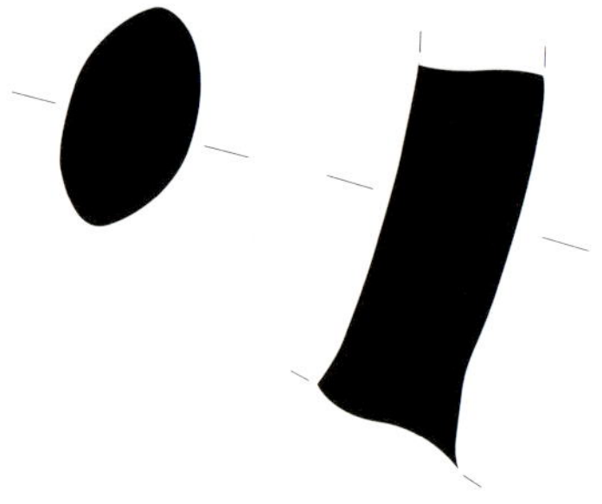

Corinthian Type A Amphora Series

The Corinthian Type A amphora, traditionally interpreted as an oil container and studied by Koehler, is, alongside Type B, the longest-lived series in southern Italy and Sicily, with production beginning in the second half of the 8th c. BC. Eleven examples were recovered at Sant'Angelo Vecchio, the earliest of which—two flat brim rims—date only to the 6th c. BC. The A′ type of the 5th c. BC is attested by vertical handles with circular sections (verging on oval near the upper attachment point) attached immediately beneath the rim and on the rounded shoulder. Two handles that can be securely ascribed to the Corinthian Type A because of their characteristic shape and fabric are the only examples dating to the Hellenistic period.

The clays are typical of Corinthian Type A production. The color varies from pale yellow/roseate to a deep red. There are two fabrics, distinguished by their internal structure: one is harder and more compact, with small reddish and white inclusions; the other is grittier, with frequent large triangular inclusions of variable color.

Corinthian Type A Amphora Series

SAV Amp 12 Corinthian Type A′ Amphora, Type A4.1a

Provenance: Surface Slope 4, W extension. Surface find. Lot: SA 79-175P.

Dimensions and technical features: h 5.7, rim 14.5. Clay Munsell 7.5YR 8/4, with numerous large reddish and grayish rock fragments and few white inclusions (Whitebread Fabric Class 2).

Condition: Fragment of the rim and neck.

Description: Horizontal brim rim with flat lip slightly inclined toward the exterior. It cannot be determined whether the specimen is wheel-formed or hand-made.

Comparisons and comment: Koehler 1979, 104, no. 35, pl. 6; Cavalier 1985, 26–27, fig. 2b (from Lipari, beginning of the 5th c. BC); Di Sandro 1986, pl. 4, no. 54 (from Pithekoussai, beginning of the 5th c. BC); Semeraro 1997, 61, no. 74, fig. 17 (from Cavallino, end of the 6th–first half of the 5th c. BC); *Necropoleis*, 741–42, T 206-7 (from Pantanello necropolis, 475–450 BC), T 266-1 (from Pantanello necropolis, 480–450 BC); Sourisseau 2006, fig. 5, type 4c, 139 (from Camarina, Rifriscolaro necropolis, end of the 6th–first quarter of the 5th c. BC).

Date: Late 6th–first half of the 5th c. BC.

1:2

SAV Amp 13 Corinthian Type A′ Amphora, Type A4.1b

1:2

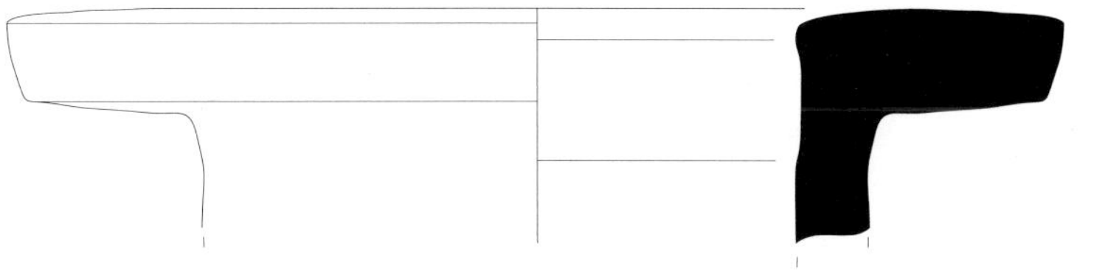

Provenance: Retaining Wall, US 9/18.
Dimensions and technical features: h 4.2, rim 14.5. Clay Munsell 7.5YR 8/4, with numerous large and small stone and mudstone inclusions.
Condition: Fragment of the rim and neck.
Description: Horizontal brim rim with flat lip.
Comparisons and comment: Koehler 1979, 104, no. 35, pl. 6; Cavalier 1985, 26–27, fig. 2b (from Lipari, beginning of the 5th c. BC); Di Sandro 1986, pl. 4, no. 54 (from Pithekoussai, beginning of the 5th c. BC); Semeraro 1997, 250, no. 941a, fig. 209 (from Otranto, end of the 6th–beginning of the 5th c. BC); Sourisseau 2006, fig. 5, type 4a, 139 (Camarina, Rifriscolaro necropolis, mid-6th–beginning of the 5th c. BC).
Date: Late 6th–first half of the 5th c. BC.

SAV Amp 14 Corinthian Type A Amphora, Type A4.2a

1:2

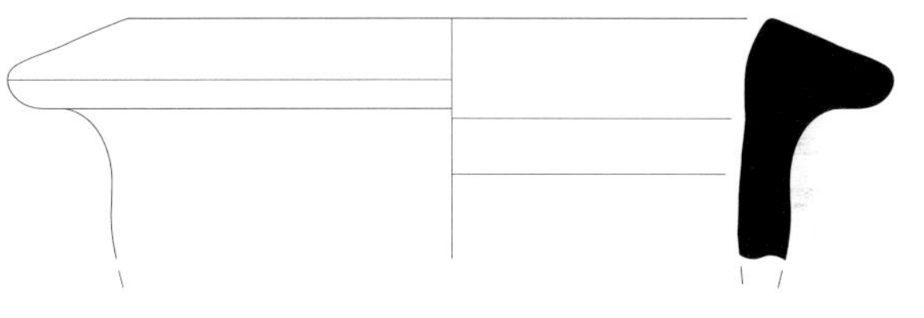

Provenance: Sounding C, Pottery Deposit. Lot: SA 79-988PL. Assemblage 15.
Dimensions and technical features: h 4.0, rim: 11.0. Clay Munsell 5YR 7/6, compact, with frequent red, brown, and black argillaceous rock fragments.
Condition: Fragment of the rim and neck.
Description: Rim with triangular section protruding from the rounded tip and inclined upper surface.
Comparisons and comment: Koehler 1979, no. 43 (second quarter of the 5th c. BC); Swift in *Survey*, 461, no. 19 (from survey, beginning of the 5th c. BC).
Date: First half of the 5th c. BC.

SAV Amp 15 Corinthian Type A Amphora, Type A4.1b

1:2

Provenance: Sounding C, Pottery Deposit. Lot: SA 79-1071P. Assemblage 15.
Dimensions and technical features: h 11.4, th 3.25. Clay Munsell 5YR 7/4, hard, with large brown and red and small white inclusions.
Condition: Fragment of a handle.
Description: Handle with circular section.
Comparisons and comment: Edlund 1982a, 30, fig. III, 9; Koehler 1979, 9–32, 89–129; *Necropoleis*, 745–46, T 324-1 (from Pantanello necropolis, 460–440 BC).
Date: Middle of the 5th c. BC.

Corinthian type A series cont.

SAV Amp 16 Corinthian Type A Amphora, Type A4.1
Provenance: Square F5. Level 1, batt. 1. Lot: SA 79-455PL.
Dimensions and technical features: h 4.5, th 3.5. Clay Munsell 10R 5/6, hard, with large medium-brown, red, and white inclusions.
Condition: Fragment of a handle.
Description: Handle with circular section.
Comparisons and comment: Koehler 1979, 9–32, 89–129 (second half of the 4th c. BC).
Date: Second half of the 4th c. BC.

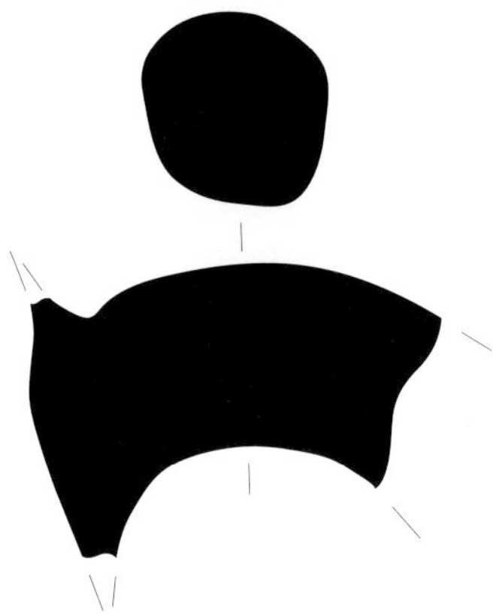

1:2

SAV Amp 17 Corinthian Type A Amphora, Type A4.2
Provenance: Sounding D. Level 2, batt. 6. Lot: SA 80-56PL.
Dimensions and technical features: h 10.0, th 2.6. Clay Munsell 5YR 6/4, hard, with a few white inclusions.
Condition: Fragment of a handle.
Description: Handle with circular section.
Comparisons and comment: Koehler 1979, 9–32, 89–129.
Date: Second half of the 4th c. BC.

1:2

SAV Amp 18 Corinthian Type A Amphora, Type A4.2
Provenance: Square G5. Level 2, batt. 1. Lot: SA 79-636PL. Assemblage 3.
Dimensions and technical features: h 11.26, th 2.5. Clay Munsell 2.5YR 6/4, compact, with black and white inclusions. Slip Munsell 2.5YR 6/8.
Condition: Fragment of a handle.
Description: Handle with circular section.
Comparisons and comment: Koehler 1979, 9–32, 89–129.
Date: Second half of the 4th c. BC.

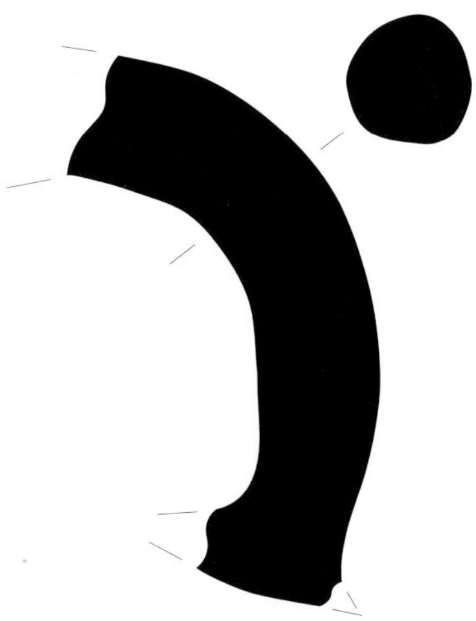

1:2

Corinthian Type B Amphora Series

The Corinthian Type B amphora was likely used for transporting wine, as suggested by the presence of a resinous coating on the interior of numerous examples. It was produced between the middle of the 6th and the second half of the 3rd c. BC mainly on Corcyra, where kilns were discovered in the vicinity of the polis' two ports. The latest research indicates the existence in the central-southern Adriatic of more than one center producing imitations of this vessel, especially in Magna Grecia and Sicily between the second half of the 4th and the beginning of the 3rd c. BC. The majority of the examples found at Sant'Angelo Vecchio come from these production centers.

At Sant'Angelo Vecchio, this type is better attested than Type A, with 12 examples dating from the 5th c. BC onward. The majority of these fragments date to the second half of the 5th c. BC, in particular five echinus rims with the upper surface slightly inclined toward the interior. An inclined rim with triangular section, a feature that would become a trademark of the Greco-Italic group, belongs to the first half of the 3rd c. BC.

The color of the clays in the Corinthian Type B series is generally pale yellow. The fabrics in Type B are more thoroughly purified and porous or soft than that of the Type A series, with small black and white inclusions.

Corinthian Type B Amphora Series

SAV Amp 19 Corinthian Type B Amphora, Type A5.1a

1:2

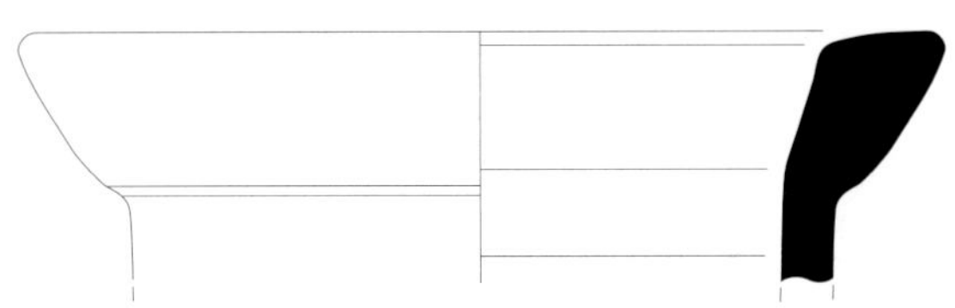

Provenance: Sounding C, Pottery Deposit. Lot: SA 79-920P. Assemblage 15.

Dimensions and technical features: h 3.9, rim 14.0. Clay Munsell 10YR 8/3, compact and hard, with (more) white and (fewer) black inclusions (Whitebread Fabric Class 1).

Condition: Fragment of the rim and neck.

Description: Echinus rim with flat upper section turned slightly toward the interior; on the neck there are two small raised areas.

Comparisons and comment: Corcyrean production (?). Edlund 1982a, 30; Koehler 1979 (from Corinth, 460–420 BC); Semeraro 1997, 63, no. 95, fig. 18 (from Cavallino, mid-5th c. BC); Albanese Procelli 2003, 38, pl. I, no. 4 (from Ramacca, mid-5th c. BC).

Date: Second half of the 5th c. BC.

SAV Amp 20 Corinthian Type B Amphora, Type A5.2a

1:2

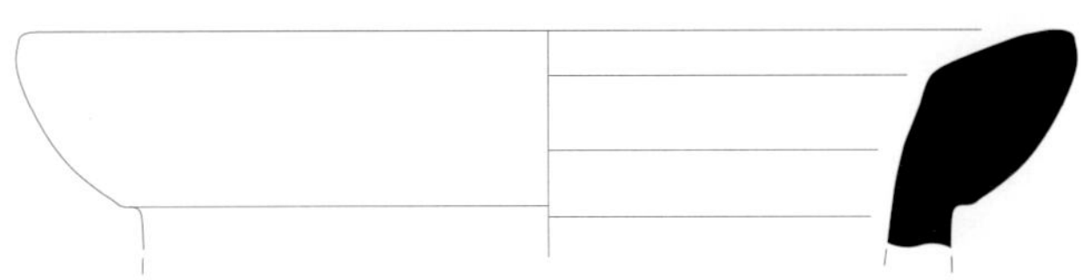

Provenance: Sounding C, Pottery Deposit. Lot: SA 79-1069P. Assemblage 15.

Dimensions and technical features: h 3.35, rim 15.5. Clay Munsell 10YR 8/4, hard, with dark and occasional white inclusions.

Condition: Fragment of the rim and neck.

Description: Echinus rim with upper section turned very sharply toward the interior.

Comparisons and comment: Calabrian production (?). Edlund 1982a, 30, fig. III, 8; Vandermersch 1989, 99, no. 496, fig. 65 (from Caulonia, 450–400 BC); *Necropoleis*, 738, CD 37 (from Pantanello necropolis, mid-5th c. BC), 744, T 256-1 (from Pantanello necropolis, 460–440 BC).

Date: Middle of the 5th c. BC.

Corinthian type B series cont.

SAV Amp 21 Corinthian Type B Amphora, Type A5.1b
Provenance: Sounding B (from SA 79-362PL). Level 3, batt. 6. Lot: SA 79-982P (see *Fig. 2.35*).
Dimensions and technical features: h 3.64, rim 10.0. Clay Munsell 10YR 8/3, hard, with black inclusions.
Condition: Fragment of the rim and neck.
Description: Echinus rim with upper section turned toward the interior; there is a groove at the juncture between rim and neck.

1:2

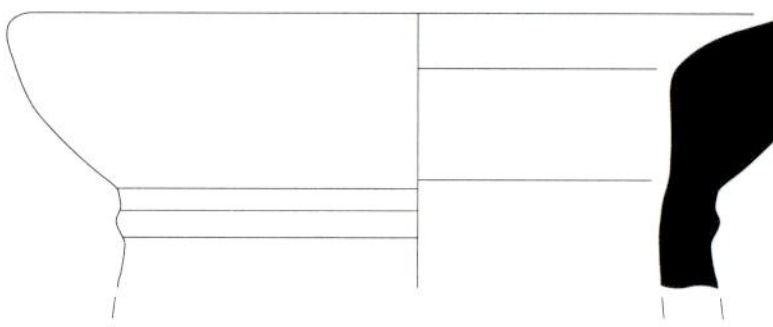

Comparisons and comment: Edlund 1982a, 49, fig. V, 18; Cavalier 1985, no. 63, cat. no. 55 (from Lipari); Desy and De Paepe 1990, 191–92, pl. 85, no. 18 (from Torre San Giovanni di Ugento, end of the 4th–first half of the 3rd c. BC).
Date: End of the 4th c. BC.

SAV Amp 22 Corinthian Type B Amphora, Type A5.2b
Provenance: Square F7. Level 3, batt. 3. Lot: SA 80-24PL (see *Fig. 2.20*).
Dimensions and technical features: h 4.37, rim 14.0. Clay Munsell 7.5YR8/4, compact and powdery on the surface, micaceous, with feldspars and occasional vughs.
Condition: Fragment of the rim and neck.
Description: Echinus rim with flat upper section turned slightly toward the center; double ribbing on the neck.
Comparisons and comment: South Italian? Desy and De Paepe 1990, 191–92, pl. 85, no. 17 (from Torre San Giovanni di Ugento, end of the 4th–first half of the 3rd c. BC).
Date: End of the 4th c. BC.

1:2

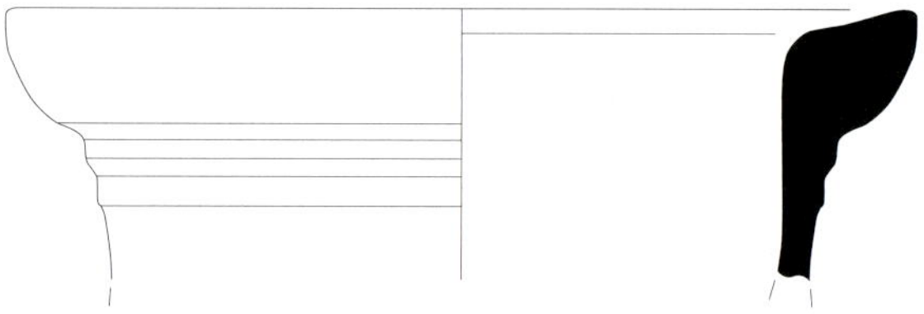

SAV Amp 23 Corinthian Type B Amphora, Type A5.2c
Provenance: Slope 3, W extension, N slope. Level 2, batt. 1. Lot: SA 79-387P. Assemblage 13.
Dimensions and technical features: h 4.0, rim 11.2. Clay Munsell 10YR 8/3, hard, with few black inclusions and occasional vughs.
Condition: Fragment of the rim and neck.
Description: Echinus rim with flat upper section becoming slightly concave at the center.
Comparisons and comment: Desy and De Paepe 1990, pl. 85, no. 9 (from Torre San Giovanni di Ugento, end of the 4th–first half of the 3rd c. BC).
Date: End of the 4th c. BC.

1:2

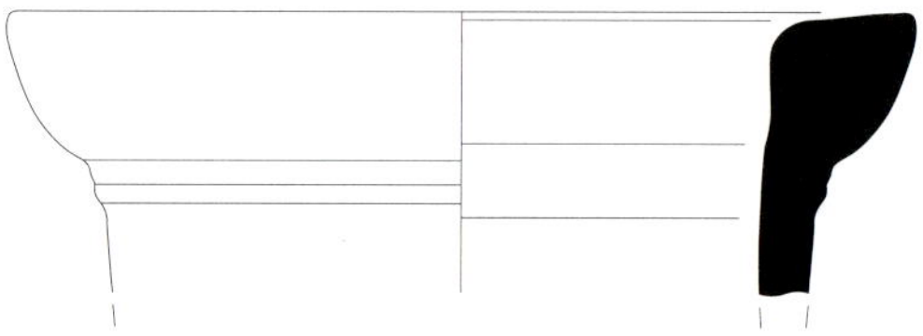

SAV Amp 24 Corinthian Type B Amphora, Type A5

Provenance: Square G7. Level 2, batt. 5. Lot: SA 79-999P. Assemblage 4.

Dimensions and technical features: h 17.5, th 2.3. Clay Munsell 10YR 6/6, hard and compact, with rare white inclusions.

Condition: Fragment of a handle.

Description: Ribbon handle with oval section.

Comparisons and comment: Locri IV, pl. LXVIII, type G (from Locri, second half of the 4th c. BC); Koehler 1979, no. 277 (from Athens, first quarter of the 3rd c. BC).

Date: End of the first quarter of the 3rd c. BC.

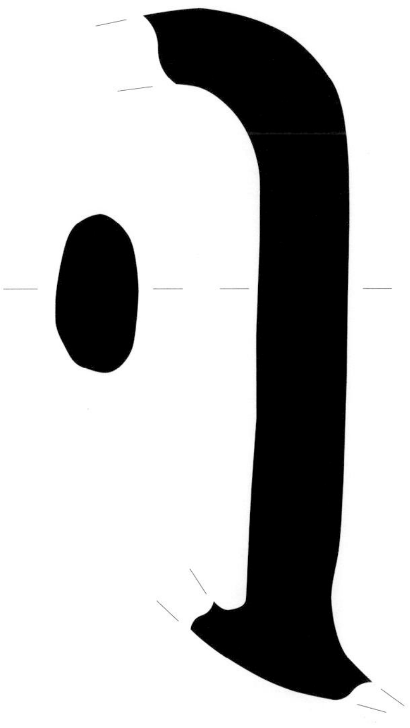

1:2

SAV Amp 25 Corinthian Type B Amphora, Type A5.3a

Provenance: Square F7. Level 3, batt. 5. Lot: SA 80-27PL.

Dimensions and technical features: h 9.0, neck diam. 10.0, th 2.0. Clay Munsell 7.5YR 8/4, hard and compact, with rare black and white inclusions.

Condition: Fragment of neck and handle with oval section.

Description: Handle of an amphora with saw tooth rim.

Comparisons and comment: Cavalier 1985, fig. 23, no. 125b (from Filicudi, Capo Graziano).

Date: Middle of the 3rd c. BC.

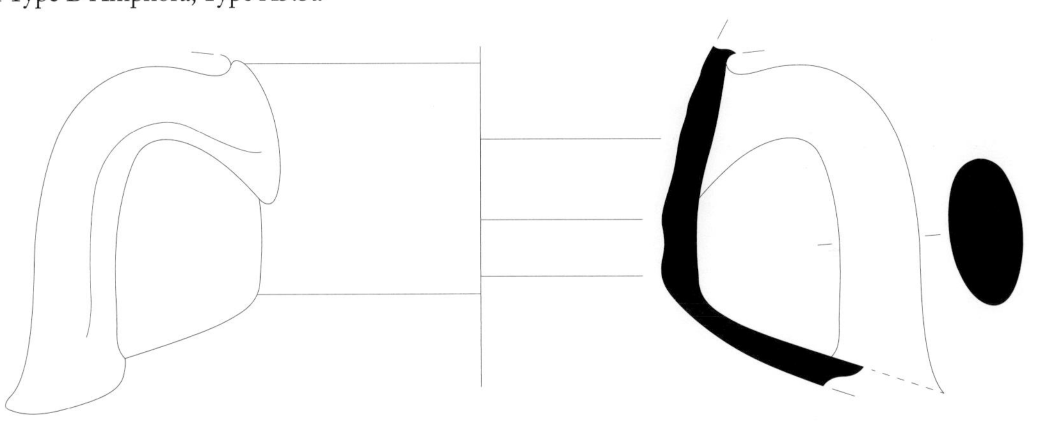

1:2

Corinthian type B series cont.

SAV Amp 26 Corinthian Type B Amphora, Type A5

Provenance: Slope 3. Level 1, batt. 2. Lot: SA 79-980P (see *Fig. 2.35*).

Dimensions and technical features: h 4.3, th 2.1. Clay Munsell 10YR 8/3, hard and compact, with rare black inclusions.

Condition: Fragment of a handle.

Description: Ribbon handle with pinched oval section.

Comparisons and comment: Edlund 1982a, 52; *Locri IV*, 236, pl. LXV, no. 228 (from Locri, beginning of the 3rd c. BC).

Date: Early 3rd c. BC.

1:2

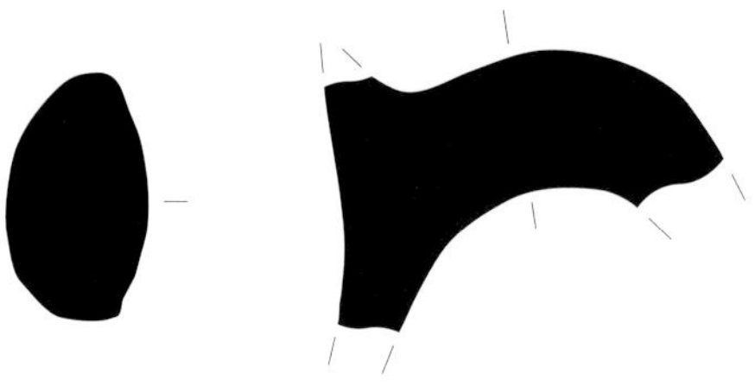

SAV Amp 27 Corinthian Type B Amphora, Type A5

Provenance: Square H3. Level 1, batt. 1. Lot: SA 79-677PL (see *Fig. 2.22*).

Dimensions and technical features: h 4.2, foot diam. 3.2. Clay Munsell 2.5YR 5/6, hard, with black and white inclusions.

Condition: Fragment of the toe.

Description: Complete toe.

Comparisons and comment: *Locri IV* (from Locri, end of the 4th–beginning of the 3rd c. BC).

Date: Second half of the 4th c. BC.

1:2

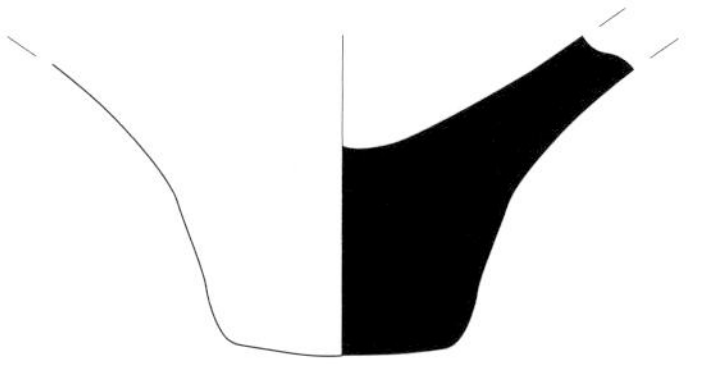

Greco-Italic Amphora Series

Greco-Italic amphorae, considered principally wine containers, circulated widely in the western Mediterranean. The most recent research suggests a highly fragmentary production of the vessels beginning at the end of the 4th c. BC in various local centers. These local series can occasionally be distinguished typologically, but they cannot be grouped under a single label. Production continued even during the Romanization of southern Italy, beginning in the middle of the 3rd c. BC and continuing until the introduction of the late Greco-Italic types in the last quarter of the 3rd and early 2nd c. BC. These latter types shared morphological features with the preceding series, but they are considered to represent the first typically Italic and Romanized vessels, forerunners of the Dressel 1 and Lamboglia 2.

To the same category belong the amphorae Vandermersch called "Italiote," widely held to be the link between the western Greek amphorae of the Classical period and the true Greco-Italic amphorae of the Hellenistic period. Lyding Will grouped them under her "type a," while Vandermersch placed them in his MGS III and MGS IV groups. There is a morphological evolution from an echinus rim or a rim slightly inclined toward the exterior, to a gradually thinner and more vertically oriented rim on the late Greco-Italic amphorae (Lyding Will's types a1, c, d, and e; Vandermersch's groups MGS V and MGS VI). Ten of the examples from Sant'Angelo Vecchio belong to the Greco-Italic group with downward-turned triangular rim. This is the most commonly attested type (94 total fragments).

The fabrics used in these vessels exhibit a certain variety in their technical features and color, probably due to the diversity of the production zones—from beige-hazelnut and very compact, with very small black inclusions, to orange (verging on beige on the exterior) and rough and compact, with small white inclusions. A large group of body fragments is noteworthy for their reddish clay with small white inclusions, which can be attributed to a local Metapontine production.

Greco-Italic Amphora Series

SAV Amp 28 Greco-Italic Amphora, Type A6.1a
Provenance: Square F2, N Kiln interior. Level 2. Lot: SA 79-809P. Inv. no. 319962. Assemblage 8.
Dimensions and technical features: h 4.2, rim: 13.7. Clay Munsell 5YR 7/6, rough, with few stone and frequent black inclusions. Exterior cream-colored slip Munsell 7.5YR 8/4.
Condition: Fragment of the rim and neck.
Description: Rim with triangular section and oblique lip.
Comparisons and comment: Olcese 2012, 543, 548, no. 5 (from the Punta Lazzaretto B wreck, second half of the 2nd c. BC); Swift in *Survey*, no. 42, 484 (Vandermersch type VI, Metaponto field survey, 260–210 BC).
Date: Late 3rd–first half of the 2nd c. BC.

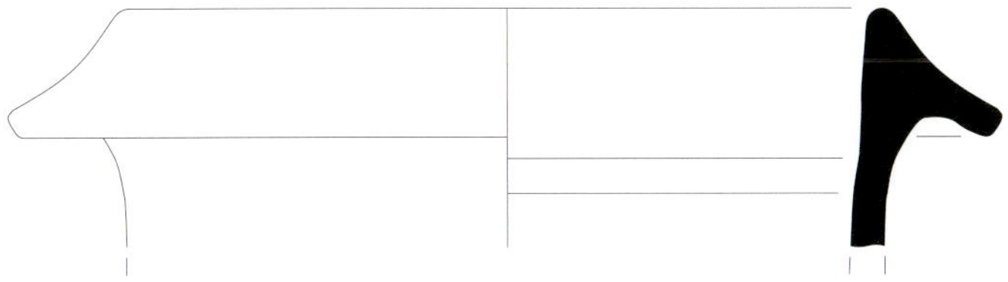
1:2

SAV Amp 29 Greco-Italic Amphora, Type A6.2a
Provenance: Square E2. Lot: SA 80-63PL. Assemblage 7.

Dimensions and technical features: h 4.45, rim: 14.7. Clay Munsell 5YR 7/6, compact, with rare black inclusions.
Condition: Fragment of the rim and neck.
Description: Rim with vertical triangular section and downward-turned exterior surface.
Comparisons and comment: Desy and De Paepe 1990, 219, no. 167 (from Torre San Giovanni di Ugento, 2nd c. BC); Lattara 6, no. A-GR-ITAbd5 (200–150 BC); Olcese 2012, 572, 580, no. 1 (from the Filicudi A wreck, first half of the 2nd c. BC); Swift in *Survey*, 463, no. P281-347 (from Pantanello, end of the 2nd c. BC).
Date: 2nd c. BC.

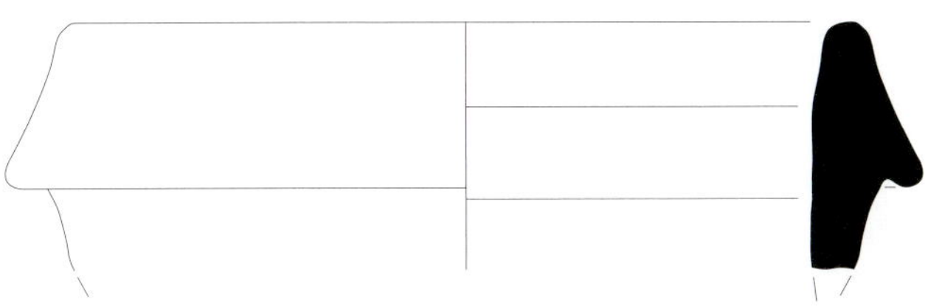
1:2

SAV Amp 30 Greco-Italic Amphora, Type A6 (Metapontine?)
Provenance: Square G5. Level 2, batt. 1. Lot: SA 79-777PL. Assemblage 3.
Dimensions and technical features: h 9.5, th 2.0. Clay Munsell 2.5YR 5/6, gritty, with white and (fewer) black inclusions.
Condition: Fragment of a handle.
Description: Handle with oval section and double ribbing; there is a fingerprint at the shoulder attachment.
Comparisons and comment: Olcese 2010, 320, n IA.34 (Vandermersch type V/VI, from Pithekoussai, end of the 4th–first half of the 3rd c. BC).
Date: First half of the 3rd c. BC.

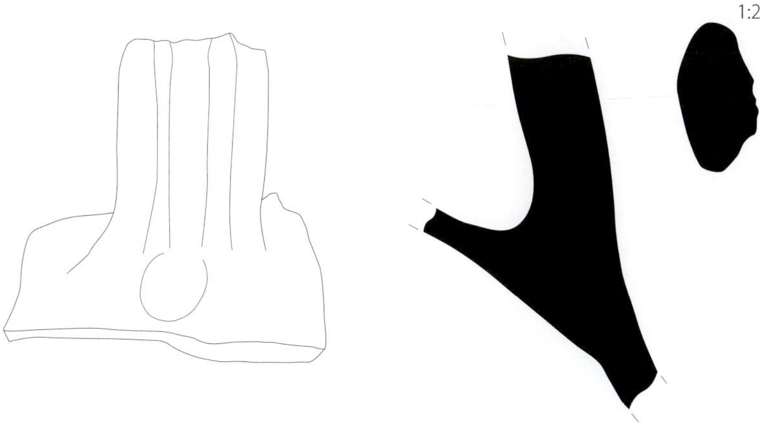
1:2

Greco-Italic series cont.

SAV Amp 31 Greco-Italic Amphora, Type A6

Provenance: Slope 4, surface cleaning. Lot: SA 79-421PL. Assemblage 8.

Dimensions and technical features: h 5.4, th 2.3. Clay Munsell 2.5YR 4/3, gritty, with numerous black and white inclusions.

Condition: Fragment of a handle.

Description: Handle with oval section; there is a fingerprint at the shoulder attachment.

Comparisons and comment: Olcese 2010, 320, no. IA.15 (Vandermersch type IV, from Pithekoussai, end of the 4th–beginning of the 3rd c. BC).

Date: Beginning of the 3rd c. BC.

1:2

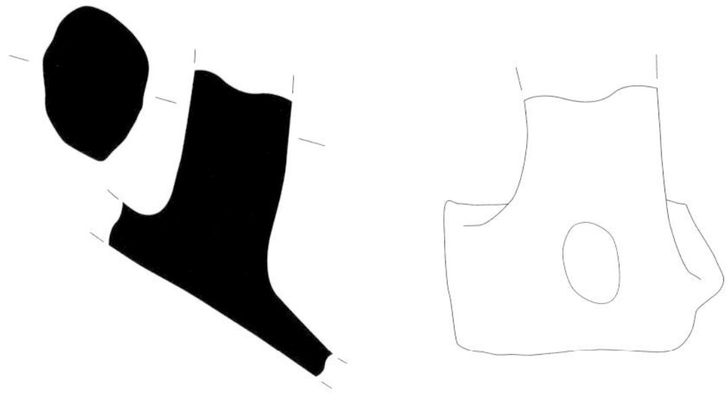

SAV Amp 32 Greco-Italic Amphora, Type A6

Provenance: Sounding A, S extension. Level 1, batt. 5. Lot: SA 79-22PL (see *Fig. 2.30*).

Dimensions and technical features: h 4.0. Clay Munsell 2.5YR 5/4, gritty, with numerous black and white inclusions. Light-colored slip Munsell 5YR 8/3.

Condition: Fragment of the shoulder.

Description: Downward-sloping shoulder.

Comparisons and comment: Metapontine production. Olcese 2010, 62–63, pls. 13–14 (Vandermersch type V, from Pithekoussai, end of the 4th–first half of the 3rd c. BC).

Date: First half of the 3rd c. BC.

1:2

SAV Amp 33 Greco-Italic Amphora, Type A6

Provenance: Square F3, N Kiln interior. Level 2, batt. 2. Lot: SA 79-875PL. Assemblage 8.

Dimensions and technical features: h 10.3, th 2.4. Clay Munsell 7.5YR 7/6, hard, with occasional black and white inclusions. Slightly lighter-colored slip Munsell 7.5YR 6/6.

Condition: Fragment of the shoulder.

Description: Shoulder fragment preserving the handle attachment, beneath which there is a fingerprint.

Comparisons and comment: Olcese 2010, 62–63, pls. 13–14 (Vandermersch type V, from Pithekoussai, end of the 4th–first half of the 3rd c. BC).

Date: First half of the 3rd c. BC.

1:2

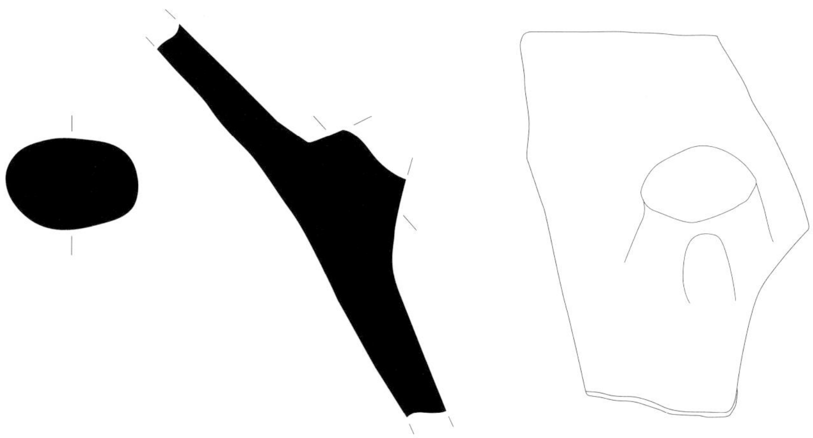

Metapontine Amphora Series

This type of amphora appears to have developed from the Greco-Italic type. Typical features include a collar rim of limited height, relatively short neck and handles, and relatively small capacity. Production in the region begins in the 2nd c. BC and is attested in the destruction levels of the stoa of Metaponto, datable to the beginning of the 1st c. BC (Giardino 2005, 423). Besides Metaponto, the type is frequently encountered in the 2nd- and 1st-c. BC levels of Herakleia, as well as in all of the contemporary agricultural installations so far identified in the territory between the two centers (De Siena and Giardino 1994, 201). Production of the type is attested at Pantanello after the middle of the 2nd c. BC (Carter 1983a, 40), justifying its identification as "Metapontine." The product(s) it transported are still unknown, and its area of circulation appears not to exceed the limits indicated above, although several examples come from Salento (Desy and De Paepe 1990, 217–18) and the area of Syracuse (Pelagatti 1970, 493–94). The single example found at Sant'Angelo Vecchio has the typical reddish-brown clay coated with a lighter-colored slip.

SAV Amp 34 Metapontine Amphora, Type A7.1a

Provenance: Sounding D. Level 1, batt. 3. Lot: SA 80-45PL.

Dimensions and technical features: h 2.5, rim 13.4. Clay Munsell 5YR 6/6, with red and small white inclusions, as well as rare black inclusions. Some vacuoles.

Condition: Rim fragment.

Description: Collar rim with two swellings in the upper and lower sections.

Comparisons and comment: Desy and De Paepe 1990, 218, no. 160, fig. 93 (from Ugento, 2nd–1st c. BC); De Siena and Giardino 1994, 200, fig. 3 (from Herakleia, 2nd–1st c. BC); Swift in *Survey*, 463, no. PZ81-871 (from Pantanello, mid-2nd c. BC).

Date: 2nd c. BC.

1:2

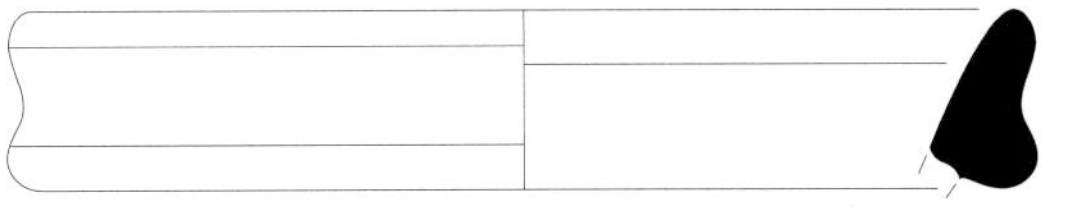

28
Opus Doliare

Annalisa Concilio

Cat. No.	Provenance	Description	Date	NFr	MNI
SAV OD 01	SA 79-879P Square F3, N Kiln interior, Level 2, batt. 2	rim	4th–3rd c. BC	1	1
SAV OD 02	US 9, 18 Slope 4, cleaning of the Retaining Wall	rim body		2	1
SAV OD 03	SA 79-829P Square F3, N Kiln exterior, Level 2, batt. 1	base	4th–3rd c. BC	1	
SAV OD 04	SA 79-956T + SA 79-997PL Square G7, E of wall, Level 2, batt. 5	lid	4th c. BC	2	
SAV OD 05	SA 79-420PL Sounding B, Level 3, batt. 10	body		2	
Uncataloged	US 42+9, 18+35 Slope 4	body		5	
Uncataloged	SA 79-3PL Surface finds from newly cut road	body		1	
Uncataloged	US 35 Slope 4	body		1	
Uncataloged	SA 79-42PL Sounding A, Level 3, batt. 4	body		1	
Uncataloged	Square F7, Level 3, batt. 8	body		1	

Table 28.1 Quantification of opus doliare fragments from Sant'Angelo Vecchio.

Large containers typically functioned as long-term storage vessels for solid or liquid foodstuffs. Among the most common of these were pithoi (opus doliare), which featured a rather thick body wall, wide mouth, mostly flat base, and spherical or ovoid body with a massive lid that guaranteed maximum protection for the contents. Compared to other major ceramic classes, pithoi required a longer, more complex, and more difficult production process that also resulted in higher costs. Consequently, pithoi were afforded special attention and care in order to extend their use-life as long as possible. A solid, inclusion-rich fabric and the common practice of burying the vessels partially or entirely both served to prevent damage to the large, costly vessels. If damage did occur, the vessels were repaired if possible, evidenced by holes bored through the body wall for the insertion of metal clamps, as in SAV OD 05, and by the clamps themselves, usually made of lead (such as those found at Sant'Angelo Vecchio, SAV M 08 and SAV M 09).

Pithoi are recovered mainly in domestic contexts, many of which were equipped with dedicated storerooms or clusters of rooms designed to accommodate the vessels.[1] They are also documented at sacred areas,[2] necropoleis,[3] and production facilities.

The pithoi from Sant'Angelo Vecchio (17 fragments from at least two individuals,[4] *Table 28.1*), all highly fragmentary with the exception of the lid SAV OD 04, were partly recovered in association with material belonging to the phase of the terracotta production facility attested between the end of the 4th and the 3rd century BC. The rim SAV OD 01 and the base SAV OD 03, both found in the North Kiln (Kiln 5) of the Kiln Complex (square F3), also date to the same period (see Ch. 2, "Site Phasing, Stratigraphy, and Site Assemblage"). Both fragments also have a hard and thick layer of concretion over the entire surface, suggesting that they may have been reused as building material.

1 For the attestations at the Lucanian sites of Serra di Vaglio, Oppido Lucano, Cancellara, Tolve-Valle di Chirico, Baragiano, Tolve-Moltone, and Pomarico, see Russo Tagliente 1992, 158–92. On Pomarico, see also Maturo 1997, 250; on Serra di Vaglio, see also G. Greco 1991, 60ff. The prevalence of the domestic sphere was corroborated by the Bradano-Basento field survey: the majority of the sites with pithos fragments were identified as farmhouses (Vittoria in *Survey*, 349–50).

2 Pithoi are attested at many sacred sites in Lucania for ritual functions or for the storage of offerings: Russo Tagliente 1995, 93–94 (Armento); Barra Bagnasco 2001, 227, La Rocca 2000, 14 (Chiaromonte); on the sanctuary at Ascoli Satriano see Osanna 2003, 357–66.

3 On the use of pithoi as urns, see Russo Tagliente 1992, 200; Lo Porto 1973, 219. The recovery of pithoi in necropoleis, however, is not necessarily an indicator of cremation or *enchytrismos* burials; it may also be correlated with the use of pithoi as covers or markers, or connected to funerary rites.

4 The count of individual vessels was calculated from the rims. The subdivision of the body fragments as proposed in *Table 28.1* is based on the fabrics.

Catalog of Opus Doliare

SAV OD 01 Rim

Provenance: Square F3, N Kiln interior. Level 2, batt. 2. Lot: SA 79-879P. Assemblage 8.

Dimensions and technical features: rim 50.0, h 5.2. Clay very hard and compact, with small and medium-sized red and black stone inclusions, diffuse calcite fragments, and rare mica.

Condition: One fragment of the rim.

Description: Rim with rectangular section and flat upper surface.

Comparisons and comment: Different interior rim profile but similar to Ricciardini 2002, 178, pl. 40, no. 39 (Ascoli Satriano, end of the 4th–early 3rd c. BC), similar to Peluso 1992, 261, pl. LXXIX, no. 249 (Locri, first half of the 3rd c. BC).

Date: 4th–3rd c. BC.

1:3

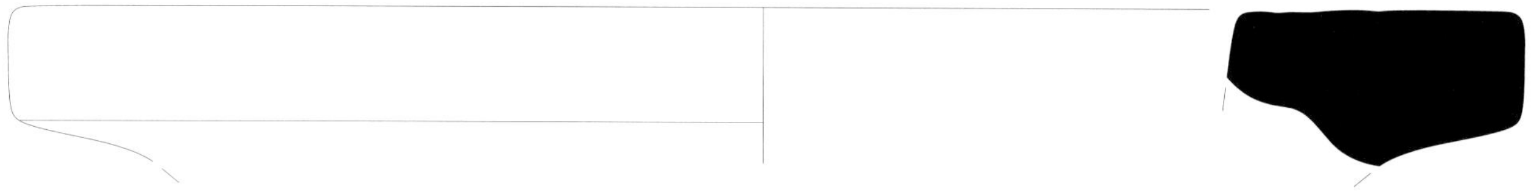

SAV OD 02 Rim

Provenance: Slope 4, Retaining Wall; US 9, 18.

Dimensions and technical features: rim 36.0, h 7.8. Clay Munsell 5YR 6/6 (reddish yellow) on the exterior surface, Munsell Gley 1 3/N (very dark gray) at the core, with abundant large and medium-sized grog fragments, frequent small and medium-sized opaque white inclusions, and pointed silver mica.

Condition: Two fragments of the rim and body.

Description: Oblique lip with slightly convex upper surface, squared rim, and two shallow parallel grooves following the rim; short truncated conical neck.

Comparisons and comment: Cotton 1977, 372, fig. 142, no. 428 (Cozzo Presepe).

1:3

SAV OD 03 Base

Provenance: Square F3, N Kiln exterior. Level 2, batt. 1. Lot: SA 79-829P (see *Fig. 2.30*).

Dimensions and technical features: base 24.0, h 9.8. Clay Munsell 5YR 6/6 (reddish yellow), very hard and compact, with elongated vacuoles, numerous small and medium-sized black and red-brown clasts, abundant small and medium-sized quartzite, and frequent small silver and gold mica. The interior surface bears large and often heavily encrusted red-brown stone inclusions.

Condition: One fragment of the base.

Description: Massive disc base with flat resting surface and modeled exterior profile. The interior surface is heavily encrusted.

Comparisons and comment: Russo Tagliente 1995, 94, fig. 94, no. 250 (Armento, 4th–3rd c. BC); Vittoria in *Survey*, type Pi8b1, nos. 244–245.

Date: 4th–3rd c. BC.

1:3

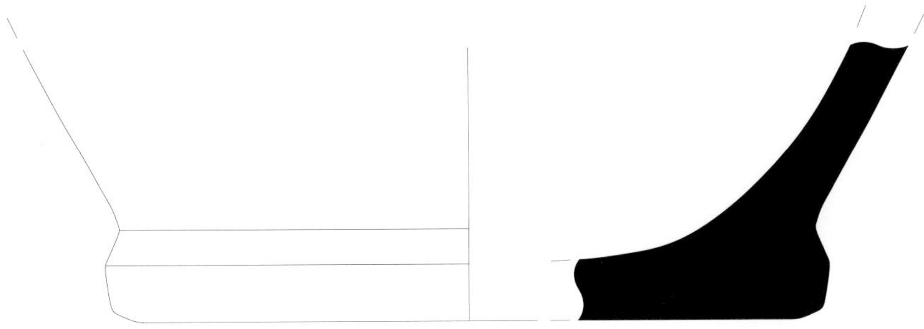

SAV OD 04 Lid

Provenance: Square G7, E of wall. Level 2, batt. 5. Lot: SA 79-956T + SA 79-997PL. Assemblage 4.

Dimensions and technical features: rim 46.4, h 8.0. Clay Munsell 5YR 6/6 (reddish yellow), hard and compact, with numerous small red and brown clasts, frequent small quartzite, and small and medium-sized calcite.

Condition: Two fragments of the knob and rim.

Description: Flat disc continuing straight to the rim; truncated conical knob with rounded top featuring a shallow central depression.

Comparisons and comment: Similar to De Miro 2000, fig. 116, no. 1292 (Agrigento, 5th–4th c. BC).

Date: 4th c. BC.

1:3

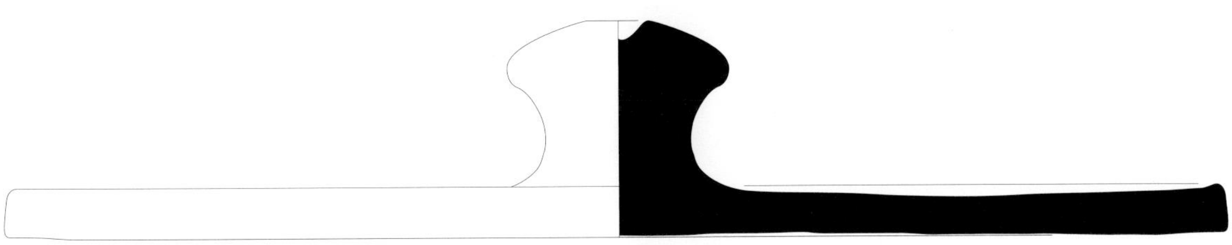

Opus doliare cont.

SAV OD 05 Body

1:2

Provenance: Sounding B. Level 3, batt. 10. Lot: SA 79-420PL (see *Fig. 2.35*).

Dimensions and technical features: th 2.2. Clay Munsell 5YR 6/6 (reddish yellow), hard and porous, with very numerous calcite and quartzite inclusions; very frequent small, medium-sized, and large fresh biotite flakes, more abundant than is usual in Metapontine fabrics; and numerous small and medium-sized shiny black inclusions. The fabric, distinct from the others found at Sant'Angelo Vecchio in terms of the composition and dimensions of the inclusions, may indicate that the fragment was imported from Calabria.

Condition: Fragment of the body with repair hole.

Description: Body wall of uniform thickness and unclear location within the vessel due to its small size, but probably originally belonging to the upper part; a hole was bored through the wall for a repair.

Date: 4th or 3rd c. BC.

29

The Architectural Terracottas: Some Considerations on Production

Anna Lucia Tempesta

Among the many fragments of architectural terracottas that came to light during the excavation of the complex at Sant'Angelo Vecchio,[1] there are a few fragments of the antefixes that capped the terminal cover tiles on the long sides of the roof as well as acroteria from the ridge of the roof. All were very fragmentary and presented clear traces of overexposure to heat, indicating that they would not have actually been used for roofing in antiquity. The excavation also produced five fragments of terracotta antefix molds, three of which are Archaic types—an antefix with palmette and one with a Silenos head.[2] The remaining two, the Artemis Bendis type and an Aphrodite or (more likely) a young satyr (SAV AT 09 and SAV AT 06), date to the 4th century BC.

Two antefix fragments bearing the head of Dionysos (SAV AT 01 and SAV AT 02) came from the collapse of cover tiles located to the west of the north-south wall in square G6, in Levels 2 and 3, just above the virgin soil (Tile Fall 2). A mold fragment (SAV AT 06), chronologically consistent with the antefixes, was found in the same collapse, just slightly to the northeast (square G7). A fragment of a Late Archaic mold (SAV AT 07) was recovered from the small bank of pan and cover tiles in square H3, used to delimit the clay deposit in the southernmost part of the excavation.

The molds SAV AT 08 and SAV AT 09, found in the area of the Kiln Complex, were much less fragmentary. The Late Archaic mold with a Silenos head was among the material from layer 2 in Sounding A, defined by a substantial collapse of pan and cover tiles covering the entire area and delimited to the east by the perimeter wall of the production complex. The mold depicting a female head with lion-skin covering, generally attributable to the 4th century BC, was among the materials in the collapse of the South Kiln (Kiln 4).

With the exception of several Etruscan contexts,[3] no concentrations of similar antefix molds are currently known in any part of the ancient world, even though they cover a fairly wide chronological span between the end of the Archaic period and the Early Hellenistic period. For this reason, the finds from Sant'Angelo Vecchio, which are very closely associated with an area where terracottas were produced, are of great importance.

The Production of Molds

The terracotta molds represent a particular form of artistic creation and pose specific questions pertinent to many specialists.[4] In fact, most forms of ancient production involved the use of molds—from statuettes to reliefs, from architectural terracottas to lamps, from relief-vases to cooking stands. Transcending differences in materials, geography, and chronology, mold-made manufacturing affected the processes of artistic creativity, production in series, and the circulation of the resulting artifacts.[5]

The widespread adoption of molds for producing large numbers of artifacts in series, such as antefixes, greatly affected the products' quality.[6] In fact, it is not difficult to identify differing levels of quality even within the same type series. Beyond the variable behavior of clay during the drying and firing phases, these differences were also due to the state of preservation of the

[1] See Ch. 32, "Roof Tiles and Bricks."

[2] SAV AT 05 and SAV AT 08. The specific type of fragment SAV AT 07 could not be identified, though its semicircular profile belies its Archaic date. For more on molds, see Ch. 30, "The Molds from Sant'Angelo Vecchio."

[3] Examples include the villa with atelier at Selvasecca near Blera, the artisans' quarter of Vigna Parrocchiale by the Sanctuary of Hera at Caere, and the great temple at Falerii (for a corpus of the published molds from these and other excavations, see Billot 1997, 76ff).

[4] This was confirmed by the success of the international conference held at Lille in December 1995 (see *Moulage*). The terminology adopted here is taken from this volume (Muller 1997).

[5] For techniques and production methods, see Cuomo di Caprio 1985; Bonghi Jovino 1990; Mannoni and Giannichedda 1996; for the Metapontine material in particular, see Cracolici 2003.

[6] The relative scarcity of antefix molds has led many scholars to believe that they were made of perishable material such as gypsum. However, use of gypsum is mainly documented from the 2nd c. BC onward. Furthermore, the products from gypsum molds would be recognizable by the presence of bubbles on the surface (for this phenomenon, see Muller 1996, 27–61). For the advantages of the use of clay or gypsum in the making of molds, see Karivieri 1996, 21–30.

Figure 29.1 Antefix with head of Dionysos from Metaponto. Metaponto, National Archaeological Museum inv. no. 317569; scale: 1:2. (Aut. MBAC-SBA-BAS UPProt n. 0003131 31.05.2012, Cl. 28.13.10/1.12)

molds—examples with duller edges and blurred details are almost always attributable to worn molds. This phenomenon, widely attested at Metaponto, is due to the ancient craftsman's habit of exploiting the molds until they were completely worn out and then correcting any defects by retouching or painting. The wide use of sticks in retouching Metapontine antefixes reflects the craftsman's attempts to obtain a higher-quality product with greater detail.[7]

As for many other terracotta products, there are also numerous examples, or copies, obtained from parallel molds and derivative molds taken from casts,[8] as well as from secondary prototypes. The decoration on the plaque is of a lower relief than that on the original or first-generation examples. This phenomenon, called "contraction" or "shrinkage," is due to the clay's loss of weight and volume first during drying and then during firing.[9] The shrinkage varies greatly, from a minimum of 6 percent to a maximum of 12 percent of the overall volume. In fact, many elements affect this process, including the nature of the clay, the quantity of water absorbed during preparation, the quality and the quantity of the inclusions,[10] the thickness of the walls, the firing, and even the position in which the piece is placed during drying.[11]

The "reduction" (or shrinkage)[12] that affected every object made with a derived mold must also be taken into account. The reduction is clearly subject to the same variable parameters of contraction, and therefore in this case, too, it is not always easy to establish a fixed numerical coefficient on which to base the identification of the subsequent generations. It may oscillate between 6–12 percent and 20 percent.[13]

In order to identify the technical and manufacturing characteristics distinguishing a particular production in a specific locality—and thus, ultimately, a workshop's production methods—the availability of a wide range of terracotta samples representing the same subject is essential, so as to create a broad view of the variants and the different solutions adopted in the production process. Study of the Metapontine antefixes indicates that generally a 10–12 percent reduction is sufficient to identify a successive generation.[14] In many cases, however, despite the immediate recognition of different generations, the state of preservation limits the possibility of identifying dimensions and measurements at the same points. The limited number of examples within the same type, as well as the presence of small differences and details, intentional

[7] Compare the Metapontine terracottas with those from Capua, where the mold is generally retouched, rather than the products (Bonghi Jovino 1965, 16–19).

[8] These are, respectively, molds of the same generation reproducing the same type but distinguished by several accidental details, and molds taken from a cast that was not the prototype but rather a replica or a secondary prototype. The difference is substantial, in that the former process produced variants which, according to Muller (1997, 454), differ unintentionally from the originals, while the latter produced versions of a type.

[9] During the drying phase, the clay loses the water absorbed during the cycles of purification, washing, preparation, and manipulation through evaporation. During firing, the water present within the clay's molecular structure is eliminated. The latter is a chemical process and therefore irreversible.

[10] Degreasing agents, or non-plastic materials, served to avoid undesired effects, in particular severe shrinkage, difficulties in the drying phase, and the need for very high firing temperatures. These agents were mineral components from volcanic (*pozzolana*, mica) and metamorphic (quartz and silica) rocks. Grog, on the contrary, is baked clay, ground to varying grades of fineness depending on its use, which increases the porosity of the clay fabric and at the same time reduces the contraction caused by drying and firing.

[11] See Jastrow 1941, 3; Muller 1996, 42.

[12] The reduction is the difference in size and volume between two successive generations of objects, and therefore the sum of the two shrinkages (the shrinkage of the clay used for the mold plus the shrinkage of the clay used for the replica).

[13] With the use of plaster for derivative molds, for example, one of the stages of shrinkage can be avoided and, therefore, the overall reduction of the examples of the next generation.

[14] The same as that established for the protomes at Locri: see Barra Bagnasco 1986, 22.

and otherwise, creates further obstacles. In the case of antefixes decorated with heads, the measurable elements are obviously few (distance between the eyes, distance between the forehead and chin, etc.), and a certain imprecision may be caused by surface wear, making the identification of specific workshop characteristics especially difficult.

Figure 29.2 Palmette antefix from San Biagio. Metaponto, National Archaeological Museum inv. no. 19721; scale: 1:2. (Aut. MBAC-SBA-BAS UPProt n. 0003131 31.05.2012, Cl. 28.13.10/1.12)

Local Production and Workshops at Sant'Angelo Vecchio

While the central role of northwest Greece in the development of the antefix decorated with human faces is well known,[15] in Magna Grecia the rival colonies of Taranto and Metaponto can both claim to have invented them. That is to say, these settlements were the first in the Italiote world to employ terracottas for the covering and decoration of roofs, beginning with the first attestations in the Orientalizing period. The earliest plaques have been found at Taranto and Metaponto. They were more or less semicircular, decorated with human heads (male and female, respectively), and equipped with an attachment for a curved cover tile following the tradition of the earliest Laconian roofs in the second half of the 7th century BC.[16]

Very close to the Corinthian types and comparable to the antefixes of Calydon in the shape of the face and the Daedalic hairstyle, the examples from Metaponto are completely different in form. They constitute, at the close of the 7th century BC, an autonomous and original reworking of Peloponnesian models, with some Tarantine influences.[17] In this rather hazy cultural sphere, which acquired its own character only later through close contact with northwest Greece, and which was possibly influenced by the Laconian sphere, the dialog was almost never one-sided.[18]

The vitality of this dialog, which was certainly not occasional, is attested not only by the experimentation widely seen in the nascent colonial architecture, but also by a series of perfectly recognizable parallels in architectural solutions reconstructed from archaeological documentation. The period between the second half of the 6th to the early part of the 5th century BC saw a surprising number of types and variants of roof decoration—sure proof, on the one hand, of the conspicuous possibilities expressed in this field by a lively and proud sociopolitical elite, and, on the other hand, of the continuous exchange between the Achaian

15 Above all, the areas centered around the Gulf of Corinth, including Corcyra, Aetolia, central Greece, and Corinth, along with the Argolid. Pliny the Elder's claim that a Sicyonian potter named Boutades invented the terracotta masks decorating the ends of wooden roof beams (*NH* XXXV.151–52) seems to suggest that the use of antefixes originated there, although this has not been confirmed by archaeological finds. For a synthesis of the question with ample bibliography, see Aversa 2002, 231ff.

16 The example from Taranto, no longer considered female after comparison with contemporary terracotta figures that certainly are female, is housed in the Trieste museum: Van Buren 1923, 81, 147, no. 49, pl. XVI, fig. 66; Laviosa 1954, 229, n. 2; Winter 1978, 31, pl. 8, 4; Mertens Horn 1991, 12. The antefixes with female heads found at Metaponto came from the area later occupied by the castrum (Adamesteanu 1982, 309, fig. 10 regarding the find; also Orlandini 1983, 336, fig. 313) and the urban sanctuary (unpublished). They were probably part of a roofing system with sima with cavetto profile decorated with a Doric *kymation*, painted or in low relief (Viola 1996, 165–66, and no. 116; Aversa 2002, 235–38).

17 Furthermore, the same eclectic character emerges in the numerous votive terracottas found both in the urban sanctuary of the Achaian colony and in the extra-urban sanctuaries at San Biagio alla Venella and at the so-called Tavole Palatine. It is also visible in the contemporary products of other Achaian colonies from Sybaris to Poseidonia, suggesting the existence of a koinè, a common stylistic language between settlements of the same origin.

18 In this sense, the slightly later adoption of the decorative system with antefixes with human heads in the indigenous center at Serra di Vaglio, geographically and culturally under the influence of Metaponto, is extremely significant. See G. Greco 1991, 36–37, figs. 91–93.

Figure 29.3 "Artemis Bendis" antefix type from Metaponto, urban Sanctuary of Apollo. Metaponto, National Archaeological Museum inv. no. 125202; scale: 1:2. (Aut. MBAC-SBA-BAS UPProt n. 0003131 31.05.2012, Cl. 28.13.10/1.12)

area in which Metaponto was situated and the area of Taranto, culturally linked to her mother city. The singularity and creativity of certain terracotta decorative elements and of some architectural creations (simae, revetment plaques, and antefixes) adopted in both settlements certainly suggest a circulation of molds between the two places.[19] Therefore it is not surprising to find among the items from Sant'Angelo Vecchio a mold for an antefix (SAV AT 08), the first generation of which is attested at Taranto and in Apulian spheres, and of which no casts have been found to date at Metaponto or in the territory under the direct control of the Achaian colony.[20]

During the course of the 6th century BC, Italiote production of antefixes seems to have abandoned male and female heads and adopted exclusively palmettes and the heads of Gorgons and Silenoi. The antefix with Silenos head found at Sant'Angelo Vecchio, although in very low relief, heavily worn, and bearing small fractures and imperfections, is rather well made and was clearly taken from a cast. The ratio of contraction or retraction, calculated on the almost complete example in the Taranto museum, is 13 percent. The lack of casts among the finds from the Metapontine sanctuaries does not diminish the value of this production center, which must be considered a derivative and not a creative atelier. It could, nevertheless, have made products, at least during the second half of the 6th century BC, for the nearby sanctuary at San Biagio.

This hypothesis, attractive and rich in consequence, might be definitively confirmed by a second Archaic mold, unfortunately very fragmentary, recovered among the collapsed tile against the walled structures of Sant'Angelo Vecchio (SAV AT 05).

This mold produced a type of antefix with a palmette springing from a pair of volutes that was very common on Late Archaic roofs at Metaponto. This type alone constitutes 84 percent of all the antefixes decorated with vegetal motifs (*Fig. 29.2*). Known from over 100 examples of casts or replicas, all very fragmentary, the type presents numerous variants and six different versions that demonstrate the use of parallel molds and derivative molds (*sourmolues*)—not always recognizable due to the state of preservation—as well as a peculiar pattern of circulation. In fact, 85 percent of the known fragments come from the urban sanctuary at Metaponto, 10 percent from the extra-urban sanctuary at San Biagio, and the remaining 5 percent from the extra-urban *Heraion* of the Tavole Palatine, the site of Incoronata, and farmhouses scattered across the Metapontine chora. The fragments from San Biagio all belong to the version without the large apotropaic eye within the volute and with a herringbone pattern in very low relief framing the semicircular plaque. According to current research, this version appears to be exclusive to this sanctuary. Thus the identification of the variant or version of the palmette type from which the Sant'Angelo mold was created would be of great importance for the recognition of one of the possible

[19] In this regard, the reconstruction at Taranto of an exact replica of roof G of the sanctuary at San Biagio alla Venella, dated to the end of the 6th c. BC, is exemplary (Viola 1996, 168 with bibliography for the individual finds).

[20] It must be stressed, however, that the repertoire of antefixes known from the centers of Metaponto and Taranto is constantly growing.

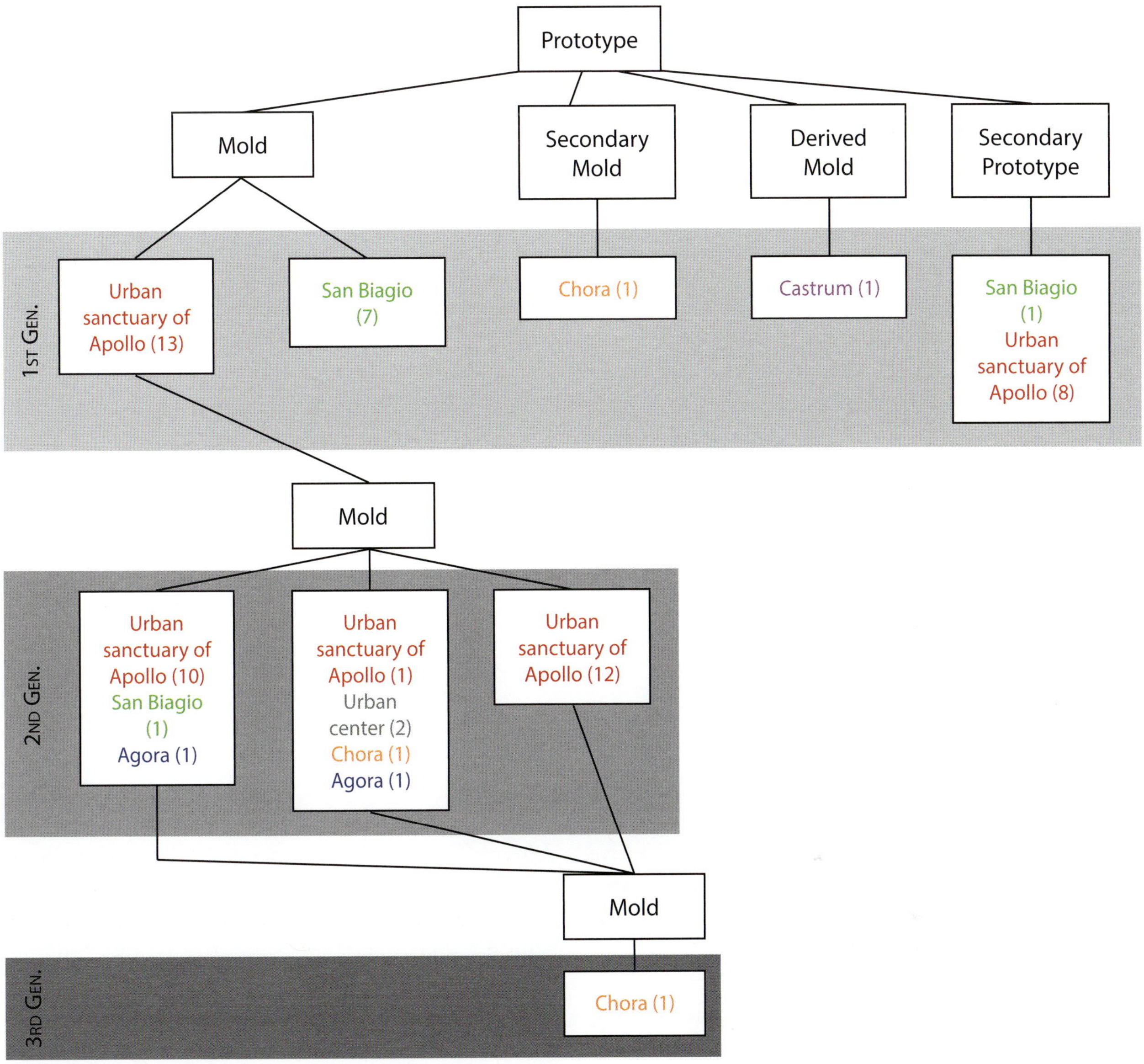

Figure 29.4 "Artemis Bendis" antefix type. Diagram of the two prototypes, the relationships between generations, and the locations of derived productions based on the examples from Metaponto and the Metapontino. (LMJ/ALT/ICA)

clients/recipients of the products from the atelier of Sant'Angelo Vecchio.

The fragment preserves only the central button and part of the curl of one of the volutes, but the dimensions of the button make it possible to exclude the presence of the eye. The three known versions with the eye have a larger volute and, consequently, a button of larger diameter. No hypothesis can be proposed regarding the presence or absence of the herringbone-pattern frame.

Two other molds belong to Late Classical or Hellenistic antefixes: SAV AT 06 and SAV AT 09. The first, perhaps of the type with the head of a young satyr, is also known from an almost complete example from the sanctuary at San Biagio.[21] The second belongs to the Artemis Bendis type, very probably the most widely circulated type among the Greek colonies of the Ionian coast from Taranto to Rhegion, encountered also in Apulian territory between Peucetia and Messapia (*Fig. 29.3*).[22] Artemis Bendis alone represents 37 percent of the Metapontine antefixes with female head. It is known in three different versions

[21] See Tempesta 1995, 56–57, fig. 11 (inv. no. 125209).

[22] See, for example, a number of antefixes from Brindisi (Sciarra 1976, 40, no. 266) and from the Messapian sites of Rudiae and Cavallino, in the Provincial Museum at Lecce (unpublished).

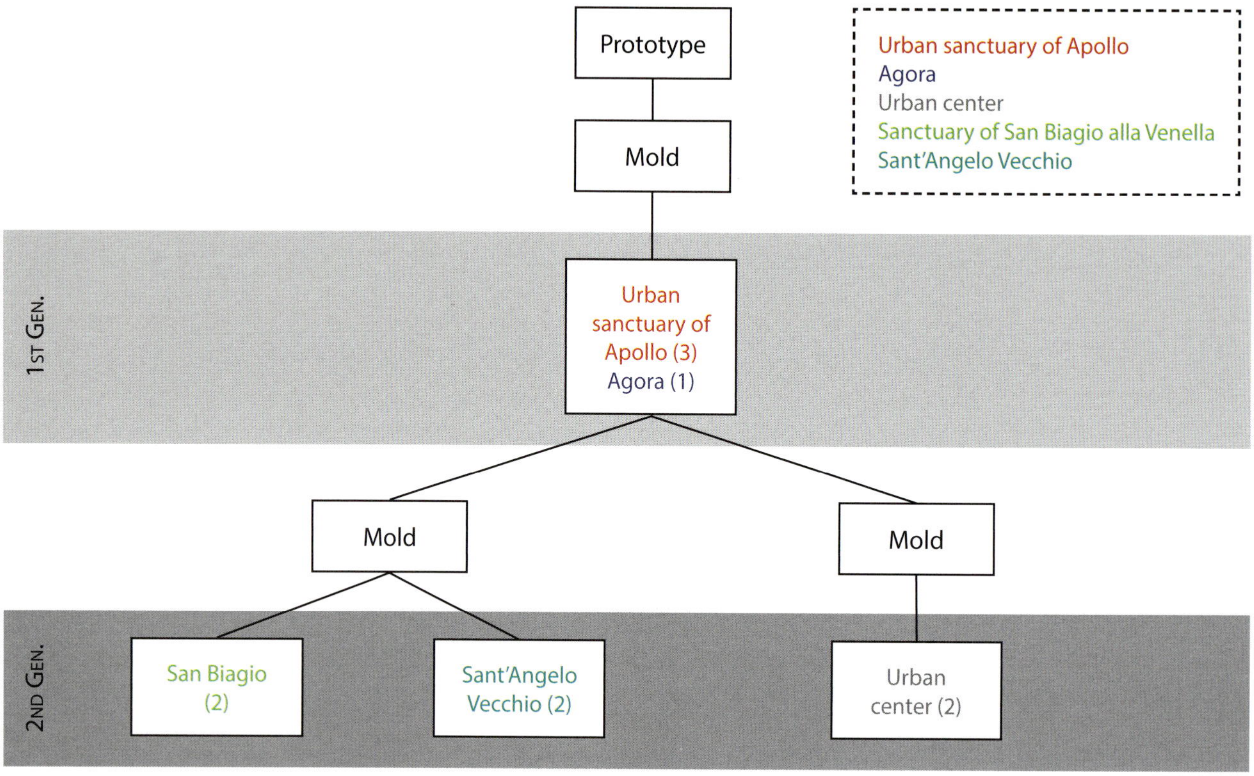

Figure 29.5 Antefix with head of Dionysos. Diagram of the prototype, the relationships between generations, and the locations of derived productions from Metaponto and the Metapontino. (LMJ/ALT/ICA)

and a number of variants, some due to the excessive wear on the molds used and the habit of the Metapontine craftsmen of directly retouching the casts (*Fig. 29.5*), discussed above.[23]

The Artemis Bendis mold from Sant'Angelo Vecchio, SAV AT 09, was certainly obtained from a cast, as suggested by the reduced dimensions and the cut at the base of the neck, and recognizable for its rather fuzzy relief and imperfections or slight chipping of the surfaces (*Fig. 29.4*). Given the difficulty of calculating the ratio of retraction (that is to say, the sum of the two retractions) directly, due to both the lack of parallels with analogous materials and the fragment's poor state of preservation, the ratio of reduction was measured on a purpose-made gypsum cast,[24] resulting in a ratio of just over 10 percent with respect to the examples of the second generation. Our mold could therefore have produced third-generation examples of the Artemis Bendis type, known at present only among the materials from the Roman villa at Termitito a few kilometers to the west,[25] where the antefix decorations from the inner courtyard and one of its two *impluvia* were preserved.[26]

Only two casts were found at Sant'Angelo Vecchio (SAV AT 01 and SAV AT 02). They are fragmentary, and their surfaces vitrified from excessive firing. Both are of the type bearing a head of Dionysos on a plaque with a pointed arch and small cover tile with an arched section attached to the back, probably reused as construction material in the building of the walls in the western part of the excavation. The representation of the god belongs to the canonical type of the end of the 4th century BC, but with the interesting addition at the sides of the face of two rearing griffins with bird's heads and lion's bodies in very low relief that was orig-

[23] For the type and its description, see Tempesta 1995, 46–48, fig. 3.

[24] A terracotta positive was also made and then discarded, because the omission of inclusions (or, more correctly, the use of unknown quantities of inclusions in the clay fabric) and firing in an electric kiln led to excessive reduction of the piece.

[25] Orlandini 1983, fig. 561.

[26] The context of the Roman villa of Termitito is being studied by a team from the Department of Archaeology of the Università del Salento under the direction of Liliana Giardino, whom the author wishes to thank for providing the information about the finds. Figure 29.4 shows the mold seriation for this type and illustrates the generational development of the Metapontine materials.

inally painted. They allude to a particular connotation assumed by the cult of the god and give the image a chthonic and mysterious character.[27]

The type is present in the agora at Metaponto, in the colony's residential quarters, in the urban sanctuary, and in the extra-urban sanctuary at San Biagio alla Venella. To date, no antefixes of the same type are known outside of the Metapontino. A completely different version, with more decorative traits and later in date, is present at Herakleia. Two different reductions are known, taken from a single prototype, with a reduction coefficient of just over 12 percent (*Fig. 29.5*).

The examples from the urban sanctuary, all first-generation, were found in the area between the Temples of Hera (A) and Apollo (B) intended for votive hoards and occupied, between the end of the 4th and the first half of the 3rd century BC, by the small *sacellum* H. The plaque found in the agora comes from a tile collapse close to the eastern *parodos* of the theater, below which were sealed materials datable to the beginning of the last quarter of the 4th century BC.

The plaques decorating the portico of a metalworking structure that faced the east-west stenopos delimiting the southern side of the agora and the atrium of a house in Insula III immediately south of the sacred area were of the second generation. Also belonging to this generation were the examples found at San Biagio, which, together with a group of antefixes with male heads (Dionysos, Pan, satyrs, Helios) from secondary molds, decorated the roofs of the buildings dating to the sanctuary's final phase. This last phase occurred between the last decade of the 4th and the beginning of the 3rd century BC, before the area's radical change in function with the construction of a farmhouse and the agricultural reorganization of the territory.[28]

This type therefore does not seem to have been widely circulated—that is, with the exchange of molds and/or casts between various specialized workshops located at some distance from each other. On the contrary, it would seem a typical example of a product that had a strictly local circulation, with the exchange of molds or (more likely) casts between workshops that were geographically close within a single territorial district.

In fact, according to the present state of the research, there are 14 fragments belonging to 10 examples, of which four are complete or almost completely reconstructable, while six are fragmentary. All of the casts found between the sanctuary and the agora appear to have been made with the same mold—some of the technical and manufacturing characteristics are identical (such as the clay fabric with abundant ceramic inclusions, the thickness and shape of the plaques, the use of a stick to refine and incise the details of the vine leaves forming the wreath).[29] Some differences in the shape of the cover tile probably distinguish the productions from the same workshop intended for different buildings in the urban center.

For the second-generation antefixes, on the contrary, it is possible to identify the presence of at least two parallel molds. One, sharper and without imperfections, was used to produce the examples found at San Biagio and, very probably, the fragments from Sant'Angelo Vecchio. The second, much more abraded and bearing small lesions (noted from the imperfections on the casts), was used to produce the examples in the urban area, from Insulae III and IV. The essential completeness of the example from Insula IV makes it possible to identify the gradual wear on the mold and thus establish the temporal succession of the copies within the production sequence. The antefix from Insula III is "sharper," while the one from the metalworking installation is much more worn and has more accentuated imperfections from the casting process.[30] The chronological difference between the use of the two derivative molds is quite considerable, as the sacred buildings on the terrace at San Biagio were certainly abandoned in the first years of the 3rd century BC, while the metalworking structure facing onto the agora was undoubtedly in use during the 2nd century BC.[31]

Again, there are small but nevertheless readily visible differences in the clay composition and in the shapes of (and attachment to) the cover tiles between the very fragmentary examples from Sant'Angelo Vecchio and those from San Biagio. The differences seem to suggest the activity of two different workshops operating at the same time and clearly using the same mold. Given the few examples available for the reconstruction of the mold seriation and their very fragmentary state,

[27] For this association and information on Dionysiac iconography at Metaponto, see Tempesta 2005, 87–89.

[28] See Tempesta in *San Biagio*.

[29] The only difference of any significance is in the shape of the attached cover tile, which varies from semi-cylindrical to only very slightly curved, and the presence, in one case, of mica among the inclusions.

[30] The defect visible on the upper leaves of the wreath is a clear sign of breakage of the mold, which apparently did not prevent its continued use.

[31] A second antefix of the type with a Silenos head also comes from here. This type was previously unknown and suggests a continuity of production of these terracottas until the mid-2nd c. BC.

however, the interesting hypothesis that the kilns at Sant'Angelo Vecchio also produced for the sanctuary at San Biagio cannot be completely excluded.

Finally, the present state of the research suggests that the creative atelier was located within the asty of Metaponto and that it produced for the urban sanctuary and the public buildings of the asty. A circulation of parallel molds from a derivative atelier can be identified only for the later production used in the asty.

General examination of the terracotta architectural fragments from Sant'Angelo Vecchio leaves no doubt about the existence of an extensive local production. The rural workshop was not limited to copying models created elsewhere, but also adopted iconographical themes and variations in at least two different historical moments, the second half of the 6th century BC and the end of the 4th–beginning of the 3rd century BC. These two periods clearly saw a strong demand for antefixes for monuments in the asty and for the sanctuaries located at crucial points in the chora, which led to both the adoption of types and molds from Taranto and the creation of new types.

Even in the absence of certainties, the finds from Sant'Angelo Vecchio demonstrate a decentralization of the production of architectural terracottas in peripheral territories, as well as a rapid and diffused circulation of the derivative products within the territorial sphere that was more or less under the control of the coastal towns. For the moment, the workshops at Sant'Angelo Vecchio produced the greatest number and variety of types produced from the second half of the 6th century BC onward in the Greek world.[32]

[32] In fact, almost all of the antefix molds found in Greece are of types with palmettes and vegetal motifs. Furthermore, a third fragment, no. 3.3, can be added to the two Late Archaic molds discussed here. The type cannot be identified, but it has a semicircular plaque, like those used in Magna Grecia exclusively in the Archaic period.

Catalog of Architectural Terracottas
Antefixes

SAV AT 01 Plaque with Head of Dionysos

1:2

Provenance: G6, SW side of wall. Level 2, batt. 2. Lot: SA 79-1050T. Inv. no. 320042 (see *Fig. 2.20*).

Dimensions and technical features: h 9.5, w 10.7, th plaque 1.7. Clay: surface Munsell 5YR between 7/4 (pale yellow) and 6/4 (pale olive), fracture between Munsell 7.5YR 7/6 (reddish yellow) and 5Y 6/4 (pale olive); fine-grained fabric, hard, compact, and harsh/rough, with abundant inclusions. Mold-made; surfaces slipped.

Condition: One fragment. Worn surfaces, covered with encrustations; some parts are vitrified from over-firing.

Description: Part of the lower right-hand side of the face is preserved, from the cheekbone to the chin and neck. The lower end of a small bunch of grapes—part of the divinity's crown—can be seen at the height of the ear. On the back, a curved cover tile is preserved for a maximum length of 12 cm, attached to the upper half of the plaque.

Comparisons and comment: Well-attested among the Metapontine antefixes, this type shows the face of a young male, tilted to the right, with large eyes and gaze full of pathos, wide nose, fleshy lips of the mouth half-closed, and pronounced chin. The long curly locks of hair hang down on the strong, powerful neck. At the sides of the face are two rearing griffins, with bird's heads and lion's bodies, in very low relief (for the discussion and interpretation see Tempesta 1995, 59, fig. 14). The type is present at Metaponto in two different reductions, but without substantial variations. First-generation examples come from the urban sanctuary and agora, more precisely from the area between Temples A and B intended for votive hoards and occupied by the small sacellum H, as well as from the eastern parodos of the theater, below a tile collapse sealing materials dating to the beginning of the final third of the 4th c. BC. The type was produced throughout the 3rd c. and is attested well into the 2nd c. by the second-generation examples found at the nearby extra-urban sanctuary at San Biagio alla Venella (see Adamesteanu 1975, 254) and those from Insulae III and IV at Metaponto, which faced onto the sanctuary and the agora, respectively (Tempesta 2005, 83ff). Compare to *Figure 29.1*.

Date: Last quarter of the 4th c. BC.

SAV AT 02 Plaque with Head of Dionysos

1:2

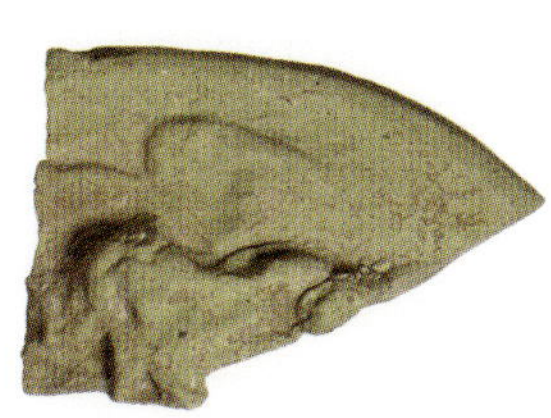

Provenance: G6, W of S wall corner. Level 3, batt. 1. Lot: SA 79-1018T. Inv. no. 319989. Assemblage 4.

Dimensions and technical features: h 4.7, w 7.3, th plaque 1.0. Clay: surface Munsell 5Y 7/3 (pale yellow), fracture GLEY 1 between 6/1 and 4/1 (dark greenish gray); fine-grained fabric, very hard, compact, and harsh/rough, with small inclusions. Mold-made with retouches by stick; slipped surfaces. On the back there is a manufacturing fault along the edge.

Condition: One fragment. Surfaces are vitrified by over-firing and covered with thin encrustations.

Description: Part of the upper, sharply-arched edge of an antefix with the remains of a crown of vine leaves and grape clusters in heavily-modeled relief. The housing for a curved cover tile is preserved on the back.

Comparisons and comment: For the type, see SAV AT 01.

Date: Last quarter of the 4th c. BC.

Acroteria

1:2

SAV AT 03 Acroterion Disc?

Provenance: G6, E of wall. Level 2, batt. 1. July 3, 1979. Lot: SA 79-1064P. Assemblage 3.

Dimensions and technical features: h 11.1, w 6.4, th plaque 1.1; max th 2.2. Clay: surface Munsell 7.5YR 7/6 (reddish yellow), fracture 5YR 6/6 (reddish yellow); fine-grained fabric, hard, compact, and powdery, with inclusions and small vughs. Mold-made; slipped surfaces.

Condition: One fragment. Worn surfaces are covered with encrustations.

Description: A small part of the disc is preserved, broken at the edge, with a decoration of concentric modeled listels of varying thickness. From the surviving part a minimum diameter of 22.5 cm can be reconstructed. The rear face is completely smooth, with the exception of a thin raised listel along the outer edge. The surfaces show traces of the original slip.

Comparisons and comment: Characteristic elements of traditional Peloponnesian roofing systems based on the Laconian model, disc acroteria were used to close and protect the top of the roof ridge. In the Laconian tradition figurative motifs were absent, the preference being for geometric elements, sometimes only painted, arranged in concentric patterns. In the west, their circulation was limited to a few cases (the most well-known being that of the Marasà temple at Locri) and, after initially following the Peloponnesian aniconic archetypes, the artisans of Magna Grecia came to prefer figurative decorations (gorgons, Achelöos, heraldic sphinxes, mythological scenes), solutions already known and circulated in Arcadian contexts. In any case, the greater diversity demonstrated in the west with respect to the mainland Greek prototypes originated in the decidedly limited dimensions that these architectural terracottas could have in Magna Grecia, to the point that in German archaeological publications such elements are specified as "ridge antefixes" (*Firstantefixe*). A second substantial difference, which they share with the Aetolian examples, is the circular form: the canonical Peloponnesian types, on the contrary, formed three quarters of a circle, terminating in correspondence with the apex of the tympanum. However, the western examples are still too few to be able to reconstruct the circulation history, but there is no doubt about the link between mainland Greece and the colonial world in the experimentation with, and fusion of, architectural experiences and solutions of different origins (an exemplary instance is the aristocratic building at Serra di Vaglio, datable to the first quarter of the 6th c. BC, with a mixed roofing system using both disc acroteria and continuous frieze acroteria with scenes of horsemen that were the work of Metapontine workshops: G. Greco 1996, 270–71, figs. 11–3). A possible parallel for this example, taking due note of the differences in context and material, is a marble fragment found in the area of Temple B dedicated to Apollo in the urban sanctuary of Metaponto (*Metaponto I*, 137, fig. 141).

Date: 4th c. BC.

SAV AT 04 Acroterion Disc?

Provenance: H4. Level 2, batt. 1. June 26, 1979. Lot: SA 79-538PL (see *Fig. 2.21*).

Dimensions and technical features: h 7.9, w 7.3, th plaque 2.1, max th 2.9. Clay: surface Munsell 5YR 6/1 (gray), fracture 5YR 7/6 (reddish yellow); fine-grained fabric, quite hard and harsh/rough, with abundant small inclusions. Mold-made.

Condition: One fragment. Surfaces are covered with heavy lime and sandy encrustations; some areas are blackish due to excessive exposure to heat.

Description: A small section of the disc's edge is preserved, decorated with concentric modeled listels of varying thickness. The diameter of the acroterion can be reconstructed from the surviving part as ca. 18.5 cm. The back face is completely smooth.

Comparisons and comment: See SAV AT 03.

Date: 4th c. BC.

1:2

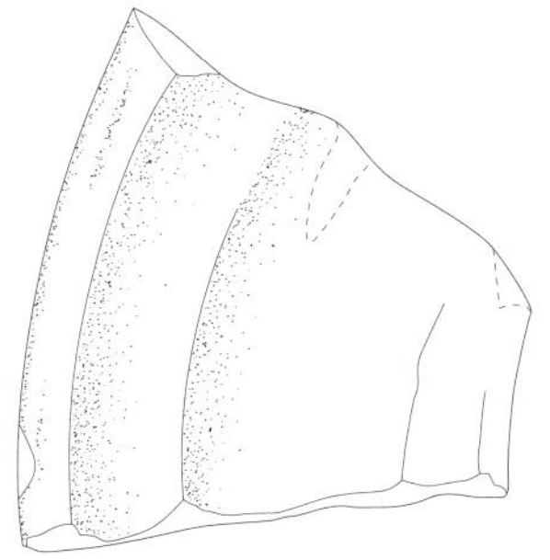

Molds

SAV AT 05 Palmette Antefix Mold

Provenance: G5. Level 1, batt. 2. June 25, 1979. Lot: SA 79-600PL. Assemblage 3.

Dimensions and technical features: h 3.9, w 4.1, th plaque 1.3. Clay: surface and fracture Munsell 5YR 7/6 (reddish yellow); fine-grained fabric, soft and powdery, with occasional inclusions and vughs.

Condition: One fragment. Worn surfaces are covered with encrustations.

Description: Central part of the mold; edges missing; shallow. The central button of one of the volutes at the base of the palmette is preserved. The back face is completely smooth.

Comparisons and comment: This is the earliest and probably the most numerous of the antefixes of the so-called Tarantine type. The cast has a semi-elliptical form decorated with a seven-petal palmette bordered by a thin, raised listel emerging from a pair of volutes, the interior of which are sometimes decorated with two large apotropaic eyes. Originally the petals were painted alternately red and black; the cream color was used only for the listel and volutes. The type was present in five different variants in the northern part of Metaponto's urban sanctuary, later occupied by Temple D, and by the temenos along the western side of the sacred area, which was later closed by the monumental stoa. Numerous examples have also been found in the chora, from the extra-urban sanctuaries at the Tavole Palatine and San Biagio alla Venella (Adamesteanu 1964, 360–61) to the Archaic structure at Incoronata, only 3 km south of Sant'Angelo Vecchio, which had a roof with sima-casing decorated with a plaited motif in very low relief (Carter 2006, 157–73). Compare to *Figure 29.2*.

Date: 6th c. BC.

1:2

Molds cont.

SAV AT 06 Antefix Mold with Male Head

1:2

Provenance: G7. Level 2, batt. 2. July 4, 1979. Lot: SA 79-786PL (see *Fig. 2.20*).

Dimensions and technical features: h 6.5, w 8.5, th 1.5. Clay: surface Munsell 7.5YR 8/4 (pink), fracture 7.5 YR 7/6 (reddish yellow); fine-grained fabric, moderately hard, compact, and granular, with occasional small inclusions and vughs.

Condition: One fragment. Worn and chipped surfaces are covered with a strong, heavy layer of lime encrustations.

Description: The lower right corner of the mold is preserved, along with the attachment of the left shoulder of an unidentifiable figure. The relief is shallow, the back face completely smooth.

Comparisons and comment: The mold may have been used for the creation of an antefix with a head, as suggested by its dimensions and the form of the edge, which has a semi-elliptical profile. The horizontal cut of the shoulder and lack of jewelry in the surviving design exclude a female type representing Aphrodite or Io; it may instead represent the head of a young satyr (Tempesta 1995, 56–57, fig. 11).

Date: 4th c. BC.

SAV AT 07 Antefix Mold?

1:2

Provenance: H3, cleaning of Level 1, batt. 3. July 20, 1979. Lot: SA 79-1024PL. Assemblage 2.

Dimensions and technical features: h 8.0, w 5.5, th 1.3. Clay: surface and fracture Munsell 2.5YR 6/8 (light red); fine-grained fabric, hard and porous, with minute inclusions and occasional vughs.

Condition: One fragment. Surfaces are very worn and heavily encrusted.

Description: The fragment is the lower left corner of a mold with a semi-circular plaque. The subject cannot be identified. The back face is completely smooth.

Comparisons and comment: A useful indicator for the dating is provided by the plaque's profile, which excludes identification as a Late Classical or Hellenistic antefix.

Date: 6th c. BC.

SAV AT 08 Antefix Mold with Head of Silenos

1:2

Provenance: Sounding A, subdivision of Room 1 to the S. Level 2, batt. 1. Lot: SA 79-20T. Inv. no. 319902. Assemblage 5.

Dimensions and technical features: h 17.0, w 12.2, reconstructed w 27.0, th 1.2. Clay: surface Munsell 5YR 7/6 (reddish yellow), fracture 2.5YR 6/6 (light red); fine-grained fabric, moderately hard, compact, and well-levigated.

Condition: Several joining fragments; incomplete. Surfaces are worn, with heavy encrustation.

Description: The fragment is the right side of a semicircular plaque with the face of a Silenos. It has a low, triangular forehead, large oval eyes, a snub nose, thin moustache, and long feline ears set vertically on the temples. The beard is separated into four rows of concentric locks, while the hair, arranged in thin wavy sections on the top of the head, falls in a series of triangular locks (like "wolf's teeth") on either side of the face. The curvilinear edge of the plaque is bordered by a double listel. The relief is shallow, the surfaces worn by prolonged use. The back surface is rough, with the remains of incomplete fingerprints in several points.

Comparisons and comment: See Edlund 1984, 287–88, fig. 8; Billot 1997, 86 (but dated to the 4th c. BC). For the moment the type has no parallels at Metaponto but is found at Taranto (see Laviosa 1954, 244, fig. 76.1, with bibliography pertinent to examples housed in other European museums; Herdejürgen 1978, passim) and at Bari (unpublished).

Date: Second half of the 6th c. BC.

Molds cont.

SAV AT 09 Antefix Mold with Female Head wearing a Lion Skin

1:2

Provenance: F2. Level 2, batt. 1. June 24, 1979. Lot: SA 79-715T. Inv. no. 319933 (see *Fig. 2.30*).

Dimensions and technical features: h 19.8, w 9.7, th 2.2. Clay: surface and fracture Munsell 2.5 YR 6/8 (light red); fine-grained fabric, hard and compact, with occasional inclusions and vughs.

Condition: One fragment. Surfaces are worn and chipped, covered in part with lime and sandy encrustations.

Description: The fragment is the left half of a semi-elliptical plaque with a representation of the so-called Artemis Bendis. Part of the front-facing female face is preserved, from the low, triangular forehead to the left cheek, along with half of the lion skin headdress; a small wing appears among the curls beside the temple and the hair, which is in small, wavy curls. The rear of the plaque is convex.

Comparisons and comment: See Edlund 1984, 287; Edlund 1986, 121, fig. 4; Billot 1997, 86. The mold is for a type of antefix that is widely documented in Magna Grecia and which seems to find its model in a terracotta head of Medusa from Taranto, now housed in the Allan Pierson Museum in Amsterdam (Mertens Horn 1994, 246, fig. 287). However, the identification of this female face is the subject of a long-running debate among scholars (for a synthesis of the various identifications, see Tempesta 1995, 46–48). One of the more widely accepted hypotheses identifies it as Artemis Bendis, a divinity of Thracian origin assimilated with Artemis, whose cult is attested by archaeological finds at numerous sites in Magna Grecia (for the iconography of the goddess, see *LIMC II.1*, 690–92, nos. 927–34). A recently reproposed second reading of this and other types produced in Taranto identifies it as the head of Medusa, characterized by the use of an oriental headdress (*petasos*, Phrygian cap, or lion skin). However, this type of antefix occurs in all of the settlements of the Ionian coast (Taranto, Metaponto, Herakleia), where it seems to have been used in the urban centers and, to a different extent, in the necropoleis (Laviosa 1954; Herdejürgen 1982; Tempesta 1996b; Neutsch 1967, pl. 37). It is known through a series of versions and variants which are distinguished mainly by their stylistic rendering, the set of the head, and the presence or absence of the neck. Lastly, this type's long production period runs from the last years of the 5th c. BC until the mid-3rd c. BC, with maximum circulation in the second half of the 4th c. BC. Three reductions have been identified within this long chronological span, produced from the creation of new molds—like this fragment—taken from casts. Compare to *Figure 29.3*.

Date: 4th c. BC.

30

The Molds from Sant'Angelo Vecchio

Ingrid E. M. Edlund-Berry

Studies of kilns and terracotta production have provided much valuable insight into the local economy of workshops and distribution of such essential commodities as pottery and architectural terracottas, including roof tiles.[1] In addition, the presence of molds in kiln or workshop contexts emphasizes the role of artisans who were responsible for developing techniques of manufacture and for establishing the different types of objects to be made.[2]

As can be expected, kilns were created to facilitate the production of specific types of terracottas and pottery, and their lifespan was related to the history and use of a site, whether as a sanctuary or as a settlement. What is remarkable about the modest site of Sant'Angelo Vecchio is that it has produced a surprisingly large number of molds of different types and dates, and that some of them may have been used to produce terracottas used at neighboring sites.[3]

The history of Sant'Angelo Vecchio includes an Archaic phase (House on the Hill); a late 6th/early 5th-century BC sacred enclosure marked by an inscribed boundary stone[4]; a 5th-century BC Retaining Wall, Pottery Deposit, and a cluster of burials; a late 4th- to early 3rd-century BC production of votive plaques in several kilns with a corresponding workshop shed; and finally a Kiln Complex with two main kilns used in the 2nd century BC in conjunction with a workshop shed.[5] While the different features fit well into the pattern known from other sites in the chora of Metaponto, with evidence of habitation and a rural cemetery, a sacred precinct, and terracotta production, it is the kilns that mark Sant'Angelo Vecchio as a site of importance.[6] The kiln production thus spans a period of up to two hundred years, from the late 4th into the 2nd century BC, and took place in two main locations, in the area of the deposit of votive plaques, perhaps reaching as far as a small kiln (later incorporated into the Kiln Complex), and in the Kiln Complex consisting of a small kiln and two larger ones, the North and South Kilns (Kilns 5 and 4).

The types of molds represented include a silenos antefix (SAV AT 08), an Artemis Bendis antefix (SAV AT 09), fragments of a palmette antefix (SAV AT 05), a male head antefix (SAV AT 06), an unidentified antefix (SAV AT 07), and two votive plaques, a seated female (SA 79-47T) and a standing female (SA 79-1100T),[7] a rosette mold with a Medusa head (SA 79-1019T), and a loom weight mold with an Eros (SAV LW 25; *Fig. 30.1*). Both the findspots and the different types of molds suggest a complicated history of production and distribution where Sant' Angelo Vecchio formed part of a network within Metaponto and the surrounding territory already in the 6th century BC.

Findspots of the Molds

The molds come from different areas of the site, including:

I. The Kiln Complex, Sounding A (SAV AT 08, SA 79-47T) and square F2 (SAV AT 09);

II. The workshop sheds, square G4 (SAV LW 25), square G5 (SAV AT 05), square G7 (SAV AT 06), and square H3 (SA 79-1019T, SAV AT 07); and

III. The deposit of votive plaques (SA 79-1100T) (*Fig. 31.2*).

---

[1] See Ch. 29, "The Architectural Terracottas: Some Considerations on Production," Ch. 32, "Roof Tiles and Bricks," and Ch. 33, "Kiln and Workshop Furniture." For a detailed analysis of manufacturing techniques of both mold-made and hand-made terracottas, see most recently, Winter 2009, 505–38; for a catalogue and discussion of molds found in the ancient Mediterranean, see Billot 1997, 39–129.

[2] See, for example, Smith 1998, 31–51; Edlund-Berry 2011, 353–55. Although many of the studies of terracottas and molds pertain primarily to Etruria and central Italy, the main topics apply also to southern Italy, including the Greek colonies.

[3] For other sites with which the material from Sant'Angelo Vecchio seems to have connections in terms of the production of terracottas, see Ch. 29, "The Architectural Terracottas: Some Considerations on Production."

[4] See p. 68 in Ch. 3, "The Inscribed Stone."

[5] See Ch. 32, "Roof Tiles and Bricks," for the possible identification of the owners of the kiln workshop with the Aufidii family in the 2nd c. BC, and, for the documentation of the family, Tagliamonte 2011, 232–35.

[6] For the chora of Metaponto, see *Necropoleis*; Carter 2006; *Survey*.

[7] Both molds will be studied in detail by R. Ammerman in the forthcoming volume on coroplastic from Sant'Angelo Vecchio.

Figure 30.1 Assembly of molds from Sant'Angelo Vecchio: (1) SAV AT 08; (2) SA 79-47T; (3) SAV LW 25; (4) SAV AT 05; (5) SAV AT 09; (6) SAV AT 06; (7) SA 79-1019T; (8) SAV AT 07; and (9) SA 79-1100T. (CW/ICA)

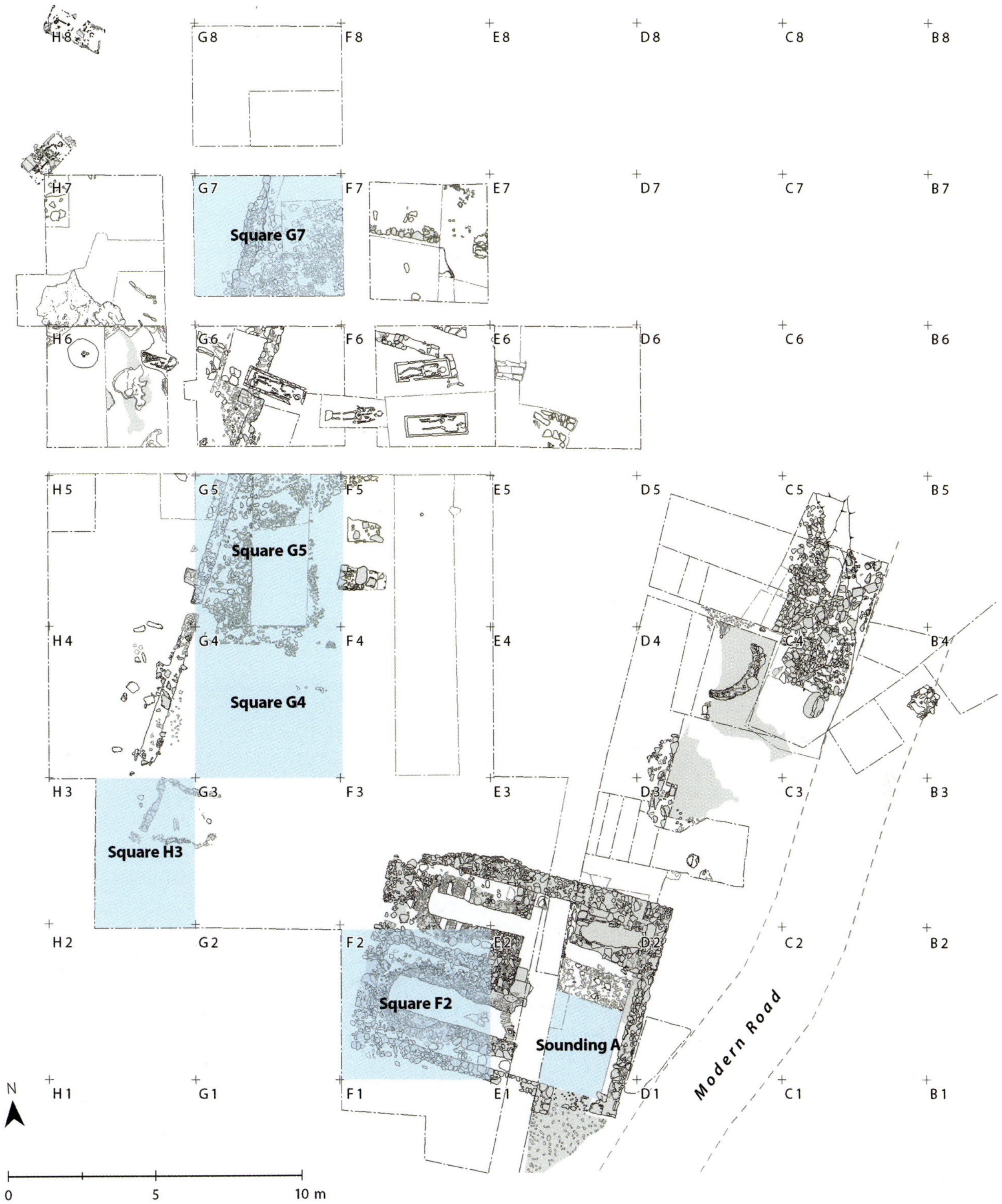

Figure 30.2 Sant'Angelo Vecchio, site plan with findspots of molds indicated by squares or soundings. The deposit of votive plaques corresponds with the area along the modern road where votive plaques were discovered as part of a salvage operation. (JT/ICA)

Lot No.	Cat. No.	Inv. No.	Provenance	Description	Chronology
SA 79-20T	SAV AT 08	319902	Sounding A	Silenos antefix mold	550–500 BC
SA 79-47T		319931	Sounding A	Seated female votive plaque mold	4th/3rd c. BC
SA 79-435T	SAV LW 25	320001	Square G4	Eros loom weight mold	4th/3rd c. BC
SA 79-600PL	SAV AT 05		Square G5	Palmette antefix mold	6th c. BC
SA 79-715T	SAV AT 09	319933	Square F2	Artemis Bendis antefix mold	4th c. BC
SA 79-786PL	SAV AT 06		Square G7	Male head antefix mold	4th c. BC
SA 79-1019T		319929	Square H3	Medusa rosette mold	4th/3rd c. BC
SA 79-1024PL	SAV AT 07		Square H3	Unidentified antefix mold	6th c. BC
SA 79-1100T		319930	Deposit of votive plaques	Standing female votive plaque mold	4th/3rd c. BC

Table 30.1 The molds from Sant'Angelo Vecchio.

Kiln Complex, Sounding A

Within the Kiln Complex, Sounding A represents a square within the open courtyard south of the small kiln, and east of the larger North and South Kilns. The second stratum of the area consisted of a tile fall that included tiles with stamped impressions (crown, SAV RT 67; wheel, SAV RT 93), and a sima fragment (SAV RT 129). Mixed in with the tile fragments in the tile fall was a fragment of a mold showing a seated female belonging to the reclining Dionysos/Hades type of the votive plaques (SA 79-47T).[8] Since the fragment is broken, and no positives (casts) of this plaque or any others were found within the tile fall, it is possible that the mold had been used for the construction of one of the kilns in squares F2–F3 (South or North Kiln), probably the South Kiln. Also found in this stratum was a section of a silenos antefix mold (SAV AT 08) of 6th-century BC date, for which, like all the other mold fragments with the exception of the votive plaque molds, no positive (cast) exists at the site.[9]

Kiln Complex, Square F2

Square F2 includes the west section of the South Kiln in the Kiln Complex. A female antefix mold, interpreted to represent Artemis Bendis (SAV AT 09), was found here in the first stratum, second level—that is, above and not within the preserved section of the kiln. It is therefore not possible to ascertain its relationship to the production of terracottas in the South Kiln.

Workshop Sheds, Squares G4–G7, and Square H3

The workshop shed in squares F6 and G6 represents an important stratigraphic point at the site, since its long western wall is partly placed directly above some of the 5th-century BC tombs (Tomb 1 in square G6; Tomb 2 in square F6).[10] Since it seems unlikely that the original 5th-century BC inhabitants would have built a workshop shed directly over the family cemetery, it is probable that the area changed hands, and that the production shifted from farming in the earlier period, beginning with the House on the Hill in the 6th century BC, to an industrial production of terracottas and pottery in the 4th–2nd centuries BC.

To the first phase of kiln production belongs a workshop shed documented in squares F7 and G7. Based on the presence of smoke hole tiles and loom weights, this structure may have been used as a workshop, but also as living space, inhabited by the owner of the kilns for production of votive plaques in the area labeled the deposit of votive plaques (for which, see below). An inscribed Herakas tile (SAV RT 02) and fragments of a bearded head plaque (SA 79-565T),[11] as well as a well-preserved maenad-satyr plaque (SA 80-10T),[12] and antefix fragments found in the southeast corner of square G6 (SAV AT 02 and SAV AT 01) suggest a connection between the structure and the production of terracottas.

The mold fragments from this area of the site, are, however, all connected with the second phase of the workshop shed, the features of which included a clay deposit (squares H3–H4) and Wall 1, a long wall stretching from square H4 through square G6 with square pilaster supports on the outside facing west. Two clusters of stones preserved in square F5 may have served as foundations for wooden posts that supported the roof. A tile fall to the east of the long wall suggests that the structure served as a roofed

[8] See Ammerman forthcoming.

[9] See Ch. 29, "The Architectural Terracottas: Some Considerations on Production."

[10] See Ch. 4, "The Tombs at Sant'Angelo Vecchio."

[11] See Ammerman forthcoming.

[12] See Ammerman forthcoming.

Figure 30.3 Sant'Angelo Vecchio, detail of site plan with second phase of workshop shed. (JT/ICA)

Figure 30.4 Sant'Angelo Vecchio, detail of site plan with area between Soundings A and C. (JT/ICA)

workshop and storage area for the material produced in the kilns (*Fig. 30.3*).

Two molds come from the area interpreted as a clay deposit, at the southern end of the workshop shed, that is, the Medusa rosette mold (SA 79-1019T), and the 6th-century BC mold tentatively interpreted as representing an antefix (SAV AT 07), both from Stratum 1. Noteworthy finds in this area were tile fragments with stamped designs (SAV RT 55, SAV RT 74), a Herakas tile fragment (SAV RT 01), and a horseshoe-shaped tile (SAV RT 104).

The Eros loom weight (SAV LW 25) was found further north, as an isolated find from Stratum 1 in square G4. The palmette antefix mold (SAV AT 05) appeared in square G5, also in Stratum 1 together with a fragment of the standing Zeus votive plaque (SA 79-466T).[13] Finally, the male head antefix mold (SAV AT 06) came from square G7, west of the wall in Stratum 2 together with a few curved tiles and coarse ware pottery.

Although the find context links these mold fragments to the area of the later workshop shed, their distribution cannot be directly tied to any particular kiln or sequence of production. Rather, the chronological range from the 6th-century BC (silenos antefix SAV AT 08, antefix fragments SAV AT 05 and SAV AT 07) to the 4th/3rd-century BC molds (Medusa rosette mold SA 79-1019T, Eros loom weight SAV LW 25, and male head antefix mold SAV AT 06) suggests that the presence of molds was not related to the structures in which they appeared, and that the production of terracottas at the site should be evaluated primarily on the examples of positives (casts).

Deposit of Votive Plaques

The best evidence of terracotta production at Sant'Angelo Vecchio comes from the large assembly of votive plaques. Since, unfortunately, the plaques were discovered through roadwork and the material was gathered as part of a salvage effort, it was not possible to establish any kind of stratigraphy or context. Thus the term "deposit" simply indicates that a concentration of plaques was discovered, some in excellent condition, others less so.[14]

While the deposit of votive plaques has provided good evidence for the different types of plaques, less is known about the molds used and the actual production process. Unfortunately, there is only one small mold fragment from this area, showing a standing female (SA 79-1100T),[15] the second of the plaque molds from the whole site; the other portraying a seated female found in Sounding A (SA 79-47T).[16] In spite of the scarcity of plaque molds, the number of plaques and their findspots throughout the site suggest that such plaques constituted one of the main types of objects produced at Sant'Angelo Vecchio. Remains of two or more kilns along the slope between Sounding A and Sounding C indicate that the plaques were produced here in the late 4th to early 3rd centuries BC (*Fig. 30.4*). In addition to the kilns, the earlier workshop shed may have served as a combination of dwelling and storage area for the production, whereas the later, larger workshop shed consisted primarily of a storage space.

Types of Molds

Considering the types of terracotta objects produced at Sant'Angelo Vecchio it is remarkable that there are only two mold fragments preserved for the votive plaques (SA 79-47T and SA 79-1100T), which are by far the most numerous of the terracottas found at the site. The lack of more and better preserved plaque molds may perhaps be attributed to the craftsmen who were needing to use such molds at other production sites in the area as they moved from one site to the next (*Fig. 30.5*).[17]

The largest number of molds is represented by antefixes identified by Tempesta as a silenos (SAV AT 08), a palmette (SAV AT 05), Artemis Bendis (SAV AT 09), a male head (SAV AT 06), and an unknown type (SAV AT 07). Since we do not have any examples of antefixes made from these molds, it is possible that, like the Dionysos antefixes (SAV AT 02, SAV AT 01) and acroteria fragments (SAV AT 04, SAV AT 03), the antefixes produced here were used at other sites.[18]

The Eros loom weight mold SAV LW 25 and Medusa rosette mold SA 79-1019T are particularly well preserved, but there is no evidence that they were actually used for production at the site. With the exception of the plaque molds, they, like the other molds, document the presence of artisans at the site, but not necessarily actual use.[19]

Chronology of the Molds

The dates of the molds range from the 6th century BC (palmette antefix SAV AT 05, unidentified antefix SAV AT 07, and silenos antefix SAV AT 08) to the 4th century BC (male head antefix SAV AT 06, Ar-

13 See Ammerman forthcoming.

14 See Ch. 1, "The Many Lives of a Rural Site: Sant'Angelo Vecchio," and Ammerman forthcoming.

15 See Ammerman forthcoming.

16 See Ammerman forthcoming.

17 For a discussion of resident and itinerant artisans, see Winter 2009, 535–38.

18 See Ch. 29, "The Architectural Terracottas: Some Considerations on Production," and SAV LW 25 in Ch. 31, "Loom Weights."

19 Because of the widespread use of terracotta roof decoration, it is possible that the acroterial terracottas from Sant'Angelo Vecchio were used at one or more of the buildings at the site. For reconstructions of the different roofing systems, see Ch. 32, "Roof Tiles and Bricks," in particular the roof that includes the silenos antefix (*Fig. 32.2*).

Figure 30.5 Composite of molds and positives (casts): (1) SA 79-47T mold (*left*) and SA 79-417T cast (*right*); (2) SA 79-1100T mold (*left*) and SA 79-411T cast (*right*).

temis Bendis antefix SAV AT 09), and late 4th/early 3rd century BC (votive plaques, seated female SA 79-47T, standing female SA 79-1100T, Eros loom weight SAV LW 25 and Medusa rosette SA 79-1019T).

Most surprising, because of their Archaic date, are the three antefix molds, of which the silenos SAV AT 08 is well preserved. Unless we are missing a whole chapter of the site's early history, where silenos and other antefixes were produced in kilns no longer preserved and perhaps used on the House on the Hill or another important early building, here or at another site, these molds remain isolated elements.[20]

Interpretation

In many ways the presence of molds at the site of Sant'Angelo Vecchio is not surprising considering the kilns that existed there. On the other hand, the variety of molds suggests different facets of the site that may

[20] For comparanda from other sites in the Metaponto area, including S. Biagio and Incoronata, see Ch. 29, "The Architectural Terracottas: Some Considerations on Production."

be difficult to interpret. First of all, combined with the fragments of acroteria (SAV AT 04, SAV AT 03) and antefixes (SAV AT 02, SAV AT 01),[21] the molds that represent architectural terracottas such as the silenos (SAV AT 08), palmette antefix (SAV AT 05), Artemis Bendis (SAV AT 09), male head (SAV AT 06), and unidentified antefix (SAV AT 07) suggest the presence of architectural features that unfortunately does not correspond well with the actual remains of structures at the site. Unless we postulate that the Archaic House on the Hill was adorned with silenos and palmette antefixes, there is no further indication of a 6th-century BC building at the site. Likewise, the 4th-century BC antefixes and acroteria, as well as the antefix molds, do not correspond to any preserved structure. It is therefore possible that the production at Sant'Angelo Vecchio, both during the 6th and 4th centuries BC, supplied terracottas to other sites, perhaps even the city of Metaponto.[22]

As for the molds for votive plaques, it is perhaps surprising that we do not have more examples of the different types. Again, it is possible that in this case the artisans removed the molds after a sufficient number of plaques had been made, and continued to the next production site.

The best-preserved molds, the Eros loom weight and the Medusa rosette, are not represented by any examples produced at the site. Because of their distinct iconography, well represented in the Metaponto area, it is again possible that the finished product was taken to sites in the area.

Importance

Although many sites in ancient Italy preserve examples of terracotta molds, their numbers are small in relation to the masses of mold- and handmade terracottas of all kinds, including the architectural decoration.[23] The study of molds can provide important information about terracotta production and mold sequences, as shown by Tempesta for Sant'Angelo Vecchio.[24] Likewise, the typology of molds allows us to follow the distribution of motifs as molds were reused and copied in different areas, including the Greek colonies around Metaponto.[25]

Other subjects of discussion pertain to the creation of workshops and the role of permanent and traveling artisans in varying types of settings, including habitation sites, commercial areas, and sanctuaries.[26] Although it is not possible to link any particular type of mold to specific artisans, the care with which kilns and workshops were established and maintained suggests that terracotta production was an important part of the economy, and that it involved large groups of people, ranging from the artisans to the kiln builders and workers.

The distribution of molds can further suggest commercial exchange between neighboring sites such as Metaponto and Taranto as well as the local distribution within smaller areas such as Sant'Angelo Vecchio and nearby San Biagio. As shown by Tempesta, a careful comparison of mold sequences and actual terracottas enables us to suggest production centers (such as Sant'Angelo Vecchio) as well as recipients (such as San Biagio and Incoronata).[27]

Although the preservation of molds no doubt in the past has been subject to chance, studies of old material combined with ever-increasing new discoveries provide examples of sites with a wealth of molds. According to the study by Billot, molds have been found primarily at sites in Etruria and central Italy (for example, at Caere, Falerii, Orvieto, Poggio Civitate, and Selvasecca,) and Campania (for example, at Capua), whereas southern Italy and Magna Grecia are represented by fewer examples, including sites such as Medma and Metaponto (*Fig. 30.6*).[28]

Considering the size, location, and function of Sant'Angelo Vecchio in comparison with the Greek cities in the south and the Etruscan and Faliscan cities in the north, it is remarkable that this site has produced not only a greater number of molds but also a greater variety than any other site. This fact can be attributed to a number of causes but primarily to the long history of concentrated kiln production at the

21 See Ch. 29, "The Architectural Terracottas: Some Considerations on Production."

22 The best parallel for a site where molds have been found which may or may not have been used to produce terracottas for the site itself is Selvasecca in southern Etruria, for which see Söderlind 2006, 116–21; Klynne 2006–2007, 29–57.

23 See Billot 1997.

24 See Ch. 29, "The Architectural Terracottas: Some Considerations on Production."

25 See Ch. 29, "The Architectural Terracottas: Some Considerations on Production."

26 Recent petrographic analyses of different clays have documented distributions of terracotta objects within Etruria and Latium, for which see Winter 2009, 537–38.

27 See Ch. 29, "The Architectural Terracottas: Some Considerations on Production."

28 See Billot 1997.

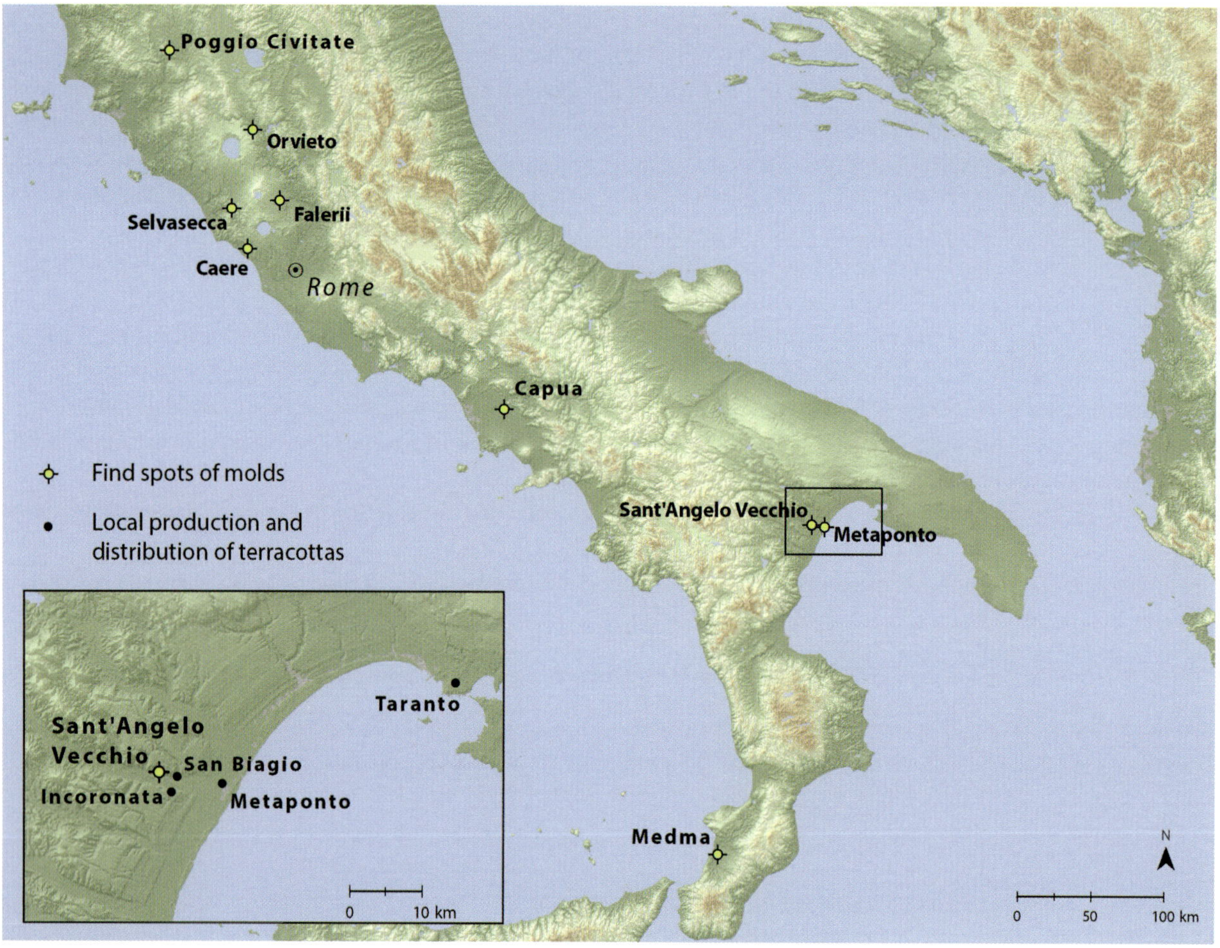

Figure 30.6 Map of central and southern Italy: terracotta production, molds, and findspots.

site. Seen in terms of the distribution within the site, and the chronological framework, each of the different molds represents a phase in the history of the site itself, but also within the immediate territory and the city of Metaponto. While other sites may preserve individual molds designed for local production, often in connection with sanctuaries and kilns, it is rare to find examples of molds that may not even have been used at the site itself. Within the period of some three hundred years (6th to early 3rd century BC), this area of southern Italy witnessed much growth and development, displayed in architecture as well as in trade, and it is here that Sant'Angelo Vecchio plays its role in defining the significance of local production seen in relation to the larger picture of communication and expansion.

31
Loom Weights

Lin Foxhall and Alessandro Quercia

Most of the loom weights from Sant'Angelo Vecchio were found in contexts of unusual and interesting use.[1] Of the four contexts where loom weights appeared, two (possibly three) appear to be closely related.

House on the Hill

The three pyramidal loom weights discovered (SAV LW 01, SAV LW 02, and one uncataloged) in the House on the Hill are consistent with the Archaic date of the site. This is the only context in which the loom weights discovered are likely to have been used for making textiles, because of the probable domestic character of the site. Loom weights from other contexts were probably not in use in domestic settings.

Area of the 3rd-century BC Votive Plaque Production and Kilns

The area of the 3rd-century BC kilns used for the production of votive plaques was located on the slopes on the easternmost sector of the main site. Fifteen loom weights were found in this area, within the large deposit of votive plaques. Except for two pyramidal loom weights (SAV LW 05 and SAV LW 10), the examples found here belong to the disc-shaped type (*oscilla*). A few are incomplete, and even some of the complete examples are quite worn.

Shed/Workshop Area

The area in the center of the site (squares F4–F7, G3–G7, and H3–H7) appears to have consisted of a shed or workshop that was built over the 5th-century BC tombs and was associated with the kilns. Of the eleven loom weights found (SAV LW 13–SAV LW 23), two are pyramidal (SAV LW 15, one uncataloged) and the remaining nine are disc-shaped. One of the more remarkable finds from this area was a mold for making disc-shaped loom weights with the relief decoration of the figure of Eros (SAV LW 25). An example of a loom weight made from this mold or one very similar was included in the assemblage from the Pantanello sanctuary (PZ80.0180). The combination of the loom weights in the area of the kilns and the mold suggests that at least some loom weights (probably *oscilla* or disc-shaped) were made in the kilns that were mainly devoted to the manufacture of votive plaques. Some may have been kept as samples for potential customers to view. The presence of a random selection of loom weights, however—many quite worn—in the area of the kilns as well as in the Workshop Area suggests that some of this random assortment played a role in the manufacturing process, probably as kiln spacers. It seems unlikely that the loom weights were being used for textile manufacture in these contexts, and their use as kiln spacers would have been an additional phase of their life after they had ceased to be used for weaving.

Kiln Complex and Soundings A and B

Ten loom weights (SAV LW 26–SAV LW 31) were found inside the Kiln Complex (squares E2–E3, F1–F4). It is somewhat surprising that eight of these were pyramidal, with only two disc-shaped loom weights (SAV LW 28 and SAV LW 29), as in Metapontine contexts pyramidal loom weights are generally associated with contexts of the 5th century BC and earlier. Again it is likely, given the position of the finds, that these were selectively collected for use as kiln spacers, and were neither manufactured in this workshop nor used for textile manufacture in the context of their findspots. Some of the pyramidal loom weights may have been considerably older than the Late Republican context in which they were found. In Metapontine farmhouse sites of the Bradano to Basento field survey, disc-shaped loom weights were the most common type, while pyramidal loom weights were rare in sites of the later 4th and 3rd centuries BC.[2]

[1] For an overview on types and chronology of the loom weights from the Metapontine countryside, see Foxhall in *Survey*.

[2] For a similar loom weight, see Firmati 2012, which could be consistent with the context. Similar pyramidal objects were also found in kilns where Hispanic terra sigillata was fired, interpreted as test pieces; see Mezquìriz 1985, 118 pl. LII,1.

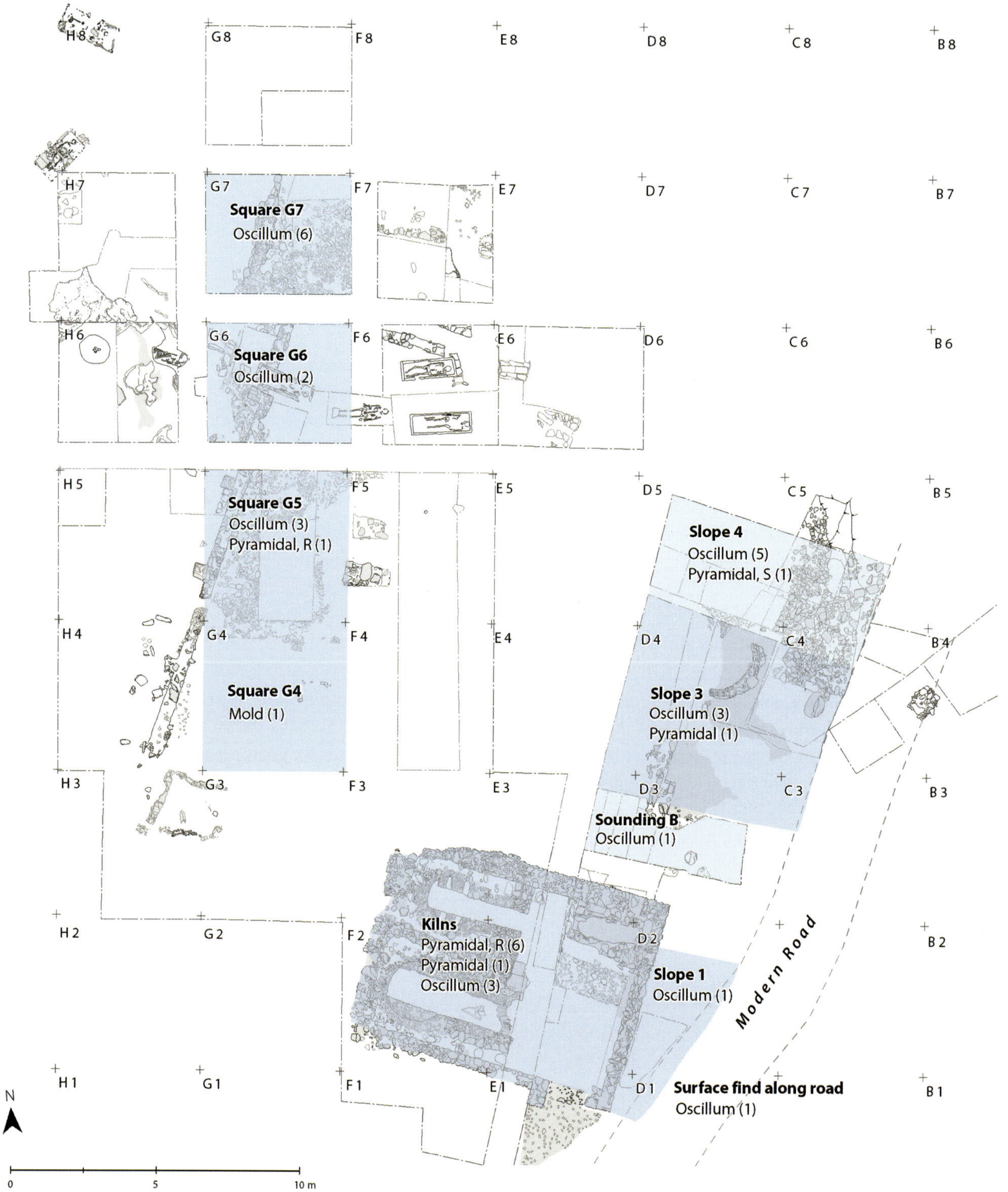

Figure 31.1 Distribution map of loom weights; areas highlighted in blue indicate general locations of listed finds. (JT/ICA)

Catalog of Loom Weights

1:2

SAV LW 01 Pyramidal Loom Weights

Provenance: House on the Hill, Test Trench 3. Level 1, batt. 1. Lot: SA 79-1014T A–B (see *Fig. 2.39*).

Dimensions and technical features: A—h 6.3, base major 2.8 x 3.7, base minor 1.1, hole diam. 0.6, weight 60 g. B—h 4.8, base major 2.5, weight 40 g.

Condition: A—Entire. B—Damaged on the top.

Description: Pyramidal, square variant. A—Quite fine and hard pale yellow clay (Munsell 5Y 7.2), overfired, sharp and smooth fracture of the section, frequent small black inclusions, some small quartz. B—Quite fine soft pale brown clay (Munsell 10YR 8.4), sharp powdery fracture of the section.

Comparisons and comment: Pyramidal loom weights probably dated to the Archaic and Classical periods are documented in Metaponto, in particular from the countryside (Foxhall in *Survey*, 541) and from the extramural votive deposit at Crucinia locality (Lo Porto 1981, 316, nos. 315–34). Productions of pyramidal loom weights are also occasionally recorded for the Archaic period in some indigenous settlements near the territory controlled by Metaponto, for example in Pisticci in the first half of the 7th c. BC (Lo Porto 1973, 156) and in Montescaglioso. In this last site, the workshop produced local Matt-painted pottery, imitations of Ionic-type B2 cups, as well as loom weights of the 6th c. BC (Lo Porto 1988–1989, 384–93). Archaic examples are sporadically attested at Cozzo Presepe, from a context in association with spindle whorls (du Plat Taylor/Macnamara/Ward Perkins 1983, 381–83). Recently, two concentrations of loom weights containing over 400 pyramidal specimens were found in the *anaktoron* of Torre di Satriano, dated from the second half of the 6th c. to 480 BC (Quercia and Foxhall 2012; Quercia and Foxhall forthcoming a).

Date: Late Archaic?

1:2

SAV LW 02 Pyramidal Loom Weight

Provenance: House on the Hill, Test Trench 2. Level 1, batt. 4. Lot: SA 79-673T (see *Fig. 2.39*).

Dimensions and technical features: h 7.2, base major 3.7, base minor 1.2, hole diam. 0.7, weight 84 g.

Condition: Chipped.

Description: Pyramidal, square variant. Fine and hard reddish yellow clay (Munsell 5YR 6.8), sharp smooth frequent small- and medium-sized black and gray inclusions, a few small chunks of quartz and mica.

Comparisons and comment: See SAV LW 01.

Date: Late Archaic?

Loom weights cont.

SAV LW 03 Disc-shaped Loom Weight

1:2

Provenance: Slope 1, surface find from Persephone plaque. Lot: SA 79-281T. Inv. no. 319944 (see *Fig. 2.35*).

Dimensions and technical features: diam. 6.4 cm, th. 2.2, hole diam. 0.5, weight 117 g.

Condition: Complete.

Description: Disc with two holes. Palmette in relief (from circular disc stamp?). Fine hard reddish yellow clay (Munsell 5YR 6.8), sharp smooth fracture, a few small black inclusions and sporadic mica. Smoothed surface with parallel lines.

Comparisons and comment: This decorative motif is broadly attested on *oscilla* and pyramidal loom weights from Metaponto (Lo Porto 1966, 153, pl. IV.4; Foxhall in *Survey*, 553, no. 34) and mainly from Herakleia (Lo Porto 1961, fig. 19d; Neutsch 1967, 174, pl. 39.1–2; Osanna, Prandi and Siciliano 2008, pl. XVII). The same decoration in relief is documented sporadically in loom weights from indigenous contexts of Apulia, such as Ascoli Satriano (Lo Monaco 2002, 298, no. 14, pl. 78). Palmettes, as well as lotus flowers and meanders in relief were impressed mainly by using the *stampi per focacce*, large circular disks documented at Metaponto and Herakleia (Lo Porto 1966, 156–57, fig. 17, pl. VI: Quercia and Foxhall 2012, 375–76): these impressions were in most cases not perfectly centered.

Date: 4th–3rd c. BC.

SAV LW 04 Disc-shaped Loom Weight

1:2

Provenance: Sounding B. Level 3, batt. 3. Lot: SA 79-303T. Inv. no. 319945 (see *Fig. 2.35*).

Dimensions and technical features: diam. 6.4, th. 2.2, hole diam. 0.5, weight 120 g.

Condition: Complete.

Description: Disc with two holes. Palmette in relief (from circular stamp?). Fine soft reddish yellow clay (Munsell 7.5YR 8.6), sharp powdery fracture, sporadic black inclusions.

Comparisons and comment: See SAV LW 03.

Date: 4th–3rd c. BC.

SAV LW 05 Pyramidal Loom Weight

Provenance: Slope 3. Level 3, batt. 1. Lot: SA 79-296T. Assemblage 11.

Dimensions and technical features: h 5.5, base major 6.0, base minor 3.5, hole diam. 0.5, weight 60 g.

Condition: Broken around hole and chipped at bottom.

Description: Pyramidal. One hole. Some wear at corners. Hard, pinkish beige black glaze to plain ware–type fabric.

Comparisons and comment: See above.

SAV LW 06 Disc-shaped Loom Weight

1:2

Provenance: Slope 3, W extension. Level 2, batt. 1. Lot: SA 79-101T. Inv. no. 319869. Assemblage 11.
Dimensions and technical features: diam. 7.2, diam. holes 0.4, weight 130 g.
Condition: Complete.
Description: Disc with two holes. Circular stamp, illegible; fingerprint on the other side. Fine soft pale brown clay (Munsell 10YR 8/4), sharp smooth fracture, sporadic black inclusions.
Comparisons and comment: Oscilla with oval and circular stamps are frequently found at sites in the chora of Metaponto from the 4th c. BC onward (Foxhall in *Survey*, 545–47, 549–53).
Date: 4th–3rd c. BC.

SAV LW 07 Disc-shaped Loom Weight

1:2

Provenance: Slope 3, W extension, continued N (D4). Level 1, batt. 4. Lot: SA 79-86T. Inv. no. 319888 (see *Fig. 2.35*).
Dimensions and technical features: diam. 7.0, th. 2.2, hole diam. 0.6, weight 130 g.
Condition: Chipped.
Description: Disc with two holes. Fine hard reddish yellow clay (Munsell 7.5YR 6.6), sharp smooth fracture. Occasional inclusions (mica?). White slip (?) on one of the surfaces: one surface is irregular, because perhaps it was in contact with a spacer or because it originally had relief decoration, no longer preserved.
Comparisons and comment: See above.
Date: 4th–3rd c. BC.

SAV LW 08 Disc-shaped Loom Weight

1:2

Provenance: Slope 4, gully wall N of Sounding C. Lot: SA 80-18T.
Dimensions and technical features: diam. 6.9, th. 2.3, hole diam. 0.7, weight 139 g.
Condition: Complete.
Description: Disc with two holes. Edges folded on themselves on one surface; the other surface is smoothed. Quite fine hard reddish yellow clay (Munsell 5YR 7.8), sharp smooth fracture, frequent small black and gray inclusions (quartz?), some mica.
Comparisons and comment: See above.
Date: 4th–3rd c. BC.

Loom weights cont.

SAV LW 09 Disc-shaped Loom Weight

1:2

Provenance: Slope 4, loose soil. Level 1, batt. 1. Lot: SA 79-51T (A). Inv. no. 319891 (A) (see *Fig. 2.35*).
Dimensions and technical features: diam. 5.5, th. 1.9, hole diam. 0.4 to 0.6, weight 73 g.
Condition: Complete?
Description: Disc with two holes. Fine soft reddish yellow clay (Munsell 5YR 6.8), sharp powdery fracture. Sporadic mica.
Comparisons and comment: See above.
Date: 4th c. BC.

SAV LW 10 Pyramidal Loom Weight

1:2

Provenance: Slope 4, loose soil. Level 1, batt. 1. Lot: SA 79-51T (C). Inv. no. 319892 (see *Fig. 2.35*).
Dimensions and technical features: h 5.7, base major 2.5, weight 48 g.
Condition: Damaged on the top, hole missing.
Description: Pyramidal, square variant. Coarse hard pink clay (Munsell 7.5YR 7.4), harsh fracture, numerous small- and medium-sized gray and black inclusions (sand or quartz?), sporadic mica.
Comparisons and comment: See above.

SAV LW 11 Disc-shaped Loom Weight
Provenance: Slope 4, Sounding C. Level 2, batt. 7. Lot: SA 79-363T (from SA 79-205PL) (see *Fig. 2.35*).
Dimensions and technical features: diam. 6.5, th. 2.2, hole diam. 0.4, weight 124 g.
Condition: Concreted all over.
Description: Disc, two holes. Hard, pale orange beige plain ware–type fabric.
Comparisons and comment: See above.
Date: 4th–3rd c. BC.

SAV LW 12 Disc-shaped Loom Weight
Provenance: Slope 4, Sounding C, E of D and including division. Level 2, batt. 3. Lot: SA 79-316T (A). Assemblage 13.
Dimensions and technical features: diam. 7.2, th. 2.8, hole diam. 0.6, weight 205 g.
Description: Disc, two holes. Faint stamp on front, round but slightly pointed at one end, illegible. Buff to beige hard plain ware.
Comparisons and comment: See SAV LW 06.
Date: 4th–3rd c. BC.

SAV LW 13 Disc-shaped Loom Weight
Provenance: Square G5. Level 2, batt. 1. Lot: SA 79-547T (A). Assemblage 3.
Dimensions and technical features: diam. 7.5, th. 2.4, hole diam. 0.4, weight 174 g.
Condition: Complete.
Description: Disc, two holes. Half-circle or sigma points up facing holes. Beige, slightly sandy, black-gloss to plain ware–type fabric.
Comparisons and comment: The impression of a sigma or a half circle motif is already documented in the territory of Metaponto (e.g., from Fattoria Stefan), as well as in other Greek and indigenous sites of southern Italy from the Archaic period onward. A similar motif occurred on a pyramidal example from a 7th-c. BC context at Francavilla Marittima (Kleibrink 2010, fig. 186), although this decoration appears to be more common in the Hellenistic period, as some *oscilla* from Ginosa (Capurso 1985, 60, pl. 21, no. 69) and Tiriolo (Ferri 1927, 341, fig. 7) show.
Date: 4th–3rd c. BC.

SAV LW 14 Disc-shaped Loom Weight
Provenance: Square G5. Level 2, batt. 1. Lot: SA 79-547T(B). Assemblage 3.
Dimensions and technical features: th. 2.2, weight 49 g.
Condition: Very broken, only small piece of top survives.
Description: Disc, had two holes originally, only one partially surviving. Gray, overfired, slightly sandy plain ware–type fabric.
Comparisons and comment: See above.
Date: 4th–3rd c. BC.

SAV LW 15 Pyramidal Loom Weight
Provenance: Square G5. Level 2, batt. 1. Lot: SA 79-787T. Assemblage 3.
Dimensions and technical features: h 9.2, base 3.0 x 4.5, hole diam. 0.6–0.9, weight 206 g.
Condition: Complete.
Description: Pyramidal. Rectangular with one hole. Quite fine soft reddish yellow clay (Munsell 5YR 6.8), sharp powdery fracture, frequent small black inclusions (quartz?), sporadic medium-sized quartz inclusions.
Comparisons and comment: See above.
Date: 4th–3rd c. BC.

1:2

SAV LW 16 Disc-shaped Loom Weight
Provenance: Square G6, E of wall. Level 2, batt. 2. Lot: SA 79-738T(A). Assemblage 3.
Dimensions and technical features: diam. 8.6, th. 1.9, hole diam. 0.5, weight 129 g.
Condition: Half is missing, totally concreted.
Description: Disc, two holes originally, only one survives. Orange brown plain ware–type fabric.
Comparisons and comment: See above.
Date: 4th–3rd c. BC.

SAV LW 17 Disc-shaped Loom Weight
Provenance: Square G7. Level 1, batt. 1. Lot: SA 79-767T (A) (see *Fig. 2.20*).
Dimensions and technical features: diam. 7.6, th. 1.9, hole diam. 0.4, weight 153 g.
Condition: Complete.
Description: Disc with two holes. Quite fine soft pale yellow clay (Munsell 2.5Y 8.4), sharp powdery fracture, frequent small- and medium- sized black inclusions (quartz?), sporadic chamotte.
Comparisons and comment: See above.
Date: 4th–3rd c. BC.

1:2

SAV LW 18 Disc-shaped Loom Weight
Provenance: Square G6, E of wall. Level 2, batt. 2. Lot: SA 79-738T (B). Assemblage 3.
Dimensions and technical features: diam. 6.5, th. 2.1, hole diam. 0.3, weight 100 g.
Condition: Broken on top right-hand side, completely concreted.
Description: Disc, two holes, one survives only partially. Orange brown plain ware–type fabric.
Comparisons and comment: See above.
Date: 4th–3rd c. BC.

Loom weights cont.

SAV LW 19 Disc-shaped Loom Weight

1:2

Provenance: Square G7. Level 1, batt. 1. Lot: SA 79-767T (B) (see *Fig. 2.20*).

Dimensions and technical features: diam. 5.3, th. 1.9, hole diam. 0.4, weight 65 g.

Condition: Complete.

Description: Disc with two holes. Decoration: circular stamp, illegible. Fine soft pink clay (Munsell 7.5YR 8.4), surface Munsell 10YR 8.4. Sharp powdery fracture, frequent small black inclusions (quartz?).

Comparisons and comment: See SAV LW 06.

Date: 4th–3rd c. BC.

SAV LW 20 Disc-shaped Loom Weight

1:2

Provenance: Square G7, cleaning around wall 4 to E. Lot: SA 79-911T. Inv. no. 319946.

Dimensions and technical features: diam. 7.5, th. 1.7, hole diam. 0.5, weight 120 g.

Condition: Damaged.

Description: Disc with two holes. Fine soft reddish yellow clay (Munsell 7.5YR 8.6), sharp powdery fracture, sporadic black inclusions.

Comparisons and comment: See above.

Date: 4th–3rd c. BC.

SAV LW 21 Disc-shaped Loom Weight

1:2

Provenance: Square G7, cleaning operation N-S wall 4, SE corner. Lot: SA 80-31T.

Dimensions and technical features: diam. 5.6, th. 1.9, hole diam. 0.4, stamp diam. 0.7, weight 62 g.

Condition: Chipped.

Description: Disc with two holes. Decoration: circular stamp with swan, goose or duck (water bird) in profile. Fine hard pale brown clay (Munsell 10YR 7.4), sharp smooth fracture, occasional small inclusions (quartz and mica).

Comparisons and comment: Oval and circular impressions of water birds occur frequently in Metapontine loom weights. A ring stamp of a bird is on an *oscillum* from the Metapontine chora (Foxhall in *Survey*, 549, no. 9), while a dove is impressed on two specimens from Favale deposit: (M. Liseno 2004, 68, pl. XXXI.e). Various birds (swan, dove) are impressed on some cubic and rectangular loom weights from Segesta (Biagini 2008, 648–50, 652, nos. 3–4, 9, 19, 33).

Date: 4th–3rd c. BC.

SAV LW 22 Disc-shaped Loom Weight

1:2

Provenance: Square G7, E of wall. Level 2, batt. 1. Lot: SA 79-800T. Assemblage 17.

Dimensions and technical features: diam. 7.1, th. 1.9, hole diam. 0.4, weight 126 g.

Condition: Chipped.

Description: Disc with two holes. Quite fine hard reddish yellow clay (Munsell 5YR 6.8), surface Munsell 10YR 8.4, sharp smooth fracture, frequent small inclusions of mica or quartz, a few medium-sized quartz, sporadic white inclusions (calcite?).

Comparisons and comment: See above.

Date: 4th–3rd c. BC.

SAV LW 23 Disc-shaped Loom Weight

1:2

Provenance: Square G7, E of wall. Level 2, batt. 5. Lot: SA 79-985T. Assemblage 4.

Dimensions and technical features: diam. 6.8, th. 2.2, hole diam. 0.4, weight 148 g.

Condition: Chipped.

Description: Disc with two holes. Quite fine hard reddish yellow clay (Munsell 7.5YR 6.8), sharp rough fracture of the section. Numerous inclusions of small and medium size (quartz?), some foraminifera and mica.

Comparisons and comment: See above.

Date: 4th–3rd c. BC.

SAV LW 24 Disc-shaped Loom Weight

1:2

Provenance: Chance find (2008), Sant'Angelo Vecchio.

Dimensions and technical features: diam. 7.5, th. 1.8, hole diam. 0.8, stamp diam. 1.9, weight 127 g.

Condition: Damaged.

Description: Disc with two holes. Circular stamp with rosette, 10 petals. Fine hard pink clay (Munsell 5YR 8/4), sharp and smooth fracture, a few small quartz inclusions.

Comparisons and comment: Rosettes are a very common motif that decorated Metapontine loom weights and numerous other examples from southern Italy since the 5th c. BC, if not earlier. This decoration appears to have close parallels with gemstones of the 7th and 6th c. BC. Impressions of identical rosettes are documented in chronologically and topographically different household locations in the Metapontino (Foxhall in *Survey*, 546–47; Foxhall 2012, 203), which raises questions on the presence of "heirloom/heritage" stamps and on the mobility and links between the women who lived in the Metaponto territory. The rosette impression, with slightly different variants given by size, shape, and number of petals, also occurs in other Greek cities of the Ionian coast, such as Herakleia (Meo 2011, 2–3, fig. 4) and in a few disc-shaped loom weights from indigenous sites close to the Metapontine chora (Cozzo Presepe: Morel 1970, 105, fig. 29a, c, and d). The same decorative motif occurred also in pyramidal examples from inland Lucanian settlements, such as Pomarico (Sartoris 1997, 229, no. 9, pl. 91.) and Oppido Lucano (Lissi Caronna 1980, 219, nos. 4–5, fig. 134; Lissi Caronna 1983, fig. 89, tr. 30 sud str. 1, 295, nos. 7–8, fig. 90, 329, no. 9, fig. 125, 332, nos. 13–14, fig. 125; Lissi Caronna/Armignacco Alidori/Panciera 1992, 205, no. 22, fig. 25, fig. 219, no. 13, fig. 41, 280, no. 5, fig. 105, 281, no. 22, fig. 109, 284, nos. 36–37, fig. 108, 461, no. 2, fig. 245, 465, no. 46, fig. 247.

Date: 4th–3rd c. BC.

Loom weights cont.

SAV LW 25 Mold for Disc-shaped Loom Weight
Provenance: Square G4. Level 1, batt. 1. Lot: SA 79-435T. Inv. no. 320001 (see *Fig. 2.22*).
Dimensions and technical features: diam. 7.5, th. 1.8, hole diam. 0.8, stamp diam. 1.9, weight 127 g.
Condition: Complete.
Description: Mold for relief-decorated loom weight, which represents an Eros running and holding perhaps a bow and a quiver.
Comparisons and comment: A loom weight possibly made from this mold appears in the Pantanello sanctuary assemblage (PZ80.0180). The same motif, which appears to be typical of Metaponto, is documented on other loom weights from the urban and suburban area (Pesce 1936, 446, fig. 9; Lo Porto 1996, 153, pl. V.3; Adamesteanu 1980, 287, fig. 301). Molds for decorating specifically *oscilla* in relief are not frequently found in archaeological contexts: two inscribed examples are stored at the Archaeological Museum of Taranto (Ferrandini Troisi/Buccoliero/Ventrelli 2012, nos. 11 and 88) and some *oscilla* molds were identified at the Hellenistic site of Monte di San Fratello near Messina, the ancient Apollonia (Bonanno 2008, 61, nos. 20–21).
Date: 4th–3rd c. BC.

1:2

SAV LW 26 Pyramidal Loom Weight
Provenance: Square E2, cleaning of S Kiln praefurnium. Lot: SA 80-64T. Assemblage 7.
Dimensions and technical features: h 10.9, base major 5.9 x 3.5, base minor 4.1 x 3.2, hole diam. 0.8–0.9, weight 352 g.
Condition: Complete.
Description: Pyramidal, rectangular variant. Holes on the shorter side. Quite fine hard pale yellow clay (Munsell 5Y 8.2), overfired, sharp smooth fracture, frequent small- and medium-sized black and reddish brown inclusions (probably quartz).
Comparisons and comment: This is one of the four almost identical loom weights ranging in weight from 340 to 352 g found in the same context. For a similar loom weight, see Firmati 2012, 206, pl. XXVI.4 (second half of the 2nd–first half of the 1st c. BC).

1:2

SAV LW 27 Pyramidal Loom Weight
Provenance: Square E2, cleaning of S Kiln praefurnium. Lot: SA 80-66T. Assemblage 7.
Dimensions and technical features: h 10.5, base major 5.9 x 4.2, base minor 4.4 x 3.2, hole diam. 0.7, weight 334 g.
Condition: Damaged on the bottom.
Description: Pyramidal, rectangular variant. Hole on the shorter side. Coarse very hard pale olive clay (Munsell 5Y 6.3), overfired, sharp smooth fracture, frequent small- and medium-sized black, white and red brown inclusions (probably quartz).
Comparisons and comment: See above.

1:2

Loom weights cont.

SAV LW 28 Disc-shaped Loom Weight 1:2

Provenance: Square F1. Level 1, batt. 2. Lot: SA 79-784T (A). Inv. no. 319947 (A) (see *Fig. 2.30*).

Dimensions and technical features: diam. 11.0, th. 2.5, hole diam. 0.9, weight (preserved half) 168 g.

Condition: Broken.

Description: Disc with two holes, the preserved one with wear traces. Fine hard light brown clay (Munsell 7.5YR 6.4), sharp smooth fracture, sporadic black inclusions and mica.

Comparisons and comment: See above.

Date: 4th–3rd c. BC.

SAV LW 29 Disc-shaped Loom Weight 1:2

Provenance: Square F1. Level 1, batt. 2. Lot: SA 79-784T (B). Inv. no. 319947 (B) (scc *Fig. 2.30*).

Dimensions and technical features: diam. 8.0, th. 2.5, hole diam. 0.5, weight 146 g.

Condition: More than half preserved.

Description: Disc with two holes. Illegible circular stamp? Fine soft reddish yellow clay (Munsell 5YR 7.8), powdery fracture. Sporadic mica and black inclusions.

Comparisons and comment: See above.

Date: 4th–3rd c. BC.

SAV LW 30 Pyramidal Loom Weight 1:2

Provenance: Square F3, N Kiln interior. Level 2, batt. 2. Lot: SA 79-870T. Assemblage 8.

Dimensions and technical features: h 10.2, base major 5.8 x 3.8, base minor 3.2 x 2.8, hole diam. 0.8, weight 300 g.

Condition: Chipped.

Description: Pyramidal, rectangular variant. Fine hard reddish yellow clay (Munsell 7.5YR 7.6), surface Munsell 2.5Y 8/2, irregular granulated fracture, a few small black inclusions.

Comparisons and comment: See introduction for further details.

Date: 4th–3rd c. BC.

SAV LW 31 Pyramidal Loom Weight

1:2

Provenance: Square F3, N Kiln interior, SE corner of kiln near baulk. Level 2, batt. 4. Lot: SA 79-881T. Inv. no. 320046. Assemblage 8.

Condition: Complete.

Description: Pyramidal.

Comparisons and comment: See above.

32

Roof Tiles and Bricks

Carlo Rescigno, Francesco Perugino, and Eliana Vollaro

Among the materials from the site of Sant'Angelo Vecchio, there is a substantial number of tile fragments. They come from diverse stratified contexts, collapses, and dumps, as well as from contexts in which they had been reused (*Fig. 32.1*); in fact, both flat and curved tiles were commonly used in the construction of walls, where they functioned as bricks.

This study concentrated on a morphological-typological analysis of both the pan and cover tiles as well as the creation of a typology. The majority of the fragments could be attributed to two different roofing systems: the first, which is here called "A," used curved pan (*tegulae*) and cover (*imbrices*) tiles in the so-called Laconian model, while the second, "B," used flat tiles in two different systems of assemblage, interlocking and overlapping. In the interlocking system, a tile is inserted between the raised flanges of the ones below it, and thus is tapered by means of notching at the front corners. In the second system, called "overlapping," the upper tile covers the lower one and bears two notches on its lower surface to allow it to fit onto the flanges of the lower tile. Neither system is typical of a specific period; rather, each one is attested at various times in the areas where they are encountered.[1]

In this study, types are identified via a code with three elements. The letter identifying the system is followed by the abbreviation of the form; the final element, a number, corresponds to the individual type, indicated by a serial number beginning with 1. The typological codes are summarized in *Table 32.1*.

In the absence of mineralogical analyses, a detailed classification of the fabrics was not created. In this study, the problem was addressed preliminarily by limiting analysis to the visual and empirical identification of macro-groups. Samples were taken from each identified group as reference material, as described below. Within a group, subdivisions can be identified, but they are of limited value due to the empirical analysis. In the case of fabrics with similar characteristics but slight differences in color, texture, and inclusions, a further subdivision was made using letters in alphabetical sequence, which follow the reference number.

T1: orange-pink clay, hard and compact, with rare insertions of lighter color; occasional vacuoles, chamotte, some large, yellowish, spongy, porous inclusions.
- a: color varying between straw-yellow and greenish-gray, particularly compact.
- b: dark gray nucleus, vitrified; very compact and sharp. Surface orange–light pink in color.

T2: yellow-beige clay, moderately compact, powdery to the touch, clean fracture, frequent large quartz/feldspar inclusions, occasional moderately large biotite inclusions.

T3: clay differing from T1 only in the presence of larger inclusions, among which there appears to be straw.

T4: compact brick-red clay, hackly fracture, moderately hard, abundant milk-white, reddish, and gray quartz inclusions.

The most common clay type attested at Sant'Angelo Vecchio is T1, of local origin. The few fragments of tiles and cover tiles belonging to system Ba are different, characterized by a fabric of the T2 type, perhaps a regional production, as the abundant fairly large quartz-feldspar inclusions and occasional microscopic biotite inclusions seem to suggest. For this fabric it may even be suggested that the clays came from geological deposits that are more typically Calabrian.[2] The catalog contains only the fragments that are significant for their type or state of preservation. For all others, a typological table is presented at the end of the study (*Table 32.3*).

The Roofing Systems

Carlo Rescigno

Using the typological analysis, the site report, and the stratigraphic sequence, the identified roofing systems

[1] Rescigno 2001, 201–2.

[2] Observation under the microscope was kindly undertaken by K. Swift.

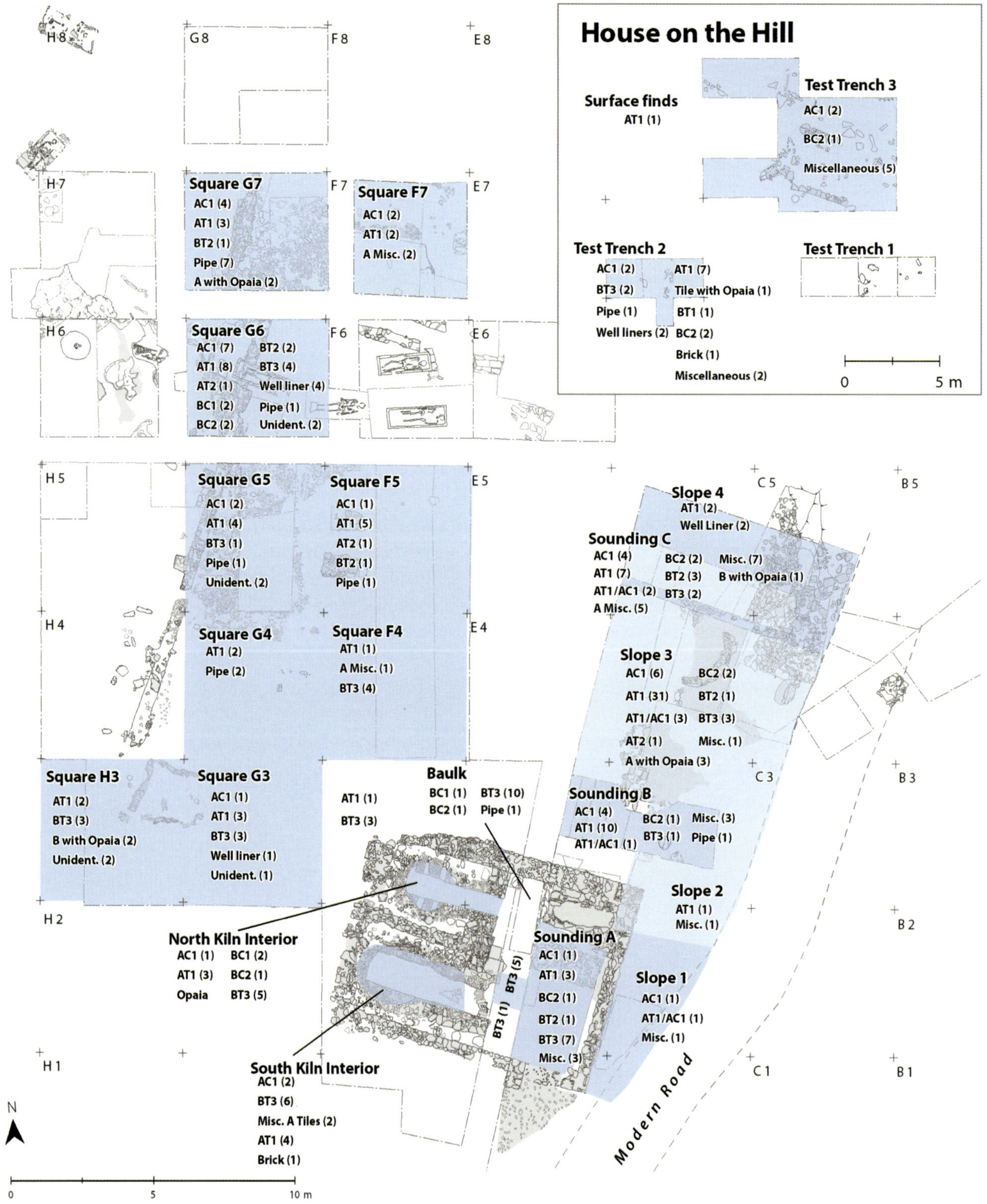

Figure 32.1 Distribution map of pan tiles and cover tiles across the excavation areas; areas highlighted in blue indicate general locations of listed finds. The distribution map includes only the materials collected during the excavations of 1979–1981 (all of which were diagnostic fragments), and therefore this map illustrates the distribution of functional and morphological types, and not the total quantity encountered. (JT/CR/ICA)

Figure 32.2 Proposed reconstruction of the roofing system using sima with channel and silenos-head antefixes. (PF/ICA)

Figure 32.3 Proposed reconstruction of the roofing system using sima with channel. (PF/ICA)

System A	Laconian System
Aa	*Archaic system*
Type AT0	Tile with concave profile
Opaia	Tile with opaion
Ab	*Hellenistic system*
Type AT1	Tile with concave profile
Type AT2	Tile with concave-convex profile
Type AC1	Cover tile with depressed arch
Opaia	Tile with opaion
System B	**System with flat tiles**
Ba	*Archaic/Classical system*
Type BT1	Flat overlapping tile
Type BC1	Horseshoe-shaped cover tile
Type BS1	Sima
Bb	*Hellenistic/Roman system*
Type BT2	Flat overlapping tegula
Type BC2	Imbrex with depressed arch
Bc	*Roman system*
Type BT3	Flat tegula for insertion
Opaia 1	Tile with opaion
Opaia 2	Tile with opaion
Tubuli (gutter pipes)	

Table 32.1 Types of roof tiles from Sant'Angelo Vecchio.

can be associated with the site's individual periods of occupation (*Fig. 32.2–Fig. 32.7*).

System Ba (tiles BT1, BC1, BS1). The roof with flat pan tiles and semicircular cover tiles is associated with the site's earliest occupation period. This system was found in the House on the Hill context, and the same tiles were used in the small cemetery on the hill slope, which predates the Early Hellenistic phases and thus provides an important terminus ante quem. The dating is supported by various parallels. A fragment of a sima (SAV RT 129), which may be interpreted as a sima with drip channel,[3] probably in secondary deposition, was found in the small courtyard in front of the Late Republican Kiln Complex. The sima preserves traces of a painted red line, the attachment point of the open channel, the form of the side, and the impression of the flange of the tile which seems similar to the flattened type documented in system Ba. The same courtyard produced a fragment of a mold for a silenos-type antefix (SAV AT 08),[4] which, together with a few other ante-

[3] G. Greco 1991, 42, figs. 99–100.

[4] For more on molds, see Ch. 29, "The Architectural Terracottas: Some Considerations on Production," and Ch. 30, "The Molds from Sant'Angelo Vecchio."

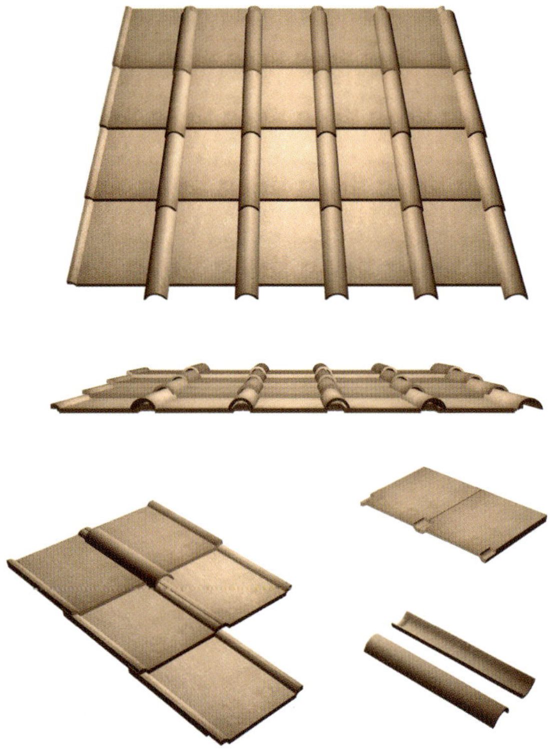

Figure 32.4 Reconstruction of flat tile system Ba attested at Sant'Angelo Vecchio. (PF/ICA)

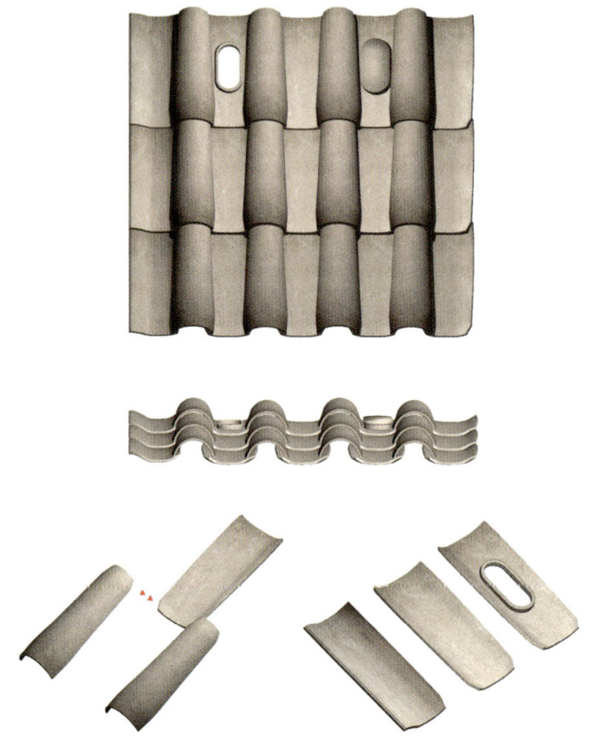

Figure 32.5 Reconstruction of the Laconian roofing system Ab attested at Sant'Angelo Vecchio. (PF/ICA)

fix fragments dating to the 6th–5th centuries BC, could indicate the nearby presence of a building of a certain architectural pretension (*Fig. 32.2* and *Fig. 32.3*).

System Aa (tile AT0, opaion). The use of Laconian tiles is attested contemporaneously with other systems. The small side edge of tile SAV RT 130, with traces of red paint on the inner surface, is the only fragment that can be securely attributed to the Archaic period.

System Ab (tiles AT1, AT2, AC1). The majority of the Laconian tiles can be attributed to the Early Hellenistic period, which was the period of the kilns and the terracotta relief plaque dumps. The excavation determined their use as construction material for the first kilns and as roofing for the structures that must have preceded the Late Republican workshop shed. Two overlapping collapses were exposed in the western sector, the earliest of which consisted entirely of curved tiles. The variations in shape and technical characteristics among the fragments from Sant'Angelo Vecchio enable comparison with the better-preserved examples from Fattoria Fabrizio, and thus reconstruction of their standard dimensions. Two stamps with Greek letters were found on tiles of this type.

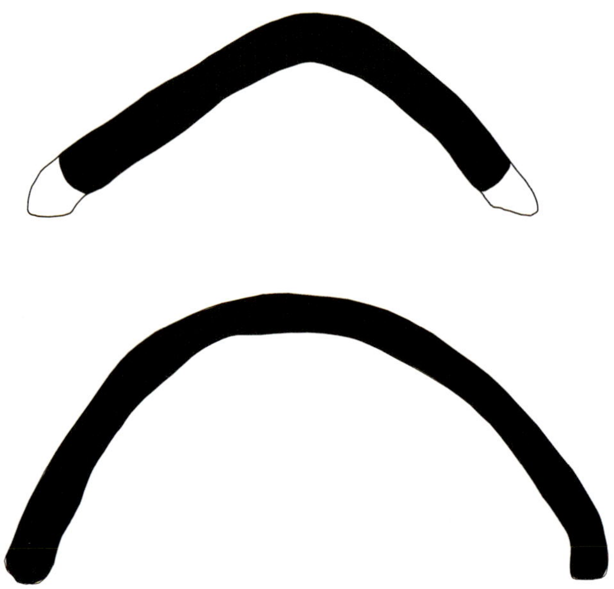

Figure 32.6 Laconian cover tiles from Pantanello, PZ 78-614, examples of standard size; scale 1:3.

System Bb (tile BT2). The area of Soundings A and C produced a few fragments of a type of flat tile, which was also employed using an overlapping roofing sys-

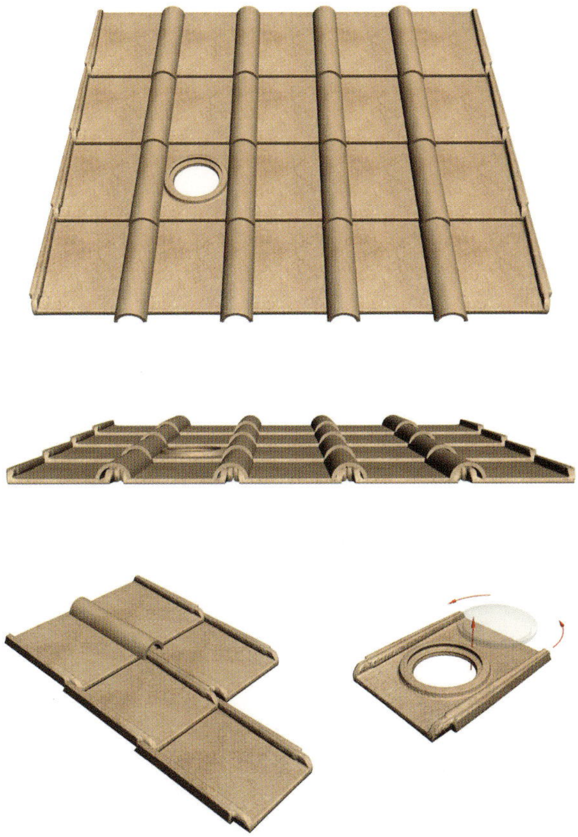

Figure 32.7 Reconstruction of flat tile system Bc attested at Sant'Angelo Vecchio. (PF/ICA)

Figure 32.8 Cover tile BC1, SA 79-1020, example of standard size; scale 1:3.

tem. Some of the fragments, however, are similar to the tiles of type Bc.

System Bc (tile BT3). This system can be attributed to the final occupation phase of the area. It consists of well-made and well-finished pan tiles assembled by interlocking square flanges, often with a longitudinal groove on the upper surface. These types of tile were found in the small courtyard of the Kiln Complex and in the area of Wall 1, a workshop shed, to the roof of which these tiles belonged. The distance between the pilasters corresponds to a multiple of the tiles (ca. 3.1 m, corresponding to eight times the width of a tile of this type, 39.0 cm). Therefore the function of the pilasters present on the external face of the wall was probably to strengthen the structure, marking the point on which the sloped roof-beams, tied in with side beams, must have rested and dispersed their weight. This hypothesis is supported by the fact that the scant traces of the workshop shed's column bases are aligned with the pilasters, thus constituting the point where the sloped beam must have rested on the architrave in correspondence with the columns and dispersed its weight. The agreement of dimensions indicates a sharing of standards between *fictores* and architects. This type of tile is associated with a substantial number of stamps, both letters and symbols.

Tile stamps, thus, can be associated with two of the five roofing systems identified. A small number of impressed marks and a single stamp with Greek letters belong to the Laconian type of roofing system (Ab), dating to the second occupation period. The first group comprises four marks of uncertain identification, all traced on the underside of the tiles, mostly curvilinear in form and made with an implement that had a rounded point: in two cases a sort of O,[5] on another a straight line joining a half circle,[6] and lastly a small impressed oval mark, which may be accidental and therefore insignificant. These were probably workshop marks, part of a production monitoring process, perhaps a rudimentary numbering system for tallying the lots of completed materials. Beside this system, there were identifying stamps that carried another sort of information. Only one such stamp is documented, with two incomplete examples of the same stamp, both on the underside of tiles.[7] The Greek letters are inside a rectangular cartouche. The inscription is damaged but can be reconstruct-

[5] SAV RT 05; stamp h 3.0, w 3.0. SAV RT 17; stamp h 5.0, w 4.0.

[6] SAV RT 19; h 11.0 cm. A similar mark composed of two joining elements, a flattened O intersected by a long and straight line, also appears on the curved, unworked—and therefore lower—side of a fragment from Sant'Angelo Vecchio, inv. no. 316619, placed among the unidentified examples, perhaps as part of an imbrex: l 8.0, h 3.0. Similar impressed marks were also observed on tiles from Fattoria Musillo, occupied in the 6th and 4th–3rd c. BC (Amore 1992–1993, 111, fig. 24).

[7] SAV RT 01; stamp w 4.0, h 2.0. SAV RT 02; stamp w 5.0, h 2.0.

Figure 32.9 Distribution of *Herakas* stamps in the chora of Metaponto. (JT/ICA)

ed as hHPAK[ΑΣ] by joining the two examples from Sant'Angelo Vecchio and examining other Metapontine stamps. The eta is large and nearly square, the rho is characterized by a small eye, the alpha by a straight horizontal cross-stroke, and the kappa by short, oblique strokes. The presence of the aspirate, as noted by Lo Porto, is a characteristic trait of the alphabet at Metaponto, Taranto, and Herakleia (*Fig. 32.10*).[8] As mentioned above, stamps bearing the same content, but with variations in the letter shape, are also known in the Metapontine area and the immediate surroundings (*Fig. 32.9*).[9]

The following list includes the attestations known to the author, including the two "new" examples.

Variant a: Lucignano between Matera and Ginosa, an example on a tile fragment from a votive deposit which, judging from the published record, seems to be of the Laconian

[8] Lo Porto 1966, 152 n. 2.

[9] The published attestations from Sybaris (*NSc* 1972 Suppl., 316, no. 20, fig. 349 on p. 326, this is a stamp in Latin) and from Lipari (*Meligunis-Lipara II*, Palermo 1965, 331–32, the name documented here is Heràkleios) should be removed from the list of stamps belonging to the *Herakas* group, thus restricting the distribution of the type to the Metapontino and environs, the furthest findspot being the Lucignano forest (see above).

However, given the latest evidence on the extent of the Metapontine chora (De Siena 2010, 15), this may be considered an area exposed to Metapontine influence. Small (2006, 329–30) assigns this site to the Peucetian area and, for reasons of topographic continuity, the district dominated by the settlement of Montescaglioso; although he acknowledges that the stamp came from a specific mold, he suggests a Metapontine provenance and thus reads the stamp as Herakas, at the same time proposing a different solution. In fact, in the same volume he discusses another example from the Peucetian area, a stamp from San Felice in the orbit of Botromagno with a double register of letters which he interprets as a single name, Herakleidas or Herakleides, with parallels from Herakleia and Taranto; this piece could represent a further parallel for the stamp from Lucignano. A mold with the inscription HRAK written from right to left was recently found in layers of fill in the southern sector of the residential quarter of Herakleia at Policoro (De Siena 2011, 643).

type; the stamp, perhaps impressed on the underside, missing the second half, is within a rectangular cartouche and lacks the annotation of the aspirate; the alpha has a straight horizontal cross-stroke.[10]

Variant b: Sant'Angelo Vecchio, two fragmentary examples on Laconian tiles with annotation of the aspirate; the alpha has a straight and horizontal cross-stroke.

Variant c: Contrada Sansone, complete but damaged tile fragment bearing a stamp in a rectangular cartouche with annotation of the aspirate; the sigma is lunate; the cross-stroke of the alpha is bent; the stamp is known only from the transcription;[11] and Pantanello, tile stamp in rectangular cartouche with missing front part that perhaps bore the aspirate; the lunate sigma is damaged, the alpha probably had a bent cross-stroke.[12]

The distribution therefore appears to be concentrated around Metaponto. With the exception of Contrada Sansone and Pantanello, where the identification of the stamp should be verified, each site has its own variants. For Lo Porto, the annotation of the aspirate and the presence of the alpha with bent cross-stroke establish a chronological variant which would limit the distribution of the attestations to the 4th century BC (variant a) or push it slightly into the following century (variant c). The examples from Sant'Angelo Vecchio would fit between the two groups, as they share features encountered in both (presence of the aspirate, alpha with horizontal cross-stroke). Therefore, the formation of the letters cannot provide a relative chronology by itself, and the variants can only be observed, with no information as to whether they are due to chronological factors.[13]

Wreath within a square cartouche, type 1, single stamp	SAV RT 66–SAV RT 85
Wreath within a square cartouche, type 1, double stamp	SAV RT 55
Wreath within a rectangular cartouche, type 2	SAV RT 90; SAV RT 91
Wheel with eight spokes	SAV RT 93; SAV RT 94
Anchor with ring	SAV RT 92
AVF in rectangular cartouche	SAV RT 86–SAV RT 89

Table 32.2 Stamps with symbols and letters.

Despite the difficulty created by the small number of attestations,[14] Herakas can be considered a name in the genitive, and thus the stamps from Sant'Angelo Vecchio fall within the well-attested and widespread group of nominal marks.[15] These names probably served to certify ownership of the workshops. This interpretation seems to be supported by contrast and comparison with other stamps showing the specific indication of the commission for which the tiles were intended to be used: building or structure, references to the city or sacred space of the sanctuaries.[16] Occasionally this second type of stamp also appears with onomastic marks, suggesting that private workshops worked on public commissions,[17] but the material is notoriously *sub iudicio*, given that the system of stamping and marking is not always legible in all contexts, and the reference to the workshop owner's name is subject to dispute. The Metapontine stamps represent the simplest solution, limited to the name. The onomastic mark denoted monitoring of the production cycle and the quality of the product. The area within which the stamp was distributed has been substantially narrowed down, to the wider territorial context of the Metapontine chora and the immediately surrounding district, and thus, in this specific case, there is no room for hypotheses on long-distance export.[18] Based on a

10 Lo Porto 1973, 224, no. 16, pl. LXIX.4.

11 Lo Porto 1966, 152, no. 11.

12 Carter 1983a, 450, fig. 21; 454, no. 21.

13 The letters with bent horizontal cross-strokes occur within the general development of an alphabet with curvilinear elements that became fully established during the course of the 3rd c. BC (Guarducci 1995, I, 371). Examples of alphas with bent cross-strokes, however, are known in Magna Grecia, at Taranto, from as early as the 4th c. BC (Guarducci 1995, I, 290–91, n. 2). The lunate letters, including the sigma, perhaps documented in our stamps, also had early beginnings despite being widespread in the Late Hellenistic period (Guarducci 1995, I, 377).

14 *LGPN III.A*, 193, Ἡρακᾶς, which documents the stamp from Pizzica; *LGPN IV*, 153, again, Ἡρακᾶς, from the Cimmerian Bosphorus, although this is an example from the Roman Imperial period.

15 For Sicily, see the initial observations in Wilson 1979, in particular p. 23, sometimes more than one name in the genitive. On Hellenistic stamps in southern Italy, the possible significance of the names, and the many variants, see Small 2006.

16 For a discussion of public workshops or private workshops which undertook public commissions, see Wilson 1979, 23–24; Small 2006.

17 See the reinterpretation of the stamp of the *veri reggini* in Lazzarini 1982, identified as the city's ethnic (indicating the public commission) and the potter's name (indicating the private ownership of the workshop which received the commission).

18 For a detailed study of stamps, names, and political control of territories, see Simonetti (2001), who observes onomastic similarities with Locri and

Figure 32.10 Tile stamps; scale 1:3.

Figure 32.11 Example used as a burial cover, SA 79-1074, (BT1); scale 1:6.

Figure 32.12 Example used as a burial cover, SA 79-1028, (BT1); scale 1:6.

comparison with other models,[19] a more likely scenario is that of artisan mobility—that is, production on different sites by *figuli* from a single (urban?) workshop. In this case, the workshop would have specialized in roof tiles, relocating as needed according to the commissions offered to them, such as the construction of a sanctuary or a farmhouse. The workshops may also have had rural satellite facilities located at strategic points within the territory. This model, among others, could explain the variants in the stamps and the use of local clays.[20] In these cases, the onomastic stamp may have served to certify the quality of the production and the provenance of the tiles, while the impressed marks may instead relate to the monitoring of the production cycle and the numbering of the production lots.[21]

The stamps with letters and symbols, the largest group of stamps from Sant'Angelo Vecchio, belong to a third structural period. The two types occur only on tiles (where enough of the stamp is preserved for identification) that belong to a single functional type, and in any case to a single production group, Bc (*Table 32.2*).

Stamps with symbols (*Fig. 32.10*):

Wreath within a square cartouche, type 1, circular, nine leaves formed by short, straight strokes with bulging ends, closed at the bottom by two short and narrow ties (h 3.0, w 3.0). The stamp appears on the upper surface of the tile and close to the front edge (where preserved). Twenty single stamps; one double example (i.e., the same stamp is repeated);

Wreath within a rectangular cartouche, type 2, circular, six leaves with lines crossing the circle of the wreath,

Reggio and interprets them in relation to the political struggles for control of the territory of Caulonia (the destruction of the city and its annexation to Locri by Dionysios I at the beginning of the 4th c. BC; the occupation of the town by the Mamertines of Rhegium during the 3rd c. BC). On some marks, a similarity with the demotic marks in the accounts of the archive of Olympian Zeus at Locri can be observed.

[19] Peacock 1979, where his second model seems particularly well suited to explain the Metapontine situation: small productions for export covering a limited distance, with the possibility of artisan mobility and on-site recreation of the tile-making workshop. The recreation of the *atelier* at the site of the architectural workshop appears to be a much simpler and logical hypothesis, unlike pottery workshops, which required more equipment and specialized skills and enjoyed greater freedom in commercial distribution. Also regarding the wide circulation of some stamps, which went beyond the political districts where they were made, Simonetti (2001, 452) prefers the solution of artisan mobility, in which specialized workers were sent out from a workshop (for example, the Sosenos stamp, the genitive of the name of the owner of a workshop located perhaps at Reggio, dating to the 3rd c. BC, found at Caulonia, Reggio Calabria, Lipari, Termini Imerese, Cefalù, and Solunto).

[20] This second hypothesis requires testing through clay analysis. In material from a homogeneous geological basin, there is a low probability of recognizing the differences through direct observation.

[21] On the other hand, stamps indicating a public project or the name of the structure may have functioned as a check on the commission, as well as—according to the majority of learned opinion—a deterrent against theft by certifying the ownership of the material.

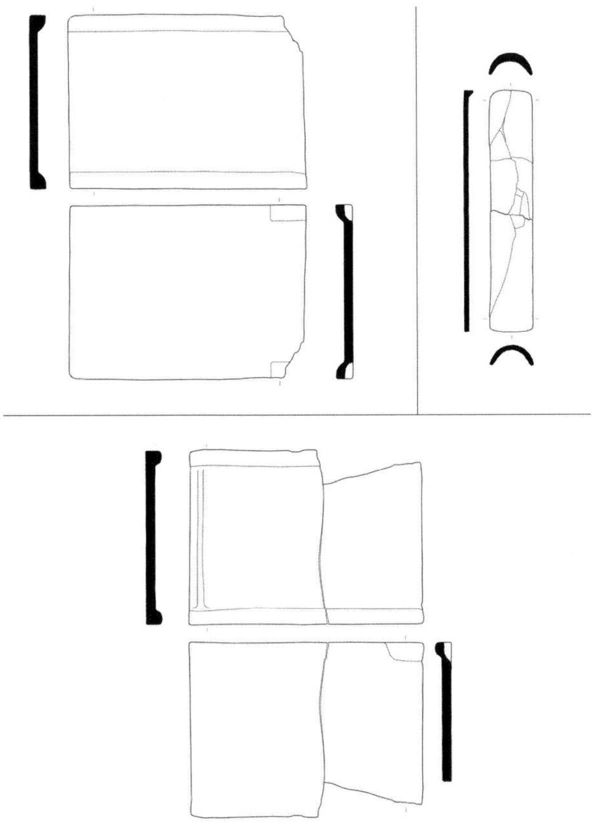

Figure 32.13 Uncertain provenience, examples of standard size from the Pantanello necropolis; scale 1:25.

closed at the bottom by two short and narrow ties (h 4.0, w 3.0). The stamp appears on the upper surface of the tile and close to the front edge (where preserved). Two examples;

Wheel with eight spokes, perhaps with central hub, within a rectangular cartouche (h 3.0, w 4.0). Two examples;

Anchor with ring, straight vertical element, and side arms, in rhomboid cartouche (max. cons. h 4.0, max. cons. w 4.0). One example;

Two unidentified examples, stamp very abraded, square cartouche (h 3.0, w 3.0).

Stamps with letters (SAV RT 86–89):

AVF in rectangular cartouche, letters in ligature, the first oblique stroke of the A has a bulging end (w 3.0). Stamp on upper surface of tile. Four examples.

To date no example of a single tile bearing both types of stamp has been found; the only example of a dou-

Figure 32.14 Example used as a burial cover, SA 79-981 (BT1); scale 1:6.

ble stamp repeats the same one, in this case the type 1 crown. No substantial conclusions, however, can be drawn from this observation. In fact, due to the fragmentary state of the finds it is impossible to evaluate whether, in some cases, the two systems—letters and symbols—coexisted on the same tile. The uniformity of the tiles suggests that the two groups of stamps belong to a single system and can probably be combined.

As *Table 32.2* demonstrates, the symbolic marks are the best-documented group, with several types: two variants of a wreath with ties, a type of wheel, and a rhomboid stamp with anchor. The other group comprises a single type in a rectangular cartouche with letters in ligature, clearly legible as AVF.

Regarding the first group, multiple parallels exist in various classes that, although differing in chronology, possibly overlap with the Late Republican period, which is the period of the fragments from Sant'Angelo Vecchio. Anchors, wreaths, and wheels occur as secondary types or countermarks, and rarely as primary types, on coins with a wide chronological range and a common period of recurrence for all known types corresponding to the 2nd–1st centuries BC.[22] The same types of stamp with symbols return with a more

[22] The diffusion of the symbols as primary and secondary types or as countermarks on coins can be followed in Crawford 1974 (to which this note refers for the numbering of the issues). The anchor, for example, appears as a secondary type after 211 BC (no. 50, Roman mint, 209–208 BC; no. 86b, mint in southeastern Italy, 211–210 BC; with a different shape, larger and more detailed, on a piece of *aes signatum*, no. 10, 260–242 BC), continuing throughout the 2nd and into the 1st c. BC in more complex compositions, associated with the rudder and in trophies (nos. 194, 258, 290, 340 etc.), often as a symbol or countermark (for example, no. 22, 3rd c. BC, and nos. 336, 343–344, 378, 385, and 405). The wreath, as a primary type, appears more rarely (for example, no. 393) and is more common as a secondary type (no. 110, 211–208 BC; no. 239, 136 BC; no. 253, 131 BC; no. 260, 128 BC); it is a frequent mint countermark throughout the 1st c. BC. The wheel, rare as a primary type (no. 24, 265–242 BC), returns as a secondary type (no. 79, 209–208 BC) and as a mint mark during the 1st c. BC, but it does not seem to be a particularly widespread type. It should be noted that a member of the Aufidii was a magistrate who minted coins and was active in 140 BC (no. 227, M. Aufidius Rusticus, whose abbreviated nomen AVF in ligature appears on the coins); he was perhaps the brother of T. Ofidius M. F. Pop., one of the senators associated with the *senatus consultum de Ambraciotibus et Athamanibus*. Circular stamps with rosettes or wheels were also observed on unidentified types of tiles recovered at the farmhouses in contrada Pagliarone, occupied from the 7th c. BC through the Roman period: Amore 1992–1993, 147, fig. 46.

varied repertoire as amphora stamps,[23] among which the Cosan amphorae of the Sesti group stand out for their similarity and associations. Here, the name stamp appears beside the stamp with symbols.[24] The link between these groups of stamps and those from Sant'Angelo Vecchio is clearly not a direct derivation, but (despite the problems associated with reading and interpretation) a common system of marking was used as a check which in many cases can be only vaguely detected.[25] Despite the variability of the materials on which they appear, the marks and countermarks seem to be intended as a check on the quantity of raw material used, on its transformation into a finished product, and on the quality of the production process.[26] Therefore, some explanation of this sort can be imagined for the production cycle of the Metapontine tiles: the symbols may indicate a form of qualitative and also quantitative monitoring of the production cycle.[27]

The AVF stamp is also known on amphorae, for example in Callander's catalog, inasmuch as this parallel is valid.[28] The suggested reading is Auf(idii). It is not, however, the parallel with the amphora stamp that supports this easy interpretation, but rather a passage from Varro well-known to archaeologists who study the Metapontine territory:

> Publius Aufidius Pontianus Amiterninus cum greges ovium emisset in Umbria ultima, quibus gregibus sine pastoribus canes accessissent, pastores ut deducerent in Metapontinos saltus et Heracleae emporium, inde cum domum redissent qui ad locum deduxerant, e desiderio hominum diebus paucis postea canes sua sponte cum dierum multorum via interesset, sibi ex agris cibaria praebuerunt atque in Umbriam ad pastores redierunt.
>
> Varro, r.r., II, 9, 6

Cicero also mentions a certain Pontianus, a friend of Atticus (*Ad Att.* XII.44.2). It has been suggested, but without reliable documentation, that P. Aufidius Pontianus could have been a member of the equestrian order.[29] The Aufidii were a family of plebeian origin, attested in Rome as early as the end of the 5th century BC. During the period in question, a member of the family served as tribune of the plebs in 170 BC (Gn. Aufidius), and another was a currency magistrate, as mentioned above (M. Aufidius Rusticus).[30] If he really did make use of personal memories or firsthand stories, the facts narrated by Varro must have occurred between the end of the 2nd century and the 30s of the 1st century BC.

This well-known passage is often quoted in relation to the history of the territory between Herakleia and Metaponto in the Late Republican period. Although a Roman garrison was stationed at Metaponto in the aftermath of the Pyrrhic War, it was only with the conclusion of the Second Punic War that Rome took full possession of the territory between Metaponto and Herakleia.[31] The situation in the 1st century BC described by Varro is considered valid also for the preceding century, with new artisan activities and trade and an

[23] There are several examples from Segesta of the use of shared stamps on tiles and amphorae, both those bearing a name (in Greek) and those bearing symbols. A type is also attested in the Metapontine repertoire: a four-spoked wheel is present on amphorae and tiles, as well as on loom weights, for which there are parallels in Lucania (Garrozzo 1995). On a Late Republican eastern Greek amphora noted in Small 2006 (332), the name, in Greek characters within a rectangular cartouche, is divided into two lines separated by an anchor motif.

[24] See Lyding Will 1979 for the Sesti amphorae, where the stamps, usually on the rim, become common toward the end of the 2nd c. BC. Production is widely attested on the amphorae from Bibracte with an ample repertoire of stamps (Laubenheimer and Rodriguez 1991; Olmer 2003), including the anchor, the wreath, and the star, which recalls a simplified version of the wheel from Sant'Angelo Vecchio. Olmer provides a synthesis of the possible meanings attributed to these stamps in the literature: a name, a chronological marker within the production sequence, and a numerical indicator of the quantity of amphorae produced.

[25] Crawford (1974, 725) notes that the phenomenon became widespread and pervasive from 170 BC onward. For the system of stamps in the production of tiles in the Roman period as a direct reflection of the norms regulating contracts and economic agreements, see Aubert 2005.

[26] See Garlan 1986 (271–73) for observations on amphora marks, beginning with the stamps from Thasos. He observes that this system, though originating in different situations, did not certify the contents (which were indicated instead by the shape of the container), but rather was a bureaucratic instrument used by the workshop for monitoring, and therefore unintelligible outside of that context. Moreover, the use of the same system of stamps on tiles—a circumstance obviously of particular interest for this site—demonstrates that it was in no way linked to the export process. Having to look for parallels elsewhere, Garlan cites the system for marking coins, which is important for the comprehension of the stamp system from Sant'Angelo Vecchio. In conclusion, Garlan prefers to leave the question open but probably within the fiscal-bureaucratic sphere.

[27] For example, the repetition of stamps could be a form of numeration: if a single stamp indicated the hundreds, a double stamp could indicate a multiple, and thus the tally of the lots produced, as has often been suggested; production was then probably certified by the letter stamp.

[28] For amphora type Dressel 38 of unspecified production, see Callander 1965, no. 157, fig. 4.5. The lines in this circular stamp differ from the example from Sant'Angelo Vecchio. The proposed chronology is the 1st c. BC. The stamp AV, in a rectangular cartouche, is documented on the rim of a Dressel 1 amphora from Bibracte (Laubenheimer 1991, 59, no. 47).

[29] Giardina 1997, 146.

[30] RE, s.v. Aufidius.

[31] Giardino 2005, 411: grey ware and relief cups are clear indicators of the change in the material culture.

important role for pastoralism.[32] A senatorial, or perhaps equestrian interest in the territories of Metaponto and Herakleia can be identified, centered around the raising of stock and the exploitation of *ager publicus*.[33] In order for this to occur, the territory was reorganized: the ratio of site distribution changed, and some important centers lost their original functions and acquired new ones.[34] The stamps from Sant'Angelo Vecchio provide fortuitous support to the descriptions in the literary sources. If they indicate a property of the Aufidii, then this senatorial and equestrian interest must be recognized as a more stable occupation of the territory.[35]

[32] Giardino 2005, 420. Finds of Gallic-type fibulae, for example, from the acropolis and the necropolis of Herakleia seem to support the hypothesis of links with the middle Adriatic area associated with the transhumance routes and Varro's description. Moreover, the presence of loom weights (often inscribed) in the urban center suggests that weaving was part of a much larger commercial enterprise. Furthermore, the presence of a merchant class has been recognized in one type of house on the acropolis (in the light of Herakleia's function as an emporium documented by Varro).

[33] Giardina (1997, 139–41) suggests patterns of stock-raising independent of the landed property, based on the exploitation of *ager publicus*.

[34] De Siena 2005.

[35] Crawford (2003) attributes to the territory and the villas, which became the new "public" spaces, the capacity to absorb the urban functions lost with the destruction of the old poleis of Magna Grecia. With regard to the stamping of these materials in southern Italy, Small (2005) notes the distribution of rural kilns in relation to the villas, recognizing them as necessary for the maintenance and service to the structure, with production activity integrated into the agricultural cycle. He also notes exceptions to the model for areas located on the coast or near important roads and crossroads, which increased the vitality and autonomy of these production centers. Moreover, emphasizing the small number of known stamps from southern Italy, he proposes various explanations for the paucity of material.

It is worth repeating that what is indicated by the association of names and symbols is impossible to identify with any certainty. The same problem exists also for earlier periods.[36] In this phase at Sant'Angelo Vecchio there is a possible break with earlier stamping systems, and the excavation data also support this interpretation: the site was transformed, and the new organization appears to be reflected in the adoption of a new system of production and stamping well-suited to Roman or Romanized contexts. The name very probably indicates the ownership of the property, which served as an alternative production area to the city. Apparently there was no longer space for the old system of the *figuli* who traveled from the central or satellite workshops to the sites where the work had been commissioned. The system of ownership changed, as did the centralized system of resource management.

[36] Stamps with symbols—sometimes enlarged versions of monetary types—appear on 4th-c. BC *tegulae* and *imbrices* in several cities of Magna Grecia, such as Croton. They have been interpreted in various ways: as decorative symbols (Spadea 1984, 155; Sabbione 1984, 299–300, who discusses a possible function for some stamps as a countermark for a public workshop and the partiality of the phenomenon of stamping) and as trademarks (Guzzo 1987, 471, as public production or a public commission for a private workshop), in some cases possibly as exceptional commissions (Spadea 1993, 25, whose reasoning is very different from the preceding ideas, instead associating a number of stamps with specific cities belonging to the Italiote League that constructed reception buildings in Croton; the circulation of such stamped materials beyond the dissolution of the League was due to reuse; Corrado 2010 also differs, attributing new values to the attested symbols).

Catalog of Roof Tiles

Laconian Systems

At least two systems (Aa and Ab) of the Laconian type were identified, which used curved tiles and semicircular cover tiles.

System Aa. The information about the Archaic Laconian roofing system at Sant'Angelo Vecchio is very scarce. The only secure documentation is a single example recovered in the lower area during the recent excavations of 2010–2011, a small fragment of the side edge of a tile (SAV RT 130, type AT0) preserving traces of red paint on the upper surface.

The upper site (House on the Hill), with attested occupation in the 6th and 5th c. BC, yielded several fragments of Laconian roofing systems (pan tiles, cover tiles, and an opaion), but these are unfortunately too fragmentary to permit secure attribution to the Archaic or Hellenistic versions. The presence of paint on the upper surface is the determining factor in recognizing Archaic examples. The fragments from the House on the Hill are completely devoid of paint, but their chronology cannot be established conclusively by this fact because Archaic tiles were only partially painted, and therefore it cannot be excluded that unpainted fragments belonged to painted tiles.

Likewise the shape of the opaion (SAV RT 36), the only whole part of a tile recovered in the House on the Hill, is not chronologically determinant. Similar examples are documented in Archaic and Hellenistic contexts (Pantanello, with traces of paint, in publication; for Hellenistic examples see Wikander type I B, Wikander 1983, 84, 6-7-9, figs. 6, 7b—A 428, 7b—A429).

AT0. Given the lack of complete examples, the full shape of the Archaic tile AT0 of the Aa system cannot be described accurately. Only one small fragment of this system remains (SAV RT 130).

System Ab (*Fig. 32.5*). The Ab system is the best documented of all thanks to the large quantity of fragments recovered from the various excavation areas. The system used two types of tile. The first, AT1 (a useful comparison can be made with the better-preserved examples from Fattoria Fabrizio: FF RT 02, FF80-226PL; FF RT 03, FF80-127T), with concave profile, was associated with the curved cover tile with depressed arch, AC1. The elements show characteristic notches at the front (curved tiles) and rear (cover tiles) edges, which allowed the single rows to be fitted together on the roof. Curved tiles are well documented in residential contexts and necropoleis of the 4th and 3rd c. BC in southern Italy, in Lucania, Daunia, and Messapia (see Perugino and Vollaro in *Fattoria Fabrizio*). The *opaion* SAV RT 36 may be compared to Wikander's type I B. The Laconian tiles with oval opaion from Nemea, Athens, and Pella (Wikander 1983, 84, 6-7-9, figs. 6, 7b (A 428), 7b (A429), however, offer parallels from later, 4th-c. BC contexts that, in this case, contrast with the excavation data (Edlund 1982a, 15–22) (see Ch. 3 "The Structures at Sant'Angelo Vecchio").

The second tile type, AT2, is a special form in which the concave tile is associated with a half cover tile. A complete example of the form could not be reconstructed from the collected fragments. One hypothesis is that these elements functioned as terminals along one of the linear edges of one of the roof slopes, joining the terracotta roof elements either to a wall or to the first element of the timber roof frame. A less probable suggestion, considering the limited numbers, is that these tiles were modular elements of another type of roof that served to join flat tiles and cover tiles as a single element, as documented, for example, in the earliest roofs in Greece (Winter 1993, 12–18) and used in modern roofing. Parallels for this type of tile include an example found in the Late Roman farmhouse at San Biagio (*San Biagio*, 40, no. 2.1.1; the tile probably dates to a 4th–3rd-c. BC occupation phase predating the Roman period) and another from Ascoli Satriano (*Ausculum I*, 302, type 1, 1.1, fig. 81, no. 1).

AT1. Type AT1 of the Ab system is well made, more or less rectangular, with thickened edges and a concave profile. It tapers close to the front edge in order to fit over the next tile. The outer sides of the front edge have two oblique notches, often unequal in length (SAV RT 04, SAV RT 06, SAV RT 09, SAV RT 11, SAV RT 12, and SAV RT 25), used to fit the cover tiles. The rims of the side edges have a range of profiles: thickened flange (SAV RT 06); offset, pointed at the top and thickened on the interior (SAV RT 13); offset, rounded upper surface, slightly concave and thickened interior (SAV RT 14); offset, forked (SAV RT 15). The concave surface of the tiles is smooth, the convex surface irregular and roughly worked; the reverse is true for the cover tiles. These features facilitate distinction of the two forms. A completely reconstructed example from Fattoria Fabrizio (FF RT 02, FF80-226PL), of similar morphology, clay, and surface treatment, suggests that the tiles from Sant'Angelo Vecchio were also of a similar (standard) size. The unit of measurement adopted for the elements in the roofing system was likely the Doric foot, 27.3 cm. On the basis of the example from Fattoria Fabrizio, the tile length would correspond to 3.5 Doric feet, equivalent to 95.55 cm, while the maximum width would be 41.0 cm, or 1.5 Doric feet. The fabric of all of the tiles is type T1, and all fragments have a light beige slip. Average dimensions (reconstructed example from Fattoria Fabrizio FF RT 02, FF80-226PL) are: l 96.0 cm; w 41.0–24.0 cm, th compressed tile edge 2.5 cm; l cut 7.5–10.0 cm; weight of complete example ca. 14.0 kg.

SAV RT 01 Laconian Tile
Provenance: Square H3. Level 1, batt. 1. Lot: SA 79-705T (see *Fig. 2.22*).
Dimensions and technical features: th 1.8.
Condition: Wall fragment; partially preserved stamp with Greek letters on the convex surface.

1:3

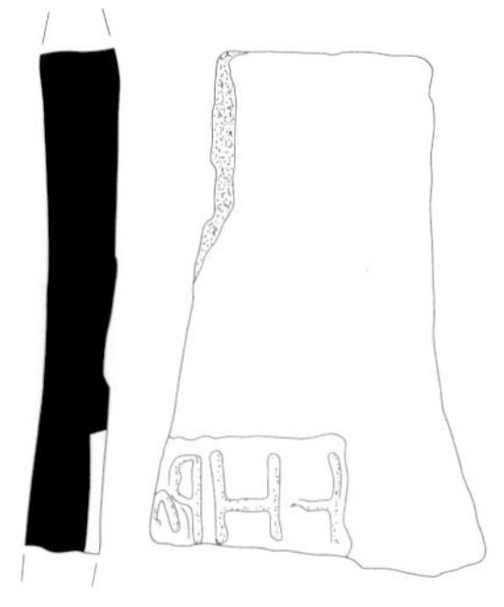

SAV RT 02 Laconian Tile
Provenance: Square G7. Level 2, batt. 6. Lot: SA 79-957T. Assemblage 4.
Dimensions and technical features: th 1.5.
Condition: Fragment of curved pan tile; partially preserved stamp with Greek letters on the lower surface.

1:3

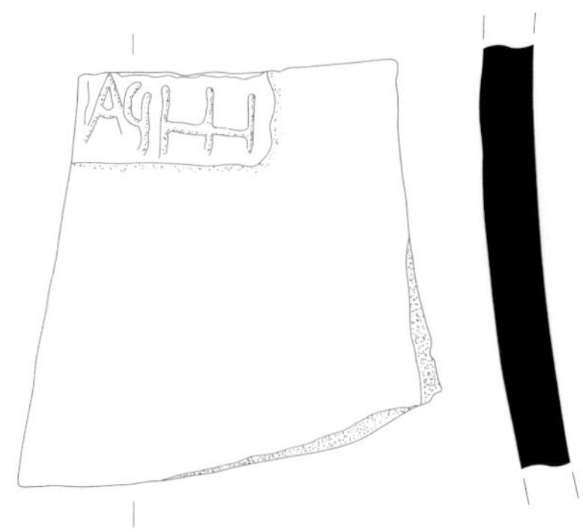

SAV RT 03 Laconian Tile
Provenance: Surface find on grassy slope above Slope 3. Lot: SA 79-64T (from SA 79-61PL).
Dimensions and technical features: th 1.2–2.3.
Condition: Side edge.

1:3

SAV RT 04 Laconian Tile
Provenance: Square D4. Level 3, batt. 6. Lot: SA 79-953T. Assemblage 13.
Dimensions and technical features: th 1.8, l cut 9.
Condition: Front edge with oblique notch.

1:3

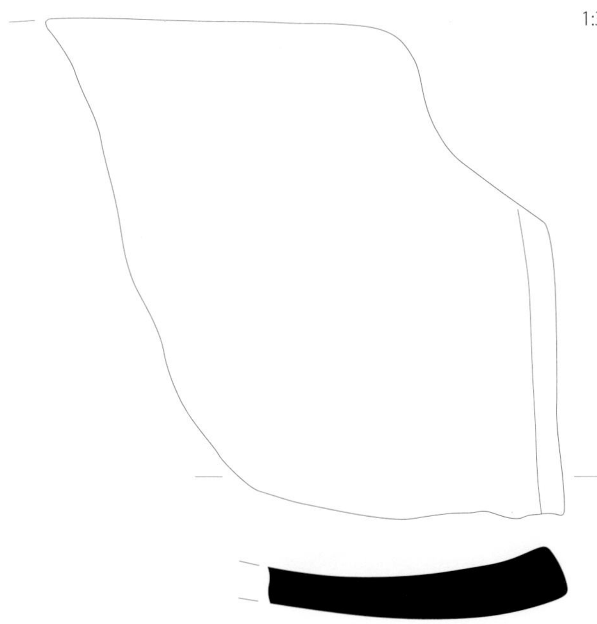

Laconian system Ab, type AT1 tiles cont.

SAV RT 05 Laconian Tile
Provenance: Square G5. Level 2, batt. 1–2. Lot: SA 79-789T. Assemblage 3.
Dimensions and technical features: th 1.3.
Condition: Fragment of flat pan with impressed mark on convex side.

1:3

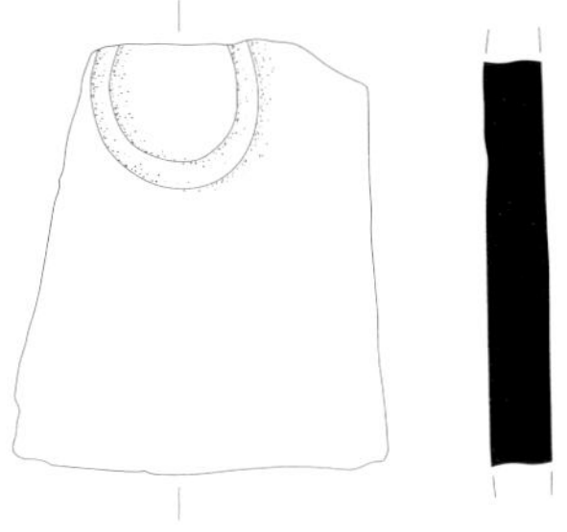

SAV RT 06 Laconian Tile
Provenance: Slope 3, W extension. Level 2. Lot: SA 79-1051T (see *Fig. 2.35*).
Dimensions and technical features: l max 37.0, w 38.0–40.0, w front edge 25.0, l right notch 8.4–10.3, edge th 2.2.
Condition: Composed of two large joining fragments; a quarter of total length preserved.

1:3

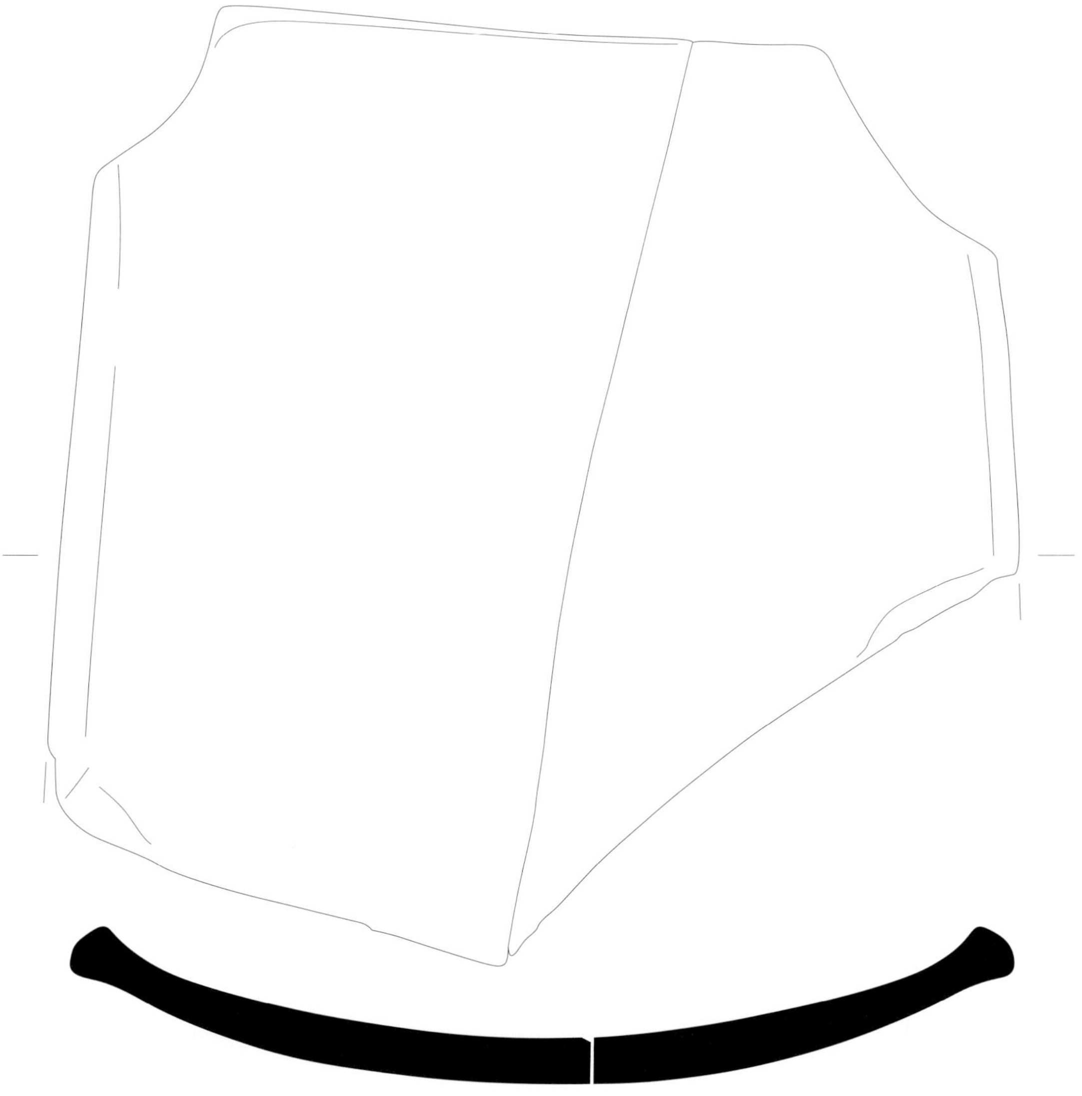

SAV RT 07 Laconian Tile
Provenance: Square G7. Level 2, batt. 4. Lot: SA 79-813T (see *Fig. 2.20*).
Dimensions and technical features: front edge th 2.2, side edge th 2.4, notch l 8.0.
Condition: Short section of front edge with oblique notch and small section of right side.

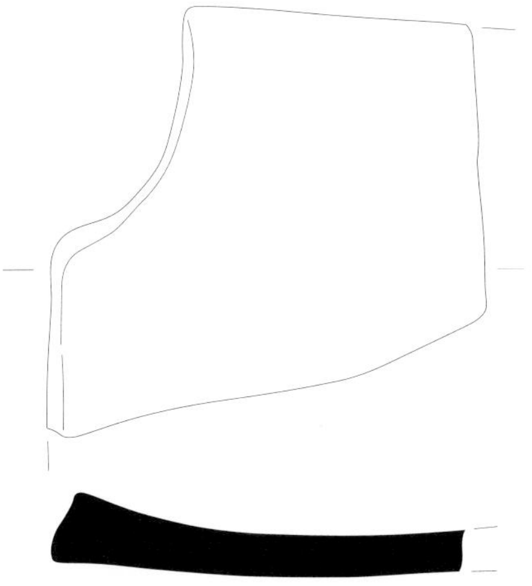

1:3

SAV RT 08 Laconian Tile
Provenance: Sounding B. Level 3, batt. 2. Lot: SA 79-186T. Assemblage 11.
Dimensions and technical features: th 1.9.
Condition: Side edge and beginning of flat pan.

1:3

SAV RT 09 Laconian Tile
Provenance: Sounding B. Level 3, batt. 2. Lot: SA 79-186T. Assemblage 11.
Dimensions and technical features: th 1.9, notch l 5.9.
Condition: Side edge and oblique notch.

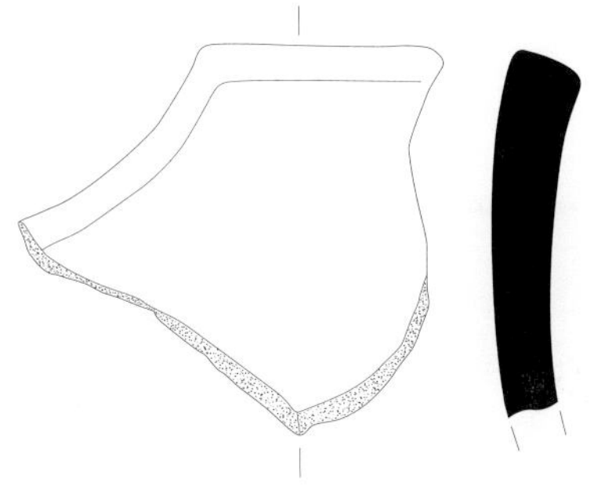

1:3

SAV RT 10 Laconian Tile
Provenance: Sounding B. Level 3, batt. 2. Lot: SA 79-186T. Assemblage 11.
Dimensions and technical features: th 2.2.
Condition: Front edge with oblique notch.

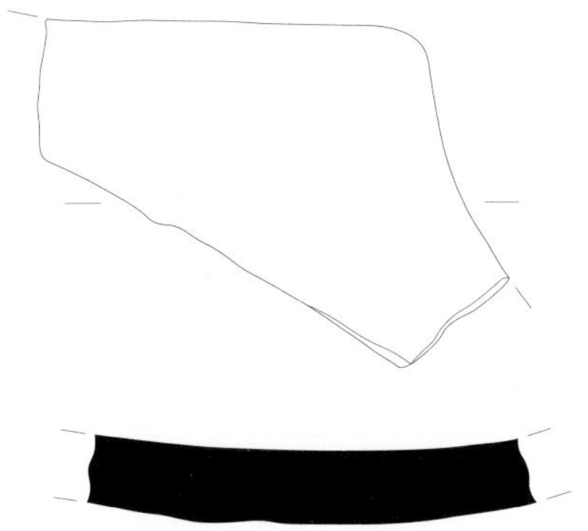

1:3

Laconian system Ab, type AT1 tiles cont.

SAV RT 11 Laconian Tile

1:3

Provenance: Sounding B. Level 3, batt. 2. Lot: SA 79-186T. Assemblage 11.
Dimensions and technical features: front edge th 1.8, notch l 5.5, side edge th 2.1.
Condition: Very small part of the front edge with oblique notch and a short section of the side edge.

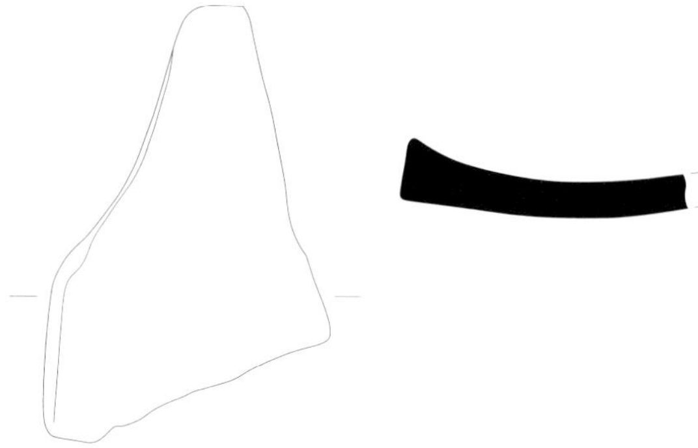

SAV RT 12 Laconian Tile

1:3

Provenance: Square G6 (from SA 79-707T). Level 1, batt. 1. Lot: SA 79-708T.
Dimensions and technical features: front edge th 2.7, notch l 6.7.
Condition: Front edge with oblique notch and short section of side edge.

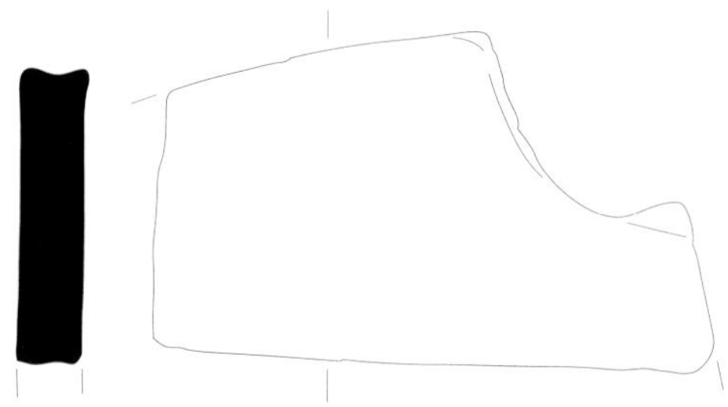

SAV RT 13 Laconian Tile

1:3

Provenance: Square F5 (possibly G5?). Level 2, batt. 1. Lot: SA 79-556T. Assemblage 3.
Dimensions and technical features: edge th 2.4.
Condition: Side edge with beginning of curved pan tile.

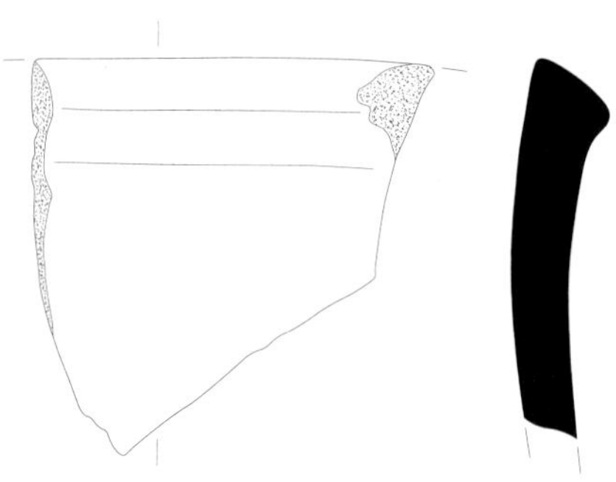

SAV RT 14 Laconian Tile

1:3

Provenance: House on the Hill, Test Trench 2. Level 1, batt. 2. Lot: SA 79-622T (see *Fig. 2.39*).
Dimensions and technical features: edge th 2.3.
Condition: Composed of three joining fragments. Side edge with beginning of curved pan tile.

SAV RT 15 Laconian Tile
Provenance: House on the Hill, Test Trench 2. Level 1, batt. 2. Lot: SA 79-622T (see *Fig. 2.39*).
Dimensions and technical features: edge th 1.5–2.0.
Condition: Side edge with beginning of curved pan tile.

1:3

SAV RT 16 Laconian Tile
Provenance: Sounding C, E of division. Level 2, batt. 3. Lot: SA 79-315T. Assemblage 13.
Dimensions and technical features: th 1.4.
Condition: Fragment of flat pan with semicircular impression on convex side.

1:3

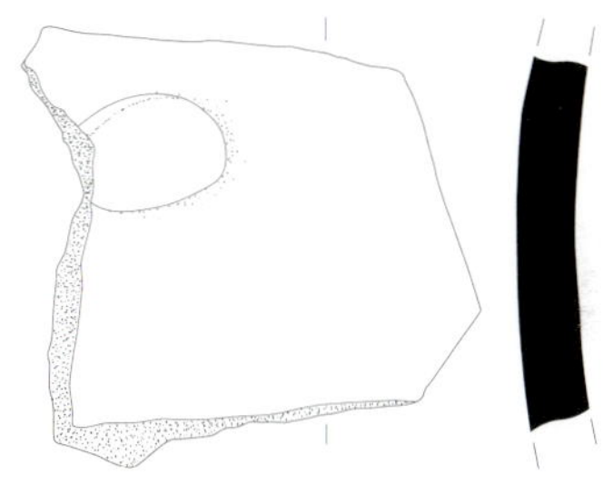

SAV RT 17 Laconian Tile
Provenance: Square F3. Level 2, batt. 1. Lot: SA 79-830T.
Dimensions and technical features: th 1.3 (see *Fig. 2.30*).
Condition: Fragment of flat pan with impressed mark on convex side.

1:3

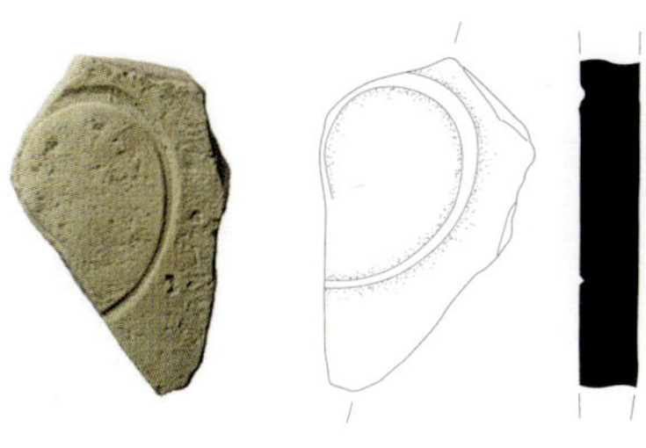

Laconian system Ab, type AT1 tiles cont.

SAV RT 18 Laconian Tile

Provenance: Square F3, SE end of N Kiln near baulk. Level 2. Lot: SA 79-819T. Assemblage 8.

Dimensions and technical features: th 2.5.

Condition: Side edge and part of flat pan. On the upper surface, deep grooves made with a stick.

1:3

SAV RT 19 Laconian Tile

Provenance: Square D4. Level 3, batt. 6. Lot: SA 79-954T. Assemblage 13.

Dimensions and technical features: th 2.2.

Condition: Fragment of flat pan with semicircular impression on convex side. On the edges, small lumps of baked-on clay, perhaps due to reuse of the fragment in a kiln.

1:3

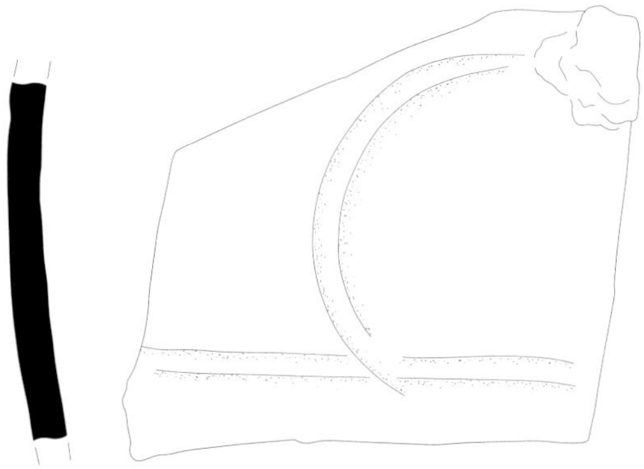

SAV RT 20 Laconian Tile 1:3

Provenance: Slope 3, W extension. Level 3, batt. 1. Lot: SA 79-318T. Assemblage 11.
Dimensions and technical features: edge th 1.6.
Condition: Short section of side edge and curved pan tile.

SAV RT 21 Laconian Tile 1:3

Provenance: Slope 3, W extension. Level 3, batt. 1. Lot: SA 79-318T. Assemblage 11.
Dimensions and technical features: edge th 2.2.
Condition: Rear edge.

SAV RT 22 Laconian Tile 1:3

Provenance: Slope 3, W extension. Level 3, batt. 1. Lot: SA 79-318T. Assemblage 11.
Dimensions and technical features: edge th 2.2.
Condition: short section of side edge and curved pan tile.

SAV RT 23 Laconian Tile 1:3

Provenance: Slope 3, W extension. Level 3, batt. 1. Lot: SA 79-318T. Assemblage 11.
Dimensions and technical features: edge th 2.0.
Condition: Short section of edge and curved pan tile.

SAV RT 24 Laconian Tile 1:3

Provenance: Slope 3, W extension. Level 3, batt. 1. Lot: SA 79-318T. Assemblage 11.
Dimensions and technical features: edge th 2.4.
Condition: Part of edge and short section of left side edge.

SAV RT 25 Laconian Tile 1:3

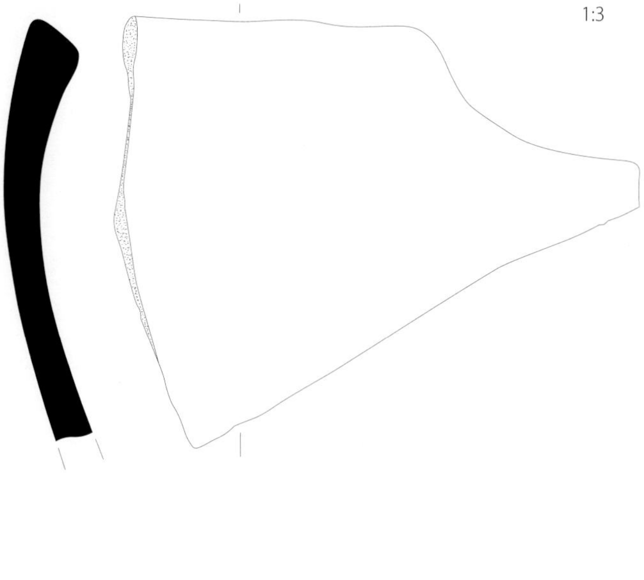

Provenance: Slope 3, W extension. Level 3, batt. 1. Lot: SA 79-318T. Assemblage 11.
Dimensions and technical features: side edge th 2.2, notch l 10.0.
Condition: Side edge with oblique notch and part of curved pan.

Laconian system Ab, type AT1 tiles cont.

SAV RT 26 Laconian Tile
Provenance: Slope 3, W extension. Level 3, batt. 1. Lot: SA 79-318T. Assemblage 11.
Dimensions and technical features: edge th 2.4.
Condition: Section of side edge and curved pan tile.

1:3

SAV RT 27 Laconian Tile
Provenance: Slope 3, W extension. Level 3, batt. 1. Lot: SA 79-318T. Assemblage 11.
Dimensions and technical features: th 2.1.
Condition: Small section of side edge and curved pan tile.

1:3

SAV RT 28 Laconian Tile
Provenance: Slope 3, W extension, Stipe S. Level 3, batt. 1. Lot: SA 79-372T. Assemblage 11.
Dimensions and technical features: front edge th 1.9, side edge th 2.1, notch l 5.8.
Condition: Short section of front edge with part of oblique notch and section of side edge and curved pan tile.

1:3

AC1. Type AT1 was associated with a cover tile with depressed arch, tapering toward the rear edge where there are two oblique notches to facilitate the fitting together of the elements, known as type AC1. The convex surface is smooth, the concave surface irregular with rough stick marks. The front edge is perfectly flat and very slightly thickened at the top, while the side edges are slightly beveled. In several cases ("forked"), a continuous groove runs along the side edges. Due to the fragmentary nature of the finds, the standard dimensions could not be reconstructed. Here, too, nearly complete reconstructed examples from Fattoria Fabrizio FF RT 03 (FF80-127T) and from Pantanello (PZ78-614T, *Fig. 32.6*, albeit slightly smaller), similar in morphology, clay, and surface treatment, serve as valid models for the dimensions. Average measurements for a cover tile (example from Fattoria Fabrizio FF80-127T) are: w front edge 28.5, w rear edge 24.0, l ca. 96.0, arc 5.0–8.0 x 23.0, l notch 5.5–7.5, th 1.7–2.5, weight ca. 4.0 kg.

SAV RT 29 Laconian Cover Tile
Provenance: Square G5. Level 2, batt. 1–2. Lot: SA 79-789T. Assemblage 3.
Dimensions and technical features: edge th 1.7–2.5.
Condition: Front section.

1:3

SAV RT 30 Laconian Cover Tile
Provenance: Sounding B. Level 3, batt. 2. Lot: SA 79-186T. Assemblage 11.
Dimensions and technical features: edge th 1.7–2.2.
Condition: Section of side edge and beginning of the pan tile.

1:3

SAV RT 31 Laconian Cover Tile
Provenance: Sounding B. Level 3, batt. 2. Lot: SA 79-186T. Assemblage 11.
Dimensions and technical features: th 1.7–2.3.
Condition: Section of side edge.

1:3

SAV RT 32 Laconian Cover Tile
Provenance: Sounding C. Level 2, batt. 2. Lot: SA 79-192T.
Dimensions and technical features: edge th 1.4–2.3.
Condition: Section of side edge.

1:3

SAV RT 33 Laconian Cover Tile
Provenance: Sounding C. Level 2, batt. 7. Lot: SA 79-379T (see *Fig. 2.35*).
Dimensions and technical features: arc 24.0, edge th 1.4–2.0.
Condition: Composed of two joining fragments from the central section of a cover tile, with both sides preserved.

1:3

Opaia (Fig. 32.5). An elongated, oval-shaped *opaion* with raised edge with rectangular profile, flattened on the top, can be attributed to the Aa or Ab system. The average measurements of *opaion* aperture are: w 21.0 cm, l 40.0 cm, h raised edge including thickness of the flat pan 4.0 cm.

SAV RT 34 Opaion
Provenance: Square F3, N Kiln interior. Level 2, batt. 2. Lot: SA 79-872T. Assemblage 8.
Dimensions and technical features: max. l 15.0, max. w 10.0, opaion th 2.8, opaion h 3.9.
Condition: Short section of edge with beginning of curved pan tile.

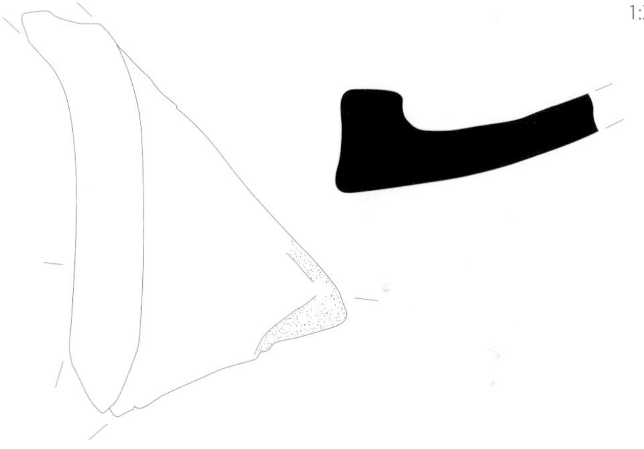

1:3

SAV RT 35 Opaion
Provenance: Square G7, from top of wall 4. Level 2, batt. 1. Lot: SA 79-771T.
Dimensions and technical features: max. l 14.0, max. w 13.0, opaion th 1.9, opaion h 5.0.
Condition: Short side of aperture, nearly complete.

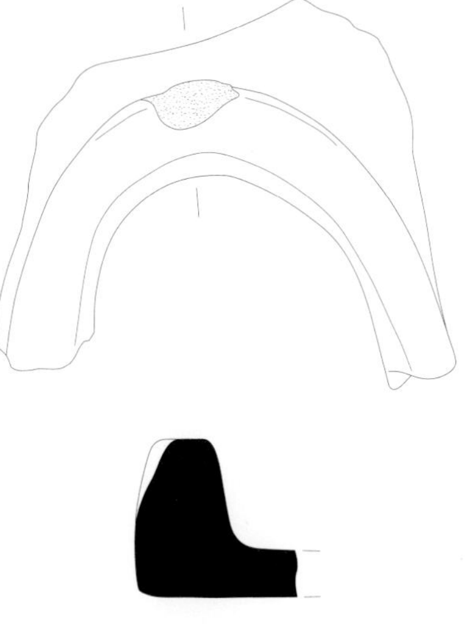

1:3

Laconian system Ab, opaia cont.

SAV RT 36 Opaion
Provenance: House on the Hill, Test Trench 2. Level 1, batt. 1. Lot: SA 79-666T (see *Fig. 2.39*).
Dimensions and technical features: max. l 42.0, max. w 35.0, opaion max. l 37.0, opaion w 21.0, opaion edge th 1.8–2.2, opaion h 2.6–4.2.
Condition: Eight joining fragments. Nearly complete except for one of the short sides. Small sections of curved pan tile.

1:4

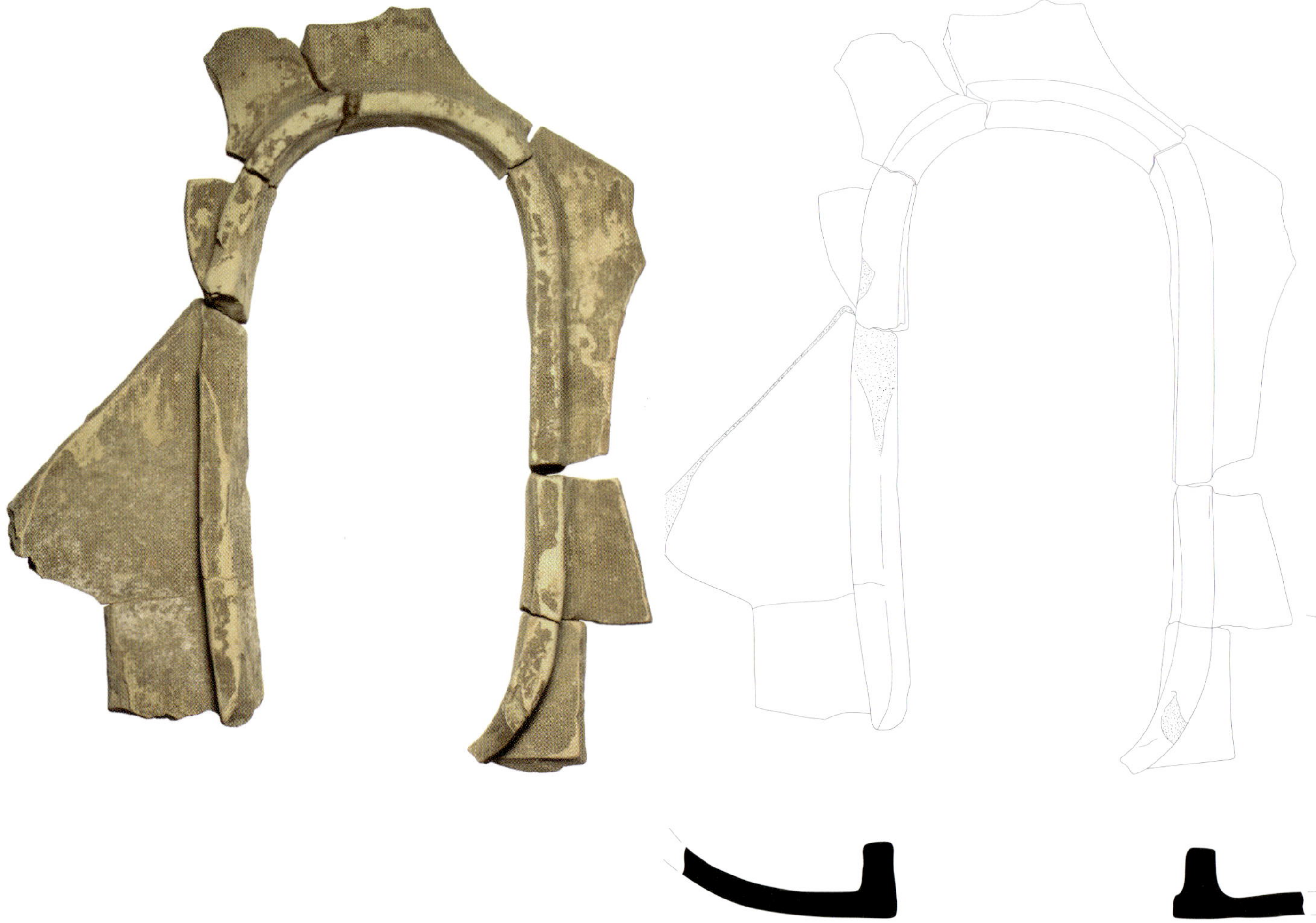

SAV RT 37 Opaion
Provenance: Square G7, E of wall. Level 2, batt. 1. Lot: SA 79-803T. Assemblage 17.
Dimensions and technical features: l max 13.0, opaion th 1.8.
Condition: Fragment of tile with section of the edge of an aperture, perhaps of an opaion. The profile of the edge is approximately rectangular.

1:4

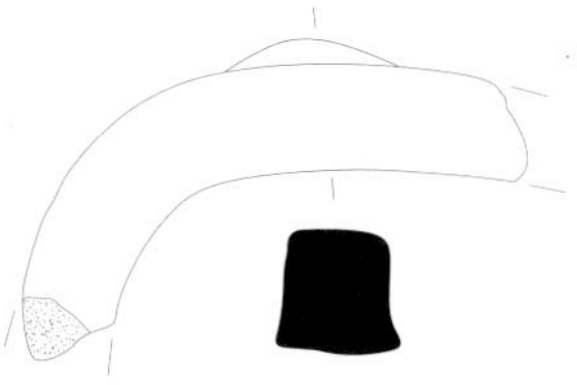

Miscellaneous. This section describes fragments which, because of their small size or poor state of preservation, could not easily be attributed to a specific Laconian form, whether curved pan tile or cover tile.

SAV RT 38 Pan or Cover Tile?
Provenance: Sounding A. Level 3, batt. 5. Lot: SA 79-44T (see *Fig. 2.30*).
Dimensions and technical features: th 2.0.
Condition: Short section of edge.

1:4

SAV RT 39 Pan or Cover Tile?
Provenance: S area of Room 1. Level 2, batt. 2. Lot: SA 79-25T. Assemblage 5.
Dimensions and technical features: th 2.0–2.6.
Condition: Short section of edge.

1:4

SAV RT 40 Pan Tile?
Provenance: S area of Room 1. Level 2, batt. 2. Lot: SA 79-25T. Assemblage 5.
Dimensions and technical features: th 1.7–2.3.
Condition: Short section of edge.

1:4

SAV RT 41 Pan or Cover Tile?
Provenance: Sounding B. Level 3, batt. 2. Lot: SA 79-186T. Assemblage 11.
Dimensions and technical features: th 1.8.
Condition: Short section of edge.

1:4

SAV RT 42 Pan or Cover Tile?
Provenance: Sounding C. Level 2, batt. 2. Lot: SA 79-192T. Assemblage 13.
Dimensions and technical features: th 1.9.
Condition: Short section of edge.

1:4

Laconian system Ab, type AT2 tiles cont.

SAV RT 43 Pan Tile?
Provenance: Sounding C. Level 2, batt. 2. Lot: SA 79-192T. Assemblage 13.
Dimensions and technical features: th 1.2.
Condition: Fragment of body.

1:3

AT2. In this type, the concave tile is associated with a half cover tile, creating an undulating, concave-convex profile.

SAV RT 44 Laconian Tile
Provenance: Square H4. Level 1, batt. 1. Lot: SA 79-667T (see *Fig. 2.21*).
Dimensions and technical features: edge th 1.8.
Condition: Section of side edge. Surface has short, impressed notches made before firing.

1:3

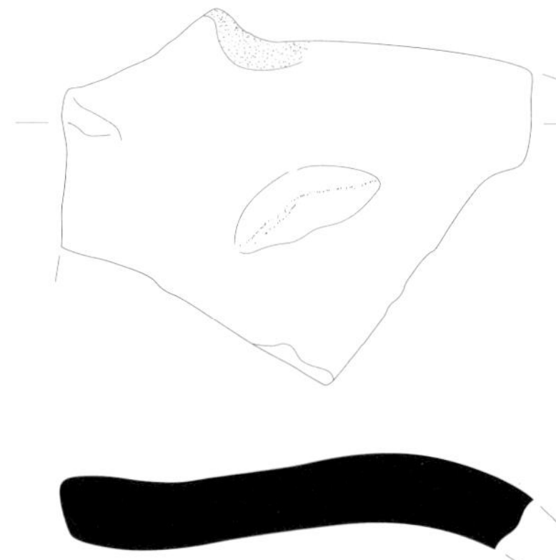

SAV RT 45 Laconian Tile
Provenance: Square F5 (possibly G5?). Level 2, batt. 1. Lot: SA 79-556T. Assemblage 3.
Dimensions and technical features: edge th 1.8–3.2, notch l 6.8.
Condition: Side edge with oblique notch.

1:3

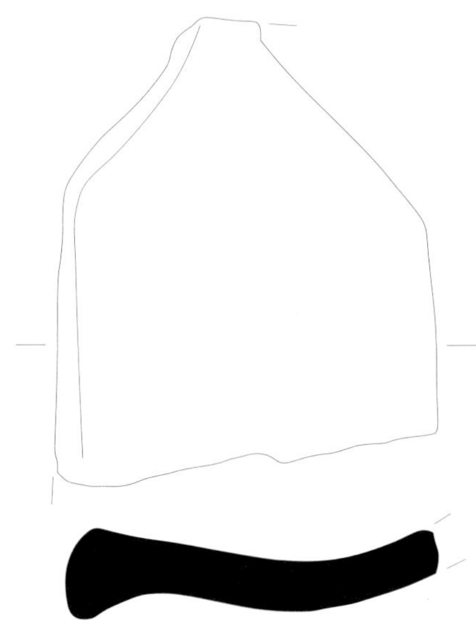

SAV RT 46 Laconian Tile
Provenance: Square G6, W of wall. Level 2, batt. 1. Lot: SA 79-733 (see *Fig. 2.20*)
Dimensions and technical features: edge th 2.4.
Condition: Two joining fragments of a side edge.

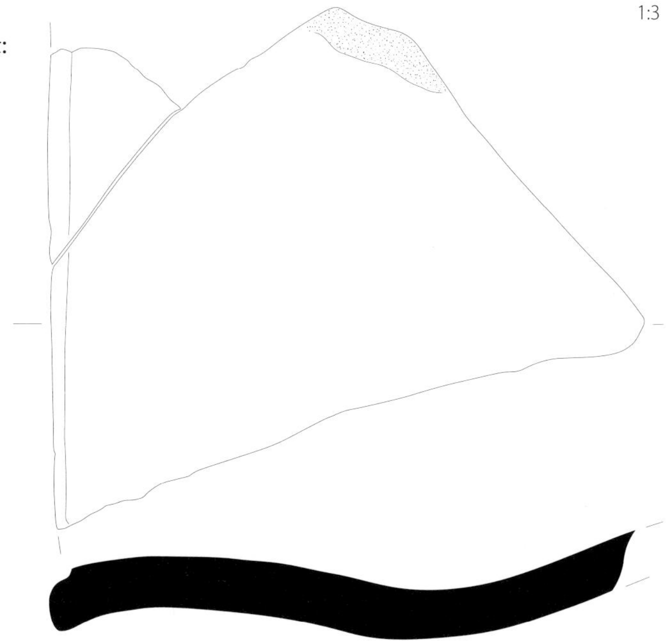

Flat Tile Systems

Elements attributable to two different types of assembly were identified among the flat tile fragments. In the overlapping system, the tiles have rectangular notches on the lower surface, into which fit the flanges of the lower tiles. In the interlocking system, each element narrows at the front in order to permit its insertion between the flanges of the tile below.

System Ba (*Fig. 32.4*)

BT1. A few fragments with a specific form of flange, a flattened semicircle (BT1), can be attributed to the overlapping system. A tile from a tomb covering at Sant'Angelo Vecchio (SA 79-1074T, *Fig. 32.11*) clearly shows how a raised border of this type can coexist on the same element with another border with a quarter-circle profile, slightly less flattened and of a smaller standard size. The lower housings for the overlapping of the flanges on the corresponding elements of the lower tile are rectangular, with a slightly beveled inner corner. The inner surface is smooth.

The tile from the House on the Hill (SAV RT 47) and other examples from tomb contexts (SA 79-1028T, *Fig. 32.12*), two examples *sine inventario* of uncertain provenience (*Fig. 32.13*), and the tile SA 79-981T (*Fig. 32.14*) document the presence of a ridge with rectangular section, a few millimeters thick, running across the tile's rear edge or slightly inside of it, between the two flanges. This transverse ridge, which served to hook the tile onto the one below or to prevent the infiltration of water, is a variant that is also present on tiles in Greece and Asia Minor in Archaic and Hellenistic contexts (Wikander 1988, 208–9, fig. 3, C2a), and in Magna Grecia in the same periods. An example of a flat tile with a ridge on the short side dating to the Archaic period comes from Parco del Cavallo at Sybaris (Zancani Montuoro 1961, 34–35, n. 1, fig. 11). The same element recurs, although raised slightly higher, on the tiles of the Heroon of Poseidonia (*Poseidonia-Paestum II*, 30–33; Mertens 1992, 554–55) and at Locri (*Locri IV*, 321, pl. XCIV, 9–10) in contexts of the mid-4th and 3rd c. BC.

At Sant'Angelo Vecchio, judging from the technical characteristics and the associations in tomb contexts (SA 79-1020), this type of tile can be ascribed to a specific type of narrow cover tile with a horseshoe-shaped arch (type BC1). The find context suggests a date within the Archaic and Classical periods. The comparison with complete tiles from tomb coverings is also useful for the creation of a size prototype for reference purposes: the measurements correspond to 3.0 Doric feet (27.3 cm). In Lucanian and Apulian tile productions, a more or less uniform standard size is encountered (*Torre di Satriano I*, 122–23), between 68.0 and 71.0 cm long and 48.5 to 54.0 cm wide. The only visible difference in the type BT1, attested at Sant'Angelo Vecchio, is in the first measurement (see *Necropoleis*, 91 n. 107). The tile fabrics are types T1 and T2. Average measurements are as follows: l 80.0 cm, w 53.4 cm, l lower notch 11.5 cm, w notch 5.0 cm, w raised border 5.0, th 2.2.

Flat system Bb, type BT1 tiles cont.

SAV RT 47a–b Pan Tile
Provenance: House on the Hill. Test Trench 2. Level 1, batt. 2. Lot: SA 79 550TA, SA 79-550TB (see *Fig. 2.39*).
Dimensions and technical features: flat pan th 2.3, flange w 6.3. The flange reaches the rear edge, a continuous ridge (w 0.2, h 0.4) runs between the flanges at 5.0 cm from the rear edge. Clay type T1, but with more vacuoles, uniform lower surface, calcareous encrustations present all over.
Condition: Two fragments from rear edge. No tapering or notches in lower surface.

1:3

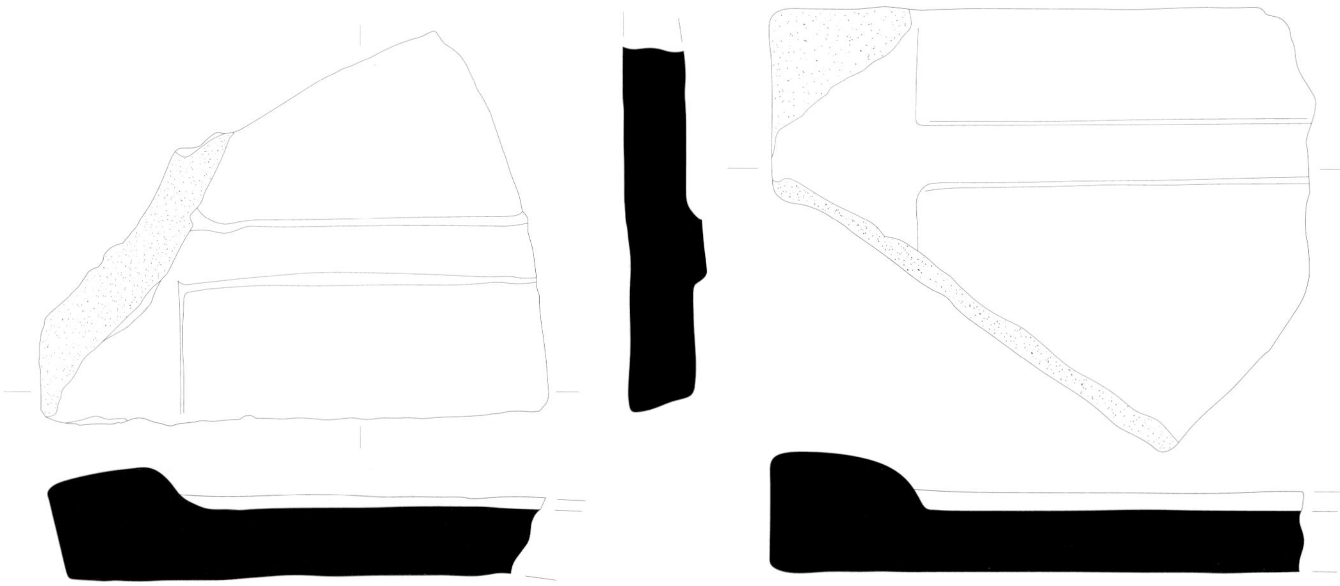

System Bb
BT2. A second type of tegula belonging to the overlapping system of assembly was identified among the material examined. This tile, documented by a few examples, has raised flanges with a sub-rectangular profile, the inner surface often thickened at the point where it overlapped with the next tile. A slight depression, the width of a finger, ran the length of the tile at the joint between the raised flange and the tile's flat surface. The latter element is more frequent in tiles from Lucania (cf., Rescigno 2001, 201–2) and can be considered the result of one of the phases of the production process: the depressions are the traces of sticks arranged on the flat part of the tile, along the long sides, during the drying phase. Where the form of the depression is particularly accentuated, its function may have been to facilitate the overlapping of the imbrex, thus forming a slot into which the edges of the cover tile were inserted. The dimensions, however, do not always support this hypothesis (*Torre di Satriano I*, 126). Due to the fragmentary nature of the finds, the standard size of the flat pan could not be reconstructed, and the stamps could not be identified. Surfaces are usually slipped, fabric T1 type.

SAV RT 48 Pan Tile
Provenance: Sounding B. Level 2, batt. 2. Lot: SA 79-186T-H. Assemblage 11.
Dimensions and technical features: flange w 4.8, h 4.0, flat pan th 2.0. Clay type T1.
Condition: Right side of front edge. Sub-rectangular flange. Uniformly impressed depression at point where flat surface joins flange.

1:3

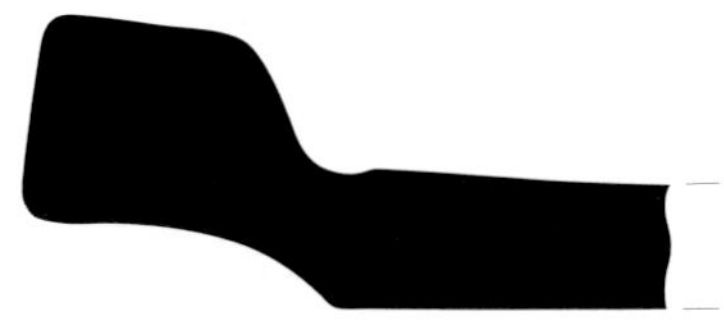

SAV RT 49 Pan Tile
Provenance: Sounding A, S end of Room 1. Level 2, batt. 2. Lot: SA 79-25T-B. Assemblage 5.
Dimensions and technical features: flange w 4.2, h 6.0, flat pan th 3.1, housing l 6.7. Clay type T1 but more refined and with orange tones.
Condition: Right side of front edge. Rectangular flange. Uniformly impressed depression at point where flat surface joins flange. Very irregular lower surface.

1:3

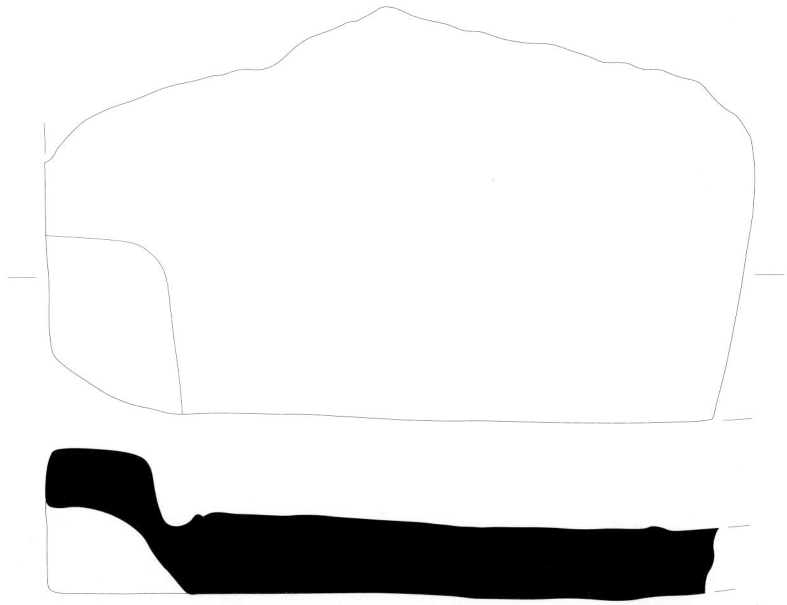

SAV RT 50 Pan Tile
Provenance: Square F5. Level 3, batt. 2. Lot: SA 79-556T-B. Assemblage 3.
Dimensions and technical features: flange w 3.2, h 4.5. Clay type T1.
Condition: Fragment of flange with upper projection.

1:3

SAV RT 51 Pan Tile
Provenance: Sounding C. Level 2, batt. 7. Lot: SA 79-378T (see *Fig. 2.35*).
Dimensions and technical features: flange w 4.0, h 4.6, flat pan th 2.5. Clay type T1.
Condition: Fragment of flange with beginning of flat pan.

1:3

SAV RT 52 Pan Tile
Provenance: Sounding C. Level 2, batt. 7. Lot: SA 79-210T (see *Fig. 2.35*).
Dimensions and technical features: w 4.2, flat pan th 2.4, w of overlapping section 4.0. Clay type T1.
Condition: Left side of front edge. Flange and beginning of flat pan. Traces of fingerprints on lower surface.

1:3

Flat system Bc, type BT2 tiles cont.

SAV RT 53 Pan Tile

1:4

Provenance: Sounding C. Level 2, batt. 2. Lot: SA 79-192. Assemblage 13.
Dimensions and technical features: flange h 4.0, flat pan th 3.0. Clay type T1.
Condition: Fragment of flange and beginning of flat pan.

A fragment was identified as the front edge of another tile belonging to a morphologically similar system with an apparently smaller standard size.

SAV RT 54 Pan Tile

1:4

Provenance: Square D4, within stone wall. Level 3, batt. 5. Lot: SA 79-955T-A. Assemblage 13.
Dimensions and technical features: flat pan th 2.2, housing l 9.2, w 4.6. Clay type T1.
Condition: Front edge, semicircular depressed flange with upper projection. Encrustations on the surfaces.

System Bc

BT3. The second group of the interlocking type comprises tegulae with raised flanges along the entire longitudinal surface of the flat pan, ending in correspondence with the tile's rear edge or terminating just short of it. The flanges gradually taper and decrease in width toward the interlock, thus ensuring the correct assembly of the elements. The raised flanges have a quarter-circle profile with a sort of groove running the length of the upper surface. As on the BT2 type, some examples have a continuous shallow depression at the junction with the tile's flat surface.

This type is well attested in the Sinni River valley (Rescigno 2001, 201–2, 40, fig. 35, 43–44, fig. 41, 80–85, fig. 76) and in Calabria. In fact, the profile of the raised border has interesting parallels among the 4th–3rd-c. BC materials from the Calabrian site of Blanda Julia, contexts dating to between the end of the 4th and the 1st c. BC (La Torre and Mollo 2006, 387–88, 606, pl. CVIII, 7–8).

This group contains the examples of rectangular stamps with the letters AVF and square stamps with a nine-point wreath closed by a pair of ties below. The tile length seems to correspond to 2.0 feet (one foot equaling 29.6 cm) and 1.0 Roman palm (7.3 cm), while the width was approximately 1.0 foot and 1.0 palm.

The surfaces are usually slipped. All fragments are clay type T1. The majority of the fragments have thick gray-beige calcareous encrustations on their surfaces. Average measurements are: l 65.0 cm, w 39.0 cm, flange h 5.2–7.0 cm, average th of flat pan 2.0–2.8 cm, l of interlock 6.2 cm.

SAV RT 55 Pan Tile

1:4

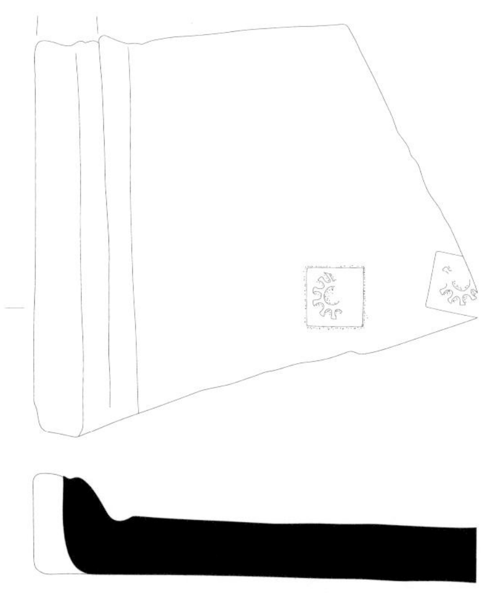

Provenance: Square H3. Level 1, batt. 2. Lot: SA 79-704T. (see *Fig. 2.22*).
Dimensions and technical features: flat pan th 2.8, flange w 3.4–4.1, h 5.3. Clay type T1 compact and with a few inclusions.
Condition: Side and front edge. On the upper surface of the flat pan two impressed stamps in rectangular cartouches, close together but not aligned, one fragmentary. Stamps show nine-point crown with two ties. Flange with rounded inner surface. Lower surface fairly regular but with irregular impressions, which are also visible on the side edge.

SAV RT 56 Pan Tile

1:4

Provenance: Square F4. Level 1, batt. 1. Lot: SA 79-458T.
Dimensions and technical features: th 2.8, flange h 5.2, w 4.5, interlock l 7.5. Clay type similar to T1, but with occasional small black and gray sandy inclusions.
Condition: Fragment of flat pan with flange preserved at interlock. Right corner.

SAV RT 57 Pan Tile

1:4

Provenance: Sounding A. Level 1, batt. 5. Lot: SA 79-23T (see *Fig. 2.30*).
Dimensions and technical features: th 2.3, flange h 4.4. Clay type T1.
Condition: Fragment of flat pan with flange.

SAV RT 58 Pan Tile

1:4

Provenance: Sounding A, 50 m from N end. Level 1, batt. 2. Lot: SA 79-12T (see *Fig. 2.30*).
Dimensions and technical features: th 2.7, flange w 3.5, h 7.0. Clay type similar to preceding examples.
Condition: Fragment of flat pan with flange.

SAV RT 59 Pan Tile

1:4

Provenance: Sounding A, S end of Room 1. Level 2, batt. 1. Lot: SA 79-21T. Assemblage 5.
Dimensions and technical features: h 5.0, w 3.3. Clay type T1.
Condition: Flange fragment.

SAV RT 60 Pan Tile

1:4

Provenance: Square H3. Level 1, batt. 1. Lot: SA 79-1065T (see *Fig. 2.22*).
Dimensions and technical features: th 2.8, flange h 5.3, w 3.5. Clay type T1.
Condition: Partially reconstructed tile, three fragments. Calcareous encrustations all over.

Flat system Bc, type BT3 tiles cont.

SAV RT 61 Pan Tile
Provenance: Square F3, N Kiln interior. Lot: SA 79-861T-A. Assemblage 8.
Dimensions and technical features: h 5.4, w 4.3, th 2.5. Clay type T1.
Condition: Flange fragment with interlock and section of flat pan.

1:4

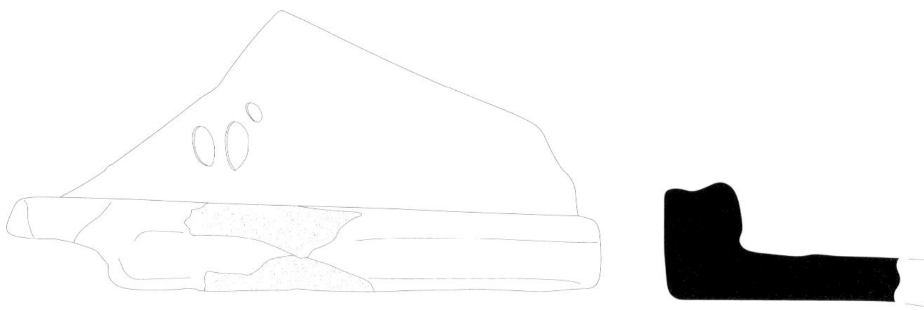

SAV RT 62 Pan Tile
Provenance: Sounding A, S area of Room 1. Level 2, batt. 2. Lot: SA 79-25T (see *Fig. 2.30*).
Dimensions and technical features: th 2.2, h 5.2, w 3.5. Clay type similar to preceding examples.
Condition: Fragment of flat pan with flange.

1:4

SAV RT 63 Pan Tile
Provenance: Sounding A, S area of Room 1. Level 2, batt. 2. Lot: SA 79-25T. Assemblage 5.
Dimensions and technical features: h 5.2, th 2.0. Clay type similar to preceding examples.
Condition: Fragment of flat pan with flange.

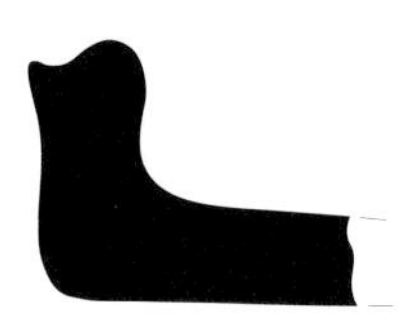

1:4

SAV RT 64 Pan Tile
Provenance: Sounding A, S area of Room 1. Level 2, batt. 2. Lot: SA 79-25T. Assemblage 5.
Dimensions and technical features: flange th 2.5, h 5.2. Clay type similar to preceding examples.
Condition: Fragment of flat pan preserved in vicinity of interlock. Left corner.

1:4

SAV RT 65 Pan Tile
Provenance: Sounding C. Level 2, batt. 2. Lot: SA 79-129. Assemblage 13.
Dimensions and technical features: Clay type T1.
Condition: Fragment of narrowest part of flange.

1:4

BT3: Fragments with Square Stamp of Nine-point Crown and Pair of Ties

SAV RT 66 Pan Tile
Provenance: Square E2/E3. Level 2, batt. 2. Lot: SA 81-4T. Assemblage 5.
Dimensions and technical features: w 44.5, th 2.8, flange h 6.2, interlock l 6.2. Clay type T1.

Condition: Two joining fragments from approximately the central section of the flat pan, ca. 11.0 cm from the front edge. The stamp with nine-point crown and pair of ties is barely legible.

1:4

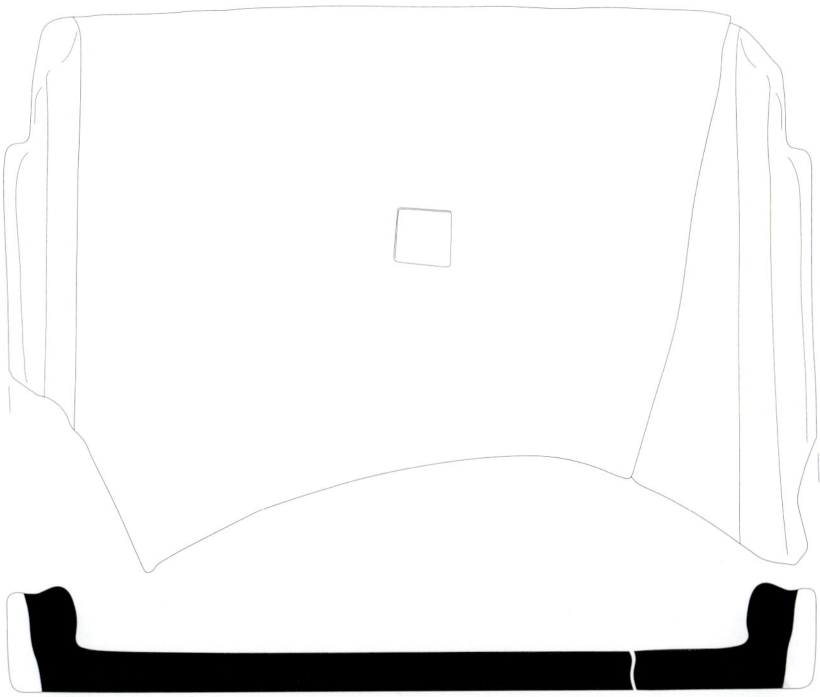

SAV RT 67 Pan Tile
Provenance: Sounding A, S area of Room 1. Level 2, batt. 1. Lot: SA 79-13T. Assemblage 5.
Dimensions and technical features: flat pan th 2.7. Clay type T1.

Condition: Fragment from terminal section of flat pan with short section of flange. At ca. 11.0 cm from front edge stamp with nine-point crown and pair of ties.

1:4

Flat system Bc, stamped type BT3 tiles cont.

SAV RT 68 Pan Tile
Provenance: Square F3, N Kiln interior. Level 2, batt. 2. Lot: SA 79-1059T. Assemblage 8.
Dimensions and technical features: flat pan th 3.0, flange l 2.4. Covered with calcareous encrustations. Clay type T1.
Condition: Front part of tile with small section of flange preserved at narrowest point. Three joining fragments. At ca. 9.5 cm from front edge a square stamp with nine-point crown and pair of ties, barely legible.

1:3

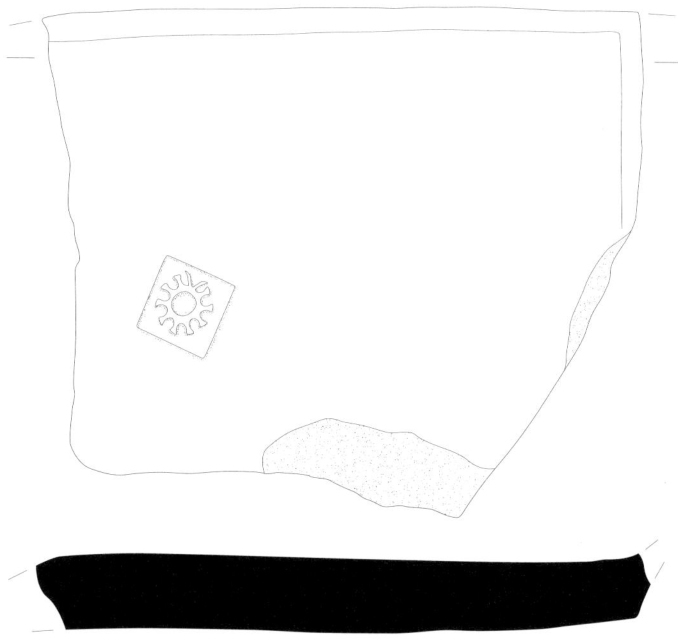

SAV RT 69 Pan Tile
Provenance: Baulk at entrance to N and S Kilns. Level 2, batt. 3. Lot: SA 81-22T. Assemblage 5.
Dimensions and technical features: th 3.0, flange h 5.4. Clay type T1.
Condition: Fragment from terminal section of flat pan with section of flange. At ca. 12.8 cm from flange square stamp with nine-point crown and pair of ties, incomplete and abraded.

1:3

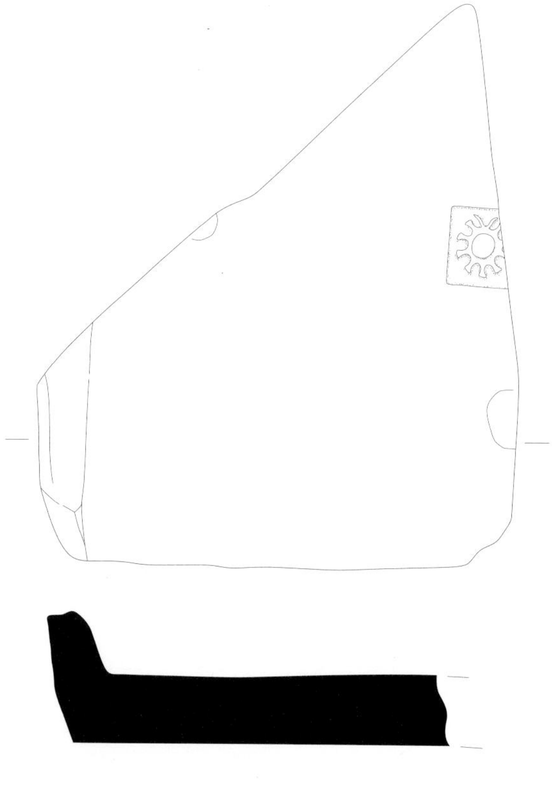

SAV RT 70 Pan Tile
Provenance: Surface find.
Dimensions and technical features: th 2.2. Clay type T1. Lot: SA 81-1058T.
Condition: Fragment of flat pan with square stamp with nine-point crown and pair of ties, incomplete and very abraded, barely legible. Calcareous encrustation on surface.

1:3

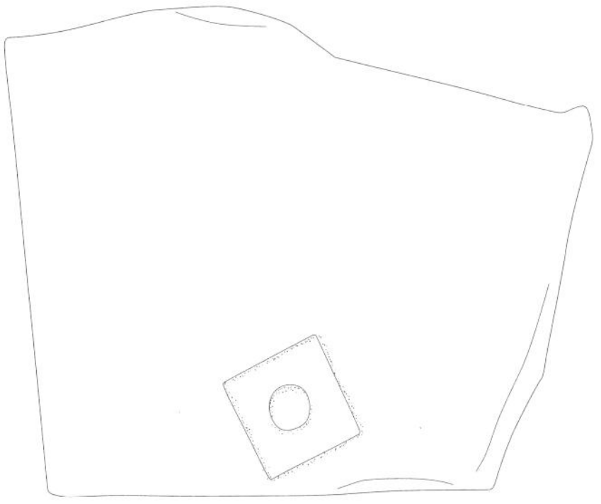

SAV RT 71 Pan Tile
Provenance: Cut in between N and S Kilns. Level 2, batt. 1/2. Lot: SA 81-16T. Assemblage 5.
Dimensions and technical features: th 2.8. Clay type T1.
Condition: Fragment of flat pan with square stamp with nine-point crown and pair of ties, incomplete and very abraded, barely legible.

1:3

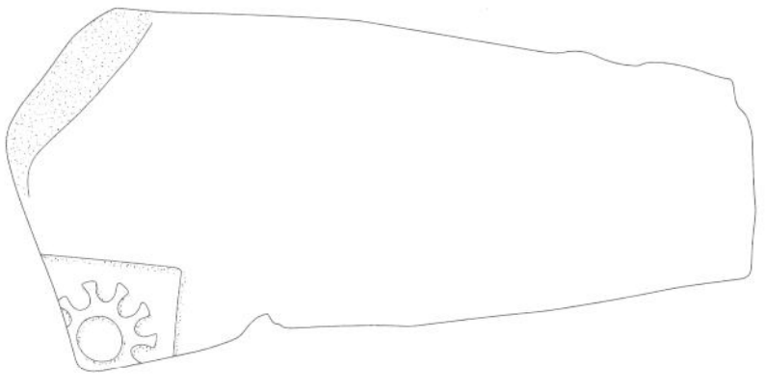

SAV RT 72 Pan Tile
Provenance: Cut in baulk, between N and S Kilns. Level 2, batt. 1/2. Lot: SA 81-15T. Assemblage 5.
Dimensions and technical features: th 2.2. Clay type T1.
Condition: Fragment of flat pan with square stamp with nine-point crown and pair of ties, very abraded, barely legible. Calcareous encrustation on surface.

1:3

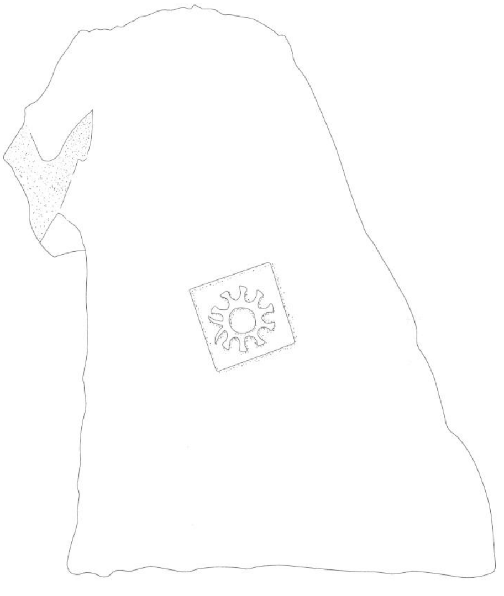

SAV RT 73 Pan Tile
Provenance: Cut in between N and S Kilns. Level 2, batt. 1/2. Lot: SA 81-18T. Assemblage 5.
Dimensions and technical features: th 2.2. Clay type T1.
Condition: Fragment of flat pan with square stamp with nine-point crown and pair of ties, incomplete and very abraded, barely legible. Calcareous encrustation on surface.

1:3

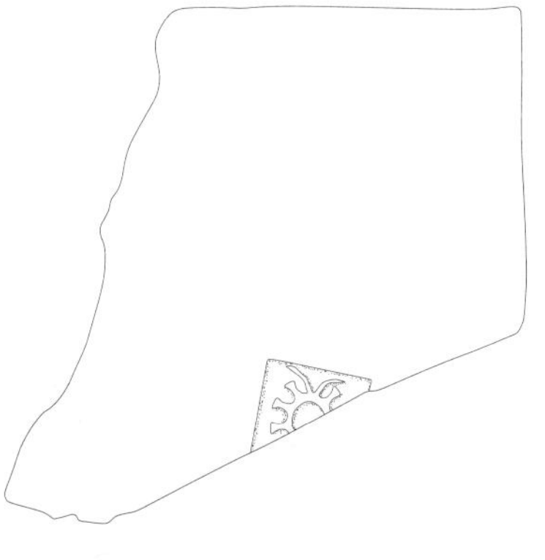

Flat system Bc, stamped type BT3 tiles cont.

SAV RT 74 Pan Tile
Provenance: Square H3. Level 1, batt. 1. Lot: SA 79-743T (see *Fig. 2.22*).
Dimensions and technical features: th 2.2. Clay type T1.
Condition: Fragment of front edge with square stamp with nine-point crown and pair of ties.

1:3

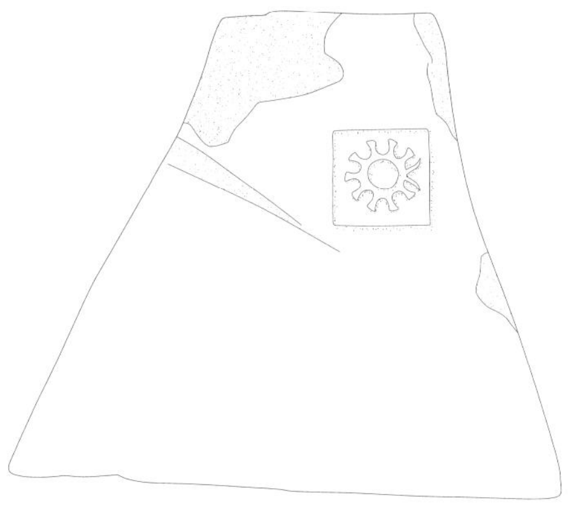

SAV RT 75 Pan Tile
Provenance: Unknown. Lot: SA 81-11T. Assemblage 5.
Dimensions and technical features: th 2.5, flange h 5.7. Clay type T1.
Condition: Fragment of flat pan with flange. At ca. 18.0 cm from flange stamp as on preceding example.

1:3

SAV RT 76 Pan Tile

Provenance: Baulk at entrance to N and S Kilns. Level 2, batt. 3. Lot: SA 81-21T. Assemblage 5.

Dimensions and technical features: th 2.2. Clay type T1.

Condition: Fragment of flat pan with square stamp with nine-point crown and pair of ties. Calcareous encrustation on surface.

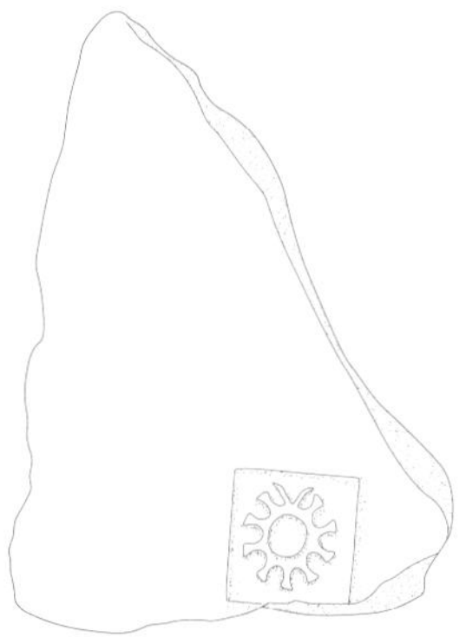

1:3

SAV RT 77 Pan Tile

Provenance: Square E2. Level 1, batt. 1. Lot: SA 79-764T.

Dimensions and technical features: Clay type T1 (see *Fig. 2.30*).

Condition: Small fragment with square stamp with nine-point crown and pair of ties, incomplete. Calcareous encrustation on surface.

1:3

SAV RT 78 Pan Tile

Provenance: Cut in baulk, between N and S Kilns. Level 2, batt. 1/2. Lot: SA 81-17T. Assemblage 5.

Dimensions and technical features: th 2.2. Clay type T1.

Condition: Fragment of flat pan with square stamp with nine-point crown and pair of ties, very abraded, barely legible. Calcareous encrustation on surface.

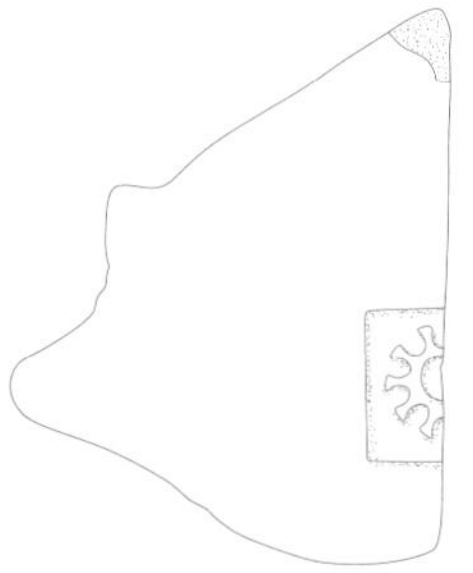

1:3

Flat system Bc, stamped type BT3 tiles cont.

SAV RT 79 Pan Tile
Provenance: Cut in baulk between N and S kilns. Lot: SA 81-13T. Assemblage 5.
Dimensions and technical features: interlock l 6.6. Clay type T1.
Condition: Fragment of terminal section of flat pan with trace of flange. Square stamp with nine-point crown and pair of ties.

1:3

SAV RT 80 Pan Tile
Provenance: Square E2/E3, N extension of the cut of the baulk. Lot: SA 81-14T. Assemblage 5.
Dimensions and technical features: th 2.2. Clay type T1.
Condition: Fragment of front edge with square stamp with nine-point crown and pair of ties, very abraded, barely legible.

1:3

SAV RT 81 Pan Tile

Provenance: Square E2/E3. Lot: SA 81-5T. Assemblage 5.

Dimensions and technical features: th 2.2. Clay type T1.

Condition: Fragment of front edge with square stamp with nine-point wreath and pair of ties, very abraded, barely legible.

1:3

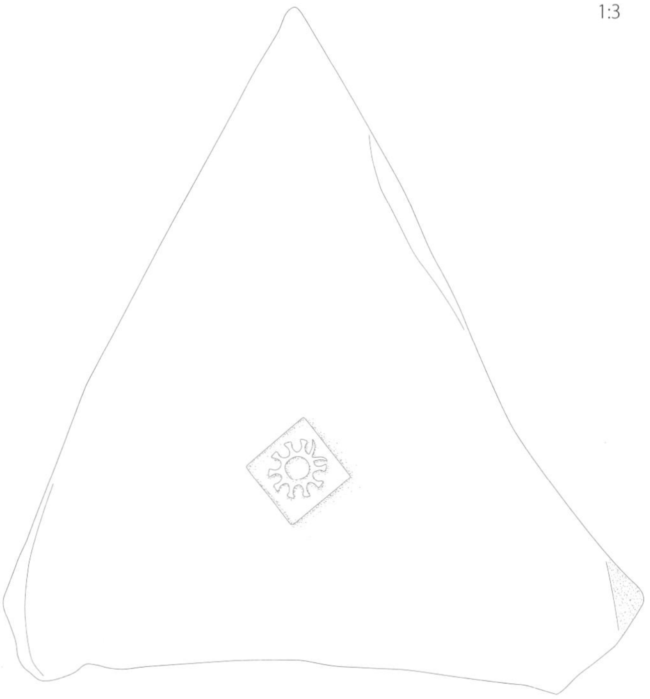

SAV RT 82 Pan Tile

Provenance: Square F3. Level 1, batt. 2. Lot: SA 79-821T. Assemblage 8.

Dimensions and technical features: th 2.2. Clay type T1.

Condition: Fragment of flat pan with square stamp with nine-point crown and pair of ties, very abraded, barely legible. Calcareous encrustation on surface.

1:3

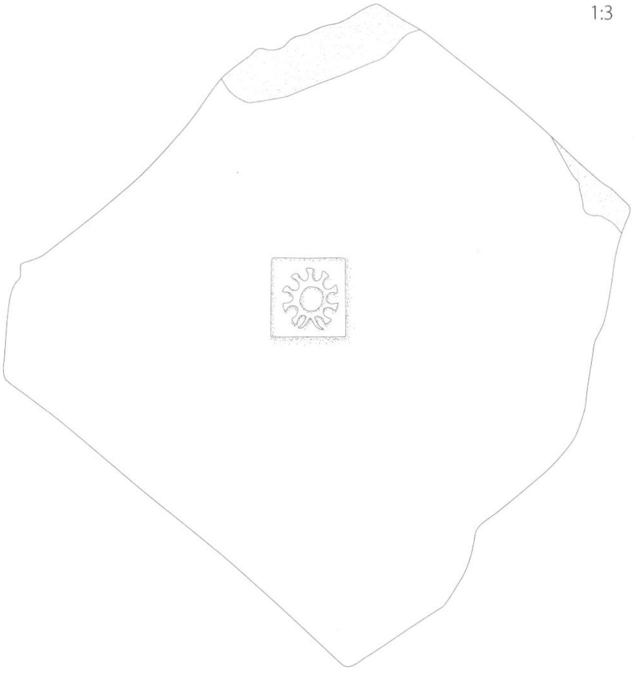

Flat system Bc, stamped type BT3 tiles cont.

SAV RT 83 Pan Tile

Provenance: Square E2/E3, N extension of the cut of the baulk. Lot: SA 81-10T. Assemblage 5.

Dimensions and technical features: th 2.5. Clay type T1.

Condition: Fragment of front edge. At ca. 14.5 cm square stamp with nine-point crown and pair of ties, very abraded, barely legible.

1:3

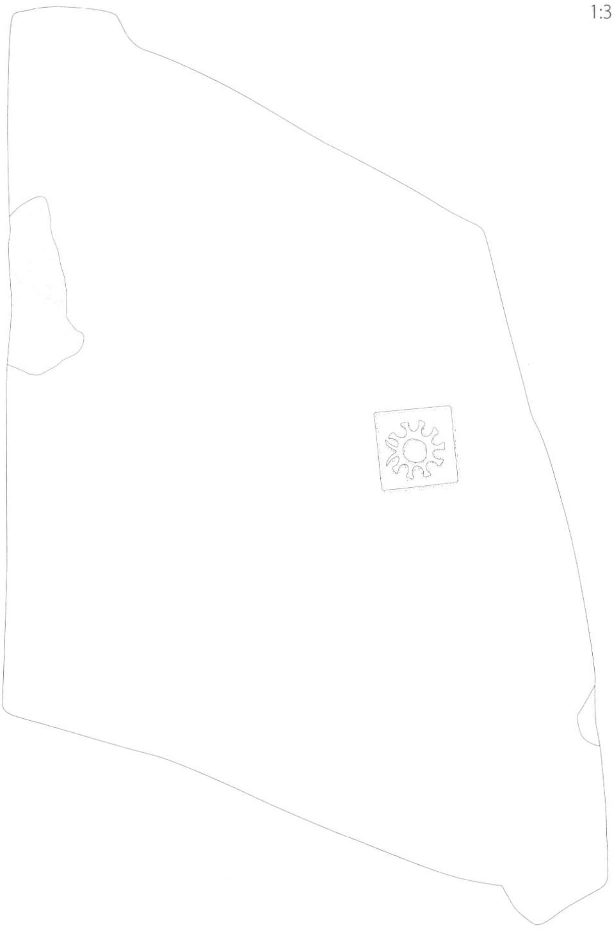

SAV RT 84 Pan Tile

Provenance: Sounding A, S area of Room 1. Level 2, batt. 1. Lot: SA 79-13T. Assemblage 5.

Dimensions and technical features: th 2.4.

Condition: Fragment of flat pan with square stamp with nine-point crown and pair of ties, very abraded, barely legible.

1:3

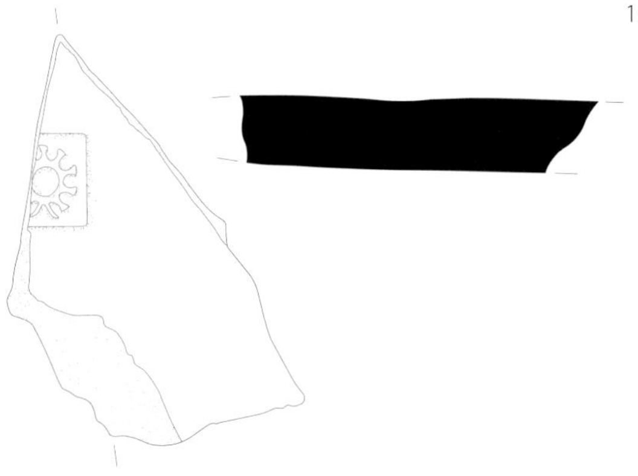

SAV RT 85 Pan Tile
Provenance: Square E2. Level 1, batt. 1. Lot: SA 79-936T (see *Fig. 2.30*).
Dimensions and technical features: th 2.5. Clay type T1.
Condition: Front edge. No trace of flanges preserved. A short distance from the edge and parallel to it, square stamp with nine-point crown and pair of ties.

1:3

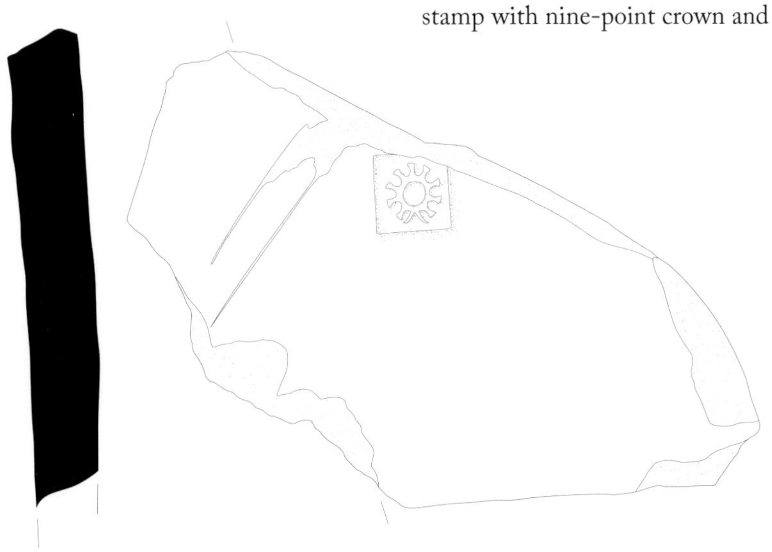

BT3: Fragments with AVF Rectangular Stamp

SAV RT 86 Pan Tile
Provenance: Square E2, baulk at S Kiln. Level 2, batt. 1. Lot: SA 81-6. Assemblage 5.
Dimensions and technical features: th 3.8, flange h 6.2. Clay type T1.
Condition: Fragment of flat pan with flange. At ca. 17.5 cm from flange part of rectangular stamp [AVF].

1:3

Flat system Bc, stamped type BT3 tiles cont.

SAV RT 87 Pan Tile
Provenance: Cut in baulk between N and S Kilns. Level 2, batt. 1/2. Lot: SA 81-9T. Assemblage 5.
Dimensions and technical features: Clay type T1.
Condition: Fragment of flat pan, left side of stamp missing.

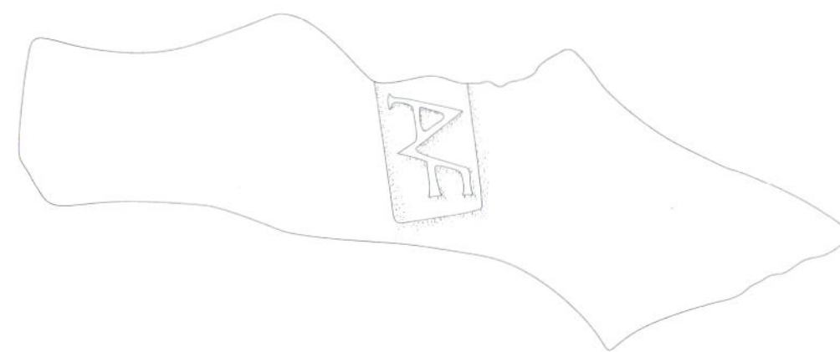

1:3

SAV RT 88 Pan Tile
Provenance: Square F2. Level 2, batt. 2. Assemblage 7.
Dimensions and technical features: Clay type T1.
Condition: Fragment of flat pan, right side of stamp missing.

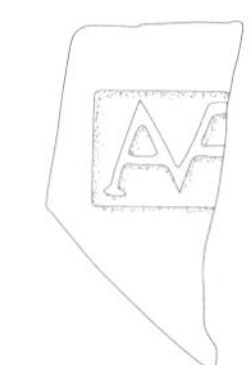

1:3

SAV RT 89 Pan Tile
Provenance: Baulk at entrance to N and S Kilns. Level 2, batt. 3. Lot: SA 81-20T. Assemblage 5.
Dimensions and technical features: Clay type T1.
Condition: Fragment of flat pan, right side of stamp missing.

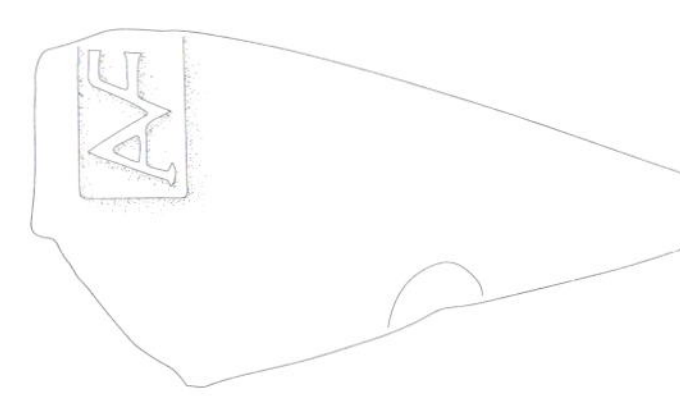

1:3

Miscellaneous stamped tile fragments. In the absence of characterizing elements, the following stamped fragments cannot be attributed to a single type of tile or roofing technique. They do, however, appear very similar to the preceding examples.

Rectangular Stamp, Seven-point Wreath and Pair of Ties

SAV RT 90 Pan Tile
Provenance: Square F2, S Kiln interior. Level 2. Lot: SA 79-794T. Assemblage 7.
Dimensions and technical features: th 2.5. Clay type T1.
Condition: Fragment of flat pan with rectangular stamp, seven-point crown, and pair of ties, partially preserved.

1:3

SAV RT 91 Pan Tile
Provenance: Square F2, S Kiln interior. Level 2. Lot: SA 79-812T. Assemblage 7.
Dimensions and technical features: th 2.5. Clay type T1.
Condition: Fragment of front edge. At ca. 7.5 cm from edge there is a rectangular stamp with seven-point crown and pair of ties.

1:3

Square Stamp with Anchor

SAV RT 92 Pan Tile
Provenance: Square F3. Level 2, batt. 1. Lot: SA 79-818T. Assemblage 8.
Dimensions and technical features: th 2.5. Clay type T1.
Condition: Fragment of flat pan with square stamp of an anchor.

Square Stamp with Eight-spoked Wheel

SAV RT 93 Pan Tile
Provenance: Sounding A, S area of Room 1. Level 2, batt. 2. Lot: SA 79-24T. Inv. no. 319908. Assemblage 5.
Dimensions and technical features: th 2.6. Clay type T1.
Condition: Fragment without edges.

SAV RT 94 Pan Tile
Provenance: Square E2 (from SA 79-772PL). Level 2, batt. 2. Lot: SA 79-774T (see *Fig. 2.30*).
Dimensions and technical features: th 2.5. Clay type T1.
Condition: Fragment with incomplete stamp.

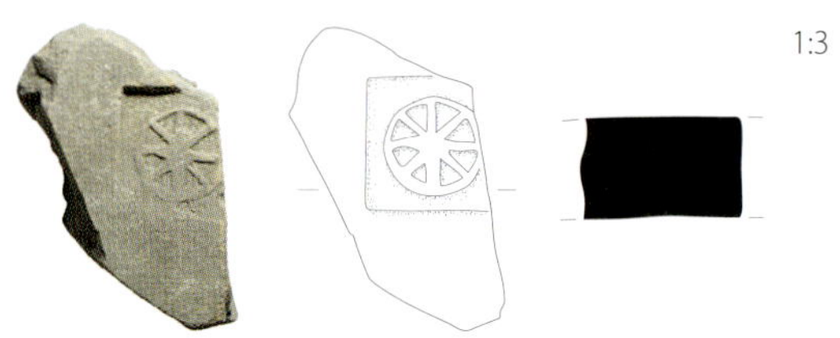

BC1. A group of fragments attests the presence of cover tiles with an arched profile, smaller than those attributable to the Laconian roofing system. There are two distinct types within the group. In the first type, BC1, documented by only a few fragments, the arch has a rectangular, horseshoe-shaped profile. The resting edges are oblique, the side edges can be defined as slightly forked. Using the evidence in the tomb contexts (*Fig. 32.13*) and one partially restored but complete example (SA 79-1020) composed of several fragments, with smoothed upper surface, irregular underside, and flat resting-edges, made of clay type T2 (*Fig. 32.7*), a standard size can be suggested for, and associated with, a specific tile in the overlapping system of assemblage (BT1), in use from the Archaic period onward. The clay types are T1 and T2. Average dimensions are: l 80.0 cm, front w 15.5 cm, rear w 15.0 cm, h front 7.5 cm, h rear 6.0 cm.

SAV RT 95 Cover Tile
Provenance: Square F3, N Kiln interior. Level 2, batt. 2. Lot: SA 79-820T. Assemblage 8.
Dimensions and technical features: th 1.8. Clay type T2.
Condition: Fragment with short sections of side edges.

1:3

BC2. The second type, BC2, is characterized by a depressed arch with irregular profile. The cover tile seems to taper toward the highest part and, like the Laconian types, rests on oblique edges. The form of this cover tile finds a close parallel among those in use at the Italic settlement of Oppido Mamertina from the 3rd through the 1st c. BC (*Oppido Mamertina*, 360, fig. 362) and among the "Y" type documented from the mid-3rd c. BC onward at Locri (*Locri IV*, 324, pl. XCV, 14). Due to the poor state of preservation of the fragments, neither the length nor the complete profile can be reconstructed. The cover tile type BC2 has an interesting, although chronologically anomalous, parallel among those from the Late Roman farmhouse at San Biagio (inv. SB 80-418T, see *San Biagio*, 41, 3.1.1). The fabric types in this group are T1 and T3. Average measurements are: front w 22.0 cm, opening 6.5 x 17.0 cm, th 2.0–2.2 cm.

SAV RT 96 Cover Tile
Provenance: Sounding A. Level 2, batt. 3. Lot: SA 79-36T (see *Fig. 2.30*).
Dimensions and technical features: th 2.5. Clay type T1.
Condition: Front section.

1:3

SAV RT 97 Cover Tile
Provenance: Square G6, W of wall. Level 2, batt. 1. Lot: SA 79-751PL (see *Fig. 2.20*).
Dimensions and technical features: th 1.5. Clay type T1.
Condition: Front section of imbrex; slightly forked edges. Irregular walls.

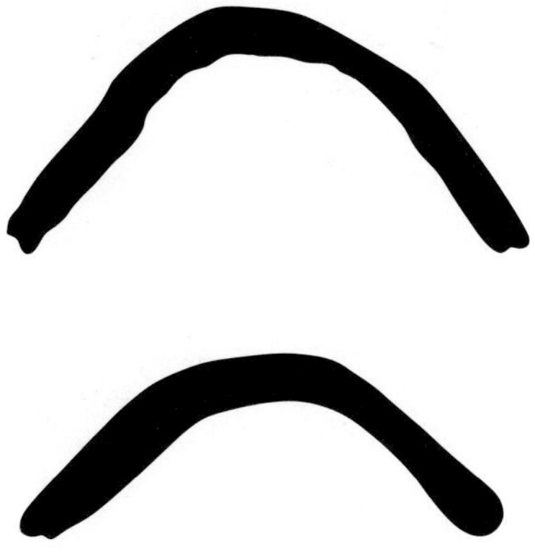

1:3

The following fragment may be attributed to a similar shape with an apparently smaller standard size (probably associated with a BT2 tile from an overlapping system) that rested on a flat edge.

SAV RT 98 Cover Tile
Provenance: Square D4. Level 3, batt. 4. Lot: SA 79-823T.
Dimensions and technical features: th 1.8. Clay type T1.
Condition: Rear section of flat area.

1:3

SAV RT 99 Cover Tile
Provenance: Sounding C, E of division. Level 2, batt. 3. Lot: SA 79-195T.
Dimensions and technical features: th 1.5. Clay type T1.
Condition: Fragment of side edge, slightly forked.

1:3

SAV RT 100 Cover Tile
Provenance: Square F3, Level 2, end of kiln. Lot: SA 79-819T.
Dimensions and technical features: th 2.0. Clay type T1.
Condition: Front section.

1:3

SAV RT 101 Cover Tile
Provenance: Square D4, within stone wall. Lot: SA 79-955.
Dimensions and technical features: th 2.0. Clay type T1.
Condition: Cover tile fragment.

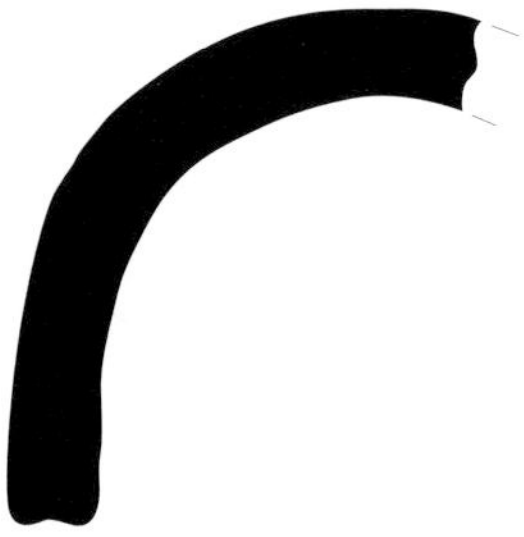
1:3

Opaia 1. Two types of tile with opaion can be associated with one of the flat tile roofing systems. These perforated tiles let out smoke and at the same time let in air and light (Wikander 1983). Type 1 has a sub-circular aperture with a raised border, located ca. 2.0 cm from the edge of the tile. The tile itself has a sub-rectangular profile with a flat upper edge, thickened on the interior. Due to the state of preservation of this example, it can only be suggested that close to the aperture there was a hole used for lifting the tile. It is likely that this group of *opaia keramis* was associated with a fragment, perhaps a cover, with circular flange, externally thickened offset, and flat base (cf., SAV RT 102). The function of the piece might be confirmed by an unpublished item from Fratte di Salerno that is completely identical to the Metapontine example (the author wishes to thank P. Monda for bringing this item to her attention). All fragments are clay type T1. Average measurements: diam. 28.0 cm, diam. aperture 19.0 cm.

SAV RT 102 Opaion
Provenance: Slope 3, W extension. Level 3, batt. 1. Lot: SA 79-380P. Assemblage 11.
Dimensions and technical features: th 2.4, h 2.5. Clay type T1.
Condition: Opaion cover. Section of the flat pan and part of the insertion point preserved.

1:3

SAV RT 103 Opaion
Provenance: Square D4. Level 3, batt. 4; Level 3, batt. 5. Lot: SA 79-897T, SA 79-825PL, SA 79-897T. Assemblage 13.
Dimensions and technical features: th 3.2, h 2.7, w 3.0. Clay type T1.
Condition: Tile with opaion. Three joining fragments, including section of an edge with beginning of flat pan.

1:3

Opaia 2. Among the analyzed materials only one fragment documents the presence of a skylight-tile differing from type 1. The aperture is sub-circular, as in the previous type, but the raised border has a rectangular section with a flat upper border lacking an internal offset. This example has numerous parallels in Lucanian contexts (*Pomarico Vecchio I*, 250, pl. 98, 9; *Roccagloriosa I*, 304, fig. 200, 551; *Torre di Satriano I*, 128–29, pl. II, 18; La Torre and Mollo 2006, 397.400, 610, pls. CXI, 1–3, CXII 1–2) of the 4th and 3rd c. BC and is widely attested in Roman (Adam 1989, 230, figs. 500–501) and Greek (Wikander 1983) contexts.

The examples SAV RT 103 and SAV RT 104 are both fragmentary, so that the piece cannot be reconstructed in its entirety. Moreover, it is also possible that these items had other functions. Terracotta discs with central hole have been found at Syracuse (Cultrera 1938, 289–90, fig. 17) and in the Sanctuary of Demeter at Herakleia (Nava 2001, 953, pl. XXXIII); in the latter context, the examples were used as a well cover. The presence of wells is in fact confirmed at Sant'Angelo Vecchio by the recovery of fragments of cylindrical terracotta well-liners, some of which had been reused as a cover for Tomb 10 (see p. 121 in Ch. 4, "Tomb 10"). Nonetheless, other items with similar shape come from the Athenian Agora, where they have been interpreted as covers for pithoi (*Agora XXXIII*, 98, nos. 175–77, fig. 29, pl. 23).

SAV RT 104 Opaion
Provenance: Square H3. Level 1, batt. 3. Lot: SA 79-1025T. Assemblage 2.
Dimensions and technical features: w 3.4, h 5.2, th 2.8. Clay type 1.
Condition: Fragment with section of raised border and beginning of flat pan.

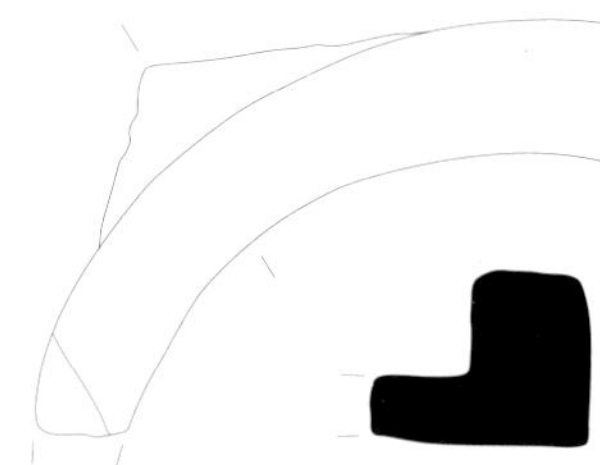
1:4

Tubuli (gutter pipes). A group of tubuli, perhaps gutter pipes, have cylindrical bodies that taper slightly toward the top. The lower edge terminates with a molded band, barely marked on the exterior by a thickening at the edge and parallel grooves, on the interior by a thinning of the wall to facilitate insertion onto the flange of the next tubulus. In fact, the upper border terminates with a depressed flange. The standard size is ca. 17.0 cm in length with a projecting flange of 5.0 cm. All fragments are fabric type T1. Water-removal systems using terracotta pipes are well documented in Vesuvian contexts, where the use of these elements is significant and widespread even during the years of the eruption (Adam 1989, 284–88). Average measurements are: l 17.5 cm, flange l 5.0 cm, max. diam. 9.0 cm, diam. 7.0 cm, th 1.8 cm.

SAV RT 105 Gutter Pipe
Provenance: House on the Hill, W side of Test Trench 2. Level 1, batt. 2. Lot: SA 79-583PL (see *Fig. 2.39*).
Dimensions and technical features: th 1.8. Clay type 1.
Condition: Section of upper border: rim marked by slight thickening on the exterior and a narrow groove, in correspondence the inner surface of the body wall thins where one pipe inserts onto next one below.

1:4

SAV RT 106 Gutter Pipe
Provenance: Square G7. Level 2, batt. 2. Lot: SA 79-1073T. Inv. no. 319943. Assemblage 17.
Dimensions and technical features: diam. 8.5–7.5. Clay type T1.
Condition: Preserved at height of flange, visible wheel marks on body.

1:4

SAV RT 107 Gutter Pipe
Provenance: Square G7. Level 2, batt. 2. Lot: SA 79-1073. Inv. no. 319942. Assemblage 17.
Dimensions and technical features: th 2.2, diam. 7.0–9.0. Clay type T1.
Condition: Reconstructed from two fragments, almost complete. Missing part of upper flange. Calcareous encrustation on all surfaces. Wheel marks visible on body.

1:4

Unidentified Tiles

The first small group is comprised of flat pan fragments with a very slightly curved profile, the convex side smoothed and the other irregular. It is unlikely that they are parts of cover tiles due to their thickness (3.0 cm) and, in some cases, the curvature that, if reconstructed, would form a very wide span with depressed arch. The second group of flat pan fragments has a slightly undulating profile. The side edges are rounded or have a sort of low flange. Fragment SA 79-195T(M) has traces of baked clay on both surfaces attesting to its reuse in a kiln. The third group contains an assortment of unidentifiable types.

SAV RT 108 Pan Tile
Provenance: Sounding B. Level 3, batt. 9. Lot: SA 79-398P (see *Fig. 2.35*).
Dimensions and technical features: th 2.0. Clay type 1.
Condition: Fragment of flat pan with a hole made before firing.

1:4

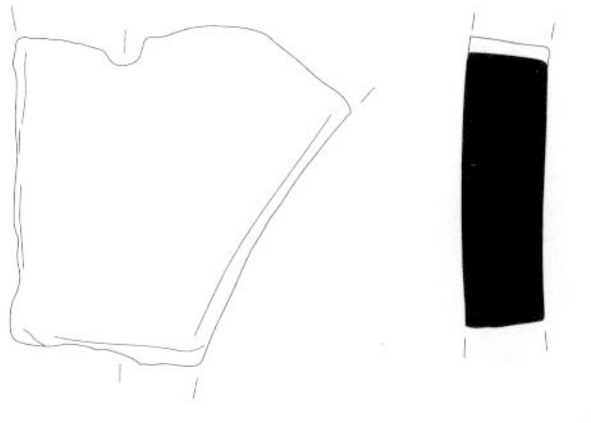

SAV RT 109 Pan Tile
Provenance: Square F3, N Kiln interior. Lot: SA 79-766T. Assemblage 8.
Dimensions and technical features: edge th 1.6–2.6. Clay type T1.
Condition: Two joining fragments from side edge with low "flange."

1:4

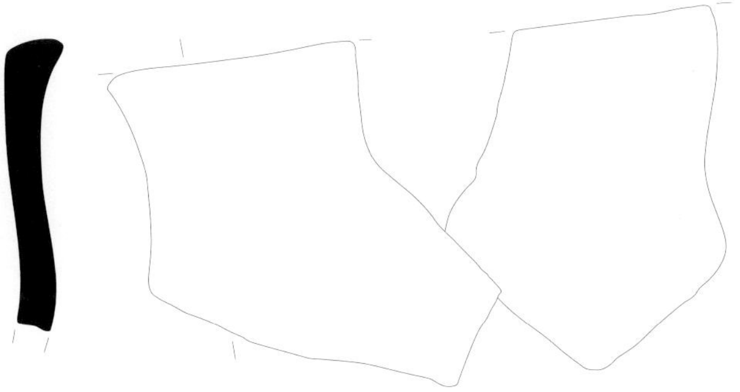

Unidentified tiles cont.

SAV RT 110a–b Pan Tile
Provenance: Square G7. Level 2, batt. 4. Lot: SA 79-813T-B; SA 79-733PL. Assemblage 17.
Dimensions and technical features: th 1.0–1.6. Clay type T1a but with greenish core.
Condition: Two fragments from edge and part of flat pan. Over-fired and deformed.

1:3

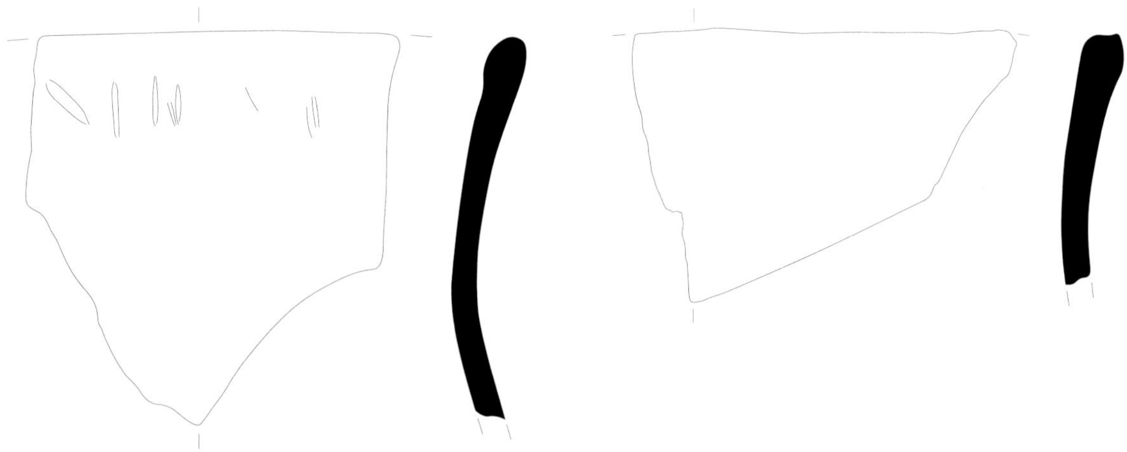

SAV RT 111 Pan Tile
Provenance: Square E2. Level 1, batt. 2. Lot: SA 79-798PL. Assemblage 5.
Dimensions and technical features: th 1.6–2.5. Clay type T1.
Condition: Fragment of flat pan with convex profile with section of a border closed by a raised listel with rectangular section, narrowing toward the bottom where the tile thickens.

1:3

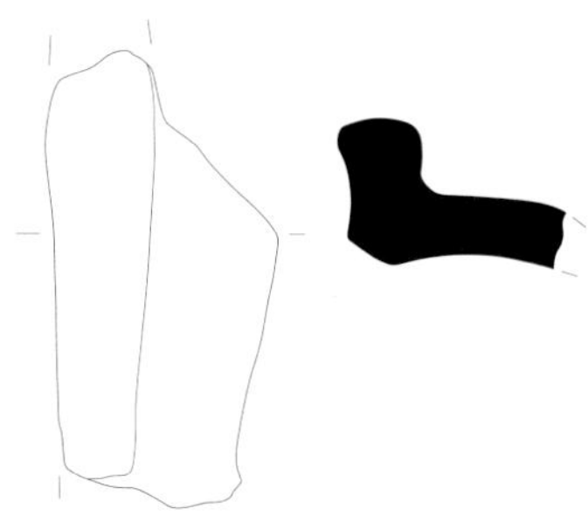

SAV RT 112 Cover Tile?
Provenance: Slope 3, W extension. Level 2, batt. 1. Lot: SA 79-313. Assemblage 11.
Dimensions and technical features: th 2.5. Clay type T1.
Condition: Kalypter hegemon? Border fragment, perhaps from a cover tile with a high arch that has no parallels in the materials from the same context.

1:3

SAV RT 113 Pan Tile
Provenance: Slope 3, W extension. Level 2, batt. 1. Lot: SA 79-131PL. Assemblage 13.
Dimensions and technical features: th 3.5.
Condition: Very thick fragment of flat pan with concave profile with part of an arched element in relief on the upper surface.

1:3

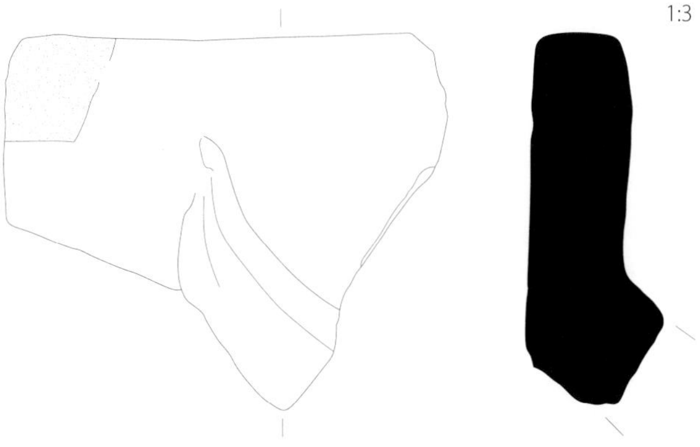

SAV RT 114 Concave Fragment
Provenance: House on the Hill, W side of Test Trench 2. Level 1, batt. 2. Lot: SA 79-569T (see *Fig. 2.39*).
Dimensions and technical features: th 1.6–5.0. Clay type T1.
Condition: Fragment with concave profile. Corner and part of two sides. One of the sides closed by a listel with rectangular section, the thickness of which increases from the corner toward the center. At the fracture the edge expands into an ample curve.

1:3

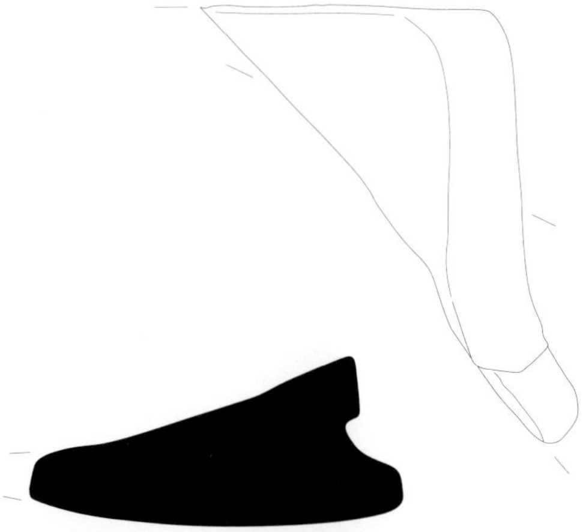

SAV RT 115 Pan Tile
Provenance: Square G3. Level 1, batt. 1. Lot: SA 79-670T.
Dimensions and technical features: th 2.4. Clay type T1.
Condition: Fragment of flat pan. Corner and section of two sides. Thickness increases from the corner toward the center. At ca. 7.5 cm from vertical edge there is a trace of a projection.

1:3

Unidentified tiles cont.

SAV RT 116 Pan Tile

Provenance: Westward Hummock. Level 1, batt. 1–2. Lot: SA 79-1031T.

Dimensions and technical features: th 3.5. Clay type T1.

Condition: Fragment of flat pan with trace of join to what seems to be a flange or listel. Groove at join. On the lower surface join marked by slight step.

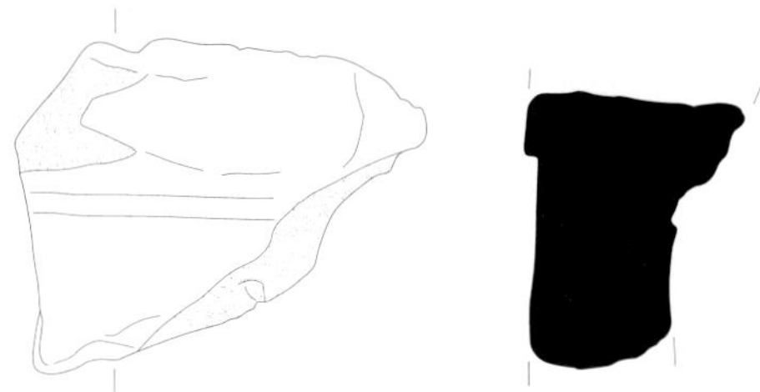

1:3

SAV RT 117 Sima?

Provenance: Square G6. Level 1, batt. 2. Lot: SA 79-747T.

Dimensions and technical features: h 16.5, w 21.0, th 4.0–2.0. Clay type T1.

Condition: Sima? Terminal section of the right side with a flat band which narrows toward the center then thickens again at the point where tile would have inserted into another horizontal tile or listel.

1:3

SAV RT 118 Pan Tile

Provenance: Sounding B. Level 2, batt. 2. Lot: SA 79-186T-F. Assemblage 11.

Dimensions and technical features: th 1.4, h 2.2.

Condition: Section of side edge and section of flat tile. The edge folds gently upward to form a round, slightly flattened border.

1:3

SAV RT 119 Cover Tile?

Provenance: Unknown. Inv. no. 316619.

Dimensions and technical features: th 1.0. Clay type T1, but more compact and bright orange in color.

Condition: Cover tile? Fragment without borders. Upper surface smoothed. On the lower surface impressed oval cut by a straight line.

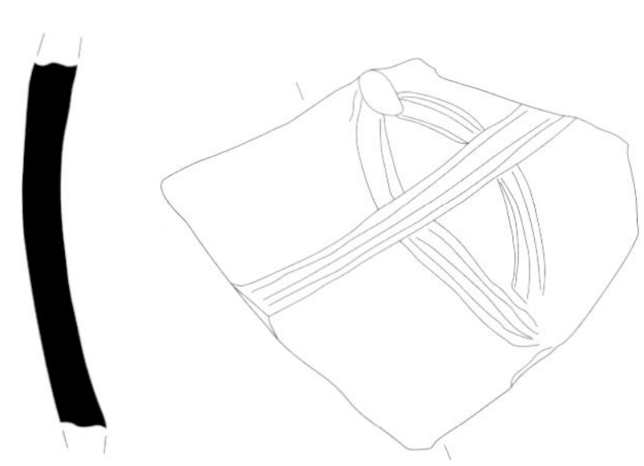

1:3

Tiles reused in the construction of the kilns. This is a small group of pan and cover tiles belonging to different roofing systems that were reused in the construction of the kilns, as suggested by the presence of baked clay, used as bonding material, on the walls of the fragments. There are some examples of fragments from different tile types bonded together in one block. Rare fragments of over-fired cover and pan tiles are also present, presumably wasters.

AT1/AT2/AC1 Laconian Tiles and Laconian Cover Tiles

SAV RT 120a–b Laconian Tiles
Provenance: Sounding A, S end of Room 1. Level 2, batt. 1. Lot: SA 79-15T. Assemblage 5.
Dimensions and technical features: th 1.0–2.2. Clay type T1.
Condition: Fragments of Laconian tiles with a layer of baked clay on the surfaces.

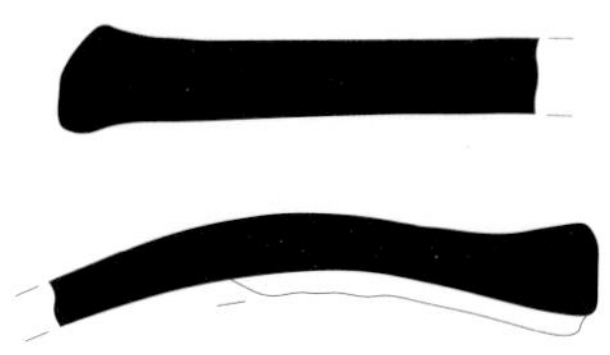

1:3

SAV RT 121 Laconian Pan Tile
Provenance: Slope 3, W extension. Level 2, batt. 2. Lot: SA 79-166T (see *Fig. 2.35*).
Dimensions and technical features: th 1.9–2.3. Clay type T1.
Condition: Side edge with beginning of curved pan tile.

1:3

SAV RT 122 Laconian Pan Tile
Provenance: Slope 3, W extension. Level 2, batt. 2. Lot: SA 79-166T (see *Fig. 2.35*).
Dimensions and technical features: th 2.2. Clay type T1.
Condition: Side edge with section of curved pan tile.

1:3

SAV RT 123 Laconian Pan Tile
Provenance: Slope 3, W extension. Level 2, batt. 2. Lot: SA 79-166T (see *Fig. 2.35*).
Dimensions and technical features: th 2.0. Clay type T1.
Condition: Small section of side edge with beginning of curved pan tile.

1:3

SAV RT 124 Laconian Pan Tile
Provenance: Slope 3, W extension. Level 3, batt. 1. Lot: SA 79-318.
Dimensions and technical features: th 2.0–3.0. Clay type T1.
Condition: Side edge with section of curved pan tile.

1:3

SAV RT 125 Laconian Pan Tile
Provenance: Slope 3, W extension. Level 2, batt. 2. Lot: SA 79-166T (see *Fig. 2.35*).
Dimensions and technical features: th 2.0. Clay type T1.
Condition: Fragment of curved pan, probably type AT2.

1:3

Kiln Wasters

SAV RT 126 Laconian Kiln Waster?
Provenance: Square H3. Level 1, batt. 1. Lot: SA 79-677PL (see *Fig. 2.22*).
Dimensions and technical features: th 1.4–2.2.
Condition: Fragment of edge, particularly deformed. Perhaps a Laconian tile waster.

1:3

SAV RT 127 Laconian Kiln Waster?
Provenance: Slope 3, W extension, S Stipe. Level 3, batt. 1. Lot: SA 79-374P. Assemblage 11.
Dimensions and technical features: th 1.3–2.0.
Condition: Very deformed fragment of flat pan with section of edge, perhaps a Laconian tile waster.

1:3

Floor tiles. Approximately square tiles were certainly used for paving floors.

SAV RT 128 Floor Tile
Provenance: House on the Hill, Test Trench 2. Level 1, batt. 2. Lot: SA 79-622T (see *Fig. 2.39*).
Dimensions and technical features: th 2.7.
Condition: Two joining fragments, completely vitrified.

1:3

Sima

SAV RT 129 Sima

Provenance: Sounding A, S area of Room 1. Level 2, batt. 2. Lot: SA 79-27T. Assemblage 5.

Dimensions and technical features: h 7.7, l 20.0, th 4.3. Clay type T1.

Condition: Small fragment comprising high side joining to rear tile slab at a right angle. On the flat part of the tile, in correspondence with the right edge, is an attachment for a flange-like element (l 5.0) with a flattened quarter-circle profile. The lower surface has no notches but is slightly damaged. On the thickened side there appears to be a slight step, projecting just a few mm (ca. 0.2 cm). On the front face of the side closest to the break there is a trace of detachment (beginning at ca. 15.0 cm from the edge). At the rear, the side seems to fold, and therefore it may have been open from above as well—perhaps the attachment for a drip-spout with open channel. Heavy encrustation on the surfaces. On the front face of the side, perhaps traces of a painted red line on a light-colored slip.

1:3

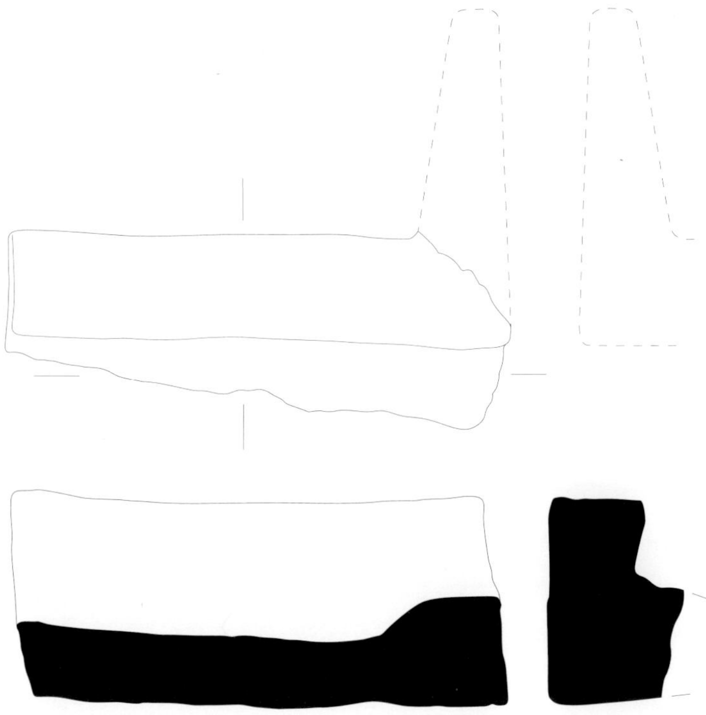

Appendix

Fragments Found in the 2010–2011 Cleaning

Cleaning operations performed in the area of Sant'Angelo Vecchio in 2010–2011 yielded new data about the tiles used at the site, most of which were Laconian and attributable to types AT1–AC1, dated 4th–3rd c. BC. Flat tiles of type BT2, with overlapping assembly and square flanges, were also found, although to a lesser extent. The following items are included in the catalog to broaden the general picture of the archaeological remains recovered at Sant'Angelo Vecchio.

System Aa
ATO

SAV RT 130 Laconian Tile
Provenance: Square D4. US 32.
Dimensions and technical features: max. l 6.3, edge th 2.2. Clay type T1.
Condition: Fragment of the side edge and part of the flat section. Traces of red coloring on the inner wall surface and the side edge, with faint remains also on the outer wall.

1:3

System Ab

The corpus of fragments bearing incised decorations was enriched with these examples bearing patterns/letters made with the fingers on the rough inner wall of Laconian cover tiles of type AC1 (for a thorough discussion of the function of these patterns, see above, p. 469, "The Roofing Systems"). Within this small sample, the patterns that can be identified are crescent moons or circles, in some cases circles with cross-bars (SAV RT 131 and SAV RT 132) documented among the materials recovered in 1979, as well as a zigzag which could be identified as a letter, perhaps a sigma or nasal letter (SAV RT 133).

AC1

SAV RT 131 Laconian Cover Tile
Provenance: Square D4. US 32.
Dimensions and technical features: max. l 8.3, flat section th 1.0. Clay type T1.
Condition: Wall fragment. On the inner surface, a pattern incised with the fingers, perhaps a circle with crossbar.

1:3

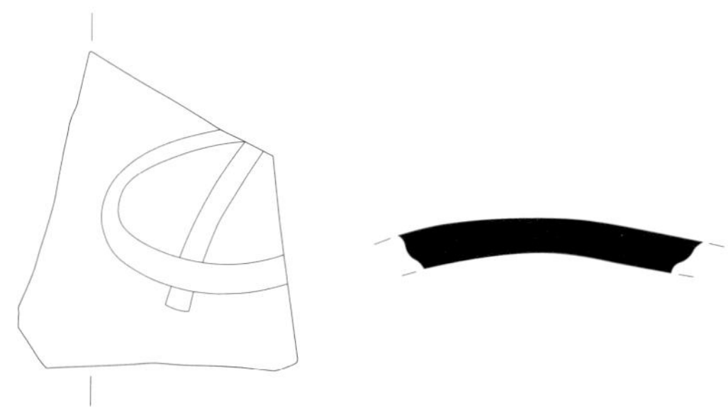

SAV RT 132 Laconian Cover Tile
Provenance: Square D4. US 32.
Dimensions and technical features: max. l 8.8, flat section th 1.5. Clay type T1.
Condition: Wall fragment. On the inner surface, traces of a pattern, perhaps a crescent moon or circle.

1:3

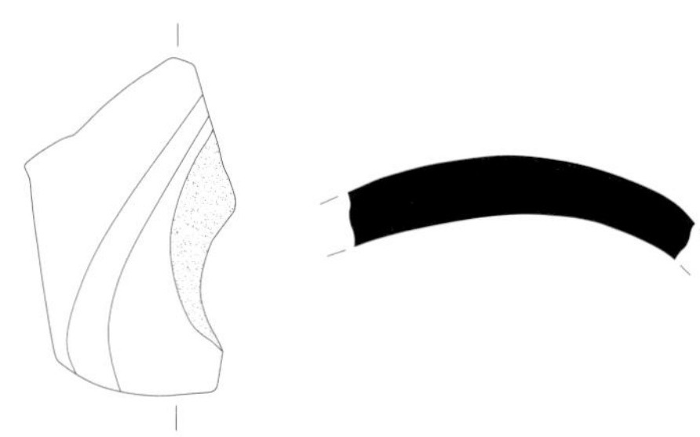

SAV RT 133 Laconian Cover Tile
Provenance: Square D4. US 32.
Dimensions and technical features: max. l 6.8, th flat section 1.2. Clay type T1.
Condition: Wall fragment. On the inner surface, pattern incised with the fingers, perhaps a zigzag.

1:3

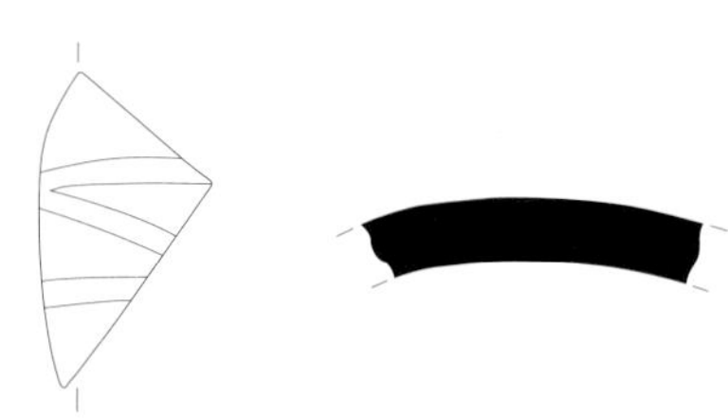

Among the items recovered in the area of the Retaining Wall is a cover tile fragment with a profile that is either triangular or "pentagonal," with two sloping segments joining another two segments that served as resting surfaces. This type was not present among the materials recovered in 1979, but it circulated rather widely from the Archaic through the Roman period. The type recalls several examples of roofing materials recovered at the Pantanello sanctuary (Rescigno forthcoming), currently under study for publication.

"Pentagonal" Cover Tile

SAV RT 134 "Pentagonal" Cover Tile

1:3

Provenance: Square D5. US 9, 18.

Dimensions and technical features: max. l 7.0, flat section th 2.0. Clay type T1.

Condition: Side edge.

System B

BT2. SAV RT 135 has a large flange with rectangular section that links it to type BT2. However, the height of the flange also suggests an Archaic sima with channel.

SAV RT 135 Pan Tile

1:3

Provenance: Square C5, US 42.

Dimensions and technical features: max. l 28.0, flat section th 2.7, flange h 4.5. Clay type T2.

Condition: Part of the large flange and a small portion of the flat section.

Form	Type	Area	Square/Slope	L.	b.	Lot	Chronology	NFr
Cover	BC2	HoH	Test Trench 2	1	2	SA 79-622T		2
			HoH			SA 79-1062PL		1
	AC1						Archaic?	2
			W side of Test Trench 2	1	2	SA 79-569	4th–3rd c. BC	2
	BC2		Westward Hummock	1	1–2	SA 79-1030T		5
	AC1	Kilns	F2, S Kiln interior	1	2	SA 79-837T	4th–3rd c. BC	1
	BC2		Baulk at entrance to N and S Kilns			SA 81-19		1
	AC1		E2	1	2	SA 79-753T-B	4th–3rd c. BC	1
			F2, S Kiln interior	2	4	SA 79-837T		1
			F3, N Kiln interior			SA 79-861		1
	BC1							1
	AC1	Slope 2	Sounding B	2	1	SA 79-306T	4th–3rd c. BC	1
				3	10	SA 79-420		1
		Slope 3	D4	3	1	SA 79-953T		1
				1	4	SA 79-83		2
			W extension	2		SA 79-1052T		1
					1	SA 79-313T		1
					3	SA 79-108PL		1
	BC2	Slope 4	Sounding C	2	2	SA 79-192T		1
	AC1		Sounding C, E of division		3	SA 79-195T	4th–3rd c. BC	2

Table 32.3 Materials not presented in the catalog.

Form	Type	Area	Square/Slope	L.	b.	Lot	Chronology	NFr
Cover	AC1	Wall and Tombs	F5 (possibly G5?)	2	1	SA 79-556T	4th–3rd c. BC	1
			F7	2	2	SA 80-5		1
			F7, NW Trench?	1		SA 80-4?		1
			G3		1	SA 79-670T (from SA 79-669T)		1
			G5	2	2	SA 79-789T		1
	BC1		G6	1	1	SA 79-708T (from SA 79-707T)		1
	AC1			1	2	SA 79-785T	4th–3rd c. BC	1
				1	5	SA 79-707T		1
						SA 79-765T		1
			G6 (from SA 79-707T)	1	1	SA 79-708T-C (from SA 79-707T)		1
			G6, E of wall	2	2	SA 79-739PL		1
			G6, W of wall	2	1	SA 79-748T		1
	BC2							1
	AC1					SA 79-750T	4th–3rd c. BC	1
	BC1				2	SA 79-742T		1
	AC1		G7	1	2	SA 79-796 (PL?)	4th–3rd c. BC	1
			G7, top of wall	2	1	SA 79-771T		1
			G7	2	2	SA 79-816T		5
				2	7	SA 79-839T		1
	BC1					SA 81-2T		1
	BC2							2
Unknown		HoH	W side of Test Trench 2	1	2	SA 79-583PL		1
		Kilns	F2, S Kiln interior			SA 79-837T		1
		Slope 4	Sounding C, E of division	2	3	SA 79-195T		1
		Wall and Tombs	G5	1	2	SA 79-600PL		1
			G6, E of wall	2	2	SA 79-737T-E		1
						SA 79-741T (from SA 79-737T)		1
			H3	1	2	SA 79-560PL		1
Brick	Brick	Kilns	E2	1	2	SA 79-754T (from SA 79-753T)		1
Plaque	Waste	Wall and Tombs	G7	1	2	SA 79-6P		1
Pan	AT1		G5 (from SA 79-600PL)	1	2	SA 79-600PL	4th–3rd c. BC	1
		HoH	Test Trench 2	1	2	SA 79-622T		5
	BT3							1
			W side of Test Trench 2	1	2	SA 79-569T		1
	AT0		HoH			SA 79-1062PL	Archaic?	2
	AT1		Surface finds from area around abandoned farmhouse on hill			SA 79-1PL	4th–3rd c. BC	1
	BT3		Westward Hummock	1	1–2	SA 79-1031T		1
								2
		Kilns	E2	1	1	SA 79-766T		1
				2	1	SA 79-753T-A		1
			E2, baulk at S Kiln	2	1	SA 81-6T		1
	AT1		E2, cleaning of kiln praefurnium			SA 80-63PL	4th–3rd c. BC	1

Table 32.3 cont. Materials not presented in the catalog.

Form	Type	Area	Square/Slope	L.	b.	Lot	Chronology	NFr
Pan	BT3	Kilns	F2	2	2	SA 79-792T		1
			F2, S Kiln	2		SA 79-790T		1
			F2, S Kiln interior			SA 79-836T		4
						SA 79-837T		1
Pan or Cover	AT1/ AC1			2		SA 79-794T		1
						SA 79-801PL		1
						SA 79-836T		3
						SA 79-837T		1
Pan	AT1		F3, N Kiln interior			SA 79-861	4th–3rd c. BC	1
	BT3					SA 79-861		2
						SA 79-861T-C		1
	AT1		G4	2	1–2	SA 79-545T		1
			Sounding A; Slope Area 4, N of Sounding A, found during clearing off of loose soil left by bulldozer			SA 79-52PL	4th–3rd c. BC	4
			Sounding A, S area of Room 1	2	2	SA 79-25TA		1
						SA 79-26PL		1
Pan or Cover	AT1/ AC1		Sounding A, S area of Room 1	2	2	SA 79-15T		
Pan	BT3	Kilns; Sounding D	Sounding D	1	1	SA 80-42PL		1
	AT1						4th–3rd c. BC	3
					2	SA 80-44PL	4th–3rd c. BC	3
	BT3				3	SA 80-45PL		1
Pan or Cover	AT1/ AC1			2	7	SA 80-58PL		1
Pan	AT1	Slope 1	Surface cleaning			SA 79-359PL	4th–3rd c. BC	1
Pan or Cover	AT1/ AC1							1
Pan	AT1	Slope 2	W extension, topsoil	1	1	SA 79-77PL	4th–3rd c. BC	1
			Sounding B	2	1	SA 79-306T		3
Pan or Cover	AT1/ AC1							1
Pan	AT1			3	2	SA 79-186T	4th–3rd c. BC	2
						SA 79-300PL		1
	BT3				7	SA 79-358PL		1
	AT1	Slope 3	D4	3	4	SA 79-824T; from SA 79-823T	4th–3rd c. BC	1
					6	SA 79-954T		1
Pan or Cover	AT1/ AC1							1
Pan	AT1			3	1	SA 79-321T	4th–3rd c. BC	1
			Clearing of grass and topsoil		1	SA 79-76PL		1
			NE extension	2	2	SA 79-142PL		1
	BT3		NW extension		3	SA 79-135PL		1
	AT1		Stipe S, W extension	3	1	SA 79-376PL	4th–3rd c. BC	1
Pan or Cover	AT1/ AC1		Surface cleaning	2	1	SA 79-313T		1

Table 32.3 cont. Materials not presented in the catalog.

Form	Type	Area	Square/Slope	L.	b.	Lot	Chronology	NFr
Pan	AT1	Slope 3	W extension	2	1	SA 79-313T	4th–3rd c. BC	5
						SA 79-314T		1
						SA 79-424PL		1
	BT3					SA 79-313T-B		1
	AT1				2	SA 79-166T	4th–3rd c. BC	1
	AT2							1
	AT1			3	1	SA 79-131	4th–3rd c. BC	1
						SA 79-424PL		1
Pan or Cover	AT1/ AC1		W extension, Stipe N	2	1	SA 79-193T		1
Pan	AT1					SA 79-385PL	4th–3rd c. BC	1
Pan or Cover	AT1/ AC1		Sounding C	2	2	SA 79-192T		1
Pan	AT1		Surface find from grassy slope above Slope 3			SA 79-61PL	4th–3rd c. BC	1
	BT3							1
	AT1	Slope 4			2	SA 79-60PL	4th–3rd c. BC	2
			Sounding B	3	7	SA 79-358PL	4th–3rd c. BC	1
	BT3		Sounding C	2	2	SA 79-192T (A)		1
	AT1					SA 79-193T	4th–3rd c. BC	1
					7	SA 79-377T (B)		1
	Opaia 2		Sounding C, E of division	2	1–3	SA 79-205, SA 79-322PL, SA 79-765		1
	AT1				3	SA 79-195T	4th–3rd c. BC	4
Pan or Cover	AT1/ AC1							2
Pan	AT1	Wall and Tombs	F4	1	1	SA 79-457PL	4th–3rd c. BC	1
			F5 (possibly G5?)	2	1	SA 79-556T		4
			F7	2	2	SA 80-5		2
			G3	1	1	SA 79-669T		1
	BT3							3
	AT1					SA 79-670T (from SA 79-669T)	4th–3rd c. BC	2
			G4	2	1	SA 79-545T		2
	BT3		G5	1	1	SA 79-600PL		1
	AT1				2		4th–3rd c. BC	1
						SA 79-789T		1
				2	1			1
	BT3		G6	1	1	SA 79-708T (from SA 79-707T)		1
						SA 79-708T-C (from SA 79-707T)		1
	AT1				2	SA 79-785T	4th–3rd c. BC	2
Pan or Cover	AT1/ AC1							2
Pan	AT1				4	SA 79-679T	4th–3rd c. BC	1
	BT3				5	SA 79-684PL		1

Table 32.3 cont. Materials not presented in the catalog.

Form	Type	Area	Square/Slope	L.	b.	Lot	Chronology	NFr
Pan	AT1	Wall and Tombs	G6, E of wall	2	1	SA 79-765T	4th–3rd c. BC	2
	BT2					SA 79-737T-C		1
	BT3					SA 79-765		1
	BT2				2	SA 79-741T (F) (from SA 79-737T)		1
	BT3					SA 79-741T (from SA 79-737T)		2
	AT1		G6, W of wall	2	1	SA 79-750T	4th–3rd c. BC	2
			G6, W of wall (from SA 79-737T)	1	2	SA 79-741T (from SA 79-737T)		2
			G7	2	4	SA 79-813T		1
	BT2							1
	AT1		H3	1	1	SA 79-560PL	4th–3rd c. BC	1
			H tile				4th–3rd c. BC	2
	BT3		Surface find			SA 79-11T		1
	AT1		Surface finds from cut of road			SA 79-3PL	4th–3rd c. BC	1
Pan						SA 79-839T	4th–3rd c. BC	4
Pan or Cover	AT1/ AC1							1
Tomb		Wall and Tombs	F6, Tomb 2	3	1	SA 79-921T (C)		1
			F6, Tomb 3	4	1	SA 79-1020T	6th–5th c. BC	1
			F6, W baulk			SA 79-1020T		1
			G6, E of wall	2	2	SA 79-922T		1
Gutter pipe		Slope 4	Sounding B	3	9	SA 79-398PL		1
		Wall and Tombs	F5	2	1	SA 79-556T		1
			G4	1	4	SA 79-461T		1
			G5	1	2	SA 79-600PL		1
			G6, E of wall	2	2	SA 79-737T		1
			G6, E of wall ; G7, cleaning around wall to E	2	2	SA 79-1012T; SA 79-737T		1
			G7, E of wall	2	1	SA 79-802PL		3
Misc.		Kiln	Sounding A, deep one in Room 1	3	2	SA 79-40PL		1
		Slope 2	Sounding B	2	2	SA 79-190PL		1
				3	2	SA 79-198PL		1
		Slope 4	Sounding C	2	3	SA 79-323PL		3
				2	4	SA 79-329PL		1
			Sounding C, E of division	2	3	SA 79-195T		1
		Kilns	Sounding D	1	1	SA 80-42PL		4
					2	SA 80-43PL		2
					3	SA 80-45PL		2
				2	7	SA 80-57PL		1
		Wall and Tombs	F4	1	1	SA 79-457PL		1
			F7	2	2	SA 80-5PL		3
			F7, E of wall	2	4	SA 80-6PL		1
		Slope 1	Surface cleaning			SA 79-359PL		1
		Slope 2	W extension, topsoil	1	1	SA 79-77PL		1

Table 32.3 cont. Materials not presented in the catalog.

Form	Type	Area	Square/Slope	L.	b.	Lot	Chronology	NFr
Misc.		Slope 3	W extension, Stipe N	2	1	SA 79-385PL		1
			Surface find from grassy slope above Slope 3			SA 79-61PL		1
			W extension	2	3	SA 79-108PL		1
			NW extension		3	SA 79-135PL		1
			W extension, Stipe N	2	1	SA 79-385PL		1
		Kilns	E2, cleaning of kiln praefurnium	2	2	SA 80-63PL		1
					2	SA 80-63PL		1
		HoH	Test Trench 2	1	2	SA 79-622T		2
		Wall and Tombs	G1			SA 80-14PL		1
			G6	1	5	SA 79-684PL		1

Table 32.3 cont. Materials not presented in the catalog.

33

Kiln and Workshop Furniture

Alessandro Quercia

The presence of kiln and workshop furniture and tools is a basic factor in identifying ceramic ateliers because they represent conclusive evidence for production. Nevertheless, the archaeological indicators of craft production have received less attention in the literature than the structures themselves (kilns and relevant installations) or their products.[1] Moreover, owing to a lack of proper understanding, this type of evidence has not always been correctly identified, and scholars have paid little attention to these objects or to their relevant functional systems.[2]

A more comprehensive approach to these finds has recently been gaining ground in archaeological research, focusing on their relationship to and function within their contexts.[3] Better knowledge of the workshop furniture and the relevant issues creates a more detailed picture of an atelier's organization and productive processes.

The function of kiln firing supports, the published evidence for which is still far from satisfactory,[4] has been discussed in a number of studies.[5] These spacers played a decisive role in stacking, separating, and stabilizing products during firing. They facilitated the correct circulation of the hot gasses passing through the materials placed in the firing chamber, kept the surfaces of large vessels from touching, and prevented unwanted effects such as over-firing or blemishing. The numerous shapes and types of kiln supports attested in pottery workshops appear to be associated with different degrees of quality in the technological process of firing.[6]

[1] There are a few concise studies, such as Kaloyeropoulou 1970, Biers 1971, and Papadopoulos 1992.

[2] One of the main issues is the difficulty of identifying the correct function of the wedge-shaped kiln supports, which in the past have been interpreted as religious objects, as well as representations of tongues or theater-tickets: see Papadopoulos 1992, 213 n. 35.

[3] See, e.g., Cracolici's (2003) monograph on the kiln spacers from the *kerameikos* of Metaponto. See also Dell'Aglio 1996a; Papadopoulos 2003, 216–17; Merker 2006, 20–22; Cuomo di Caprio 2007, 528–31.

[4] See the list of South Italian finds in Cracolici 2003, 19–24.

[5] Papadopoulos 1992, 216 ff.; Cuomo di Caprio 2007, 528–31; Cracolici 2003, *passim*.

[6] For a typology of these finds, see Cracolici 2003, 25–32 (with references to previous classifications).

The workshop furniture from the site of Sant'Angelo Vecchio consists mainly of kiln firing supports (SAV KF 01–SAV KF 13) and a few tools (SAV KF 14), all made of clay. Kiln wasters (SAV KF 15–SAV KF 16), which are likewise important indicators of production, were also found in the workshop dumps. The most attested type at Sant'Angelo Vecchio is the wedge-shaped support, with the straight edge raised to form a distinctive lip (type 1); the overall number of relevant finds is 41 out of 70 identified kiln firing supports (*Fig. 33.1*). They are small, rough, squared objects with considerable variation in size: length and height can range from 3.2 to 7.4 cm, with most specimens falling between 3.7 and 4.7 cm. Four variants have been identified on the basis of the presence or absence of a projecting lip (SAV KF 01–SAV KF 04), which probably had a practical role in preventing the material from slipping during stacking in the firing chamber, as some hypothetical reconstructions assume.[7] The lips project from 0.2 to 0.9 cm. Variant d (SAV KF 04), which occurs in a large number of finds, is consistent with the group of wedge-shaped supports, although it is distinguished by the lack of the lip. Wedge-shaped supports were quite widespread in Greece (Athens, Corinth, Thasos, Elis, and Locris) and southern Italy (Taranto, Locri, Entella, Metaponto, and some southern Lucanian sites such as Montescaglioso, Pisticci, and Gravina di Puglia)[8] from the Archaic through the Hellenistic, Roman, and Byzantine periods. The shape, which was entirely consistent with its practical function, was probably very conservative, and it is impossible to detect a typological evolution through time.[9] It has been assumed that this shape had precursors in the Early Iron Age on the basis of some triangular pieces of clay, discovered in the Athenian Agora, that are consistent with this period.[10]

In comparison to the wedge-shaped spacers, there are few attestations of the elliptical (type 2; SAV KF

[7] Kaloyeropoulou 1970, 433, fig. 10; Karagiorga 1971, 29, fig. 2.

[8] See the full references in the catalog below.

[9] For the chronological range, see Papadopoulos 1992, 211.

[10] Papadopoulos 2003, 216, no. 86.

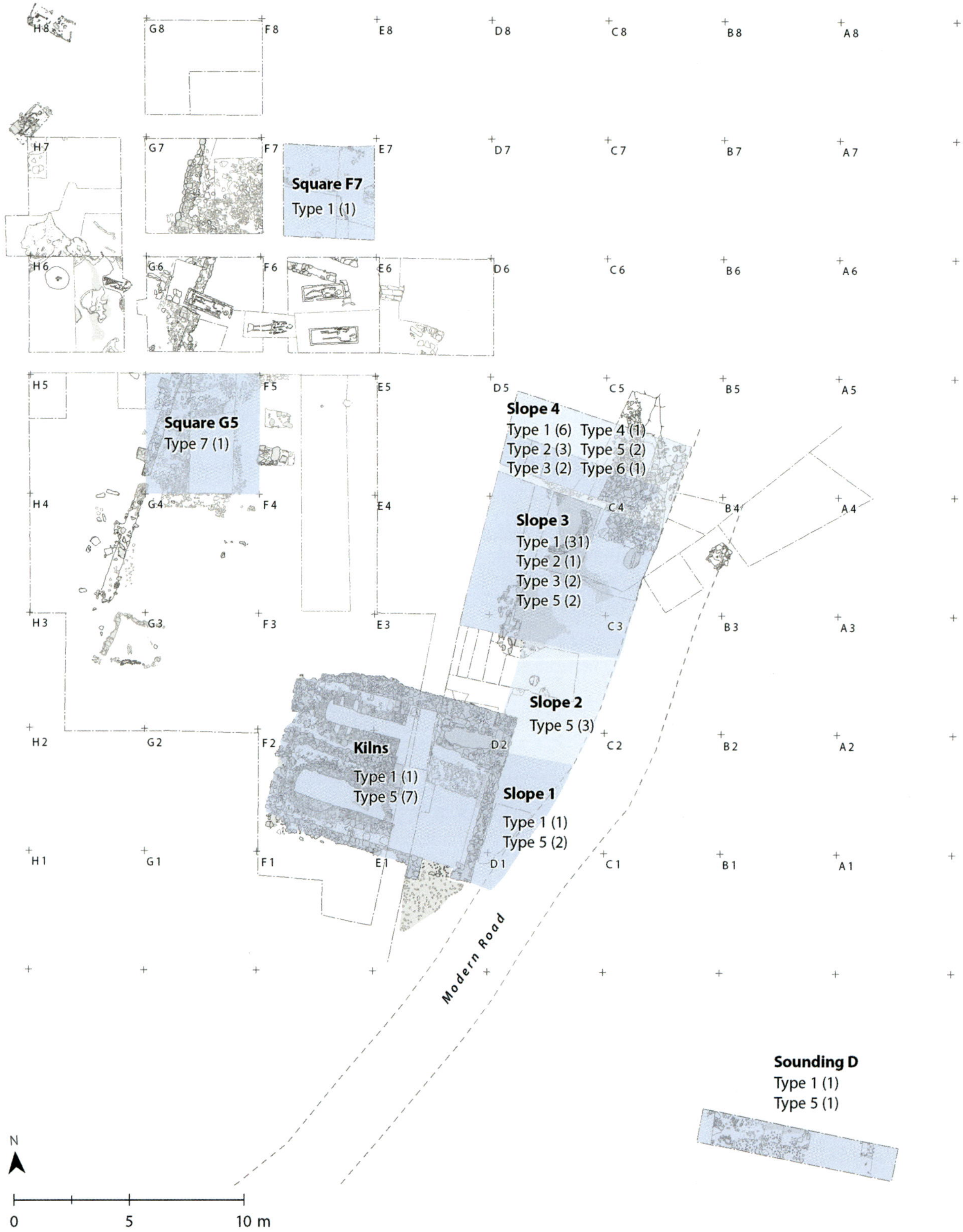

Figure 33.1 Distribution map of kiln spacers; areas highlighted in blue indicate general locations of listed finds. (JT/ICA)

Form	Type	MNI
Kiln firing support	1.a	12
	1.b	10
	1.c	2
	1.d	17
	2.a	2
	2.b	2
	3	4
	4	1
Kiln firing support?	5	17
	6	1
	7	1
	Unknown	1
Pestles		6
Wasters		3
Total		**79**

Table 33.1 Quantification of the workshop materials (furniture and wasters).

05 and SAV KF 06) and truncated pyramidal supports (type 3; SAV KF 07): eight examples in all. These types were also present at an Archaic production site located in the Metapontine chora (Pisticci),[11] and in Classical and Hellenistic Tarentine pottery workshops.[12] Notably, the ring-shaped stacking support, which was one of the most common kiln firing spacers in Classical and Hellenistic production contexts and occurred widely in the *kerameikos* of Metaponto,[13] is almost completely absent in the Sant'Angelo assemblage: just one example relevant to type III-C of Cracolici's classification is attested (type 4; SAV KF 08).

A large number of parallelepiped bars (17 examples; type 5) is also attested in Sant'Angelo Vecchio. Their surfaces are regular and partially smoothed, and at least two sizes may be identified on the basis of their dimensions. One of them (SAV KF 09), which is fragmentary, might have surpassed approximately 14–15 cm in length; it is 7–8 cm wide and 4–5 cm thick. The second variant (SAV KF 10–SAV KF 11), which is much better attested, is approximately half the size of the previous one. Its length is 8.9–9.7 cm, its height 4.8–6.1 cm, and its thickness 2.2–2.8 cm. These parallelepiped bars can be interpreted as kiln firing supports on the basis of evidence from the archaeological literature: similar items were discovered in kilns and ceramic ateliers in Greece (e.g., Thasos) and Roman Gaul (Avenches, Lousonna, Savignac, and Bourzanel) that produced both pottery and tiles.[14] Unfired and fired rectangular solids of clay were also sometimes used as supports during the firing of *pitharia* in many traditional pottery workshops.[15] A sub-rectangular and tapered bar (SAV KF 12; type 6) is approximately consistent with the previous type.

A curvilinear or horseshoe-shaped bar with rectangular section and concave profile, which is not completely preserved (SAV KF 13; type 7), can also be interpreted as a kiln firing support. The item recalls the description of some objects (*boudins*) discovered in a Hellenistic ceramic workshop at Thasos, which have been understood as such.[16]

At Sant'Angelo Vecchio, there are few tools that can be identified as relevant to the furniture of the workshop. Six pestles were found in the excavations; they are all of the same type, consisting of a truncated conical base and a curvilinear and tapered upper part (SAV KF 14). This kind of tool, which occurs frequently in domestic and production contexts, was used for grinding solid substances, such as food (cereals, legumes, etc.), into powder. The presence of clay pestles in several ceramic workshops, such as the *kerameikos* of Metaponto[17] and Tarentine workshops,[18] in association with kiln wasters and spacers, suggests a role somewhere in the manufacturing process, for example in pressing and kneading the clay or other substances (pigments). Moreover, the recovery of a mold for pestles and the evidence for the production of mortaria and basins in the Tarentine atelier discovered in Via Leonida indicate that pestles might also have been one of the products of the workshop. Their association with mortaria and basins, the function of which is strongly related to the kneading and pressing of solid substances, does not appear to be merely incidental.

A probable mold for loom weights (SA 79-435T) is also relevant to the workshop furniture, as it provides important evidence for production of these artifacts at the site. Two over-fired items with surfaces rich in vughs and blisters might have been kiln wasters or deformed products of the workshop (SAV KF 15–

[11] Cracolici 2003, 97–98.

[12] The workshop is located in Via Leonida, 52: Dell'Aglio 1996a, 56–57.

[13] Cracolici 2003, *passim*.

[14] See n. 10 for the full reference.

[15] Blitzer 1990, 696.

[16] Garlan 1986, 213.

[17] Silvestrelli forthcoming.

[18] In Via Leonida and near the Caserma Mezzacapo: Dell'Aglio 1996a, 56–57.

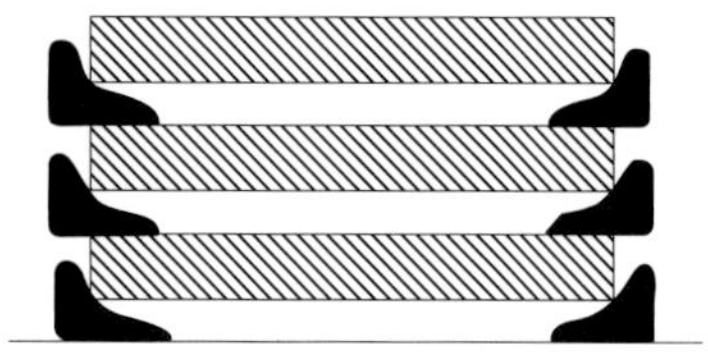

Figure 33.2 Hypothetical reconstruction of a kiln stacking system for tiles. (After Karagiorga 1971, fig. 1)

SAV KF 16). As their irregular shapes prevent precise identification, their interpretation as tile wasters must remain hypothetical. There is no clear evidence for test pieces at Sant'Angelo Vecchio: none of the items appears to have been used as a tool for checking the temperature of the firing chamber.[19]

One of the most important issues raised by the kiln furniture found in the atelier of Sant'Angelo Vecchio is the useful information that the objects might provide about the production processes. In particular, it would be interesting to determine whether the various types of kiln firing supports were used individually, for the firing of specific products, or if their role was generic and not technologically specialized.

Two-thirds of the kiln supports were wedge-shaped spacers (counting the elliptical and truncated pyramidal types, which are close to them in the generic similarities of their shapes). The meager evidence available in the archaeological literature suggests that the wedge-shaped supports were used in kilns that produced a range of products, such as vessels (slipped and plain wares, amphorae), loom weights, tiles, and bricks.[20] It has been suggested that, in some cases, supports with a projecting lip were specifically associated with the production of tiles (*Fig. 33.2*)[21] and pithoi (*Fig. 33.3*).[22] According to the two idealized reconstructions, the supports might have been used to separate and stack large objects. Furthermore, Merker has suggested that the wedges discovered in the tile kilns at Corinth were not used for ordinary pottery, but rather they would have supported larger items, such as the roof tiles and decorated architectural revetments that were the actual products of the atelier.[23] A stamped wedge-shaped spacer found at Phlius, close to Corinth, provides another connection between this type of kiln spacer and tile production.[24] The name on the support (Πρειμος) is the same as the signature found on lamps and tiles at Corinth and Athens—the objects probably belonged to the same craftsman, who owned a workshop in Corinth in the 3rd century BC, where tiles were also produced.

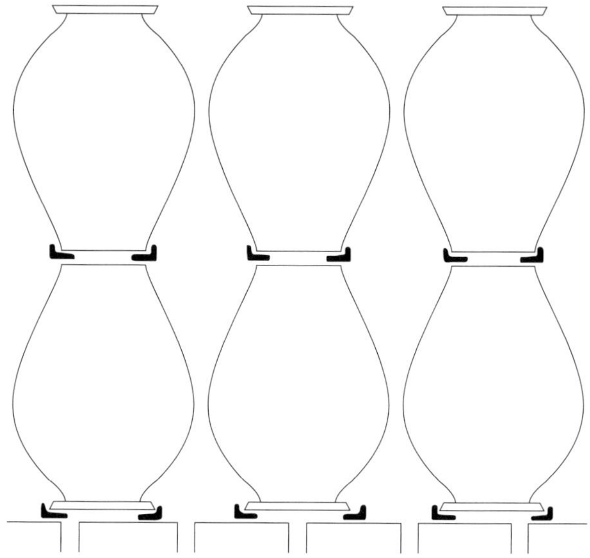

Figure 33.3 Hypothetical reconstruction of a kiln stacking system for pithoi. (After Kaloyeropoulou 1970, fig. 10)

Rectangular bars, the second most common type of support attested at Sant'Angelo Vecchio (if considered as spacers), also appear to have been used in kilns that produced small and large vessels in Archaic fine and plain wares,[25] Roman slipped and plain wares,[26] and even tiles.[27] An experimental firing of curvilinear and rectangular tiles conducted in a reconstructed

[19] For test pieces in production contexts, see *Metaponto I, passim* (*kerameikos* of Metaponto) and Papadopoulos 2003, 225–69 (Athens).

[20] Plain wares, amphorae, loom weights, and perhaps tiles (as indicated by the discovery of numerous stamped examples) were produced in a Hellenistic atelier at Thasos where at least 100 wedge-shaped supports were discovered (Garlan 1986, 213–15). Large and small vessels were produced in a workshop discovered in Atalanta (Lampropoulou 1983, 78–79). Black-gloss and red-figure pottery, large plain vessels, and terracottas were produced at Taranto in the atelier of Via Leonida and the atelier next to the Caserma Mezzacapo (Dell'Aglio 1996a, 56–57, 71, who suggests that the wedge-shaped spacers separated the stacks of vessels within the kiln). Sub-geometric indigenous pottery and B2 cups were certainly produced at Montescaglioso, although production of black-gloss and plain wares cannot be excluded, as well as tiles (Lo Porto 1988–1989, 389). Slipped and plain bowls, basins, and large closed shapes were fired in the Archaic kiln of Entella (Guglielmino 2000, 703). Kaloyeropoulou (1970, 434) suggests the use of this type of support in the production of pithoi, plates, Megarian bowls, cups, tiles, and other domestic vessels.

[21] Karagiorga 1971, 29, fig. 1.

[22] Kaloyeropoulou 1970, 433, fig. 10.

[23] Merker 2006, 22.

[24] Biers 1971.

[25] At Phari, on Thasos: Peristere et al. 1985, 34.

[26] From Gallo-Roman contexts: Luginbühl 2001, 338 (Avenches and Lousonna); Bénévent 1997, 132 (Savignac).

[27] The atelier of Bourzanel, 1st c. AD: Gruat et al. 1999.

kiln, carried out according to the methods adopted by modern potters, confirmed that the insertion of broken rectangular pieces of tiles or brick between the objects allows for better circulation of the hot gasses.[28]

Significant also is the effective absence of ring-shaped stacking supports in the atelier of Sant'Angelo Vecchio. These were widespread in ancient ceramic ateliers, as well as in the urban *kerameikos* of Metaponto. It has been assumed that this type of support was associated mainly with the firing of fine wares and figured pottery.[29] The frequent presence on the body of one or more pierced holes perhaps facilitated the correct circulation of reducing gasses all along the surfaces of the vases, a feature essential for both glossed and figured pottery.[30] The absence of these objects in the workshop furniture of Sant'Angelo Vecchio does not appear to be coincidental, instead possibly reflecting the absence of these more specialized productions in the atelier. This hypothesis is apparently confirmed by the lack of evidence for wasters pertinent to these ware classes. The absence of this type of spacer, however, could be equally well explained by chronology—perhaps ceramic firing techniques underwent a radical but as yet undocumented change in the Hellenistic period. This problem will have to be revisited after an analysis of the provenience of this material, with the aim of determining its chronology.

Nevertheless, a significant difference between the atelier of Sant'Angelo Vecchio and the urban *kerameikos* of Metaponto is clear on the basis of the kiln firing supports: the ratio of wedge-shaped and ring-shaped spacers is completely inverted at the two production sites. Cracolici has suggested that the wedge-shaped supports appeared at Metaponto in the Archaic period and that they gradually decreased in number and size during the Classical and Hellenistic phases of the urban *kerameikos*, when the stacking ring spacers became predominant.[31] In the Hellenistic atelier of Sant'Angelo Vecchio, the wedge-shaped spacers were much more numerous than the other shape, and therefore it can be assumed that an "evolutionary" change from plain spacers (i.e., the wedge-shaped type) to more specialized supports (i.e., the stacking ring-shaped type) did not occur there.

Analysis of the kiln furniture thus does not lead to secure identification of the atelier's products, but it does provide several hypotheses. Tiles and large plain vessels might have been produced there, as indicated by the predominance of wedge-shaped supports and parallelepiped bars, as well as the possible wasters. Production of black-gloss and slipped pottery is arguably less probable. The wedge-shaped and parallelepiped supports, which apparently did not have a high degree of technological specialization like the stacking ring-shaped spacers, were likely used in the firing of various products. Their shape probably made them fairly adaptable to practical needs "in any way they were found to be useful at a given moment."[32] The hypothetical existence of different sizes of rectangular bars could be explained in this way.

The workshop furniture from Sant'Angelo Vecchio is closer to that of the Tarentine atelier located in Via Leonida rather than that of the urban *kerameikos* of Metaponto. In the Via Leonida atelier, there is an abundance of the same types of supports as at Sant'Angelo Vecchio (wedge-shaped, elliptical, and truncated pyramidal), but the more specialized shapes that were probably associated with the firing of black-gloss and red-figured pottery are also well attested.

[28] Rostoker and Gebhard 1980, 218, fig. 11. See also Blitzer 1990, 696, pl. 109.

[29] Cracolici 2003, 109–12.

[30] Cuomo di Caprio 2007, 528–29.

[31] Cracolici 2003, 109.

[32] Merker 2006, 22.

Catalog of Kiln Furniture

SAV KF 01 Kiln Firing Support Type 1.a

1:2

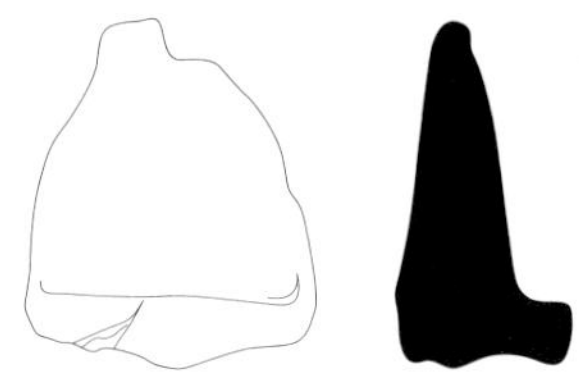

Provenance: Slope 4, surface cleaning (Sounding C, E extension). Lot: SA 79-422T.

Dimensions and technical features: h 4.6, w 4.0, th 0.7–1.4; projecting lip 0.9. Very fine and soft, red fabric, Munsell 5YR 7/1; sporadic small calcareous inclusions; sharp, smooth fractures. Both surfaces have been smoothed.

Condition: Partly chipped.

Description: Wedge-shaped body with sharp, thick, and long projecting ledge at the bottom (type 1, variant a).

Comparisons and comment: The variant is well attested in ceramic ateliers of Greece and South Italy; Papadopoulos 1992, 209–11, fig. 5 (from Athens and Corinth, 4th c. BC through the Roman and Byzantine periods); Merker 2006, 32, no. 24, fig. 22 (from Corinth, tile works area, 4th c. BC or earlier); Kaloyeropoulou 1970, 433–35, fig. 6 (from Epitalion, 1st c. BC–3rd c. AD), figs. 7–8 (from Athens), fig. 9 (from Corinth, 4th c. BC–2nd. C. AD or later); Biers 1971, 414, fig. 1(from Phlius, near Corinth, 3rd c. AD); Garlan 1986, 213, fig. 14 (from Thasos, from an atelier dated end of the 4th–3rd c. BC); Karagiorga 1971, 29, fig. 7, pl. 1 (from Elis, Hellenistic and Roman periods); Lo Porto 1988–1989, 389, fig. 101 (from Montescaglioso, Matera, from a workshop area dated 6th–4th c. BC); Cracolici 2003, 97–98, type VI-A, figs. 12 and 53, cat. nos. 338–339 (from Pisticci, Matera, contrada Torretta, from a kiln dated second half of the 6th–second half of the 5th c. BC); Dell'Aglio 1996a, 68–70, nos. 31 and 51 (from Taranto, 5th–4th c. BC).

SAV KF 02 Kiln Firing Support Type 1.b

1:2

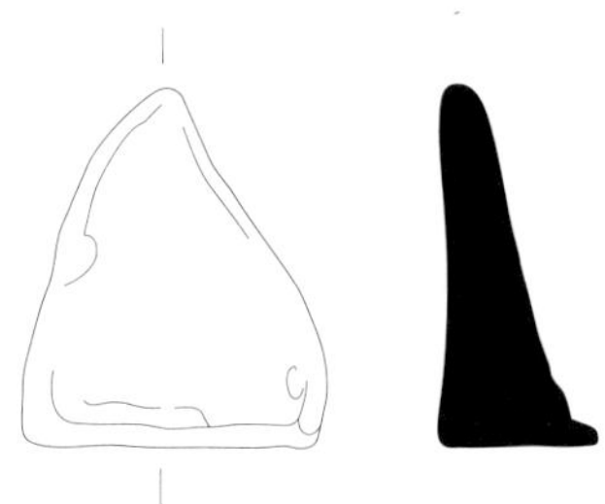

Provenance: Slope 3, W extension. Level 2, batt. 2. Lot: SA 79-142PL (see *Fig. 2.35*).

Dimensions and technical features: h 5.0, w 4.1, th 0.7–1.7; projecting lip 0.3. Very fine and soft fabric, Munsell 2.5Y 8/3; sporadic small calcareous inclusions; sharp, smooth fractures. Two of the three surfaces have been smoothed; the third one bears shallow parallel grooves.

Condition: Complete.

Description: Wedge-shaped body with sharp, thin, and short projecting lip at the bottom (type 1, variant b).

Comparisons and comment: The variant is attested in an atelier discovered on Thasos, which produced amphorae and plain wares from the end of the 4th to the 3rd c. BC: Garlan 1986, 213–15, fig. 14 (three items below).

SAV KF 03 Kiln Firing Support Type 1.c

1:2

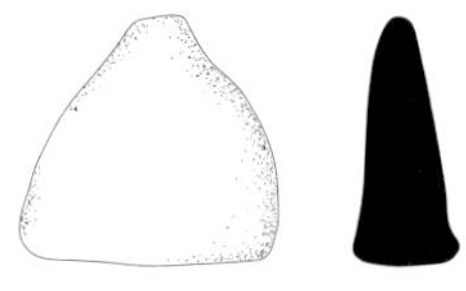

Provenance: Sounding C. Level 2, batt. 2. Lot: SA 79-328T. Assemblage 13.

Dimensions and technical features: h 3.2, w 3.7, th 0.6–1.2; projecting lip 0.2. Very fine and soft fabric, Munsell 10YR 8/4; sporadic small calcareous grains; sharp, smooth fractures. The surface of the lip and the inner surfaces have been smoothed.

Condition: Complete.

Description: Wedge-shaped body with sharp, thin, and short projecting lips on both sides of the bottom (type 1, variant c).

Comparisons and comment: The variant with double border appears to be less common than the other wedge-shaped supports: see Lampropoulou 1983, 78, pl. 3 (Atalanta, Locris, from a ceramic atelier functioning between the end of the 4th and the 1st c. BC); Papadopoulos 1992, 213, fig. 6, no. MC 947, pl. 51:b, c (Athens, from a deposit of the 4th c. BC or slightly later); Merker 2006, 31, no. 19 (Corinth, tile works area, 4th c. BC or earlier).

SAV KF 04 Kiln Firing Support Type 1.d

1:2

Provenance: Slope 3, W extension, Stipe N. Level 2, batt. 1. Lot: SA 79-389TB. Inv. no. 319992. Assemblage 13.

Dimensions and technical features: h 3.7, w 4.2, th 0.6–1.7. Fine and soft fabric, Munsell 2.5Y 7/2; sporadic small inclusions; sharp, smooth fractures. The surfaces have been smoothed.

Condition: Complete.

Description: Wedge-shaped body, without projecting lip at bottom (type 1, variant d).

Comparisons and comment: The occurrences of this variant appear to be as frequent as those of the other spacers with projecting lip. The variant seems to be attested also at the *kerameikos* of Metaponto (*Metaponto I*, 418, n. 263, fig. 58, from a waste dump of the second quarter of the 4th c. BC) and in the ceramic workshop of Locri (*Locri III*, pl. V.1). See Merker 2006, 31, no. 18 (from Corinth, tile works area, 4th c. BC or earlier); Garlan 1986, 213–15, fig. 14 (from Thasos, end of the 4th–3rd c. BC); Guglielmino 2000, 703, pl. CXXIII, 4 (Entella, Sicily, from a Late Archaic kiln).

SAV KF 05 Kiln Firing Support Type 2.a

1:2

Provenance: Sounding C, E of division. Level 2, batt. 3. Lot: SA 79-316T C. Assemblage 13.

Dimensions and technical features: h 7.7, w 9.4, th 0.5–1.9. Very fine and soft fabric, Munsell 2.5Y 8/2; sporadic small calcareous inclusions; sharp, powdery fractures. One of the surfaces has been smoothed.

Condition: Complete.

Description: Semi-elliptical body with depressed edges (type 2, variant a).

Comparisons and comment: The type can be compared with a kiln support from an atelier located in Taranto (Via Leonida), which produced vessels from the end of the 5th to the middle of the 3rd c. BC (Dell'Aglio 1996a, 68, no. 24).

Kiln furniture cont.

SAV KF 06 Kiln Firing Support Type 2.b
Provenance: Slope 3, W extension. Stipe N. Level 2, batt. 1. Lot: SA 79-390T. Assemblage 13.
Dimensions and technical features: h 10.4, w 11.8, th 0.4–5.1. Very fine and soft fabric, Munsell 7.5YR 7/4; sporadic small calcareous inclusions; sharp, powdery fractures. The surfaces are rough.
Condition: Complete.
Description: Semi-elliptical body without depressed edges; the lower part is horseshoe-shaped (type 2, variant b).
Comparisons and comment: The type may be compared to Cracolici's type VI-B, which occurs in a Late Archaic kiln at Pisticci, near Metaponto (Cracolici 2003, 97–98, no. 337, fig. 53). A similar kiln spacer is attested in a production area at Taranto, in Via Leonida (Dell'Aglio 1996a, 68, no. 25, 5th–4th c. BC).

1:2

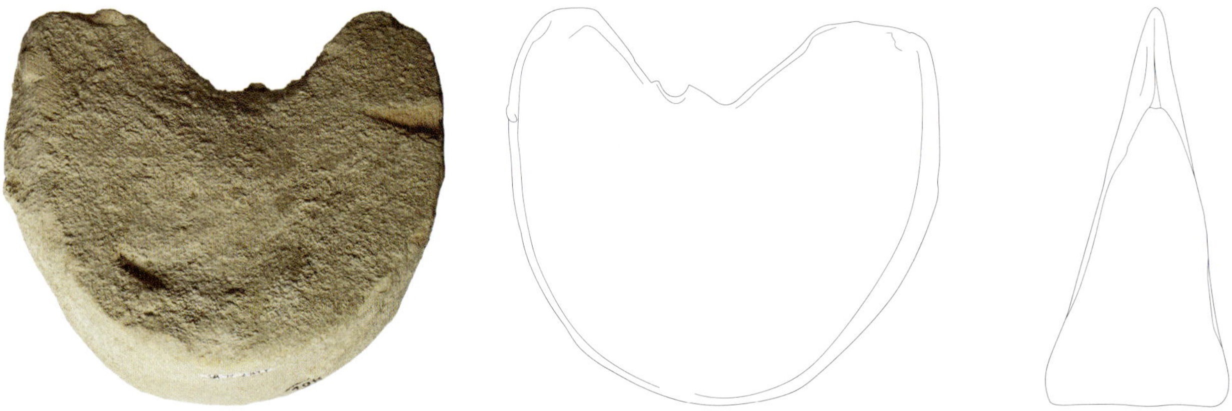

SAV KF 07 Kiln Firing Support Type 3
Provenance: Sounding C, E of division. Level 2, batt. 3. Lot: SA 79-316T B.
Dimensions and technical features: h 7.2, w 9.0, th 1.2–2.9. Very fine and soft fabric, Munsell 7.5YR 7/6; sporadic small calcareous inclusions, with some vughs; sharp, smooth fractures. The surfaces are rough.
Condition: Complete.
Description: Truncated pyramidal body (type 3).
Comparisons and comment: The type is well attested in a ceramic atelier located in Taranto (Via Leonida, 52), which produced black-gloss and plain wares, as well as mortaria and louteria, starting at the end of the 5th c. BC (Dell'Aglio 1996a, 68, nos. 24–26 and 30, 5th–4th c. BC). See also the Romano-British kiln bars from west-central Leicestershire (Pollard 2005, 151, fig. 2, 25–28).

1:2

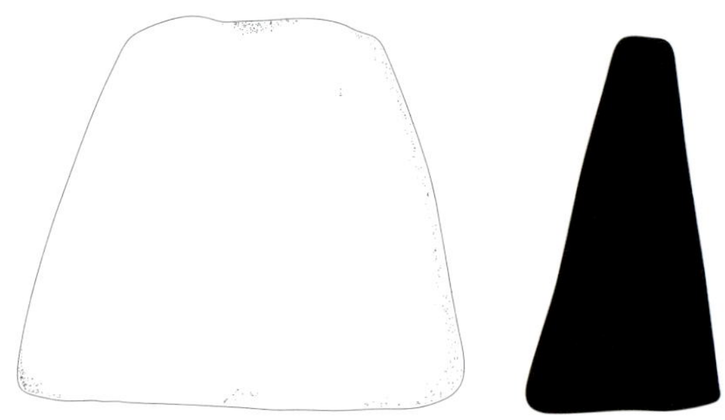

SAV KF 08 Kiln Firing Support Type 4

Provenance: D4. Level 3, batt. 4. Lot: SA 79-825PL. Assemblage 13.

Dimensions and technical features: h 2.6, th 3.1. Very fine and soft fabric, Munsell 7.5 YR 7/6; sporadic small calcareous inclusions, with some vughs; sharp, smooth fractures.

Condition: Only a small part is preserved.

Description: Stacking ring with rectangular section and L-shaped profile (type 4).

Comparisons and comment: The ring-shaped support, which corresponds to type III-C in Cracolici's classification, is one of the most common types of kiln firing support in the Greek world during the Classical and Hellenistic periods; Papadopoulos 1992, 215, fig. 7, in particular nos. MC 973, MC 579, MC 577, MC 858 (from the Athenian Agora); Cracolici 2003, 41, fig. 8 (Metaponto, from waste dumps of the end of the 5th–4th c. BC).

SAV KF 09 Kiln Firing Support? Type 5

Provenance: Slope 1. Surface find from area above Persephone plaque. Lot: SA 79-282T (A) (see *Fig. 2.35*).

Dimensions and technical features: h 7.0, l 8.2, th 4.2. Very fine and hard fabric, Munsell 10YR 7/4; sporadic small calcareous inclusions, with some vughs; sharp, smooth fractures. The surfaces have not been smoothed; possible traces of mortar are present.

Condition: Incomplete—approximately half of the item is probably preserved.

Description: Parallelepiped bar (type 5).

Comparisons and comment: See SAV KF 10.

1:2

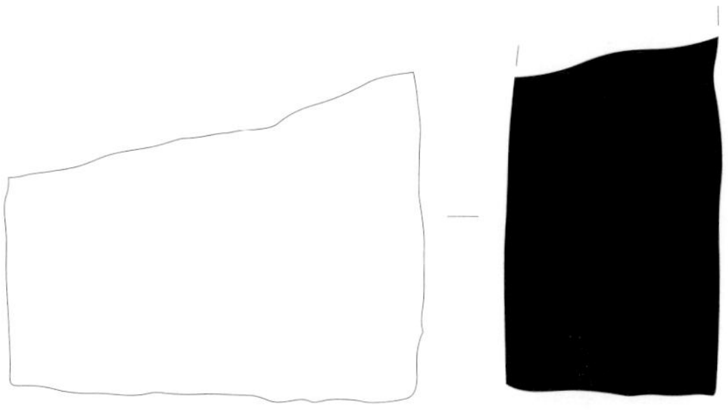

SAV KF 10 Kiln Firing Support? Type 5

Provenance: F1. Level 1, batt. 1. Lot: SA 79-769T (see *Fig. 2.30*).

Dimensions and technical features: h 4.8, l 9.3, th 2.4. Soft and rough fabric, Munsell 5YR 7/8; frequent small calcareous inclusions, with some grog fragments. The surfaces have been smoothed.

Condition: Complete.

Description: Parallelepiped bar (type 5); one of the edges is depressed.

Comparisons and comment: Similar bars are attested at a Thasian atelier, which produced pottery from the 8th to the 6th c. BC (Peristere et al. 1985, 31, fig. 4) and in a Tarentine ceramic workshop dated end of the 5th–middle of the 3rd c. BC (Dell'Aglio 1996a, 68, no. 29). Small rectangular clay bars encountered in several ceramic workshops of Gallo-Roman areas have been identified as kiln firing supports: Luginbühl 2001, 338, fig. VI, 25.1 (western Switzerland); Bénévent 1997, 132, fig. 24 (Savignac, France, 1st c. BC–3rd c. AD); Gruat et al. 1999, 85, fig. 32, 8 (Bourzanel, from a tile workshop of the 1st c. AD).

1:2

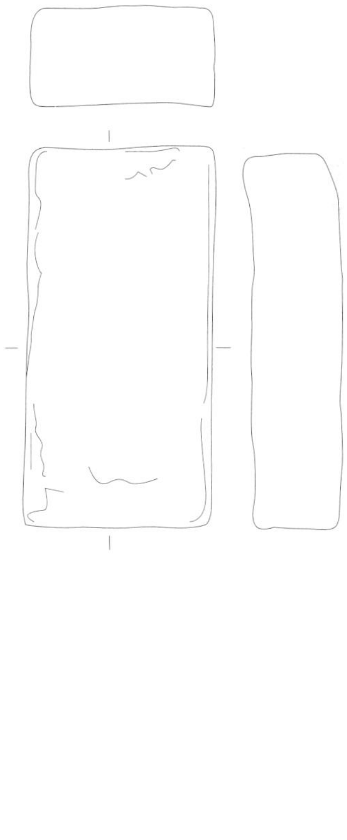

Kiln furniture cont.

SAV KF 11 Kiln Firing Support? Type 5

Provenance: Sounding D. Level 1, batt. 1. Lot: SA 80-41PL.

Dimensions and technical features: h 4.6, l 8.9, th 2.2. Very fine and hard fabric, Munsell 5Y 5/1; sporadic small calcareous inclusions; smooth, sharp fractures. The surfaces of the short sides are smoothed, those of the long ones are rougher. Two doubtful and illegible letters are carved on one surface.

Condition: Complete.

Description: Parallelepiped bar (type 5) with slightly curvilinear profile.

1:2

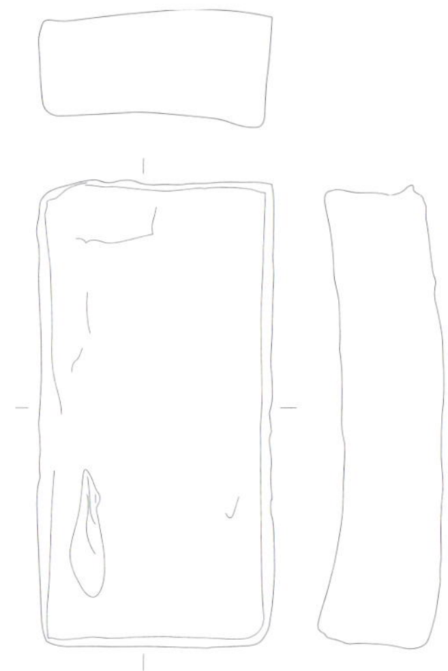

SAV KF 12 Kiln Firing Support? Type 6

Provenance: Slope 4. Level 1, batt. 1. Lot: SA 79-51T (C) (see *Fig. 2.35*).

Dimensions and technical features: h 2.7, l max. 5.0, th 2.3, 1.5. Fine and hard fabric, overfired, Munsell 5Y 6/3 (pale olive); sporadic small calcareous inclusions; smooth, sharp fractures.

Condition: The top is missing.

Description: Sub-rectangular and tapered bar (type 6).

1:2

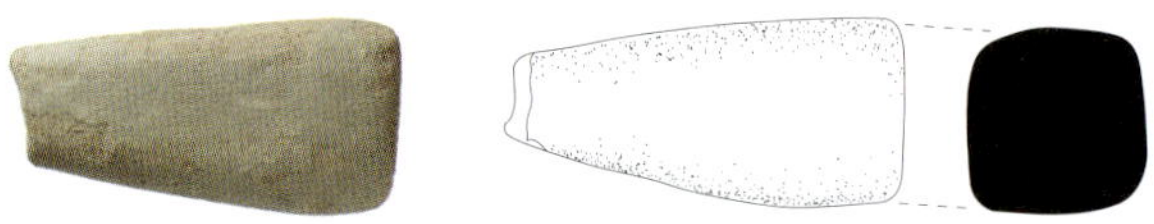

SAV KF 13 Kiln Firing Support? Type 7

Provenance: G5. Level 1, batt. 2. Lot: SA 79-635TA (from SA 79-600PL). Assemblage 3.

Dimensions and technical features: h 4.1, l 7.1, th 5.4. Very fine and hard fabric, Munsell 5YR 7/6; sporadic small micaceous inclusions, with some vughs; smooth, sharp fractures. The surfaces have not been smoothed.

Condition: Incomplete.

Description: Curvilinear (perhaps horseshoe-shaped?) bar with rectangular section and concave profile (type 7).

Comparisons and comment: The object is similar to the "boudins légèrement incurvés, longs d'une dizaine de cm et de la grosseur d'une anse d'amphore" recovered in a Hellenistic ceramic workshop on Thasos and identified as kiln firing supports (Garlan 1986, 213).

SAV KF 14 Pestle

Provenance: Slope 3, W extension. From SA 79-108T. Level 3, batt. 1. Lot: SA 79-138T. Assemblage 13.

Dimensions and technical features: h 10.7, base 7.3, th 3.7. Soft fabric, Munsell 5YR 7/8 (reddish yellow). Sporadic calcareous inclusions.

Condition: Incomplete.

Description: Truncated conical base with curvilinear and tapered upper part.

Comparisons and comment: This type of tool is widely attested in Classical and Hellenistic domestic and production contexts, as at Taranto: Dell'Aglio 1996a, 70–71, nos. 37–40 (from an atelier located in Via Leonida, 5th–4th c. BC; nos. 39–40 are molds for pestles) and 75–76 (from an atelier located near the Caserma Mezzacapo, end of the 4th c. BC).

1:2

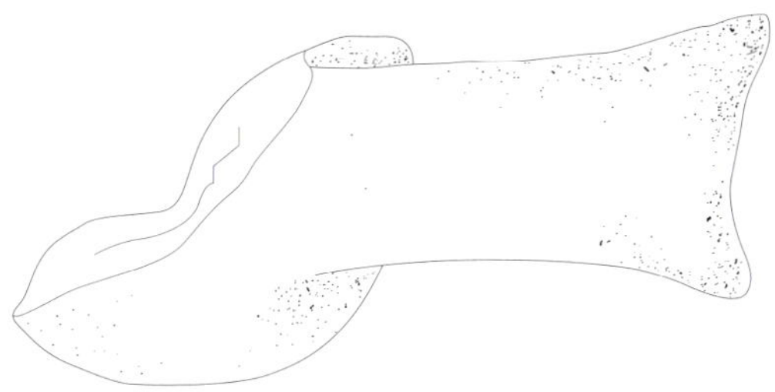

SAV KF 15 Kiln Waster?

Provenance: Sounding D. Level 1, batt. 1. Lot: SA 80-40S.

Dimensions and technical features: h 8.5, l 5.0, th 2.3. Hard and rough fabric, Munsell 5Y 5/1; hackly fractures. The surfaces are irregular, with several vughs and blisters; one of these is slightly curvilinear.

Condition: Incomplete.

Description: Approximately rectangular shape; possibly a fragment of a tile.

SAV KF 16 Kiln Waster?

Provenance: Slope 4, W extension. Level 1, batt. 1. Lot: SA 79-120SS (see *Fig. 2.35*).

Dimensions and technical features: h 17.0, l 10.5, th 4.0. Very hard and rough fabric (because of over-firing), Munsell 5Y 4/2; frequent inclusions ranging in size from small to large.

Condition: Incomplete, composed of three fragments.

Description: Curvilinear shape; the object might be a tile or a kiln pillar. The upper surface is smoothed, the lower one irregular.

34

Evidence for Pottery Production: Fabric Analysis

Keith Swift

Introduction/Discussion

The corpus of plain and banded pottery from Sant'Angelo Vecchio frequently included misfired pieces. Although fired very highly, with a characteristically brittle consistency and entirely reduced grayish color, these over-fired pieces do not in themselves constitute direct evidence for pottery production in the vicinity since they may still have been used—they are not warped or deformed sufficiently to be rendered completely useless.

The objective of this study was therefore to identify over-fired pieces that were incontrovertibly not usable—"wasters." The presence of such pieces would indicate pottery production at some point in the general vicinity (the precise findspot determined by nature of their disposal and post-depositional processes at the site). Once wasters are identified, the frequency of over-fires can provide important supporting evidence for pottery production.

Wasters were indeed present, but as a production site for terracottas and tiles, Sant'Angelo Vecchio presents a more complicated case than usual. At known production sites, there is always the possibility that sherds of pottery (already fired elsewhere, used, and broken) may have ended up incidentally in the kilns (as detritus, or perhaps incorporated into the structure, etc.; see Ch. 3) and thus "over-fired."

To rule out this possibility, a very real one at Sant'Angelo Vecchio, the wasters were examined for evidence of over-firing during the primary (i.e., the "real") firing, during which the plastic-clay vessels were transformed into fired ceramics. Such pieces would provide direct evidence of pottery production at, or very near, the site.

The key piece is a waster in a coarse fabric from the body of a closed form (no. 1, SA 79-600PL). The interior of the sherd has a large expansion void parallel to the vessel walls, which was created by the build-up of gasses within the fabric during (over-) firing. A flow structure is visible within the large vugh, creating a "stalagmite"-like feature that was formed when the matrix was plastic. The exterior is deformed, undulating, but does not exhibit the same degree of vitrification (which might be expected had this piece been over-fired during a second firing). It therefore seems very likely that the vugh on the interior was created when the clay matrix was still plastic, and thus that the over-firing occurred during primary firing, producing a piece that was deformed, unusable, and unlikely to have travelled far from its point of production.

Other examples of over-fired sherds include the rim of a banded one-handler (no. 2, SA 79-1037P). This example was fairly typical of the over-fires with their reduced dark gray fabric, frequent fine rounded voids, and, in this case, the pale streaky trace of an over-fired slip, originally intended to be dark against the pale background of the unslipped parts of the body, characteristic of banded ware (see Ch. 20).

<table>
<tr><th>Class</th><th>Form/Type</th><th>Provenance</th><th>Lot</th><th>Cat. No.</th><th>Chronology</th><th>R</th><th>B</th><th>H</th><th>S</th><th>NR</th></tr>
<tr><td>GW</td><td>Dish</td><td>Sounding B, L. 3 b. 10</td><td>SA 79-433P</td><td>SAV GW 07</td><td>Second half of the 2nd–1st c. BC</td><td></td><td>1</td><td></td><td></td><td>1</td></tr>
<tr><td rowspan="5">BW</td><td rowspan="2">One-handler</td><td>House on the Hill, Test Trench 3, L. 1, b. 1</td><td>SA 79-1037P</td><td>SAV GW 02</td><td rowspan="3">Late Archaic–Early Classical</td><td>1</td><td></td><td></td><td></td><td>1</td></tr>
<tr><td rowspan="2">House on the Hill, topsoil, L. 2, b. 1</td><td rowspan="2">SA 79-1062PL</td><td></td><td></td><td>1</td><td></td><td></td><td>1</td></tr>
<tr><td>Closed form</td><td></td><td></td><td></td><td></td><td>1</td><td>1</td></tr>
<tr><td>Bowl</td><td>Sounding C, Pottery Deposit</td><td>SA 79-988PL-5</td><td>SAV BW 48</td><td></td><td></td><td>1</td><td></td><td></td><td>1</td></tr>
<tr><td>Lekane (Lne 1)</td><td>G6, W of wall, L. 2, b. 1</td><td>SA 79-733PL</td><td></td><td>5th–first half of the 4th c. BC</td><td>1</td><td></td><td></td><td></td><td>1</td></tr>
</table>

Table 34.1 Quantification of over-fired examples from Sant'Angelo Vecchio.

Class	Form/Type	Provenance	Lot	Cat. No.	Chronology	R	B	H	S	NR
PCW	One-handler	Surface finds 10 m from SW corner of abandoned farm-house, N of Sounding A	SA 79-39PL		Late Archaic–Early Classical			1		1
		Slope 3, W extension, L. 2, b. 2	SA 79-84PL				1			1
	Jug (J1)	Slope 3, surface clearing	SA 79-426PL			1				1
	Jug/amphora	G5, L. 1, b. 2	SA 79-636PL					1		1
	Lekane/basin (Ba2)	Sounding C, L. 2, b. 6	SA 79-201PL	SAV PCW 13	3rd–2nd c. BC	1				1
	Lekane/basin	Surface finds, gully wall N of Sounding C, L. 1, b. 3	SA 80-16PL			1				1
		Sounding C, Pottery Deposit	SA 79-988PL						2	2
		House on the Hill, Test Trench 3, L. 1, b. 1	SA 79-1062PL	SAV PCW 03		1				1
	Closed form	House on the Hill, S half of Test Trench 2 on top of hill, L. 1, b. 1	SA 79-557PL		Late Archaic–Early Classical				1	1
		House on the Hill, Test Trench 3, Topsoil, L. 2, b. 1	SA 79-1062PL						10	10
		Surface find (plowed furrow parallel to road to farm)	SA 79-752PL						1	1
		H4, L. 1, b. 1	SA 79-664PL						2	2
			SA 79-538PL						1	1
		G5, L. 1, b. 2	SA 79-600PL	SAV PCW 01					1	1
		G6, W of wall, L. 2, b. 1	SA 79-733PL						1	1
		G7, cleaning around wall to W, L. 2	SA 79-1010PL						1	1
		F6, L. 1, b. 2	SA 79-600PL						1	1
		Slope 3, W extension, L. 2, b. 1	SA 79-79PL						1	1
		Slope 3, W extension, surface find from grassy slope above Slope 3, L. 1, b. 1	SA 79-61PL						1	1
		Sounding D, L. 1, b. 2	SA 80-45PL						1	1
		Sounding D, L. 2, b. 3	SA 80-51PL				1			1
	Open form	E3, between Soundings A and B, L. 2, b. 1	SA 79-772PL		1st c. BC–1st c. AD		1		1	2
		Sounding D, L. 2, b. 3	SA 80-51PL					1		1
CkW	Pan (PD5)	F3, inside kiln, L. 2, b. 2	SA 79-924P	SAV CkW 07	2nd–1st c. BC	1				1
	Klibanos (Kl1)		SA 79-923P; SA 79-1048PL	SAV CkW 25		1				1
RT	Floor tile	House on the Hill, W side of Test Trench 2, L. 1, b. 2	SA 79-622T	SAV RT 128						1
	Laconian tile (AC1/AT1?)	H3, L. 1, b. 1	SA 79-677PL	SAV RT 126	Hellenistic					1
		G6, E of wall, L. 2, b. 2	SA 79-739PL							1
		Slope 3, W extension, S Stipe, L. 3, b. 1	SA 79-374P	SAV RT 127						1
	Pan tile	E2/E3, S Kiln interior, L. 2, b. 2	SA 79-770T		Roman					1
Totals						**8**	**6**	**3**	**26**	**48**

Table 34.1 cont. Quantification of over-fired examples from Sant'Angelo Vecchio.

The very over-fired rim and upper body of a lekane (no. 3, SA 79-1062PL) shows similarities in hardness and firing color to examples of tile wasters from Sant'Angelo Vecchio. The upper surface of the flange is undulated, the piece very hard and brittle, and it seems very likely that this piece is a genuine waster, rather than from an over-fired but usable vessel.

Although, as mentioned above, over-fires alone do not directly demonstrate local pottery production, when combined with wasters they do provide supporting evidence. The three pieces described above stand out in the small corpus of over-fired pieces (for samples examined, see *Table 34.1*) as scant, but nonetheless present evidence for pottery production at Sant'Angelo Vecchio. Wasters are, in general, among the more abundant testaments to pottery production in the archaeological record, and writing off even a limited presence at any site would be a mistake.

Fabric Samples

A)

B)

1. Over-fired Coarse Ware Body Sherd from a Closed Form
SA 79-600PL.

Although semi-vitrified, the other parts of the fabric do not seem to show flow structures, i.e., to be sufficiently wasted as the stalagmite. This would indicate that the matrix was over-fired when plastic, not the result of re-firing (as would be the case if, e.g., the piece formed part of a kiln lining or structure).

A. Showing section with large expansion void. Microscopy, magnification x10, width of field ca. 20 mm.

B. Showing another part of the section with smaller, elongated expansion voids. The whitish crystalline inclusions are probably quartz, the well-rounded yellow crater-like features are over-fired calcium carbonates. Microscopy, magnification x20, width of field ca. 10 mm.

Fabric samples cont.

A) 

B) 

2. Over-fired Banded One-handler Rim
SA 79-1037P.

A. Microscopy showing section, magnification x10, width of field ca. 20 mm. A hard, gray matrix with occasional very fine rounded voids and elongated vughs.

B. External surface showing pale streak of slip on a darker background of slip. Microscopy, magnification x20, width of field ca. 10 mm.

A) 

B) 

3. Over-fired Lekane Rim
SA 79-1062PL.

A. Microscopy showing the upper surface of the flange, the banded slip producing a pale streaky slip as in no. 2. (NB: Borders of the fractures show modern adhesive used to reconstruct the piece.) Magnification x10, width of field ca. 20 mm.

B. Microscopy showing the typical texture of the broken section, with abundant fine, irregular expansion voids, most less than 2 mm. Magnification x10, width of field ca. 20 mm.

35
Metal Objects

Marta Mazzoli

The excavations at Sant'Angelo Vecchio yielded a small number of metal finds that, despite their scarcity, nevertheless round out the picture of the material life of the area dedicated to production, widely documented by other and more abundant classes of materials. Most of the metal finds, in fact, were recovered inside the Kiln Complex (in Soundings A and A-1) or in its immediate vicinity (in Sounding C north of Kiln 1). Though none of the examples can be conclusively assigned to a manufacturing function, they were certainly used in this manufacturing area.

An axe-head and an iron block (SAV M 05 and 06) came from the topsoil layer of the Kiln Complex. They were found mixed with bricks dating to the Late Republican phase along with earlier materials probably reused in the construction of the kilns. A second axe-head (SAV M 04) and a lead strip (SAV M 07) were found on the courtyard floor level (see *Fig. 2.35*). An iron nail (SAV M 01) came from a layer of collapse belonging to a Late Republican structure (especially the wall in square G6 built directly above Tomb 1). It was found mixed with tiles forming a very compact layer, and was probably part of the structure's roof. Given its size and the parallels with other sites of similar date, it was likely a nail used for attaching the rafters to the beams.

Two lead clamps (SAV M 08 and SAV M 09) were found in Sounding C, located immediately north of the kilns. Lead clamps were generally used for repairing large ceramic containers or architectural elements,[1] and their clearly utilitarian function provides significant information on the optimization of labor in the manufacturing area.[2] The clamps were mixed with other pottery material datable to the first phase of the workshops producing votive terracottas in this area, and they therefore can be dated between the end of the 4th and the 3rd century BC.

It is difficult to determine the chronology of these objects, as they were constantly in use over a prolonged period of time, and parallels are found in contexts of various dates. The proposed dating is based on the find contexts and associations with other classes of materials. Furthermore, two previously published objects from the cemetery area have been included in the present work, with new interpretations based on further study.[3] The iron strip in particular (previously identified as a belt hook) is reinterpreted as a fragment of a strigil handle (see p. 118 in Ch. 4, "T03-03").

[1] Colivicchi 2004, 63.

[2] Lead has a low melting point (327.4°C), making it a very suitable material for making repairs, and at the same time it is possible to melt it down on a simple open fire (Giumlia Mair 2002, 108–9).

[3] Carter 1998, 443–47. See above, Ch. 4.

Catalog of Metal Objects
Iron Objects

SAV M 01 Nail

1:2

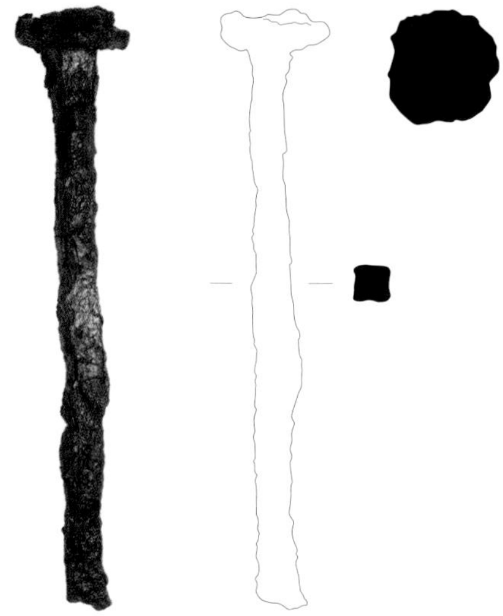

Provenance: G6, E of wall, NE corner near baulk. Level 2, batt. 3. Lot: SA 79-995M. Inv. no. 319830. Assemblage 3.
Dimensions: l 13.4, head diam. 2.5, th 0.88, weight 54 g.
Material: Iron.
Condition: Lower part of shank incomplete; oxidized.
Description: Almost complete nail, circular, slightly convex head, fragmentary long square-sectioned shank, lower part missing where the section narrows, probably due to oxidization. Given the dimensions and comparisons with other contexts, this seems to be a nail used for fixing rafters to roof beams.
Comparisons and comment: Edlund 1982a, 83; Carandini and Ricci 1985, pl. 4.5 (Settefinestre); Busanna 2000, type B2CP, 114 (Asolo); Fiori 2005, type A3 (Monte Bibele, 4th–2nd c. BC); Ciampoltrini and Andreotti 2001, type E3 (Pisa).
Date: Late Republican.

SAV M 02 Rod

1:2

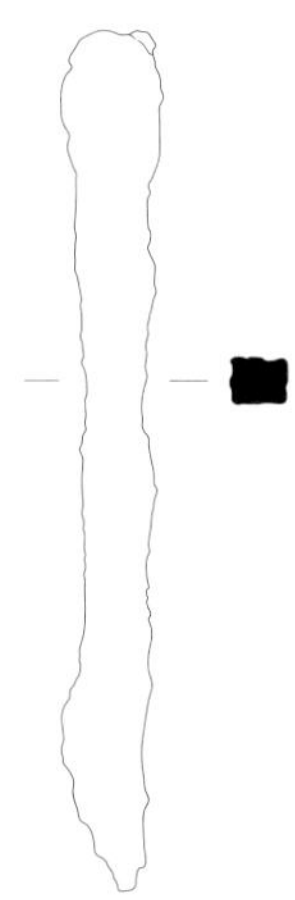

Provenance: Sounding A-1, more or less in E2. Lot: SA 79-706M. Inv. no. 319829. Assemblage 5.
Dimensions: l 11.4, w 0.8, th 0.6, weight 33 g.
Material: Iron.
Condition: Fragmentary; oxidized.
Description: Long rod with slightly rounded head, rectangular section, terminating in a point. Given its fragmentary state, this may have been a carpentry nail. However, considering its provenance (Sounding A-1 is the kiln area), it may be a multifunctional object used for making incisions and perforations in leather, wood, and pottery.
Comparisons and comment: Macnamara 1992, 239, no. 1911, fig. 108 (Gravina); Fiori 2005, 179, nos. 12–13, type A3 (Monte Bibele, 4th–2nd c. BC); Carandini and Ricci 1985, pl. 4.5 (Settefinestre); Busanna 2000, type B2CP, 114 (Asolo); Ciampoltrini and Andreotti 2001, type E3 (Pisa).
Date: Late Republican.

1:2

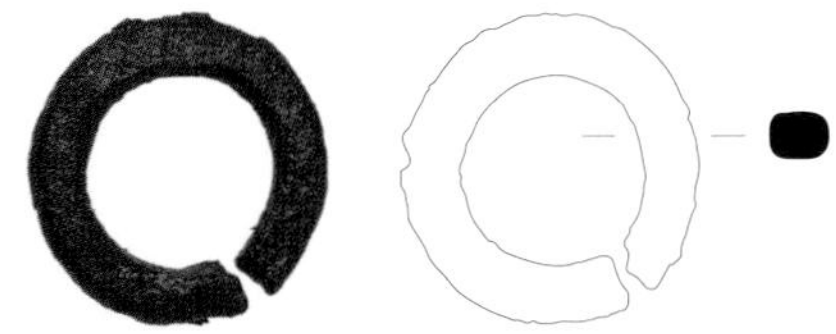

SAV M 03 Ring
Provenance: Slope 3, surface cleaning. Lot: SA 79-428M. Inv. no. 319828.
Dimensions: diam. 4.12, w 0.82, w 0.67, weight 32 g.
Material: Iron.
Condition: Small section missing; slightly oxidized; slight traces of wear on opposite side to fracture.
Description: Ring with rectangular section, with indication of ribbing, slightly rounded edges. Identifiable as a linking element between metal parts (chains, or joining or articulating element between objects).
Comparisons and comment: Edlund 1982a, 52; Macnamara 1992, 241, no. 1937a, 1937b, 1939, fig. 111 (Gravina); Meirano 1997, 239, fig. 108 (Pomarico Vecchio).

1:2

SAV M 04 Axe
Provenance: Sounding A. Level 2, batt. 3. Lot: SA 79-31M (see *Fig. 2.30*).
Dimensions: l 8.1, w 4.2, th 1.5, weight 116 g.
Material: Iron.
Condition: Incomplete, heel missing; heavily oxidized.
Description: Axe-head with trapezoidal profile, blade has a convex cutting edge.
Comparisons and comment: Edlund 1982a, 97; *Greci, Enotri e Lucani*, 147, no. 2.10.64; 151, no. 2.11.54.
Date: Late Republican.

1:2

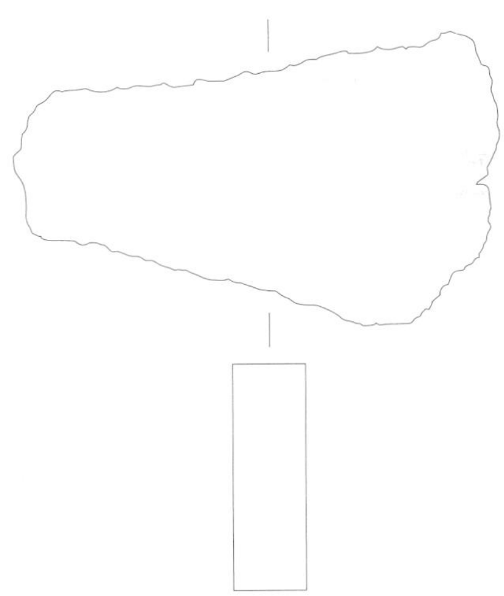

SAV M 05 Axe
Provenance: Sounding A. Level 1, batt. 4. Lot: SA 79-68M. Inv. no. 319826 (see *Fig. 2.30*).
Dimensions: l 10.7, w 6.47, th 1.9, weight 297 g.
Material: Iron.
Condition: Incomplete, heel missing; oxidized.
Description: Axe-head with trapezoidal profile, wide blade has a convex cutting edge.
Comparisons and comment: Edlund 1982a, 93; *Greci, Enotri e Lucani*, 147, no. 2.10.64; 151, no. 2.11.54.
Date: Late Republican.

Iron objects cont.

SAV M 06 Block

1:2

Provenance: Sounding A. Level 1, batt. 4. Lot: SA 79-68Mbis. Inv. no. 319826 (see *Fig. 2.30*).
Dimensions: l 4.5, w 4.4, th 1.2, weight 148 g.
Material: Iron.
Condition: Incomplete; oxidized.
Description: Thick, sub-rectangular plaque with rectangular section, at least one side has missing parts. At the center of the lower face is a blind hole which corresponds on the upper face to hammer marks. Found in association with the axe (inv. no. 319826) and detached during conservation. If associated with the axe, it could be interpreted as a head-fastener from a craft-working tool, of which the central pivot is not preserved.
Comparisons and comment: Edlund 1982a, 93.
Date: Late Republican.

Lead Objects

SAV M 07 Strip

1:2

Provenance: Sounding A. Level 2, batt. 3. Lot: SA 79-33M. Inv. no. 319876 (see *Fig. 2.30*).
Dimensions: l 6.2, w 5.9, th 0.27, weight 156 g.
Material: Lead.
Condition: Missing part of inner edge.
Description: Wide strip folded in half with irregular edges. One face shows hammer marks. Given that it is incomplete it is difficult to define its precise function. However, on the basis of comparisons with material not of precisely the same date, this may be a fishing-net weight or a plaque used for repairing terracotta or wooden objects.
Comparisons and comment: Edlund 1982a, 97; Hicks and Hicks 1992, 309, nos. 177–179 and 182, fig. 10 (Otranto); Macnamara 1992, 247, nos. 2000 and 2002, fig. 120, pl. XXI (Gravina).
Date: Late Republican.

SAV M 08 Mending Clamp

1:2

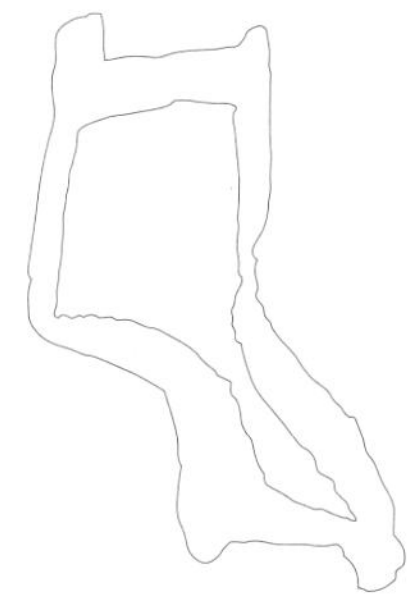

Provenance: Sounding C. Level 2, batt. 5. Lot: SA 79-298M. Inv. no. 319827 (see *Fig. 2.35*).
Dimensions: l 7.6; w 5.5; th 0.5; weight 102 g.
Material: Lead.
Condition: Complete.
Description: Mending clamp, presumably for large pottery containers, comprising two twisted rods with hemispherical sections joined at the extremities by two hooks with circular sections. Grooves visible on the convex surfaces of the two lateral rods.
Comparisons and comment: Edlund 1982a, 71; Visonà 1999, 371, nos. 846–48, fig. 382 (Oppido Mamertina); Macnamara 1992, 246–47, nos. 1987–1990, fig. 119 (Gravina); Colivicchi 2004, 66, no. 230, pl. 11 (Gravisca).
Date: End of the 4th–beginning of the 3rd c. BC.

SAV M 09 Mending Clamp

1:2

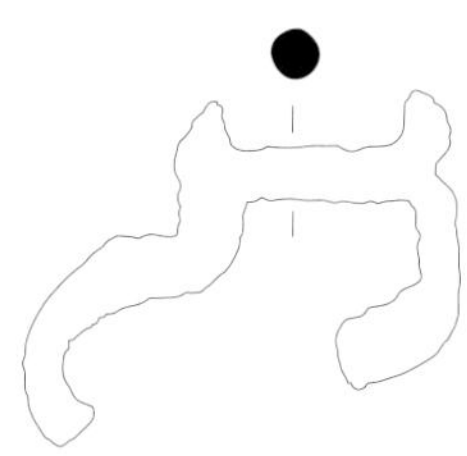

Provenance: Sounding C. Level 3, batt. 3. Lot: SA 79-1023M. Inv. no. 319831.
Dimensions: l 4.2; w 5.9; th 0.6; weight 54 g.
Material: Lead.
Condition: Half of the object is missing.
Description: Mending clamp, presumably for large pottery containers, the same as the previous artifact, but fragmentary. The two twisted rods with hemispherical section, grooves on the convex surface, and one hook at one end with semicircular section are preserved.
Comparisons and comment: Edlund 1982a, 71; Visonà 1999, 371, nos. 846–48, fig. 382 (Oppido Mamertina); Macnamara 1992, 246–47, nos. 1987–1990, fig. 119 (Gravina); Colivicchi 2004, 66, no. 230, pl. 11 (Gravisca).
Date: End of the 4th–beginning of the 3rd c. BC.

36
Greek Lithic Material

Cesare D'Annibale

The use of lithic material (stone) as a tool source occurs infrequently at Greek sites in the Metapontino. During the Greek period, manufacturing lithics required specialized industries, most notably millstones of primarily volcanic origin. This trade in particular was tied to a few specific production and distribution centers located beyond the region. Local lithic material instead was used sporadically and idiosyncratically, likely for expedient needs or personal expression as opposed to a formal production. Seven of the eight lithic artifacts from Sant'Angelo Vecchio attributed to the Greek period were locally procured, whereas one volcanic grinding stone is indicative of a broader Mediterranean commercial network. The lithic objects recovered from the site pertain to both domestic functions and ritual connected to the funerary aspects of the site.

Millstones

Milling is one of the key activities in processing agricultural products on a farmhouse site. Hand-operated grinding stones, the majority of which are lava stone, occur frequently on many of the farmhouse sites identified in the chora of Metaponto.[1] The most common grinding system, consisting of a saddle quern and hand grinder, has a long prehistoric tradition and was used as early as the Neolithic period. The manufacture and commercial distribution of volcanic millstones became standardized by the Archaic period, and the mechanics of this system remained in use virtually unchanged until the Late Classical period.[2] The two elements adhere to a regimented size and form. The quern stone is rectangular with beveled edges, likely to be inserted or wedged in a box. The hand-operated grinder that is pushed across the face of the quern is ellipsoidal in shape. In its most sophisticated form, the grinder exhibits an eared downward projection at the ends, intended to stabilize or guide its movement across the quern, and a keeled dorsal ridge with a triangular cross-section.[3]

The commercially manufactured millstone from Sant'Angelo Vecchio is an andesite hand grinder (SAV Li 01). Although heavily worn from use, it retains its original ellipsoidal shape. In cross-section, the back of the grinder is convex, but severely reduced in thickness, a sign of prolonged and extensive use. The general form is widespread in Magna Grecia, finding many parallels throughout southern Italy.[4] In the Metapontino, an abundant number of such finds indicate that the distribution of these millstones is far more common than the limited number of documented occurrences elsewhere would suggest.[5]

The preferential use of andesite and dacite as the raw material of choice for millstones, especially from the Archaic and into the Classical period, is an indication of a specialized industry operating in the Mediterranean. The virtual exclusion of other stone types suggests a controlled production and commercial distribution emanating from a few centers. These production centers can be tracked to several localities in the Aegean basin, especially around the Saronic Gulf and

---

1 Volcanic millstones were recovered from 107 farmhouse sites identified through the Institute of Classical Archaeology's field survey of the chora of Metaponto. These were also recovered from the excavated farmhouse sites of San Biagio, Fattoria Fabrizio, and Sant'Angelo Grieco (*San Biagio*, 168; D'Annibale in *Fattoria Fabrizio*, 370–71); other excavated farmhouse sites in the chora producing volcanic millstones include Fattoria Pacciano, Pascucci-Torraca, Risimini, Pagliarone 2, and Musillo (Amore 1992–1993, tab. 13).

2 For a discussion on the saddle quern during the Archaic and Classical periods, see Amouretti 1986, 138, and Kardulias and Runnels 1995, 116, 123.

3 For a hand-operated grinder retaining these features, see example from Fattoria Fabrizio (D'Annibale in *Fattoria Fabrizio*, 376).

4 Similar forms can be seen at Serra di Vaglio (G. Greco 1991, 75, fig. 175); Roccagloriosa from 5th-c. BC contexts (Gualtieri 1990, 83, 309, fig. 202, n. 590); an Archaic house at Elea (Bencivenga Trillmich 1983, 426, 443, fig. 11); Sibari (Guzzo et al. 1974, 116, fig. 124); Francavilla Marittima (Kleibrink 1970–1971, 78–80, tavv. 31–32); and Morgantina in Archaic contexts (White 1963, 201). A number of farmhouse sites identified by ICA's field survey of the chora of Croton also had this type in their assemblage.

5 The form is found on 15 farmhouse sites from the Metaponto field survey, including sites 58, 283, 549, 647, 850, 912, 927, 950, 961, 974, 977, 1051, 1061, 1077, and 1081. From site excavations, the form is found at Fattoria Fabrizio and Sant'Angelo Grieco (D'Annibale in *Fattoria Fabrizio*); Fattoria Musillo and Fattoria Pacciano in the Saldone locality (Amore 1992–1993, photos 16 and 23). See also from Incoronata (Castoldi 1995, 110, fig. 221; Castoldi 2000, 47, fig. 50–51; Lambrugo 2003, 91–92, fig. 85, 156–57).

outlying islands.[6] Preliminary mineralogical analysis of volcanic grinding stones from Metapontine farmhouse sites has revealed that some of these do originate from the Aegean.[7] That Archaic- to Classical-period saddle quern millstones tend to have an Aegean origin has been confirmed by other petrographical studies, such as that from the Monte Castellazzo area in Sicily.[8]

Despite the presence of a dominant commercial network for distributing volcanic grinding stones in Magna Grecia and supplying standardized equipment to farmhouse sites, other material does occasionally enter the monopoly of this formalized industry. At Sant'Angelo Vecchio, two thin slabs of fine-grained sandstone are good candidates for grinding stones as they display classic signs of wear from grinding on one face (SAV Li 02 and SAV Li 03). Sandstone slabs and fragments used in wall construction are abundant on site, but these two particular fragments were evidently the two elements of a saddle quern grinding system. SAV Li 02, with a convex dorsal surface, most likely functioned as a hand grinder or upper stone, and SAV Li 03, with a rectangular cross-section, could have been a quern or lower grinding stone fragment. The locally available stones could have been used without needing much alteration or modification.[9] Although millstones are generally indicative of cereal processing, their use in association with the ceramic industry for grinding clay and temper cannot be excluded, especially given the presence of kilns at Sant'Angelo Vecchio.[10] It is certainly plausible that the two grinding stones of locally available sandstone from Sant'Angelo Vecchio were employed in this way. These objects document an adaptive strategy either to overcome deficiencies in the formal tool assemblage or as a quick response to an immediate necessity.

Weight/Whetstone

The use of local lithic material on Greek sites in the Metapontino was generally limited to atypical and often unique objects. These materials were for the most part procured from within the immediate environs of the site and were rarely altered or modified for the production of tools. One such object from Sant'Angelo Vecchio is an elongated nodule of limonite (SAV Li 04). The stone is perforated for suspension and likely functioned as a weight. The nodule shows some slight modification, as the end opposite the perforation has been shaved. This feature may hint that the artifact was also used as a honing or sharpening stone. Its precise function is difficult to determine, as this example was produced arbitrarily out of necessity rather than as part of standard manufacture. Several other functions may be proposed, such as a plummet in construction-related activity, a weight for a fishing or birding net.[11] Interestingly, the material is identical to another nodule recovered from the kiln deposit at Pantanello.[12]

Spherical Pebbles

Two spherical sandstone pebbles were recovered in diverse contexts that suggest both a functional use and a role in funerary ritual for this artifact type. The pebble included as a burial offering comes from Tomb 2, the burial of a young male and dated to the middle of the 5th century BC (SAV Li 06).[13] Although inclusion of this object as a grave good is somewhat atypical, it does have parallels from the chora of Metaponto and beyond.[14] The presence of stone objects of this type is an extremely rare occurrence in Metapontine burials and in Magna Grecia in general. The infrequent incidence could be a result of misidentification of such pebbles as just natural inclusions found accidentally within the grave.[15] Nevertheless, the occurrence is purposeful, as these recoveries demonstrate.

A clear functional role for pebbles from Hellenistic and Roman burial contexts is evident from the recovery of pebbles belonging to *tintinnabuli* or ceramic

[6] For a broader analysis of volcanic grinding stone provenance in Greece, see Williams-Thorpe and Thorpe 1993, and Kardulias and Runnels 1995.

[7] Recent analysis of volcanic grinding stones from the Metaponto field survey was carried out by A. Maraudino and N. Palumbo.

[8] Ferla et al. 1984.

[9] For the use of local sandstone as grinding stones at Incoronata, see Castoldi 1986, 115, and D'Annibale forthcoming.

[10] At Sant'Angelo Grieco, a similar sandstone slab with a ground surface was found in association with the kiln (SG 81-20).

[11] At Cozzo Presepe, a perforated stone recovered in a mid-6th–c. BC context suggests that these objects are occasionally recovered, though lacking a specific function other than as a generic suspended weight (Macnamara 1983b, 230, fig. 36 no. 29).

[12] PZ93 context 159.170. The nodule is incomplete, missing roughly 2/3 of its original size. Nevertheless, its proportions and weight approximate that of the Sant'Angelo Vecchio example.

[13] Carter and Hall in *Necropoleis*, 444. See p. 113 in Ch. 4, "Tomb 2."

[14] A burial excavated at field survey site 635 in the Pizzica locality of Metaponto also included a spherical sandstone pebble—a grave offering. A similar offering of pebbles in a burial (burial 29) occurs at Muro Tenente, where five pebbles were offered to a female between the ages of 20 and 23, dated from the end of the 4th to the mid-3rd c. BC (three pebbles in right hand and two near the left; Tetteroo and Waagen 2010, 130–40). At Agrigento, 25 pebbles were recovered from a 6th-c. BC tomb, and two from a tomb dating to the first half of the 5th c. BC (Franchi Dell'Orto and Franchi 1988, 32, tomb 1116, tav. XII; 54, tomb 1181, tav. XLIV; 371, tomb 11).

[15] The pebble from Tomb 2 was not included among the grave offerings in the publication of the burials at Sant'Angelo Vecchio (Carter and Hall in *Necropoleis*, 444).

toys used as rattles.[16] Other pebble recoveries in burials, however, are especially problematic as any number of interpretations can be proposed for their presence as grave goods. A wide range of uses, such as tokens, marbles, or simply personal keepsakes, is easily conceivable. The pebbles may, though, represent symbolic markers from which any number of perspectives on their meaning in social behavior and afterlife beliefs can be inferred. Ethnicity may also be postulated from the preponderance of these pebbles in indigenous contexts.[17] Could they hint at an indigenous presence at Sant'Angelo Vecchio? In addition, the presence of a *Cypraea* shell from Tomb 4 (SAV Sh 01), a somewhat uncommon shell type in Greek burials but very much present in indigenous burials, is perhaps another subtle connection to Italic customs.

[16] For examples of such use at Brindisi, see Cocchiaro 1988, 201, tomb 178, nos. 359–60; 92, tomb 153, no. 60. From Cavallino see Pancrazi 1979, 214, fig. 86, no. 12.

[17] See spherical pebble finds from Incoronata (De Faveri 1999, 154, 512, tav. XXXVI, nos. 151–52). Two other stone pebbles were recovered from the 1977–78 University of Texas excavations at Incoronata (D'Annibale forthcoming). See also Cozzo Presepe (Macnamara 1983b, 228, fig. 28).

The other pebble found in a domestic context offers several possibilities for its use as a tool, or an object associated with personal and recreational functions (SAV Li 05). The more obvious recreational interpretations include a marble or gaming object, and a personal collectible is also plausible. Still other uses can be proposed, such as weights[18] or sling stones for hunting or gaming purposes.[19] In prehistoric contexts, spheroid stones are usually associated with polishers or rubbers/grinders—this function could have continued on a sporadic basis during the Greek period.[20] Two sandstone items (SAV Li 07 and SAV Li 08) are also linked to this type of function, since their natural tubular form is predisposed for use as a pestle or a rolling grinder.

[18] See examples from Sibari (Guzzo/Rota/Daddi 1973, 451, fig. 412) and from Oppido Mamertina (Brizzi 1999c, 342, fig. 333).

[19] See *Cozzo Presepe* for interpretation of pebbles as sling stones; at Sant'Angelo Vecchio, they were not necessarily related to combat activities (Macnamara 1983b, 230, fig. 28).

[20] Two local (chert?) pestles from Fattoria Pascucci-Torraca, a predominantly Hellenistic farmhouse (Saldone locality; Amore 1992–1993, 115, photo 25); a local chert pestle from Cozzo Presepe (Macnamara 1983b, 230, fig. 37); one oblong and one ovoid calcarenitic grinder (likely local) from villa del Moltone, Tolve, dated to the end of the 4th to the beginning of the 3rd c. BC (P. Bottini 1992, 80, 85, nos. 69928, 70034).

Catalog of Greek Lithic Material

SAV Li 01 Grinding Stone

Provenance: G5, Level 2, batt. 1. SA 79-659S. Assemblage 3.
Material: Volcanic; andesite.
Dimensions and technical features: l 12.2, w 10.85, th 3.5, weight 634 g.
Condition: Medial fragment; just over one third of original.
Description: Hand-grinder or upper stone from a saddle quern grinding system. Ellipsoidal form; convex cross section.
Comparison and comment: For general form, see FF Li 02 (D'Annibale in *Fattoria Fabrizio*); Metaponto field survey sites 44, 58, 283, 549, 647, 850, 912, 927, 950, 961, 974, 977, 1051, 1061, 1077, and 1081; see also example from Fattoria Musillo (Amore 1992–1993, photo 23). See notes 4 and 5 for other comparisons.
Date: Archaic to Classical.

1:2

SAV Li 02 Grinding Stone

Provenance: C5, US 19; SAV 2010.
Material: Sedimentary; sandstone.
Dimensions and technical features: l 10.5, w 9.4, th 2.6, weight 363 g.
Condition: End fragment; just over one third to nearly half of original; edge damage on one lateral margin; exposed to heat.
Description: Hand-grinder or upper stone from a saddle quern grinding system. Ellipsoidal form; convex cross section.
Comparison and comment: For the form, general comparisons can be made with volcanic grinding stone SAV Li 01. Sandstone as the material for millstones in the Greek period is somewhat rare; however, other instances do occur such as at Incoronata and Sant'Angelo Grieco (Castoldi 1986, 115; D'Annibale forthcoming).
Date: Greek.

1:2

SAV Li 03 Grinding Stone

Provenance: C5, US 35; SAV 2011.
Material: Sedimentary; sandstone.
Dimensions and technical features: l 10.5, w 7.5, th 2.7, weight 235 g.
Condition: Incomplete; side fragment.
Description: Saddle quern stone; rectangular cross section.
Comparison and comment: General comparisons for the quern or lower stone can be made with volcanic millstones recovered from the Metaponto field survey at sites 49, 54, 579C, 876, 937, 950, 961, 1011, 1072, 1083, 1092, and 1101. However, the use of sandstone as millstones is comparatively less frequent. For general comparisons, see comment for SAV Li 02, above.
Date: Greek.

1:2

SAV Li 04 Weight/Whetstone

Provenance: Sounding B. Level 3, batt. 4. SA-79-369T. Assemblage 11.
Material: Limonite.
Dimensions and technical features: l 12.5, w 6.0, th 2.5–3.3, weight 158 g.
Condition: Complete.
Description: Flat elongated nodule; perforated with biconical drilling that is slightly off-center, perforation aperture 1.3 exterior to 0.7 interior; lateral edges and distal end have been shaved down to shape.
Comparison and comment: Perforated stones as suspended weights is the most common functional use assigned to this object type. Not an unusual occurrence but nevertheless few documented references for these in Greek contexts. One such find from nearby Cozzo Presepe (Macnamara 1983b, 230, fig. 36 no. 29).
Date: Greek.

1:2

SAV Li 05 Pebble

1:2

Provenance: Quadrant D4, US 32; SAV 2011.
Material: Sandstone.
Dimensions and technical features: l 2.0, w 1.9, th 1.9, weight 8.5 g.
Condition: Complete with some encrustation.
Description: Spherical pebble.
Comparison and comment: See pebble SAV Li 06, below, from Tomb 2. Spherical pebbles from domestic contexts are found at Pantanello and at Incoronata (De Faveri 1999, 154, 512, tav. XXXVI, nos. 151–52); see also sling stones recovered at Cozzo Presepe (Macnamara 1983b, 228, fig. 28).
Date: Greek.

SAV Li 06 Pebble

1:2

Provenance: F6, Tomb 2; SA 79-1029.
Material: Sandstone.
Dimensions and technical features: l 2.8, w 2.2, th 2.2, weight 17 g.
Condition: Complete.
Description: Spherical pebble, smoothed all around.
Comparison and comment: Although the pebble was slightly larger, parallels for this occurrence are to be seen at field survey site 635 (a burial). Pebbles in burial contexts can be seen from a late 4th- to mid-3rd-c. BC burial at Muro Tenente where five pebbles were found (Tetteroo and Waagen 2010, 130–40). See also discussion on spherical pebbles above for further references.
Date: Mid-5th c. BC (the date for this object is linked to the pottery found in the tomb).

SAV Li 07 Grinder/Pestle?

1:2

Provenance: House on the Hill; SA 79-1062PLa. Assemblage 1.
Material: Sandstone.
Dimensions and technical features: l 4.2, th 0.8–1.0, weight 6.3 g.
Condition: Complete, worn.
Description: Tubular form with convex wider end.
Comparison and comment: Likely utilized as a pestle.
Date: Greek.

SAV Li 08 Grinder?

1:2

Provenance: House on the Hill; SA 79-1062PLb. Assemblage 1.
Material: Sandstone.
Dimensions and technical features: l 3.0, th 1.3, weight 9.1 g.
Condition: Fragment; fractured at both ends.
Description: Tubular form, evenly rounded and smooth surface.
Comparison and comment: Possibly utilized as a rolling grinder.
Date: Greek.

37
Post-Medieval Pottery

Erminia Lapadula

The excavations at Sant'Angelo Vecchio yielded no Medieval fragments. A single example, probably a basin attested by one base and two body fragments, dates to the Post-Medieval period. The technical features of the item, especially the type of glazed coating, as well as the form, indicate modern production. The find context—the surface of the kilns area—confirms the proposed chronology and links the item to the masseria and the use of the area by later inhabitants (see Ch. 6).

Catalog of Post-Medieval Pottery
Green and Yellow Glazed Pottery

SAV PMP 01 Basin

Provenance: Square F2 (kilns area). Level 1, batt. 1. Lot: SA 79-443Pa, b, c (see *Fig. 2.30*).

Dimensions and technical features: h 7.9, base diam. 16.8, th 1.0. Clay Munsell 5YR 6/6 (reddish yellow), slip pale yellow (Munsell 2.5Y 8/2), rare small dark and large red inclusions distributed randomly.

Condition: One fragment of the base and two non-joining fragments of the body. The exterior surface of the base is missing; the interior and exterior surfaces of the walls are marked by large chips.

Description: Flat base with straight and slightly flaring wall. Slip applied on both interior and exterior surfaces; glaze applied over the slip: yellow on the interior, green on the exterior.

Date: 19th c. AD.

1:3

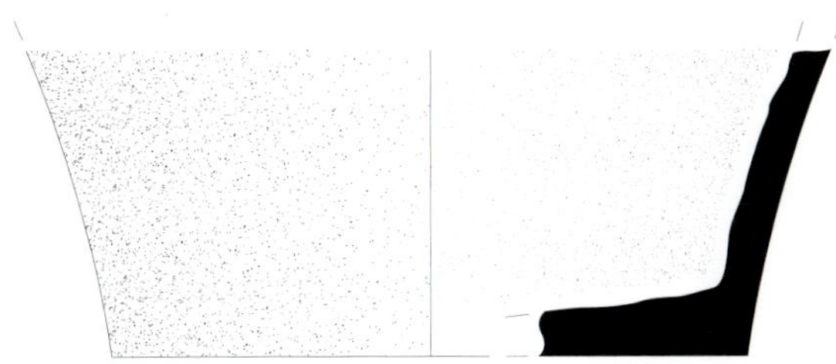

Appendices

Appendix A
Assemblage Tables

Francesca Silvestrelli

Ware	Form	Type	Cat. No.	Chronology	R	B	H	S	MNI	Total
Archaic Fine Ware?	n. id.							2	1	2
BG	Olpe			Late Archaic–Classical			1			1
	Closed forms							10		10
	B3b cup		SAV BG 07	550–500 BC	1				1	1
	Cup-skyphos		SAV BG 15	530–480 BC		1				1
	Cup-skyphos		SAV BG 13	500–450 BC	1				1	1
	Kylix	C		470–440 BC		6	3		1	9
				5th c. BC			1			1
	A-type skyphos			500–450 BC	2				1	2
			SAV BG 17	500–450 BC	1				1	1
			SAV BG 18	475–450 BC	1				1	1
			SAV BG 19	450–400 BC	1				1	1
		n. id.						1		1
	Skyphos with offset rim			Late Archaic	1			3	1	4
	Ionic-type B2 cup		SAV BG 06	520–480 BC		3				3
				Late Archaic	15			8	12	23
	Open forms			Late Archaic				21		21
					2		2	44	1	48
BW	Jug	J 1.a	SAV BW 32	5th c. BC	1				1	1
		J 3.a	SAV BW 37	5th–4th c. BC	2				1	2
	Hydria	H 4.a	SAV BW 40	perhaps 5th c. BC	1				1	1
	Jug/amphora						1		1	1
	Olpe		SAV BW 43		1				1	1
								1		1
	Closed forms						2	18	2	20
	Lekane				1		1		2	2
	One-handler	SB/O 2.a	SAV BW 05	5th–4th c. BC	1				1	1
		SB/O 3		Late Archaic–Early Classical	1				1	1
		SB/O 3		Late Archaic–Early Classical	1				1	1
				Late Archaic–Early Classical	4	2			6	6
	Open forms						2		2	2

Table A1 Assemblage 1: House on the Hill, SA 79-1062PL; Phase 2.

Ware	Form	Type	Cat. No.	Chronology	R	B	H	S	MNI	Total
PCW	Oinochoe?						1		1	1
	Jug	J 1	SAV PCW 51		2				1	2
		J 2	SAV PCW 55		1				1	1
			SAV PCW 70			1			1	1
			SAV PCW 71			1			1	1
						3		1	3	4
	Jug/amphora						4		2	4
	Closed forms					4	2	123	6	129
	Large basin				1				1	1
	Lekane/basin	Ba 2			3				2	3
	One-handler	SB/O 3	SAV PCW 01	End of the 6th–beginning of the 5th c. BC	1			1	1	2
					3				3	3
	Open forms							1		1
	n. id.							83		83
CkW	Chytra				1				1	1
		P1.2	SAV CkW 31	End of the 6th c. BC	1				1	3
		P1.5		600–400 BC	10		9	1	10	22
							2		2	2
	Closed forms			580–500 BC		1		28		29
							3		2	3
	Open forms			580–500 BC				49	0	49
	n. id.			580–500 BC	1			44	1	45
Amp	Archaic	A3.1b	SAV Amp 04	500–450 BC	1				1	
	Amphora	Archaic Type B		500–450 BC				2	2	
		Corinthian A		550–500 BC				1	1	
				500–450 BC				4	4	
		Type B		450 BC				4	4	
	n. id.							6	6	
Pottery Total					**63**	**22**	**34**	**456**	**100**	**561**
RT	Cover tile	AC1		Hellenistic system						1
				Archaic or Hellenistic system						1
		BC2		Hellenistic/Roman system						1
	Tile	AT0		Archaic system						2
	n. id.	System Aa–Ab		Archaic/Hellenistic system						5
Li	Grinder/ pestle?		SAV Li 07							1
	Grinder?		SAV Li 08							1
Total					**63**	**22**	**34**	**456**	**100**	**573**

Table A1 cont. Assemblage 1: House on the Hill, SA 79-1062PL; Phase 2.

Ware	Form	Type	Cat. No.	Chronology	R	B	H	S	MNI	Total
Fig	Cup			500–460 BC	1				1	1
BG	Amphora			5th c. BC	2			3	1	5
	Oinochoe		SAV BG 33	430–400 BC			1	4		5
	Jug			End of the 5th c. BC		1			1	1
	Jug		SAV BG 30	420–350 BC		1			1	1
	Closed forms									12
	Ionic-type B2 cup		SAV BG 04	500–470 BC	7			23	6	30
	Skyphos with offset rim		SAV BG 08	500–480 BC	8	2		19	2	29
			SAV BG 09	500–450 BC	5		1	4	3	10
			SAV BG 10	5th c. BC					1	1
	Skyphos with concave rim		SAV BG 11–12	470–440 BC	14	3	1	15	4	33
				5th c. BC	2			1	1	3
				Mid-5th c. BC					2	2
	Kylix		SAV BG 01	475–450 BC			2		1	4
	A-type skyphos		SAV BG 20	470–450 BC		5				5
			SAV BG 22	460–370 BC	3	6		6	1	15
			SAV BG 21	430–400 BC		1				1
	Cup-skyphos		SAV BG 14	430–400 BC	1			1	1	2
	C-type skyphos		SAV BG 23	450–425 BC	2			3	1	5
			SAV BG 26	400–370 BC	4			2	3	6
				370–325 BC	2				2	2
	Skyphos									5
	Open forms			Late Archaic				3		3
										37
BW	Jug	J 1.a	SAV BW 29	5th c. BC	3				2	3
		J 3.b	SAV BW 38	5th c. BC	1				1	1
	Jug/amphora						4			4
	Lebes	Le 1	SAV BW 45	End of the 6th–5th c. BC	1				1	1
		Le 2	SAV BW 46		1				1	1
	Closed forms					10				10
	Lekane	Lne 3	SAV BW 23	5th–4th c. BC	1					1
		Lne 1	SAV BW 17–18	5th–first half of the 4th c. BC	3				2	3
		Lne 1			6				1	6
		Lne 2	SAV BW 21	Second half of the 6th–5th c. BC	1				1	1
		Lne 3	SAV BW 22	5th c. BC	8				2	8

Table A2 Assemblage 15: Pottery Deposit; Phase 3.

Ware	Form	Type	Cat. No.	Chronology	R	B	H	S	MNI	Total
BW	One-handler	SB/O 1.a	SAV BW 01	5th–4th c. BC	2				1	2
		SB/O 1.b	SAV BW 04	Second half of the 5th–4th c. BC	2			1	1	2
		SB/O 2.a	SAV BW 06–07	5th–4th c. BC	5	1		2	2	8
		SB/O 2.b	SAV BW 10	5th–4th c. BC	1	3		3	1	6
		SB/O 2.b	SAV BW 11	5th–4th c. BC	2			1	1	2
		SB/O 2.b	SAV BW 09		4	2			1	6
		SB/O 3.b	SAV BW 13	5th–4th c. BC	1	1			1	2
		SB/O 3.b	SAV BW 14	Second half of the 6th–first half of the 4th c. BC	3				1	3
					1			5	4	6
	Open forms					6		6	2	12
			SAV BW 48			1			1	1
	n. id.					4			2	4
PCW	Dinos	Dn 1	SAV PCW 45	4th c. BC?	2				1	2
	Jug					4				4
		J 1	SAV PCW 47	5th c. BC	3	35		2	1	40
		J 1	SAV PCW 49–50		3	12	2		3	17
		J 2	SAV PCW 54	Classical	1				1	1
	Jug/amphora						4		3	4
	Oinochoe		SAV PCW 73		1				1	1
	Closed forms					42	3	8	7	53
	Bowl/lekane							2	1	2
	Large basin				1			1	2	2
	Lekane/basin	Ba 1	SAV PCW 26		1				1	1
		Ba 2	SAV PCW 10		12	1	1		6	14
		Ba 8	SAV PCW 19		1				1	1
			SAV PCW 26			3	1	8	6	12
	Flanged lekane			500–400 BC	1				1	1
	Open forms					13	1		1	14
	n. id.					2				2
Min	Lebes		SAV Min 02	520–400 BC	1				1	1
	n. id.							1	1	1
Amp	Corinthian A			450–400 BC		3				3
		A4.1b	SAV Amp 15	450 BC			1		1	1
		A4.2a	SAV Amp 14	450 BC	1	2			1	3
	Corinthian B	A5.1a	SAV Amp 19	450–400 BC	1				1	1
	Type B	A5.2a	SAV Amp 20	450 BC	1				1	1
CkW	Chytra	P1.5			2	1			2	3
					1				1	1
	Casserole	C3	SAV CkW 10	End of the 6th–beginning of the 4th c. BC	2		1		1	3
Pottery Total					**132**	**161**	**23**	**123**	**107**	**496**
LA	Lamp	Circular	SAV LA 44	480–325 BC	1	1		1	1	3
Total					**133**	**162**	**23**	**124**	**108**	**499**

Table A2 cont. Assemblage 15: Pottery Deposit; Phase 3.

Ware	Form	Type	Cat. No.	Chronology	R	B	H	S	MNI	Total
BG	C-type skyphos			370–325 BC	1				1	1
	Kantharos		SAV BG 40	350–300 BC				4	0	4
	Open forms							1		1
PCW	Bowl							2		2
	Closed forms							6		6
Mor	Mortarium	MO4	SAV Mor 01	4th c. BC	2				1	2
Amp	Type B	A5	SAV Amp 24	275 BC			1		1	1
OD	Pithos lid		SAV OD 04	4th c. BC	1		1			2
	n. id.							1		1
Pottery Total					**4**		**2**	**14**	**3**	**20**
KF	Kiln firing support	1								1
LW	Oscillum	Oscillum	SAV LW 23	4th–3rd c. BC						1
TC	Plaque			Hellenistic						5
AT	Antefix		SAV AT 02	Last quarter of the 4th c. BC						1
RT		AT0–AT1		Archaic/Hellenistic system						1
	Cover tile	AT1	SAV RT 02	Hellenistic system						1
										1
	Pipe?				2				1	2
	Opaion?			Hellenistic system					1	1
										1
	Tiles counted and discarded									779
Total					**6**	**0**	**2**	**14**	**5**	**814**

Table A3 Assemblage 4: Tile Fall 2; Phase 4.

Ware	Form	Type	Cat. No.	Chronology	R	B	H	S	MNI	Total
RF	Hydria			450–400 BC	1				1	1
BG	Open forms							3		3
BW	Closed forms							1		1
	Large bowl	LB 1		4th–3rd c. BC	2				1	2
PCW	Large bowl	LB 2	SAV PCW 23	4th–3rd c. BC	1				1	1
	Jug	J 3	SAV PCW 57	4th–3rd c. BC	1				1	1
		J 4			1				1	1
	Closed forms					1		1	1	2
	n. id.							3		3
CkW	Casserole	C10.2	SAV CkW 24	First half of the 3rd c. BC	1				1	1
		n. id.						1		1
	Chytra			Classical–Hellenistic				1		1
Amp	Greco-Italic			350–300 BC				2	1	2
Pottery Total					**7**	**1**		**12**	**8**	**20**
KF	Clay amorphous fragment									1
TC	Relief plaque			End of the 4th–3rd c. BC						2
RT	Tile	AT1/AC1		Hellenistic system						2
Total					**7**	**1**		**12**	**8**	**25**

Table A4 Assemblage 16: Destruction Layer in G6, G7, and H7; Phase 4.

Ware	Form	Type	Cat. No.	Chronology	R	B	H	S	MNI	Total
BG	Dish		SAV BG 37	320–250 BC	1	2			1	3
	Mug		SAV BG 29	470–430 BC	3			2	1	5
	Ionic-type B2 cup			Late Archaic				2		2
BW								1		1
	Olpe		SAV BW 44			1			1	1
	One-handler	SB/O 1.a	SAV BW 02	4th–beginning of the 3rd c. BC	1				1	1
PCW	Jug					6		1	3	7
	Closed forms					1	1	9	1	11
	Lekane/basin	Ba 2			1				1	1
	n. id.							2		2
Min						1				1
CkW	Closed forms			310–250 BC				6		6
	Casserole				1		1	4		6
	Pan	P3.1		Hellenistic	5	1		1	5	7
Amp	n. id.							1		1
Pottery Total					**12**	**12**	**2**	**29**	**14**	**55**
KF	Mudbrick									16
LW	Oscillum									1
TC	Plaque			Early Hellenistic						1302
Total					**12**	**12**	**2**	**29**	**14**	**1374**

Table A5 Assemblage 9: Slope Area, Bulldozed Stipe; Phase 4.

Ware	Form	Type	Cat. No.	Chronology	R	B	H	S	MNI	Total
PCW	Plate/lid	8a	SAV PCW 20		4	2		9	1	15
	Jug							1		1
		J 4		Early Hellenistic	1				1	1
CkW	Open forms					1				1
Pottery Total					**5**	**3**		**10**	**2**	**18**
KF	Kiln firing support	1								2
		1								2
TC				Early Hellenistic						1
RT	Tile	AT1		Hellenistic system						1
										1
	Tiles									70
Total					**5**	**3**		**10**	**2**	**95**

Table A6 Assemblage 14: Slope Area, Grey Stipe; Phase 4.

Ware	Form	Type	Cat. No.	Chronology	R	B	H	S	MNI	Total
Fig	Amphora				1				1	1
	Cup-skyphos			Late Archaic	2			1	1	3
	Pelike			400–300 BC			1		1	1
	Open forms							1		1
BG	Squat lekythos							1		1
	Cup			Late Archaic				3		3
	Open forms							1		1
BW	Jug	J 2	SAV BW 36	325–250 BC	5				4	4
	Jug/amphora						1			1
	Lebes	Le 1		500–400 BC	1				1	1
	Closed forms					1		3		4
	Dish			4th–3rd c. BC	1					1
	n. id.							2		2
PCW	Jug	J 1			3				3	3
		J 4	SAV PCW 60–61	Late 4th–3rd c. BC	4				2	4
			SAV PCW 68			1		1	1	2
	Closed forms					3		10	2	13
	Deep basin	DB 1	SAV PCW 39	4th–3rd c. BC	2				1	2
	Deep basin					1			1	1
	Open forms							1		1
	n. id.							5		5
CkW	Chytra	P8.3		End 4th–3rd c. BC	3			1	1	4
							2	3		5
	Closed forms					1			1	1

Table A7 Assemblage 11: Slope Area, terracotta and tile concentration, Sounding B, and Slope 3; Phase 4.

Ware	Form	Type	Cat. No.	Chronology	R	B	H	S	MNI	Total
CkW	Casserole	C1.1	SAV CkW 09	End of the 4th–3rd c. BC	1				1	1
		C10.1c	SAV CkW 23	340–300 BC	5		1	1	1	7
		C6.2			2				2	2
		C6.3		350–270 BC	3		1		2	4
		C9.2	SAV CkW 20	End of the 4th–3rd c. BC	1			1	1	2
							3			3
	Pan	PD2			2				2	2
		PD3.1			2				2	2
		PD3.2	SAV CkW 04	End of the 4th–3rd c. BC	1				1	1
	Open forms					1		3		4
	Klibanos	Kl2	SAV CkW 27	End of the 4th–3rd c. BC	1	1		4	2	6
	Lid							1		1
Amp	Greco-Italic			350–300 BC				1		1
Amp?								1		1
Pottery Total					**40**	**9**	**9**	**45**	**34**	**102**
KF	Kiln firing support	1a		Early Hellenistic						3
		2		Early Hellenistic						1
		5		Early Hellenistic						1
	Kiln mudbrick		Ch. 9, Sample 1					4		4
										1
	Clay amorphous fragment									2
LW	Oscillum		SAV LW 06	4th–3rd c. BC						3
	Pyramidal		SAV LW 05	4th–3rd c. BC						1
TC				Early Hellenistic						177
RT	Tile		SAV RT 118							1
		AT0/AT1		Archaic/Hellenistic system						1
			SAV RT 41							1
	Cover tile?	n. id.	SAV RT 112							1
	Opaion lid	Opaia 1		Roman system						1
		n. id.								2
		Opaia 1	SAV RT 102	Roman system						1
	Pan tile	BT3		Roman system						1
	Tile	AT1/AC1	SAV RT 8–11, 20–28, 30–31, 127	Hellenistic system						35
		BT2	SAV RT 48	Hellenistic/ Roman system						1
		n. id.								317
Fau	Shell		SAV Sh 12							1
Li	Weight		SAV Li 04							1
Total					**40**	**9**	**9**	**49**	**34**	**659**

Table A7 cont. Assemblage 11: Slope Area, terracotta and tile concentration, Sounding B, and Slope 3; Phase 4.

Ware	Form	Type	Cat. No.	Chronology	R	B	H	S	MNI	Total
BG	Squat lekythos			4th c. BC				2	0	2
	Closed forms							2	2	2
	Ionic-type B2 cup		SAV BG 05	540–500 BC		1				1
	C-type kylix			5th c. BC	1				1	1
	A-type skyphos			450–400 BC	1				1	1
	C-type skyphos			470–430 BC		1				1
			SAV BG 25	420–380 BC		1				1
	Bolsal		SAV BG 28	425–400 BC		2		3	1	5
	Dish		SAV BG 36	325–275 BC		2				2
	Stemmed cup		SAV BG 39	320–250 BC	2				1	2
	Open forms			Late Archaic			1	3		4
								1		1
U	Unguentarium		SAV U 02	End of the 4th–3rd c. BC					1	1
BW	Jug	J 2	SAV BW 34	300 BC	1				1	1
	Jug/amphora				1		1		2	2
	Closed forms					1		2	1	3
	One-handler	SB/O 3.a	SAV BW 12	5th– 4th c. BC	6				2	6
	Lekane	Lne 1	SAV BW 19	Second half of the 5th–first half of the 4th c. BC	1				1	1
	Lid				1				1	1
	n. id.					1				1
PCW	Jar	Ja 1	SAV PCW 74	End of the 4th–3rd c. BC	1				1	1
	Jug	J 4	SAV PCW 59	4th–3rd c. BC	3				3	3
		J 5	SAV PCW 62	4th–3rd c. BC	1				1	1
			SAV PCW 67	3rd c. BC		3		1	3	4
	Jug/amphora					2	1	1	3	4
	Closed forms							6		6
	Bowl				1				1	1
	Carinated lekane		SAV PCW 05	4th–3rd c. BC	1				1	1
	Deep basin	DB 1	SAV PCW 40	4th–3rd c. BC	2				1	2
		DB 3b	SAV PCW 44		2				1	2
	Large basin					1			1	1
	Lekane/basin	Ba 1	SAV PCW 08	5th c. BC?	1				1	1
	Basin	Ba 8			1					1
	Lid		SAV PCW 78		2		1		3	3
	n. id.							5		5
Lou	Louterion					1			1	1

Table A8 Assemblage 13: Slope Area, terracotta and tile concentration, debris from the collapse of Kiln 1; Phase 4.

Ware	Form	Type	Cat. No.	Chronology	R	B	H	S	MNI	Total
CkW	Chytra	P2.2	SAV CkW 35	End of the 4th–3rd c. BC	3	1	1		1	5
		P2.3	SAV CkW 36	End of the 4th–3rd c. BC	1				1	1
		P3.2		End of the 4th–3rd c. BC	1				1	1
		P4	SAV CkW 42	End of the 4th–3rd c. BC	1				1	1
						1	2	3	4	6
	Kakkabé	P6.1	SAV CkW 44	End of the 4th–3rd c. BC	1				1	1
	Closed forms	F2	SAV CkW 50	End of the 4th–3rd c. BC		1		1	1	3
							1		1	1
	Pan	PD2		End of the 4th–3rd c. BC	2		1		2	3
		PD3.1		End of the 4th–3rd c. BC	2				2	2
		PD3.2a	SAV CkW 03	End of the 4th–3rd c. BC	1				1	1
	Casserole	C6.2		End of the 4th–3rd c. BC	3	3	1	1	3	12
		C6.3	SAV CkW 14	End of the 4th–3rd c. BC			1		1	1
		C7		End of the 4th–3rd c. BC	1			1	1	2
		C7a	SAV CkW 15	End of the 4th–3rd c. BC	1	1	1	1	1	4
		C8.2	SAV CkW 18	End of the 4th–3rd c. BC	1				1	1
		C9.3		End of the 4th–3rd c. BC	1				1	1
		C10.1	SAV CkW 21	End of the 4th–3rd c. BC	2		1		2	3
		C10.1c		End of the 4th–3rd c. BC	4		1	1	1	6
		n. id.		End of the 4th–3rd c. BC			3	2	1	5
	Open forms							1		1
	Klibanos	Kl2	SAV CkW 26	End of the 4th–3rd c. BC	6	1		1	4	10
								1		1
	Lid	Li1	SAV CkW 28	End of the 4th–3rd c. BC	3		2	1	4	6
		Li1/L2		End of the 4th–3rd c. BC	3		1		4	14
		Li2	SAV CkW 29	End of the 4th–3rd c. BC	5		1		5	9
		Li4		2nd–1st c. BC	1				1	1
								1		1
Amp	Archaic Type B	A3.1c	SAV Amp 05	500–450 BC	1		1		1	2
	Type B	A5.2c	SAV Amp 23	End of the 4th c. BC	1			1	1	2
	Type B			280–250 BC			1		1	1
	Greco-Italic			350–300 BC				1	1	1
	n. id.							1		1
Pottery Total					**74**	**24**	**23**	**44**	**86**	**186**

Table A8 cont. Assemblage 13: Slope Area, terracotta and tile concentration, debris from the collapse of Kiln 1; Phase 4.

Ware	Form	Type	Cat. No.	Chronology	R	B	H	S	MNI	Total
KF	Clay amorphous fragment									20
	Kiln firing support	1	SAV KF 03–04	End of the 4th–3rd c. BC						23
		2	SAV KF 05–06	End of the 4th–3rd c. BC						3
		3								3
		4	SAV KF 08	End of the 4th–3rd c. BC						1
		5		End of the 4th–3rd c. BC						2
				End of the 4th–3rd c. BC						1
	Kiln mudbrick									1
			Counted							20
	Mold									1
	Pestle		SAV KF 14	4th–3rd c. BC						5
LW	Oscillum		SAV LW 12	End of the 4th–3rd c. BC						1
TC	Plaque									158
RT	Floor tile									3
	n. id.									3
	Opaion lid?									1
	Pan tile	BT3	SAV RT 65	Roman system						2
			Counted							4
	Cover tile	Miscellanea								3
	Tile	AT0-AT1		Laconian Archaic/ Hellenistic system						6
		AT1		Hellenistic system						47
		AT1/AC1	SAV RT 04, 16, 19, 42–43	Hellenistic system						24
		BT2/AC2	SAV RT 53–54	Hellenistic/Roman system						6
		BT3	SAV RT 113	Roman system						1
		Opaia							1	2
		Opaia 1	SAV RT 103	Roman system						1
		Opaia 2		Roman system						1
		Miscellanea								3
			Counted							566
	Tile, small cube-like fragment		Counted							370
	Stones used in the construction of the kiln		Counted							6
	Brick and TC slag		Counted							40
	Miscellanea	Miscellanea								59
n. id.					1				1	1
Fau	Shell		SAV Sh 09, 10, 22							9
Total					**75**	**24**	**23**	**44**	**88**	**1583**

Table A8 cont. Assemblage 13: Slope Area, terracotta and tile concentration, debris from the collapse of Kiln 1; Phase 4.

Ware	Form	Type	Cat. No.	Chronology	R	B	H	S	MNI	Total
BG	Shouldered lekythos			Mid-5th c. BC		1			1	1
BG	Skyphos with offset rim			500–450 BC	1				1	1
BG	Open forms							1		1
GW	Open forms		SAV GW 11	Late Republican		1			1	1
GW	Open forms			Late Republican				1		1
BW	Jug		SAV BW 42			1			1	1
BW	Jug	J 2.a	SAV BW 33	Perhaps 4th–beginning of the 3rd c. BC	1				1	1
BW	Closed forms							1		1
BW	One-handler	SB/O 1.a	SAV BW 03	4th c. BC–beginning of 3rd c. BC	1				1	1
BW	Skyphoid cup		SAV BW 15	Second half of the 3rd–first half of the 2nd c. BC	1				1	1
BW	Dish		SAV BW 27		2				1	2
PCW	Large basin		SAV PCW 29	Early Hellenistic	1				1	1
PCW	Large basin		SAV PCW 30		1				1	1
PCW	Large basin?				1				1	1
PCW	Plate/lid	8b		Early Hellenistic	2					2
PCW	Jar				2			1	1	3
PCW	Jug/amphora						10		1	10
PCW	Small table amphora?				1				1	1
PCW	Closed forms						2	3	3	5
PCW	n. id.				1			3	1	4
CkW	Chytra	P3.3a		Classical	1				1	1
CkW	Chytra	P3.3b	SAV CkW 41	2nd–1st c. BC	1	1		8	1	10
CkW	Casserole	C8a	SAV CkW 17	4th–3rd c. BC	1				1	1
Mor	Mortarium		SAV Mor 05	4th c. BC		1			1	1
Mor	Mortarium		SAV Mor 03	4th–second half of the 3rd c. BC	1				1	1
Lou	Louterion		SAV Lou 04	End of the 5th–4th c. BC		1			1	1
Lou	Louterion		SAV Lou 05	End of the 5th–4th c. BC		1			1	1
Amp	Corinthian A	A4.2	SAV Amp 18	350–300 BC			1		1	1
Amp	Greco-Italic	A6	SAV Amp 30	300–250 BC			1		1	1
Amp	Greco-Italic			350–300 BC				3		3
Amp	Greco-Italic			300–250 BC			1		1	1
Pottery Total					**19**	**7**	**15**	**21**	**28**	**2**

Table A9 Assemblage 3: Tile Fall 1; Phase 5.

Ware	Form	Type	Cat. No.	Chronology	R	B	H	S	MNI	Total
KF	Kiln firing support	7	SAV KF 13							1
LW	Pyramidal, rectangular variant		SAV LW 15	4th–3rd c. BC						1
	Oscillum		SAV LW 13	4th–3rd c. BC						1
			SAV LW 14	4th–3rd c. BC						1
			SAV LW 16	4th–3rd c. BC						1
			SAV LW 18	4th–3rd c. BC						1
TC	Plaque			Early Hellenistic						1
										11
AT	Antefix mold		SAV AT 05	6th c. BC						1
	Acroterion disc		SAV AT 03	4th c. BC						1
AT?						1				1
RT	Tile	AT1	SAV RT 05, 13	Hellenistic system						14
		AT2	SAV RT 45	Hellenistic system						1
		BT2	SAV RT 50	Hellenistic/Roman system						3
	Cover tile	AC1	SAV RT 29	Hellenistic system						5
	Pan tile	BT3		Roman system						3
	Gutter pipe			Roman system						3
	Opaion lid?			250–200 BC					1	1
	Well liner									2
	n. id.									3
	Tiles counted and discarded									1670
M	Iron nail		SAV M 01	Late Republican						1
Li	Grinding stone		SAV Li 01	Archaic to Classical						1
Fau	Bone									4
Total					**19**	**8**	**15**	**21**	**29**	**1795**

Table A9 cont. Assemblage 3: Tile Fall 1; Phase 5.

Ware	Form	Type	Cat. No.	Chronology	R	B	H	S	MNI	Total
BG	Skyphos with concave rim			470–440 BC		1				1
	C-type skyphos			370–325 BC	3				3	3
	Open forms							4		4
BW	Jug					1			1	1
	Large bowl	LB 1	SAV BW 25	4th–3rd c. BC	1				1	1
PCW	Closed forms							1		1
	Skyphoid cup		SAV PCW 02	320–250 BC	1				1	1
CkW	Chytra							1		1
Pottery Total					**5**	**2**		**6**	**6**	**13**
LW	Oscillum		SAV LW 22	4th–3rd c. BC						1
RT	Cover tile	AC1		Hellenistic system						4
	Tile	Opaia	SAV RT 37	Hellenistic system						1
	Gutter pipe		SAV RT 106–107	Roman system						5
	Flat pan tile		SAV RT 110							1
	Thick pan tile and curved tile		Counted							220
Fau	Bone									3
Total					**5**	**2**		**6**	**6**	**261**

Table A10 Assemblage 17: tile concentration in G7; Phase 5.

Ware	Form	Type	Cat. No.	Chronology	R	B	H	S	MNI	Total
BG	Open forms							1		1
PCW	Large basin		SAV PCW 33		1	1			1	2
	n. id.							3		3
CkW	Casserole	C6.3		End of the 4th–3rd c. BC	1				1	1
	Chytra	P1.3	SAV CkW 32	6th–1st c. BC	1			1	2	2
Pottery Total					**3**	**1**		**5**	**4**	**9**
KF	Mold			Second half of the 4th–3rd c. BC						1
TC	Plaque									1
AT	Antefix mold		SAV AT 07	Second half of the 6th c. BC						1
RT	Tile	Opaia 2	SAV RT 104	Roman system						1
Total					**3**	**1**		**5**	**4**	**13**

Table A11 Assemblage 2: tile concentration inside the clay-settling basin; Phase 5.

Ware	Form	Type	Cat. No.	Chronology	R	B	H	S	MNI	Total
BG	Ionic-type B2 cup			Late Archaic				1		1
	Ring-handled cup			225–150 BC	1				1	1
	Bowl		SAV BG 43	Late 3rd–2nd c. BC	1				1	1
	C-type skyphos			370–325 BC	1				1	1
	Open forms							7		7
	Closed forms							1		1
GW	Open forms			Late Republican				1		1
		GW C1?	SAV GW 09	150–100 BC	1				1	1
RW	Dish		SAV RW 01	Late Hellenistic–Early Imperial (late 2nd c. BC–early 1st c. AD)	1				1	1
	Juglet		SAV RW 03	Late Hellenistic–Early Imperial (late 2nd c. BC–early 1st c. AD)		1			1	1
	Closed forms		SAV RW 04	Late Hellenistic–Early Imperial (late 2nd c. BC–early 1st c. AD)		2		1	1	3
ITS	Plate		SAV Imp 01	Early 1st c. AD	1					1
Lag	Lagynos		SAV Lag 03	End of the 2nd–1st c. BC	1			1	1	2
PCW	Closed forms						3	12	2	15
	Large jar	LJ 1	SAV PCW 76		1				1	1
						1			1	1
	Large basin		SAV PCW 38			1			1	1
					1				1	1
	n. id.							5		5
Rom	Bottle		SAV Imp 12	1st c. BC–1st c. AD				2		2
	Closed form		SAV Imp 18	1st c. AD		1			1	1
	Closed form		SAV Imp 17	1st c. BC–1st c. AD		1			1	1
	Olla		SAV Imp 16	Late Republican–Early Imperial	1				1	1
		J 1	SAV Imp 13	1st c. BC–1st c. AD	2				1	2
	Basin	Ba 7	SAV Imp 06	Late 1st c. BC–1st c. AD	4	3		4	1	11
	Bowl	LB 2	SAV Imp 02	1st c. AD	1				1	1
Lou	Louterion		SAV Lou 01	5th–4th c. BC	3				1	3
	Louterion		SAV Lou 06	2nd–1st c. BC		1			1	1
CkW	Chytra							4		4
	Lid			300–200 BC	1				1	1
	Open forms							1		1
	Pan	PD4	SAV CkW 06	200–0 BC	8			1	1	9
Amp	Archaic Type B			500–450 BC				2		2
				500–450 BC				1	1	1
	Greco-Italic			350–300 BC				10		10
		A6		350–300 BC				2		2
	Greco-Italic						1		1	1
Pottery Total					**29**	**11**	**4**	**56**	**26**	**100**

Table A12 Assemblage 5: Kiln Complex, courtyard tile fall; Phase 5.

Ware	Form	Type	Cat. No.	Chronology	R	B	H	S	MNI	Total
KF	Kiln firing support	5								1
	Mud-brick									8
TC				Hellenistic						2
	Mold			Hellenistic						1
	Plaque			Hellenistic						1
AT	Antefix mold		SAV AT 08	550–500 BC						1
RT	Sima		SAV RT 129	Archaic						1
	n. id.									2
		Ab	SAV RT 39	Ab system, Hellenistic period						1
		Ab	SAV RT 40							1
	Cover tile	AC1		Hellenistic system						1
		BC1		Archaic/Classical system						1
		BC2		Hellenistic/Roman system						1
	Laconian tile	AT1/AC1	SAV RT 120	Hellenistic system						1
	Tile	AT1		Hellenistic system						2
		AT1/AC1								1
		BT2	SAV RT 49	Hellenistic/Roman system						1
	Gutter pipe	BC		Roman system						1
	Pan tile	BT3								38
			SAV RT 59							1
			SAV RT 63							1
			SAV RT 64							1
			SAV RT 66							1
			SAV RT 67, 69							2
			SAV RT 71							1
			SAV RT 72							1
			SAV RT 73							1
			SAV RT 75, 76							2
			SAV RT 78							1
			SAV RT 79							1
			SAV RT 80							1
			SAV RT 81							1
			SAV RT 83							1
			SAV RT 84							1
			SAV RT 86							1
			SAV RT 87, 89							2
			SAV RT 93							1
	Cover tile	n. id.	SAV RT 111							1
	Tile (counted and discarded									3059
M	Iron nail		SAV M 02							
Total					**29**	**13**	**4**	**56**	**27**	**3249**

Table A12 cont. Assemblage 5: Kiln Complex, courtyard tile fall; Phase 5.

Ware	Form	Type	Cat. No.	Chronology	R	B	H	S	MNI	Total
PCW	Jug	J 3			1				1	1
					1				1	1
	Open forms							1		1
Rom	Jug/amphora		SAV Imp 11	Early Imperial			1		1	1
CkW	n. id.						1		1	1
Pottery Total					**2**		**2**	**1**	**4**	**5**
KF	Kiln mudbrick		Ch. 9, Samples 2–5							12
RT	Pan tile	BT3		Roman system						1
Total					**2**		**2**	**1**	**4**	**18**

Table A13 Assemblage 6: deposit within Kiln 3 (squares E2 and E3); Phase 5.

Ware	Form	Type	Cat. No.	Chronology	R	B	H	S	MNI	Total
RW	Closed form		SAV RW 04	Late Hellenistic–Early Roman (late 2nd c. BC–early 1st c. AD)		2		1	1	3
PCW	Basin?							2		2
Amp	Greco-Italic	A6.2a	SAV Amp 29	2nd c. BC	1				1	1
				350–300 BC				1	1	1
	Greco-Italic			350–300 BC				1	1	1
Pottery Total					**1**	**2**		**5**	**4**	**8**
KF	Kiln element									11
	Mudbrick									32
			Counted							25
LW	Pyramidal, rectangular variant		SAV LW 26							1
			SAV LW 27							1
										3
RT	Tile/cover tile	AT1/AC1		Hellenistic system						9
		Miscellanea AT0–AT1		Archaic/Hellenistic system						2
			Counted							186
	Pan tile	BT3		Roman system						5
			SAV RT 88	Roman system						1
			SAV RT 90	Roman system						1
			SAV RT 91	Roman system						1
			Counted							60
	Brick		Counted							20
	n. id.									2
Fau	Animal bone									1
	Bone									1
Total					**1**	**2**		**5**	**4**	**370**

Table A14 Assemblage 7: Kiln Complex, collapse of Kiln 4; Phase 5.

Ware	Form	Type	Cat. No.	Chronology	R	B	H	S	MNI	Total
BG	Ring-handled cup		SAV BG 42	225–150 BC	1				1	1
	Open forms							1		1
GW	Mug	GW Bo1	SAV GW 01	Second half of the 1st c. BC	10	1	1	0	1	12
			SAV GW 02	Second half of the 1st c. BC	6	0	0	0	1	6
	n. id.		SAV GW 15	Late Republican	0	0	1	0	0	1
RW	Juglet		SAV RW 02	Late Hellenistic–Early Imperial (late 2nd c. BC–early 1st c. AD)		2			1	2
Lag	Lagynos		SAV Lag 02		1				1	1
			SAV Lag 03	End of the 2nd–1st c. BC	1				1	1
PCW	Large container				1				1	1
	Jar	Jar 1	SAV PCW 75		1				1	1
	Jug	J 1			2				1	2
	Large basin		SAV PCW 32		2				1	3
	Large basin?				2				2	2
	Small table amphora?		SAV PCW 53		2			2	1	4
	n. id.				1			3		4
Rom	Jug	-	SAV Imp 09	1st c. AD		2		12		14
Lou	Louterion		SAV Lou 02	End of the 5th–early 4th c. BC	1				1	1
Mor	Mortarium				1	1		1	1	3
CkW	Pot	P5	SAV CkW 43	150–0 BC	1		1	2	1	4
	Chytra	P8.1	SAV CkW 48		1		1	4	1	6
		P8.2	SAV CkW 49	2nd–1st c. BC	1		1	1	1	3
	Lid	Li1.1		400–200 BC	1				1	1
		Li1.3		500–300 BC	1				1	1
		Li4		2nd–1st c. BC	1			1	1	2
	Pan	C10.1		300–100 BC	1			1	1	2
		PD5	SAV CkW 07	2nd–1st c. BC	5	1			1	6
		PD6	SAV CkW 08	2nd–1st c. BC	2				1	2
	Open forms			Second half of the 1st c. BC				1		1
	Klibanos	Kl1	SAV CkW 25	2nd–1st c. BC	5	1	1		1	7
		n. id.		300–180 BC				1		1
Amp	Archaic Type B	A3	SAV Amp 10	Second half of the 6th c. BC		1			1	1
	Corinthian A			Late 6th–early 5th c. BC	1					1
	Greco-Italic	A6.1a	SAV Amp 28	200–150 BC	1				1	1
				350–300 BC				1	1	1
		A6	SAV Amp 33	300–250 BC				4	1	4
	Metapontine			200–100 BC			1	4	2	5
	n. id.				1			1	1	2
OD	Pithos		SAV OD 01	4th–3rd c. BC	1					1
Pottery Total					**55**	**10**	**7**	**40**	**32**	**112**

Table A15 Assemblage 8: Kiln Complex, collapse of Kiln 5; Phase 5.

Ware	Form	Type	Cat. No.	Chronology	R	B	H	S	MNI	Total
KF	Kiln element									7
	Kiln firing support	5								3
LW	Pyramidal		SAV LW 31							1
	Pyramidal, rectangular variant		SAV LW 30							1
AT?								2		2
RT	Sima, waterspout									1
		AB		Hellenistic system						1
	Cover tile	AC1		Hellenistic system						1
		BC1		Archaic/Classical system						1
			SAV RT 95	Archaic/Classical system						1
		BC2		Hellenistic/Roman system						1
	Tile	AT1		Hellenistic system						5
			SAV RT 18	Hellenistic system						1
		Opaia	SAV RT 34	Hellenistic system						1
		BT3		Roman system						4
			SAV RT 68	Roman system						1
			SAV RT 82	Roman system						1
			SAV RT 92	Roman system						1
	Brick		Counted							70
	Curved tiles		Counted							210
	Thick pan tile		Counted							60
	Mudbrick		Counted							20
Construction Material	Stones		Counted							50
Fau	Bone									3
	Shell	*Pectinidae and Ostrea edulis*	SAV Sh 11, 24							4
n. id.								3		3
Total					**55**	**10**	**7**	**45**	**32**	**566**

Table A15 cont. Assemblage 8: Kiln Complex, collapse of Kiln 5; Phase 5.

Appendix B
Census of Production Sites

Francesca Silvestrelli

	Location	Kilns	Indicators of Production	Types of Production	Chronology	Observations	Bibliography
1A	*Kerameikos* (*Fig. 5.3*).	Only the E half of the circular combustion chamber of an Archaic clay kiln is preserved (diam. 0.90 m, max. h. 0.50 m)	Deposit with production discards, lumps of clay (kiln lining?)	Ionic-type cups, skyphoi with offset rim, lekanides	Second half of the 6th c. BC	The excavated part of the *ergasterion* occupied an area of 150 m^2. The workshop consists of an open-air area with a well and a roofed room (Room D), to which is annexed the kiln placed in a small open-air space. To the E of the kiln, a deposit with production discards was discovered.	D'Andria 1980a, 370–75; Fischer-Hansen 2000, 101–2; Silvestrelli 2004, 111–12; Di Giuseppe 2012, 101; Silvestrelli forthcoming
1B	*Kerameikos*		Deposit with production discards, (no. 12)	BG, BW, PW, CkW	Second quarter of the 5th c. BC	The workshop is composed of 2 rooms (H and D). Deposit no. 12 was found inside Room D.	D'Andria 1980b; Silvestrelli 2004, 111–12; Di Giuseppe 2012, 101; Silvestrelli forthcoming
1C	*Kerameikos*		Deposit with production discards, (nos. 16, 17, 19, and 22), lumps of clay (kiln lining?), and kiln firing supports	BG, BW, PW, CkW	Second half of the 5th c. BC		Cracolici 2003, 79–84; Silvestrelli forthcoming
1D	*Kerameikos*	Part of the combustion chamber and central pillar (diam. 0.40 m) of a circular clay kiln (max. l. 1.05 m, max w. 0.75 m)	Deposits with production discards, (nos. 1, 4, 7, 8, 23, and 24), lumps of clay (kiln lining?), test pieces, and kiln firing supports	RF (Dolon, Creusa, and Anabates Painters), Xenon Group, BG, BW, PW, CkW	First quarter of the 4th c. BC		D'Andria 1980a, 375–418, 434–35, 440–44; Silvestrelli 1996; Cracolici 2003, 60–79; Silvestrelli 2004, 113–14; Silvestrelli 2005, 114; Di Giuseppe 2012, 42–44, 101

Table B1 Production sites from the asty of Metaponto.

	Location	Kilns	Indicators of Production	Types of Production	Chronology	Observations	Bibliography
1E	*Kerameikos*	**Kiln A** *(Cuomo di Caprio Type Ib)*: stocking channel (l. 2.00 m; w. 0.48 m; h. max. 0.22) built with mudbricks 0.11 m long. Circular combustion chamber (diam. 1.20 m; max. h. 0.33 m; h. 0.25 m), built with curvilinear mudbricks (0.26 m x 0.10 m; h. 0.06 m) lined with clay (th. 0.05 m), partly vitrified. The perforated floor, now lost, was supported by a central column; diam. 0.18 m). **Kiln B** *(Cuomo di Caprio Type Ib)*: stocking chamber (l. 1.60 m; int. w. 0.40 m; ext. w. 0.54 m) and combustion chamber (diam. 1.17 m; max. h. 0.27 m) with central pillar (diam. 0.20 m) supporting the perforated floor, not preserved. The combustion chamber is built with circular mudbricks (w. 0.11 m).	Deposit with production discards (nos. 3, 5, and 6), lumps of clay (kiln lining?), test pieces, and kiln firing supports	RF, Gnathia ware, Xenon Group, BG, BW, PW, CkW	Second half of the 4th c. BC		D'Andria 1980a, 362–67, figs. 7–12, 422–34, 435–40; Cracolici 2003, 87–94; Silvestrelli 2004, 114–15; Di Giuseppe 2012, 101
1F	*Kerameikos*	**Kiln C**: stocking channel (l. 1.70 m; w. 0.60 m, not excavated) and circular combustion chamber (diam. 1.80 m; max. h. 0.60 m) built with mudbricks.	Deposit with production discards (no. 2)	Terracotta plaques	End of the 4th–3rd c. BC		D'Andria 1980a, 367–70, 418–22; Signore 1996; Silvestrelli 2004, 115; Di Giuseppe 2012, 101
1G	*Kerameikos*		Molds for Megarian bowls		2nd c. BC	The molds were found in the alluvial layer covering the *kerameikos* area. Molds for Megarian bowls were also found in the area of the North Gate.	Silvestrelli 2004, 115; Silvestrelli forthcoming; Adamesteanu 1975, 249, 251, fig. 266; Puppo 1995, 102, no. TA25
2	Plateia I/Plateia A (*Fig. 5.4*, no. 2).	Kiln	Deposit with production discards	RF (Amykos Painter)	Last quarter of the 5th c. BC	Excavated in 1966, now disappeared.	Adamesteanu, 1969, 168–69; Adamesteanu 1971, 485, fig. 481; Adamesteanu 1973b, 27–28; D'Andria 1980a, 447–52, fig. 77; Silvestrelli 2005, 115–16
3	Plateia I/Plateia A (*Fig. 5.4*, no. 3)		Kiln firing supports (Types II-C2 and II-D2)	BG?	Late 4th–3rd c. BC	Excavated in 1966.	Cracolici 2003, 37; Silvestrelli 2004, 110; Silvestrelli 2005, 116
4	Plateia A, Sounding G (*Fig. 5.4*, no. 4)		Mudbricks			The road pavement fill was built with mudbricks from destroyed kilns.	Adamesteanu 1980, 297, fig. 313

Table B1 cont. Production sites from the asty of Metaponto.

	Location	Kilns	Indicators of Production	Types of Production	Chronology	Observations	Bibliography
5	Plateia III/Plateia B (*Fig. 5.4*, no. 5)		Lumps of clay (kiln lining?), vitrified clay fragments and iron slags	Pottery (kantharoi of indigenous type)	First half of the 6th c. BC	Ceramic production discards were found in secondary deposition in the 6th-c. BC road pavement fill, suggesting the presence of workshops possibly located in the nearby insulae.	A. Bottini 1998, 700, fig. 3 (location of the excavation) and pl. LXXXVIII.3 (kantharoi)
6	Insula I, W of the Sanctuary of Apollo Lykeios (*Fig. 5.4*, no. 6)	3 kilns of unknown type	Deposit with production discards	Unguentaria	3rd c. BC		Adamesteanu 1980, 59, n. 168; Mertens 1985, fig. 1a; Mertens 1999, 286
7	Sanctuary of Apollo Lykeios, between eastern side of the temenos and the temples (square D6c7, (*Fig. 5.4*, no. 7).	Two quadrangular kilns and stocking channel opened to the W. The best-preserved kiln has a central wall.	Wasters of architectural terracottas "belonging to one of the Archaic Metapontine roofs"	Architectural terracottas?	Archaic period	Excavated in 1982 (A. De Siena). The wasters were found inside the kilns.	A. Bottini 1983, 467–68; Mertens 1985, fig. 1a–2; Adamesteanu 1987, 123; Cuomo di Caprio 1971–1972, Type II/a; Cuomo di Caprio 1992, 76; Lecce 2010–2011, 40
8	Sounding S/96 (*Fig. 5.4*, no. 8)		Kiln firing supports, wasters, lumps of clay (kiln lining?)	Pottery	Second half of the 3rd c. BC	Ceramic production discards were found in secondary deposition in the road pavement fill suggesting the presence of workshops possibly located nearby.	Nava 1997, 462–63, fig. 1; Cracolici 2003, 97, fig. 1.E; Di Giuseppe 2012, 101
9	Sounding A/94. Stenopos delimitating the S side of the agora (*Fig. 5.4*, no. 9)		Vitrified clay, Ionic cup wasters, kiln firing supports	Pottery	Second half of the 6th c. BC; 5th and late 4th c. BC	Ceramic production discards were found in secondary deposition in the 6th-c. BC road pavement fill, suggesting the presence of workshops possibly located in the nearby insulae.	A. Bottini 1995, 763–64
10	Area W of the Castrum (between the *fossatum* and the modern road leading to Masseria Sansone)		Deposit with production discards, lumps of clay (kiln lining?), kiln firing supports, grey-ware wasters, transport amphorae wasters, molds for Megarian-type bowls	GW, Megarian-type bowls, Amp	2nd–1st c. BC	The deposit was identified during agricultural works.	D'Andria 1976, 541–42, n. 9; Giannotta 1980, 62
11	Castrum, *fossatum* (*Fig. 5.4*, no. 11)		Kiln firing supports		2nd–1st c. BC		D'Andria 1976, 541, pl. XXXIII,b and XXXV

Table B1 cont. Production sites from the asty of Metaponto.

	Site	Type of Settlement	Kilns	Indicators of Production	Types of Production	Chronology	Observations	Bibliography
12	C.da Saldone, Fattoria Musillo	Farmhouse		Wasters of Ionic-type cups		6th c. BC	Excavated in 1967–1969 (Uggeri)	Uggeri 1969, 60–61; Amore 1992–1993, 60–62; Van Compernolle 1996, 299; Lanza Catti in *Fattoria Fabrizio*, 409
13	San Marco (H2), site 411	Greek farmhouse and kiln		Slag, mudbricks				Carter in *Survey*, 610, n. 132; Prieto et al. in *Survey*, 75
14	Giardinetto (F2), site 521	Farmhouse, tombs, and kiln		Slag	Unknown	Uncertain. The farmhouse was active in the Archaic period and after 450 BC.		Carter in *Survey*, 646; Prieto et al. in *Survey*, 80
15	Cozzo Presepe	Greek settlement		Waster of a transport amphora handle		Unknown	The waster is a surface find.	Morel 1970, 110–11; Morel BCGT 453–54
16	Cozzo Presepe, site 1020	Farmhouse and kiln	Kiln				Identified during the 2007 Basento-Cavone survey	D'Annibale 2007, 64
17	Site 962	Kiln					Identified in the survey conducted in the highland between Bradano and Basento Rivers (2005); unpublished	
18	C.da Saldone, Cappa d'Amore	Farmhouse	Circular combustion chamber (diam. around 4 m). A wall with three sections supported the perforated floor; stocking channel (l. 2 m)			4th c. BC	The kiln is located some 30 m from the farmhouse.	Adamesteanu 1973a, 55, fig. 1; Adamesteanu 1974, 86–87, fig. at p. 86; Carter 2006, 150; Lanza Catti in *Fattoria Fabrizio*, 409
19	Demanio Campagnolo (G7), site 95	Farmhouse and kiln		Slags		2nd–1st c. BC		Carter in *Survey*, 895; Prieto et al. in *Survey*, 60; Swift with Devereux in *Survey, Vol. IV*, 92
20	Campagnolo, Fattoria Casamassima	Farmhouse		Kiln firing support	Ceramic	3rd–2nd c. BC	Excavated in 1967–1969 (Uggeri)	Adamesteanu 1973a, 57; Amore 1992–1993, 66–67, 116, fig. 48; Lanza Catti in *Fattoria Fabrizio*, 410
21	Lago del Lupo, Fattoria Venezia	Farmhouse		Kiln firing supports	Ceramic	4th c. BC	Excavated in 1967 (A. Schnapp)	Adamesteanu 1973a, 57, fig. 2; Amore 1992–1993, 75–76, 124, fig. 48; Lanza Catti in *Fattoria Fabrizio*, 409
22	Lago del Lupo, Casa Preti Cluster (G6), site 248	Greek farmhouse and kiln		Slag		6th c. BC or second half of the 4th–3rd c. BC		Carter in *Survey*, 689, 828, 874; Prieto et al. in *Survey*, 67

Table B2 Production sites from the chora of Metaponto (see also *Fig. 5.2*).

	Site	Type of Settlement	Kilns	Indicators of Production	Types of Production	Chronology	Observations	Bibliography
23	Cogno del Prete (H4), village of Capistretti, site 266	Greek farm-house and sanctuary		Wasters of transport amphorae	Metapontine Type B and Metapontine Greco-Italic amphorae, PW	Late 4th–3rd c. BC		Swift in *Survey*, 463, fig. 12.2, 487–88, nos. 52–58; Swift in *Survey*, 37; Prieto et al. in *Survey*, 68; Swift with Devereux in *Survey, Vol. IV*, 98
24	Cugno del Pero, c.da Bufalara	Unknown	3 kilns of unknown type		Ceramic? Amp? Arch. TC?	4th–3rd c. BC	Excavations of the 19th c. brought to light 3 kilns. The location of this workshop has never been identified.	Lacava 1891, 327–28; Adamesteanu 1973a, 49, n. 2, 54–55, fig. 1
25	Cugno del Pero, c.da Bufalara	Farmhouse				Second half of the 6th c. BC	According to Adamesteanu, Ionic cups found on the site are of local production and the existence of a kiln should be hypothesized.	Adamesteanu 1973a, 54; Adamesteanu 1974, 82–85; Carter 2006, 135; Lecce 2010–2011, 39; Lanza Catti in *Fattoria Fabrizio*, 409
26	Bufalara, site 489	Greek farm-house		Mold fragment for antefix of Io	Arch. TC?	First half of the 4th c. BC		Carter in *Survey*, 709, n. 77; Ammerman in *Survey*, 538, no. 341
27	Avinella (G9), site 399	Greek farm-house and kiln		Slag		Classical–Hellenistic period		Carter in *Survey*, 610, n. 132; Prieto et al. in *Survey*, 74; Swift with Devereux in *Survey, Vol. IV*, 100
28	Avinella (G9), site 404	Greek farm-house		Mold for relief plaque of reclining banqueter	TC relief plaque?	4th c. BC		Ammerman in *Survey*, 537, no. 340
29	Pantanello	Sanctuary		3 kiln firing supports	BG?	End of the 4th–beginning of the 3rd c. BC	The 3 kiln firing supports were found in residual deposits excavated in 2008 in the area of the Late Hellenistic ceramic workshop.	Swift in *Survey*, 144

Table B2 cont. Production sites from the chora of Metaponto (see also *Fig. 5.2*).

	Site	Type of Settlement	Kilns	Indicators of Production	Types of Production	Chronology	Observations	Bibliography
30	Pantanello	Roman villa rustica?	Large Kiln: circular combustion chamber (int. diam. 3.9 m, l. 5 m) built of fragmentary tiles held by mortar (l. 0.20/0.25 m) and short stocking chamber. The kiln, built into the hillside, was set in an enclosure. Small kiln: circular combustion chamber and perforated floor supported by three arches (ext. diam. 1.5 m), short stocking channel to the S; total l. 2 m.		Tiles, Greco-Italic amphorae, GW	2nd c. BC–Early Imperial period		Carter 1979, 53–63; Carter 1983a, 421–39, 447; De Siena 2005, 448–49; Carter in *Survey*, 463–64, fig. 12.2, 910, fig. 27.11
31	Pisticci, Destra Basento, site 77 Oleodotto, prop. Viggiano	Roman villa	Kiln			2nd–1st c. BC	Excavated by Soprintendenza, 2001 (A. De Siena). The excavations brought to light structures belonging to a farmhouse and of a kiln located nearby.	Nava 2002, 673–75, pl. XXXVIII; De Siena 2005, 448–51, pl. XLV
32	San Biagio	Sanctuary? Farmhouse?	Kiln of unknown type		TC figurines?	Adamesteanu suggests that the kiln should be dated to the Archaic period. Leone (1998) suggests a Hellenistic-Roman chronology. The kiln was cleaned in 1999; no conclusive elements to establish the chronology were detected (A. De Siena, pers. comm.).	The kiln is located 50 m N of the sanctuary.	Adamesteanu 1965, 131; Adamesteanu 1967, 271; Cuomo di Caprio 1971–1972, 456, n. 13; Adamesteanu 1973a, 54; Adamesteanu 1974, 60; San Pietro 1991, 14; Osanna 1992, 80; R. Leone 1998, 123–24; Olbrich 1979, 15; Fischer-Hansen 2000, 105; Lecce 2010–2011, 39
33	Early Cluster Village, Basento valley, F10, site 530	Greek farmhouse and kiln		Slag		Archaic period	The presence of the slag might indicate the presence of a kiln.	Carter in *Survey*, 881; Carter 2012, 1119

Table B2 cont. Production sites from the chora of Metaponto (see also *Fig. 5.2*).

	Site	Type of Settlement	Kilns	Indicators of Production	Types of Production	Chronology	Observations	Bibliography
34	Pisticci, Destra Basento, La Cappella	Roman rural villa	The presence of two kilns is reported.	Tiles, dolia, and transport amphorae		2nd–1st c. BC. The site was occupied until the 4th–5th c. AD	Excavated in 1983 (A. De Siena). The kilns are separated from the nucleus of the villa.	A. Bottini 1983, 470–71; De Siena and Giardino 1994, 208; De Siena 2005, 450–51, pl. XLVI,2
35	Sant'Angelo, Basento valley (E10), site 105	Farmhouse		BG test piece fragment		End of the 4th–beginning of the 3rd c. BC. Ceramic production possibly continued also in the Late Hellenistic/Early Roman period.		Carter in *Survey*, 788; Prieto et al. in *Survey*, 60; Silvestrelli and Swift in *Survey*, 143–44, 213, no. 7; Swift in *Survey*, 37; Swift with Devereux in *Survey, Vol. IV*, 92–93
36	Sant'Angelo, Basento valley (E10), site 163	Greek kiln		Slag		Greek period	The presence of the slag might indicate the presence of a kiln. A clay bed was noted 40 m NW of the site.	Prieto et al. in *Survey*, 63
37	Sant'Angelo, Basento valley (D10), site 213	Greek farmhouse		Slags	Ceramic production?	5th–3rd c. BC	The presence of the slag might indicate the presence of a kiln.	Carter in *Survey*, 682; Prieto et al. in *Survey*, 65
38	Basento valley (D10), site 153	Greek farmhouse	Waster		Metapontine Greco-Italic amphorae	4th–3rd c. BC		Swift in *Survey* a, 463–64, 488, no. 59; Prieto et al. in *Survey*, 63
39	Sant'Angelo Grieco-Nuovo (D10), site 1	Greek farmhouse, Late Republican–Early Imperial farmhouse	Circular combustion chamber with central wall supporting the perforated floor, not preserved (Cuomo di Caprio 1971–1972, Type 1/b)			Unknown; De Siena (2005, 449) suggests that the kiln should date to the 2nd–1st c. BC	Excavated in 1981 (University of Texas at Austin and Soprintendenza for Basilicata)	Carter 1983b, 16–18; Carter in *Survey*, 611, fig. 18.25; Prieto et al. in *Survey*, 56; Swift with Devereux in *Survey, Vol. IV*, 88; Lanza Catti in *Fattoria Fabrizio*, 410
40	Sant'Angelo (D10), site 9	Greek kiln		Slag		Classical period	The presence of the slag might indicate the presence of a kiln.	Prieto et al. in *Survey*, 57
41	Sant'Angelo (C9), site 12	Greek farmhouse and kiln		Slags	Ceramic production?	Uncertain, possibly middle of the 6th c. BC	A spring providing water is present near the site.	Carter in *Survey*, 611, fig. 18.24, 646; Swift with Devereux in *Survey, Vol. IV*, 89
42	La Cannala, site 1045	Farmhouse, kiln, and tomb	Kiln			Roman period	Identified during the 2007 Basento-Cavone survey.	D'Annibale 2007, 63
43	La Cannala, site 1062		Kiln			Greek period	Identified in the 2007 Basento-Cavone survey.	D'Annibale 2007, 63

Table B2 cont. Production sites from the chora of Metaponto (see also *Fig. 5.2*).

	Site	Type of Settlement	Kilns	Indicators of Production	Types of Production	Chronology	Observations	Bibliography
44	San Vito, site 26	Ceramic kiln				Unknown	Identified during the 1981 Basento-Cavone survey; unpublished	
45	San Vito, site 28	Kiln				Unknown	Identified during the 1981 Basento-Cavone survey; unpublished	
46	Pisticci, loc. Torretta	Rural settlement	The kiln, partially destroyed by the bulldozer at the moment of discovery, was set in a natural cavity of the bedrock. A rectangular combustion chamber (l. 4.30 m, w. 3.40 m) and double stocking channel built in mudbricks (l. 2.50 m) were preserved.	Production discards, kiln firing supports from the deposit filling the kiln	BG, BW, PW	First half of the 5th c. BC	Excavated in 1992 (A. De Siena)	A. Bottini 1988, 1–18; A. Bottini 1993, 705–6; Cracolici 2003, 97–98; Lecce 2010–2011
47	Fosso del Lavandaio, near Castello di San Basilio, site 829, Area C	Farmhouse	Kiln (l. 2.20 m)?	Lumps of vitrified clay		The site was active from the Classical to the Hellenistic Period.	The site is adjacent to a tributary of the Cavone River.	Prieto 2003, 72; D. Davis 2003, 78–80, fig. 4; D. Davis et al. 2003, 11–15, fig. 9–11, 14; Lecce 2010–2011, 39–40
48	Masseria Durante	Rural complex		Loom weight wasters		The complex was active between the end of the 3rd c. BC and the early decades of the 3rd c. AD. The wasters should be dated to the 3rd–2nd c. BC.		Giardino in *San Biagio*, 5–7

Table B2 cont. Production sites from the chora of Metaponto (see also *Fig. 5.2*).

Appendix C
Archaeobotanical Analysis: Pollen and NPPs

Assunta Florenzano

The results of microscopic analyses from Sant'Angelo Vecchio are extraordinarily rich, and a long list of pollen and Non-Pollen Palynomorphs (NPPs) types has been identified. The data have been organized into four tables that are presented in this appendix. *Table C1* to *Table C4* present pollen data while *Table C5* and *Table C6* report detailed NPPs data. They were the analytical basis of the palaeoenvironmental inferences reported above.[1]

Tables C1 and C2. Pollen spectra of samples collected from modern surface soil and from Pollen Sequence 2 are reported in *Table C1*, while those from Pollen Sequences 8, 9, and 10 are reported in *Table C2*. In the first two columns, tables show taxa (botanical family and genus/species or pollen type) in alphabetical order. Then, each column reports the pollen concentration (p/g) of the sample, and the percentage values per each taxon. This is repeated in the columns per all the analyzed samples. Concentrations vary from 1,537 p/g (P22) to 19,871 p/g (modern surface soil P0). The number of taxa is high, consisting of 152 different types, ranging from 22 (P21) to 69 (P12) taxa per sample. Arboreal taxa are reported first, in the upper part of the tables. Among them, the trees oak, *Quercus* (mean: 2.2%; up to 3%), olive tree, *Olea* (2.1%; up to 13%), and pine, *Pinus* (1.7%; up to 7%) prevailed, as well as the shrub flax-leaved daphne, *Daphne* cf. *gnidium* (2.4%; up to 13%). Next are the herbaceous plants, which are dominated by Cichorieae (21%; up to 36%) and the wild grass group, Poaceae (14%; up to 22%). In general, the high values of herbs in the pollen spectra testify that an open herb-dominated environment was prevalent in the site and its surroundings. Pollen sum is reported at the bottom of the tables; the number of counted pollen grains varies from 127 (P21) to 680 (P12), depending on the pollen concentration (p/g) and the organic matter content of each sample. The last rows show the number of the observed clusters of pollen (clumps or anthers); in almost all samples pollen grains grouped in clumps have been found, while traces of anthers are observed only in the Pollen Sequence 2 (*Table C1*). They are evidence of the deposition of flowers—harvested or ingested—that became parts of the archaeological layers.

Tables C3 and C4. According to the literature, the main pollen sums useful for environmental and past economy reconstructions have been calculated in *Table C3* and *Table C4*. The first column shows the name/label of the sum, and the second column lists the taxa included in that sum. For each sample, percentage sums indicate natural or human environments. As examples of natural environments, there are mixed oakwood (mean 3%, up to 4% in P18), Mediterranean woods and maquis (6%, up to 16% in P17), and wet environments (7%, up to 10% in P18). The following sums are indicators of increased human activity: LPPI (Local Pastoral Pollen Indicators), 35% and up to 58% in P24; cereals (oat/wheat and barley), 1% and up to 3% in P27; wAI (wild Anthropogenic Indicator), 8% and up to 37% in P23; RHAPI (Regional Human Activities Pollen Indicators), 10% and up to 37% in P14 and P23; and *Juniperus* + Chenopodiaceae, 2% and up to 6% in P19. Altogether, pollen shows that a puzzle of natural and human habitats was strictly interconnected during the lifespan of the Sant'Angelo Vecchio settlement. The "wild" vegetation featured a very low forest cover, presence of Mediterranean shrubland/maquis, and fairly extended wet environments. The data suggest that the natural environment had been already shaped by the human presence in the area. In fact, the pollen sums indicate that intense pastoral activities and broad agrarian practices have highly influenced the landscape surrounding this site since ancient times.

Tables C5 and C6. Concentrations of the main Non-Pollen Palynomorphs (npp/g) are reported in *Table C5* and *Table C6*. The morphotypes (spores of fungi, algae elements, and parasite eggs), identified according to van Geel,[2] are listed in alphabetical order in

[1] See Ch. 8, "Archaeobotanical Analysis."

[2] van Geel 1978, 2001.

the first two columns. Among the 34 types observed, the fungal elements prevailed with the dominance of *Glomus* (mean: 600 npp/g) and coprophilous fungi (216 npp/g). The algal traces are also recurrent findings, and are mainly represented by HdV-181 (173 npp/g). The latter two NPP types give further confirmation that the area was affected by pastoral activities (spores of coprophilous fungi), and that wet habitats were fairly common (algal elements).

Sample Number		**P0**	**P1**	**P2**	**P3**	**P4**	**P5**	**P6**	**P7**	**P8**	**P9**	**P10**	**P11**	**P12**
Archaeological Context		**Surface Soil**	**Rural necropolis, square G6 Pollen Sequence 2**											
Chronology		recent	Post-1st c. BC	3rd c. BC	3rd c. BC	3rd c. BC	3rd c. BC	3rd c. BC	3rd c. BC	3rd c. BC	3rd c. BC	4th–3rd c. BC	5th–4th c. BC	5th–4th c. BC
Depth (cm)		*0*	*12*	*18*	*27*	*34*	*40*	*50*	*58*	*68*	*76*	*83*	*90*	*100*
Concentration (p/g)		*19871*	*18914*	*7664*	*4870*	*2383*	*4134*	*3614*	*4803*	*3746*	*2323*	*2751*	*4303*	*4090*
Trees+shrubs %		*21.4*	*18.6*	*18.3*	*12.1*	*5.9*	*9.1*	*11.1*	*4.9*	*4.7*	*6.9*	*4.4*	*17.9*	*13.7*
Herbaceous plants %		*78.6*	*81.4*	*81.7*	*87.9*	*94.1*	*90.9*	*88.9*	*95.1*	*95.3*	*93.1*	*95.6*	*82.1*	*86.3*
TREES and SHRUBS														
ANACARDIACEAE	*Pistacia lentiscus*	0.2		0.4		0.2	0.4	0.4	0.3		0.2	0.2	0.7	0.1
BETULACEAE	*Alnus*	0.4	0.7			0.2		0.2		0.2	0.2			
CAPRIFOLIACEAE	*Sambucus nigra*											0.5		
CISTACEAE	*Helianthemum*						0.2	2.6		0.2		0.2	0.5	
CORYLACEAE	*Carpinus betulus*	0.2	0.6		0.2	0.7	0.6	0.7		0.7	0.2		0.4	0.4
	Corylus	0.9	0.4	0.4	0.6	0.4	0.2	0.2		0.4	0.4	0.5	0.7	0.4
	Ostrya/Carpinus or. type													0.1
CUPRESSACEAE	*Cupressus*						0.4							
	Juniperus type		0.4		0.4			0.6	0.2					0.6
ELEAGNACEAE	*Hippophae* cf.													0.1
ERICACEAE	*Erica*													0.1
FAGACEAE	*Quercus* deciduous	2.6	2.4	0.4	1.4	1.9	2.3	1.1	1.2	2.0	3.0	1.9	2.3	1.2
	Quercus cerris cf.		0.2	0.6	0.2	0.7		0.4		0.2				0.3
	Q ilex type	2.7	0.4		0.4		0.2		0.2	0.4	0.4		0.9	
MYRTACEAE	*Myrtus*												0.2	0.1
OLEACEAE	*Fraxinus excelsior* type		0.7	0.2									0.4	1.0
	F. ornus type	0.4												
	Fraxinus undiff.				0.6		0.4	0.4						
	Olea	5.1	7.4	13.1	5.5	0.6	0.6	2.0	2.2	0.4	1.5	0.5	4.2	1.8
	Oleaceae undiff.					0.2							0.4	0.4
	Phillyrea	0.2	1.1	1.3	0.6			0.2						

Table C1 Pollen spectra of samples collected from different contexts of Sant'Angelo Vecchio (modern surface soil and Pollen Sequence 2).

Sample Number		P0	P1	P2	P3	P4	P5	P6	P7	P8	P9	P10	P11	P12
Archaeological Context		Surface Soil	Rural necropolis, square G6 Pollen Sequence 2											
PINACEAE	*Pinus halepensis*	0.4	0.2										0.5	0.3
	Pinus undiff.	6.4	2.4	0.7	0.8	0.6	0.4	2.0	0.7		0.4	0.2	3.0	3.2
RHAMNACEAE	*Rhamnus*	0.2												0.1
SALICACEAE	*Populus*	0.2	0.2	0.2	0.4	0.2	0.6		0.2	0.4	0.4	0.2		0.3
	Salix	0.4	0.2	0.7	0.4		0.4	0.2					0.5	
THYMELAEACEAE	*Daphne* cf. *gnidium*	1.3	1.3	0.4	0.4		1.9				0.2	0.2	2.6	2.6
ULMACEAE	*Ulmus*					0.4		0.2			0.2		0.7	0.1
VITACEAE	*Vitis*						0.8							
HERBACEOUS PLANTS														
ALISMATACEAE	*Sagittaria*					0.4	0.6							
APIACEAE	Apiaceae undiff.		0.2			0.6	0.4	0.2	0.3	0.4	0.6	0.5	0.9	0.6
	Anethum cf.			0.4										
	Apium			0.4							0.2			
	Bupleurum						0.2	0.2						0.3
	Eryngium type		0.2											
	Peucedanum			0.2										
	Torilis						0.2		0.2					
	Pimpinella													0.1
ASTERACEAE	*Achillea*							0.4						
	Ambrosia	0.2										0.2		
	Artemisia	2.7	1.1	0.4	0.4	0.4	0.4	0.6	0.3		0.4		0.9	1.0
	Aster type	2.7	1.9	4.5	4.3	4.5	2.1	2.8	3.5	5.1	4.5	4.2	3.5	0.5
	Asteroideae undiff.	3.5	3.9	4.9	7.6	7.4	6.3	6.1	4.4	5.1	5.6	4.7	4.9	3.2
	Bellis perennis			0.2										
	Carduus						0.4	1.8	0.3	0.2			0.4	
	Carthamus / Carduncellus						0.2							
	Centaurea undiff.	0.7	0.6		0.6		0.4	0.2	0.3		1.5			0.3
	Centaurea nigra type	1.5	0.4	0.7	1.2	1.3	1.1	2.0	0.7	0.7	0.9	1.6	1.1	0.4
	Cichorieae	2.9	20.6	21.5	13.3	26.0	26.3	22.9	31.6	35.8	27.0	20.3	15.6	6.3
	Cirsium	0.4	1.1	0.6	0.6	1.3	0.8	1.7	0.3	0.7	0.8		1.4	0.3
	Erigeron									0.2				
	Matricaria type	0.7	0.6		0.4	0.9	0.6		0.2	0.2	0.4		0.4	
BORAGINACEAE	Boraginaceae undiff.				0.2			0.4	0.3		0.4			
	Echium									0.4				

Table C1 cont. Pollen spectra of samples collected from different contexts of Sant'Angelo Vecchio (modern surface soil and Pollen Sequence 2).

Sample Number		P0	P1	P2	P3	P4	P5	P6	P7	P8	P9	P10	P11	P12
Archaeological Context		**Surface Soil**	**Rural necropolis, square G6 Pollen Sequence 2**											
BRASSICACEAE	*Brassica* type	0.9	1.7	2.6	1.0	0.4	0.4	2.6	1.5	0.4	0.6		2.1	0.4
	Brassicaceae undiff.	3.1	4.5	6.5	3.7	0.6	1.3	3.0	1.3	0.7	1.3	0.7	4.2	5.6
	Cardamine		0.2	0.2	0.4									
	Hornungia type	0.9	1.7	1.7	0.8	0.2		0.2		0.2	0.9	0.5		0.1
	Matthiola	3.5	5.8	0.4	1.2	0.2	1.1	0.9	0.7	0.4	0.6	0.7	1.4	0.6
	Sinapis type		0.6											
BUTOMACEAE	*Butomus*			1.1										
CAMPANULACEAE	*Campanula*	0.2												0.1
CANNABACEAE	*Cannabis*									0.2	0.2			
CARYOPHYLLACEAE	Caryophyllaceae undiff.	0.5	0.6	2.2	0.2	0.2	0.9	0.4	0.3	0.7	0.2			0.1
	Herniaria glabra type					0.2								
	Paronychia	0.7	0.2	0.2		0.2	0.2	0.2	0.3				0.5	0.3
	Silene dioica type	0.4		0.4	0.2		0.2							
CHENOPODIACEAE	*Beta* type	0.5	0.2	0.6		0.6			0.3	0.2	0.2		0.5	0.3
	Chenopodiaceae undiff.	0.7	0.6	0.2		0.2	0.4				0.6	0.2	0.2	0.6
	Chenopodium type	1.6	1.5	0.4	1.0	0.4	0.2	0.2	0.3		0.4		0.9	0.7
CISTACEAE	Cistaceae undiff.									0.2		0.5	0.2	
	Cistus	0.5									0.4			0.1
CONVOLVULACEAE	*Convolvulus*	0.2				0.2							0.2	0.1
CRASSULACEAE	*Sedum*										0.2			
CYPERACEAE	Cyperaceae	4.6	3.3	2.1	5.7	5.0	5.9	3.7	4.2	2.4	3.0	5.6	4.6	3.2
	Scirpus												0.2	
DIPSACACEAE	*Scabiosa/ Cephalaria*		0.2									0.2	0.2	
EUPHORBIACEAE	*Euphorbia*		0.4			0.2								
	Euphorbiaceae undiff.	0.2		0.2					0.2		0.2			

Table C1 cont. Pollen spectra of samples collected from different contexts of Sant'Angelo Vecchio (modern surface soil and Pollen Sequence 2).

Sample Number		P0	P1	P2	P3	P4	P5	P6	P7	P8	P9	P10	P11	P12
Archaeological Context		Surface Soil	Rural necropolis, square G6 Pollen Sequence 2											
FABACEAE	*Astragalus*	0.4					0.4	0.6	0.5	0.4	0.6		0.4	0.6
	Dorycnium type						0.2							
	Fabaceae undiff.	2.6	0.6	0.7	1.2	0.9	1.7	0.4	1.0	0.2	2.1	2.1	0.9	1.8
	Lathyrus type									0.2				
	Lotus	0.7	1.1.	0.4	0.6	1.5	0.8	0.7		1.1	0.4	1.4	1.2	0.4
	Medicago cf.						0.2						0.2	
	Melilotus	0.4	0.4		0.2		0.2		0.2	0.4		0.2		
	Ononis type													0.3
	Tetragonolobus purpureus cf.		0.2	0.2										
	Trifolium undiff.	4.6	1.7	1.1	0.4	0.9	0.6	1.7	1.3	1.5	1.9	2.3	0.5	2.9
	Trifolium pratense type	1.6	2.8	4.5	2.3	0.6	0.8		1.7	2.0	1.3	2.6	1.4	0.4
LAMIACEAE	Lamiaceae undiff.	0.4			0.4		0.2	0.4		0.5	0.4			
	Mentha type		0.7	0.4	0.2	0.2		1.1		0.2			0.7	0.7
	Stachys type	0.2								0.2				
LEMNACEAE	*Lemna*							0.2		0.2		0.2		
LILIACEAE	*Allium* type	0.2	0.4	0.2	0.2	0.6				0.2		0.5	0.2	0.3
	Asphodelus				0.2	0.4								
	Liliaceae undiff.	0.2		0.2	0.4		0.2	0.6		0.4	0.6		0.2	
	Scilla cf.				0.2									
NYMPHAEACEAE	*Nuphar*									0.2				
	Nymphaea cf. *alba*	0.2		0.4	1.2	2.4	2.3	0.6	0.5	0.4	0.4	0.5	0.4	0.3
PAPAVERACEAE	*Glaucium*									0.2				
	Papaver argemone cf.		0.2											
	Papaver rhoeas cf.	0.5							0.2					
	Papaveraceae undiff.		0.2		0.2	0.2					0.2			
PLANTAGINACEAE	*Plantago* undiff.	2.9		0.2	0.4	0.9	0.6	0.7	1.2	0.9	0.4	1.6	1.1	1.9
	Plantago lanceolata		0.7	0.2		1.7	1.9	1.8	1.2	0.4	1.5	1.4	1.2	0.9

Table C1 cont. Pollen spectra of samples collected from different contexts of Sant'Angelo Vecchio (modern surface soil and Pollen Sequence 2).

Sample Number		P0	P1	P2	P3	P4	P5	P6	P7	P8	P9	P10	P11	P12
Archaeological Context		Surface Soil	Rural necropolis, square G6 Pollen Sequence 2											
POACEAE	*Avena/Triticum* group	0.4	0.4	0.7	0.8	1.1	1.5	1.1	0.7	1.5	1.3	0.9	0.2	0.3
	Hordeum group	0.4				0.4								
	Phragmites australis cf.	0.5									0.9	0.2	1.6	0.3
	Poaceae wild grass group	11.1	8.9	8.0	17.8	17.6	14.4	13.8	19.0	16.9	12.8	21.7	13.5	15.7
POLYGONACEAE	*Polygonum persicaria* cf.		0.2											
	Rumex	1.5	0.2	0.2		0.2	0.8		0.2		0.4	0.5	0.2	
PRIMULACEAE	*Anagallis*												0.2	
RANUNCULACEAE	*Adonis*						0.2							
	Ranuncula-ceae undiff.	3.6	1.9	1.7	2.5	2.8	1.7	2.2	3.4	2.2	2.3	2.6	2.	1.8
	Ranunculus cf. *fluitans*					0.2				0.5				
	Ranunculus type	0.7	0.4			0.9	0.4	0.4	0.5	0.7	0.2		0.2	0.4
ROSACEAE	*Alchemilla/ Aphanes* type	0.9	0.4	0.2		0.2		1.3	0.3		0.2			
	Filipendula	0.5					0.2	0.4			1.7		0.2	0.1
	Rosaceae undiff.	0.7	0.2		0.2	0.7	0.4	0.6	0.3	0.4	1.1	0.7	0.2	0.1
RUBIACEAE	*Galium* type		0.2					0.2	0.5	0.5			0.4	
SAXIFRAGACEAE	*Saxifraga* cf.			0.2									0.2	
SCROPHULARIA-CEAE	*Digitalis* type						0.2						0.2	
	Euphrasia type						0.2		0.2	0.2				0.1
	Linaria type	0.5	0.2	0.2	0.2		0.2	0.4			0.2	1.2	0.4	1.3
	Scrophularia type	0.9	0.2						1.0		0.8	0.5		1.0
	Scrophularia-ceae undiff.	2.0	1.5	1.7	2.3	0.6	1.1	1.5	1.2	0.7	0.4	1.4	1.8	13.4
	Verbascum cf.	1.1	0.7	0.7	0.6	0.4	0.4	0.2	0.2	0.2	0.4	0.2	1.4	1.0

Table C1 cont. Pollen spectra of samples collected from different contexts of Sant'Angelo Vecchio (modern surface soil and Pollen Sequence 2).

Sample Number		P0	P1	P2	P3	P4	P5	P6	P7	P8	P9	P10	P11	P12
Archaeological Context		**Surface Soil**	**Rural necropolis, square G6 Pollen Sequence 2**											
SOLANACEAE	*Solanum*										0.2			
TYPHACEAE	*Typha/Sparganium* type			0.4	0.4	0.4			0.3	0.2		0.7	0.5	0.7
	Typha latifolia type									0.2				
URTICACEAE	*Urtica dioica* type	0.9	0.6	0.2	1.4	0.7	0.9	0.9	0.5	2.0	1.9	0.5		1.2
	Urtica membranacea		0.4		0.4	0.6		0.2	0.2	0.2		0.2	0.2	0.1
	Urtica cf. *pilulifera*			0.2	1.6	0.2	1.3	0.4	1.2	1.5	1.1	1.4	0.2	1.5
Unknown		2.2	2.2	2.2	3.9	2.6	3.6	3.5	3.7	2.7	3.9	6.1	3.5	6.6
Undeterminable pollen		1.8	2.6	4.3	4.5	3.3	3.2	3.1	2.2	1.5	3.0	4.2	2.6	3.2
Pollen Sum		**548**	**539**	**535**	**488**	**539**	**528**	**542**	**594**	**550**	**533**	**429**	**569**	**680**
Number of Taxa		**66**	**65**	**58**	**56**	**59**	**65**	**59**	**53**	**61**	**62**	**46**	**66**	**69**
Number of Pollen Clumps														
< 5 pollen grains		2	2	9	3	1	5		3	3	4	2	7	12
between 5 and 10 pollen grains				4				1			2	1	2	
>10 pollen grains								1				1	1	2
cluster of broken grains								3			1			5
Number of anthers			**1**	**4**	**2**	**1**						**1**		

Table C1 cont. Pollen spectra of samples collected from different contexts of Sant'Angelo Vecchio (modern surface soil and Pollen Sequence 2).

Sample Number		P13	P14	P15	P16	P17	P18	P19	P20	P21	P22	P23	P24	P25	P26	P27	P28
Archaeological Context		**Retaining Wall, square C4 Pollen Sequence 10**						**Slope 3, square D4 Pollen Sequence 8**					**Slope 3, square D4 Pollen Sequence 9**				
Chronology		6th–5th c. BC	6th–5th c. BC	6th–5th c. BC	6th–5th c. BC	6th–5th c. BC	6th–5th c. BC	4th–3rd c. BC/2nd–1st c. BC	4th–3rd c. BC/2nd–1st c. BC	4th–3rd c. BC/2nd–1st c. BC	End of the 4th–first half of the 3rd c. BC	End of the 4th–first half of the 3rd c. BC	4th–3rd c. BC/2nd–1st c. BC	4th–3rd c. BC/2nd–1st c. BC	4th–3rd c. BC/2nd–1st c. BC	End of the 4th–first half of the 3rd c. BC	End of the 4th–first half of the 3rd c. BC
Depth (cm)		*8*	*15*	*20*	*25*	*35*	*42*	*30*	*50*	*70*	*90*	*115*	*30*	*50*	*70*	*90*	*115*
Concentration (p/g)		*5430*	*14156*	*5679*	*14083*	*9846*	*4743*	*12185*	*3112*	*2045*	*1537*	*5153*	*1555*	*3376*	*2120*	*1816*	*3423*
Trees+shrubs %		*25*	*10.7*	*18.6*	*15*	*28.4*	*17.6*	*11.7*	*6.3*	*4.7*	*8.9*	*11*	*4.3*	*5.2*	*9.3*	*9.3*	*22.7*
Herbaceous plants %		*75*	*89.3*	*81.4*	*85*	*71.6*	*82.4*	*88.3*	*93.7*	*95.3*	*91.1*	*89*	*95.7*	*94.8*	*90.7*	*90.7*	*77.3*
TREES and SHRUBS																	
ANACARDIACEAE	*Pistacia lentiscus*		0.3					0.3				0.5				0.4	
	Schinus cf.			0.6											0.5		
BETULACEAE	*Alnus*	1.0		0.6							0.5			0.5		0.4	0.7
CAPRIFOLIACEAE	*Sambucus nigra*			0.3													
CISTACEAE	*Helianthemum*	0.3					0.3					0.3					
CORYLACEAE	*Carpinus betulus*	0.5	0.3		0.3		0.7		0.4							0.4	0.3
	Corylus	0.5	1.0	0.9		0.3	1.0	0.6	1.1	0.8	0.9	0.5		0.5		0.4	
	Ostrya/Carpinus or. type	0.3	0.3	0.3			0.3				0.5			0.5	0.5		
CUPRESSACEAE	*Cupressus*			0.3													
	Juniperus type	1.6	1.0	1.9	0.3		1.4	2.2			0.5						
ELEAGNACEAE	*Hippophae* cf.					0.3											
ERICACEAE	*Erica*	0.3									0.5						
FAGACEAE	*Quercus* deciduous	0.8	1.6	2.2	1.9	1.5	2.4	2.2	0.7	2.4	2.3	1.0	0.9	2.1	2.8	1.6	2.1
	Q. ilex type	1.0	0.3	0.3			1.0		0.7		0.5	0.8	0.9		0.5		1.0
JUGLANDACEAE	*Juglans*				0.3	0.3											
MYRTACEAE	*Myrtus*	0.8									0.9	3.9					0.3

Table C2 Pollen spectra of samples collected from Pollen Sequences 10, 8, and 9.

Sample Number		P13	P14	P15	P16	P17	P18	P19	P20	P21	P22	P23	P24	P25	P26	P27	P28
Archaeological Context		**Retaining Wall, square C4 Pollen Sequence 10**						**Slope 3, square D4 Pollen Sequence 8**					**Slope 3, square D4 Pollen Sequence 9**				
OLEACEAE	*Fraxinus excelsior* type	1.3	1.0	0.9	1.9	2.1	1.7	0.3	0.4								
	F. ornus type				0.6												
	Fraxinus undiff.	1.8									0.5						0.7
	Olea	3.1		0.6	0.6	2.7	0.3		0.4		0.5	0.3	0.9	1.0	2.8	2.0	3.5
	Oleaceae undiff.	1.0			0.6												
	Phillyrea	0.5															
PINACEAE	*Pinus*	6.0	2.3	3.1	2.2	7.0	2.0	1.3	0.4	0.8		0.8	0.5	0.5	0.5	1.2	3.8
RANUNCULA-CEAE	*Clematis*	0.3															
RHAMNACEAE	*Rhamnus*								0.4								
SALICACEAE	*Populus*			0.6	0.3	0.6	1.0		1.5	0.8	0.9	0.5	0.5		0.5		
	Salix				0.9	0.6	0.3					0.3	0.5				
TAXACEAE	*Taxus*		0.3	0.3													
THYMELAEA-CEAE	*Daphne* cf. *gnidium*	3.9	2.3	5.6	5.0	12.8	5.1	4.7			0.5	2.1			1.4	2.8	10.1
TILIACEAE	*Tilia*								0.4								
HERBACEOUS PLANTS																	
APIACEAE	Apiaceae undiff.	0.5	0.3	0.6	0.6	0.9	0.3	0.6				0.3				0.4	0.3
	Apium							0.3	0.7	0.8		0.3					0.3
	Bupleurum	0.3				0.3		0.3								0.4	0.3
	Echinophora tenuifolia cf.					0.6											
	Eryngium type	0.3			0.3	0.3	0.7										
	Oenanthe cf.			0.3													
	Peucedanum		0.6				0.7		0.4								
	Torilis	1.0			0.3												0.3

Table C2 cont. Pollen spectra of samples collected from Pollen Sequences 10, 8, and 9.

Sample Number		P13	P14	P15	P16	P17	P18	P19	P20	P21	P22	P23	P24	P25	P26	P27	P28
Archaeological Context		**Retaining Wall, square C4 Pollen Sequence 10**						**Slope 3, square D4 Pollen Sequence 8**					**Slope 3, square D4 Pollen Sequence 9**				
ASTERACEAE	*Ambrosia*		0.3														
	Artemisia	1.8	1.0	0.3	2.5	1.8		0.6				0.3			1.9	2.8	1.7
	Aster type	1.0	1.0	2.2	0.3	0.3	0.7	1.6	3.3		2.3		9.0	5.7	3.7	2.8	3.8
	Asteroideae undiff.	3.6	2.6	3.7	4.1	1.5	2.7	3.8	4.8	7.9	4.2	3.9	8.5	7.3	8.8	5.7	4.2
	Bellis perennis									0.8		3.1					
	Carduus										0.5						
	Centaurea undiff.																0.3
	Centaurea nigra type	0.3		1.2							0.5	1.0	0.9		0.5	0.8	
	Cichorieae	4.9	5.2	10.6	9.4	7.9	7.5	16.7	29.7	36.2	35.2	11.8	32.2	34.4	30.2	23.1	26.6
	Cirsium	0.5	0.3	1.2	0.6	1.2	0.3	0.6	0.4	0.8			1.9		0.9		0.7
BORAGINACEAE	Boraginaceae undiff.						0.7										
	Symphytum								0.4								0.3
BRASSICACEAE	*Brassica* type		1.3	1.9	0.6		1.4						0.5				2.1
	Brassicaceae undiff.	4.7	3.9	3.7	6.9	4.9	7.1	4.1	2.2	2.4	0.9	2.1	2.4	1.6	2.3	1.2	4.2
	Hornungia type	0.8		0.3	0.3	0.6		0.3			0.5			0.5	2.3		
	Matthiola	1.8	1.6	3.4	1.3	2.4	0.7	0.6	1.1	0.8	0.5	0.3			0.5		
BUTOMACEAE	*Butomus*				0.3												
CAMPANULACEAE	*Campanula*								0.4								
CARYOPHYLLACEAE	Caryophyllaceae undiff.	0.5			0.6		1.0	0.6			0.5	0.5		0.5	0.5	1.2	
	Paronychia				0.6	0.3		0.3									
CHENOPODIACEAE	*Beta* type	0.8		0.9			0.3	0.6			1.4			0.5		0.4	0.3
	Chenopodiaceae undiff.	0.5		0.9	0.6	1.5	0.3	1.9	1.5	3.1	1.9			3.1	1.4		
	Chenopodium type	1.0	1.3	2.2	1.3	0.6	0.3	1.3	0.4		0.9	0.5		0.5	0.9	0.4	0.3
CISTACEAE	*Cistus*						0.7	0.3									
CONVOLVULACEAE	*Convolvulus*			0.6			0.3										0.3
	Cuscuta cf.						0.3										

Table C2 cont. Pollen spectra of samples collected from Pollen Sequences 10, 8, and 9.

Sample Number		P13	P14	P15	P16	P17	P18	P19	P20	P21	P22	P23	P24	P25	P26	P27	P28
Archaeological Context		Retaining Wall, square C4 Pollen Sequence 10						Slope 3, square D4 Pollen Sequence 8					Slope 3, square D4 Pollen Sequence 9				
CYPERACEAE	Cyperaceae	5.2	5.2	4.3	4.4	4.3	8.1	6.6	5.2	3.9	5.6	3.1	4.7	4.2	4.7	5.3	2.4
	Scirpus			0.3	0.3												
DIPSACACEAE	Dipsacaceae undiff.	0.3															
	Scabiosa/ Cephalaria						0.3									0.4	
EUPHORBIACEAE	*Euphorbia*			0.6	0.6	0.6						0.3					0.3
	Euphorbiaceae undiff.					0.3		1.3									
	Mercurialis							0.3									
FABACEAE	*Astragalus*	0.3									0.5						
	Fabaceae undiff.	1.3	1.9	0.6	0.6	2.1	1.0	1.3	1.9	2.4	1.9	1.3			1.4	0.8	1.0
	Lathyrus type						0.3			0.8							
	Lotus	1.0	1.0	1.6	1.3	0.6	1.7	0.6		1.6	1.4	0.3	0.5			0.8	0.3
	Melilotus	0.3			0.3							0.3			0.5		0.3
	Tetragonolobus purpureus cf.				0.3												
	Trifolium undiff.	1.3	1.0	2.5	0.9	2.1	1.0	2.8	2.2	1.6	0.9	1.3				0.4	0.7
	Trifolium pratense type	1.0	1.6	0.9	0.6			0.6	1.1			0.5	1.4	1.6	1.4	0.4	1.0
GERANIACEAE	Geraniaceae		0.6														
LAMIACEAE	Lamiaceae undiff.				0.3		0.7									0.4	0.3
	Mentha type					0.3	0.3	0.6			0.5			0.5	0.5		
	Sideritis						0.3										
LEMNACEAE	*Lemna*					0.6	0.3						0.5			0.4	
LILIACEAE	*Allium* type		0.3		0.3				0.4			0.5					
	Asphodelus											0.3					
	Liliaceae undiff.	0.3										0.3					0.3
NYMPHAEACEAE	*Nymphaea* cf. *alba*		0.6	1.6	0.3								0.5	0.5	0.9		0.3
PAPAVERACEAE	Papaveraceae					0.9		0.6	0.4			0.3					

Table C2 cont. Pollen spectra of samples collected from Pollen Sequences 10, 8, and 9.

Sample Number		P13	P14	P15	P16	P17	P18	P19	P20	P21	P22	P23	P24	P25	P26	P27	P28
Archaeological Context		Retaining Wall, square C4 Pollen Sequence 10						Slope 3, square D4 Pollen Sequence 8					Slope 3, square D4 Pollen Sequence 9				
PLANTAGINACEAE	*Plantago* undiff.	1.3	1.3	0.6	0.9	0.3		1.6					1.4		0.		
	Plantago lanceolata	1.0	1.3	0.3	0.9	0.3		1.3	1.5			1.0		1.0	0.5	3.6	3.5
POACEAE	*Avena/ Triticum* group	0.5		0.3	0.9	0.3	0.3	1.3				1.0	0.5			2.8	0.7
	Hordeum group															0.4	
	Phragmites australis cf.		0.3	1.2	0.9	0.6		0.9	0.7			0.8		1.6	1.4		0.7
	Poaceae wild grass group	16.4	11.0	16.5	18.8	7.9	18.0	11.4	14.9	15.0	15.5	11.8	16.6	13.0	11.2	20.6	6.6
POLYGONACEAE	*Polygonum aviculare* type	0.3	0.3						0.4								
	Rumex	0.5			0.3	0.9	0.3										
POTAMOGETONACEAE	*Potamogeton*						0.3										
PRIMULACEAE	Primulaceae				0.3												
RANUNCULACEAE	Ranunculaceae undiff.	4.4	2.3		3.1	1.5	4.4	3.2	1.9	5.5	1.9	1.3	5.2	5.7	1.4	1.2	1.0
	Ranunculus type	0.8					0.3				0.9		0.5		0.5	0.8	
ROSACEAE	*Alchemilla/ Aphanes* type		0.3			0.6									0.5		
	Filipendula		0.3		0.3	0.3			0.4		0.5	1.0					
	Potentilla	0.3	0.3					0.3									
	Rosaceae undiff.	0.5				0.9	0.3	0.6			0.5	0.3					
RUBIACEAE	*Galium* type	0.3	0.3					0.6	0.4					0.5		0.4	0.3

Table C2 cont. Pollen spectra of samples collected from Pollen Sequences 10, 8, and 9.

Sample Number		P13	P14	P15	P16	P17	P18	P19	P20	P21	P22	P23	P24	P25	P26	P27	P28
Archaeological Context		**Retaining Wall, square C4 Pollen Sequence 10**						**Slope 3, square D4 Pollen Sequence 8**					**Slope 3, square D4 Pollen Sequence 9**				
SCROPHULARIACEAE	*Euphrasia* type	0.3	0.3														
	Linaria type	0.3				0.3		0.3		0.8			0.5		0.9		
	Scrophularia type		1.3										0.5	1.6			0.3
	Scrophulariaceae undiff.	2.6		2.8	1.9	4.0	2.0	2.8	1.5		0.5	0.8		3.1	1.4	1.6	1.7
	Verbascum cf.	0.5	0.6	1.2	0.3	1.2	0.3									0.8	
SOLANACEAE	*Solanum*																0.3
TYPHACEAE	*Typha/Sparganium* type	0.3	1.3	0.9	0.6	1.2			1.1		0.5				0.5		1.0
	Typha latifolia type														0.9		
URTICACEAE	*Urtica dioica* type	0.5	0.3	0.6	0.3		0.3		0.7		0.5	0.3				0.4	
	Urtica membranacea	2.6	31.7	3.7	7.2	4.3	5.4	0.9	0.4	0.8	0.5	34.1				0.4	0.3
	Urtica cf. *pilulifera*	1.0	0.3	1.2	0.6	1.8	0.7	0.6	0.4	0.8	1.4		0.9	1.0			
VALERIANACEAE	Valerianaceae	0.3					0.3										
Unknown		2.9	2.3	3.7	4.7	4.9	6.8	9.5	7.8	5.5	5.2	2.6	3.8	4.2	4.2	5.2	3.8
Undeterminable pollen		2.1	1.6	1.6	1.6	3.0	2.0	3.5	5.2	3.9	3.3	1.3	2.8	2.1	3.3	4.0	2.8
Pollen Sum		**384**	**309**	**322**	**319**	**328**	**295**	**317**	**269**	**127**	**213**	**381**	**211**	**192**	**215**	**247**	**286**
Number of Taxa		**64**	**48**	**50**	**56**	**50**	**54**	**46**	**39**	**22**	**40**	**43**	**26**	**27**	**37**	**38**	**46**
Number of Pollen Clumps																	
< 5 pollen grains		2		1	1	5	4	1						1	1	2	1
>10 pollen grains																	1

Table C2 cont. Pollen spectra of samples collected from Pollen Sequences 10, 8, and 9.

Sample Number		P0	P1	P2	P3	P4	P5	P6	P7	P8	P9	P10	P11	P12
Archaeological Context		**Surface Soil**	**Rural necropolis, square G6 Pollen Sequence 2**											
Concentration (p/g)		*19871*	*18914*	*7664*	*4870*	*2383*	*4134*	*3614*	*4803*	*3746*	*2323*	*2751*	*4303*	*4090*
Trees %		*18.6*	*15.4*	*15.9*	*10.0*	*5.2*	*5.7*	*7.2*	*4.4*	*4.2*	*6.2*	*2.8*	*12.8*	*8.8*
Shrubs + lianas %		*2.7*	*3.1*	*2.4*	*2.1*	*0.7*	*3.4*	*3.9*	*0.5*	*0.5*	*0.7*	*1.6*	*5.1*	*4.9*
Herbaceous plants %		*78.7*	*81.5*	*81.7*	*87.9*	*94.1*	*90.9*	*88.9*	*95.1*	*95.3*	*93.1*	*95.6*	*82.1*	*86.3*
Pollen Sum		*548*	*539*	*535*	*488*	*539*	*528*	*542*	*594*	*550*	*533*	*429*	*569*	*680*
Number of Taxa		*66*	*65*	*58*	*56*	*59*	*65*	*59*	*53*	*61*	*62*	*46*	*66*	*69*
Mixed oakwood	*Carpinus betulus*, *Corylus*, *Ostrya/Carpinus orientalis* type, deciduous *Quercus*	3.6	3.5	1.3	2.5	3.7	3.0	2.4	1.2	3.3	3.6	2.3	3.3	2.5
Mediterranean plants	*Cistus*, *Daphne* cf. *gnidium*, *Erica*, *Helianthemum*, *Juniperus* type, *Myrtus*, *Olea*, *Phillyrea*, *Pinus halepensis*, *Pistacia* cf. *lentiscus*, *Q. ilex* type, *Rhamnus*	10.6	10.8	15.1	7.4	0.7	3.2	5.7	2.9	0.9	2.6	1.2	9.7	6.0
Juniperus + Chenopodiaceae	*Beta* type, Chenopodiaceae undiff., *Chenopodium* type, *Juniperus* type	2.9	2.6	1.1	1.4	1.1	0.6	0.7	0.8	0.2	1.1	0.2	1.6	2.2
Wet environments	*Alnus*, *Butomus*, Cyperaceae, *Lemna*, *Nuphar*, *Nymphaea*, *Phragmites australis* cf., *Populus*, *Potamogeton*, *Ranunculus* cf. *fluitans*, *Sagittaria*, *Salix*, *Typha/Sparganium*, *Typha latifolia* type, *Ulmus*	6.2	4.5	4.9	8.2	9.1	9.7	5.0	5.2	4.5	5.1	7.5	8.4	5.0
LPPI—Local Pastoral Pollen Indicators	Asteraceae (including Cichorieae), *Galium* type, *Potentilla* type, Ranunculaceae	19.7	32.5	34.4	30.9	45.5	40.7	41.1	46.1	51.5	43.5	33.6	30.9	15.3
Cereals	*Avena/Triticum* group, *Hordeum* group	0.7	0.4	0.7	0.8	1.5	1.5	1.1	0.7	1.5	1.3	0.9	0.2	0.3
wAI—wild Anthropogenic Indicators	*Alchemilla/Aphanes* type, *Anagallis*, *Artemisia*, *Centaurea nigra* type, *Chenopodium* type, *Convolvulus*, *Galium* type, *Mercurialis*, *Papaver rhoeas* type, *Plantago lanceolata* type, *Rumex*, *Urtica dioica* type, *Urtica membranacea*, *Urtica pilulifera*	9.7	5.4	2.4	6.1	5.6	6.6	7.6	5.6	5.3	6.8	5.6	5.1	5.9
RHAPI—Regional Human Activities Pollen Indicators	*Artemisia*, *Avena/Triticum* group, *Hordeum* group, Chenopodiaceae, *Plantago lanceolata* type, *Rumex*, *Urtica dioica* type, *Urtica membranacea*, *Urtica pilulifera*	8.8	5.6	3.2	6.1	7.2	8.0	5.9	6.1	6.5	8.1	6.8	5.4	8.5

Table C3 Main pollen sums useful for environmental and past economy reconstructions at Sant'Angelo Vecchio, surface soil and Pollen Sequence 2.

Sample Number		P13	P14	P15	P16	P17	P18	P19	P20	P21	P22	P23	P24	P25	P26	P27	P28
Archaeological Context		**Retaining Wall, square C4 Pollen Sequence 10**						**Slope 3, square D4 Pollen Sequence 8**					**Slope 3, square D4 Pollen Sequence 9**				
Concentration (p/g)		*5430*	*14156*	*5679*	*14083*	*9846*	*4743*	*12185*	*3112*	*2045*	*1537*	*5153*	*1555*	*3376*	*2120*	*1816*	*3423*
Trees %		*15.9*	*6.1*	*9.3*	*9.1*	*14.9*	*9.8*	*3.8*	*4.8*	*3.9*	*5.6*	*3.7*	*4.3*	*4.7*	*7.4*	*5.7*	*12.2*
Shrubs + lianas %		*9.1*	*4.5*	*9.3*	*6.0*	*13.5*	*7.8*	*7.9*	*1.5*	*0.8*	*3.3*	*7.3*	*0.0*	*0.5*	*1.9*	*3.6*	*10.5*
Herbaceous plants %		75.0	89.4	81.4	84.9	71.6	82.4	88.3	93.7	95.3	91.1	89.0	95.7	94.8	90.7	90.7	77.3
Pollen Sum		384	309	322	319	328	295	317	269	127	213	381	211	192	215	247	286
Number of Taxa		64	48	50	56	50	54	46	39	22	40	43	26	27	37	38	46
Mixed oak-wood	*Carpinus betulus*, *Corylus*, *Ostrya/Carpinus orientalis* type, deciduous *Quercus*	2.1	3.2	3.4	2.2	1.8	4.4	2.8	2.2	3.1	3.8	1.6	0.9	3.1	3.3	2.4	2.4
Mediterranean plants	*Cistus*, *Daphne* cf. *gnidium*, *Erica*, *Helianthemum*, *Juniperus* type, *Myrtus*, *Olea*, *Phillyrea*, *Pinus halepensis*, *Pistacia* cf. *lentiscus*, *Q. ilex* type, *Rhamnus*	11.5	3.9	8.4	6.0	15.5	8.8	7.6	1.5	0.0	3.3	7.9	1.9	1.0	4.7	5.3	15.0
Juniperus + Chenopodiaceae	*Beta* type, Chenopodiaceae undiff., *Chenopodium* type, *Juniperus* type	3.9	2.3	5.9	2.2	2.1	2.4	6.0	1.9	3.1	4.7	0.5	0.0	4.2	2.3	0.8	0.7
Wet environments	*Alnus*, *Butomus*, Cyperaceae, *Lemna*, *Nuphar*, *Nymphaea*, *Phragmites australis* cf., *Populus*, *Potamogeton*, *Ranunculus* cf. *fluitans*, *Sagittaria*, *Salix*, *Typha/Sparganium*, *Typha latifolia* type, *Ulmus*	6.5	7.4	9.6	8.2	7.9	10.2	7.6	8.6	4.7	7.5	4.7	6.6	6.8	8.8	6.1	5.2

Table C4 Main pollen sums useful for environmental and past economy reconstructions at Sant'Angelo Vecchio, Pollen Sequences 10, 8, and 9.

Sample Number		P13	P14	P15	P16	P17	P18	P19	P20	P21	P22	P23	P24	P25	P26	P27	P28
Archaeological Context		Retaining Wall, square C4 Pollen Sequence 10						Slope 3, square D4 Pollen Sequence 8					Slope 3, square D4 Pollen Sequence 9				
LPPI—Local Pastoral Pollen Indicators	Asteraceae (including Cichorieae), *Galium* type, *Potentilla* type, Ranunculaceae	18.0	13.3	19.3	20.1	14.3	15.9	27.4	40.5	51.2	45.5	21.5	58.3	53.6	47.9	37.7	38.8
Cereals	*Avena/Triticum* group, *Hordeum* group	0.5	0.0	0.3	0.9	0.3	0.3	1.3	0.0	0.0	0.0	1.0	0.5	0.0	0.0	3.2	0.7
wAI—wild Anthropogenic Indicators	*Alchemilla/Aphanes* type, *Anagallis*, *Artemisia*, *Centaurea nigra* type, *Chenopodium* type, *Convolvulus*, *Galium* type, *Mercurialis*, *Papaver rhoeas* type, *Plantago lanceolata* type, *Rumex*, *Urtica dioica* type, *Urtica membranacea*, *Urtica pilulifera*	9.1	36.6	9.6	13.2	10.4	7.1	5.7	3.7	1.6	3.8	37.3	1.9	3.1	4.2	8.9	6.3
RHAPI—Regional Human Activities Pollen Indicators	*Artemisia*, *Avena/Triticum* group, *Hordeum* group, Chenopodiaceae, *Plantago lanceolata* type, *Rumex*, *Urtica dioica* type, *Urtica membranacea*, *Urtica pilulifera*	11.7	37.2	11.2	15.7	11.9	8.1	10.1	4.8	4.7	6.6	37.3	2.8	6.3	5.6	11.3	7.0

Table C4 cont. Main pollen sums useful for environmental and past economy reconstructions at Sant'Angelo Vecchio, Pollen Sequences 10, 8, and 9.

Sample Number		P0	P1	P2	P3	P4	P5	P6	P7	P8	P9	P10	P11	P12
Archaeological Context		Surface Soil	Rural necropolis, square G6 Pollen Sequence 2											
Taxa	Types													
Fungal Spores														
Cercophora type	HdV-112	0	0	29	0	13	0	20	16	0	4	6	0	12
Chaetomium	HdV-7B	0	211	344	70	22	117	80	186	14	52	64	106	54
Coniochaeta cf. *lignaria*	HdV-172	73	211	14	40	9	78	100	178	143	48	32	61	36
Delitschia type		0	35	0	0	0	0	0	0	7	4	0	15	0
Diplocladiella type		0	35	0	0	0	0	0	0	0	0	0	23	24
Diporotheca	HdV-143	0	0	0	0	0	0	0	0	7	13	0	0	24
Gelasinospora	HdV-1	0	70	29	0	0	0	13	0	7	0	0	0	0
Glomus	HdV-207	1813	1263	602	788	500	665	447	154	272	427	628	779	439
Pleospora	HdV-3B	36	35	0	0	0	0	0	0	0	0	0	0	0
Podospora type	HdV-368	36	140	143	40	62	55	60	113	54	57	83	68	90
Sordaria type	HdV-55	145	842	1074	309	150	188	233	315	123	192	269	310	192
Spegazzina		0	35	14	0	0	0	0	0	0	0	0	0	0
Sporomiella type	HdV-113	36	105	115	70	31	0	20	32	20	52	71	23	48
Tilletia		0	0	14	20	13	16	0	0	27	13	6	0	0
Trichocladium type		0	0	0	30	0	0	0	0	0	0	0	0	0
Urocystis		0	0	329	100	35	39	20	8	41	4	32	0	6
Ustulina deusta	HdV-44	0	0	0	0	0	8	0	0	0	0	0	0	0
Valsaria variospora type	HdV-140	0	0	14	40	4	16	53	57	89	44	122	53	54
	HdV-16	0	456	14	50	22	0	27	32	0	9	0	129	12
	HdV-18	218	140	272	80	4	47	53	73	27	17	26	15	42
	HdV-20	0	105	229	0	40	86	33	65	82	100	45	61	205
	HdV-124	0	0	43	0	0	8	0	8	0	4	0	8	12
	HdV-201 (fragm.)	0	0	43	0	0	0	0	0	0	0	0	0	0
	HdV-359	0	0	0	0	4	0	0	0	0	0	0	0	361
Algae														
Arcella cf. *vulgaris*	HdV-352	0	0	0	0	4	0	0	0	0	0	0	0	6
Botryococcus colony	HdV-901	0	0	0	0	0	0	7	8	0	0	0	0	0
Concentricystis		0	0	0	0	0	0	0	0	0	0	0	0	0
Desmidiales		36	0	0	0	0	0	0	0	0	0	0	8	0
Spirogyra		0	0	0	10	0	0	7	0	0	0	0	0	0
Zygnema		145	70	43	120	49	31	20	16	14	4	13	53	42
	HdV-181	326	421	172	140	172	509	340	348	82	161	160	68	180
	HdV-221	73	70	14	0	0	0	0	0	0	0	32	0	12
	HdV-492 cf.	0	0	0	0	0	0	0	0	7	4	26	15	0
Zoomicroremains														
Dicrocoelium		0	0	0	0	4	0	0	0	0	0	0	0	0
Number of Taxa		**11**	**17**	**20**	**15**	**17**	**14**	**17**	**16**	**17**	**19**	**16**	**17**	**19**

Table C5 Table of concentrations of the main Non-Pollen Palynomorphs (npp/g) found in slides examined for pollen at Sant'Angelo Vecchio, surface soil and Pollen Sequence 2.

Sample Number		P13	P14	P15	P16	P17	P18	P19	P20	P21	P22	P23	P24	P25	P26	P27	P28
Archaeological Context		Retaining Wall, square C4 Pollen Sequence 10						Slope 3, square D4 Pollen Sequence 8					Slope 3, square D4 Pollen Sequence 9				
Taxa	Types																
Fungal Spores																	
Cercophora type	HdV-112	42	92	18	0	0	0	0	0	32	14	14	0	0	20	15	12
Chaetomium	HdV-7B	198	0	88	132	360	64	1038	0	0	0	0	0	0	0	22	287
Coniochaeta cf. *lignaria*	HdV-172	42	92	35	132	30	32	807	58	0	7	54	15	0	0	0	24
Delitschia type		113	1787	176	1236	720	370	38	0	0	7	14	0	0	10	0	36
Diplocladiella type		14	137	0	88	30	0	0	12	0	0	0	0	0	0	0	0
Diporotheca	HdV-143	0	0	0	0	0	0	0	23	0	7	0	0	0	10	22	36
Gelasinospora	HdV-1	0	0	0	0	0	0	38	12	0	0	0	0	0	0	0	0
Glomus	HdV-207	806	1695	847	1015	630	418	730	463	113	281	284	265	229	89	221	491
Pleospora	HdV-3B	0	1054	0	177	150	32	0	0	0	0	0	0	0	0	0	12
Podospora type	HdV-368	0	92	0	88	0	16	115	46	32	7	0	29	35	59	51	60
Sordaria type	HdV-55	141	550	212	486	360	161	615	116	48	22	68	29	141	89	51	48
Spegazzina		0	46	0	0	0	0	0	0	0	7	0	7	0	0	0	0
Sporomiella type	HdV-113	14	0	0	44	150	0	77	12	48	7	27	0	0	20	22	72
Tilletia		0	46	0	0	0	0	0	0	0	0	0	7	0	10	0	0
Trichocladium type		0	0	18	0	0	0	0	46	0	0	0	0	0	10	0	24
Urocystis		127	46	511	0	150	177	1691	289	145	58	189	118	53	39	88	0
Ustulina deusta	HdV-44	0	0	0	0	0	0	0	23	0	0	0	0	0	0	0	0
Valsaria variospora type	HdV-140	0	46	0	0	30	0	77	81	16	22	54	52	53	79	118	72
	HdV-16	0	0	0	0	0	0	0	0	0	0	0	0	0	0	0	0
	HdV-18	14	916	88	486	210	161	231	0	64	29	27	214	70	49	0	36
	HdV-20	71	137	71	0	30	48	807	463	354	159	379	125	246	217	860	215
	HdV-124	0	0	0	0	0	0	0	0	0	0	0	0	0	0	0	0
	HdV-201 (fragm.)	28	0	0	0	0	0	77	0	0	0	14	0	0	0	0	0
	HdV-359	0	0	0	0	0	0	77	0	0	7	27	0	0	0	7	0
Algae																	
Arcella cf. *vulgaris*	HdV-352	0	46	0	0	0	0	0	0	0	0	0	0	0	0	0	0
Botryococcus colony	HdV-901	0	0	0	0	0	0	0	12	0	0	0	0	0	0	0	0
Concentricystis		42	0	18	0	0	0	0	0	0	22	0	7	0	0	0	12
Desmidiales		14	0	0	0	210	0	0	12	0	0	0	0	0	0	0	0
Spirogyra		0	0	0	0	60	0	38	12	16	0	0	0	0	10	7	24
Zygnema		28	0	35	177	150	80	154	46	0	14	27	0	35	0	15	0
	HdV-181	57	137	388	132	120	209	0	127	64	79	135	44	211	0	125	120
	HdV-221	0	0	0	0	0	0	0	0	0	0	0	0	0	0	0	12
	HdV-492 cf.	14	0	0	0	0	0	0	0	0	0	0	0	0	0	0	0
Zoomicroremains																	
Dicrocoelium		0	0	0	0	0	16	0	0	16	0	0	0	0	0	0	0
Number of Taxa		**17**	**16**	**13**	**12**	**16**	**13**	**16**	**18**	**12**	**17**	**14**	**12**	**9**	**14**	**14**	**18**

Table C6 Table of concentrations of the main Non-Pollen Palynomorphs (npp/g) found in slides examined for pollen at Sant'Angelo Vecchio, Pollen Sequences 10, 8, and 9.

Reference Material

References

Periodicals/Series/Compendia

ActaPraehistA	*Acta praehistorica et archaeologica*
AJA	*American Journal of Archaeology*
ASNSP	*Annali della Scuola Normale Superiore di Pisa*
ASTSN	*Atti della Società Toscana di Scienze Naturali, Mem. Ser. A*
AttiMGrecia	*Atti e Memorie della Società Magna Grecia*
Atti Taranto	*Atti del Convegno di Studi sulla Magna Grecia (Taranto)*
BdA	*Bollettino d'Arte*
ClRh	*Clara Rhodos*
FOLD&R	*Fasti On Line Documents & Research*
IstMitt	*Istanbuler Mitteilungen*
JAS	*Journal of Archaeological Science*
JFA	*Journal of Field Archaeology*
JHS	*Journal of Hellenic Studies*
JRS	*Journal of Roman Studies*
MEFRA	*Mélanges de l'École Française de Rome, Antiquité*
NSc	*Notizie degli Scavi di Antichità*
PBSR	*Papers of the British School at Rome*
RasArch	*Rassegna di Archeologia*
RM	*Mitteilungen des Deutschen Archäologischen Instituts, Römische Abteilung*
StAnt	*Studi di Antichità*

Short Title Abbreviations

3rd Hellenistic Pottery Conference 1994
1994. Γ' επιστημονική συνάντηση για την ελληνιστική κεραμική: χρονολογημένα σύνολα—εργαστήρια (24–27 Σεπτεμβρίου 1991, Θεσσαλονίκη). Κείμενα. (Βιβλιοθήκη της εν Αθήναις Αρχαιολογικής Εταιρείας, 137.) 2 vols. Athinai: Archaiologiki Etaireia.

5th Hellenistic Pottery Conference 2000
2000. Ε' επιστημονική συνάντηση για την ελληνιστική κεραμική. Χρονολογικά προβλήματα—κλειστά σύνολα—εργαστήρια [Χανιά, 1997]. Πρακτικά: κείμενα. Athinai: Ypourgeio Politismou, KE' Ephoreia Proïstorikon kai Klasikon Archaiotiton/Tameio Archaiologikon Poron kai Apallotrioseon.

6th Hellenistic Pottery Conference 2004
2004. ΣΤ' επιστημονική συνάντηση για την ελληνιστική κεραμική: χρονολογικά προβλήματα—κλειστά σύνολα—εργαστήρια (Βόλος, 12–23 Απριλίου 2000). Πρακτικά. Athina: Ypourgeio Politismou, IG Ephoreia Proïstorikon kai Klasikon Archaiotiton/Tameio Archaiologikon Poron kai Apallotrioseon.

Agora IV
Howland, R. H. 1958. *Athenian Agora*. Vol. 4, *Greek Lamps and their Survivals*. Princeton: The American School of Classical Studies at Athens.

Agora V
Robinson, H. 1959. *Athenian Agora*. Vol. 5, *Pottery of the Roman Period. Chronology*. Princeton: The American School of Classical Studies at Athens.

Agora XII
Sparkes, B. A., and L. Talcott. 1970. *Athenian Agora*. Vol. 12, *Black and Plain Pottery of the 6th, 5th, and 4th Centuries BC*. Princeton: The American School of Classical Studies at Athens.

Agora XXIX
Rotroff, S. I. 1997. *Athenian Agora*. Vol. 29, *Hellenistic pottery: Athenian and imported wheelmade table ware and related material*. Princeton: The American School of Classical Studies at Athens.

Agora XXXII
Hayes, J. W. 2008. *Athenian Agora*. Vol. 32, *Roman Pottery, Fine-Ware Imports*. Princeton: The American School of Classical Studies at Athens.

Agora XXXIII
Rotroff, S. I. 2008. *Athenian Agora*. Vol. 33, *Hellenistic Pottery: The Plain Wares*. Princeton: The American School of Classical Studies at Athens.

Amphi Sirios
Osanna, M., and G. Zuchtriegel, eds. 2012. ΑΜΦΙ ΣΙΡΙΟΣ ΡΟΑΣ: *Nuove ricerche su Eraclea e la Siritide*. Venosa: Osanna Edizioni.

Amphores Romaines
Università degli Studi di Siena, ed. 1989. *Amphores Romaines et histoire économique: dix ans de recherche. Actus du Colloque* (Sienne 22–24 mai 1986). Collection de l'École Française de Rome, 114. Rome/Paris: École Française de Rome/Diffusion de Boccard.

Antichi Peucezi
Riccardi, A., ed. 2003. Gli antichi Peucezi a Bitonto. *Documenti ed immagini dalla necropolis di via Traiana*. Bari: Edipuglia.

Archeologia dei Messapi
D'Andria, F., ed. 1990. *Archeologia dei Messapi. Le mostre, i cataloghi* 1. Bari: Edipuglia.

Archeologia di una città
Andreassi, G., and F. Radina, eds. 1988. *Archeologia di una città: Bari dalle origini al X secolo.* Bari: Edipuglia.

Armi
Bottini, A. 1993. *Armi. Gli strumenti della guerra in Lucania.* Bari: Edipuglia.

Arpi
Mazzei, M., ed. 1995. *Arpi: L'Ipogeo della Medusa e la necropoli.* Bari.

Art in the Ancient World XXI
Royal Athena Galleries. 2010. *Art in the Ancient World XXI: Greek, Etruscan, Roman, Byzantine, Egyptian, & Near Eastern Antiquities.* New York/London: Royal Athena Galleries.

Arte e artigianato
Lippolis, E., ed. 1996. *I Greci in Occidente: Arte e artigianato in Magna Grecia.* Naples: Electa.

Arti di Efesto
Giumlia-Mair, A., and M. Rubinich, eds. 2002. *Le arti di Efesto: capolavori in metallo dalla Magna Grecia.* Cinisello Balsamo. Milan: Silvano.

Atlante II
Istituto della Enciclopedia Italiana. 1985. *Atlante delle Forme Ceramiche* II: *Ceramica fine romana nel Bacino Mediterraneo (tardo ellenismo e primo impero).* Suppl. to *Enciclopedia dell'Arte Antica.* Rome.

Attività archeologica
Lattanzi, E., ed. 1980. *Attività archeologica in Basilicata 1964–1977. Scritti in onore di Dinu Adamesteanu.* Matera: Meta.

Ausculum I
Fabbri, M., and M. Osanna, eds. 2002. *Ausculum* I. *L'abitato sulla collina del Serpente di Ascoli Satriano.* Foggia: Claudio Grenzi.

Brateís datas
Battiloro, I., and M. Osanna, eds. 2011. *Brateís datas. Pratiche rituali, votivi e strumenti del culto dai santuari della Lucania antica. Atti delle giornate di studio sui santuari Lucani (Matera, 19–20 febbraio 2010).* Venosa: Edizioni Osanna.

Cairano
Bailo Modesti, G. 1980. *Cairano nell'età arcaica. L'abitato e la necropoli.* AION 1. Naples: Istituto Universitario Orientale di Napoli.

Campagna e paesaggio
Quilici, L., and S. Quilici Gigli, eds. 2000. *Campagna e paesaggio nell'Italia antica.* Rome: L'Erma di Bretschneider.

Caracausi
Grasso, L. 1989. "Caracausi. Un insediamento rupestre nel territorio di Lentini." *Cronache di Archeologia* 28: 3–172.

Carta archeologica Sinni 4
Quilici, L., and S. Quilici Gigli, eds. 2001. *Carta archeologica della valle del Sinni* 4. *Zoni di Senise.* Atlante Tematico di Topografia Antica Suppl. X/2000–2002. Rome: L'Erma di Bretschneider.

Ceramica in Italia
Saguì, L., ed. 1998. *La ceramica in Italia: VI–VII secolo. Atti del Convegno in onore di John W. Hayes (Roma, 11–13 maggio 1995).* 2 vols. Florence.

Ceramiche fini
Frère, D., ed. 2007. *Ceramiche fini a decoro subgeometrico nel VI sec. a.C. in Etruria meridionale e in Campania.* Collection de l'École Française de Rome 389. Rome: l'École Française de Rome.

Civita di Tricarico
De Cazanove, O. 2008. *Civita di Tricarico* 1. *Quartier de la maison du monolithe et l'enceinte intermédiaire.* Collection de l'École Française de Rome 409. Rome: École Française.

CNR, Direzione Generale del Catasto, 1956–1960
Carta della utilizzazione del suolo d'Italia in scala 1:200.000. Fogli n. 14, 15, 16, 17 e 19. Milan: Touring Club.

Con il fuso
Russo, A., ed. 2006. *Con il fuso e la conocchia: la fattoria lucana di Montemurro e l'edilizia domestica nel IV secolo a.C.* Potenza: Soprintendenza per i Beni Archeologici della Basilicata.

Consiglio d'Europa, 2000
Convenzione europea del paesaggio. Florence, 20 October 2000. <http://www.sinanet.isprambiente.it/it/gelso/buone-pratiche-paesaggio/convenzione-europea-paesaggio/convenzione_europea_paesaggio.pdf>

Corinth VII.3
Edwards, G. R. 1975. *Corinthian Hellenistic Pottery. Corinth* 7.3. Princeton: The American School of Classical Studies at Athens.

Cozzo Presepe
du Plat Taylor, J., E. Macnamara, J. Ward-Perkins, P. G. Dorell, A. Small, A. Johnston, A. J. N. W. Prag, et al. 1983. "The Excavations at Cozzo Presepe (1969–72)." In *Metaponto II*, 191–406.

Cuma 2
Cuozzo, M., B. d'Agostino, and L. Del Verme, eds. 2006. *Cuma: le fortificazioni* 2: *I materiali dai terrapieni arcaici.* Naples: Università degli Studi di Napoli L'Orientale.

Da Leukania a Lucania
Torelli, M., and L. de Lachenal, eds. 1993. *Da Leukania a Lucania: La Lucania centro-orientale fra Pirro e i Giulio-Claudii.* Venosa: Istituto Poligrafico e Zecca dello Stato.

Da Zancle 1
Bacci, G. M., and G. Tigano, eds. 2000. *Da Zancle a Messina, un percorso archeologico attraverso gli scavi*, 1. Messina: Sicania.

Emergenze e problemi
Regione Puglia. 1990. *Emergenze e problemi archeologici: Manduria-Taranto-Heraclea*. Manduria: Centro Regionale Servizi Educativi e Culturali.

European Environment Agency, 2000
CORINE Land Cover Technical Guide. Addendum 2000. Brussels, 2000.

Excavations at Otranto
D'Andria, F., and D. Whitehouse, eds. 1992. *The Excavation at Otranto*. II. *The Finds*. Galatina: Congedo Editore.

Fattoria Fabrizio
Lanza Catti, E., and K. Swift. 2014. *The Chora of Metaponto* 5: *A Greek Farmhouse at Ponte Fabrizio*. Ed. J. C. Carter. Austin: University of Texas Press.

Felicitas Temporum
Russo, A., and H. Di Giuseppe, eds. 2008. *Felicitas Temporum. Dalla terra alle genti: la Basilicata settentrionale tra archeologia e storia*. Muro Lucano.

Forentum I
Giorgi, M., S. Martinelli, M. Osanna, and A. Russo. 1988. *Forentum* I. *Le necropolis di Lavello*. Venosa: Osanna.

Forentum II
Bottini, A., and M. P. Fresa. 1991. *Forentum* II: *L'acropoli in età classica*. Venosa: Osanna.

Fratte
Greco, G., and A. Pontrandolfo, eds. 1990. *Fratte: Un insediamento etrusco-campano*. Modena: Panini.

Gli ori
De Juliis, E. M., A. Alessio, A. D'Amicis, A. Dell'Aglio, D. Graepler, E. Lippolis, L. Masiello, and T. Schojer, eds. 1986. *Gli ori di Taranto in età ellenistica*. Taranto: Scorpione.

Grammichele
Barra Bagnasco, M., ed. 2006. *Da Terravecchia di Grammichele a Occhiolà: archeologia di un insediamento della Sicilia centro-orientale: campagne di scavo 2000-2001*. Alessandria: Dell'Orso.

Gravina II
Small, A. M., ed. 1992. *An Iron Age and Roman Republican Settlement on Botromagno, Gravina di Puglia. Excavations of 1965–1974*. Vol. II, *Artifacts*. Archaeological Monographs of the British School at Rome 5. London: British School at Rome.

Gravina III
du Plat Taylor, J., P. G. Dorrell, and A. M. Small. 1977. "Gravina di Puglia III: Houses and a Cemetery of the Iron Age and Classical Periods. Part Two." *PBSR* 45: 69–137.

Gravisca 4
Boldrini, S. 1994. *Gravisca. Scavi nel santuario Greco*. Vol. 4, *Le ceramiche ioniche*. Bari: Edipuglia.

Gravisca 9
Valentini, V. 1993. *Gravisca. Scavi nel santuario Greco*. Vol. 9, *Le ceramiche a vernice nera*. Bari: Edipuglia.

Gravisca 12
Gori, B., and T. Pierini. 2001. *Gravisca. Scavi nel santuario Greco*. Vol 12, Bk. 2. *La ceramica comune*. Bari: Edipuglia.

Gravisca 16
Colivicchi, F., ed. 2004. *Gravisca. Scavi nel santuario Greco*. Vol. 16, *I materiali minori*. Bari: Edipuglia.

Greci, Enotri e Lucani
Bianco, S., A. Bottini, A. Pontrandolfo, A. Russo Tagliente, and E. Setari, eds. 1996. *I Greci in Occidente. Greci, Enotri e Lucani nella Basilicata meridionale*. Naples: Electra.

Kaulonia I
Tréziny, H., ed. 1989. *Kaulonia* I. *Sondages sur la fortification nord (1982–1985)*. Cahiers du Centre Jean Bérard 13. Naples: Centre Jean Bérard.

Laos 1
Greco, E., S. Luppino, and A. Schnapp, eds. 1989. *Laos* I: *Scavi a Marcellina 1973–1985*. Taranto: Istituto per la storia e l'archeologia della Magna Grecia.

Lattara 6
Py, M., ed. 1993. Dictionnaire des céramiques antiques (VII s. av. n. è.–VII s. de n. n. è.) en Méditerranée nord-occidentale (Provence, Languedoc, Ampurdan). Lattes: Associacion pour la Recherche Archéologique en Languedoc Oriental.

LCRW I
Gurt i Esparraguera, J. M., J. Buxeda i Garrigós, and M. A. Cau Ontiveros, eds. 2005. *Late Roman Coarse Wares, Cooking Wares and Amphorae in the Mediterranean: Archaeology and Archaeometry. International Conference (Barcelona 2002). British Archaeological Reports International Series* 1340. Oxford: Archaeopress.

LCS
Trendall, A. D. 1967. *The Red-figured Vases of Lucania, Campania and Sicily*. 2 vols. Oxford: Oxford University Press.

LCS Suppl. 1
Trendall, A. D. 1970. *The Red-figured Vases of Lucania, Campania and Sicily. First Supplement*. Supplement to the *Bulletin of the Institute of Classical Studies of the University of London* 26. London: Institute of Classical Studies of the University of London.

LCS Suppl. 3
Trendall, A. D. 1983. *The Red-figured Vases of Lucania, Campania and Sicily. Third Supplement*. Supplement to the *Bulletin of the Institute of Classical Studies of the University of London* 41. London: Institute of Classical Studies of the University of London.

Les Céramiques communes
Bats, M., ed. 1996. *Les Céramiques communes de Campanie et de Narbonnaise (I s. av. J.-C.-II s. ap. J.C.). La vaisselle de cuisine et de table. Actes des Journées d'étude organisées par le Centre Jean Bérard et la Soprintendenza Archeologica per le Province di Napoli e Caserta, Napoli 1994.* Naples: Centre Jean Bérard.

Leuca
van Compernolle, R., ed. 1978. *Leuca.* Galatina: Congedo.

LGPN III.A
Fraser, P. M., and E. Matthews. 1997. *Lexicon of Greek Personal Names.* Vol. III. A. *The Peloponnese, Western Greece, Sicily and Magna Graecia.* Oxford: Clarendon Press.

LGPN IV
Fraser, P. M., E. Matthews, and R. W. V. Catling. 2005. *Lexicon of Greek Personal Names, vol.* IV. *Macedonia, Thrace, northern regions of the Black Sea.* Oxford: Clarendon Press.

LIMC
1981–1997. *Lexicon Iconographicum Mythologiae Classicae.* Zurich: Artemis Verlag.

Locri I
Barra Bagnasco, M., C. Sabbione, G. Bacci, U. Spigo, and G. Molli Boffa. 1977. *Locri Epizefiri* I. *Ricerche nella zona di Centocamere.* Florence: Sansoni.

Locri II
Barra Bagnasco, M., ed. 1989. *Locri Epizefiri* II. *Gli isolati I2 e I3 dell'area di Centocamere.* Florence: Le Lettere.

Locri III
Barra Bagnasco, M., ed. 1989. *Locri Epizefiri* III. *Cultura materiale e vita quotidiana.* Florence: Le Lettere.

Locri IV
Barra Bagnasco, M., ed. 1992. *Locri Epizefiri* IV. *Lo scavo di Marasà sud. Il sacello tardo arcaico e la "casa dei leoni."* Florence: Le Lettere.

Lo spazio del potere
Osanna, M., L. Colangelo, and G. Carollo, eds. 2009. *Lo spazio del potere. La residenza ad abside, l'anaktoron, l'episcopio a Torre di Satriano.* Venosa: Osanna.

Lo spazio del rito
Nava, M. L., and M. Osanna, eds. 2005. *Lo spazio del rito: santuari e culti in Italia meridionale tra indigeni e Greci. Atti delle giornate di studio, Matera, 28 e 29 giugno 2002.* Bari: Edipuglia.

LSAG[2]
Jeffery, L. H. 1990. *The local scripts of archaic Greece: a study of the origin of the Greek alphabet and its development from the eighth to the fifth centuries B.C.* Rev. ed. Oxford: Clarendon Press.

Megara Hyblaea 2
Vallet, G., ed. 1978. *Megara Hyblaea 2. La ceramique archaique.* Paris: Boccard.

Meligunìs Lipára II
Bernabò Brea, L., and M. Cavalier. 1965. *Meligunìs Lipára.* 2. Palermo: Flaccovio.

Metaponto I
Adamesteanu, D., D. Mertens, and F. D'Andria. 1980. *Metaponto* I. *NSc* 8th ser., 29, suppl. Rome: Accademia Nazionale dei Lincei.

Metaponto II
Adamesteanu, D., B. Chiartano, J. C. Carter, and E. Macnamara. 1983. *Metaponto* II. *NSc* 8th ser., 31. Rome: Accademia Nazionale dei Lincei.

Monte Irsi
Small, A., ed. 1977. *Monte Irsi, Southern Italy: The Canadian Excavations in the Iron Age and Roman Sites, 1971–1972.* BAR Supplementary Series 20. Oxford: BARS.

Monte Sannace
Ciancio, A., E. M. De Juliis, A. Riccardi, and F. Rossi, eds. 1989. *Monte Sannace: Gli scavi dell'acropoli (1978–1983).* Galatina: Congedo.

Moulage
Muller, A., ed. 1997. Le moulage en terre cuite dans l'Antiquité. Creation et production dérivée, Fabrication et diffusion. (Actes du XVIIIe Colloque du Centre de Recherches Archéologiques, Lille III 1995). Lille: Presses Universitaires du Septentrion.

Mozia
Famà, M. L., ed. 2002. *Mozia: gli scavi nella "zona A" dell'abitato.* Bari: Edipuglia.

Mozia XII
Nigro, L., ed. 2007. *Mozia* XII. *Zona D. La "Casa del sacello domestico," il "Basamento meridionale" e il sondaggio stratigrafico I.* Rome: Missione Archeologica a Mozia.

Museo di Taranto
Alessio, A., ed. 1988. *Museo di Taranto: Cento anni di archeologia.* Taranto: Scorpione.

Museo Taranto I.3
Dell'Aglio, A., ed. 1997. *Catalogo del Museo Nazionale Archeologico di Taranto.* I:3, *Atleti e guerrieri. Tradizioni aristocratiche a Taranto tra VI e V sec. a.C. Catalogo della mostra, Taranto 1994.* Taranto: Scorpione.

Museo Taranto II.1
Dell'Aglio, A., and E. Lippolis. 1992. *Catalogo del Museo Nazionale Archeologico di Taranto.* II:1, *Ginosa e Laterza: La documentazione archeologica dal VII al III sec. a.C. Scavi 1900–1980.* Taranto: Scorpione.

Museo Taranto II.2
De Juliis, E. M., ed. 2006. *Catalogo del Museo Nazionale Archeologico di Taranto.* II:2, *Rutigliano I: la necropolis di contrada Purgatorio, scavo 1978.* Taranto: Scorpione.

Museo Taranto III.1
Lippolis, E., ed. 1994. *Catalogo del Museo Nazionale Archeologico di Taranto.* III:1, *Taranto. La Necropoli: aspetti e problemi della documentazione archeologica tra VII e I secolo a.C.* Taranto: Scorpione.

Necropoleis
Carter, J. C., ed. 1998. *The Chora of Metaponto: The Necropoleis*. 2 vols. Austin: University of Texas Press.

Nella terra degli Enotri
La Torre, G. F., and A. Colicelli, eds. 1999. *Nella terra degli Enotri, Atti del Convegno di studi, Tortora, 18–19 aprile 1998*. Paestum: Pandemos.

Oppido Mamertina
Costamagna, L., and P. Visonà, eds. 1999. *Oppido Mamertina: Ricerche archeologiche nel territorio e in contrada Mella*. Rome: Gangemi.

Ordona III
Mertens, J. 1974. *Ordona* III. *Rapports et etudes*. Brussels/Rome.

Ordona VII/2
Iker, R. 1986. *Ordona* VII/2. *Les tombes dauniennes*. Etudes de philologie, d'archéologie et d'histoire anciennes 24. Brussels/Rome: Academia Belgica.

Ordona X
Volpe, G., ed. 2000. *Ordona* X. *Ricerche archeologiche a Herdonia*. Bari.

Ordona XI
Volpe, G., and D. Leone, eds. 2008. *Ordona* XI. *Ricerche archeologiche a Herdonia*. Bari.

Palinuro II
Naumann, R., B. Neutsch, F. Hiller, and E. Naumann, eds. 1960. *Palinuro: Ergebnisse der Ausgrabungen, II. Mitteilungen des Deutschen Archeologischen Instituts. Römische Abteilung* 3–4. Heidelberg: F. H. Kerle.

Panskoye I
Hannestad, L., V. F. Stolba, and A. N. Ščeglov, eds. 2002. *Panskoye* I. *The Monumental Building U6*. Vol. I. Aarhus: Aarhus University Press.

Paphos III
Hayes, J. W., I. K. Nicolaou, and L. L. Neuru. 1991. *Paphos Vol.* III, *The Hellenistic and Roman Pottery*. Nicosia: Department of Antiquities.

Pomarico Vecchio I
Barra Bagnasco, M., ed. 1997. *Pomarico Vecchio* I. *Abitato, mura, necropolis, materiali*. Galatina: Congedo.

Pontecagnano II.1
De Natale, S., ed. 1992. *Pontecagnano. II, La necropoli di S. Antonio: ECI. 2, Tombe della Prima Età del Ferro*. Naples: Istituto universitario orientale.

Popoli anellenici
Adamesteanu, D., E. Lissi Caronna, R. R. Holloway, J.-P. Morel, and G. Tocco. 1971. *Popoli anellenici in Basilicata*. Soprintendenza alle Antichità della Basilicata–Potenza. Naples: La Buona Stampa.

Poseidonia-Paestum II
Greco, E., and D. Theorescu. 1983. *Poseidonia-Paestum* II. *L'Agora*. Rome: École Française de Rome.

Principi Imperatori Vescovi
Cassano, R., ed. 1992. *Principi Imperatori Vescovi: Duemila anni di storia a Canosa*. Venice: Marsilio.

Problèmes de la terre
Finley, M. I., ed. 1973. *Problèmes de la terre en Grèce ancienne. Civilisations et sociétés* 33. Paris/La Haye: Mouton.

Ravitaillement en blé
Centre Jean Bérard and Centre national de la recherché scientifique. 1994. *Le ravitaillement en blé de Rome et des centres urbains des débuts de la République jusqu'au haut empire. (Actes du colloque international, Naples, 14–16 février 1991.)* Collection Centre Jean Bérard 11. Collection de l'École française de Rome 196. Naples/Rome: Centre Jean Bérard/L'École française de Rome.

Regione Basilicata, 2009
Sistema Ecologico funzionale territoriale. Regione Basilicata: Dipartimento Ambiente, Territorio e Politiche della Sostenibilità. 3 vols.

Ricerche sulla casa
D'Andria, F., and K. Mannino, eds. 1996. *Ricerche sulla casa in Magna Grecia e in Sicilia. Atti del Colloquio–Lecce, 23–24 giugno 1992. Università di Lecce, Scuola di Specializzazione in Archeologia Classica e Medioevale*. Archeologia e Storia 5. Galatina: Congedo.

Rituali
Nava, M. L., and M. Osanna, eds. 2001. *Rituali per una dea lucana. Il santuario di Torre di Satriano*. Potenza: Soprintendenza Archeologica della Basilicata.

Roccagloriosa I
Gualtieri, M., and H. Fracchia, eds. 1990. *Roccagloriosa* I. *L'abitato: scavo e ricognizione topografica (1976–1986)*. Bibliothèque de l'Institut Français de Naples. Série 2.8. Naples: Centre Jean Bérard.

Roccagloriosa II
Gualtieri, M., and H. Fracchia, eds. 2002. *Roccagloriosa* II. *L'oppidum lucano e il territorio*. Collection du Centre Jean Bérard 20. Naples: Centre Jean Bérard.

Roman Brick and Tile
McWhirr, A. 1979. *Roman Brick and Tile*. British Archaeological Reports 68. Oxford: Oxford University Press.

RVAp I
Cambitoglou, A., and A. D. Trendall. 1978. *The Red-figured vases of Apulia, I*. Oxford: Oxford University Press.

Samos III
Furtwängler, A., and H. J. Kienast. 1989. *Samos* III. *Der Nordbau im Heraion von Samos*. Bonn: Habelt.

Samos IV
Isler, H. P. 1978. *Samos* IV. *Das Archaische Nordtor und seine Umgebung im Heraion von Samos*. Bonn: Habelt.

San Biagio
Lapadula, E. 2012. *The Chora of Metaponto 4: The Late Roman Farmhouse at San Biagio*. Austin: University of Texas Press.

San Giovanni di Ruoti
Small, A. M., and R. J. Buck, eds. 1994. *The Excavations of San Giovanni di Ruoti* I. *The Villas and Their Environment.* Toronto.
Santuario di Hera
de la Genière, J., and G. Greco, eds. 2010. *Il santuario di Hera alla foce del Sele. Indagini e studi 1987–2006, Atti e Memorie della Società Magna Grecia IV, 1–2.* Rome: Società Magna Grecia.
Sibari I
Accademia Nazionale dei Lincei. 1969. *Sibari 1. Saggi di Parco del Cavallo (1969) e agli Stombi (1969–70). NSc* 8th series, 23, Suppl. No. 3. Rome.
Sibari II
Accademia Nazionale dei Lincei. 1970. *Sibari 2. Sibari: Scavi al Parco del Cavallo (1960–62; 1969–70) e agli Stombi (1969–70). NSc* 8th series, 24, Suppl. No. 3. Rome.
Sibari III
Accademia Nazionale dei Lincei. 1972. *Sibari 3. Rapporto preliminare della campagna di scavo: Stombi, Casa Bianca, Parco del Cavallo, San Mauro (1971). NSc* 8th series, 26, Suppl. No. 3. Rome.
Sibari IV
Accademia Nazionale dei Lincei. 1974. *Sibari 4. Relazione preliminare della campagna di scavo: Stombi, Parco del Cavallo, Prolungamento Strada, Casa Bianca (1972). NSc* 8th series, 28, Suppl. No. 3. Rome.
Sibari V
Accademia nazionale dei Lincei. 1974. *Sibari 5. Relazione preliminare delle campagne di scavo 1973 (Parco del Cavallo; Casa Bianca) e 1974 (Stombi; Incrocio; Parco del Cavallo; Prolungamento Strada; Casa Bianca). NSc* 8th series, 28, Suppl. No. 3. Rome.
Siris 3
Studi e ricerche della scuola di specializzazione di Matera (2000–2001). Bari: Adda.
Survey
Carter, J. C., and A. Prieto, eds. 2011. *The Chora of Metaponto* 3. *Archaeological Field Survey: Bradano to Basento.* Vols. I–IV. Austin: University of Texas Press.
Tarchna 2
Chiaromonte Treré, C., ed. 1997. *Tarquinia, Scavi sistematici nell'abitato, Campagne 1982–1988. I materiali, 1. Tarchna* II. Rome: L'Erma di Bretschneider.
Testimonianze
Tocco, G., P. Bottini, E. Pica, and P. G. Moles, eds. 1982. *Testimonianze archeologiche nel territorio di Tolve.* Matera: BMG.
ThesCRA
Thesaurus Cultus et Rituum Antiquorum. Los Angeles: The J. Paul Getty Museum.
Tocra
Boardman, J., and J. Hayes. 1966. *Excavations at Tocra, 1963–1965, The Archaic Deposits* I. British School of Archaeology at Athens, Suppl. 4. Oxford: Thames and Hudson.
Torre di Satriano I
Osanna, M., and M. M. Sica. 2005. *Torre di Satriano* I: *Il santuario lucano.* Venosa: Osanna.
Tyndaris 1
Leone, R., and U. Spigo. 2008. *Tyndaris 1. Ricerche nel settore occidentale: campagne di scavo 1993–2004.* Palermo: Regione Siciliana, Assessorato dei Beni Culturali, Ambientali e della Pubblica Istruzione.
Valesio I
Yntema, D. 2001. *Pre-Roman Valesio: Excavations of the Amsterdam Free University at Valesio, Province of Brindisi, Southern Italy.* Vol. 1, *The Pottery.* Amsterdam: Institute of Archaeology, Free University.
Vases from Magna Graecia
Mayo, M. E. 1982. *The Art of South Italy. Vases from Magna Graecia.* Richmond: Virginia Museum of Fine Arts.

Works Cited

Abbott, J. T. 2011. "Geomorphology and Geoarchaeology of the Metapontino." In *Survey*, 31–68.
Accorsi, C. A., M. Bandini Mazzanti, and A. M. Mercuri. 1995. "Analisi palinologiche." In *L'insediamento preistorico di Terragne (Manduria-Taranto).* Eds. M. A. Gorgoglione, S. di Lernia, and G. Fiorentino. 185–98. Manduria: C.R.S.E.C. TA/55.
Acsádi, Gy., and J. Nemeskéri. 1970. *History of human life span and mortality.* Trans. K. Balas. Budapest: Akadémiai Kiadó.
Adam, J. P. 1989. *L'arte di costruire presso i Romani. Materiali e tecniche.* Bergamo: Longanesi & C.
Adamesteanu, D. 1964. "Santuario di San Biagio alla Venella." *Bollettino di Archeologia* 49: 360–61.
———. 1965. "La documentazione archeologica in Basilicata." *Atti Taranto* 4: 121–43.
———. 1967a. "Problèmes de la zone archéologique de Métaponte." *Revue Archéologique* n.s.: 3–38.
———. 1967b. "L'attività archeologica in Basilicata." *Atti Taranto* 6: 255–76.
———. 1969. "L'attività archeologica in Basilicata." *Atti Taranto* 8: 163–77.
———. 1971. "L'attività archeologica in Basilicata." *Atti Taranto* 10: 467–85.
———. 1973a. "Le suddivisioni di terra nel Metapontino." In *Problèmes de la terre*, 49–61.
———. 1973b. *Metaponto.* Naples: La Buona Stampa.
———. 1974. *La Basilicata antica.* Cava dei Tirreni: Di Mauro.
———. 1975. "L'attività archeologica in Basilicata." *Atti Taranto* 14: 247–60.
———. 1979. "Metaponto." *EAA* Suppl. 1970: 481–86.
———. 1980. "Il santuario di Apollo e urbanistica generale." In *Metaponto I*, 15–311.

———. 1982. "Siris e Metaponto alla luce delle nuove scoperte archeologiche." *Annuario della Scuola Archeologica di Atene* 60: 301–12.

———. 1987. "Poleis italiote e comunità indigene." In *Magna Grecia. Lo sviluppo politico, sociale ed economico*. Ed. G. Pugliese Carratelli. 115–34. Milan: Electa.

Adamesteanu, D., and H. Dilthey. 1978. "Siris. Nuovi contributi archeologici." *MEFRA* 90 (2): 515–65.

Adamesteanu, D., D. Mertens, and A. De Siena. 1975. "Metaponto: Santuario di Apollo, Tempio D (Tempio Ionico): Rapporto preliminare." *BdA* 60: 26–49.

Albanese Procelli, R. M. 1996. "Appunti sulla distribuzione delle anfore commerciali nella Sicilia arcaica." *Kokalos* 42: 91–137.

———. 2003. "Anfore commerciali dal centro indigeno della montagna di Ramacca (Catania)." In *Archeologia del Mediterraneo. Studi in onore di Ernesto De Miro*. Eds. G. Fiorentini, M. Caltabiano, and A. Calderone. 37–50. Rome: L'Erma di Bretschneider.

Alessio, A. 1988. "La necropoli di Piazza d'Armi. Lo scavo del 1911–1914." In *Museo di Taranto*, 325–69.

Allegro, N. 1982. "Louteria a rilievo da Himera." In *Secondo Quaderno Imerese*. Eds. N. Allegro et al. 115–66. Rome: L'Erma di Bretschneider.

———. 2008. *Himera 5. L'abitato. Isolato II. I blocchi 1-4 della zona 1*. Palermo: Università di Palermo, Dipartimento di Beni Culturali.

AlQahtani, S. J. 2009. *Atlas of Human Tooth Development and Eruption*. London: Barts and The London School of Medicine and Dentistry, Westfield College, Queen Mary University of London. <http://www.atlas.dentistry.qmul.ac.uk/>

Amato, A. 2000. "Estimating Pleistocene tectonic uplift rates in the Southeastern Apennines (Italy) from erosional land surfaces and marine terraces." In *Geomorphology, Human Activity and Global Environmental Change*. Ed. O. Slaymaker. 67–87. New York: Wiley.

Ammerman, R. M. 2011a. "Coroplastic material, San Angelo Vecchio, Metaponto: Preliminary Report." Unpublished report, Institute of Classical Archaeology.

———. 2011b. "Terracotta Objects." In *Survey*, 489–538.

Amore, M. G. 1992–1993. *Aspetti di vita quotidiana nella chora metapontina. La documentazione delle fattorie*. Tesi di laurea, University of Turin.

Amouretti, M. C. 1986. *Le pain et l'huile dans la Grèce antique. De l'araire au moulin*, Annales Litteraires de l'Université de Besançon 328. Paris: Les Belles Lettres.

———. 2000. "L'artisant indispensable au functionnement de l'agricolture." In *L'artisanat en Grèce ancienne. Les productions, les diffusions. Actes Lyon 1998*. Eds. F. Blondé and A. Muller. 147–64. Lille: SEGES.

Amyx, D. A. 1958. "The Attic Stelai: Part III. Vases and Other Containers." *Hesperia* 27.3–4: 163–310.

Andreiomenou, A. K. 1971. "Ephoreia Archaiopoleion kai idiotipo Archaiologikon Syllogon." *Archaiologikon Deltion* 26: 561–63.

Andriani, R., and F. Laricchia. 2007. *Gravina-Botromagno: La necropoli di Accurso, scavi 1972*. Bari: M. Adda.

Annese, C. 2000. "Ceramiche della prima e media età imperiale dai riempimenti delle latrine della *domus A*." In *Ordona X*, 251–66.

Antinucci, F. 2007. *Musei virtuali: come non fare innovazione tecnologia*. Bari: Laterza.

Aprosio, M. 2008. *Archeologia dei paesaggi a Brindisi dalla romanizzazione al Medioevo*. Bari: Edipuglia.

Aprosio, M., and A. Pizzo. 2003. "La rozza terracotta." In *Manifattura Ceramica Etrusco-Romana a Chiusi. Il complesso produttivo di Marcianella*. Eds. G. Pucci and C. Mascione. 225–50. Bari: Edipuglia.

Arias, P. E. 1979. "Contributo allo studio delle tombe di Cavallino." In *Cavalino* I. *Scavi e ricerche 1964–1967*. Ed. O. Pancrazzi. 217–26. Galatina: Università di Lecce, Istituto di Archeologia e Storia Antica, Congedo Editore.

Arthur, P. 1996. "'Masseria Quattro Macine.' A deserted medieval village and its territory in southern Apulia: An interim report on field survey, excavation and document analysis." *PBSR* 44: 181–235.

Arthur, P., and B. Bruno, eds. 2009. *Apigliano. Un villaggio bizantino e medioevale in Terra d'Otranto. L'ambiente, il villaggio, la popolazione*. Galatina: Arti Grafiche Panico.

Aubert, J. J. 2005. "L'estampillage des briques et des tuiles: une explication juridique fondée sur une approche globale." In *Interpretare i bolli laterizi di Roma e della Valle del Tevere: produzione, storia, economia e topografia. Atti del convegno all'École Française de Rome e all'Institutum Romanum Finlandiae, 31 marzo e 1 aprile 2000*. Eds. C. Brunn and F. Chausson. 52–59. Rome: Bardi Eds.

Auriemma, R. 2004. *Salentum a Salo. Forma maris antiqui*. 2 vols. Galatina: Congedo.

Aversa, G. 2002. "Decorazioni architettoniche fittili di età arcaica tra Grecia propria e Occidente coloniale: realtà a confronto." *Annuario della Scuola Archeologica di Atene* 80: 231–79.

———. 2012. *I tetti achei. Terrecotte architettoniche di età arcaica in Magna Grecia. Tekmeria* 15. Paestum: Pandemos.

Bacci, G. 1977. "Ceramica protocorinzia, subgeometrica ed orientalizzante." In *Locri I*, 73–82.

Bagnasco Gianni, G. 1997. "La ceramica etrusca depurata acroma e a bande." In *Tarchna 2*, 100–162.

———. 2007. "Le ricerche sulle ceramiche depurate a Tarquinia: tra importazioni e produzioni locali." In *Ceramiche fini*, 93–99.

Baldoni, V. 2009. *La ceramica attica dagli scavi ottocenteschi di Marzabotto*. Bologna: Ante Quem.

Barberis, V., and V. Meirano. 1997. "La ceramica a vernice nera dalle fortificazioni." In *Pomarico Vecchio I*, 92–97.

Barbieri, F. 2002. Ceramica a vernice rossa o bruna. In *Ausculum I*, 263–67.

Barbieri, F., and T. Virtuoso. 2002. "Ceramica a vernice nera." In *Ausculum I*, 232–62.

Barello, F. 1995. *Architettura greca a Caulonia. Edilizia monumentale e decorazione architettonica in una città della Magna Grecia*. Florence: Le Lettere.

Barra Bagnasco, M. 1986. *Protomi in terracotta da Locri Epizefiri. Contributo allo studio della scultura arcaica in Magna Grecia*. Florence-Turin: Il Quadrante.

———. 1997. "Osservazioni su officine e produzione a Locri Epizefiri. L'esempio delle terrecotte con Zeus fulminate." In *Moulage*, 207–32.

———. 2001. "Il santuario indigeno di Chiaromonte." In *Carta archeologica Sinni 5*, 215–35.

Bats, M. 1988. *Vaisselle et alimentation à Olbia de Provence (v. 350-v. 50 av. J.-C.). Modèles culturels et categories céramiques*. Paris: Éditions du Centre National de la Recherche Scientifique (CNRS).

———. 1993. "Céramique commune italique." In *Lattara 6*, 357–62.

Bechtold, B. 1999. *La Necropoli di Lilybaeum*. Rome: L'Erma di Bretschneider.

Becker, M. J. 1979. "Preliminary Report on the Sant'Angelo Vecchio Tombs Excavated in 1979. Inhumations of the Fourth Century B.C." Unpublished report, Institute of Classical Archaeology.

———. 1980a. "The Saldone tombs: An analysis of the burial pattern and human skeletal material from a fifth century B.C. rural cemetery near Metaponto, Italy." Unpublished report, Institute of Classical Archaeology.

———. 1980b. "Field notes 1980. Measurements and observations of the Sant'Angelo Vecchio tomb remains. Metaponto, Italy, June." Unpublished notes, Institute of Classical Archaeology.

———. 1981. "San Angelo Vecchio Tombs of the Fourth Century B.C.: An Analysis of the Inhumations Recovered by the University of Texas, Austin." Unpublished report, Institute of Classical Archaeology.

———. 1988. "An analysis of the human skeletal remains from [the] necropolis at İkiztepe. Appendix V." In Bahadir Alkim, U. *Ikiztepe I, Birinci ve ikinci dönem kazıları*. Ed. Ö. Bilgi. 261–75. Ankara: Türk Tarih Kurumu Basimevi.

———. 1996–1997. "Roman Amphora Burials of Young Children Dating to the Third Century CE at Metaponto (Basilicata) Italy." *Archaeological News* 21–22: 20–26.

———. 1999. "Calculating stature from in situ measurements of skeletons and from long bone lengths: An historical perspective leading to a test of Formicola's hypothesis at 5th century BCE Satricum, Lazio, Italy." *Rivista di Antropologia* (Rome) 77: 225–47.

———. 2007. "Childhood among the Etruscans: mortuary programs at Tarquinia as indicators of the transition to adult status." In *Constructions of Childhood in Ancient Greece and Italy*. Eds. A. Cohen and J. B. Rutter. *Hesperia Supplement* 41. 281–92. Princeton: The American School of Classical Studies at Athens.

———. 2011. "Etruscan infants: children's cemeteries at Tarquinia, Italy, as indicators of an age of transition." In *(Re)Thinking the Little Ancestor: New Perspectives on the Archaeology of Infancy and Childhood*. Eds. M. Lally and A. Moore. *British Archaeological Reports International Series* 2271. 24–36. Oxford: Archaeopress.

———. 2012. "Coming of Age in Etruria: Etruscan Children's Cemeteries at Tarquinia, Italy." *International Journal of Anthropology* 27 (1–2): 63–86.

Becker, M. J. and L. Salvadei. 1992. "Analysis of the Human Skeletal Remains from the Cemetery of Osteria dell'Osa." In *La Necropoli Laziale di Osteria dell'Osa*. Ed. A. M. Bietti Sestieri. 53–191. Rome: Quasar.

Behre, K. E. 1981. "The interpretation of anthropogenic indicators in pollen diagrams." *Pollen Spores* 23: 225–45.

Bencivenga Trillmich, C. 1983. "Resti di casa greca di età arcaica sull'acropoli di Elea." *MEFRA* 95: 417–88.

Bénévent, C. 1997. "Un atelier de potiers gallo-romains à l'Auzéral (Savignac)." *Vivre en Rouergue. Cahiers d'Archéologie Aveyronnaise* 11: 119–53.

Bentivenga, M., M. Coltorti, G. Prosser, and E. Tavernelli. 2004. "A new interpretation of terraces in the Taranto Gulf: The role of extensional faulting." *Geomorphology* 60: 383–402.

Berggren, G. 1969. *Atlas of seeds and small fruits of North-west-European plant species with morphological descriptions*. Stockholm: Swedish Natural Science Research Council.

Berlin, A. M. 2006. *Gamla I: the pottery of the Second Temple Period. The Shmarya Gutmann excavations, 1976–1989. Israel Antiquities Authority Reports* 29. Jerusalem: Israel Antiquities Authority.

Berloco, V., and G. Schiavariello. 2012. "Archeologia e ceramica romana. Il contesto di Egnazia. La ceramica d'uso comune." In *Ceramica romana nella Puglia adriatica*. Ed. C. S. Fioriello. 388. Bari: Sedit.

Bernasconi, M. P. 2007. "Kaulonia-Casa nei pressi della Casamatta: studio dei molluschi." In *Caulonia tra Crotone e Locri. Tomo* I. *Atti del Convegno Internazionale. Firenze, 30 maggio–1 giugno 2007*. Eds. L. Lepore and P. Turi. 153–54. Florence: University Press.

Bertelli, G., and D. Roubis. 2002. *Torre di Mare* I. *Ricerche archeologiche nell'insediamento medievale di Metaponto (1995–1999)*. Bari: Adda Editore.

Berti, F., and P. De Santis. 2005. "Nuovi spunti di riflessione sui culti di Spina." In *Depositi votivi e culti dell'Italia antica dall'età arcaica a quella tardo-repubblicana. Atti del convegno di studi (Perugia, 1–4 giugno 2000)*. Eds. A. Comella and S. Mele. 437–44. Bari: Edipuglia.

Bertsch, K. 1940. *Früchte und Samen. Ein Bestimmungsbuch zur Pflanzenkunde der vorgeschichtlichen Zeit.* Stuttgart: Enke.

Bes, P. M. 2015. *Once Upon a Time in the East. The Chronological and Geographical Distribution of Terra Sigillata and Red Slip Ware in the Roman East. Roman and Late Antique Mediterranean Pottery* 6. Oxford: Oxford University Press.

Biagini, C. 2008. "Pesi da telaio e oscilla." In *Segesta* III. *Il sistema difensivo di Porta di Valle (scavi 1990–1993).* Ed. R. Camerata Scovazzo. 645–63. Mantua: SAP Edizioni.

Bianco, O., and A. Deodato. 1997. "La ceramica commune." In *Pomarico Vecchio I*, 175–95.

Bianco, S. 1983. "Il neolitico di contrada Cetrangolo nella valle del Cavone (Matera)." In *Studi in onore di Dinu Adamesteanu.* 1–16. Galatina: Congedo Editore.

———. 1993a. "Herakleia, necropoli meridionale. Via Umbria-Tomba 119." In *Da Leukania a Lucania*, 182–89.

———. 1993b. "La *chora* di Herakleia: Tursi, S. Maria d'Anglona." In *Da Leukania a Lucania*, 197–207.

———. 2011. *Enotria. Processi formativi e comunità locali: la necropoli di Guardia Perticara.* Lagonegro: Tipografia Zaccara.

———. 2012. "Guardia Perticara (PZ). I modelli fittili in forma di cofanetto della necropoli enotria di contrada San Vito." In *Lo spazio del potere*, 205–62.

Biers, W. 1971. "From the Furnace." *Archaiologika Analekta ex Athenon* 4: 414–16.

Biffino, A. 2012. "La ceramica comune e la ceramica da fuoco." In *Le fornaci romane di Le fornaci romane di Giancola (Brindisi). Bibliotheca archaeologica* 27. Eds. D. Manacorda and S. Pallecchi. 227–80. Bari: Edipuglia.

Bigi, S., G. Cantalamessa, E. Centamore, P. Didaskalou, A. Micarelli, S. Nisio, T. Pennesi, and M. Potetti. 1996. "The Periadriatic basin (Marche-Abruzzi sector, Central Italy) during the Plio-Pleistocene." *Giornale di Geologia* 59 (1–2), 3rd series: 245–59.

Bilde, P. 1993. "Mouldmade Bowls, Centres and Peripheries in the Hellenistic World." In *Centre and Periphery in the Hellenistic World.* Eds. P. Bilde, P. Engberg-Pedersen, L. Hannestad, J. Zahle, and K. Randsborg. 192–209. Aarhus: Aarhus University Press.

Billot, M.-F. 1997. "Moules et moulage du décor des toits grecs." In *Moulage*, 39–129.

———. 2000. "Centres de production et diffusion des tuiles dans le monde grec." In *L'artisanat en Grèce ancienne. Les productions, les diffusions. Actes Lyon 1998.* Eds. F. Blondé and A. Muller. 193–240. Lille: SEGES.

Bitti, M. C. 1989. "La ceramica a vernice nera. Il vasellame del II strato: vasi da mensa." In *Locri II*, 141–88.

Blasi, M. 2001. "La ceramica comune." In *Rituali*, 63–74.

Blitzer, H. 1990. "Κορωνεϊκά. Storage-jar Production and Trade in the Traditional Aegean." *Hesperia* 59: 675–711.

Blondé, F., and M. Picon. 2000. "Artisanat et histoire des techniques: le cas des céramiques." In *L'artisanat en Grèce ancienne. Les productions, les diffusions. Actes du Colloque de Lyon 1998.* Eds. F. Blondé and A. Muller. 13–26. Lille: SEGES.

Blondé, F., J. Y. Perreault, and C. Péristéri. 1992. "Un atelier de potier archaïque à Phari (Thasos)." In *Les ateliers de potiers dans le monde grec aux époques géometrique, archaïque et classique. (Actes de la Table Ronde organisée à l'École Française d'Athènes, 2 et 3 octobre 1987).* Eds. F. Blondé and J. Y. Perreault. 12–40. Athens: l'École Française d'Athènes.

Blotti, M. F. 2014. "Cooking Ware." In *Fattoria Fabrizio*, 297–319.

Boenzi, F., B. Radina, G. Ricchetti, and A. Valduga. 1971. "Note illustrative della Carta Geologica d'Italia alla scala 1:100,000 del F. 201 'Matera'." Ministero dell'Industria, del Commercio e dell'Artigianato, Servizio Geologico d'Italia. Rome: Nuova Tecnica Grafica.

Boersma, J., and D. Yntema. 1987. *Valesio. Storia di un insediamento apulo dall'Età del Ferro all'epoca tardoromana. Bilancio delle ricerche dopo tre campagne di scavo.* Brindisi: Montedipe.

Boitani, F. 1990. "Le ceramiche laconiche a Gravisca." In Lakonikà. *Ricerche e nuovi materiali di ceramica laconica.* Eds. P. Pelagatti and C. M. Stibbe. *BdA* Suppl. 64: 19–67.

Bojňanský, V., and A. Fargašová. 2007. *Atlas of Seeds and Fruits of Central and East-European Flora: The Carpathian Mountains Region.* Dordrecht: Springer.

Bökönyi, S., and E. Gál. 2010. *The Chora of Metaponto 2: Archaeozoology at Pantanello and Five Other Sites.* Ed. László Bartosiewicz. Austin: University of Texas Press.

Bonanno, C., ed. 2008a. *Apollonia. Indagini archeologiche sul Monte di San Fratello, Messina, 2003–2005.* Rome: L'Erma di Bretschneider.

———. 2008b. "Gli oscilla discoidali." In *Apollonia. Indagini archeologiche sul Monte di San Fratello, Messina, 2003–2005.* Ed. C. Bonanno. 57–61. Rome: L'Erma di Bretschneider.

Bonghi Jovino, M. 1965. *Capua preromana. Terracotte votive. Catalogo del Museo Provinciale Romano*, I. Florence: Sansoni.

———. 1982. *La necropoli preromana di Vico Equense.* Cava dei Tirreni: Di Mauro.

———. 1990. "Artigiani e botteghe nell'Italia preromana. Appunti e riflessioni per un sistema di analisi." In *Artigiani e botteghe nell'Italia preromana. Studi sulla coroplastica di area etrusco-laziale campana. Studia archeologica* 56. Ed. M. Bonghi Jovino. 19–59. Rome: L'Erma di Bretschneider.

Bottini, A. 1983. "L'attività archeologica in Basilicata." *Atti Taranto* 22: 451–73.

———. 1988. "Elena in Occidente: una tomba dalla chora di Metaponto." *BdA* 73: 1–18.

———. 1994. "L'attività archeologica in Basilicata." *Atti Taranto* 33: 695–709.

———. 1995. "L'attività archeologica in Basilicata nel 1994." *Atti Taranto* 34: 759–70.

Bottini, A., and L. Lecce. 2013. "La mesogaia lucana e il caso di Pisticci." In *La comunicazione verbale tra Greci e indigeni in Apulia nel V–IV secolo a.C.: quali elementi? (Atti del seminario di studi linguistici, archeologici e storici, Bari 2012).* Quaderni di Ostraka XV. Ed. L. Todisco. 45–60. Naples: Loffredo Editore.

Bottini, A., and E. Setari. 2003. *La necropoli italica di Braida di Vaglio in Basilicata. Materiali dallo scavo del 1994. Monumenti Antichi* 60. Rome: Giorgio Bretschneider.

Bottini, P. 1988. *Archeologia, arte e storia alle sorgenti del Lao: catalogo della mostra. Castelluccio: un centro minore tra beni culturali e memoria storica.* Matera: BMG.

———. 1992. "I materiali della villa di Monte Moltone." In *Testimonianze archeologiche nel territorio di Tolve.* 67–108. Matera/Ferrara: Edizioni La Bautta.

———, ed. 1997. *Il Museo Archeologico Nazionale dell'Alta Val d'Agri.* Lavello: Alfagrafica Volonnino.

Boulter, C. 1953. "Pottery of the Mid-fifth Century from a Well in the Athenian Agora." *Hesperia* 22.2: 59–115.

Brizzi, M. 1999a. "Le anfore." In *Oppido Mamertina*, 299–309.

———. 1999b. "La ceramica comune." In *Oppido Mamertina*, 310–27.

———. 1999c. "*Instrumentu domesticum*-Miscellanea." In *Oppido Mamertina*, 337–43.

Brückner, H. 1980. *Marine terrassen in Süditalien. Eine quartärmorphologische Studie über das Kustentiefland von Metapont.* Düsseldorfer Geographische Schriften 14. Düsseldorf: Selbstverlag des Geographischen Institutes der Universität Düsseldorf.

Bruscella, A., and T. Virtuoso. 2005. "Ceramica a vernice nera." In *Torre di Satriano I*, 261–98.

Buckberry, J. L., and A. T. Chamberlain. 2002. "Age estimation from the auricular surface of the ilium: a revised method." *American Journal of Physical Anthropology* 119 (3): 231–39.

Burgers, G.-J. 1991. *Pizzica-Pantanello: Kiln Deposit. Grey Ware.* Unpublished report, Institute of Classical Archaeology.

Burn, L. 1998. "Figured Vases." In *Necropoleis*, 592–640.

Busanna, M. S. 2000. "Il materiale metallico da carpenteria." In *Il teatro romano di Asolo. Valore e funzione di un complesso architettonico urbano sulla scene del paesaggio.* Ed. G. Rosada. 113–18. Treviso: Canova.

Buzzi Auxilia, M. C. 1989. "Vasi per bere." In *Locri II*, 107–41.

Calamita, F., M. Coltorti, P. Pieruccini, and A. Pizzi. 1999. "Evoluzione strutturale e morfogenesi Plio-Quaternaria dell'Appennino Umbro-Marchegiano tra il Pedappennino Umbro e la costa Adriatica." *Bollettino della Società Geologica Italiana* 118: 125–39.

Callander, M. H. 1965. *Roman Amphorae with an Index of Stamps.* Oxford: University Press.

Campagna, L. 1995. "Cisterna e buca di scarico di età repubblicana a Vaste (LE), scavi di Fondo S. Antonio." *StAnt* 8.2: 215–88.

———. 2000. "Le anfore della necropoli in Contrada Portinenti (Proprietà Leone)." In *Meligunìs Lipára. 10. Scoperte e scavi archeologici nell'area suburbana di Lipari.* Eds. L. Bernabò Brea and M. Cavalier. 443–78. Palermo: Flaccovio.

Camporeale, G., ed. 1997. *L'abitato etrusco del'Accesa: il quartiere B. Archaeologica* 122. Rome: Bretschneider.

Campus, L. 1981. *Materiali del Museo Archeologico Nazionale di Tarquinia II. Ceramica attica a figure nere. Piccoli vasi e vasi plastici.* Rome: Giorgio Bretschneider.

Cannata, L. 2004. "Lo scavo in proprietà Zaffino (saggi Tomasello e Iannelli)." In *Kaulonía, Caulonia, Stilida (e oltri). Contributi storici, archeologici e topografici* II. *AnnPisa, s. IV, Quaderni* 17–18, 2004 (2007). Ed. M. C. Parra. 535–88. Pisa: Scuola normale di Pisa, Classe di lettere e filosofia.

Cappers, R. T. J., R. M. Bekker, and J. E. A. Jans. 2006. *Digital Seed Atlas of the Netherlands.* Eelde: Barkhuis Publishing.

Capurso, A. 1985. *Ginosa antica.* Bari: Dedalo.

Caputo, F. and L. Bubbico. 1983. *Insediamente benedettini nell'area metapontina. L'abbazia di S. Michele a Montescaglioso.* Montescaglioso: Motola s.n.c.

Caraccia, F. 2006. *Metodi di modellazione NURBS con Rhinoceros.* L'Aquila: Janotek.

Caramiello, R., and C. Siniscalco. 1997. "Studio archeobotanico nell'abitato di Pomarico vecchio (Matera)." In *Pomarico Vecchio*, 253–68.

Carandini, A., and A. Ricci, eds. 1985. *Settefinestre. Una villa schiavistica nell'Etruria romana* III. *La villa e i suoi reperti.* Modena: Panini.

Caravale, A., and I. Toffoletti. 1997. *Anfore antiche: conoscerle ed identificarle.* Rome: Ireco.

Caravelli, A. M. 2008. "Vasellame da mensa e da dispensa in ceramica comune." In *Civita di Tricarico*, 451–82.

Carè, B., and A. Cavallo. 2006. "Ceramica comune (CC)." In *Grammichele*, 297–98.

Carrara Ronzani, M. 1978. "Ceramica apula a vernice bruna e rossa." *Rivista di Studi Liguri* 44: 219–32.

Carter, J. C. 1979. "Rural Architecture and Ceramic Industry at Metaponto, Italy, 350–50 BC." In *Roman Brick and Tile*, 45–64.

———. 1980a. *Excavations in the Territory, Metaponto, 1980.* Annual Report, University of Texas at Austin.

———. 1980b. "A Classical Landscape. Rural Archaeology at Metaponto." *Archaeology* 33: 23–32.

———. 1982. "Ricerca archeologica nel territorio di Metaponto: nuovi risultati." *Atti Taranto* 22: 476–87.

———. 1983a. "Preliminary Report of the Excavation at Pizzica Pantanello (1974–1976)." In *Metaponto* II, 407–90.

———. 1983b. "Farmhouse: Sant'Angelo Grieco." In *The Territory of Metaponto 1981–1982*. 16–18. Austin: Institute of Classical Archaeology.

———. 1994. "Sanctuaries in the chora of Metaponto." In *Placing the Gods: Sanctuaries and Sacred Space in Ancient Greece*. Eds. S. E. Alcock and R. Osborne. 161–98. New York: Clarendon.

———. 1998a. "Burial Rites and Tomb Types." In *Necropoleis*, 57–111.

———. 1998b. "Rites at the Tomb and Grave Markers." In *Necropoleis*, 115–40.

———. 1998c. "Miniatures." In *Necropoleis*, 691.

———. 2006. *Discovering the Greek Countryside at Metaponto*. Ann Arbor: University of Michigan Press.

———. 2011. "The Historic Development of the Chora, 625–25 BC." In *Survey*, 641–922.

———. 2012. "Forme e processi di territorializzazione a Metaponto." *Atti Taranto* 50: 1103–30.

Carter, J. C., L. Burn, A. R. Parente, F. Silvestrelli, K. Swift, and E. Vittoria. 2011. "Catalog of the Objects from the 1999 Pizzica Excavations." In *Survey*, 1053–1105.

Carter, J. C., and J. Hall. 1998. "Burial Descriptions." In *Necropoleis*, 237–448.

Carter, J. C., and A. Parmly Toxey. 1998a. "Banded and Dipped Pottery." In *Necropoleis*, 695–715.

———. 1998b. "Unglazed Pottery." In *Necropoleis*, 721–28.

———. 1998c. "Alabastra." In *Necropoleis*, 757–69.

Casagrande, M. 2002. "I materiali." In *L'agorà di Eraclea Lucana*. Ed. G. Pianu. 113–224. Rome: Carocci.

Casalicchio, A. 2012. "Appendice: i materiali ceramici." In *Amphi Sirios*, 327–38.

Casero, P., F. Roure, L. Edignoux, I. Moretti, C. Muller, L. Sage, and R. Vially. 1988. "Neogene geodynamic evolution of the Southern Apennines." *Bollettino della Società Geologica Italiana* 41: 109–20.

Cassi, F., and L. Viviano, eds. 2006. *I Suoli della Basilicata. Carta pedologica della Regione Basilicata in scala 1:250,000.* Regione Basilicata: Dipartimento Agricoltura e Sviluppo Rurale, Direzione Generale. <http://www.basilicatanet.il/suoli/index.htm>

Castiglione, M. T. 1997. "Ricerche a Montagna dei Cavalli. La ceramica a vernice nera." In *Archeologia e territorio*. Ed. C. A. Di Stefano. 307–14. Palermo: G. B. Palumbo.

Castoldi, M. 1986. "Manufatti di pietra." In *I Greci sul Basento*. 115. Como: Edizioni New Press.

———. 1995. "Oggetti diversi." In *Incoronata* 3: *Ricerche archeologiche all'Incoronata di Metaponto*. 3, *L'oikos Greco del saggio S: Lo scavo e i reperti*. Scavi dell'Università degli studi di Milano, Dipartimento di Scienze dell'Antichità Sezione Archeologica. 109–11. Milan: Comune di Milano, Settore cultura e spettacolo.

———. 2000. "Macine dell'area dell'*oikos*." In *Incoronata* 4: *Ricerche archeologiche all'Incoronata di Metaponto.* 4, *L'oikos del grande perirrhanterion nel contesto del saggio G.* Scavi dell'Università degli studi di Milano, Dipartimento di Scienze dell'Antichità Sezione Archeologica. 47–48. Milan: Comune di Milano, Settore cultura e spettacolo.

Catmull, E., and J. Clark. 1978. "Recursively generated B-spline surfaces on arbitrary topological meshes." *Computer-Aided Design* 10 (6): 350–55.

Cavagnera, L. 1995. "Anfore commerciali." In *Ricerche archeologiche all'Incoronata di Metaponto.* 3: *L'oikos greco del saggio S. Lo scavo e i reperti.* Eds P. Orlandini and M. Castoldi. 41–55. Milan: Edizioni Et.

Cavalier, M. 1985. *Les amphores du VIe au IVe siècle dans le fouilles de Lipari. Cahiers des amphores archaïques et classiques* 1. Naples: Centre Jean Bérard.

Cavallo, A. 2014a. "Banded Ware." In *Fattoria Fabrizio*, 231–58.

———. 2014b. "Louteria." In *Fattoria Fabrizio*, 289–90.

Cerchiai, L. 2007. "Introduzione." In *Ceramiche fini*, 1–5.

Cerri, L., M. R. Ciuccarelli, and V. Lani. 2012. "Nuovi dati sul complesso produttivo di età romana a Pesaro." *Ocnus* 20: 51–60.

Chiartano, B. 1983. "La necropolis dell'età del ferro dell'Incoronata e di S. Teodoro (Scavi 1970–1974)." In *Metaponto II*, 9–190.

Ciampoltrini, G., and A. Andreotti. 2001. "Boscaioli, carrettieri, cacciatore. Il ponte di Botronchio di Orientano (Castelfranco di Sotto, Pisa)." *RasArch* 18B: 145–73.

Ciancio, A. 1985. "Tombe arcaico-classiche nei territori di Noicattaro e Valenzano (Bari). Scavi 1979–1981." *Taras* 5: 45–107.

———. 1997. *Silbíon: Una città tra Greci e indigeni; La documentazione archeologica dal territorio di Gravina in Puglia dall'ottavo al quinto secolo a.C.* Bari: Levante.

Ciaranfi, N., P. Pieri, and G. Ricchetti. 1988. "Note alla carta geologica delle Murge e del Salento (Puglia centro-meridionale)." *Memorie della Società Geologica Italiana* 41: 449–60.

Ciaranfi, N., M. Marino, L. Sabato, A. D'Alessandro, and R. De Rosa. 1996. "Studio geologico stratigrafico di una successione infra e mesopleistocenica nella parte sudoccidentale della Fossa Bradanica (Montalbano Ionico Basilicata)." *Bollettino della Società Geologica Italiana* 115: 379–91.

Cipriani, M. 1989. "Morire a Poseidonia nel V secolo. Qualche riflessione a proposito della necropoli meridionale." *DialArch* 3rd ser., 7.2: 71–91.

Cocchiaro, A. 1988. "La necropoli." In *La necropoli di via Cappuccini a Brindisi*. Eds. A. Cocchiaro and G. Andreassi. 63–230. Brindisi: Schena Editore.

Cocchiaro, A., and G. Andreassi, eds. 1988. *La necropoli di via Cappuccini a Brindisi.* Fasano: Schena Editore.

Colangelo, L. 2005a. "Ceramica a bande." In *Torre di Satriano I*, 251–60.

———. 2005b. "Ceramica a vernice rossa o bruna." In *Torre di Satriano I*, 299–310.

———. 2009. "Le necropoli arcaiche di Torre di Satriano. Distribuzione delle tombe e rituale funerario." *Siris* 10: 7–19.

Colivicchi, F. 2001. *Catalogo del Museo Nazionale Archeologico di Taranto* III, 2: *Alabastra Tardo-Ellenistici e Romani dalla necropoli di Taranto.* Materiali e Contesti, Ministero per i Beni Culturali e Ambientali, Soprintendenza Archeologica della Puglia. Taranto: Scorpione Editrice.

———. 2002. *La necropoli di Ancona (IV–I sec. a.C.). Una comunità italica fra ellenismo e romanizzazione. Quaderni di Ostraka* 7. Naples: Loffredo.

———. 2004. "I materiali minori." In *Gravisca 16*, 28–69.

Colucci, R. 2002. "Catalogo." In *Immagine e mito nella Basilicata antica. (Potenza, Museo Provinciale, dicembre 2002–marzo 2003).* Eds. M. Osanna and M. L. Nava. 141–78. Venosa: Osanna Edizioni.

Conti, M. C. 1983. "La ceramica comune." In *Locri Epizefiri: Ricerche archeologiche su un abitato della Magna Grecia, Mostra documentaria, Locri, ottobre 1983.* Soprintendenza Archeologica della Calabria. 34–38. Milan: Università di Torino, Istituto di Archeologia.

———. 1989. "La ceramica comune." In *Locri II*, 257–346.

———. 1992. "La ceramica comune." In *Locri IV*, 241–56.

Cook, R. M., and P. Dupont. 1998. *East Greek Pottery.* London: Routledge.

Corchia, R., O. Pancrazzi, and M. Tagliente. 1982. "Cavallino. Settore C dell'abitato. Fondo 'Aiera Vecchia.' Relazione preliminare." *StAnt* 3: 5–61.

Cordano, F. 2007. "Perché, dove e quando si scriveva sulle anfore commerciali arcaiche." *Annali della facoltà di Lettere e Filosofia dell'Università degli Studi di Milano*, Vol. 60, Fasc. 3: 16–21.

Corrado, M. 2010. "Bolli figurati impressi su tegole di V e IV secolo a.C. da Kroton: sulle tracce di Apollo «Delfiniere»." *Polis* 3: 35–58.

Corrente, M. 1997. "Monili a Minervino Murge tra V e IV secolo a.C." In *Atti del 15° Convegno Nazionale sulla Preistoria-Protostoria-Storia della Daunia, San Severo 1994.* Ed. A. Gravina. 145–78. San Severo: Archeoclub d'Italia.

Corretti, A., and C. Capelli. 2003. "Entella. Il granaio ellenistico (SAS 3). Le anfore." In *Quarte giornate internazionali di studi sull'area elima.* Ed. A. Corretti. 287–351. Pisa: Scuola Normale Superiore.

Costamagna, L. 1983. "Taranto: Indagini archeologiche nell'area di Piazza Garibaldi." *Taras* 3 (1–2): 101–23.

Costantini, L. 1980. "Semi e impronte da Sant'Angelo Vecchio." Unpublished report, Institute of Classical Archaeology.

———. 1983. "Piante spontanee e piante cultivate a S. Giovanni di Ruoti, Potenza." In *Lo scavo di S. Giovanni di Ruoti ed il period tardoantico in Basilicata: atti della tavola rotunda. Roma, 4 luglio 1981.* Eds. M. Gualtieri, M. Salvatore, and A. Small. 85–90. Bari: Adriatica.

———. 2002. "Italia centro-meridionale." In *Storia dell'agricoltura italiana, vol. I, L'età antica, tomo 1, Preistoria.* Eds. G. Forni and A. Marcone. 221–34. Florence: Ed. Polistampa.

Costantini, L., and L. Costantini Biasini. 2003. "Agriculture and diet in the chora of Metaponto: The paleobotanical evidence from Pantanello." In *Living off the chora: Diet and nutrition at Metaponto.* Ed. J. C. Carter. 3–12. Austin: Institute of Classical Archaeology.

Cotecchia, V., and G. Magri. 1967. "Gli spostamenti della linea di costa quaternaria del mar Ionico tra Capo Spulico e Taranto." *Geologia Applicata e Idrogeologia* 2: 1–27.

Cotton, M. A. 1977. "Plain and Cooking Ware." In *Gravina III*, 125–32.

———. 1979. *The Late Republican Villa at Posto, Francolise.* (Supplementary Publications of the British School at Rome). London.

———. 1983. "Coarse Ware." In *Cozzo Presepe*, 364–379.

———. 1992a. "Plain light-surfaced wares." In *Gravina II*, 162–78.

———. 1992b. "Dark-Surfaced or Cooking Ware." In *Gravina II*, 179–94.

———. 1992c. "Mortaria." In *Gravina II*, 194–97.

Coudin, F. 2009. *Les Laconiens et la Méditerranée à l'époque archaïque.* Collection du Centre Jean Bérard 33. Naples: Centre Jean Bérard.

Cracolici, V. 2001. "Le ceramiche. Tecniche e modi di produzione." In *Metaponto. Archeologia di una colonia greca.* Ed. A. De Siena. 103–14. Taranto: Scorpione.

———. 2003. *I sostegni di fornace dal* kerameikos *di Metaponto.* Beni Archeologici–Conoscenza e Tecnologie. Quaderno 3. Bari: Edipuglia.

Crawford, M. H. 1974. *Roman Republican Coinage.* 2 vols. Cambridge: University Press.

———. 2003. "Brave New World: Metapontum after Metapontum." In M. Cébeillac-Gervasoni, *Les élites et leurs facettes.* Collection de l'École Française de Rome 309. 15–30. Rome: L'École Française de Rome.

Cremonesi, G. 1984. "Latronico: La Grotta no. 3." In *Testimonianze archeologiche nel territorio di Latronico.* Ed. S. Bianco. 27–32. Galatina: Congedo Editore.

Cubberley, A. L., J. A. Lloyd, and P. C. Roberts. 1988. "*Testa* and *clibani*: the baking covers of classical Italy." *PBSR* 55: 99–119.

Cultrera, G. 1938. "Siracusa. Rovine di un antico stabilimento idraulico in contrada Zappalà." *NSc* 6: 261–301.

Cuomo di Caprio, N. 1971–1972. "Proposta di classificazione delle fornaci per ceramica e laterizi nell'area italiana." *Sibrium* 11: 371–464.

———. 1979. "Pottery and Tile Kilns in South Italy and Sicily." In *Roman Brick and Tile*, 45–64.

———. 1985. *La ceramica in archeologia. Antiche tecniche di lavorazione e moderni metodi di indagine.* Rome: L'Erma di Bretschneider.

———. 1992. "Les ateliers de potiers en Grande Grèce." In *Les ateliers de potiers dans le monde grec aux époques géometrique, archaïque et classique.* Actes de la Table Ronde organisée à l'École Française d'Athènes, les 2 et 3 octobre 1987. Eds. F. Blondé and J. Y. Perreault. 69–85. Paris: De Boccard.

———. 2007. *Ceramica in archeologia. Antiche tecniche di lavorazione e moderni metodi di indagine.* 2nd ed. Studia Archaeologica 144. Rome: L'Erma di Bretschneider.

Cuozzo, M. 2006. "La ceramica in argilla depurata a decorazione lineare." In *Cuma 2*, 88–91.

Cuozzo, M. and A. D'Andrea. 1991. "Proposta di periodizzazione del repertorio locale di Pontecagnano tra la fine del VII e la metà del V sec. a.C. alla luce della stratigrafia della necropolis." *Annali dell'Istituto Orientale di Napoli* 13: 47–114.

d'Agostino, B. 1968. "Marcina?" *Dialoghi di Archeologia* 2: 139–51.

D'Amicis, A. 1988. "La necropoli di S. Lucia." In *Museo di Taranto*, 123–72.

D'Andria, F. 1976. "Metaponto romana." *Atti Taranto* 15: 539–44.

———. 1977a. "Scavi a Metaponto (1976)." *Atti Taranto* 16: 855–59.

———. 1977b. "Ceramografi metapontini." *Magna Graecia* 12: 3–4, 9–10.

———. 1978. "Metaponto fornaci (1977)." *Atti Taranto* 17: 409–12.

———. 1980a. "Scavi nella zona del *kerameikos* (1973)." In *Metaponto I*, 355–452.

———. 1980b. "I materiali del V sec. a.C. nel ceramico di Metaponto e alcuni risultati delle analisi delle argille." In *Attività archeologica*, 117–45.

———. 1980c. "Les potiers de Metaponte." *Archeologia* 147: 42–50.

———. 1997a. "Fingering Ancient Greek Potters." *Science* 275, 7/3/1997: 1425.

———. 1997b. "Detectives a Metaponto." *Archeo. Attualità del passato* 13.5 (147), maggio 1997: 34–39.

———., ed. 2009. *Castrum Minervae*. Galatina: Congedo.

D'Angella, D. 2006. *Storia civile e religiosa di Santa Maria del Casale di Pisticci*. Pisticci: Tipografia I.M.D.

D'Annibale, C. 2007. "Metaponto Survey 2007." In *The Study of Ancient Territories: Chersonesos & Metaponto. 2006–2007 Annual Report*. 63–65. Austin: Institute of Classical Archaeology.

———. 2012. "Pottery and Other Neolithic Artifacts." In *San Biagio*, 55–65.

———. 2014. "Marine Shells." In *Fattoria Fabrizio*, 143–48.

———. Forthcoming. "Lithic Material." In *The Chora of Metaponto: Incoronata*.

D'Argenio, B., and W. Alvarez. 1980. "Stratigraphic evidence for crustal thickness changes on the southern Tethyan margin during the Alpine cycle." *Geological Society of America Bulletin* 91: 681–89.

Dai Pra, G., and P. J. Hearty. 1988. "I livelli marini pleistocenici del Golfo di Taranto. Sintesi geocronologica e tettonica." *Memorie della Società Geologica Italiana* 41: 637–44.

Dai Pra, G., and S. Stearn. 1977. "Sul Tirreniano di Taranto. Datazione sui coralli con il metodo del Th330/U 234." *Geologica Romana* 16: 231–42.

Dal Sasso, C. 1995. "Reperti faunistici." In *Ricerche archeologiche all'Incoronata di Metaponto* 3: *L'oikos greco del saggio S. Lo scavo e i reperti*. 129–39. Milan: Scavi dell'Università degli studi Milano Istituto di Archeologia.

Damgaard Andersen, H., and H. W. Horsnaes. 2007. "Common Ware." In *Hellenistic and Roman Pontecagnano. The Danish Excavations in proprietà Avallone 1986–1990*. Ed. B. Tang. 94–105. Naples: Centre Jean Bérard.

Daremberg, C., and E. Saglio. 1877–1919. *Dictionnaire des antiquités grecques et romaines, d'après les textes et les monuments*. Paris: Hachette.

Davis, D. 2003. "2003 Gradiometer Survey in the Chora of Metaponto." In *The Study of Ancient Territories: Chersonesos & Metaponto. 2003 Annual Report*. 77–80. Austin: Institute of Classical Archaeology.

Davis, D., et al. 2003. *Report on Four Gradiometric Surveys in the Chora of Metaponto*. Unpublished report, Institute of Classical Archaeology.

Davis, O. K. 1987. "Spores of the dung fungus *Sporormiella*: increased abundance in historic sediments and before Pleistocene megafaunal extinction." *Quaternary Research* 28: 290–94.

de Cazanove, O. 1996. "Un edifice de repas communautaires en Lucanie interne." *MEFRA* 108.2: 901–41.

———, et al. 2008. "Terre cuites figurées, vases miniatures, thymiateria, pesons, opus doliare." In *Civita di Tricarico*, 521–42.

De Faveri, C. 1999. *Incoronata di Pisticci: Abitato e materiali dell'eta del Ferro: tipologia delle forme ceramiche*. Tesi di Laurea, Facolta di Lettere e Filosofia, Dipartimento di Scienze dell'Antichità, Università degli studi di Padova.

De Juliis, E. 1990. *L'ipogeo dei vimini di Canosa*. Bari: Laterza.

de la Genière, J. 1968. *Recherches sur l'age du fer en Italie Méridionale. Sala Consilina*. Bibliothèque de l'Institut Français de Naples. Série 2.1,1. Naples: Institut Français de Naples.

de la Genière, J., and G. Greco. 2010. "Area del tempio." In *Santuario di Hera*, 130–260.

De Leo, R. 2008. *Riforma agraria e politiche di sviluppo. L'esperienza in Puglia, Lucania e Molise (1951–1976).* Matera: Antezza.

De Marco, A. 1990. "Rapporti tra geodinamica e sedimentazione della Fossa Bradanica durante il Pleistocene: testimonianze mineralogiche." *Bollettino della Società Geologica Italiana* 109: 313–24.

De Miro, E. 2000. *Agrigento I. L'area sacra tra il tempio di Zeus e Porta V.* Rome: L'Erma di Bretschneider.

De Mitri, A. 1978. "I materiali della Grotta del Diavolo." In *Leuca: Archeologia, storia e preistoria.* 13–25. Galatina: Congedo Editore.

De Mitri, C. 2009a. "Castro. Lo scavo in via IV Novembre." In *Castrum Minervae.* Ed. F. D'Andria. 199–214. Galatina: Congedo.

———. 2009b. "Castro. Lo scavo in località 'Muraglie': nuovi dati sul circuito murario di età messapica." In *Castrum Minervae.* Ed. F. D'Andria. 121–97. Galatina: Congedo.

De Philippis, A. 1937. "Classificazioni ed indici del clima in rapporto alla vegetazione forestale italiana." *Nuovo Giornale Botanico Italiano* 44: 1–169.

De Siena, A. 1975. "Ceramica. Saggio 'Y/75'." *BdA* 60: 39–45.

———. 1980. "Note stratigrafiche sul santuario di Apollo Licio a Metaponto." In *Attività archeologica*, 83–116.

———. 1993. "Metaponto e il metapontino." In *Da Leukania a Lucania*, 114–35.

———. 1998. "Metaponto: problemi urbanistici e scoperte recenti." In *Siritide e metapontino. Storia di due territori coloniali. Atti dell'incontro di studio (Policoro, 31 ottobre-2 novembre 1991).* Cahiers du Centre Jean Bérard 20. 141–71. Naples-Paestum: Centre Jean Bérard.

———. 1999. "La colonizzazione achea del Metapontino." In *Storia della Basilicata 1. L'antichità.* Ed. D. Adamesteanu. 211–45. Rome/Bari: Laterza.

———. 2005. "Tramonto della Magna Grecia: la documentazione archeologica dai territori delle colonie greche di Metaponto e Herakleia." *Atti Taranto* 44: 433–58.

———. 2007a. "Il santuario extraurbano di S. Biagio alla Venella nel territorio di Metaponto." In *Il santuario extraurbano di San Biagio alla Venella.* Ed. J. Carella. 9–17. Bernalda: La Spiga d'Oro.

———. 2007b. "L'attività archeologica in Basilicata nel 2006." *Atti Taranto* 46: 405–63.

———. 2010. "Il santuario extraurbano delle Tavole Palatine: ruolo e tradizione." In *Il tempio di Hera (Tavole Palatine) di Metaponto. Archeologia, archeometria, conservazione.* Ed. L. Lazzarini. 13–15. *Marmora* 5/2009, Suppl. 2. Pisa/Rome: Fabrizio Serra.

———. 2011. "Attività della Soprintendenza per i Beni Archeologici della Basilicata." *Atti Taranto* 49: 609–55.

De Siena, A., and L. Giardino. 1994. "*Herakleia* e Metaponto. Trasformazioni urbanistiche e produzione agricola tra tarda repubblica e primo impero: i nuovi dati archeologici." In *Le ravitaillement en blé de Rome et des centres urbains des débuts de la République jusqu' Haut Empire.* Actes du colloque international (Naples 14–16 Février 1991). 197–211. Naples-Rome.

———. 2001. "Trasformazioni delle aree urbane e del paesaggio agrario in età romana nella Basilicata sudorientale." In *Modalità insediative e strutture agrarie nell'Italia meridionale in età romana.* Eds. E. Lo Cascio and A. Storchi Marino. 129–67. Bari: Edipuglia.

Dell'Aglio, A. 1996a. "Taranto." In *Arte e artigianato*, 51–80.

———. 1996b. "La ceramica a vernice nera. Taranto." In *Arte e artigianato*, 322–32.

Della Calce, R. 2004. *Maya 5: elaborazione e animazione grafica 3D.* Milan: Apogeo.

Delli Ponti, G. 1996. "Vaste–Poggiardo (LE). La necropoli di Fondo Aia." *StAnt* 9: 99–214.

Denoyelle, M. 2002. "Il mito greco in Occidente nel V. secolo: Metaponto ed Herakleia," In *Immagine e mito nella Basilicata antica. (Potenza, Museo Provinciale, dicembre 2002–marzo 2003).* Eds. M. Osanna and M. L. Nava. 104–12. Venosa: Osanna Edizioni.

De Stefano, A. 2008. "Un contesto ceramico di età repubblicana e primo/imperiale dall'area delle due *domus.*" In *Ordona XI*, 45–144.

De Stefano, F. 2014. "Il repertorio iconografico del santuario di S. Biagio alla Venella (Metaponto) all'alba della colonia." *Antesteria* 3: 157–69.

Desy, P., and P. De Paepe. 1990. "Torre San Giovanni (Ugento): Les amphores commerciales hellenistiques et republicaines." *StAnt* 6: 187–234.

Di Fraia, T. 1970. "Resti di un Villaggio della Cultura Tipo Serra d'Alto a Saldone presso Metaponto (Lucania)." *ASTSN* 77: 54–77.

Di Giuseppe, H. 1998. "La furnace di Calle (Tricarico): produzione e diffusione." In *Ceramica in Italia*, 735–52.

———. 2012. *Black-Gloss Ware in Italy. Production, Management and Local History. BAR-IS* 2335. Oxford: Oxford University Press.

Di Giuseppe, H., and C. Capelli. 2005. "Produzioni urbane e rurali di ceramica comune dipinta nella Lucania tardo-antica e altomedievale." In *LCRW I*, 395–411.

Dilthey, H. 1980. "Sorgenti, acque, luoghi sacre in Basilicata." In *Attività archeologica*, 83–116.

Di Meo, A. 1801. *Annali Critico-Diplomatici della Mezzana Età (a. 568–1202).* Naples 1795–1819, IX. Edizione anastatica.

Dimbleby, G. W. 1985. *The palynology of archaeological sites.* London: Academic Press.

Di Sandro, N. 1986. *Le anfore archaiche dello scarico Goesttì, Pithecusa.* Cahiers du Centre Jean Bérard 13. Naples: Centre Jean Bérard.

———. 1990. "Aspetti quantitative e tipologici delle anfore marsigliesi in Magna Grecia e Sicilia." In *Les amphores de Marseille grecque. Chronologie et diffusion (VI-I s. av. J.-C.).* Atti della tavola rotonda (Lattes 1986). Ed. M. Bats. 235–39. Aix-en-Provence: A.D.A.M. edition.

Distasi, V., and T. Perretti. 2006. "Catalogo." In *Con il fuso,* 219–21.

Doepner, D. 2002. *Steine und Pfeiler für die Götter Weihgeschenkgattungen in westgriechischen Stadtheiligtümern.* Wiesbaden.

Donati, L. 1994. *La casa dell'Impluvium. Architettura etrusca a Roselle.* Rome: Giorgio Bretschneider.

Donnarumma, R., and L. Tomay. 1999. "La necropoli di San Brancato di Tortora." In *Nella terra degli Enotri: Atti del convegno di studi (Tortora, 18–19 aprile 1998).* Eds. F. La Torre and A. Colicelli. 49–59. Paestum: Pandemos.

Doulgeri-Intzesiloglou, A. 1997. "Un atelier de produits céramiques moulés à Phères." In *Moulage,* 295–314.

Dubois, L. 2002. *Inscriptions grecques dialectales de Grande Grèce* II. *Colonies achéennes.* Geneva.

Duday, H. 2006. *Lezioni di archeotanatologia: archeologia funeraria e antropologia di campo. Testo del corso specialistico intensive di Roma (18 ottobre–6 novembre 2004).* Rome: Soprintendenza archeologica.

Duncan, G. C. 1964. "A Roman Pottery near Sutri." *PBSR* 32: 38–88.

du Plat Taylor, J. 1983. "Wheelmade Painted Pottery." In *Metaponto II,* 331–46.

Dyson, S. L. 1976. *Cosa: The Utilitarian Pottery.* Rome: American Academy in Rome.

Edlund, I. E. M. 1979. *The University of Texas Excavations at Metaponto, 1979.* Austin: University of Texas at Austin.

———. 1982a. *The University of Texas Excavations at Metaponto: S. Angelo Vecchio 1979–1981.* Unpublished report, Institute of Classical Archaeology.

———. 1982b. *The Rural Cemetery at Sant'Angelo Vecchio in the Territory of Metaponto.* Unpublished report, Institute of Classical Archaeology.

———. 1984. "Sacred and Secular: Evidence of Rural Shrines and Industry Among Greeks and Etruscans." In *Crossroads of the Mediterranean.* Eds. T. Hackens, N. D. Holloway, and R. Ross Holloway. Archeologia Transatlantica II. 277–90. Louvain la Neuve: Université Catholique de Louvain.

———. 1986. "Scavi nella zona di Metaponto: Sant'Angelo Vecchio." *BdA* 71, Ser. 6, Vols. 39–40: 119–22.

———. 1987. *The Gods and the Place. Location and Function of Sanctuaries in the Countryside of Etruria and Magna Graecia (700–400 B.C.).* Acta Instituti Romani Regni Sueciae 4°, XLIII. Stockholm: Svenska Institutet i Rom.

Edlund-Berry, I. E. M. 2011. "Artisans and trade in the Etruscan World." In *ThesCRA VI,* 353–55.

Elia, D. 2006. "Ceramica fine dei secoli IV e III a.C. (FI.)." In *Grammichele,* 157–214.

Elliott, M. 1998. "Black-Glazed Pottery." In *Necropoleis,* 643–93.

Empereur, J.-Y., and Y. Garlan, eds. 1986. "Recherches sur les amphores grecques." *Bulletin de Correspondance Hellénique* Suppl. 13: 581–86.

Ersoy, Y. 1993. *Clazomenae: the Archaic Settlement.* Doctoral dissertation, Bryn Mawr College.

Esposito, A. M. 1996. "Ceramica arcaica figurata." In *Arte e artigianato,* 306–12.

Esposito, A., and G. M. Sanidas. 2012. "La question des regroupments des activités économiques et le concept des 'quartier d'artisans': quelle approche?" In *"Quartiers" artisanaux en Grèce ancienne: une perspective méditerranéenne.* Eds. A. Esposito and G. M. Sanidas. 11–21. Lille: Presses Universitaires du Septentrion.

Ettlinger, E. 1990. *Conspectus formarum terrae sigillatae Italico modo confectae.* Bonn: R. Habelt.

Euwe, J. 1985. "The Shape of the Early Nolan Amphorae (490–480 BC): the Origin and Relation with Other Small Neck Amphorae." In *Ancient Greek and Related Pottery. Proceedings of the International Vase Symposium in Amsterdam, 12–15 April 1984.* Ed. H. A. G. Brijder. 144–50. Amsterdam: Allard Pierson Museum.

Faegri, K., P. E. Kaland, and K. Krzywinski. 1989. *Textbook of pollen analysis.* 4th ed. London: Wyley & Son.

Falcinelli, F., and M. Moraldi. 2001. "Il pero mandorlino (Pyrus Amygdaliformis Villars)." *Sherwood, foreste e alberi oggi* 68: 39–42.

Fancelli, P. 2001. "Indagini preliminari e diagnostica. Rilievo storico-critico delle fasi costruttive." In *Il manuale del restauro architettonico.* Ed. L. Zevi. E13. Rome: Mancosu.

Ferla, P., R. Alaimo, G. Falsone, and F. Spatafora. 1984. "Studio petrografico delle macine di età arcaica e classica da Monte Castellazzo di Poggioreale (Sicilia Occidentale). Rinvenimento di materiale vulcanico in gran parte estraneo alla Sicilia ed implicazioni storico-archeologiche." *Sicilia Archeologica* 56, No. 17: 25–52.

Ferrandini Troisi, F., B. R. Buccoliero, and D. Ventrelli. 2012. *Coroplastica tarantina: le matrici iscritte.* Bari: Edipuglia.

Ferrara, B. 2010. "La ceramica e vernice nera." In *Santuario di Hera,* 443–80.

Ferri, S. 1927. "Tiriolo: trovamenti fortuiti e saggi di scavo." *NSc* 6th ser., Vol. 3: 337–59.

Fiorani, D. 2011. "Il progetto della scuola di Specializzazione in Archeologia di Matera sul santuario di Timmari, località Lamia di San Francesco. Gli oggetti del rituale. Ceramica comune." In *Brateís datas,* 123–30.

Fiori, F. 2005. "L'instrumentum metallico di Monte Bibele." In *Studi sulla media e tarda età del ferro nell'Italia settentrionale.* Studi e scavi 12. Ed. D. Vitali. 149–214. Bologna: Università di Bologna, Dipartimento di archeologia, Ante Quem.

Firmati, M. 2012. "Pesi da rete e pesi da telaio." In *Le fornaci romani di Giancola (Brindisi). Le fornaci romane di Giancola (Brindisi). Bibliotheca archaeologica* 27. Eds. D. Manacorda and S. Pallecchi. 204–6. Bari: Edipuglia.

Fischer-Hansen, T. 2000. "Ergasteria in the Western Greek World." In *Polis and Politics: Studies presented to Mogens Herman Hansen*. Eds. P. Flensted-Jense, Th. Heine Nielsen, and L. Rubinstein. 91–120. Copenhagen.

Florenzano, A. 2014. "Archaeobotany at Fattoria Fabrizio." In *Fattoria Fabrizio*, 133–38.

Florenzano, A., S. Benassi, and A. M. Mercuri. 2012. "Pioggia pollinica e qualità dell'aria: polline di Olea negli uliveti dal caso studio della regione Basilicata (sud Italia)." *Atti della Società dei Naturalisti e Matematici di Modena* 142: 171–85.

Florenzano, A., I. Massamba N'siala, A. M. Mercuri, and D. Roubis. 2009. "La ricostruzione dell'ambiente della valle del fiume Bradano in età ellenistica (Basilicata orientale)." *Atti della Società dei Naturalisti e Matematici di Modena* 139: 145–55.

Florenzano, A., A. M. Mercuri, A. Pederzoli, P. Torri, G. Bosi, L. Olmi, R. Rinaldi, and M. Bandini Mazzanti. 2012. "The significance of intestinal parasite remains in pollen samples from Medieval pits in the Piazza Garibaldi of Parma, Emilia Romagna, Northern Italy." *Geoarchaeology: An International Journal* 27: 34–47.

Folk, R. L. 2011. "Geologic Background of the Metapontino." In *Survey*, 3–29.

Foxhall, L. 2011. "Loom weights." In *Survey*, 539–54.

———. 2012. "Family time. Temporality, gender and materiality in ancient Greece." In *Greek notions of the past in the archaic and classical eras. History without historians*. Eds. J. Marincola, L. Llewellyn-Jones, and C. Maciver. 183–206. Edinburgh: University Press.

Fracchia, H. 1994. "La ceramica comune e da cucina di età repubblicana e della prima e media età imperiale." In *Il complesso archeologico di Carminiello ai Mannesi, Napoli (scavi 1983–1984)*. Ed. P. Arthur. 173–79. Galatina.

———. 2001. "L'*oppidum* di M. Capitenali. L'area di Napoli. The ceramic evidence." In *Roccagloriosa II*, 38–46.

Fracchia, M., and D. Girardot. 1986. "Roccagloriosa (SA). Materiali da un pozzo di scarico sigillato." *Klearchos* 109–12: 127–78.

Fracchia, M., and H. Keith. 1990. "Coarse Ware." In *Roccagloriosa I*, 262–77.

Franchi Dell'Orto, L., and R. Franchi. 1988. *Veder greco. Le necropoli di Agrigento (Mostra internazionale Agrigento, 2 maggio–31 luglio 1988)*. Rome: L'Erma di Bretschneider.

Freed, J. 1994. "The pottery." In *San Giovanni di Ruoti I*, 51–57, 62–63, 65–66, 68–71, 82–85, 102–6.

Fuchs, F. 1980. "Quartäre Küsten- und Flussterrassen in der Umrahmung des Golfs von Tarent (süd-Italien)." *Catena* 7: 27–50.

Gabrieli, R. S. 2011. "Greek and Roman Republican Cooking Wares." In *Survey*, 439–54.

Gagliardi, V. 2010. "La vernice nera di produzione attica a Kaulonia: dati quantitativi e novità tipologiche." In *Caulonia tra Crotone e Locri, Atti del Convegno Internazionale, Firenze 30 maggio–1 giugno 2007*. Eds. L. Lepore and P. Turi. 67–80. Florence: Firenze University Press.

Garaffa, V. 2011. "Garaguso: i depositi votivi di 'Grotta delle Fontanelle": il deposito 'Autera'." In *Brateís datas*, 39–48.

Gargano, M. P. 2001. "Catalogo." In *Un luogo della Peucezia. Le scoperte archeologiche in contrada Bigetti*. Ed. A. Damato. 65–121. Palo del Colle: Liantonio.

Garlan, Y. 1986. "Quelque nouveaux ateliers amphoriques à Thasos." In *Recherches sur les amphores grecques. (Actes du Colloque international organisé par le Centre National de la Recherche Scientifique, l'Université de Rennes II et l'Ecole Française d'Athènes, Athènes 10–12 septembre 1984; Bulletin de Correspondance Hellénique Suppl. 13.)* Ed. J.-Y. Empereur. 200–276. Paris: École Française d'Athènes.

Garrozzo, B. 1995. "Bolli su coppi ed embrici in Segesta. Parco archeologico e relazioni preliminari delle campagne di scavo 1990–1993 (parte seconda)." *ASNSP* S. 3, 25: 1187–1204.

Gassner, V. 1996. "Zur Entstehung des Typus der ionisch-massiliotischen Amphoren." In *Fremde Zeiten Festschrift für J. Borchardt zum sechzigsten Geburtstag*. Ed. F. Blakolmer. 165–76. Vienna: Phoibos-Verlag.

———. 1997. *Das Sudtor der Tetragonos-Agora: Keramik und Kleinfunde*. Vienna: Osterreichische Akademie der Wissenschaften.

———. 2003. *Materielle Kultur und kulturelle Identität in Elea in spätarchaisch-frühklassischer Zeit. Untersuchungen zur Gefäß- und Baukeramik aus Unterstadt (Grabungen 1987–1994). Velia-Studien II.* Vienna: Österreichische Akademie der Wissenschaften.

Gebhard, E. R. 1998. "Small Dedications in the Archaic Temple of Poseidon at Isthmia." In *Proceedings of the Fourth International Seminar on Ancient Greek Cult*. Ed. R. Hagg. 91–115. Athens: Swedish Institute at Athens.

Giacometti, D. 2005. *Metaponto. Gli dèi e gli eroi nella storia di una polis di Magna Grecia*. Cosenza.

Giannini, C., and R. Roani. 2000. *Dizionario del restauro e della diagnostica*. Fiesole: Nardini.

Giannotta, M. T. 1980. *Metaponto ellenistico-romana: problemi topografici*. Galatina: Congedo.

———. 1994. "Tombe messapiche da Vaste e Ortelle." *StAnt* 7: 85–110.

———. 1995. "Rinvenimenti tombali da Rocavecchia (1934) al Museo Nazionale di Taranto." *StAnt* 8.2: 39–74.

———. 1996. "Rinvenimenti tombali da Rocavecchia (1934) al Museo Provinciale di Lecce: materiali di corredo e produzioni." *StAnt* 9: 37–98.

———. 1997. "Una tomba di guerriero del IV sec. a.C. e la necropoli extra-urbana di Vaste-Fondo Aia." *StAnt* 10: 153–86.

Giardina, A. 1997. *L'Italia romana. Storie di un'identità incompiuta*. Bari: Laterza.

Giardino, L. 1978. "Le ceramiche ellenistiche e romane." In *Leuca*, 121–60.

———. 1980. "Sulla ceramica a pasta grigia di Metaponto e sulla presenza in essa di alcuni bolli iscritti: studio preliminare." *StAnt* 2: 247–87, tavv. 81–86.

———. 1990. "Heraclea, necropoli meridionale: Le sepolture di II e I sec. a.C." In *Emergenze e problemi*, 73–125.

———. 1993. "Herakleia, necropoli meridionale. Via Umbria-Tomba 92." In *Da Leukania a Lucania*, 179–82.

———. 1996. "Herakleia." In *Arte e artigianato*, 35–43.

———. 1998. "Herakleia (Policoro). Contesti e materiali arcaici dal settore della Collina del Castello." In *Siritide e metapontino. Storia di due territori coloniali. Atti dell'incontro di studio (Policoro, 31 ottobre–2 novembre 1991)*. Cahiers du Centre Jean Bérard 20. 105–22. Naples-Paestum: Centre Jean Bérard.

———. 2005. "Herakleia e Metaponto: dalla *polis* italiota all'abitato proto imperiale." *Atti Taranto* 44: 387–432.

———. 2012. "The Farmhouse at San Biagio and the Agricultural Landscape of Basilicata in the Roman Period." In *San Biagio*, 1–18.

———. 2015. "Aree urbane e territori della costa ionica della Basilicata tra Pirro e Annibale." *Atti Taranto* 52: 571–617.

Gignoux, M. 1913. *Les formations marines, pliocènes et quaternaires de l'Italie du sud et de la Sicile. Annales de l'Université de Lyon 36*. Paris: Baillière & Fils.

Gigout, M. 1960. "Sur le Quaternaire marin de Tarente (Italie)." *Comptes Rendus de l'Académie des Sciences* 250: 1094–96.

Giumlia Mair, A. 2002. "Piombo: il metallo velenoso." In *Arti di Efesto*, 108–9.

Gori, B., and T. Pierini. 2001. *Gravisca. Scavi nel santuario Greco. La ceramica comune*. Bari: Edipuglia.

Grace, V. R. 1956. "Stamped Wine Jar Fragments." In *Small objects from the Pnyx:* II. Eds. L. Talcott, B. Philippaki, G. R. Edwards, and V. R. Grace. *Hesperia* Suppl. 10: 113–78.

Graepler, D. 1997. *Tonfiguren im Grab: Fundkontexte hellenistischer Terrakotten aus der Nekropole von Tarent*. Munich: Biering & Brinkmann.

Gras, M. 1987. "Amphores commerciales et histoire archaïques." *Dialoghi di Archeologia* 5.3: 41–50.

Gras, M., H. Tréziny, and H. Broise. 2004. *Mégara Hyblaia 5. La ville archaïque. L'espace urbain d'une cité grecque de Sicile orientale*. Mélanges d'archéologie et d'histoire. Suppléments 1/5. Rome: L'École Française de Rome.

Greco, A., and E. Lapadula. 2004. "L'insediamento di età tardoromana in località Badia, Cutrofiano (Lecce). Dat I preliminari sulla ceramica dipinta." In *Museo della Ceramica di Cutrofiano. Quaderno 8-9*. Ed. S. Matteo. 11–38. Galatina: Congedo.

Greco, C. 2009. "Attività della Soprintendenza per i Beni Archeologici della Basilicata." *Atti Taranto* 48: 785–824.

Greco, G. 1991. *Serra di Vaglio. La "casa dei pithoi."* Modena: Panini.

———. 1996. "Per una definizione dell'architettura domestica di Serra di Vaglio." In *Ricerche sulla casa*, 255–99.

Green, J. R. 1972. "Oinochoe." *Bulletin of the Institute of Classical Studies* 19: 1–16.

Gruat, P., C. Bénévent, V. Le Fillatre, and G. Marty. 1999. "L'Alzou: un atelier de tuiliers rutène du Ier siècle (Bourzanel)." *Vivre en Rouergue, Cahiers d'Archéologie Aveyronnaise* 13: 65–108.

Gualtieri, M. 1990. "Terrecotte architettoniche, louteria, elementi architettonici in pietra, macine, vetro." In *Roccagloriosa I*, 300–310.

Guarducci, M. 1978. *Epigrafia Greca* IV. *Epigrafi sacre pagane e cristiane*. Rome.

———. 1995. *Epigrafia greca*. 2nd ed. Rome: Istituto Poligrafico dello Stato.

Guglielmino, R. 2000. "Entella. Un'area artigianale extraurbano di età tardo arcaica." In *Atti delle terse giornate internazionali di studi sull'area elima*. (*Ghibellina–Erice–Contessa Entellina, 23–26 ottobre 1997*.) 701–713. Pisa: Scuola Normale di Pisa.

Guzzo, P. G. 1978. "Acquappesa. Loc. Aria del Vento (Cosenza). Scavo di una struttura di epoca ellenistica." *NSc* 32: 465–479.

———. 1987. "I documenti per lo studio della produzione artigianale (VII–IV sec. a.C.)." In *Storia di Calabria antica*. Ed. S. Settis. 429–474. Rome: Gangemi.

Guzzo, P. G., A. Bedini, R. Spadea, and L. Rota. 1973. "Campagna di scavo 1970: Stombi." In *Sibari* II: *Scavi al Parco del Cavallo (1960–1962; 1969–1970)*. *NSc* 24: 216–366.

Guzzo, P. G., M. N. Pagliardi, U. Spigo, and L. Rota. 1974. "Descrizione dei materiali." In *Sibari* III. *Rapporto preliminare della campagna di scavo: Stombi, Casa Bianca, Parco del Cavallo, San Mauro (1971)*. *NSc* 8th ser., 26, Suppl. No. 3. 48–146. Rome: Accademia Nazionale dei Lincei.

Guzzo, P. G., L. Rota, and F. Daddi. 1973. "Parco del Cavallo." In *Sibari* II. *Scavi al Parco del Cavallo (1960–62; 1969–70) e agli Stombi (1969–70)*. *NSc* 8th ser., 24, Suppl. No. 3. 367–547. Rome: Accademia Nazionale dei Lincei.

Hall, J. 1998. "Grave Goods." In *Necropoleis*, 563–90.

Hänsel, B. 1973. "Policoro (Matera). Scavi eseguiti nell'area dell'acropoli di Heraclea negli anni 1965–1967." NSc 27: 400–492.

Hasaki, E. 2002. *Ceramic Kilns in Ancient Greece: Technology and Organization of Ceramic Workshops*. Doctoral dissertation, University of Cincinnati, Cincinnati, Ohio.

Hayes, J. W. 1981. *Corpus Vasorum Antiquorum. Canada. Royal Ontario Museum, Toronto*. Oxford/New York: Oxford University Press.

———. 1984. *Greek and Italian Black-Gloss Wares and Related Wares in the Royal Ontario Museum*. Toronto: Royal Ontario Museum.

———. 1985. "Sigillate orientali." In *Atlante II*, 1–96.

———. 1997. *Handbook of Mediterranean Roman Pottery*. Norman: University of Oklahoma Press.

Hearty, P., and G. Dai Pra. 1992. "The age and stratigraphy of Middle Pleistocene and younger deposits along the Gulf of Taranto (Southeast Italy)." *Journal of Coastal Research* 8: 882–905.

Hempel, K. G. 2001. *Die Nekropole von Tarent im 2. und 1. jahrhundert v. Chr.: Studien zur materiellen Kultur*. Taranto: Scorpione Editrice.

Henneberg, M., and R. J. Henneberg. 1998. "Biological Characteristics of the Population Based on Analysis of Skeletal Remains." In *Necropoleis*, 503–59.

Herdejürgen, H. 1978. *Götter, Menschen und Dämonen: Terrakotten aus Unteritalien*. Basel-Mainz: Archäologischer Verlag und Philipp von Zabern.

———. 1982. "Tarentinische Antefixe." In *Antike Kunstwerke aus der Sammlung Ludwig* II. *Terrakotten und Bronzen*. Ed. E. Berger. 32–135. Basel: Archäologischer Verlag.

Hevly, R. H. 1981. "Pollen production, transport, and preservation: potentials and limitations in archaeological palynology." *Journal of Ethnobiology* 1: 39–54.

Hicks, A. J., and M. J. Hicks. 1992. "The small objects." In *Excavations at Otranto*, 279–313.

Hill, D. K. 1962. "The Long-Beaked Bronze Jug in Greek Lands." *AJA* 66: 57–63.

Hjelmqvist, H. 1977. "Economic plants of the Italian Iron Age from Monte Irsi." In *Monte Irsi*, 274–81.

Hoffmann, A. 2002. *Grabritual und Gesellschaft. Gefässformen, Bildthemen und Funktionen unteritalisch-rotfiguriger Keramik aus der Nekropole von Tarent*. Internationale Archäologie 76. Rahden/Westf.: Leidorf.

Hubner, G. 2000. "Zur Lagynos: Aspekte und Retrospektiven." In *Ε΄ Επιστημονική Συνάντηση για την Ελληνιστική Κεραμική: Χρονολογικά Προβλήματα. Κλειστά Σύνολα - Εργαστήρια* (Χανιά, Απρίλιος 1997). 161–78. Athens: Υπουργείο Πολιτισμού. Ταμείο Αρχαιολογικών Πόρων και Απαλλοτριώσεων.

Ingoglia, C., M. Triscari, and G. Sabatino. 2008. "Archeometallurgy in Messina: Iron Slag from a Dig at Block P, Laboratory Analyses and Interpretation." *Mediterranean Archaeology and Archaeometry* Vol. 8, No. 1: 49–60.

Ingravallo, E. 1980. "L'insediamento eneolitico di Pizzica Pantanello presso Metaponto (Basilicata)." *ASTSN* 87: 317–27.

Interdonato, E. 2002. "Le tombe dall'area dell'abitato." In *Ausculum I*, 323–52.

Iozzo, M. 1981. "Louteria fittili in Calabria. Analisi e classificazione preliminare." *Archeologica Classica* 33: 143–93.

Isler, H. P. 1975. "Monte Iato (Palermo). Scavi 1972–1974." *NSc* 8th. ser., Vol. 29: 531–56.

———. 1978. "Monte Iato: l'ottava campagna di scavo." *Sicilia Archeologica* XI, 38: 7–29.

Jastrow, E. 1941. "Abformung und Typenwandel in der antiken Tonplastik." *Opuscola archaeologica* 2: 1–99.

Jentel, M.-O. 1977. "Hellenistic vases with relief decoration." In *Monte Irsi*, 164–69.

Kaloyeropoulou, A. 1970. "From the Techniques of Pottery." *Archaiologika Analekta ex Athenon* 3: 429–34.

Kanowski, M. G. 1984. *Containers of Classical Greece: A Handbook of Shapes*. St. Lucia/New York/London: University of Queensland Press.

Karagiorga, T. 1971. "Κεραμεικός κλίβανος έν Ηλιδι." *Archaiologika Analekta ex Athenon* 4: 27–31.

Kardulias, P. N., and C. Runnels. 1995. "The Lithic Artifacts: Flaked Stone and Other Nonflaked Lithics." In *Artifact and Assemblage: The Finds from a Regional Survey of the Southern Argolid, Greece. Vol. I: The Prehistoric and Early Iron Age Pottery and the Lithic Artifacts*. Eds. C. Runnels, D. Pullen, and S. Langdon. 74–139. Stanford: Stanford University Press.

Karivieri, A. 1996. *The Athenian Lamp Industry in Late Antiquity. Papers and Monographs of the Finnish Institute at Athens* 5. Helsinki: Suomen Ateenen-instituutin säätiö.

Keay, S., and D. Williams. 2005. *Roman Amphorae: A Digital Resource*. Archaeology Data Service and the University of Southampton.

Keys, A. L. 1983. *The Gray Ware from the 1980–1981 Excavations at Pizzica*. MA thesis. Austin: The University of Texas at Austin.

Kirsopp Lake, A. K. 1934–1935. "*Campana supellex*: The Pottery Deposit at Minturnae." *Bollettino dell'Associazione Internazionale di Studi Mediterranei* 5: 97–114.

Kleibrink, M. 1970–1971. "Anfora attica a figure nere e macine per grano." *AttiMGrecia* n.s. 11–12: 75–80.

———. 2010. *Guida, Parco Archeologico 'Lagaria' a Francavilla Marittima presso Sibari*. Rossano: Associazione "Lagaria" ONLUS.

Klynne, A. 2006–2007. "The Villa Selvasecca Revisited." *Opuscula Romana* 31–32: 29–57.

Koehler, C. G. 1979. *Corinthian A and B Transport Amphoras*. Ann Arbor: University of Michigan Press.

———. 1981. Corinthian Developments in the Study of Trade in the Fifth Century. *Hesperia* 50: 449–58.

Kögler, P. 2010. *Feinkeramik aus Knidos vom mittleren Hellenismus bis in die mittlere Kaiserzeit ca. 200 v. Chr. bis 150 n. Chr.* Wiesbaden: Reichert Verlag.

Krauskopf, I. 2005. "Perirrhanterion/louterion." In *ThesCRA* 5, 178–82.

Krogman, W. M. 1962. *The Human Skeleton in Forensic Medicine*. Springfield, IL: Charles C. Thomas Publisher.

Kundert Gibbs, J., and P. Lee. 2004. *Maya 5: guida completa*. Milan: Apogeo.

Kustermann Graf, A. 2002. *Selinunte: necropoli di Manicalunga: le tombe della contrada Gaggera*. Soveria Mannelli: Rubbettino.

Kyriakou, D. 1994. "Ἑλληνιστικὴ κεραμικὴ ἀπὸ τὸ νεκροταφεῖο τῆς Πάτρας." In *3rd Hellenistic Pottery Conference 1994*, 185–95, pls. 127–41.

Lacava, M. 1891. *Topografia e storia di Metaponto*. Naples.

Ladstätter, S. 2003. "Ein hellenistischer Fundkomplex in SR 12." In *Das Hanghaus 1 in Ephesos. Funde und Ausstattung. Forschungen in Ephesos VIII/4*. Eds. C. Lang-Auinger, B. Asamer, and U. Müller-Kaspar. 70–80. Vienna: Österreichische Akademie der Wissenschaften.

———. 2010. "Keramik." In *Das Prytaneion in Ephesos. Funde in Ephesos* IX/4. Ed. M. Steskal. 85–172. Vienna: Österreichische Akademie der Wissenschaften.

Lafratta, B. 1996a. "Pisticci, l'Abbazio di S. Maria del Casale." In *Monasteri Italo-Greci e Benedettini in Basilicata*. Vol. II. Eds. L. Bubbico, F. Caputo, and A. Maurano. 174–79. Matera.

———. 1996b. "Pisticci, La Grancia di San Basilio." In *Monasteri Italo-Greci e Benedettini in Basilicata*. Vol. II. Eds. L. Bubbico, F. Caputo, and A. Maurano. 180–82. Matera.

Lalonde, G. V. 1968. "A Fifth-century Hieron Southwest of the Athenian Agora." *Hesperia* 37: 123–33.

———. 1991. "Horoi." In *The Athenian Agora. Vol. XIX, Inscriptions*. Eds. M. K. Langdon, M. B. Walbank, and G. V. Lalonde. 5–51. Princeton: American School of Classical Studies at Athens.

———. 2006a. "IG I^3 1055 B and the Boundary of Melite and Kollytos." *Hesperia* 75: 83–119.

———. 2006b. *Horos Dios. An Athenian shrine and cult of Zeus*. MonGrRom, 11. Leiden.

Lamboglia, N. 1952. "Per una classificazione preliminare della ceramica campana." In *Atti del I Congresso Internazionale di Studi Liguri (1950)*. 139–206. Bordighera: Istituto Internazionale di Studi Liguri.

Lamboley, J.-L. 1996. "Vaste (Lecce). La porte est et la tombe des Pizzinaghe." *StAnt* 9: 361–430.

———. 1999. *Muro Leccese. Sondages sur la fortification nord.* Rome: École Francaise de Rome.

Lambrugo, C. 2003. "Oggetti diversi." In *Ricerche archeologiche all'Incoronata di Metaponto. 6, L'oikos Greco del saggio E. Lo scavo e i reperti*. Scavi dell'Università degli Studi di Milano, Dipartimento di Scienze dell'Antichità Sezione Archeologica. 91–92. Milan: Comune di Milano, Settore cultura e spettacolo.

———. 2004. "Ceramica magnogreca a figure rosse. I vasi miniaturistici." In *La Collezione Lagioia. Una raccolta storica dalla Magna Grecia al Museo Archeologico di Milano*. Ed. G. Sena Chiesa, C. Lambrugo, and M. Dolci. 239–47. Milan: Edizioni ET.

Lampropoulou, L. 1983. "Δύο κεραμικοί κλίβανοι στην Αταλάντη Λοκρίδος." *Archaiologika Analekta ex Athenon* 16: 74–79.

Lang-Auinger, C., B. Asamer, and U. Müller-Kaspar. 2003. *Das Hanghaus 1 in Ephesos: Funde und Ausstattung. Forschungen in Ephesos* VIII/4. Vienna: Österreichische Akademie der Wissenschaften.

Langdon, M. K. 1999. "Hymettiana III. The Boundary Markers of Alepovouni." *Hesperia* 68: 481–508.

Lanza Catti, E. 2014. "Appendix C: Census of Farmhouses." In *Fattoria Fabrizio*, 409–17.

Lanza Catti, E., F. Silvestrelli, K. Swift, A. Tubelli, and E. Vittoria. 2011. "Archaic and Black-Gloss Fine Ware." In *Survey*, 143–270.

Lapadula, E. 2007. "Le ceramiche di uso comune di produzione locale." In *La circolazione delle ceramiche nell'Adriatico tra Tarda Antichità e Altomedioevo, III Incontro di Studio CER.AM.IS.* Eds., S. Gelichi and C. Negrelli. 341–48. Mantua: SAP.

———. 2009. "Il 'castrum' di San Salvatore (Bernalda, MT). Indagine archeologica di emergenza." *FOLD&R Italia:* 137. <http://www.fastionline.org/docs/FOLDER-it-2009-137.pdf>

———. 2011a. "Imperial and Late Roman Settlement in the Metapontino." In *Survey*, 1137–45.

———. 2011b. "Medieval and Post-Medieval Ceramics." In *Survey*, 1173–1201.

———. 2012a. "The Excavation and Structures." In *San Biagio*, 19–53.

———. 2012b. "The Materials: The Roman Imperial Period." In *San Biagio*, 103–68.

La Rocca, L. 2000. "A proposito dei santuari rurali in Lucania." In *Campagna e paesaggio*, 7–18.

La Torre, G. F. 2002. *Un tempio arcaico nel territorio dell'antica Temesa. L'edificio sacro in località Imbelli di Campora S. Giovanni. Corpus delle stipi votive in Italia* 14: *Regio 3, 4. Archaeologica* 133. Rome: Giorgio Bretschneider.

La Torre, G. F., and F. Mollo. 2006. *Blanda Julia sul Palécastro di Tortora–Scavi e ricerche (1990–2005)*. Messina: Di.Sc.A.M. Eds.

Lattanzi, E. 1980. "L'attività archeologica in Basilicata nel 1979." *Atti Taranto* 19: 399–409.

———., ed. 2007. *Il Museo Nazionale di Reggio Calabria. I tesori della Magna Grecia*. Rome: Gangemi.

Laubenheimer, F., and C. Rodriguez. 1991. *Les amphores de Bibracte. Le materiel des fouilles anciennes*. Documents d'Archéologie Française 29. Paris: Maison des Sciences de l'Homme.

Laubenheimer, F., F. Widemann, and M. Attas. 1978–1979. "Atelier de potiers gallo-romain de Salléles d'Aude (Narbonne). Le chargement du four B 5." *ActaPraehistA* 9–10: 115–24.

Laurenzana, M. 2011. "Il progetto della scuola di Specializzazione in archeologia di Matera sul santuario di Rossano di Vaglio. Gli oggetti del ritual. Ceramica comune." In *Brateís datas*, 267–75.

Laviosa, C. 1954. "Le antefisse fittili di Taranto." *Archeologia Classica* 6: 217–50.

Lazzarini, M. L. 1982. "I «veri reggini»." *Klearchos* 24: 145–57.

Le Bailly, M., and F. Bouchet. 2010. "Ancient dicrocoeliasis: occurrence, distribution and migration." *Acta Tropica* 115: 175–80.

Lecce, L. 2010–2011. "Una fornace tardoarcaica nella chora di Metaponto." *Siris* 11: 15–44.

Lejsgaard Christensen, J. 2010. "The Black Glossed Pottery." In J. K. Jacobsen and S. Handberg. *Excavation on the Timpone della Motta Francavilla Marittima (1992–2004). I. The Greek Pottery*. 334–357. Bibliotheca Archaeologica 21. Bari: Edipuglia.

Leone, D. 2000. "Le ceramiche tardoantiche della fattoria di Posta Crusta." In *Ordona X*, 387–436.

———. 2002. "Aspetti della circolazione delle merci nell'Apulia tardoantica tra importazioni e produzioni locali: la ceramica comune dipinta." In *L'Africa romana*, 881–84.

Leone, R. 1998. *Luoghi di culto extraurbani d'età arcaica in Magna Grecia*. Florence: Le Lettere.

Lepre, A. 1981. "Le campagne pugliesi nell'età moderna." In *La Puglia tra Medioevo ed età moderna. Città e campagna*. Ed. C. D. Fonseca. 273–331. Milan: Electa.

Leroux, G. 1913. *Lagynos. Recherches sur la céramique et l'art ornamental hellénistique*. Paris, Univ. diss.

Licinio, R. 1981. "L'organizzazione del territorio fra XIII e XV secolo." In *La Puglia tra Medioevo ed età moderna. Città e campagna*. Ed. C. D. Fonseca. 202–71. Milan: Electa.

———. 1998. *Masserie medievali. Masserie, massari e carestie da Federico II alla Dogana delle Pecore*. Bari: Adda Editore.

Liebig, J. 1848. *Chemistry in its Application to Agriculture and Physiology*. London: Taylor and Walton.

Limoncelli, M. 2012. *Il restauro virtuale in archeologia*. Rome: Carocci.

Lippolis, E. 1984. *La necropoli del Palazzone di Perugia. Ceramiche comuni e verniciate*. Rome: Giorgio Bretschneider.

———. 1994a. "La tipologia dei *semata*." In *Museo Taranto III.1*, 108–29.

———. 1994b. "La necropoli ellenistica: problemi di classificazione e cronologia dei materiali." In *Museo Taranto III.1*, 238–81.

———. 1995. "Corinto e l'Occidente: la costa del Mar Ionio." *Atti Taranto* 34: 508–50.

———. 2005. "Taranto romana: dalla conquista all'età augustea." *Atti Taranto* 44: 235–412.

Liseno, A. 2007. *Dalla capanna alla casa: dinamiche di trasformazione nell'Italia sud-orientale (VIII–V sec. a.C.)*. Bari: Progedit.

Liseno, M. G. 2004. *Metaponto. Il deposito votivo Favale*. Corpus delle stipi votive in Italia 12. Regio III, 5. Rome: Giorgio Bretschneider.

Lissi Caronna, E. 1980. "Oppido Lucano (Potenza). Rapporto preliminare sulla seconda campagna di scavo, 1968." *NSc* 8th ser., Vol. 34: 119–297.

———. 1983. "Oppido Lucano (Potenza). Rapporto preliminare sulla terza campagna di scavo, 1969." *NSc* 8th ser., Vol. 37: 215–352.

———. 1983–1984. "Oppido Lucano: quattro case di IV–III sec. a.C." *AttiMGrecia* n.s. 24–25: 193–212.

———. 1990–1991. "Oppido Lucano (Potenza). Rapporto preliminare sulla quarta campagna di scavo (1970). Materiale archeologico rinvenuto nel territorio del comune." *NSc* n.s., 1–2: 185–344.

———. 1998–1999. "Metaponto (Matera). Scavo di parte di una fattoria in Contrada Casa Teresa, proprietà Durante." *NSc* 9th ser., 9–10: 177–205.

Lissi Caronna, E., V. Armignacco Alidori, and S. Panciera. 1992. "Oppido Lucano (Potenza). Rapporto preliminare sulla quarta campagna di scavo (1970). Materiale archeologico rinvenuto nel territorio del comune." *NSc* 9th ser., 1–2: 185–488.

Lo Monaco, A. 2002. "Pesi da telaio e fuseruole." In *Ausculum I*, 288–300.

Long, L., J. Miro, and G. Volpe. 1992. "Les èpaves archaiques de la pointe Lequin (Porquerolles, Hyères, Var). Des données nouvelles sur le commerce de Marseille à la fin du VIe et dans la première moitié du Ve s. Av. J.C." In *Marseille grecque et la Gaule. Actes du Colloque, Marseille 18–23 novembre 1990*. Ed. M. Bats. 199–234. Lattes: A.D.A.M.

Lo Porto, F. G. 1961. "Recenti scoperti di tombe archaiche in Taranto." *BdA* 46: 268–82.

———. 1966. "Metaponto. Scavi e ricerche archeologiche." *NSc* 8th ser. 20: 136–231.

———. 1970–1971. "Tomba messapica di Ugento." *AttiMGrecia*, n.s., 11–12: 99–152.

———. 1973. *Civiltà indigena e penetrazione greca nella Lucania orientale. Monumenti Antichi* 48, Serie Miscellanea I-3. 149–250. Rome.

———. 1981. "Metaponto (Matera): Nuovi scavi nella città e nella sua necropoli." *NSc* 8th ser., 35: 289–391.

———. 1988–1989. "Metaponto (Matera): Rinvenimenti nella città antica e nel suo retroterra ellenizzato." *NSc* 8th ser., 42–43: 299–441.

———. 1991. *Timmari. L'abitato, le necropoli, la stipe votiva.* Rome: G. Bretschneider.

———. 1994. "Due nuove tombe scoperte a Cavallino." *StAnt* 7: 47–84.

———. 1996. "Tombe arcaiche di Peuceti emergenti." *StAnt* 9: 7–36.

Luginbühl, T. 2001. *Imitations de sigillée et potiers du Haut-Empire en Suisse occidentale. Archéologie et histoire d'un phénomène artisanal antique.* (*Cahiers d'Archéologie Romande* 83.) Lausanne: Selbstverlag.

Lund, J. 2005. "An economy of consumption: The Eastern Sigillata A industry in the late Hellenistic period." In *Making, Moving and Managing, The New World of Ancient Economies, 323–31 BC.* Eds. Z. H. Archibald, J. K. Davies, and V. Gabrielsen. Oxford: Oxbow Books.

Lyding Will, E. 1979. "The Sestius Amphoras. A Reappraisal." *JFA* 6: 339–50.

———. 1982. "Greco-Italic amphorae." *Hesperia* 51: 338–56.

———. 1987. "The Roman Amphoras." In *The Roman Port and Fishery of Cosa.* Ed. A. McCann. 171–220. Princeton: Princeton University Press.

Macnamara, E. 1983a. "One handlers." In *Metaponto II,* 333–37.

———. 1983b. "Site A (and Site C) Native Settlement, Archaic Greek Occupation and Early Hellenistic Circuit Wall." In *Cozzo Presepe,* 207–43.

———. 1992. "The metalwork." In *Gravina II,* 230–39.

Macrì, P. 2009. "Le necropoli della Valle del Sauro: Aliano, Alianello e Guardia Perticara. Proposta per una metodologia di studio." *Siris* 10: 113–22.

Malfitana, D., J. Poblome, and J. Lund. 2005. "Eastern Sigillata A in Italy. A socio-economic evaluation." *Babesch* 80: 199–212.

Manacorda, D. 1986. "A proposito delle anfore cosiddette 'greco-italiche': una breve nota." In Recherches sur les amphores grecques. Eds. J.-Y. Empereur and Y. Garlan. *Bulletin de Correspondance Hellénique* Suppl. 13. 581–86.

———. 2012. "Le materie prime." In *Le fornaci romane di Giancola (Brindisi). Bibliotheca archaeologica* 27. Eds. D. Manacorda and S. Pallecchi. 97–101. Bari: Edipuglia.

Manacorda, D., and S. Pallecchi, eds. 2012. *Le fornaci romane di Giancola (Brindisi). Bibliotheca archaeologica* 27. Bari: Edipuglia.

Mancusi, M. 2001. "Considerazioni sulla necropoli." In *Carta archeologica Sinni 3,* 274–79.

Maniére, G. 1971. "Une officine de tuilier gallo-romaine du Haute-Emore à Coilàdere, par Cazères (Haute-Garonne)." *Gallia* 29: 191–99.

Mannino, K. 1993. "Le monete in terracotta." In *Moneta e non moneta. (Atti del Convegno Internazionale di Studi Numismatici, Milano 11–15 maggio 1992). Rivista Italiana di Numismatica e Scienze Affini* 95. 207–42. Milan: Comune di Milano.

———. 1996. "Gli ateliers attici e la nascita della produzione figurata. In *Arte e artigianato,* 363–70.

———. 1998. "Nuovi dati sulle 'monete' in terracotta." *StAnt* 9: 61–71.

Mannoni, T., and E. Giannichedda. 1996. *Archeologia della produzione.* Turin: Einaudi.

Marchegiani, M. C. 1997. "La ceramica tornita matt-painted." In *Pomarico Vecchio I,* 161–74.

Marin, M. M., et al. 1982. *Ceglie Peuceta I.* Bari: Dedalo.

Martelli, M. 1979. "Primi considerazioni sulla statistica delle importazioni greche in Etruria nel periodo arcaico." *Studi Etruschi* 47: 37–52.

Martin, R., and K. Saller. 1957. *Lehrbuch der Anthropologie in systematischer Darstellung mit besonderer Berücksichtigung der anthropologischen Methoden.* 3rd edition. 4 vols. Stuttgart: G. Fischer.

Maruggi, G. A. 1982. "Taranto. Rinvenimenti nella necropoli." *Taras* II, 1–2: 51–80.

———. 1988. "La necropoli del Regio Arsenale." In *Museo di Taranto,* 185–234.

Masiello, L. 2006. "Ceramica acroma." In *Museo Taranto II.2,* 317–34.

Mastrocinque, A., ed. 2009. *Grumentum romana. Atti del convegno di studi (Grumentum Nova, 28–29 giugno 2008).* Moliterno.

Mastronuzzi, G. 2005. "L'archeologia di un luogo di culto in Messapia: Vaste-Piazza Dante." In *Lo spazio del rito,* 235–47.

———. 2013. *Il luogo di culto di Monte Papalucio ad Oria. 1. La fase arcaica.* BACT 12. Bari: Edipuglia.

Maturo, C. 1997. "Varia." In *Pomarico Vecchio I,* 249–52.

Mazier, F. 2007. *Modélisation de la relation entre pluie pollinique actuelle, vegetation et pratiques pastorals en moyenne montagne (Pyrenées et Jura): application pour l'interprétation des données polliniques fossils.* Doctoral dissertation. Université de Franche Comté.

Mazzei, M., and L. Pietropaolo. 1995. "Ceramica a vernice nera, a pasta grigia, suddipinta e a figura rosse." In *Arpi,* 245–47.

Medri, M. 2000. "Disegno ricostruttivo." In *Dizionario di archeologia: temi, concetti e metodi.* Eds. R. Francovich and D. Manacorda. 133–38. Bari: Laterza.

Meirano, V. 1997. "I reperti metallici." In *Pomarico Vecchio I,* 235–46.

Meo, F. 2011. "Rediscovering ancient activities: textile tools in a 3rd–2nd century BC context from Herakleia, Southern Basilicata, Italy." *Archaeological Textiles Newsletter* 53: 2–11.

Mercuri, A. M., M. Bandini Mazzanti, P. Torri, L. Vigliotti, G. Bosi, A. Florenzano, L. Olmi, and I. Massamba

N'siala. 2012. "A marine/terrestrial integration for mid-late Holocene vegetation history and the development of the cultural landscape in the Po Valley as a result of human impact and climate change." *Vegetation History and Archaeobotany* 21 (4–5): 353–72.

Mercuri, A. M., A. Florenzano, I. Massamba N'siala, L. Olmi, D. Roubis, and F. Sogliani. 2010. "Pollen from archaeological layers and cultural landscape reconstruction: case studies from the Bradano Valley (Basilicata, southern Italy)." *Plant Biosystems* 144: 888–901.

Merker, G. S. 2006. *The Greek Tile Works at Corinth. The Site and the Finds. Hesperia* Suppl. 35. Princeton: American School of Classical Studies at Athens.

Mertens, D. 1985. "Metapont. Ein neuer Plan des Stadtzentrums. Ein Kurzbericht über die Arbeite des DAI Rom im Rahmen der gemeinsam mit der Soprintendenz der Basilicata durchgeführten Ausgrabung." *AA* 1985: 645–71.

———. 1992. "Note sull'architettura di Poseidonia-Paestum. Problemi e stato della ricerca." *Atti Taranto* 27: 541–74.

———. 1999. "Metaponto: l'evoluzione del centro urbano." In *Storia della Basilicata 1. L'antichità*. Ed. D. Adamesteanu. 247–94. Rome/Bari: Laterza.

Mertens Horn, M. 1991. "Un nuova 'antefissa' a testa femminile da Akrai ed alcune considerazioni sulle ninfe in Sicilia." *BdA* 66: 9–28.

———. 1994. "Antefissa." In *Enciclopedia dell'Arte Antica. Supplemento* II. 242–52. Rome: Istituto della Enciclopedia italiana.

Metzger, I. R. 1994. "Ein hellenistisches Grabmonument in Eretria." In *3rd Hellenistic Pottery Meeting 1994*, 71–79.

Meyer-Schlichtmann, C. 1988. *Die pergamenische Sigillata aus der Stadtgrabung von Pergamon: Mitte 2. Jh. v. Chr. - Mitte 2. Jh. n. Chr.* Berlin [u.a.]: De Gruyter.

Meyers, G. E. 2013. "Women and the Production of Ceremonial Textiles: A Reevaluation of Ceramic Textile Tools in Etrusco-Italic Sanctuaries." *AJA* 117.2: 247–74.

Mezquìriz, M. A. 1985. "Terra sigillata ispánica." In *Atlante delle forme ceramiche II. Ceramica fine romana nel bacino mediterraneo, tardo ellenismo e primo impero. Enciclopedia dell'arte antica, classica e orientale.* Ed. G. Pugliese Carratelli. 97–166. Rome: Istituto della Enciclopedia italiana.

Milanese, M., ed. 2006. *Vita e morte dei villaggi rurali tra Medioeva ed Età Moderna. Dallo scavo della villa de Geriti ad una pianificazione della tutela e della conoscenza dei villaggi abbandonati della Sardegna.* Quaderni dei villaggi abbandonati della Sardegna, 2. Florence: All'Insegna del Giglio.

Mitsopoulos-Leon, V., and C. Lang-Auinger. 1991. *Die Basilika am Staatsmarkt in Ephesos: Kleinfunde.* Vienna: Schindler.

Molnar, S. 1971. "Human tooth wear, tooth function and cultural variability." *American Journal of Physical Anthropology* 34 (2): 175–89.

Monaco, C., L. Tortorici, and W. Paltrinieri. 1998. "Structural evolution of the Lucanian Apennines." *Journal of Structural Geology* 20, 617–38.

Monckton, S. G. 2002. "Plant remains." In *The excavations of San Giovanni di Ruoti.* Vol. 3, *The faunal and plant remains.* Eds. M. R. MacKinnon, A. Small, and R. J. Buck. 201–15. Toronto: University of Toronto Press.

Mongiello, L. 1984. *Le masserie di Puglia. Organismi architettonici ed ambiente territoriale: analisi del fenomeno, studio del repertorio, disegno delle partiture.* Bari: Adda Editore.

Montagna Pasquinucci, M. 1972. "La ceramica a vernice nera del Museo Guarnacci di Volterra." *MEFRA* 84 (1): 269–498.

Moore, P. D., J. A. Webb, and M. E. Collins. 1991. *Pollen Analysis.* 2nd ed. Oxford: Blackwell Scientific Publications.

Morel, J.-P. 1970. "Fouilles à Cozzo Presepe prés de Metaponto." *MEFRA* 82: 73–116.

———. 1981. *Ceramique campanienne: les formes.* Rome: École Française de Rome.

Morter, J., and J. R. Leonard. 1998. "Storage Amphorae." In *Necropoleis*, 731–47.

Moscati, P. 2009. "Archeologia e musealizzazione virtuale." In *Quale futuro per l'archeologia? Workshop internazionale, Roma, 4-5 dicembre 2008.* Eds. A. L. D'Agata and S. Alaura. 189–200. Rome: Gangemi.

Muggia, A. 2000. "La sfera infantile e il simbolismo iconografico: alcuni casi dalla necropoli di Valle Trebba a Spina." *Ostraka* 9: 87–94.

———. 2004. *Impronte nella sabbia: tombe infantili e di adolescenti dalla necropoli di Valle Trebba a Spina. Quaderni di Archeologia dell'Emilia Romagna* 9. Florence: All'Insegna del Giglio.

Muller, A. 1996. "Le terres cuites votives du Thesmophorion. De l'atelier au sanctuaire." *Études Thasiennes* 17: 27–61.

———. 1997. "Description et analyse des productions moulées. Proposition de lexique multilingue, suggestions de méthode." In *Moulage*, 437–63.

———. 1999. "Les ateliers de coroplathes thasiens." In *Thasos. Matières premières et technologie de la préhistoire à nos jours. Actes du Colloque International, Thasos, Limenaria, 26–29/9/1995.* Ed. A. Muller. 279–91. Athens: L'École Française d'Athènes.

———. 2000. "Artisans, Techniques de production et diffusion." In *L'artisanat en Grèce ancienne. Les productions, les diffusions. Actes Lyon 1998.* Eds. F. Blondé and A. Muller. 92–106. Lille: SEGES.

Müller, P. 1976. "Gestempelte Ziegel." *Studia Ietina* 1: 49–77.

Munzi, P. 1999. "Laos: aspetti di vita quotidiana attraverso lo studio del materiale ceramico." In *Nella terra degli Enotri*, 91–98.

Musgrave, J. H. and N. K. Harneja. 1978. "The estimation of adult stature from metacarpal bone length." *American Journal of Physical Anthropology* 48: 113–19.

Musumeci, A. 1989. "Vasellame di destinazione varia." In *Caracausi*, 117–29.

Nafissi, M. 1989. "Distribution and Trade." In C. M. Stibbe. *Laconian Mixing Bowls. A History of the krater Lakonikos from the Seventh to the Fifth Century BC*. 68–88. Amsterdam: Allard Pierson Museum.

Nava, M. L. 1997. "L'attività archeologica in Basilicata nel 1996." *Atti Taranto* 36: 455–94.

———, ed. 1998. *Trésors d'Italie du Sud. Treasures from the South of Italy. Tesori dell'Italia del Sud.* Milan: Skira.

———. 2000. "L'attività archeologica in Basilicata nel 1999." *Atti Taranto* 39: 675–726.

———. 2001. "L'attività archeologica in Basilicata nel 2000." *Atti Taranto* 40: 939–80.

———. 2002. "L'attività archeologica in Basilicata nel 2001." *Atti Taranto* 41: 719–65.

———. 2003a. "L'attività archeologica in Basilicata nel 2002." *Atti Taranto* 42: 653–717.

———. 2003b. "La Basilicata." *ACSM* 42: 651–718.

———. 2004. "L'attività archeologica in Basilicata nel 2003." *Atti Taranto* 43: 935–1000.

Neboit, R., 1982. "Instabilité et morphogenesè au Quaternaire en Lucania orientale." *Revue de Géographie Physique et de Géologie dynamique* 23: 15–26.

Neeft, C. W. 1995. "Corinthian Pottery in Magna Graecia." *Atti Taranto* 34: 367–402.

Neutsch, B., ed. 1967. *Archäologische Forschungen in Lukanien. II. Herakleiastudien. Mitteilungen des Deutschen Archäologischen Instituts. Römische Abteilung* 11. Heidelberg: F. H. Kerle Verl.

Niccolucci, F., ed. 2002. *Virtual Archaeology: Proceedings of the VAST Euroconference (Arezzo, 24–25 November 2000). BAR-IS* 1075. Oxford: Archeopress.

Nigro, M. 2006a. "La ceramica in argilla grezza." In *Cuma 2*, 57–80.

———. 2006b. "La ceramica a vernice nera d'importazione e d'imitazione." In *Cuma 2*, 91–100.

Novellis, D. 2009. "Archeobotanica a Torre di Satriano. Aggiornamenti e novità dallo scavo 2008." In *Lo spazio del potere*, 217–25.

Nuzzo, A. N. 2003. "Kerameikos di Metaponto, campagna di scavo 2003. Analisi paleobotaniche." Unpublished report, Dipartimento di Beni Culturali, Università del Salento.

Oakley, J. H. 1988. "Attic Red-Figured Skyphoi of Corinthian Shape." *Hesperia* 57: 165–91.

Ober, J. 1995. *Greek Horoi: Artifactual Texts and the Contingency of Meaning*. In *Methods in the Mediterranean: Historical and Archaeological Views on Texts and Archaeology*. Ed. D. Small. Mnemosyne Suppl. 135. 91–123. Leiden.

Olbrich, G. 1979. *Archaische Statuetten einer Metapontiner Heiligtums. Studia Archaeologica* 23. Rome: L'Erma di Bretschneider.

Olcese, G. 1993. *Le ceramiche comuni di Albintimilium. Indagine archeologica e archeometrica sui materiali dell'area del Cardine*. Florence: All'insegna del giglio.

———. 2003. *Ceramiche comuni a Roma e in area romana: produzione, circolazione e tecnologia (tarda età repubblicana-prima età imperiale)*. Mantua: S.A.P.

———. 2004. "Anfore Greco-italiche: alcune osservazioni sull'origine e sulla circolazione alla luce di recenti ricerche archeologiche ed archeometriche." In *Archaeological Methods and Approaches: Industry and Commerce in Ancient Italy*. Eds. E. C. De Sena and H. Dessales. 173–92. BAR International Series 1262. Oxford.

———. 2010. *Le anfore Greco italiche di Ischia: archeologia e archeometria*. Rome: Edizioni Quasar.

———. 2012. *Atlante dei siti di produzioni ceramica*. Rome: Edizioni Quasar.

Olivero Ferrero, G. 1989. "La ceramica a vernice nera. Il vasellame di IV e III strato." In *Locri II*, 71–106.

Olmer, F. 2003. *Les Amphores de Bibracte. 2. Le commerce du vin chez les Édues d'après les timbres d'amphores, Collection Bibracte* 7. Paris: Éditions de la Maison des sciences de l'homme.

Orlandini, P. 1983. "Le arti figurative." In *Megale Hellas. Storia e civiltà della Magna Grecia*. Ed. G. Pugliese Carratelli. 331–554. Milan: Scheiwiller.

Orlandos, A. K. 1966. *Les matériaux de construction et la technique architecturale des anciens Grecs. Travaux et memoires des anciens membres étrangers de l'École et de divers savants* 16. Paris: École Française d'Athènes.

Orsi, P. 1990. *La necropoli di Passo Marinaro a Camarina. Campagne di scavo 1904–1909, Monumenti Antichi. Serie Miscellanea, volume IV (LIV della Serie generale)*. Ed. M. T. Lanza. Rome: Accademia Nazionale dei Lincei.

Ortner, D. J., and W. G. J. Putschar. 1981. *Identification of pathological conditions in human skeletal remains*. Washington, DC: Smithsonian Institution Press.

Osanna, M. 1992. *Chorai coloniali da Taranto a Locri. Documentazione archeologica e ricostruzione storica*. Rome: Istituto Poligrafico e Zecca dello Stato, Libreria dello Stato.

———. 2003. "Conclusioni." In *Siris 3*, 357–66.

———. 2007. "La ricostruzione del paesaggio antico attraverso l'archeologia: il caso di Torre di Satriano." In *Il paesaggio archeologico – Resti e contesti: prospettive di condivisione su tutela e valorizzazione*. 14–17. Rome: Ministero per i Beni e le Attività Culturali.

———. 2008. "L'attività archeologica in Basilicata nel 2007." *Atti Taranto* 47: 911–44.

Osanna, M. and A. Russo. 1988. "Catalogo delle tombe del settore sud della necropoli in Contrada Casino." In *Forentum* I: *Le necropoli di Lavello*. Eds. M. Giorgi,

S. Martinelli, M. Osanna, and A. Russo. 98–140. Venosa: Edizioni Osanna.

Osanna, M., M. C. Pilo, and C. Trombetti. 2009. "Ceramica attica nei santuari della costa ionica dell'Italia meridionale: colonie achee e indigeni tra *paralia* e *mesogala*." In *Ceramica attica da santuari della Grecia, della Ionia e dell'Italia. Atti convegno Perugia 14–17 marzo 2007*. Eds. S. Fortunelli and C. Masseria. 455–94. Venosa: Osanna.

Osanna, M., L. Prandi, and A. Siciliano, eds. 2008. *Eraclea. Culti greci in Occidente. Fonti scritte e documentazione archeologica, II.* Taranto: Città di Castello, Istituto per la Storia e l'Archeologia della Magna Grecia.

Özyiğit, Ö. 2000. "The chronology of Pergamene appliqué ware." In *5th Hellenistic Pottery Conference 2000*, 195–98.

Pagliara, C. 1981. "Materiali epigrafici da Vaste e Muro." *StAnt* 2: 207–20.

Pancrazi, O. 1979. "L'abitato. Il settore B." In *Cavallino I. Scavi e ricerche 1964–1967*. Ed. O. Pancrazi. Università di Lecce, Istituto di Archeologia e Storia Antica. 115–96. Galatina: Congedo.

Panella, C. 1996. "Lo studio delle ceramiche comuni in età romana: qualche riflessione." In *Les Céramiques communes*, 9–15.

Papadopoulos, J. K. 1992. "Λασανα, Tuyères, and Kiln Firing Supports." *Hesperia* 61: 203–21.

———. 2003. *Ceramicus redivivus. The Early Iron Age Potters' Field in the Area of the Classical Athenian Agora. Hesperia* Suppl. 31. Princeton: American School of Classical Studies at Athens.

Parmly Toxey, A., and J. C. Carter. 1998. "Unglazed Pottery." In *Necropoleis*, 718–30.

Patacca, E., and P. Scandone. 2001. "Late thrust propagation and sedimentary response in the thrust-belt–foredeep system of the southern Apennines (Pliocene–Pleistocene)." In *Anatomy of an Orogen: The Apennines and Adjacent Mediterranean Basins*. Eds. G. B. Vai and I. P. Martini. 401–40. Great Britain: Kluwer Academic Publishers.

Peacock, D. 1979. "An Ethnoarchaeological Approach to the Study of Roman Bricks and Tiles." In *Roman Bricks and Tiles* 5–10.

———. 1982. *Pottery in the Roman World: An Ethnoarchaeological Approach*. London/New York: Longman.

Pedio, T. 1990. *Centri scomparsi della Basilicata*. Venosa: Edizioni Osanna.

———. 1998. *Cartulario della Basilicata (476–1443)*. Vol. I. Venosa: Appia 2 Editrice.

Pelagatti, P. 1970. "Akrai (Siracusa). Ricerche nel territorio." *NSc* Ser. 8, Vol. 24: 436–99.

———. 1972. "Naxos, 2. Ricerche topografiche e scavi 1965–1970. Relazione preliminare." *BdA* 57: 211–20.

———. 1980. "L'attività della Soprintendenza alle antichità della Sicilia Orientale. Parte II." *Kokalos* 26–27: 694–730.

———. 1990. "Ceramica laconica in Sicilia e a Lipari. Materiali per una carta di distribuzione." In Lakonikà. *Ricerche e nuovi materiali di ceramica laconica*. Eds P. Pelagatti and C. M. Stibbe. *BdA* Suppl. 64: 123–92.

———. 1995. "Le anfore commerciali." *Atti Taranto* 34: 403–16.

Pellegrino, E. 2009. "La céramique commune d'époque romaine dans le department des Alpes-Maritimes (II siècles av. J.-C. au III ap. J.-C.)." In *Les Céramiques communes d'Italie et de Narbonnaise. Structures des production, typologies et contexts inédits, II s. av. J.-C.–III s. apr. J.-C. Actes de la table ronde de Naples (2–3 novembre 2006)*. 165–89. Naples.

Peluso, P. 1992. "L'opus doliare." In *Locri IV*, 257–62.

Peristere, K., F. Blondé, J. Y. Perreault, and M. Brunet. 1985. "Θάσος 1985. Πρώτη ανασκαφική έρευνα σ'ένα εργαστήρι αγγειοπλαστικής στη θέση "Φαρί" Σκάλας Μαριών." *Archaiologika Analekta ex Athenon* 18: 29–38.

Peroni, R. 1967. "Tipologia e analisi stilistica nei materiali della preistoria: breve messa a punto." *Dialoghi di Archeologia* 1: 155–58.

Perugino, F., and E. Vollaro. 2014. "Roof Tiles." In *Fattoria Fabrizio*, 353–60.

Pescatore, T., P. Renda, and M. Tramutoli. 1988. "Rapporti tra le unità lagonegresi e le unità sicilidi nella media Valle del Basento, Lucania (Appennino Meridionale)." *Memorie della Società Geologica Italiana* 41: 353–61.

Pesce, G. 1936. "Metaponto. Ritrovamenti vari." *NSc* 6th ser., Vol. 12: 439–49.

Pianu, G. 1990. *La necropoli meridionale di Eraclea* 1. *Le tombe di secolo IV e III a.C.*. Rome: Quasar Editore.

———. 2002. *L'agorà di Eraclea Lucana*. Collana del Dipartimento di storia dell'Università degli studi di Sassari 12. Rome: Carocci.

Picon, M., and G. Olcese. 1994. "Per una classificazione in laboratorio delle ceramiche comuni." In *Ceramica romana e archeometrica: lo stato degli studi. Atti delle Giornate Internazionali di Studio (Castello di Montegufoni, 1993)*. Ed. G. Olcese. 105–14. Florence: All'Insegna del Giglio.

Pieri, P., L. Sabato, and M. Tropeano. 1996. "Significato geodinamico dei caratteri deposizionali e strutturali della Fossa bradanica del Pleistocene." *Memorie della Società Geologica Italiana* 51: 501–15.

Pierobon, R. 1979. "*Lagynos*: funzione e forma." *Rivista di Studi Liguri* 45: 27–50.

Pietropaolo, L. 1997. "L'approdo di S. Sabina (Brindisi). Le ceramiche comune di età romana. Aspetti tipologici, tecnologici e distributivi." In *Atti del Convegno Nazionale di Archeologia Subacquea*, Anzio 30–31 maggio e 1 giugno 1996. 249–70. Bari: Edipuglia.

Pimpl, H. 1997. *Perirrhanteria und Louteria. Entwicklung und Verwendung grosser Marmor- und Kalksteinbecken auf figürlichem und säulenartigem Untersatz in Griechenland*. Berlin: Köster.

Pisani, M. 2012. "Impianti di produzione ceramica e coroplastica in Sicilia." In *"Quartiers" artisanaux en Grèce ancienne: une perspective méditerranéenne*. Eds. A. Esposito and G. M. Sanidas. 311–32. Lille: Presses Universitaires du Septentrion.

Poblome, J. 2004. "Comparing ordinary crafts: textile and pottery in production in Roman Asia Minor." *Journal of the Economic and Social History of the Orient* 47: 491–506.

———. 2006. "Made in Sagalassos. Modelling regional potential and constraints." In *Territorio e produzioni ceramiche. Paesaggi, economia e società in età romana. Atti del Convegno Internazionale Pisa, 20–22 ottobre 2005*. Eds. S. Menchelli and M. Pasquinucci. *Instrumenta* 2. 347–55. Pisa: Pisa University Press.

Poblome, J., O. Bounegru, P. Degryse, W. Viaene, M. Waelkens, and S. Erdemgil. 2001. "The sigillata manufactories of Pergamon and Sagalassos." *JRS* 14: 143–65.

Poblome, J., and R. Brulet. 2005. "Production Mechanisms of Sigillata Manufactories. When East Meets West." In *Trade Relations in the Eastern Mediterranean from the Late Hellenistic Period to Late Antiquity: The Ceramic Evidence. Acts from a Ph.D. seminar for young scholars*. Sandbjerg Manorhouse, 12–15 February 1998. Halicarnassian Studies III. Eds. M. Berg Briese and L. E. Vaag. 27–36. Odense.

Poblome, J. and M. Zelle. 2002. "The Table Ware Boom: a Socio-economic Perspective from Western Asia Minor." In *Patris und Imperium*. Eds. C. Berns, H. von Hesberg, L. Vendeput, and M. Waelkens. 275–87. Leuven: Peeters.

Pollard, R. 2005. "The Roman Pottery Industry of West-central Leicestershire." *Journal of Roman Pottery* 12: 149–54.

Pomar, L., and M. Tropeano. 2001. "The Calcarenite di Gravina Formation in Matera (Southern Italy): New Insights for Coarse-grained, Large-scale, Cross-bedded Bodies Encased in Offshore Deposits." *Bulletin of the American Association of Petroleum Geologists* 85.4: 661–89.

Pontrandolfo, A. 2000. "La ceramica attica di IV secolo in area tirrenica." In *La céramique attique du IVe siècle en Méditerranée occidentale. Actes du colloque international organisé par le Centre Camille Jullien, Arles, 7–9 décembre 1995*. Ed. B. Sabattini. 121–30. Naples: Centre Jean Berard.

Poppe, G. T. and Y. Goto. 1993. *European Seashells. Vol. 2 (Scaphopoda, Bivalvia, Cephalopoda)*. Wiesbaden: Verlag Christa Hemmen.

Postrioti, G. 1996. *La stipe votiva del tempio «E» di Metaponto*. Corpus delle stipi votive in Italia IX, Regio III. 2. Rome: Giorgio Bretschneider.

Prag, A. J. N. W. 1983. "Black-Glazed Pottery." In *Metaponto II*, 351–64.

———. 1992a. "Black Glaze Ware." In *Gravina II*, 67–131.

———. 1992b. "Grey Glaze Ware." In *Gravina II*, 134–55.

Preacco Ancona, M. C. 1989. "Il vasellame del I strato." In *Locri II*, 195–255.

———. 1992. "La ceramica fine della 'Casa dei Leoni.'" In *Locri IV*, 113–71.

Prieto, A. 2003. "Field Survey of the Metapontino." In *The Study of the Ancient Territories: Chersonesos & Metaponto. 2003 Annual Report*. 71–74. Austin: Institute of Classical Archaeology.

Prohászka, M. 1998. "Metal Objects and Coins." In *Necropoleis*, 787–835.

Psalti, A. 2004. "Ταφικά σύνολα ελληνιστικών χρόνων από την Ερέτρια." In *6th Hellenistic Pottery Conference 2004*, 505–12, pls. 243–46.

Pucci, G. 1985. "Terra Sigillata Italica." In *Atlante II*, 359–404.

Puppo, P. 1995. *Le coppe megaresi in Italia*. Rome: L'Erma di Bretschneider.

Py, M. 1993. "Céramique grecque orientale." In *Lattara 6*, 435–44.

Quercia, A. 2003. "La ceramica da fuoco dello scarico 1 dal *Kerameikos* di Metaponto." In *I sostegni di fornace dal* kerameikos *di Metaponto*. Ed. V. Cracolici. Beni Archeologici–Conoscenza e Tecnologie. Quaderno 3. 175–99. Bari: Edipuglia.

Quercia, A., and L. Foxhall. 2012. "Tracing Networks Project: Craft Traditions in Ancient Mediterranean. I pesi da telaio come indicatori di dinamiche produttive e culturali nelle attività tessili del Sud Italia in età preromanà." In *La lana nella Cisalpina Romana: economia e società. Studi in onore di Stefania Pesavento Mattioli. Atti del convegno (Padova-Verona, 18–20 maggio 2011)*. Eds. P. Basso and M. S. Busana. 367–81. Padua: Quasar Edizioni.

———. Forthcoming a. "Temporality, materiality and women's networks: The production and manufacture of loom weights in the Greek and indigenous communities of southern Italy." In *Material Crossovers: Knowledge networks and the movement of technological knowledge between craft traditions*. Eds. K. Rebay-Salisbury, L. Foxhall, and A. Brysbaert. Oxford: Routledge.

———. Forthcoming b. *Loving unloved objects. Production and use of the loom weights in pre-Roman southern Italy*.

Radić, I. 1991. "Three more louteria finds in the eastern Adriatic." *International Journal of Nautical Archaeology* 20.2: 155–60.

Raux, S. 1998. "Méthodes de quantification du mobilier céramique. Etat de la question et pistes de réflexion." In *La quantification des céramiques. Conditions et protocole. (Actes de la table ronde du Centre archéologique européen du Mont Beuvray, Glux-en-Glenne, avril 1998)*. Eds. P. Arcelin and M. Tuffreau-Libre. 11–16. Glux-en-Glenne: Centre archéologique européen du Mont Beuvray.

Reese, D. S. 1989. "Faunal Remains from the Altar of Aphrodite Ourania, Athens." *Hesperia* 58 (1): 63–70.

———. 1992. "The Marine and Fresh-Water Shells." In *Excavations at Otranto*. Vol. II: *The Finds*. Eds. F. D'Andria and D. Whitehouse. 347–52. Galatina: Congedo Editore.

Reille, M. 1999. *Pollen et spores d'Europe et d'Afrique du Nord. Supplement II.* URA CNRS 1152. Marseille: Laboratoire de botanique historique et palinologie.

Rescigno, C. Forthcoming. "Architectural Materials from the Pantanello Sanctuary." In *The Chora of Metaponto 7: A Greek Sanctuary at Pantanello.* Austin: University of Texas Press.

Rescigno, C., and M. Corrado. 2001. "Le colline di Senise. Il territorio tra la Fiumarella S. Arcangelo e il fosso Bomberto." In *Carta archeologica Sinni* 4, 11–223.

Reynolds, P. 2010. "Trade networks of the East, 3rd to 7th centuries: The view from Beirut (Lebanon) and Butrint (Albania) (fine wares, amphorae and kitchen wares)." In *LRCW 3. Late Roman Coarse Wares, Cooking Wares and Amphorae in the Mediterranean: Archaeology and archaeometry.* Eds. S. Menchelli, M. Pasquinucci, and S. Santoro. 89–114. Oxford: BAR.

Riccardi, A. 2007. "Ceramica a fase e di stile misto." In *Museo Taranto II.2*, 351–85.

Ricciardini, M. R. 2002. "I grandi contenitori." In *Ausculum I*, 166–97.

Riley, J. A. 1977. "The Coarse Pottery from Berenice. Cooking Ware." In *Excavations at Sidi Khrebish Benghazi (Berenice).* Supplement to Libya Antiqua 5, vol. 2. Ed. J. A. Lloyd. 237–76. Tripoli: Socialist People's Libyan Arab Jamahirya Secretariat of Education Department of Antiquities.

Rinaldi, M. 2005. "Ceramica comune." In *Torre di Satriano 1*, 222–39.

Risberg, Ch. 1992. "Metal-working in Greek sanctuaries." In *Economies of Cult in the Ancient Greek World. Proceedings of the Uppsala Symposium, 1990.* Eds. T. Linders and B. Alroth. 33–40. Uppsala: Acta Universitatis Upsaliensis.

———. 1998. "Production in a sacred place." *Topoi* 8/2: 671–79.

Rizza, G., D. Palermo, and F. Tomasello. 1992. *Mandra di Gipari. Una officina protoarcaica nel territorio di Priniàs. Priniàs* 2. Catania: Università di Catania, Centro di studi per l'archeologia greca.

Robinson, E. G. D. 2012. "New Pixe-Pigme Analyses for South Italian Pottery." *Mediterranean Archaeology* 25: 15–41.

Roncoroni, P. 2004. "Reperti archeologici nel territorio di Alessandria del Carretto (CS)." In *Il Progetto Archeologico di Raganello. Rapporto preliminare 2002–2003.* Eds. P. Attema, M. van Leusen, and P. Roncoroni. 54–73. Groningen: Groningen Institute of Archaeology.

Rossi, F. 1989. "Le fasi di occupazione: cronologie e materiali. Fase III. Dall'ellenismo alla romanizzazione (seconda metà del IV secolo a.C.-I sec. d.C.)." In *Monte Sannace*, 155–210.

Rostoker, W., and E. Gebhard. 1980. "The Reproduction of Roof Tiles for the Archaic Temple of Poseidon at Isthmia, Greece." *JFA* 8: 211–27.

Rotroff, S. I., and J. H. Oakley. 1992. *Debris from a public dining place in the Athenian Agora. Hesperia Supplement* 25. Princeton: The American School of Classical Studies at Athens.

Rotroff, S. I., A. Oliver, I. Hanfmann, and G. M. A. Hanfmann. 2003. *The Hellenistic pottery from Sardis: The Finds through 1994.* Cambridge, MA: Archaeological Exploration of Sardis.

Roubis, D. 1996. "Ricerche archeologiche nell'abitato indigeno di Difesa San Biagio (Montescaglioso)." In *Ricerche sulla casa*, 235–53.

Rubinich, M. 1992. "La ceramica fine della frequentazione arcaica e del sacello." In *Locri IV*, 85–111.

Rüdiger, U. 1969. "Santa Maria d'Anglona (Com. Tursi, Prov. Matera): Scavi nell'anno 1967." *NSc* 8th ser., Vol. 23: 171–97.

Ruga, A. 2010. "Espressioni di eusebeia domestica a Crotone." In *Caulonia tra Crotone e Locri. Tomo I. Atti del Convegno Internazionale. Firenze, 30 maggio–1 giugno 2007.* Eds. L. Lepore and P. Turi. 209–26. Florence: University Press.

Ruga, A., D. Roubis, C. Rescigno, and R. Fiorillo. 2005. "Ricerche nella chora meridionale di Crotone: prospezioni e scavi (1990–1991)." In *Kroton e il suo territorio tra VI e V secolo a.C.: Aggiornamenti e nuove ricerche, Atti del Convegno di Studi, Crotone 3–5 marzo 2000.* Eds. R. Belli Pasqua and R. Spadea. 149–206. Crotone: Grafica Seriart.

Russo, A. 1999. "Il ruolo dell'acqua nei luoghi sacri della Basilicata antica." In *Archeologia dell'acqua in Basilicata.* Ed. M. L. Nava. 103–26. Lavello.

Russo, S. 2007. *All'ombra di Murat. Studi e richerche sul Decennio francesce.* Bari: Edipuglia.

Russo Tagliente, A. 1992. *Edilizia domestica in Apulia e Lucania. Ellenizzazione e società nella tipologia abitativa indigena tra VIII e III sec. a.C.* Galatina: Congedo.

———. 1995. "Armento. Archeologia di un centro indigeno." *BA* 35–36: 1–198.

Sabbione, C. 1984. "L'artigianato artistico." *Atti Taranto* 23: 243–301.

Sacco, A. 1914. *La Certosa di Padula disegnata, descritta e narrata su documenti inediti con speciale riguardo alla topografia, alla storia e all'arte della contrada.* Edizione anastatica del 1982, a cura di Vittorio e Angelina Bracco. Salerno: Boccia Editore.

Safont, S., A. Malgosa, and M. E. Subirà. 2000. "Sex assessment on the basis of long bone circumference." *American Journal of Physical Anthropology* 113 (3): 317–28.

Salvatore, M., ed. 1991. *Il Museo Archeologico Nazionale di Venosa.* Potenza: IEM editrice.

Sanidas, G. M. 2013. *La production artisanale en Grèce. Une approche spatiale et topographique à partir des exemples de l'Attique et du Péloponnèse du VIIe au Ier siècle avant J.-C. Archéologie et Histoire de l'art* 33. Paris: CTHS.

San Pietro, A. 1991. *La ceramica a figure nere di S. Biagio (Metaponto).* Galatina: Congedo.

Sartoris, A. 1997. "I pesi da telaio." In *Pomarico Vecchio*, 227–30.

Sauer, D., S. Wagner, H. Brückner, F. Scarciglia, G. Mastronuzzi, and K. Stah. 2010. "Soil development on marine terraces near Metaponto (Gulf of Taranto, southern Italy)." *Quaternary International* 222: 48–63.

Saunders, S. L., and J. du Plat Taylor. 1992. "Wheel-made Painted Ware." In *Gravina*, 13–41.

Scafuro, M. 2009. "Grandi Contenitori." In *Fratte*, 154–58.

Scarano, G. 1992. "La necropoli di Metaponto in età ellenistica." *Bollettino storico della Basilicata*. 8: 17–24.

Scarfi, B. M. 1961. "Gioia del Colle (Bari). Scavi nella zona di Monte Sannace. Le tombe rinvenute nel 1957." *Monumenti Antichi dei Lincei* 45: 145–332.

Schäfer, J. 1968. *Hellenistische Keramik aus Pergamon*. Berlin: De Gruyter.

Schmiedt, G., and R. Chevallier. 1959. "Caulonia and Metaponto." *L'Universo* 39: 349–70.

Schojer, T. 1988. "La necropoli di Contrada Corti Vecchie." In *Museo di Taranto*, 469–521.

Sciallano, M., and P. Sibella. 1994. *Amphores. Comment les identifier?* Aix-en-Provence: Edisud.

Sciarma, A. 2005. "La diffusione delle *lagynoi* nelle necropoli etrusche tardo-ellenistiche." *Ostraka* 14.2: 209–77.

Sciarra, B. 1976. *Brindisi. Museo Archeologico Provinciale*. Bologna: Calderini.

Sconfienza, R. 1996. "Sistemi idraulici in Magna Grecia: classificazione preliminare e proposte interpretative." *Bollettino della Deputazione di Storia Patria per la Basilicata* 12: 25–66.

———. 2005. *Fortificazioni tardo classiche e ellenistiche in Magna Grecia. BAR-IS* 1341. Oxford: Oxford University Press.

Seif El-Din, M. and M.-D. Nenna. 1993. "La vaisselle en faïence du musée gréco-romain d'Alexandrie." *Bulletin de Correspondance Hellénique* 117.2: 565–602.

Sella, M., C. Turci, and A. Riva. 1988. "Sintesi geopetrolifera della Fossa Bradanica (avanfossa della catena appenninica meridionale)." *Memorie della Società Geologica Italiana* 41: 87–107.

Selli, R. 1962. "Il Paleogene nel quadro della geologia dell'Italia centro-meridionale." *Memorie della Società Geologica Italiana* 3: 737–89.

Semeraro, G. 1983. "Otranto dal VI sec. a.C. all'età ellenistica (Scavi 1977–1979)." *StAnt* 4: 125–212.

———. 1992. "La ceramica pre-romana e romana." In *Excavations at Otranto*, 13–44.

———. 1997. *En neusì. Ceramica greca e società nel Salento arcaico*. Lecce-Bari: Martano.

Sereni, E. 1961. *Storia del paesaggio agrario italiano*. Bari: Laterza.

Serritella, A. 1995. *Pontecagnano II.3. Le nuove aree di necropoli nel IV e III secolo a.C., Annali dell'Istituto Universitario Orientale di Napoli, Sezione Archeologia e Storia Antica, Quaderno n. 9*. Naples: Istituto Universitario Orientale.

Sestieri, P. C. 1940. "Metaponto. Campagna di scavi (marzo –aprile 1939)." *NSc* 7th ser., Vol. 1: 51–122.

Shefton, B. B. 1996. "The Castulo Cup: An Attic Shape in black glaze of special significance in Sicily." In *I vasi attici ed altre ceramiche coeve in Sicilia. Atti del convegno internazionale, Catania, Camarina, Gela, Vittoria, 28 marzo–1 aprile 1990*. 85–98. Palermo: Centro di studio sull'archeologia greca Catania.

Shugart, H. H. 1984. *A Theory of Forest Dynamics*. New York: Springerverlag.

Siebert, G. 1978. "Signatures d'artistes, d'artisans et de fabricants dans l'Antiquité classique." *Ktema* 3: 111–31.

Signore, G. M. 1996. "Rilievi fittili con recumbente dal ceramico di Metaponto." *StAnt* 9: 299–360.

Silvestrelli, F. 1996. "L'officina dei pittori di Creusa, Dolone e dell'anabates a Metaponto." In *Arte e artigianato*, 400–402.

———. 2000. *Ricerche sul kerameikos di Metaponto*. Doctoral dissertation, Università Cattolica del Sacro Cuore.

———. 2004. "L'archeologia della produzione in Italia meridionale. Il caso del kerameikos di Metaponto." In *Metodi e pratica della cultura materiale. Produzione e consumo dei manufatti*. Ed. E. Giannichedda. 107–16. Bordighera.

———. 2005. "Le fasi iniziali della ceramica a figure rosse nel kerameikos di Metaponto." In *La Cèramique apulienne: bilans et perspectives. Atti della Tavola Rotonda organizzata dal Centro Jean Bérard, Napoli 30 novembre e 1 dicembre 2000*. 113–23. Naples: Centre Jean Bérard.

———. 2008. "La distribuzione della ceramica a figure rosse dei Pittori di Creusa, di Dolone e dall'anabates." In *Le perle e il filo. A Mario Torelli per i suoi settanta anni*. 271–92. Venosa: Osanna Edizioni.

———. 2014. "Red-figured Vases from Metaponto: The Evidence from the Necropoleis along the Coast Road." In *The Italic People of Ancient Apulia: New Evidence from Pottery for Workshops, Markets, and Customs*. Eds. T. H. Carpenter, K. M. Lynch, and E. G. D. Robinson. 96–115. New York: Cambridge University Press.

———. Forthcoming. "Il *kerameikos* di Metaponto." In *Contestualizzare la ceramica greca. Nuove metodologie nello studio delle produzioni artigianali dell'antichità classica. Atti della Summer School, Università di Lecce in collaborazione con il King's College di Londra, Cavallino (LE), 28 giugno–7 luglio 2001*. Eds. F. D'Andria, and G. Semeraro.

Silvestrelli, F., and K. Swift. 2011. "Evidence for Black-Gloss Fine-Ware Production in the Chora." In *Survey*, 143–45.

Simonetti, M. 2001. "I bolli di Caulonia." *ASNSP* 4, Quad. 12: 413–63.

Slane, K. W. 1997. "The fine wares." In *Tel Anafa II,i: The Hellenistic and Roman pottery*. Ed. S. C. Herbert. *Journal of Roman Archaeology Supplementary Series* 10. 247–406. Ann Arbor, MI: Kelsey Museum of the University of Michigan.

Slane Wright, K., and R. E. Jones. 1980. "A Tiberian Pottery Deposit from Corinth." *Hesperia* 49, No. 2: 135–77.

Small, A. 1992. "Baking Lid (clibani)." In *Gravina II*, 191–94.

———. 2005. "The Production and Distribution of Bricks and Tiles in South Italy: The Evidence of Vagnari." In *Noctes Campanae*. Eds. W. V. Harris and E. Lo Cascio. 191–211. Naples: Luciano.

———. 2006. "Impressions of Ethnic Identity: Hellenistic Tile Stamps in South Italy." In *Across the Frontiers. Etruscans, Greeks, Phoenicians & Cypriots*. Eds. E. Herring, I. Lemos, F. Lo Schiavo, L. Vagnetti, R. Whitehouse, and J. Wilkins. 327–37. London: Accordia Research Institute.

———, ed. 2011. *Vagnari. Il villaggio, l'artigianato, la proprietà imperiale*. Bari: Edipuglia.

Small, A. M., B. Roe, J. W. Hayes, C. J. Simpson, G. Guzzetta, M. Mackinnon, and S. G. Monckton. 1994. "A Pit Group of c. 80–70 BC from Gravina di Puglia." *PBSR* 49: 197–260.

Smith, C. 1998. "Traders and artisans in archaic central Italy." In *Trade, Traders and the Ancient City*. Eds. H. Parkins and C. Smith. 31–51. London and New York: Routledge.

Söderlind, M. 2006. "The Function of the Terracottas from Selvasecca." In *Deliciae Fictiles* III. Eds. I. E. M. Edlund-Berry, G. Greco, and J. Kenfield. 116–21. Oxford: Oxbow Books.

Söldner, M. 2007. ΒΙΟΣ ΕΥΔΑΙΜΩΝ. *Zur Ikonographie des Menschen in der rotfiguren Vasenmalerei Unteritaliens. Die Bilder aus Lukanien*. Möhnesee: Bibliopolis.

Sophianou, C. 2011. "Loomweights. Use and Manufacture at Trypitos, Siteia." In Στεγα. *The Archaeology of Houses and Households in Ancient Crete. Hesperia Supplements* 4. Eds. K. T. Glowacki and N. Vogeikoff-Brogan. 421–30. Athens: American School of Classical Studies at Athens.

Soricelli, G. 1994. "'Terra Sigillata' della prima, media e tarda età imperiale." In *Il complesso archeologico di Carminiello ai Mannesi. Napoli (Scavi 1983–1984)*. Ed. P. Arthur. 109–68. Galatina.

Sourisseau, J.-C. 2006. "Les amphores commerciales de la nécropole de Rifriscolaro à Camarine. Remarques preliminaries sur les productions corinthiennes de type A." In *Camarina 2006 anni dopo la fondazione. Nuovi studi sulla città e sul territorio*. Atti Convegno Internazionale Ragusa, 7 dicembre 2002/7–9 aprile 2003. Eds. P. Pelagatti, G. Di Stefano, and L. de Lachenal. 129–47. Rome: Istituto Poligrafico dello Stato.

———. 2011. "La diffusion des vins grecs d'Occident du VIIIe au IVe s. av. J.C., sources écrites et documents archéologique." *Atti Taranto* 49: 142–252.

Spadea, R. 1984. La topografia. *Atti Taranto* 23: 119–66.

———. 1993. "Crotone tra IV e III secolo a.C.: precisazioni topografiche." In *Crotone e la sua storia tra IV e III secolo a.C.* Ed. M. L. Napoletano. 19–34. Naples: Arte Tipografica.

———, ed. 1999. *Il tesoro di Hera. Scoperte nel Santuario di Hera Lacinia a Capo Colonna di Crotone, Catalogo Mostra Museo Barracco, 28 marzo–30 giugno 1996*. Milan: Edizioni ET.

Sparkes, B. A. 1962. "The Greek Kitchen." *JHS* 82: 121–37.

Spinello, M. 1968. *Note storiche del monastero e Chiesa Benedettina di Santa Maria la Sanità sul Monte Corno di Pisticci*. Matera: BMG.

Steskal, M. 2010. *Das Prytaneion in Ephesos, Forschungen in Ephesos IX/4*. Vienna: Österreichische Akademie der Wissenschaften.

Stibbe, C. M. 1989 *Laconian Mixing Bowls. A History of the krater Lakonikos from the Seventh to the Fifth Century BC*. Amsterdam: Allard Pierson Museum.

Stolba, V. F. 2009. "Beads, Pendants and Charms: The Evil Eye Belief among the Greek and Indigenous Population of Taurica." *Vestnik Drevnej Istorii* 2: 109–28.

Surra, A., and M. C. Preacco Ancona. 1997. "La ceramica a vernice nera: forme aperte." In *Pomarico Vecchio I*, 65–82.

Swift, K. 2011a. "Archaic to Late-Republican Transport Amphorae." In *Survey*, 455–88.

———. 2011b. "Ceramic Production in the Metapontine Chora." In *The Study of Ancient Territories: Chersonesos & South Italy. 2008–2011 Annual Report*. 15–17. Austin: Institute of Classical Archaeology.

———. Forthcoming a. "Plain and Banded Ware." In *The Chora of Metaponto 7: A Greek Sanctuary at Pantanello*. Austin: University of Texas Press.

———. Forthcoming b. "Mold-made Vessels: Louteria, Basins and Pithoi." In *The Chora of Metaponto 7: A Greek Sanctuary at Pantanello*. Austin: University of Texas Press.

Swift, K., and A. Devereux. 2011. "Assemblage Tables." In *Survey*, Vol. IV, 88–119.

Tagliamonte, G. L. 2011. "Un bollo laterizio dal santuario del Monte San Nicola di Pietravairano (CE)." In *Corollari: scritti di antichità etrusche e italiche in omaggio all'opera di Giovanni Colonna. Studia erudita* 14. Ed. D. Maras. 232–35. Pisa-Rome: Fabrizio Serra Editore.

Tansi, S. 1746. *Historia Cronologica Monasterii S. Michaelis Arcangeli Montis Caveosi*. Naples: Edizione anastatica.

Tarditi, C. 2007. "Vasellame e utensili metallici." In *Museo Taranto II.2*, 561–71.

Taylor, J. 2000. "Cultural Depositional Processes and Post-depositional Problems." In *Extracting Meaning from Ploughsoil Assemblages. The Archaeology of Mediterra-*

nean Landscapes 5. Eds. R. Francovich, H. Patterson, and G. Barker. 16–26. Oxford: Oxbow.

Tempesta, A. L. 1995. "Antefisse fittili da Metaponto." *Bollettino della Basilicata* 11: 41–66.

———. 1996a. "Catalogo." In *Greci, Enotri e Lucani nella Basilicata meridionale*. Eds. S. Bianco, A. Bottini, A. Pontrandolfo, A. Russo Tagliente, and E. Setari. 152, 155–57. Naples: Electa Editore.

———. 1996b. "Scheda catalogo." In *I Greci in Occidente* (Catalogo della Mostra di Venezia). Ed. G. Pugliese Carratelli. 651–52. Milan: Bompiani.

———. 2005. "Attestazioni del mito di Io e del culto di Dioniso a Metaponto: alcune considerazioni." In *Lo spazio del rito*, 83–90.

———. 2012. "Architectural Terracottas." In *San Biagio*, 89–97.

Termini, A. 1997. "La montagnola di Marineo. Le anfore." In *Archeologia e territorio*. Ed. G. Di Stefano. 157–69. Palermo: Palumbo.

Tetteroo, C., and J. Waagen. 2010. "Le sepolture di Muro Tenente 1993–2002." In *L'insediamento Messapico di Muro Tenente: Scavi e ricerche 1998–2009*. Eds. G.-J. Burgers and Ch. Napolitano. 85–140. Mesagne: Museo del Territorio, Mesagne (BR).

Theseider Dupré, L. 1992. "Le lucerne." In *Locri IV*, 186–203.

Threipland, L. M., and M. Torelli. 1970. "A Semi-Subterranean Etruscan Building in the Casale Pian Roseto (Veii) area." *PBSR* 38: 62–121.

Tinelli, M. 2006. "Il casale medievale di Cerceto nel territorio di Cannole (LE): una nota preliminare." *Quaderni del museo della Ceramica di Cutrofiano*, 10-II: 49–78.

Tommaselli, M. 1984. "La Grancia di San Basilio." In *L'esperienza monastica benedettina e la Puglia*. Vol. II. Atti del Convegno di Studi organizzato in occasione del XV centenario della nascita di San Benedetto (Bari-Noce-Lecce-Picciano, 6–10 ottobre 1980). Ed. C. D. Fonseca. 366–72. Galatina: Congedo.

———. 1986. *Masserie fortificate del Materano*. Rome: De Luca Editore.

Tramontana, E. 2008. "Anfore di produzione greca e greco-occidentale." In *Tyndaris 1*, 257–58.

Trendall, A. D. 1989. *Red-Figure Vases of South Italy and Sicily: A Handbook*. London: Thames and Hudson.

Tropeano, M., and L. Sabato. 2000. "Response to Plio-Pleistocene mixed bioclastic-litoclastic temperate-water carbonate systems to forced regressions: The Calcarenite di gravina Formation, Puglia, SE Italy." In *Sedimentary Response to Forced Regressions. Geological Society Special Publication* 172. Eds. D. Hunt and R. L. Gawthorpe. 217–43. London: Geological Society.

Trotter, M., and G. C. Gleser. 1952. "Estimation of stature from long bones of American whites and Negroes." *American Journal of Physical Anthropology* 10: 463–514.

———. 1958. "A re-evaluation of estimation of stature based on measurements of stature taken during life and of long bones after death." *American Journal of Physical Anthropology* 16: 79–123.

Tschurtschenthaler, M. and V. Gertl. 2012. "Siris Herakleia 2011." FASTI Online. <http://www.fastionline.org/conservation/site/AIAC_4271>.

Tubelli, A. 2006. "Coppe ioniche." In *Cuma 2*, 43–50.

———. 2011. "Archaic Fine-Ware Forms." In *Survey*, 148–55.

Turchiano, M. 2000a. "La cisterna e il suo contesto. Materiali tardoantichi dalla domus B." In *Ordona X*, 343–384.

———. 2000b. "Ceramiche della prima e media età imperiale dai riempimenti delle latrine della *domus A*." In *Ordona X*, 251–66.

Tzannes, M-Ch. 2012. "Eros Pursuing a Maiden on an Early Lucanian Hydria in the Benaki Museum." In *Kerameos Paides: Studies Offered to Professor Michalis Tiverios by his Students*. Eds. E. Kefalidou and D. Tsiafaki. 218–27. Thessalonika: Etairia Andrion Epistimonon.

Uggeri, G. 1969. "Kleroi arcaici e bonifica classica nella chora di Metaponto." *La Parola del Passato* 24: 51–71.

Van Buren, E. 1923. *Archaic Fictile Revetments in Sicily and Magna Graecia*. London: Murray.

Van Compernolle, T. 1996. "Coppe di tipo ionico." In *Arte e artigianato*, 299–305.

———. 2000. "Les céramiques ioniennes en Méditerranée centrale." In *Ceràmiques jònies d'època arcaica: centres de producció i comercialització al Mediterràni occidental. Actes de la taula rodonda celebrada a Empùries, 26–28 de maig 1999*. Eds. P. Cabrera Bonet and M. Santos Retolaza. 89–100. Monografies Emporitanes 11. Barcelona: Museu d'Arqueologia de Catalunya.

Vandermersch, Ch. 1989. "Le material amphorique." In *Kaulonia I*, 90–109.

———. 1994. *Vins et amphores de Grande Grèce et de Sicilie, IVe-IIIe s. avant J.-C.* Centre Jean Bérard, Etudies I. Naples: Centre Jean Bérard.

van Geel, B. 1978. "A palaecological study of Holocene peat bog sections in Germany and the Netherlands." *Review of Palaeobotany and Palynology* 25: 1–120.

———. 2001. "Non-pollen palynomorphs." In *Tracking environmental change using lake sediments*. Vol. 3: *Terrestrial, algal and silicaceous indicators*. Eds. J. P. Smol, H. J. B. Birks, and W. M. Last. 99–119. Dordrecht/Boston: Kluwer Academic Publishers.

van Geel, B., J. Buurman, O. Brinkkemper, J. Schelvis, A. Aptroot, G. van Reenen, and T. Hakbijl. 2003. "Environmental reconstruction of a Roman Period settlement site in Uitgeest (The Netherlands), with special reference to coprophilous fungi." *JAS* 30: 873–883.

van Geel, B., D. P. Hallewas, and J. P. Pals. 1983. "A late Holocene deposit under the Westfriese Zeedjik near Enkhuizen (prov. of Noord-Holland, The Netherlands):

Palaeoecological and archaeological aspects." *Review of Palaeobotany and Palynology* 38: 269–335.

Vecchio, P. 2002. "Ceramica comune." In *Mozia* 203–74.

Vegas, M. 1973. *Cerámica común romana del Mediterráneo occidental*. Publicaciones eventuales 22. Barcelona: Universidad, Instituto de Arqueología y Prehistoria.

Verbicaro, G. 2010. "Aree produttive a Crotone tra VII e III sec. a.C." In *Caulonia tra Crotone e Locri. Tomo* I. *Atti del Convegno Internazionale. Firenze, 30 maggio–1 giugno 2007*. Eds. L. Lepore and P. Turi. 222–41. Florence: University Press.

Verrocchio, V. 1998. "Anfore." In *Ceramica in Italia*, 655–704.

Vessberg, O., and A. Westholm. 1956. *The Swedish Cyprus Expedition* Vol. IV, Part 3. *The Hellenistic and Roman Periods in Cyprus*. Stockholm: The Swedish Cyprus Expedition.

Vezzani, L. 1967. "I depositi plio-pleistocenici del litorale ionico della Lucania." *Atti della Accademia Gioenia di Catania* 18: 159–80.

Villard, F., and G. Vallet. 1955. "Megara Hyblaea. V. Lampes du VII siècle et chronologie des coupes ioniènnes." *MEFRA* 66: 7–34.

Villing, A., and E. Pemberton. 2010. "Mortaria from Ancient Corinth: Form and Function." *Hesperia* 79.4: 555–638.

Viola, M. L. 1996. "Le terrecotte architettoniche." In *Arte e artigianato*, 163–79.

Visonà, P. 1999. "Gli oggetti in metallo." In *Oppido Mamertina*, 363–71.

Vittoria, E. 2011a. "Grey Ware." In *Survey*, 271–301.

———. 2011b. "Plain and Banded Ware." In *Survey*, 337–424.

———. 2011c. "Grey Wares from the Farmhouses." In *The Study of Ancient Territories. Chersonesos & South Italy. Report for 2008–2011*. Austin: Institute of Classical Archaeology.

Vivliodetis, E. 2012. "The Lagynoi." In *The Antikythera Shipwreck: The Ship, the Treasures, the Mechanism. (National Archaeological Museum, April 2012–April 2013)*. Eds. N. E. Kaltsas, E. Vlachogianni, and P. Bouyia. 152–63. Athens: Kapon Editions.

von Mehren, M. 2007. "Coarse Wares." In *Hellenistic and Roman Pontecagnano. The Danish Excavations in proprietà Avallone 1986–1990*. Ed. B. Tang. 105–13. Naples: Centre Jean Bérard.

Voultsiadou, E. and D. Vafidis. 2007. "Marine Invertebrate diversity in Aristotle's Zoology." *Contributions to Zoology* 76 (2): 103–20.

Wallace-Hadrill, A. 2008. *Rome's Cultural Revolution*. Cambridge: Cambridge University Press.

Walter, H., and H. Lieth. 1967. *Klimadiagramm-Weltatlas*. Jena: Gustav Fischer Verlag.

Wheeler, R. E. M. 1956. *Archaeology from the Earth*. London: Penguin.

White, D. 1963. "A Survey of Millstones from Morgantina." *AJA* 67.2: 199–206.

Wikander, Ö. 1983. "OPAIA KERAMIS. Skylight-tiles in the Ancient World." *Opuscula Romana* 14: 81–99.

———. 1988. "Ancient Roof Tiles. Use and Function." *Opuscula Atheniensia. Annual of the Swedish Institute at Athens* 17: 203–16.

Williams-Thorpe, O., and R. S. Thorpe. 1993. "Geochemistry and Trade of Eastern Mediterranean Millstones from the Neolithic to Roman Periods." *JAS* 20: 263–320.

Wilson, R. J. A. 1979. "Brick and Tiles in Roman Sicily." In *Roman Brick and Tile*, 11–43.

Winter, N. 1978. "Archaic Architectural Terracottas decorated with Human Heads." *RM* 85: 27–57.

———. 1993. *Greek Architectural Terracottas: From the Prehistoric to the End of the Archaic Period*. Oxford: Clarendon Press.

———. 2009. *Symbols of Wealth and Power. Architectural Terracotta Decoration in Etruria and Central Italy, 640–510 B.C.* Ann Arbor: University of Michigan Press.

Wooldridge, H. G. 1980. *Sant'Angelo Vecchio. Faunal Analysis*. Unpublished report, Institute of Classical Archaeology.

Woolf, S. J. 1973. "La storia politica e sociale." In *Storia d'Italia*, vol. III. *Dal primo Settecento all'Unità*. 3–508. Turin: Giulio Einaudi editore.

Yntema, D. 1990a. "A Specific Group of Black-Gloss Ware Excavated at Valesio: The HFR Group and Its Connections." *StAnt* 6: 167–86.

———. 1990b. *The Matt-Painted Pottery of Southern Italy*. Galatina: Congedo.

———. 1993. *In Search of an Ancient Countryside: The Amsterdam Free University Field Survey at Oria Province of Brindisi, South Italy (1981–1983)*. Amsterdam.

———. 2001. "Basins and mortars." In *Valesio I*, 257–67.

———. 2005. *Conspectus Formarum of Apulian Grey Gloss Wares (Ceramica a pasta grigia). Inventory of Forms of the Apulian Grey Gloss Wares*. 2nd ed. Amsterdam: Institute of Archaeology, Vrije Universiteit.

Zancani Montuoro, P. 1961. "La campagna archeologica del 1932 nella piana del Crati. I ritrovamenti al «Parco del Cavallo»." *AttiMGrecia* n.s. 4: 7–63.

Index

Page numbers in *italics* refer to tables or illustrations.

B

D

E

F

L

M

N

O

P

Q

R

S

T

U

V

W

X

Z